A CENTURY OF BRITISH GEOGRAPHY

A Century of British Geography

EDITED BY
RON JOHNSTON & MICHAEL WILLIAMS

Published for THE BRITISH ACADEMY
by OXFORD UNIVERSITY PRESS

Oxford University Press, Great Clarendon Street, Oxford OX2 6DP

Oxford New York
Auckland Bangkok Bogotá Buenos Aires Cape Town Chennai
Dar es Salaam Delhi Hong Kong Istanbul Karachi Kolkata
Kuala Lumpur Madrid Melbourne Mexico City Mumbai Nairobi
São Paulo Shanghai Singapore Taipei Tokyo Toronto

First published 2003

British Library Cataloguing in Publication Data
Data available

ISBN 0–19–726286–4

Typeset in New Century Schoolbook by
Alden Bookset, Osney Mead
Printed in Great Britain
on acid-free paper by
Antony Rowe Limited
Chippenham, Wiltshire

Contents

I. Introductory

II. Environment

III. Place

IV. Space

V. Geography in action

VI. Geography moving forwards

List of Plates

Between pages 398 and 399

Notes on Contributors

Robert Bennett is Professor of Geography in the University of Cambridge and Fellow of St Catharine's College. He previously held posts at UCL, LSE and Berkeley. His research is concerned with public policy, economic development (particularly human resources skills and small firms), and local economic capacity-building. Recent books include *Enterprise and Human Resource Development: Local Capacity Building*, with Andrew McCoshan (1993), *Local and Regional Economic Development: Renegotiating Power under Labour*, with Diane Payne (2000). He has acted as adviser to many organisations in the public and private sectors, including House of Commons committees. He was a Leverhulme Research Professor 1995–2000. He is chair of the British Academy's Research Committee and chaired the Academy's Review of Graduate Studies (2001). He is a recipient of the Royal Geographical Society Murchison Award (1982) and Founder's Medal (1998), and is a corresponding member of the Austria Academy of Sciences.

Andrew Cliff is Professor of Theoretical Geography in the University of Cambridge and a Fellow of Christ's College. With Peter Haggett, he has worked on the geographical diffusion of epidemic diseases for some 20 years, and they have cooperated on a number of research projects supported by the Leverhulme Trust, the Wellcome Trust, the World Health Organization, and the US Centers for Disease Control. Their books at the interface of geography and epidemiology include *Spatial Diffusion* (1981), *Measles in the Pacific* (1985), *Spatial Aspects of Influenza Epidemics* (1987), *Atlas of Disease Distributions* (1988), *International Atlas of AIDS* (1992), *Measles: An Historical Geography* (1993), *Deciphering Epidemics* (1998) and *Island Epidemics* (2000).

Hugh Clout is Professor of Geography at University College London, and is also Dean of the Faculty of Social and Historical Sciences. His educational background is at the University of London and the Université de Paris. He has been on the academic staff of UCL since 1967. His major research focus is the historical geography of France, with books and articles devoted to agricultural conditions in the nineteenth century and rural reconstruction after the First World War. He has also published on various themes in rural geography and on aspects of regional development in Western Europe, especially in France. As a Londoner, he has edited two books on the city and also the *Times History of London*. Current research focuses on urban and rural reconstruction in France after the Second World War. The history of geography and geographers in France and in Britain has always fascinated him.

Catherine Delano-Smith graduated and gained her DPhil from the University of Oxford. She is now editor of *Imago Mundi: The International Journal for the History of Cartography* and Research Fellow at the Institute of Historical Research, University of London; she was formerly Reader in Historical Geography at the University of Nottingham, where she had been since 1967. Her work on the changing environment and landscape of Mediterranean Europe was summarised in her book *Western Mediterranean Europe: A Historical Geography of Italy, Spain and Southern France since the Neolithic* (1979). Since then she has written on prehistoric maps and on a number of aspects of the history of pre-modern cartography. She has co-authored *Maps in Bibles* (1991, with Elizabeth Ingram); *English Maps: A History* (1999, with Roger J. P. Kain) and *Plantejamentos i Objectivos d'una História Universal de la Cartografia: Approaches and Challenges in a Worldwide History of Cartography* (2001, with David Woodward and Cordell Yee). She has an essay on map signs in Volume 3 of the University of Chicago Press multi-volume, international *History of Cartography*, for which she serves on the Advisory Board.

Sally Eden first became interested in environmental research when analysing green consumers, which led her to investigate the public perceptions of the environment. Since then, she has also worked on the development of environmental policy, business responses to environmental issues, ecological restoration, and the construction of nature. She is currently a Lecturer in Environmental Management at the University of Hull.

Ken Gregory is Emeritus Professor of Geography in the University of London, Visiting Professor at the University of Southampton, and Honorary Professor at the University of Birmingham. He chose to retire from his post as Warden of Goldsmiths University of London in 1998 and from 1998 to 2001 held a Leverhulme Emeritus Fellowship to investigate patterns of river channel adjustment. He is currently President of the Global Continental Palaeohydrology Commission of INQUA and Chair of the Organising Committee for the 30th International Geographical Congress UK 2004 Glasgow. Prior to six years as Head of Goldsmiths College he was at the University of Southampton (1976–92), including posts as Head of Department, Dean of Science, and Deputy Vice-Chancellor, and at the University of Exeter (1962–76), where he was Reader in Physical Geography (1972–6). Research interests and publications have included hydrogeomorphology, palaeohydrology, and the development of physical geography. He received the Back Award in 1980 and the Founder's Medal in 1993 from the Royal Geographical Society, the Linton award of the British Geomorphological Research Group in 1999, and the Geographical Medal from the Royal Scottish Geographical Society in 2000. He holds a DSc from the University of London (1982), honorary degrees from the universities of Southampton (1997) and Greenwich (1997), is a Liveryman of the Goldsmith's Company (1998), a Fellow of Goldsmiths College (1998), and a Fellow of University College London (1999).

Peter Haggett is Emeritus Professor of Urban and Regional Geography in the University of Bristol. His lifelong research interest has been in the role of spatial processes in human geography. This has focused largely on spatial diffusion models with special reference to the spread of infectious diseases through human populations. His joint research contributions in that field (together with A. D. Cliff, J. K. Ord, and M. Smallman-Raynor) were recently summarised in the Clarendon Lectures on Geography at Oxford as *The Geographical Structure of Epidemics* (2000).

Sir Peter Hall is Professor of Planning at the Bartlett School of Architecture and Planning, University College London, and Director of the Institute of Community Studies. From 1991 to 1994 he was Special Adviser on Strategic Planning to the Secretary of State for the Environment, with special reference to issues of London and South East regional planning, including the East Thames Corridor (Thames Gateway) and the Channel Tunnel Rail Link. In 1998–9 he was a member of the Deputy Prime Minister's Urban Task Force, which reported in June 1999. He is author or editor of over 30 books on urban and regional planning and related topics, including *London 2000* (1963, 1969), *The World Cities* (1966, 1977, 1983), *Planning and Urban Growth: An Anglo-American Comparison* (1973, with M. Clawson), *Urban and Regional Planning* (1975, 1982), *Europe 2000* (editor, 1977), *Great Planning Disasters* (1980), *Growth Centres in the European Urban System* (1980, with D. Hay), *The Inner City in Context* (editor, 1981), *Silicon Landscapes* (1985, with A. Markusen), *Can Rail save the City?* (1985, with C. Hass-Klau), *High-Tech America* (1986, with A. Markusen and A. Glasmeier), *The Carrier Wave* (1988, with P. Preston), *Cities of Tomorrow* (1988), *London 2001* (1989), *The Rise of the Gunbelt* (1991, with A. Markusen, S. Campbell, and S. Deitrick), *Technopoles of the World* (1994, with M. Castells), *Sociable Cities* (1998, with C. Ward), *Cities in Civilization (*1998), and *Urban Future 21* (2000, with U. Pfeiffer). He has received the Founder's Medal of the Royal Geographical Society for distinction in research and is an Honorary Member of the Royal Town Planning Institute, a Fellow of the British Academy, and a member of the Academia Europea. He holds ten honorary doctorates from universities in the UK, Sweden, and Canada. He was knighted in 1998 for services to the Town and Country Planning Association.

Ray Hudson is Professor of Geography and Chair of the International Centre for Regional Regeneration and Development Studies at the

University of Durham. He first moved to Durham to a lectureship in 1972, after studying for the degrees of BA and PhD in Geography at the University of Bristol. He was subsequently awarded the degree of DSc from Bristol and an Honorary DSc from Roskilde University, Denmark. His research focuses on geographies of economies and territorial development strategies, especially in the context of Europe. He has published widely on these issues. Recent publications include *Digging Up Trouble: The Environment, Protest and Opencast Coal Mining* (2000, with H. Beynon and A. Cox), *Placing the Social Economy* (2002, with A. Amin and A. Cameron) and *Producing Places* (2001). He is a Member of the Academy of Learned Societies for the Social Sciences and of the Royal Geographical Society, of which he is currently a Vice-President.

Ron Johnston is a Professor in the School of Geographical Sciences at the University of Bristol. After graduating from the University of Manchester he spent 11 years in Australasia—at Monash University and the University of Canterbury—before taking up a chair at the University of Sheffield in 1974. He was Vice-Chancellor of the University of Essex from 1992 to 1995. His main research interests have been into the social geography of cities and the geography of elections, in addition to work on the history and philosophy of human geography. On the latter, he has written a major text—*Geography and Geographers* (5th edn, 1997)—plus *On Human Geography* (1986) and *A Question of Place* (1991), and edited several collections, including *The Dictionary of Human Geography* (4th edn, 2000, with Derek Gregory, Geraldine Pratt, and Michael Watts).

Roger J. P. Kain is a graduate and fellow of University College London and is Montefiore Professor of Geography and Deputy Vice-Chancellor in the University of Exeter. He has published a number of books on maps and mapping including *An Atlas and Index of the Tithe Files of the Mid-Nineteenth Century* (1986), *Cadastral Maps in the Service of the State* (1992, winner of the Newberry Library Kenneth Nebenzahl Prize), *The Tithe Maps of England and Wales* (1995, winner of the Library Association's McColvin Medal) and *Historical Atlas of South-West England* (1999). He wrote *English Maps: A History* (British Library, 1999) jointly with the co-author of this chapter, Catherine Delano-Smith, and is a contributor and adviser to the University of Chicago multi-volume *History of Cartography*. Roger Kain has served as Secretary and Chair of the Institute of British Geographers Historical Geography Research Group and was Chair of the Royal Geographical Society–Institute of British Geographers Annual Conference in 2002. He was elected a Fellow of the British Academy in 1990 and has chaired the Human Geography and Social Anthropology Section and the Academy Research Grants Committee. He was a Vice-President of the Academy and member of its Council (1999–2000) and is currently its Treasurer.

David Livingstone is Professor of Geography and Intellectual History at the Queen's University of Belfast. He has research interests in the history of geographical thought and practice and in the historical geography of scientific knowledge. He is the author of several books including *Nathaniel Southgate Shaler and the Culture of American Science* and *The Geographical Tradition*. His Hettner Lectures, delivered in 2001 at the University of Heidelberg, were published under the title *Science, Space and Hermeneutics*. He has recently held a British Academy Research Readership, during which he completed *Putting Science in its Place* for the University of Chicago Press (2003). He is now beginning work on a project to investigate the historical geography of the reception of Darwinism in several Victorian cities.

Linda McDowell is Professor of Economic Geography at University College London. She has also taught at the Open University, Cambridge, and the LSE, as well as holding research positions at the University of Kent in Canterbury and the Institute of Community Studies in Bethnal Green, London. Before that she was an undergraduate at Newnham College, Cambridge and a graduate student in the Bartlett School of Architecture and Planning at UCL. Her main research interest is the interconnections between economic restructuring, new forms of work, and the transformation of gender relations in contemporary Britain and she has published widely in this area. Her current work includes a study of the labour-market-entry behaviour of low-skilled young white men (*Redundant Masculinities?*) and an investigation of migrant women's work in Britain in the 1950s. She is also interested in feminist knowledge and methodologies and is the author of *Gender, Identity and Place: Understanding Feminist Geographies* (1999).

Doreen Massey is Professor of Geography at the Open University. She has worked for many years on issues of space and place, including research that has ranged from concerns with regional inequality and the future of cities through questions of politics and the conceptualisation of place, to the need to rethink space against the grain of dominant philosophical traditions. Her books include *Spatial Divisions of Labour* and *Space, Place and Gender*. She is co-founder and joint editor of *Soundings: A Journal of Politics and Culture*.

Ceri Peach is Professor of Social Geography at the University of Oxford and a Fellow of St Catharine's College; he has held visiting fellowships at ANU, Berkeley, Yale, UBC, and Harvard. His research interests are in patterns of migration, segregation, and intermarriage. He currently directs the Leverhulme Trust Domesday Survey of the impact of Muslim Mosques, Sikh Gurdwaras, and Hindu Mandirs on the cultural landscapes of English cities. His books include *West Indian Migrations to Britain: A Social*

Geography (1968), *Urban Social Segregation* (1975), *Ethnic Segregation in Cities* (1981, with Vaughan Robinson and Susan Smith), *Geography and Ethnic Pluralism* (1984, with Colin Clarke and David Ley), *South Asians Overseas* (1990, with Colin Clarke and Steven Vertovec), *The Ethnic Minority Populations of Britain: Vol. 2, Ethnicity in the 1991 Census* (1996), and *Islam in Europe: The Politics of Religion and Community* (with Steven Vertovec).

David Rhind has had a varied career, studying geography and geology at Bristol after changing from physics on his first day as an undergraduate. From there he obtained an Edinburgh PhD for a study of the fluvioglacial history of the Tweed Valley. A computer user and programmer from the early 1960s, he built a pioneering database of borehole information as a research fellow then joined the Experimental Cartography Unit based in the Royal College of Art and Imperial College, working on the earliest computer mapping. He was subsequently Lecturer, then Reader in Geography in the University of Durham, where he designed and built the first environmental information system for the European Commission, led a team producing widely used software and texts for (and atlases from) the Census of Population, and launched what is now the National On-line Manpower Information System. Both in Durham and later as Professor of Geography at Birkbeck College he was intimately involved in developments in geographical information systems worldwide and was a Vice-President of the International Cartographic Association. Appointed as the first-ever academic to be Director-General of the Ordnance Survey, he led the creation of the world's first national topographic database and promoted many initiatives to change that organisation in his seven-year tenure—whilst simultaneously serving as a Council member of the Economic and Social Research Council and on other public bodies. He was elected to the fellowship of both the Royal Society and the British Academy in 2002. Currently he is Vice-Chancellor and Principal of The City University.

Ian Simmons graduated in 1959 from University College London and stayed there to work for a PhD, supervised jointly in the geography and botany departments. He subsequently held appointments at the universities of Bristol and Durham (Professor Emeritus 2001–), with visiting professorships in North America: he was ACLS Postdoctoral Fellow at the University of California Berkeley (1964–5) and C. O. Sauer Memorial Lecturer in 1998. His published work in journals has been mainly on the environmental relations of later Mesolithic societies in the British uplands and his books have put this story of influence and impact into wider frameworks, including the global. His scientific background and empirical methodologies have of late been tempered by the increasing shifts within geography to engagement with the humanities and the history of ideas. He was elected to the Society of Antiquaries in 1980, Academia Europaea in 1996, and the British Academy in 1997.

David M. Smith is Emeritus Professor of Geography at Queen Mary, University of London, where he has been based since 1973. He has also held academic posts at the universities of Manchester, Southern Illinois, Florida, Natal, the Witwatersrand, and New England, as well as visiting appointments in various parts of the world. Following an early interest in industrial location and regional development, his research has been concerned with geographical perspectives on human welfare, inequality, and social justice. In recent years he has explored the interface of geography with ethics, or moral philosophy. His books include *Human Geography: A Welfare Approach* (1977), *Geography and Social Justice* (1994), and *Moral Geographies: Ethics in a World of Difference* (2000). His work has combined theory with case studies set in Eastern Europe, Israel, South Africa, and the United States.

Peter J. Taylor is Professor of Geography at Loughborough University and Associate Director of the Metropolitan Institute at Virginia Tech. Concern for the global is the hallmark of his geography. Over the last two decades he has developed a world-systems political geography, including a textbook (*Political Geography: World-Economy, Nation-State and Locality*, 4th edn, 2000, with C. Flint), monographs on world hegemony (*The Way the Modern World Works*, 1995), and ordinary modernity (*Modernities: a Geohistorical Perspective*, 1999), plus several edited collections, the latest being on Americanisation (*The American Century*, 2000, edited with D. Slater). Current work focuses upon developing quantitative measures of the world city network and he is founder and co-director of the Globalization and World Cities (GaWC) Study Group and Network, a virtual centre for international, interdisciplinary research on world cities (www.lboro.ac.uk/gawc).

John Thornes was an undergraduate at Queen Mary College, London University, obtained an MSc at McGill University and a PhD at King's College London. After 14 years at the LSE he became Reader in Geography there, before being Head of Geography at Bedford College, London, Bristol University, and King's College. Following ill health in 1996 he later retired from teaching at King's to become Research Chair in Physical Geography. His research has been mainly in geomorphology, supplemented by a strong interest in ecology. His latest work deals with dynamical systems investigation of the relationships between land degradation and ecological dynamics. He is a Founder's Medallist of the Royal Geographical Society, an Honorary Fellow of Queen Mary and Westfield

College, and a Linton Medallist of the British Geomorphological Research Group. He has supervised 34 successful PhD students. He is an Honorary Member of the Spanish Geomorphological Society. In 2002 he was Leverhulme Millennium Research Fellow and investigated a new model of sediment transfer in ephemeral channels.

Nigel Thrift is a Professor in the School of Geographical Sciences at the University of Bristol. He has taught or carried out research at the universities of Cambridge, Leeds, and Wales in the UK, plus the Australian National University, UCLA, and the National University of Singapore. He has also been a Fellow of the Netherlands Institute of Advanced Study and the Swedish Collegium for Advanced Study in the Social Sciences. He has been involved in a number of innovations in British geography, including time-geography, the geography of finance, non-representational theory, and performance. His main current research interests are in management knowledges, time and time consciousness, information and communications technology, and the performance of affect. His recent publications include *Spatial Formations* (1996), *Money/Space* (1997, with Andrew Leyshon), *Shopping, Place and Identity* (1998, with Danny Miller, Peter Jackson, Beverly Holbrook, and Mike Rowlands), *City A–Z* (2000, co-edited with Steve Pile), *Timespace* (2002, co-edited with Jon May), *Cities* (2003, with Ash Amin) and *The Cultural Geography Handbook* (2003, co-edited with Kay Anderson, Mona Domosh, and Steve Pile).

Michael Williams is Professor of Geography at the University of Oxford, and Vice-Provost of Oriel College. After graduating from the University of Wales he spent 17 years in the University of Adelaide, South Australia, before taking up his present position at Oxford; he was elected a Fellow of the British Academy in 1989. He has been a visiting professor at the universities of Wisconsin (Madison), Chicago, and California (Los Angeles). He has a long-standing research interest in initial settlement and landscape change as well as the evolution of global land use/land cover transformation and change, especially forests and wetlands. His books include *The Draining of the Somerset Levels* (1970), *The Making of the South Australian Landscape* (1974), *Australian Space, Australian Time* (1975, with J. M. Powell), *Americans and their Forests* (1989), *Wetlands: A Threatened Landscape* (1991), *Deforesting the Earth: From Prehistory to Global Crisis* (2003), and he has edited with Terry Coppock, Hugh Clout, and Hugh Prince the previously unpublished essays of H. C. Darby on *The Relations of History and Geography: Studies in England, France and the United States* (2002). He is now working on a biography of Carl Sauer.

Sir Alan Wilson is Vice-Chancellor of the University of Leeds. He graduated in mathematics from Cambridge, converting to the social sciences through research on cities in the 1960s. He was appointed Professor of Urban and Regional Geography in the University of Leeds in 1970 and has been Vice-Chancellor there since 1991. His latest book, *Complex Spatial Systems*, was published in 2000. He was awarded the Founder's Medal of the Royal Geographical Society in 1992, elected as a Member of Academia Europaea in 1992, a Fellow of the British Academy in 1994, and an Academician of the Academy of Learned Societies in the Social Sciences in 2000. He was knighted for services to higher education in 2001. He was a co-founder in the early 1980s of GMAP Ltd. He has served on many public bodies and is currently a member of the Economic and Social Research Council.

Introduction

RON JOHNSTON
MICHAEL WILLIAMS

Geography has a dual definition associated with its literal meaning as 'earth description'. In general discourse, it is understood as providing answers to the question 'what is where?', giving information, according to its *OED* definition, about 'the main physical features of an area'—although human-made features (such as cities) are often incorporated. Alongside this vernacular usage is the academic, referring to the discipline that provides accounts for what is where—and why. Very often, the nature of the academic discipline is misunderstood by those whose general appreciation of geography relates to the vernacular usage only: they believe that (academic) geographers are those who provide 'place descriptions', and may be taken aback—as indeed are some scholars in other disciplines—when they discover what it is that geographers have studied and are currently studying, plus the apparently constant changes in the nature of geographical scholarship.

Geographers are not alone in suffering from these dual and confusing definitions of their discipline, though we guess that their identity problem is somewhat greater than that of their contemporaries in most of the humanities, and perhaps many of the social sciences too. Although geography is a well-established discipline in UK universities, as well as the country's schools, it is not necessarily a well-understood one. Geography's contributions to education and scholarship are not as widely appreciated as they might be; many still see its activities as little more than information-gathering and depiction exercises.

In the vernacular sense geography is a very long-established practice. The question 'what is where?' has undoubtedly been asked in all societies at all times, and ways of answering it have long been practised, by accurate identification of locations (and their mapping) and by careful description, through text and other media. What geographers do—in the vernacular sense—is universal across time and space, as is their means of portrayal; indeed, as historians of cartography are showing, the use of maps preceded written languages in some societies. And those ways have been taught for a long time: both mapmaking and the study of geographical descriptions were on the syllabus at the UK's ancient universities several centuries before the separate discipline was 'invented' (Cormack, 1997; Withers and Mayhew, 2002).

Geography was only formulated as a distinct academic discipline a little over a century ago, as one of the new disciplines being added to university curricula. The information that it produced became valuable to an increasing proportion of the population, for whom knowledge of 'what is where?' was central to their commercial and/or political activities. Alongside this material usage, educationalists argued for the importance of such knowledge to citizens in general—so that they could know themselves better through comparison with others, and of the environments they inhabited. By the late nineteenth century, therefore, universal primary and secondary school education incorporated geography and the foundations were laid for the training of teachers at the country's universities. A few decades later, that training was being underpinned by the provision of degree courses—and by mid-twentieth century these were on offer at almost all of the UK's universities and university colleges.

Furthermore, geography was studied in many of the universities of the British Empire and Commonwealth, with a large number of the new departments both led and staffed from the UK, with some of the appointees on short-term contracts and others taking up permanent residence.[1] They laid the foundations for British geography to have a continuing strong influence there as the number of expatriate staff declined, to be replaced by those locally trained (initially at the undergraduate level only, with postgraduate degrees being obtained from UK universities until graduate schools were established late in the century). There is also a strong British presence in the United States, where geography is much weaker as a high school discipline. In the 1950s–1970s, many UK graduates moved to the United States for their graduate training, and a proportion remained to take up posts there, extending the foundations of what became known as Anglo-American geography.

Like their contemporaries in other disciplines, the small number of individuals appointed early in the century to provide geography degree courses focused most of their time and attention on their roles as teachers. Research activity was slight, and much of what was done involved 'filling in the gaps' in the maps: advancement of knowledge was little more than the acquisition of information and placing it in a classification system. Increasingly, however, and especially after 1945 as the nature of universities changed and the number of academic geographers began to increase substantially, research occupied a larger part of staff time. New approaches were adopted that were oriented more to explanation rather than just

[1] The main exception to this was in the Indian sub-continent, where expatriate appointments were few.

description—the basic question changed from 'what is where?' to 'what is where and why?'. The number of graduate students grew too, and by the end of the century there were flourishing large graduate schools in several institutions.

With a few notable exceptions—several of which are outlined in this book—a research culture only blossomed in UK university geography departments during the last third of the twentieth century. This recent change has undoubtedly contributed substantially to the confusion over the differences between the vernacular and academic understandings of the nature of geography—a confusion probably exacerbated because the geography taught at school to many people now in their 40s and older was of the former rather than the latter variety, stressing information rather than explanation. As a consequence, academic geography as currently practised in the country's universities is not well understood—even in some other parts of those universities. In part, this must be the academic geographers' own fault, for not 'selling their discipline' as well or as thoroughly as they might have, but until relatively recently they have been hampered by the discipline's small size.

Whatever the reason for its poor visibility and legibility,[2] academic geography increasingly flourished in the country's universities through the twentieth century. It was a popular discipline in schools, which provided a continuing supply of students wishing to study it at university—and then return to the schools to teach it to the next generations. For much of the century, however (as indicated in Chapter 2), the subject was grudgingly accepted within many universities. It was generally associated with the humanities, but lacked both the status and the much larger mass of scholars that characterised those disciplines. Its links to the sciences were weak, and it failed to embrace the burgeoning social sciences quickly enough when they sprang to the forefront of attention around mid-century. It was a small, generally poorly regarded discipline and it took some time after its own mid-century period of rapid growth and change (when student numbers increased markedly) for it to make a substantial mark within the wider academy.

One indicator of that marginal status is the lack of a geographical presence in both the British Academy—'the learned society for the humanities and social sciences'—and the Royal Society for most of the century. No geographer was elected to the Academy's fellowship until 1967, when Clifford Darby won the recognition that his contributions to historical geography so richly deserved. It was a further seven years before he was joined by another geographer—the Russian-born and French-trained Jean Gottmann, who held the Chair of Geography at the University

[2] And perhaps credibility too: geography appears to be the butt of more jokes about academic disciplines than any other!

of Oxford—and the complement of geographers did not reach double figures until 1990. By 2002, when the Academy was 100 years old, there were only 20 geographers in its fellowship, with a further three having died since their election (the first three to be elected: Darby, Gottmann, and Terry Coppock). This situation is fairly similar to that of other social science disciplines, but much below the level of representation in the Academy for the humanities.[3] Alongside this lack of representation in the Academy, only two geographers were elected to the Royal Society (H. J. Fleure and S. W. Wooldridge) during the twentieth century, neither of them specifically as geographers. After Wooldridge's death in 1963, there were no geographers in its fellowship (other than Donald Walker, a biogeographer who has spent most of his career at the Australian National University) until the election of David Rhind in 2002.

After more than a century with a formal presence in the UK's universities, therefore, as a discipline geography lacks the visibility and associated sense of identity that its general size warrants. It may not be *terra incognita* to the British Academy, or to the UK academic world more generally, but it is certainly not part of the familiar world of many other academics, let alone the general public. Hence the decision of the Academy to mark its own centenary by a series of volumes commemorating the contributions of its various disciplines offered geographers there an opportunity to put their discipline into this major prestigious showcase and demonstrate—to the Academy generally and to wider audiences—what their discipline has achieved through a century of effort.

In deciding how to present these achievements, as editors we were faced with devising a structure that would tell the full story in a coherent way, a task impeded by two major problems. The first is that geography straddles the main divide within academic life, with the humanities and social sciences on the one side and the natural and life sciences on the other. The discipline's roots lie in both traditions—as Chapters 1 and 2 show—but as it evolved into a fully-fledged research discipline in the second half of the twentieth century, so a split became increasingly apparent between physical geographers, whose links were with natural (especially earth and environmental) scientists, and human geographers, allied to the social sciences and humanities. In a book sponsored by and associated with the work of the British Academy, giving these two major subdivisions of the discipline equal weight seemed inappropriate. Against that argument, however, it had to be recognised that the study of physical geography is integral to the discipline's history as a whole. Furthermore, the divorce between the two parts of the discipline is far from complete: for many physical geographers,

[3] Comparisons are not straightforward. However, less than 2 per cent of geographers employed in UK universities are Fellows of the Academy, compared to some 10 per cent of those employed in the humanities and some 17 per cent of archaeologists.

understanding contemporary landscape formation and processes involves appreciation of human activity, while for many human geographers 'nature' provides not only the backdrop against which much human behaviour occurs but also the raw materials of much human activity. Thus we have included chapters on physical geography per se, as a key component of the larger discipline, and on various aspects of society–nature relationships.

The second major barrier related to divisions within human geography. As the discipline has grown, so research has become more specialised, and the sub-discipline has been fragmented by a variety of cleavages. Some of those relate to the places studied—defined as either formal (Europe, Latin America, etc.) or functional (e.g. urban, rural) regions; some relate to substantive divisions, linked to other social sciences (e.g. social, cultural, political, and economic geography); some relate to the approach adopted (for example, Marxist and feminist); and yet others to the methods adopted (qualitative and quantitative being the common binary division). Using these, we could have commissioned a wide range of chapters, each covering a segment of the discipline alone; but the result would have been a lot of small tales and no big stories.[4] Moreover, some of those stories are continuing vigorously whereas others are relatively dormant.

An alternative and, to us, better approach was to provide a series of overviews that emphasised the main disciplinary themes—and yet employed a structure with the dual advantages of being far from artificial and invented for this volume alone.[5] We decided to build the book around three main themes that dominated academic geography throughout the twentieth century—environment, place and space.[6] Although, as the various chapters make clear, interpretations of each of these have varied considerably over the century, they nevertheless remain enduring fundamentals of geographic study. So too does the basic geographic question of 'what is where?', which has been addressed with a common tool—the map; although, again, its nature has also changed substantially recently (Chapters 11 and 12). Furthermore, geography has not been isolated in an ivory tower: its knowledge—mapped and otherwise—has always been deployed by academic geographers, their graduates, and others, in myriad applied situations.

Having determined the general structure, we then decided on a bipartite division of the whole. The first part would set the larger scene, through a series of substantial chapters setting out the broad picture; the second

[4] As in a comparable book on *Geography in America* (Gaile and Willmott, 1989).

[5] The problem, as we saw it, with a further American volume on *Geography's Inner Worlds* (Abler, Marcus, and Olson, 1992).

[6] And these are not our invention—see Massey, Allen, and Sarre (1999) on *Human Geography Today*.

would illustrate some of the themes in more detail, not through case studies but rather with slightly shorter essays developing some topics in greater depth.

After an opening pair of chapters on the discipline's history—the first covering the long period before its formal institution as an academic discipline within the country's universities, and the second dealing with that process of institutionalisation—the next four sections of the book cover the major themes. There is one for each of the three basic themes (environment, place and space) plus a fourth on geography in action, on its major tool (the map) and on geography as 'useful knowledge'. The final section, comprising seven chapters, fulfils the second of our objectives. Selection of the topics to be covered was far from easy, and many other areas of geographic endeavour could have been included therein. The topics were chosen to illustrate the range and depth of contemporary geographic scholarship in the UK and our selection was confirmed by the geographers on the Section S3 committee of the British Academy in 2000.

Having obtained agreement from the geographers in the British Academy, and then from the Academy's Publications Committee, we solicited contributions. The response was enthusiastic, and although our initial timetable had to be extended somewhat because of some authors' commitments elsewhere (not least on the 2001 Research Assessment Exercise!) we were still able to deliver the manuscript to the Academy and its publishers during the centenary year.

In the original plan, we had considered writing an epilogue, not only looking back at the achievements of a century of scholarly endeavour but also looking forward. We decided not to do that, however. Any further retrospective views would be repetitive of the excellent material that had gone before; and any prospective essay would be fraught with the usual dangers of forecasting in a fast-flowing stream of intellectual progress.[7] Academic geography is vigorous and stimulating as a new century begins, and the prospects on the research front are exciting.[8] But fulfilment of that potential will not be problem-free. Academic geography has thrived in the UK (more so than in any other English-speaking country) because of the discipline's strength in the schools. Subjects are always jostling for slots in a crowded curriculum, however, and, as recent coverage of those debates has shown (Rawling, 2002), geographers have to be eternally vigilant to protect their position—and hence the wellsprings of their subject's vitality in the universities. The achievements set out in this volume provide some of the raw materials for making the case that geography remains what educational authorities accepted about the discipline in the late nineteenth

[7] As Richard Chorley (1973) discovered was the case when he invited contributions to a book on *Directions in Geography*.

[8] On which see Thrift (2002)—but also the commentaries that follow his essay.

century—a vital component in the preparation of an informed citizenry regarding the nature of the world we live in and are constantly changing, and of our place in it.

Acknowledgements

In producing this volume we have accrued a large number of substantial debts. Our initial thanks must go to our fellow geographers in the British Academy for their support when we proposed this venture, and to the Academy's officers and staff who then provided the needed support: in particular, we are grateful to Peter Brown, James Rivington, Vicky Baldwin, and Janet English at the British Academy, and, most especially, Ann Kingdom for her superb work as copy-editor and indexer, and for liaising with the authors and editors during the production process. But our greatest debt is to the authors who agreed to participate, who reacted well to our pressures over time limits, readily negotiated with our editorial foibles, and produced such excellent material. This is their book much more than ours.

The editors and publishers have made every effort to trace the copyright holders of material reproduced in this volume. If any have inadvertently been overlooked, appropriate arrangements will be made at the earliest opportunity.

References

Abler, R. F., Marcus, M. G., and Olson, S., eds (1992) *Geography's Inner Worlds: Pervasive Themes in Contemporary American Geography*. New Brunswick, NJ: Rutgers University Press.

Chorley, R. J., ed. (1973) *Directions in Geography*. London: Methuen.

Cormack, Lesley B. (1997) *Charting an Empire: Geography at the English Universities, 1580–1620*. Chicago, IL: University of Chicago Press.

Gaille, G. J. and Willmott, C. J., eds (1989) *Geography in America*. Columbus, OH: Merrill Publishing.

Massey, D., Allen, J., and Sarre, P., eds (1999) *Human Geography Today*. Cambridge: Polity Press.

Rawling, E. (2002) *Changing the Subject: The Impact of National Policy on School Geography 1980–2000*. Sheffield: The Geographical Association.

Thrift, N. J. (2002) The future of geography. *Geoforum*, 33, 291–8.

Withers, C. W. J. and Mayhew, R. J. (2002) Rethinking 'disciplinary' history: geography in British universities, c. 1580–1887. *Transactions, Institute of British Geographers*, NS27, 11–29.

I. Introductory

1

British geography, 1500–1900

an imprecise review

DAVID N. LIVINGSTONE

At least two dilemmas face writers charged with the task of providing introductory historical surveys of traditions of intellectual and practical inquiry. First, such reviews are too commonly seen as prefatory and antiquarian addenda to the 'real' disciplinary story that follows. Read in this register, the cognitive integrity of earlier regimes of knowledge is compromised either because their accomplishments are couched in the language of teleological inevitability, or because their now unfashionable fixations are relegated to the obsessions of a benighted prehistory. In large measure such judgements stem from a prevailing xenophobia towards the past and a sense that, in intellectual pursuits, as in mobile telephones, new is good, newer is better, and newest is best. Second, the definitional stability that scholarly traditions acquire over the course of time can easily mask those historical negotiations that delivered, for one reason or another, closure on hitherto contested methods of analysis, lines of inquiry, metaphysical assumptions, and the like. The settlements that history delivers can too easily be misconstrued as natural—and thus necessary—states of affairs. Contingency is mistaken for essence.

In beginning with these remarks, I do not mean to imply either that access to past discourses is easily achieved or that contemporary readers can just 'exit' from the present and readily recover the thought-forms of earlier epochs. To be sure, the rejection of presentism as historiographical naiveté is well founded. Reconstructing the past in the light of modern orthodoxy is to engage in the business of writing history backwards. As Quentin Skinner put it many years ago, presentist history can become a means of fixing 'one's own prejudices on to the most charismatic names, under the guise of innocuous historical explanation. History then becomes a pack of tricks we play on the dead' (Skinner, 1969, 13–14). Yet in an important sense, presentism is unavoidable. As Hans-Georg Gadamer (1960, 1976) has convincingly established, it is only through our present-day prejudices, not by seeking to jettison or bracket them, that we have any access to the past at all. The hermeneutic task of coming to terms with the texts and practices of temporally and cognitively distant worlds, involves, as David Linge (1976, xii) puts it, 'both the alien that we strive to understand and the familiar world that we already understand'. Such concessions, however, should not be mobilised to underwrite the cynicism about intellectual history that afflicts certain groves of the modern academy, where the past, we are told, should be relegated to the grave on the presumption that the only context that matters for scholarship is the present one.[1] Declarations of this stripe fail to take with sufficient seriousness the degree to which any academic enterprise is embedded in a tradition of inquiry. And at the same time, they curtail the possibility of disciplinary self-criticism that arises from historical investigations unveiling constitutive connections between social circumstances and cognitive commitments. Indeed, geography's historical complicity in a range of intellectual projects, discursive practices, and political performances continues to cast shadows over its current operations. For these and many other reasons, this introductory chapter cannot be dismissed as a mere prolegomenon to the main narrative that follows.

Historiographical dilemmas, of course, do not exhaust the pitfalls surrounding the reconstruction of British geography's intellectual ancestry, and four further introductory remarks are in order. First, a rudimentary distinction can be drawn between what might be called 'geography as discipline' and 'geography as discourse'. This rough-and-ready separation is intended to mark a difference between the intellectual content and practical operations of what was denominated 'geography' at specific points in time and those wider 'earth-writing' projects that did not come within the ambit of the subject's self-definition at particular times and places.[2] In point of fact, that cognitive boundary line has been anything but stable and the interweaving cartographies of 'discipline' and 'discourse' underscore the need to portray the origins of British geography on as wide a canvas as possible. This consideration presses on us the need to lay aside presumptions about what the terrain of past geographical knowledge must look like. For geographical enterprises have been intimately intertwined with a host of related discourses: natural magic, imperial politics, celestial cartography, natural theology, conjectural prehistory, mathematical astronomy, speculative anthropology, travel writing, national identity, and various species of literary endeavour. Indeed this is not even the right way to express things. For on many occasions such undertakings simply *were* geography. William Cuningham's *The Cosmographical Glass* of 1559, which surveyed numerous standard geographical principles and practices, for example, did not

[1] Within geography, scepticism about the value of tracing the subject's intellectual and institutional history is vigorously expressed in Barnett (1995).

[2] Such concerns animate Mayhew's (2001) Oakeshott-inspired, historiographical concerns.

hesitate to outline 'the Planets and signes governing every region' of the earth. Here geography was part and parcel of an astrological mentalité (see Livingstone, 1998). Or again, when the Scotsman Robert Dick published his book on *The Christian Philosopher* in 1823, he devoted considerable attention to geography as one of the 'Sciences which are related to Religion and Christian theology'. Here, to sever geography from natural theology is to forge a distinction without a difference. All of this means, to re-apply David Lowenthal's (1985) choice dictum, that the geographical past is a foreign country and things are done differently there.

A second set of preliminary observations rotates around the temporal and territorial parameters within which the following account is located. In a nutshell, the point at issue here is that the story of British geography over the last five centuries or so cannot be narrated in isolation from a much wider global and historical framework. I intend to take up the story in late-sixteenth-century Britain. But this geographical inheritance owed much to earlier forms of learning and practice (Wright, 1925; Kimble, 1938; Lindberg, 1992), with roots in such diverse traditions as Ptolemaic astronomy, Aristotelian mechanics (Grant, 1994), Plinian natural history (Grant, 1977), cartographic endeavours at various scales (Harley and Woodward, 1987), and nautical pilotage (Taylor, 1956). Neither can British geography be sequestered from its wider channels of intellectual exchange with continental European geographical traditions and others too. Patterns of trade (Brotton, 1997) and the transmission of knowledge between 'East and West' (Montgomery, 2000), notably in the form of practical mathematics and cartography, thus played a role of one kind or another in the shaping of early modern British geography, as did connections with European centres of learning like Paris, where British chart-makers came in contact with figures like Peter Apian, Gemma Frisius, Oronce Finé, and Pedro Nuñes (Taylor, 1954).

Arising immediately from such considerations is a third, closely related, suite of issues revolving around the simple fact that British geography is not coterminous with geography in Britain. By this I mean to advertise what will be a recurring theme of this essay—namely, that much British geography was made well beyond Britain's shores. The geographical knowledge that was produced and consumed in Britain was routinely acquired in distant places. Here we might think of missionary endeavours in the garnering of worldwide geographical data, or of the expeditions of figures like James Cook whose overseas excursions broadened the compass of British geographical consciousness, or the exploits of the East India Company and the Great Trigonometrical Survey of India, which delivered detailed cartographic data on the sub-continent. The list could be expanded *ad libitum*. But these instances serve to recall the significance of global circuitry and the capillary networks of communication through which the raw ingredients of British geographical knowledge were acquired, brought back home, and reassembled in archives, libraries, museums, gardens, and other depositories. That these tangled webs of intellectual transaction were at once vehicles of commodity exchange *and* conduits of political power directly implicates geography in imperial imperatives of various sorts.

Fourthly, to speak of British geography as though that label referred to a singular entity is, ironically, to fail to take with sufficient seriousness geography itself. By this I simply mean that at any specific point in time there were many different forms of geographical knowledge circulating in the British Isles. At one scale of operations, significant differences are discernible between the several sub-regions. In conspicuous ways, geography in Enlightenment Scotland (Withers, 1995, 2001), say, differed from its English counterpart (Mayhew, 1998a, 2000). At another scale, recent research has disclosed something of the different forms geography assumed in different university curricula (Withers and Mayhew, 2002). All this underscores the significance of location in the history of intellectual enterprises and redraws attention to the role of space in the generation and reception of scientific knowledge more generally (Livingstone, 1995, 2002, 2003). Attending to the divergent configurations of geography in different venues thus necessarily pluralises the origins and genealogy of what has come to be called 'Geography' in the last century or so.

Given these caveats, what follows can obviously be nothing more than an impressionistic, and thus imprecise, sketch of a half-millennium's history. At the very most it constitutes a rough guide to a vast terrain. It is, in other words, indicative rather than comprehensive. Accordingly, rather than elaborating a remorselessly temporal chronology of geographical texts or theories or practitioners, the remainder of this essay is structured around a number of interconnected themes that, in one way or another, snake their way through British geography's narratives. Where they unite is in their emphasis on the placing of geographical knowledge and practice in the conditions of their making and in connecting the subject's history to wider social, intellectual, and political affairs.

Curricular geographies and institutional settings

While we are accustomed to thinking that the 'disciplinary history' of virtually any sphere of scholarly endeavour has its roots in the academic division of labour that was settled during the late Victorian era, there are good reasons for locating the origins of British geography's disciplinarity considerably earlier. Thanks to the inquiries of Charles Withers and Robert Mayhew (2002)—building on the earlier historical investigations of J. N. L. Baker (1935, 1963) and Eva Taylor (1930, 1934)—something of geography's curricular role in late-sixteenth- and early-seventeenth-century English and Scottish universities is now beginning to be mapped.

Geography, it is now clear, was taught in British universities long before its formalisation as a departmental discipline in the late nineteenth century. In Oxford, for example, we know that 'Descriptive Geography' was taught by George Abbot and John Prideaux in the decades around 1600 and by Peter Heylyn during the first half of the seventeenth century; that 'Mathematical Geography' was taught at Magdalen by William Pemble and John Bainbridge (the Savilian Professor of Astronomy) and at Christ Church by John Hullet and Thomas Edwards at various stages during the same century; and that numerous others were also engaged in various forms of geography teaching over the following century and beyond. In Cambridge, geography's early presence was markedly less visible, with only a handful of individuals recorded as engaged in geographical teaching, though the list includes such distinguished names as Edward Wright, Laurence Echard, Isaac Newton, and William Whiston for the period between 1580 and 1710.

In Scotland too there is evidence of a wide range of individuals engaged in teaching geography during the same period. To take a very few instances, in Glasgow it was taught by the reformer Andrew Melville as a branch of mathematics during the mid-1570s; in St Andrews both George Buchanan, Principal of St Leonard's College, and William Wellwood provided geographical instruction alongside mathematics, civil history, and astronomy during the second half of the sixteenth century; in Edinburgh, the mathematician George Sinclair gave instruction in 'Speciall and Theoricall Geography' in the early 1670s. And the pattern persisted into the period conventionally known as the Scottish Enlightenment with thinkers like Thomas Reid—a central figure in the Scottish Philosophy of Common Sense—delivering a lecture course on 'The Elements of Geography' in 1752–3.

These curricular developments, of course, have to be understood against the background of educational thinking in seventeenth- and eighteenth-century Britain (Withers and Mayhew, 2002). The style of written work produced by Oxford teachers of geography like George Abbot and Nathanael Carpenter was markedly different from medieval encyclopedic geography and reflects the predominance of the system of conceptual regulation emanating from the French logician Petrus Ramus, whose anti-Aristotelian dialectic impressed itself in significant ways on English Puritanism (Morgan, 1999). In the Scottish universities during the half-century or so around 1600, geographical instruction was domiciled in an educational system that gave pride of place to theological erudition. The early history of modern geography, then, persistently discloses intimate connections not only with mathematics and astronomy and, on the human side, with history, but also with natural theology in various guises.

Matters of curricular delivery do not exhaust the scope of geography's significance within the world of the early modern university. Another—and related—marker of the subject's university presence is bibliometric.

The ownership of geographical books at Oxford and Cambridge around the turn of the seventeenth century, according to Lesley Cormack's (1997) analysis, was substantial. And the pattern of ownership reveals at least two striking trends. First, the 40 years after 1580 witnessed a progressive differentiation of the subject into three relatively distinct spheres of endeavour—namely, mathematical geography, descriptive geography, and chorography (Cormack, 1991b). At the same time it is clear that mathematical geography tended to decline during the period while works of descriptive geography grew dramatically—a shift no doubt reflecting the appetite for the new geographical information arising from overseas travel. But intellectual curiosity is not the whole story. For descriptive geography had practical advantages for aspiring merchant and gentry alike: it was profitable for trade, it enhanced the arts of diplomacy, and it helped convey an impression of worldly wisdom. Secondly, both in Oxford and Cambridge, a distinct geography of geographical book ownership is discernible. At Oxford, Corpus Christi affiliates owned by far the greatest number of geography texts, followed by those from Christ Church, St John's, and New College. At Cambridge, Peterhouse and St John's topped the list, and then Corpus Christi and Trinity. And it is therefore not surprising that a network of fellows and alumni with geographical interests was associated with these particular colleges, among them Richard Hakluyt, William Camden, William Lambarde, Thomas Lydiat, Samuel Purchas, Richard Devereaux, Edward Lively, and William Bedwell. Not all of these individuals devoted their scholarly energies exclusively to geographical matters: some were local historians, some theologians, others Bible translators, still others natural philosophers. But their interest in geographical affairs of one sort or another attests to the vitality of the subject in the early modern English university.

The university, however, was not the only institutional arena within which geography flourished in the early modern period. The royal court, we are now beginning to realise, was also an important site. For early-seventeenth-century Scotland, for example, Withers (1997) has shown how court masques and triumphal processions constituted theatrical spaces where the territorial idea of Britain as a 'united kingdom' was performed. In this environment, geographical claims were incarnated in a range of choreographed enactments. Indeed the coronation pageant of Charles II in 1661 was organised by the Scottish-born chorographer and geographer, John Ogilby, precisely for the purposes of national self-fashioning. By this stage, courtly geography was already well established in the British Isles. A range of geographical practitioners congregated at the royal court of Queen Elizabeth I and Henry, Prince of Wales, during the late sixteenth and early seventeenth centuries (Cormack, 1991a, 1997). Sometimes these 'Richmond geographers', as they have been styled, were charged with translating foreign geographical works; sometimes their advice was sought on foreign policy and trade; sometimes they placed their geographical skills at the service of Henry's virulent anti-Catholicism. Among these mention might be made of the mathematical geographers Edward Wright, Thomas Hariot, and William Barlow (who combined theory and practice for the benefit of navigation), and of descriptive geographers like Robert Dallington, Arthur Georges, and Thomas Lydiat (who were no less committed to Henry's passion for global Protestant hegemony). Whatever the particulars, the royal household provided these court geographers with an influential space for the production and consumption of geographical knowledge, and with a forum to disclose the imperial potential of the subject to their patrons. But their significance extended beyond their individual advancement at court and their ideological bolstering of the state's imperial ambitions. According to Cormack (1997, 209), not only did they collectively provide 'a powerful impetus for the creation of a seventeenth-century English psyche', their synthesis of abstract theorising and practical engagement 'helped prepare the way for a new vision of science that was to develop in the second half of the seventeenth century' (Cormack, 1997, 224).

Besides the university and royal court, the cause of geography as a scholarly enterprise in Britain was advanced during the centuries that followed in at least two other institutional arenas: schoolrooms and learned societies. In eighteenth-century England, a good deal of the geography that was taught in the grammar schools was communicated within the framework of a humanist curriculum whose radius was circumscribed by classical learning (Mayhew, 1998b). At least in part geography's curricular *raison d'être* in this environment lay in the subject's value to ancient historians as they sought to identify locations in classical texts. In order to fulfil such aims, books like Patrick Gordon's *Geography Anatomiz'd* (which had reached its eighteenth edition by 1744), John Harris's *The Description and Uses of the Celestial and Terrestrial Globes* (which appeared in 1703 and was revised at least a dozen times before the end of the century) and William Guthrie's *A New Geographical, Historical, and Commercial Grammar* (which appeared in 1770 and had attained a 24th edition by 1827) are known to have been recommended reading. Even in the dissenting academies, traditionally thought to have been allergic to the seeming irrelevance and impracticality of a classical syllabus, geography's attachment to scholastic history continued. Nevertheless, geography's presence in the school sector cannot solely be attributed to the tradition of civic humanism in which its role as an accessory to classics fitted most comfortably. In commercial schools dedicated to the mercantile classes (such as Thomas Watts's eighteenth-century educational establishment) and at naval institutions like the Chelsea Naval Academy, mathematical geography—eschewed in the grammar schools—was taught. The presence of geography at the school level was thus important in the subject's evolution, not least because of the need it created for the production of geography texts, or for the incorporation of geographical materials into larger reference works. John Newbery's *Geography Made Familiar and Easy to Young Gentlemen and Ladies*, which first made its appearance in 1748 as the sixth volume of the *Circle of the Sciences*, for example, popularised for a juvenile audience the work of Pat Gordon.

In the long run, geography's presence in the school curriculum had wider consequences too, and not least during the Victorian period when ad hoc provision of voluntary schooling by various religious societies for working-class children was supplanted by a state system regulated by statutory obligations (Maddrell, 1998). During this era geographical learning found itself handmaiden to such diverse interests as scientific method, teleological explanation, scriptural exegesis, moral philosophy, racial ideology, and imperial citizenship. An emphasis on the power of taxonomy, for example, was foremost in William C. Woodbridge's *Rudiments of Geography* (1828), which stressed the value of classification as a mnemonic device equipping pupils with both scientific knowledge and moral apparatus. Matters of biblical interpretation were central to the pedagogic activities of the Church of England's National Society for elementary education, with its publication of works like *Palestine and Other Scripture Geography* in 1852, which was intended as much to promote biblical appreciation as to convey regional information. By contrast, scientific empiricism was predominant in works like Archibald Geikie's *Physical Geography*, first published in 1873 as one of the 'Science Primers' brought out under the joint editorship of Huxley, Roscoe, and Stewart. This was a 269-paragraph pocketbook complete with a catechismal afterword designed to confirm the solidly fact-based, inductive nature of the subject. But it would be mistaken to think that all such writings were in tension with religious sensibilities. In Geikie's work, geography's service to the aims of a scientific natural theology is readily discernible (Livingstone, 1984).

As the century wore on, matters of imperial citizenship progressively obtruded (Ploszajska, 1999). Avril Maddrell (1998) has scrutinised these texts and uncovered the degree to which they were implicated in the discursive politics of race and ethnicity. Works like Thomas Milner's *Our Home Islands* of 1860, for example, valorised British climate and soil as the grounds of national superiority, while the senior geography texts published by Longman around the turn of the century portrayed various African peoples in the censorial language of stupidity, barbarity, and impurity. The promulgation of such stereotypes in geography texts suited Britain's administrative needs during the heyday of its imperial success, and not least as reformers portrayed the colonies as a safety valve for the nation's urban poor (Madrell, 1996). Indeed after 1882 the geography of the colonies was formally introduced into the national syllabus for elementary schools. And it is thus highly significant that many of the discipline's newly professionalised leaders—such as Mackinder, Herbertson, and Lyde—were the authors of key school texts. The reason was plain. As Mackinder (whose educational writings spanned the entire gamut of the school system from infant to teenage texts) put it in 1911, one of the nation's greatest needs was 'to equip the young citizen . . . with a knowledge of the chief contrasts of the political and commercial world' (1911, v). Seen in this guise, geographical education simply *was* imperial pedagogy (Ó Tuathail, 1996). In printed text, photographic illustration, and the practices of field-work alike, school geography was mobilised to heighten feelings of national patriotism (Ploszajska, 1998).

The schoolroom was not the only space where imperial geography proliferated. In a variety of formal and informal institutional arenas, like the Raleigh Travellers' Club and the African Association, and most particularly at the Royal Geographical Society (RGS), what Felix Driver (2001) has recently dubbed 'Geography Militant' was nurtured. Founded in 1830, the Royal Geographical Society was intended to provide a forum for the promotion of geographical knowledge, for gathering together maps, charts, images, and other materials of geographical importance, for servicing the needs of the British government, for providing would-be travellers with relevant information, and for advancing the cause of empire (Mill, 1930). Frequented by military personnel, administrative officialdom, and scientific practitioners, the RGS was a polysemic social space. In its various locations, from the rooms of the Horticultural Society at 21 Regent Street where it was initially constituted, through its occupancy of 3 Waterloo Place from 1839, 15 Whitehall Place from 1854, and 1 Savile Row from 1871, to its current suite of rooms at Lowther Lodge at Kensington Gore (where it has been based since the outbreak of the first world war), it has been at one and the same time a centre of calculation, a place of accumulation, a space of exhibition, a debating chamber on foreign policy, a site of conversation, a publication house, a reading room, and a theatre of communication.

By 1852, for instance, its library already housed some 4000 volumes, a figure that had risen to around 50,000 by 1893. Its map room acquired large numbers of maps, photographs, and lantern slides through which the world and its peoples were visualised for British consumption. In its own premises, and also from time to time at St James's Hall to accommodate large numbers, members of the Society heard the likes of Burton, Livingstone, Speke, and Eyre relate their travel narratives. Indeed their popularity was such, according to Mill (1930, 74), that 'the seats of the Council had to be kept by policemen to prevent them from being rushed by the crowd of Fellows and their friends'. Scientific figures like Roderick Murchison and Alfred Russel Wallace published their work in the pages of the RGS's proceedings. The Society was also a space of instruction, not least for travellers. Not only did it produce edition after edition of *Hints to*

Travellers—a kind of self-help manual on field observation that first appeared in 1854—but in 1879 it undertook to provide instruction on practical astronomy, surveying, and map-making under the tutelage of the Society's Map Curator, John Coles. Over the years its curricular interests expanded in several directions. During the mid-1880s, classes for intending travellers were held in geology, botany, and photography, and later still in zoology, meteorology, and anthropology. Indeed, as H. R. Mill (1930, 94) recollects it, a 'little astronomical observatory was fitted up on the flat roof [of the Savile Row premises], the latitude and longitude of which had to be fixed within a certain limit of accuracy by every aspirant for the Society's certificate'.

At the same time the RGS was also a major force promoting the subject in the British school system and in securing its place as a modern university discipline (Stoddart, 1975; Scargill, 1976), most particularly through the indefatigable efforts of Francis Galton, John Scott Keltie, and Halford Mackinder. That these activities were part and parcel of an irrepressibly imperial *mentalité* is clear from Keltie's (1885) conviction that the Society could usefully advance national interests by providing 'the lower classes' with a knowledge of emigration opportunities to the colonies. At the same time, the halls of the RGS were corridors of military conversation. After all, in his presidential address of 1842, W. R. Hamilton (1842, lxxxviii), diplomat and FRS, reminded his hearers that geography was 'the mainspring of all the operations of war, and of all the negotiations of a state of peace'. Predictably, there remained a strong military presence within a resolutely male membership (women were only admitted to fellowship in 1892) that rose from 460 in 1830 to 6400 by 1930 (Bell and McEwan, 1996).

In all these, and in many other ways too, the RGS presented itself as the institutional face of Victorian geography (Livingstone, 1992). As its long-time associate and chronicler H. R. Mill put it, reflecting on 40 years in Savile Row, the RGS was

> . . . the Mecca of all true geographers, the home port of every traveller. Here the men who were to wipe out 'Unexplored' from the maps of the continents were trained for their labours; and here on their return their records were tested and used to confirm or correct the map of the region they had traversed; here hundreds of foreign geographers who had studied to discover and to comprehend the Earth resorted as to a shrine. The earlier homes of the Society have been swallowed up in the transformation of London, but even when 1 Savile Row has also passed away the spirit of Geography must surely continue to haunt that corner. (Mill, 1930, 94–5)

Shrine it may have been, but only of an entirely ecumenical sort. For as Driver (2001, 47) has recently insisted, it is difficult 'to characterize a body which finds room for missionaries, anti-slavery campaigners, roving explorers, mountaineers, antiquarians, geologists and naturalists under its umbrella as a coherent "centre" at all.' Moreover, the RGS was not the only institutional face of geography throughout the Victorian era. A number of other provincial geographical institutions came into being as the century wore on. During the 1880s and 1890s, geographical societies flourished in Manchester, Liverpool, Tyneside, Southampton, and Hull, and the Royal Scottish Geographical Society, with its four branches, was born in 1884 (Adams, Crosbie, and Gordon, 1984; Mackenzie, 1995). In large measure these organisations were a function of a variety of crises that afflicted British society at the time—tariff reform, a naval scare, and unrest in Africa and India. Yet it was precisely their immediate utility to the commercial interests of municipal imperialism that denied them significant longevity, with the exception of the Manchester and Scottish societies, which from their beginnings enjoyed a rather stronger intellectual orientation.

Organisational arrangements of one sort or another are an indispensable constituent of any discipline's genealogy, and geography is no exception. In universities and schools, in learned societies and royal courts, and doubtless in other spaces too, the discipline was shaped and reshaped. While these different arenas were intimately intertwined with the cognitive content of geographical knowledge, the subject's intellectual history, however, cannot be read as a mere epiphenomenon of its institutional disposition. And so it is appropriate to turn briefly to several of the dominant strains of intellectual and practical endeavour that have woven their way through geography's narrative since the early modern period. The strands I have chosen to highlight—and others could surely be elaborated—are intended to exhibit something of the range of discourses in which geography has routinely been embedded. On occasions, geographical work has been transacted in the language of science and mathematics; on others, it has found a literary voice the most appropriate vehicle for both descriptive evocations and for prosecuting ideological causes of various stripes. At the same time, it has been profoundly implicated in the political languages of national identity and in the geopolitics of empire, not least through its expeditionary impulses. Geography's natural philosophy heritage, its textual constructions of local identity, and its imperial configurations are thus the motifs around which the remainder of this analysis is structured.

Geography, natural philosophy, and the scientific tradition

The prevalence of what was known as 'mathematical geography' was a conspicuous feature of university geography instruction from the late sixteenth century onwards. Frequently understood as competence in the 'use of globes', mathematical geography advertises the subject's scientific associations, notably with astronomy and geometry. Indeed the rubric governing the Savilian chair of astronomy at Oxford, established in 1619, stipulated that the subject should encompass teaching on 'optics, gnomonics, geography and the rules of navigation' (cited in Withers and Mayhew, 2002). And such concerns were at the heart of the teaching provided by numerous individuals, including Nathanael Carpenter and Isaac Newton. Although Carpenter's 1625 text, *Geography Delineated Forth*, was not exclusively a work of geographical mathematics, it did—despite its anti-Copernicanism—represent a move towards the liberation of geography as natural philosophy from ecclesiastical authority; to that degree it was part and parcel of the Puritan outlook that Carpenter had earlier expressed in his *Philosophia Libera* of 1622. In Newton's case, his updating of Varenius's *Geographia Generalis* (originally published in 1650) was precisely in order to advance his own version of the mechanical philosophy and to provide further refutation of Aristotelianism (Bowen, 1981, 77–90). After all, Varenius defined 'general geography' (which was to be supplemented by a second branch, 'special geography', concerned with empirical knowledge of particular locations) as an applied mathematical pursuit incorporating both terrestrial and celestial spheres. It was 'that part of mixed mathematics which explains the state of the earth and of its parts depending on quantity, viz., its figures, place, magnitude and motion, with celestial appearances' (Baker, 1955). Besides his reworking of Varenius's Cartesianism into a Newtonian key,[3] Newton added a range of diagrammatic schemes and wind tables, and corrected several computational errors. By so doing, as Warntz (1989, 182) noted, he demonstrated his continuing 'delight in the use of geometry'.

The character of early modern geography as natural philosophy and practical mathematics is confirmed in other ways too. At Gresham College, a site frequently if controversially thought of as an embryonic Royal Society, the task of teaching geography was allocated to the professor of astronomy, who also undertook instruction in the arts of navigation (Cormack, 1997, 204). Here Edmund Gunter, Henry Gellibrand, and Henry Briggs all practised mathematical geography. Such geographers, and others like them, devoted their energies to the resolution of practical navigational problems (Bennett, 1986). Matters of map projection, the determination of longitude at sea, and compass variation were thus some of the puzzles that attracted the minds of late-sixteenth-century geographers like William Bourne, William Borough, John Dee, and Edward Wright. The desire to solve questions such as these, of course, was fired by a good deal more than intellectual enthusiasm. These were problems of a maritime state seeking to extend its power over an ever-expanding empire. Mathematical geography and imperial politics went hand in hand. Geography in this guise was practical and patriotic at the same time.

[3] This Newtonian transformation continued in the 1712 Jurin edition.

These astronomical and mathematical associations clearly substantiate the scientific thrust of early modern geography. But it would be mistaken to consider such proclivities as evidence that geography had severed connections with earlier traditions of inquiry like natural magic. William Cuningham, as we have already noted, happily instructed his readers in the principles of both astrology and geography (Livingstone, 1992). John Dee, the sixteenth-century Elizabethan magus and court geographer, chronicled the overseas voyages of reconnaissance, manufactured surveying instruments, conjured angels in his domestic den at Mortlake, and regarded mathematics as the numerological key to unlocking the correspondences between the terrestrial and celestial worlds, between the human body and cosmic form (French, 1972; Clulee, 1988; Harkness, 1997). In the late sixteenth century, Thomas Blundeville incorporated within the umbrella pursuit of Cosmography the particular sciences of astronomy, geography, chorography, *and* astrology (Livingstone, 1992). And significantly, while later French and English editions of Varenius's *Geographia Generalis* expunged his inclusion of astrology as an adjunct to geography,[4] its place as a geographical corollary was retained in Newton's Cambridge Latin versions of 1672 and 1681, perhaps on account of Newton's own determined alchemical pursuits and his more general interest in hermeticism, neoplatonism, and occult forces (Dobbs, 1975; Henry, 1990).

The impact of the physical sciences on early modern British geography was clearly substantial; but it does not exhaust the range of the subject's connections with scientific enterprises. During the eighteenth and nineteenth centuries, the traditions of natural history and earth science had no less an impact on geographical thinking and theory (Bowler, 1992). The study of the biogeographical distribution of plants, for example, was conducted by the German philosopher-naturalist Johann Reinhold Foster, who came to England in 1766 and taught at the Dissenters' Academy in Warrington before joining James Cook on his second expedition. Although personal relations on the ship were not good, the published work from that expedition provided evidence for the conviction that the natural world was an integrated unity. The idea that vegetative, animal, and human life were all intimately interwoven with geology, geography, and climate did much to advance a belief in the integrity of regional identity (Browne, 1983, 34). The push to map worldwide botanical, zoological and anthropological distribution patterns provided a stimulus for many of the scientific expeditions of the eighteenth and nineteenth centuries.

[4] Baker (1955, footnote 15) notes that in the original, Varenius included Astrology along with Chronology and Geography as activities 'to which the name of Science . . . doth most properly belong'. Lukermann (1999) argues that Varenius, like Ptolemy, considered astrology a geographical science on methodological grounds.

The study of global geography stimulated by these various endeavours helped perpetuate the Linnaean metaphor of the economy of nature and the Humboldtian analogy of nations of plants. In England, the conception of nature as a polity manifested itself in works of English natural history like Gilbert White's *Natural History of Selborne.* The idea of a geography of life, much advanced by figures like Joseph Banks and James Cook, continued to find expression in the later evolutionary writings of Alfred Russel Wallace, J. D. Hooker, and Charles Darwin, for whom geographical distribution, isolation, and range were central to unlocking the mysteries of speciation and divergence. Take the case of Wallace. Building on William Swainson's *Treatise on the Geography and Classification of Animals* (1835), his voyage with H. R. Bates to the Amazon from 1848 to 1852, and his RGS-funded expedition to the Malay Archipelago between 1854 and 1862, Wallace insisted on the central importance of geographical distributions in the unfolding evolution of life. It was thus appropriate that his celebrated depiction of the line dividing Asian and Australian faunas appeared in the pages of the *Journal of the Royal Geographical Society* (Wallace, 1863; see also Camerini, 1993, 1996). The foundations of this fundamentally cartographic venture, however, had been laid many years before when, as a land surveyor on the Welsh borders in the early 1840s, he mapped the line dividing the agricultural practices of the Welsh hill farmers from their English counterparts—crucially at the very time he encountered Malthus's population principles (Moore, 1997). Moreover, just the year after his mapping of the faunal 'Wallace line', he laid out a comparable racial line separating 'the Malayan and Polynesian Races'. As he himself put it: 'On drawing the line which separates these races, it is found to come near to that which divides the zoological regions . . . a circumstance which appears to me very significant of the same causes having influenced the distribution of mankind that have determined the range of other animal forms' (Wallace, 1880, 19). Geographical distributions were clearly of fundamental importance in Wallace's project to connect up questions of origins, evolution, migration, and speciation—work for which he received the Founder's Medal of the RGS. Given these developments it was entirely suitable that Darwin's cousin, Francis Galton, should have played a prominent role in the history of the RGS. African explorer, travel writer and RGS activist, Galton went on to use the multiplication of regression coefficients in his eugenic work on ancestral heredity and experiments on directional selection (Gillham, 2001).

Geological concerns also conditioned the development of geography during the Victorian era. The work of the Scottish geologists James Hutton, especially in the version made popular by John Playfair in his *Illustrations of the Huttonian Theory of the Earth* of 1802, and John Fleming did much to confirm the dictum that, in the study of the earth, knowledge of the present is the key to the past (Porter, 1977; Laudan, 1987). Subsequently there were debates about whether this methodological principle implied that rates of change had been constant, and on this issue uniformitarians and catastrophists took up different positions. But these differences have often been presented in a much too stereotypal fashion, with all sorts of mediating positions being silenced. While uniformitarianism eventually won out, versions of catastrophism were promulgated by Roderick Murchison, who served four terms as President of the RGS, and Mary Somerville, who received the Society's Patron's Medal for her *Physical Geography* of 1848 (Stafford, 1989; Sanderson, 1974). Their work, and others like it, kept Victorian geography strongly in touch with geological developments. In the latter years of the century such affiliations frequently served to connect the discipline with the newer Darwinian currents of evolutionary thought. But this was not invariably so. An anti-Darwinian form of geology was championed by the Reverend Samuel Haughton FRS, for 30 years professor of Geology at Trinity College Dublin, President of the Royal Irish Academy, medical graduate, and author—at one and the same time—of a hangman's manual of best practice, of papers on the mathematical physics of plane waves in elastic media, and of an anatomical work on *Animal Mechanics* (Spearman, 2002). This stance clearly came through in the *Six Lectures on Physical Geography* he delivered in 1876 and published in 1880. Routinely though, geological writers tended to promote the evolutionary character of physical geography.[5] This stance was reinforced by post-Darwinians like T. H. Huxley, whose *Physiography* of 1877 sought to re-present physical geography to young readers in the language of causal explanations, and Archibald Geikie, who spoke to the RGS in 1879 on 'Geographical evolution' (Stoddart, 1996).

The evolutionary impetus that the earth and life sciences gave to Victorian geography continued to find expression in the early-twentieth-century geographical work of figures like Marion Newbigin, whose *Animal Geography* (1913) was set in a specifically Darwinian framework and designed to keep zoological matters within the arc of geographical concern.[6] To Newbigin, zoogeography necessarily encompassed questions of both distribution and adaptation. In the case of H. J. Fleure, who secured an election to Fellowship of the Royal Society on the strength of his mapping of racial types in Wales, his evolutionary inclinations took anthropological shape as he worked on the borderlands between geography, zoology, and archaeology (Campbell, 1972). Numerous other geographers operating with natural science principles could be mentioned and are well known in the annals of the discipline. What is less frequently appreciated, however, is the degree to which this tradition of British

[5] Within geography, evolution was frequently conceived in Lamarckian terms (see Campbell and Livingstone, 1983).
[6] Maddrell (1997) provides a feminist reading of Newbigin's science.

empiricism had long been domesticated within broader theological discourses.

In fact, since the period of the so-called Scientific Revolution, natural history had customarily been conducted within the framework of natural or physico-theology (Glacken, 1967). Features of natural history were understood to provide evidence of design and were thus mobilised to support arguments for the existence of God. The hydrological cycle, for example, was taken by seventeenth-century figures like John Keill as a sign of divine existence and beneficence (Tuan, 1968). In similar vein John Ray, John Graunt, and William Derham read various aspects of the physical and human geography of the globe in teleological terms (Davies, 1969).[7] This tradition continued until well into the Victorian period. Mary Somerville (1848) insisted in her *Physical Geography* that the patterns of human settlement demonstrated the arrangement of divine wisdom; Haughton urged the readers of his *Lectures on Physical Geography* (1880, 16) to 'never think for a moment of banishing the Creator from our thoughts in our contemplation of Nature'; David Thomas Ansted (1881, 362) also spoke of the 'order and Law of the Creator' in his *Physical Geography*; and Archibald Geikie insisted in 1886 that the object of physical geography was to display 'more and more of that marvellous plan after which this vast world has been framed, to gain a deeper insight into the harmony and beauty of creation, with a yet profounder reverence for Him who made and who upholds it all' (Geikie, 1886, 356). For geography, as for the sciences more generally, natural theology remained very largely the common context of discourse at least until the final decades of the nineteenth century (Young, 1980).

Textual geography, political language, and the making of identity

The cognitive content of early modern geography, as reflected in its instructional expression, is not circumscribed by the horizons of natural science. In the mid-seventeenth century, its humanist leanings were announced in Peter Heylyn's (1657, 19) insistence that history without geography was like 'a dead carkasse' having 'neither life, nor motion'. From its earliest days in the university curriculum, then, the subject was frequently taught as a supplement to history and this, as Withers and Mayhew (2002) make clear, explains why in later years geography was delivered to undergraduates by such distinguished Oxford historians as Thomas Arnold and E. A. Freeman. Later still, in 1901, H. B. George, Fellow of New College, insisted in the preface to his work on *The Relations of Geography and History*,

[7] Harrison (1998) advances the interesting thesis that changes in how nature was read reflected an earlier shift in scriptural hermeneutics.

that 'Every reader of history is aware that he must learn some geography, if he would understand what he reads' (George, 1901, v).

This persistent bond with historical scholarship reveals a textual impulse in geography's lineage and an orientation towards a humanist literary genre that most clearly manifested itself in the practice of chorography or special geography, that branch of geographical endeavour which, in contrast to universal and mathematical geography, was concerned with regional description. With roots in the tradition of the medieval chronicle, and by focusing on local description and emphasising estate survey, provincial antiquarianism, shire genealogy, and parochial chronology, chorography did a good deal to foster senses of regional identity. Because its animus lay in the specification of local *differences*, chorography encouraged the practices of provincial science and a patriotic sense of loyalty to place.[8] Eighteenth-century chorography, as might be expected, thrived during a time when political life in Britain was characterised by a widening schism between Court and Country. When taken in aggregate, geography in this guise—as Cormack (1997, 12) puts it—'provided a means of self-definition for the English people, encouraging them to categorise people from foreign locales as the "other," as well as providing "normal" English standards against which the "other" could be judged'. At the same time, the cartographic ventures that accompanied chorography could be seen as subversive when they promoted county identity over allegiance to the monarch (Helgerson, 1986, 1992). Loyalty to land might easily replace loyalty to crown.

Whatever the motive, local gentry harnessed the geographical skills of early surveyors like Leonard Digges and Edward Worsop to 'conceptualise their own world in terms of maps rather than merely words or deeds' (Cormack, 1997, 169). In the second half of the seventeenth century, the county maps of Christopher Saxton, John Speed, and John Norden were multi-functional articles, acting at once as symbols of local identity, as evidence of land-holding, and as exhibits in legal cases. But the activities of these 'artists of the shire', as chorographers have been called, were not restricted to cartographic preoccupations. Collecting natural objects and cultural artefacts, recording human customs and weather conditions, detailing shire toponymy and biotic life, and describing monuments and agricultural methods, were all features of the chorographer's craft—a craft integrating the new Baconian inductivism with Renaissance humanist ideas about an organic and hierarchical universe (Gentry, 1985). This chorographic branch of Elizabethan geographical practice most clearly surfaced in William Camden's *Britannia* of 1586, which did much to establish local history as an intellectual genre (Rockett, 1990). Camden necessarily relied

[8] Jankovic (2000) argues that senses of county identity in works of chorography were often refracted through the lens of local weather conditions.

on a widespread network of correspondents and collectors, many of whom became chorographers in their own right. Among these are numbered figures like Richard Carew, Sampson Erdeswick, and William Burton, who wrote surveys of Cornwall, Staffordshire, and Leicestershire respectively. The importance of chronology in this species of literary endeavour, moreover, was often closely associated with theological concerns (sometimes of a millennial variety), and it is in this context that Archbishop Ussher's computational estimate of the earth as created in 4004 BC must be placed.

In one way or another, chorographic endeavours were bound up with matters of identity. At the local level, the cataloguing of regional culture and nature did a good deal to foster senses of shire pride. At the national level, British chorography made its contribution to a wider geographical discourse that 'helped its scholars and practitioners develop a belief in the inherent superiority of Protestants over Catholics, in terms of both their treatment of other peoples and the climatic determinism that made temperate Protestant northerners more just, industrious, and courageous' (Cormack, 1997, 13). Chorography thus furthered, at once, a Baconian enthusiasm for particular history, an aristocratic delight in naturalistic readings of social hierarchy, and the patriotic needs of an imperial nation. These associations continued on into the eighteenth century. As Jankovic (2000, 99) concludes from his analysis of how clerical practitioners of regional geography were able to combine their local researches with the growth of a more international 'republic of letters', the 'patriotic politics of chorography was the metanarrative of Augustan natural history'.

Geography's mobilisation in the cause of identity formation was not restricted to national or regional senses of belonging. Nor was it confined to matters of natural history. In the period of the English Revolution, geography books were frequently a medium through which ecclesiastical squabbles—with all their political ramifications—were fought out. A sensitivity to the political languages of church–state relations in the period thus discloses something of the extent to which writers of geography texts throughout the 'long eighteenth century' were embroiled in political infighting (Mayhew, 2000). Because geography books routinely surveyed governmental histories, church constitutions, and religious controversies, they offered ample scope for their authors to intervene in the ecclesiastical politics of Enlightenment England. The Arminian sympathies of Peter Heylyn, for example, came through in his *Cosmographie* of 1657, while George Abbot—archbishop of Canterbury—clearly displayed his Calvinist partisanship in his *Briefe Description of the Whole World* (1599), as too did both Nathanael Carpenter and William Pemble during the first half of the seventeenth century. Geographical works thus disclosed their authors' confessional allegiances in various ways. How they described Catholic spaces, how they represented regional ecclesiologies, how they depicted religious authorities, how they interpreted the dissolution of the monasteries, and how they dealt with political insurrection in the era of the English Civil War were just some of the subjects that became the conduit through which contending politico-theological creeds took geographical form. Heylyn, for example, constructed what Mayhew calls a 'Laudian Geography' (after the anti-Puritan Archbishop William Laud) in which loyalty to the crown, a latitudinarian vision of the Anglican tradition, a defence of the English constitution, an affection for church adornment, and an antipathy to Presbyterianism are clearly apparent. With such sentiments, his patriotic dedication of *Microcosmus* of 1621 to the Prince of Wales is hardly surprising. Heylyn's geography, it is clear, was part and parcel of a wider theological and political agenda. And in the decades around 1700, according to Mayhew (2000), similarly distinctive confessional geographies can be discerned. In the case of Edmund Bohun, whose *Geographical Dictionary* first appeared in 1688, a malleable Anglican Royalism—moulded to changing circumstances—can be seen. In the geographical works of Laurence Echard and Edward Wells, Whig and Tory versions of Anglicanism plainly come to the surface.

The inescapably political character of regional description and geographical compilation was not an exclusively English phenomenon. As Withers (2001) has recently shown, chorographic endeavours were crucial to the making of identity in Scotland too. The activities of figures like Sir Robert Sibbald, Martin Martin, and John Adair in the service of the state during the late seventeenth century are important here (Withers, 1996, 1999a). Sibbald, appointed to the position of Geographer Royal in 1682, was sure that cartographic undertakings, the elucidation of medical topography, and the harvesting of natural and cultural data through a system of distributed Queries, were crucial to national self-knowing. And so he embarked on a range of projects—not all successful, nor completed—which rendered him 'a personal "centre of calculation" situated at the hub of a network of correspondents' (Withers, 2001, 80). Martin, an established geographical informant for the Royal Society and close associate of Sibbald, published a range of first-hand chorographies of the Scottish Highlands and Islands. And Adair, despite testy relationships with both Sibbald and Martin, was commissioned to undertake a geographical survey of the kingdom. In these projects, as Withers writes, whatever their different operational procedures, 'Scotland was being geographically constituted through local geographical enquiries'. These geographical practitioners were, in fundamental ways, 'civil servants in Scotland's "move to modernity"' (Withers, 2001, 82, 109).

Of course the efforts of Sibbald, Martin and Adair were, to one degree or another, continuous with earlier geographical writings that had sought to sustain the idea of an enduring Scottish identity. Various writers from as early as the thirteenth century, and including later-sixteenth-century figures like Hector Boece, John Major, and George Buchanan, deployed geographical data in their works of historical apologetic. So too did travel writers whose chorographical descriptions did much to give shape to Scotland throughout the same period (Withers, 2001). Through cartographic survey, local description, and the cataloguing of regional data, all these individuals were implicated in the construction and preservation of the idea of Scotland as a coherent entity with distinctive natural and cultural formations. And given the early, official, recognition by the state that geography was a powerful instrument in the cultivation of national self-knowledge, it is not surprising that throughout the Enlightenment period in Scotland, geographical crafts and learning were broadcast through a variety of different channels (Withers, 1999b, 2001). The publication of books, maps, and globes for polite consumption, the writing of school texts, the giving of public lectures, the provision of university instruction, and the undertaking of statistical surveys in the period all disclose geography as a distinct species of civic humanism.

Besides the immediately ideological services they rendered, humanist versions of geography also enjoyed a significant presence in wider literary circles. The writings of the eighteenth-century essayist and critic Samuel Johnson, for example, were suffused with geographical sensibility. In his preface to Macbean's *Dictionary of Ancient Geography* (1773), he insisted that to read the Ancients, a knowledge of ancient geography was fundamental; in his writings as a moralist he deployed geographical evidence to assault moral relativism and to confirm the universalism of such drives as hope, fear, and hate; as a philosophical traveller his rich local descriptions were sculpted from attentive geographical observations (Mayhew, 1997). Though with different predilections and emphases, similar geographical deployments are to be found in the writings of figures like John Locke and Oliver Goldsmith. Indeed eighteenth-century compilers of scholarly editions of William Shakespeare, such as Theobald, Warburton, Malone, and Johnson himself, routinely felt the need to provide critical commentary on the geographical accuracy of the bard's works precisely because of the important role geographical knowledge played in the tradition of humanist textual scholarship (Mayhew, 1998c; see also Roberts, 1991; Gillies, 1994). All of this serves to remind us of the literary wing of geography's history and the importance of textual dimensions in its ongoing narrative.

Chorographic rehearsal and literary association were not the only ways in which geography contributed to the construction of senses of identity. Two final moments are illustrative. First, geography's long-standing curiosity about racial difference and its deployment of cartography as a key investigative device have frequently come together in the subject's past. While this marriage of convenience was often enlisted to portray racial geography on a global scale, it also found expression in efforts to map the anthropometric features of the British people. In large part these cartographic ventures were integral to a degenerationist discourse about national decline and the presumed inferiority of certain ethnic groups, ideas that gripped the Victorian imagination. Various anthropometric surveys of Scotland, for example, sought to find in computational science a means of measuring Scottish identity very largely through the mapping of pigmentation statistics. For similar reasons, John Beddoe endeavoured to isolate what he described as incidences of 'nigresence' using a quasi-algebraic formula based on the discrimination of hair colour (Stepan, 1982; Winlow, 2001). Certainly these practices were conducted under the auspices of anthropology, but their fundamentally geographical bent, and the fact that early-twentieth-century geographers like H. J. Fleure perpetuated this tradition of inquiry, warrants alluding to them here.

Secondly, for nineteenth-century Ireland, Campbell (1996) has shown how educational provision and impulses towards modernisation conspired to facilitate the cultivation of geography as a key component in national policy. The story here rotates about the twin poles of the British government in London and the Protestant Anglo-Irish ascendancy of Ireland as they brokered agricultural and industrial change. Initially in institutions like the Royal Dublin Society (founded in 1731), which was geared to agricultural and manufacturing improvement, and in the Geological Society of Ireland (established in 1845), several of whose directors had strongly geographical leanings, and later through the immense personal influence of Horace Plunkett, whose liberal progressivism found outlet in the formation of the Irish Agricultural Organisation Society in 1894, geography played its part in the processes of national reform. The establishment right at the end of the century of the Department of Agricultural and Technical Instruction (DATI) under Plunkett's headship also favoured geographical instruction. Later still DATI funds were directed towards the teaching of commercial geography at the Rathmines School of Commerce in Dublin. Given these mainsprings, it is not surprising that in the university sector early-twentieth-century Irish geography was often taught in association with commercial subjects. Through its perceived value to economic growth, geography contributed to the fashioning of national self-perception.

Empire, travel, and geographical imaginings

With its overseas expansionist ambitions, it is understandable that mathematical geography and chorography did not exhaust the scope of geographical enterprises in Elizabethan England. When combined with a long-standing fascination for travellers' tales and stories of strange peoples and practices—of the sort famously reported in John Mandeville's narratives of around the middle of the fourteenth century—the taste for works of descriptive geography grew rapidly. Indeed what Eva Taylor called 'colonial geography' (referring to accounts of the Americas in the wake of Raleigh's Virginia experiment), together with instructions to the 'urbane traveller' on how to use any Grand Tour to good philosophical effect, flourished in the seventeenth century, even as the genre of English pilgrimage writings declined in the aftermath of the Reformation (Taylor, 1934). In both cases, descriptive geography blossomed as records of foreign climate, soil, and commodities, together with commentaries on the potential of new spaces for plantation, became available. John Smith's account of Virginia, dating from 1606, for instance, contained observations on weather conditions, vegetation, and the possibilities for silk culture. In this he continued a tradition established by Thomas Hariot whose *Briefe and True Report of the New Found Land of Virginia* (1588) was animated by a concern to determine the resource potential of Raleigh's Virginian 'enterprise'.[9] Indeed the very title, which stretches to over 150 words, confirms this intellectual direction in its reference to 'the commodities there found and to be raysed, as well as merchantable, as others for victuall, building and other necessarie vses for those that are and shalbe the planters there'.

The orientation of overseas reports, however, was not always to the west. There was a dramatic increase in works (often translated from European languages) registering conditions in the East Indies, among the Turks or in the Congo, portraying voyages of various kinds, or cataloguing England's overseas adventures. Thrilling and titillating, such books were often used, as appendages to history, for undergraduate teaching in the period. Perhaps the most conspicuous of these works was Richard Hakluyt's compilation of *The Principal Voiages and Navigations of the English Nation* (1589). Master of Christ Church Oxford, diplomat and churchman, Hakluyt amassed a wealth of material from mariners, merchants, and travellers. And his text was much more than an expeditionary catalogue. It was a work of imperial apologetic. According to Cormack (1994, 21), it awakened in the commercial classes 'an awareness of [their] role in the world and encouraged [them] to risk life and limb for the glory of queen, country, and purse'. In Hakluyt pragmatism and patriotism fitted hand in glove.

This imperial impetus, moreover, found voice not only in the chronicling of foreign voyages and in alluring descriptions of unknown territory, but also in the images that adorned the title pages of numerous works of descriptive geography. Imperial frontispieces, like the illustrative cover of John Dee's *General and rare memorials pertayning to the perfect arte of navigation* of 1577, often had royal emblems, magical symbols, and an eschatological sense of Protestant inevitability embedded in their iconographic

[9] This work was, relatively speaking, sensitive to the native people, or 'Lesser Virginians' as he called them. Different views on whether or not Hariot should be read as the agent of a colonialist project are to be found in Greenblatt (1981) and Sokol (1994).

codes (Cormack, 1998). In broadly similar terms, John Wolfe—in his translation of Jan Huygen van Linschoten's work describing Dutch and Portuguese voyages in 1598—staged the text in such a way as to inject into English consciousness a sense of the need to become more imperially proactive. That would result in the global dissemination of England's 'true Religion and Civill Conversation' (quoted in Cormack, 1998). Sir Walter Raleigh's *The History of the World* (1614), too, deployed both royal and masonic symbols on the map that dominates its frontispiece.

Taken overall, these practices did much to develop in the English psyche an imperial *mentalité* woven around a triadic set of convictions: that the world was measurable and therefore controllable; that the English had a right and duty to exercise dominion across the oceans; and that the peoples they encountered did not match up to the excellence of English culture. Imperialism, to put it another way, delivered to the English categories of interpretation that enabled them to cultivate a sense of self-definition that was hewn out of a sense of difference from other races and nations.

In other ways too, religious sentiment stimulated interest in far-away places, most notably the Holy Land, for which a variety of geographical commentaries—often intended to illuminate the pages of Scripture—became available at the time and continued a tradition stretching back at least to the fourth century AD. Several different sources combined to stimulate this curiosity: the establishment in 1581 of the Levant Company; the acquisition by British libraries of a range of Middle Eastern documentary materials; the emergence of philosophical travel literature as a feature of the European canon; and a concern, in the aftermath of the Reformation, to locate key events in the Bible. Commercial interests, new cultures of travel, and religious sentiments thus conspired to bring into renewed focus questions about the location of the Garden of Eden, Mount Ararat, Paradise, and so on (Withers, 1999c). In his 1650 portrayal of Palestine, which further developed his 1634 account of the Crusades, Thomas Fuller (Anglican Royalist, Civil War chaplain and self-designated chorographer) delineated the Holy Land's sacred topography, depicted the territories of the 12 tribes of Israel, provided ethnographic sketches of people, places, and customs, and included numerous colour maps. He also devoted space to describing a variety of natural resources—spices, dates, figs, salt, iron, and so on—as well the culinary, labour, and aesthetic potential of various local animals. In sum, Fuller's text, which synthesised both ancient and contemporary sources, was intended to serve the needs of both scriptural exegesis and commercial appetite (Butlin, 1999). In similar vein, George Sandys used his report of a 1610 journey to the Middle East both to describe the region's economic, social, and political conditions and to deliver critical commentary on the Islamic world (Butlin, 1999). Later, in the early years of the eighteenth century, the Anglican clergyman Edward Wells published a suite of historical geographies of both the Old and New Testaments. These were encyclopedic volumes explicitly concerned to identify 'all the places mention'd, or referr'd to' in the biblical record (Butlin, 1992, 34).

The whole tradition of scholarship devoted to scriptural geography continued to flourish in the nineteenth century, though in the midst of changing ideological contingencies. Significant here were those traditions of textual hermeneutics that raised questions about the reliability of the biblical documents and incorporated into the practice of scriptural interpretation perspectives from German Higher Criticism.[10] At the same time, the scientific interventions of figures like Darwin and Huxley delivered challenges to traditional understandings of the Genesis narratives. In the midst of these shifting conditions, scriptural geography was often mobilised as a conservative force to confirm the trustworthiness of the Bible. Whatever the motivations, historical geographies of the Holy Land enjoyed remarkable success during the Victorian period—notably George Adam Smith's text (Butlin, 1988)—as Palestine was opened up more and more to the scholarly gaze and travel lust of Europeans (Shepherd, 1987). These works thus had a significant, if unappreciated, role to play in the genealogy of Orientalism. As for the readership of this literary genre, no doubt it was considerably enlarged by the growing popularity of the tours to Palestine that Thomas Cook began to organise in the late 1860s, which did much to extend Palestine tourism to a humbler class of traveller than those members of high society who had accompanied Dean Stanley in his excursions to the Holy Land during the 1850s and 1860s. As Regius Professor of Ecclesiastical History at Oxford, and author of *Sinai and Palestine* (1856), Stanley even accompanied the Prince of Wales to Palestine in 1862. Cook, by contrast, was a Dissenter and in touch with a different segment of society. Armed with evangelical piety, Liverpool sardines, Gloucester cheese, and a hefty suspicion of the 'monkish legends' that attached themselves to holy sites, Cook's companions sought the awakening of spiritual emotions through the immediate sight of Palestinian landscapes (Larsen, 2000).

By now 'scientific travel' was already well established in the English scientific tradition. Captain James Cook, for example, following the precedents of Bougainville and La Perouse, had undertaken three expeditions between 1768 and 1780, the first, under the auspices of the Royal Society, to the South Pacific. Its chief aim was to record the transit of Venus; but numerous floral and faunal specimens were collected, anthropological inquiries pursued, landmarks named, and coastal waters mapped. By so doing, Cook and his collaborators, numbered among whom were the Forsters and Joseph Banks, brought antipodean spaces into cultural circulation in

[10] Something of the German tradition of writing on the historical geography of the Holy Land can be cleaned from Goren (1999).

the vocabulary of European science (Carter, 1987; Livingstone, 1992, ch. 4; Gascoigne, 1994). Mapping, naming, and collecting, moreover, were not the only practices involved in bringing foreign geography back home; Cook took professionally trained artists and illustrators with him on all his voyages, and he used them in different ways to speak to different audiences, depending on whether he wanted to appeal to aesthetic connoisseurs or to natural history savants.

The imperial shape that Cook's field science assumed was replicated elsewhere. It surfaced, for example, in what has been called the 'Orientalist geography' of James Rennell (Bravo, 1999). Rennell, a contemporary of Banks, acquired a reputation for remarkable accuracy as a maritime surveyor, and he put these skills to work in the service of empire in his cartographic labours in India, where he spent his formative years working for the East India Company's Bengal Engineers. His *Bengal Atlas* and *Atlas of Hindustan*, which charted interior littorals rather than coastlines, provided colonial administrators with tools for imperial management (Bayly, 1989). In this particular context, as later with the Great Trigonometrical Survey of India, geographical accuracy and precision mapping constituted the Newtonian face of British imperial rule (Edney, 1997). In cases like these, as with the RGS-sponsored surveying ventures of Sir Robert Schomburgk during the first half of the 1830s, which transformed the territory of Guiana into a space of determinate shape and size, British geographical knowledge was made on the imperial frontier (Burnett, 2000). Later, during mid-century, Murchison—already renowned as a fox-hunter, soldier, and geologist—took up geography's cause, not least because he considered it crucial to Britain's commercial and military interests (Stafford, 1999). Besides this, his leading position in the RGS afforded him the opportunity to orchestrate the society's overseas expeditions, and he unhesitatingly used these to test out his own geological theories (Stafford, 1988). The imperial mould in which Murchison's science was cast, moreover, found dramatic expression in the imperial vocabulary in which he depicted the diffusion of the concept of Silurian strata; it invaded continents, enlisted recruits, and engaged in the field of battle (Secord, 1982).

Throughout, it is clear that the geographical culture of Britain remained inextricably intertwined with the brute facts of colonies and commerce. For in the distant making of geographical knowledge, indigenous traditions were regularly incorporated. As Drayton (1998, 237), reflecting on eighteenth-century forebears, observes: 'on every continent there were sophisticated native natural-historical and geographical traditions which were quietly privatized into the cultural property of Western science and of individual savants'. And this was no less true of the geographical knowledge that was made at home. For courtesy of the filigree network of colonial officialdom, species and samples found their way back to Britain where they were re-ordered according to the norms of European taxonomy. Given that the global distribution patterns that geographers mapped had materialised out of an acquisitive colonial culture, it is understandable that the vocabulary of biogeographical explanation was replete with what Janet Browne (1996, 315) calls 'the muscular language of expansionist power'.

For all that, the subject's colonial engagement was not monochrome and a variety of factors acted in different ways to inflect imperial geography. Questions of gender, class, politics, religion, and much else besides all served to complicate the story (see, for example Blunt, 1994 and McEwan, 1994). The role of the missionary movement will serve as illustration. Missions sustained an altogether ambiguous relationship with imperial authority, their practices proving to be inconstant and not predictably compliant with imperial ambitions. Local political allegiances, inter-denominational feuding, shifting views about trade, irritation at colonial meddling in mission schools, and many other specific historical contingencies all serve to expose generalisation as stereotype. What is undoubtedly the case, is that missionaries played a significant scholarly role in the cultivation of geography and allied subjects. As Porter (1999, 241) has tellingly commented: 'Residence, prolonged observation, knowledge of vernaculars, and close contacts gave missionary observations depths unmatched by early armchair ethnologists and most travellers'. Small wonder that figures like A. C. Haddon and J. G. Frazier relied on missionary correspondents, as indeed did Charles Darwin. As for geographical knowledge, the importance of missionaries in the cultivation of a conservation mind-set (Grove, 1989) and the contributions made in different ways by figures like Mary Slessor (McEwan, 1995) and David Livingstone (Driver, 2001) only represent the tip of an iceberg whose dimensions are still in need of detailed inspection. Scrutiny of the ways in which missionary magazines represented distant places and peoples to a wide readership, the role they played in politically charged debates about acclimatisation, with all its implications for colonial settlement, and their significance in the initial mapping of sites remain a real desideratum.

Yet for all these modulations, the intrinsically imperial character of Victorian geography is conspicuous as it strove to occupy a distinctive niche in the professionalising division of academic labour. Nowhere, perhaps, is this more dramatically revealed than in the thought and practice of Sir Halford Mackinder, who did so much to give the subject determinate shape in the years around 1900. For him geography was, in fundamental ways, about 'thinking imperially' as he explicitly put it in 1907. Mackinder's geopolitical vision can best be thought of—to use Ó Tuathail's words (1996, 88)—as 'a scopic regime that configures the world-as-exhibition for European imperial eyes'. Such a reading is entirely in keeping with Mackinder's key role in the Colonial Office Visual Instruction Committee—a body set up precisely for the purpose of presenting the sights of empire to mass audiences in Britain and whose photographer he drilled in exactly what should be captured on lantern slides (Ryan, 1997). As geography moved towards professional birth around the turn of the twentieth century, imperial imperatives had a good deal to do with the success of that venture.

Over a half-millennium's history, British geography has assumed many different forms in many different arenas. Whether as a species of natural philosophy and mathematics, as a form of regional portraiture, as overseas lore, or expeditionary travel; whether in university curricula or at royal courts, in school texts or learned societies; whether as a vehicle of national and local identity or as a channel of imperial desire: geography has been inextricably intertwined with the social, intellectual, political, and religious history of the British Isles.

References

Adams, I. H., Crosbie, A. J., and Gordon, G. (1984) *The Making of Scottish Geography: 100 years of the R.S.G.S.* Edinburgh: Department of Geography, University of Edinburgh and the Royal Scottish Geographical Society.

Ansted, David Thomas (1881) *Physical Geography*, 6th edn. London: W. H. Allen.

Baker, J. N. L. (1935) Academic geography in the seventeenth and eighteenth centuries. *Scottish Geographical Magazine*, 51, 129–143.

Baker, J. N. L. (1955) The geography of Bernard Varenius. *Transactions and Papers, Institute of British Geographers*, 21, 51–60.

Baker, J. N. L. (1963) *The History of Geography*. Oxford: Blackwell.

Barnett, Clive (1995) Awakening the dead: who needs the history of geography? *Transactions, Institute of British Geographers*, NS20, 417–19.

Bayly, Christopher A. (1989) *Imperial Meridian: the British Empire and the World, 1780–1830*. Harlow: Longman.

Bell, Morag and McEwan, Cheryl (1996) The admission of women fellows to the Royal Geographical Society, 1892–1914. *Geographical Journal*, 162, 295–312.

Bennett, James A. (1986) The mechanic's philosophy and the mechanical philosophy. *History of Science*, 24, 1–28.

Blunt, Alison (1994) *Travel, Gender, and Imperialism: Mary Kingsley and West Africa*. New York: Guilford Press

Bowen, Margarita (1981) *Empiricism and Geographical Thought: from Francis Bacon to Alexander von Humboldt*. Cambridge: Cambridge University Press.

Bowler, Peter J. (1992) *The Fontana History of the Environmental Sciences*. London: Fontana.

Bravo, Michael T. (1999) Precision and curiosity in scientific travel: James Rennell and the orientalist geography of the new imperial age (1760–1830). In Jas Elsner and Joan-Pau Rubiés, eds, *Voyages and Visions: Towards a Cultural History of Travel*. London: Reaktion Press.

Brotton, Jerry (1997) *Trading Territories: Mapping the Early Modern World*. London: Reaktion Books.

Browne, Janet (1983) *The Secular Ark: Studies in the History of Biogeography*. New Haven, CT: Yale University Press.

Browne, Janet (1996) Biogeography and empire. In N. Jardine, J. A. Secord, and E. C. Spary, eds, *Culture of Natural History*. Chicago, IL: University of Chicago Press, 305–21.

Burnett, D. Graham (2000) *Masters of All They Surveyed: Exploration, Geography, and a British El Dorado*. Chicago, IL: University of Chicago Press.

Butlin, Robin A. (1988) George Adam Smith and the historical geography of the Holy Land: contents, contexts and connections. *Journal of Historical Geography*, 14, 381–404.

Butlin, Robin A. (1992) Ideological contexts and the reconstruction of biblical landscapes in the seventeenth and early eighteenth centuries: Dr Edward Wells and the historical geography of the Holy Land. In A. R. H. Baker and G. H. Bigger, eds, *Ideology and Landscape in Historical Perspective*. Cambridge: Cambridge University Press, 31–62.

Butlin, Robin A. (1999) English perspectives on the historical geography of the Holy Land in the seventeenth century. In Yossi Ben-Artzi, Israel Bartal, and Elchanan Reiner, eds, *Studies in Geography and History in Honour of Yehoshua Ben-Arieh*. Jerusalem: The Magnes Press, 20–32.

Camerini, Jane R. (1993) Evolution, biogeography, and maps: an early history of Wallace's line. *Isis*, 84, 700–27.

Camerini, Jane (1996) Wallace in the field. *Osiris*, 11, 44–65.

Campbell, J. A. (1972) *Some Sources of the Humanism of H. J. Fleure*. Oxford: School of Geography, University of Oxford, Research Paper No. 2.

Campbell, J. A. (1996) 'Modernisation' and the beginnings of geography in Ireland. In V. Berdoulay and J. A. van Ginkel, eds, *Geography and Professional Practice*. Utrecht: Koninklijk Nederlands Aardrijkskundig Genootschap.

Campbell, J. A. and Livingstone, D. N. (1983) Neo-Lamarckism and the development of Geography in the United States and Great Britain. *Transactions, Institute of British Geographers*, NS8, 267–94.

Carter, Paul (1987) *The Road to Botany Bay: An Essay in Spatial History*. London: Faber.

Clulee, Nicholas (1988) *John Dee's Natural Philosophy: Between Science and Religion*. London: Routledge.

Cormack, Lesley B. (1991a) Twisting the lion's tail: practice and theory at the court of Henry Prince of Wales. In Bruce Moran, ed., *Patronage and Institutions: Science, Technology, and Medicine at the European Court*. Woodbridge: Boydell Press, 67–84.

Cormack, Lesley B. (1991b) 'Good fences make good neighbours': geography as self-definition in early modern England. *Isis*, 82, 639–61.

Cormack, Lesley B. (1994) The fashioning of an empire: geography and the state in Elizabethan England. In Anne Godlewska and Neil Smith, eds, *Geography and Empire*. Oxford: Blackwell, 15–30.

Cormack, Lesley B. (1997) *Charting an Empire: Geography at the English Universities, 1580–1620*. Chicago, IL: University of Chicago Press.

Cormack, Lesley B. (1998) Britannia rules the waves? Images of empire in Elizabethan England. *Early Modern Literary Studies*, 4, Special Issue No. 3, 10, 1–20.

Cuningham, William (1559) *The Cosmographical Glasse, Conteinyng the Pleasant Principles of Cosmographie, Geographie, Hydrographie, or Nauigation*. London.

Davies, Gordon L. (1969) *The Earth in Decay: A History of British Geomorphology 1578–1878*. London: MacDonald.
Dick, Thomas (1823) *The Christian Philosopher; or, the Connection of Science and Philosophy with Religion*. Glasgow: Chalmers and Collins.
Dobbs, B. J. T. (1975) *The Foundations of Newton's Alchemy, or 'the Hunting of the Greene Lyon'*. Cambridge: Cambridge University Press.
Drayton, Richard (1998) Knowledge and empire. In P. J. Marshall, ed., *The Oxford History of the British Empire. Volume II. The Eighteenth Century*. Oxford: Oxford University Press, 231–52.
Driver, Felix (2001) *Geography Militant: Cultures of Exploration and Empire*. Oxford: Blackwell.
Edney, Matthew H. (1997) *Mapping an Empire: the Geographical Construction of British India, 1765–1843*. Chicago, IL: University of Chicago Press.
French, John (1972) *John Dee: The World of an Elizabethan Magus*. London: Routledge and Kegan Paul.
Gadamer, Hans-Georg (1960) *Truth and Method*. London: Sheed & Ward (2nd, revised edn 1989).
Gadamer, Hans-Georg (1976) *Philosophical Hermeneutics*, trans. and ed. D. E. Linge. Berkeley, CA: University of California Press.
Gascoigne, John (1994) *Joseph Banks and the English Enlightenment: Useful Knowledge and Polite Culture*. Cambridge: Cambridge University Press.
Geikie, Archibald (1873) *Physical Geography*. London: Macmillan.
Geikie, Archibald (1886) *Elementary Lessons in Physical Geography*. London: Macmillan.
Gentry, James Robert (1985) English chorographers 1656–1695: artists of the shire. Unpublished PhD thesis, University of Utah.
George, H. B. (1901) *The Relations of Geography and History*. Oxford: Clarendon Press.
Gillham, Nicholas Wright (2001) *A Life of Sir Francis Galton: From African Exploration to the Birth of Eugenics*. Oxford: Oxford University Press.
Gillies, J. (1994) *Shakespeare and the Geography of Difference*. Cambridge: Cambridge University Press.
Glacken, Clarence J. (1967) *Traces on the Rhodian Shore: Nature and Culture in Western Thought from Ancient Times to the End of the Eighteenth Century*. Berkeley, CA: University of California Press.
Goren, Haim (1999) Carl Ritter's contribution to Holy Land research. In Anne Buttimer, Stanley D. Brunn, and Ute Wardenga, eds, *Text and Image: Social Construction of Regional Knowledges*. Leipzig: Beiträge zur Regionalen Geographie, vol. 49, 28–37.
Grant, Edward (1977) *Physical Science in the Middle Ages*. Cambridge: Cambridge University Press.
Grant, Edward (1994) *Planets, Stars, and Orbs: the Medieval Cosmos, 1200–1687*. Cambridge: Cambridge University Press.
Greenblatt, Stephen (1981) Invisible bullets: Renaissance authority and its subversion, Henry IV and Henry V. *Glyph: Textual Studies*, 8, 40–60.
Grove, Richard H. (1989) Scottish missionaries, evangelical discourses and the origins of conservation thinking in South Africa, 1820–1900. *Journal of South African Studies*, 15, 163–87.

Hamilton, William R. (1842) Presidential address. *Journal of the Royal Geographical Society*, 12, lxxxviii–lxxxix.

Harkness, Deborah E. (1997) Managing an experimental household: the Dees of Mortlake and the practice of natural philosophy. *Isis*, 88, 247–62.

Harley, J. B. and Woodward, David (1987) *The History of Cartography, Volume One. Cartography in Prehistoric, Ancient, and Medieval Europe and the Mediterranean*. Chicago, IL: University of Chicago Press.

Harrison, Peter (1998) *The Bible, Protestantism and the Rise of Natural Science*. Cambridge: Cambridge University Press.

Haughton, Samuel (1880) *Six Lectures on Physical Geography*. Dublin and London: Hodges, Foster and Figgis, and Longmans, Green.

Helgerson, Richard (1986) The land speaks: cartography, chorography and subversion in Renaissance England. *Representations*, 16, 51–85.

Helgerson, Richard (1992) *Forms of Nationhood: the Elizabethan Writing of England*. Chicago, IL: University of Chicago Press.

Henry, John (1990) Magic and science in the sixteenth and seventeenth centuries. In R. C. Olby, G. N. Cantor, J. R. R. Christie, and M. J. S. Hodge, eds, *Companion to the History of Modern Science*. London: Routledge, 583–86.

Heylyn, Peter (1657) *Cosmographie in Four Books, Containing the Chorographie and Historie of the Whole World*. London, 2nd edn.

Jankovic, Vladimir (2000) The place of nature and the nature of place: the chorographic challenge to the history of British provincial science. *History of Science*, 38, 79–113.

Keltie, J. Scott (1885) *Geographical Education: Report to the Council of the Royal Geographical Society*. London: Royal Geographical Society.

Kimble, George H. T. (1938) *Geography in the Middle Ages*. London: Methuen.

Larsen, Timothy (2000) Thomas Cook, Holy Land pilgrims, and the dawn of the modern tourist industry. In R. N. Swanson, ed., *The Holy Land, Holy Lands and Christian History*. Woodbridge: The Boydell Press.

Laudan, Rachel (1987) *From Mineralogy to Geology: the Foundations of a Science, 1650–1830*. Chicago, IL: University of Chicago.

Lindberg, David (1992) *The Beginnings of Western Science: the European Scientific Tradition in Philosophical, Religious, and Institutional Context, 600* BC *to* AD *1450*. Chicago, IL: University of Chicago Press.

Linge, David (1976) Editor's Introduction. In Hans-Georg Gadamer, *Philosophical Hermeneutics*, trans. and ed. by D. E. Linge. Berkeley, CA: University of California Press, xi–lviii.

Livingstone, David N. (1984) Natural theology and Neo-Lamarckism: the changing context of nineteenth century geography in the United States and Great Britain. *Annals of the Association of American Geographers*, 74, 9–28.

Livingstone, David N. (1992) *The Geographical Tradition: Episodes in the History of a Contested Enterprise*. Oxford: Blackwell.

Livingstone, David N. (1995) The spaces of knowledge: contributions towards a historical geography of science. *Society and Space*, 13, 5–34.

Livingstone, David N. (1998) Geography and Renaissance magic. In Gregory A. Good, ed., *Sciences of the Earth: An Encyclopedia of Events, People, and Phenomena*. New York: Garland Publishing, 294–7.

Livingstone, David N. (2002) *Science, Space and Hermeneutics.* Heidelberg: University of Heidelberg, The Hettner Lectures 2001.
Livingstone, David N. (2003) *Putting Science in its Place: Geographies of Scientific Knowledge.* Chicago, IL: University of Chicago Press.
Lowenthal, David (1985) *The Past is a Foreign Country.* Cambridge: Cambridge University Press.
Lukermann, Fred (1999) The *Praecognita* of Varenius: seven ways of knowing. In Anne Buttimer, Stanley D. Brunn, and Ute Wardenga, eds, *Text and Image: Social Construction of Regional Knowledges.* Leipzig: Beiträge zur Regionalen Geographie, vol. 49, 11–27.
Mackenzie, John M. (1995) The provincial geographical societies in Britain, 1884–1914. In Morag Bell, Robin Butlin, and Michael Heffernan, eds, *Geography and Imperialism, 1820–1940.* Manchester: Manchester University Press.
Mackinder, Halford J. (1911) *The Nations of the Modern World: An Elementary Study in Geography.* London: George Philip and Son.
Maddrell, Avril M. C. (1996) Empire, emigration and school geography: changing discourses of imperial citizenship, 1880–1925. *Journal of Historical Geography*, 22, 373–87.
Maddrell, Avril M. C. (1997) Scientific discourse and the geographical work of Marion Newbigin. *Scottish Geographical Magazine*, 113, 33–41.
Maddrell, Avril M. C. (1998) Discourses of race and gender and the comparative method in geography school texts 1830–1918. *Environmental and Planning D: Society and Space*, 16, 81–103.
Mayhew, Robert J. (1997) *Geography and Literature in Historical Context: Samuel Johnson and Eighteenth-century English Conceptions of Geography.* Oxford: University of Oxford, School of Geography, Research Paper No. 54.
Mayhew, Robert J. (1998a) The character of English geography, *c.* 1660–1800: a textual approach. *Journal of Historical Geography* 24, 385–412.
Mayhew, Robert J. (1998b) Geography in eighteenth-century British education. *Paedagogica Historica* 34, 731–69.
Mayhew, Robert J. (1998c) Was William Shakespeare an eighteenth-century geographer? Constructing histories of geographical knowledge? *Transactions, Institute of British Geographers*, NS23, 21–37.
Mayhew, Robert J. (2000) *Enlightenment Geography: The Political Languages of British Geography, 1650–1850.* Basingstoke: Macmillan.
Mayhew, Robert J. (2001) The effacement of early modern geography (*c.* 1600–1850): a historiographical essay. *Progress in Human Geography*, 25, 383–401.
McEwan, Cheryl (1994) Encounters with West African women: textual representations of difference by white women abroad. In Alison Blunt and Gillian Rose, eds, *Writing Women and Space: Colonial and Postcolonial Geographies.* New York: Guilford Press, 73–100.
McEwan, Cheryl (1995) 'The mother of all peoples': geographical knowledge and the empowering of Mary Slessor. In Morag Bell, Robin Butlin, and Michael Heffernan, eds, *Geography and Imperialism, 1820–1940.* Manchester: Manchester University Press, 125–50.
Mill, Hugh Robert (1930) *The Record of the Royal Geographical Society, 1830–1930.* London: Royal Geographical Society.

Montgomery, Scott L. (2000) *Science in Translation: Movements of Knowledge through Cultures and Time.* Chicago, IL: University of Chicago Press.

Moore, James (1997) Wallace's Malthusian moment: the common context revisited. In Bernard Lightman, ed., *Victorian Science in Context.* Chicago, IL: University of Chicago Press, 290–311.

Morgan, John (1999) The Puritan thesis revisited. In D. N. Livingstone, D. G. Hart, and M. A. Noll, eds, *Evangelicals and Science in Historical Perspective.* New York: Oxford University Press, 43–74.

Newbery, John (1748) *Geography made Familiar and Easy to Young Gentlemen and Ladies.* London: printed for J. Newbery.

Newbigin, Marion I. (1913) *Animal Geography: The Faunas of the Natural Regions of the Globe.* Oxford: Clarendon Press.

Ó Tuathail, Gearóid (1996) *Critical Geopolitics: The Politics of Writing Global Space.* London: Routledge.

Ploszajska, Teresa (1998) Down to earth? Geography fieldwork in English schools, 1870–1944. *Environment and Planning D: Society and Space*, 16, 757–74.

Ploszajska, Teresa (1999) *Geographical Education, Empire and Citizenship: Geographical Teaching and Learning in English Schools, 1870–1944.* London: Historical Geography Research Group of the Royal Geographical Society.

Porter, Andrew (1999) Religion, missionary enthusiasm and empire. In Andrew Porter, ed., *The Oxford History of the British Empire. Volume III: The Nineteenth Century.* Oxford: Oxford University Press, 222–46.

Porter, Roy (1977) *The Making of Geology: Earth Sciences in Britain, 1600–1830.* Cambridge: Cambridge University Press.

Roberts, J. A. (1991) *The Shakespearean Wild: Geography, Genus and Gender.* Lincoln, NB: University of Nebraska Press.

Rockett, William (1990) Historical topography and British history in Camden's *Britannia. Renaissance and Reformation*, 14, 71–80.

Ryan, James R. (1997) *Picturing Empire: Photography and the Visualization of the British Empire.* London: Reaktion Books.

Sanderson, Marie (1974) Mary Somerville: her work in physical geography. *Geographical Review*, 64, 410–20.

Scargill, D. I. (1976) The RGS and the foundations of geography at Oxford. *Geographical Journal*, 142, 438–61.

Secord, James (1982) King of Siluria: Roderick Murchison and the imperial theme in nineteenth-century British geology. *Victorian Studies*, 25, 413–42.

Shepherd, N. (1987) *The Zealous Intruders: The Western Rediscovery of Palestine.* London: Collins.

Skinner, Quentin (1969) Meaning and understanding in the history of ideas. *History and Theory*, 8, 3–53.

Sokol, B. J. (1994) The problem of assessing Thomas Harriot's *A briefe and true report* of his discoveries in North America. *Annals of Science*, 51, 1–16.

Spearman, T. O. (2002) *Victorian Polymath: Samuel Haughton.* Dublin: Royal Irish Academy.

Stafford, Robert A. (1988) Roderick Murchison and the structure of Africa: a geological prediction and its consequences for British expansion. *Annals of Science*, 45, 1–40.

Stafford, Robert A. (1989) *Scientist of Empire: Sir Roderick Murchison, Scientific Exploration and Victorian Imperialism*. Cambridge: Cambridge University Press.

Stafford, Robert A. (1999) Scientific exploration and empire. In Andrew Porter, ed., *The Oxford History of the British Empire. Volume III: The Nineteenth Century*. Oxford: Oxford University Press, 294–319.

Stepan, Nancy (1982) *The Idea of Race in Science: Great Britain 1800–1960*. Basingstoke: Macmillan.

Stoddart, D. R. (1975) The RGS and the foundations of geography at Cambridge. *Geographical Journal*, 141, 216–39.

Stoddart, D. R. (1996) *On Geography and its History*. Oxford: Blackwell.

Swainson, William (1835) *A Treatise on the Geography and Classification of Animals*. London: Longman Rees.

Taylor, E. G. R. (1930) *Tudor Geography, 1485–1583*. London: Methuen.

Taylor, E. G. R. (1934) *Late Tudor and Early Stuart Geography, 1583–1650*. London: Methuen.

Taylor, E. G. R. (1954) *The Mathematical Practitioners of Tudor and Stuart England 1485–1714*. Cambridge: Cambridge University Press.

Taylor, E. G. R. (1956) *The Haven-finding Art: A History of Navigation from Odysseus to Captain Cook*. London: Hollis & Carter.

Tuan, Yi-Fu (1968) *The Hydrological Cycle and the Wisdom of God: A Theme in Geoteleology*. Toronto: University of Toronto Press.

Wallace, Alfred Russel (1863) On the physical geography of the Malay archipelago. *Journal of the Royal Geographical Society*, 33, 217–34.

Wallace, Alfred Russel (1880) *The Malay Archipelago*. London: Macmillan, 7th edn.

Warntz, William (1989) Newton, the Newtonians, and the *Geographia Generalis Varenii*. *Annals of the Association of American Geographers*, 79, 165–91.

Winlow, Heather (2001) Anthropometric cartography: constructing Scottish racial identity in the early twentieth century. *Journal of Historical Geography*, 27, 507–28.

Withers, Charles W. J. (1995) How Scotland came to know itself: geography, national identity and the making of a nation, 1680–1790. *Journal of Historical Geography*, 21, 371–97.

Withers, Charles W. J. (1996) Geography, science and national identity in early modern Britain: the case of Scotland and the work of Sir Robert Sibbald (1641–1722). *Annals of Science*, 53, 29–73.

Withers, Charles W. J. (1997) Geography, royalty and empire: Scotland and the making of Great Britain, 1603–1661. *Scottish Geographical Magazine*, 113, 22–32.

Withers, Charles W. J. (1999a) Reporting, mapping, trusting: practices of geographical knowledge in the late seventeenth century. *Isis*, 90, 497–521.

Withers, Charles W. J. (1999b) Towards a history of geography in the public sphere. *History of Science*, 37, 45–78.

Withers, Charles W. J. (1999c) Geography, Enlightenment, and the paradise question. In David N. Livingstone and Charles W. J. Withers, eds, *Geography and Enlightenment*. Chicago, IL: University of Chicago Press, 67–92.

Withers, Charles W. J. (2001) *Geography, Science and National Identity: Scotland since 1520*. Cambridge: Cambridge University Press.

Withers, Charles W. J. and Robert J. Mayhew (2002) Re-thinking 'disciplinary' history: geography in British universities, *c.* 1580–1887. *Transactions, Institute of British Geographers*, NS27, 1–19.

Woodbridge, William C. (1828) *Rudiments of geography on a new plan: designed to assist the memory by comparison and classification with numerous engravings of manners, customs, and curiosities: accompanied with an atlas exhibiting the prevailing religions, forms of government, degrees of civilization, and the comparative size of towns, rivers, and mountains.* London: Geo. B. Whittaker.

Wright, John Kirtland (1925) *The Geographical Lore of the Time of the Crusades: A Study in the History of Medieval Science and Tradition in Western Europe.* New York: American Geographical Society.

Young, Robert M. (1980) Natural theology, Victorian periodicals and the fragmentation of a common context. In Colin Chant and John Fauvel, eds, *Darwin to Einstein: Historical Studies on Science and Belief.* Harlow: Longman, 69–107.

2

The institutionalisation of geography as an academic discipline

RON JOHNSTON

Although, as David Livingstone has illustrated in the previous chapter, geography has a long history as an intellectual activity in the United Kingdom, with its subject matter being taught at the universities in the late sixteenth century (Baker, 1963; Cormack, 1997; Mayhew, 2000; Withers and Mayhew, 2002), it was only institutionalised as a separate academic discipline in the country's universities at the end of the nineteenth century.[1] Its growth was very slow during the first half of the twentieth century but then accelerated rapidly—despite some setbacks in the 1960s.[2] By the end of that century there were some 1600 staff in university and college departments offering geography programmes, with some 7000 places available on undergraduate degree programmes in geography each year, plus over 170 taught postgraduate programmes with nearly 3000 places available. In addition there were more than 800 full-time students registered for postgraduate research degrees, and a further 640 registered as part-time students.[3]

The process of institutionalisation involved the creation of a community of geographers within the British academic system, whose presence was accepted (and occasionally promoted) by the representatives of other disciplines. Academic disciplines, like territorial nation-states, emerge when

[1] On the teaching of geography in previous periods, see Mayhew (1998a, 1998b, 2000) and Withers (2002).

[2] Kirwan (1965, 373) records that 358 students were taking honours degrees in geography in 1938–9, 513 in 1951–2 and 826 in 1962–3.

[3] Most of these data are taken from Royal Geographical Society (2000). The data are approximations only: the staff numbers are all those listed; the undergraduate places are for honours degrees in geography (there are more than 2000 further places for named joint degrees including geography); the postgraduate places are for all degrees, whatever their titles. The numbers registered for postgraduate degrees are taken from the Higher Education Statistics' Agency website: http://www.hesa.ac.uk/holisdocs/pubinfo/student

their sovereignty over certain defined areas of scholarship is generally accepted, even if the boundaries of their separate identities become increasingly difficult to delineate (Haggett, 1989). The nature of each community's 'sovereignty' is characterised by its concepts of both subject content and teaching-research practice: disciplines are defined by what they study and how they do that (Granö, 1996). According to Geertz (1983, 157):

> . . . most effective academic communities are not that much larger than most peasant villages and just about as ingrown. . . From such units, intellectual communities if you will, convergent data can be gathered, for the relations among the inhabitants are typically not merely intellectual, but political, moral, and broadly personal (these days, increasingly, marital) as well. Laboratories and research institutes, scholarly societies, axial university departments, literary and artistic cliques, intellectual factions, all fit the same pattern: communities of multiply connected individuals in which something you find out about A tells you something about B as well, because, having known each other too long and too well, they are characters in one another's biographies.

Untangling such biographies, individual and collective, is the key to understanding geography's institutionalisation in the UK, indeed the institutionalisation of any academic discipline, during the twentieth century (on which, see also Withers, 2002). Here the focus is on the collective, while recognising that communities are usually only effective in such contexts because of the work of key individuals, many of whom are named in the following discussion.

Scholarly societies and the emergence of geography

The institutionalisation of a discipline usually involves interest groups promoting it within the university system and beyond, almost invariably pointing to the benefits that would accrue to society as a whole, and particularly to crucial sectors, as a result of the production of graduates trained in the discipline. In the UK, a central role in that promotion exercise during the late nineteenth century was undertaken by the Royal Geographical Society (RGS).

Founded in 1830 as the Geographical Society of London by individuals who had started meeting around 1825 as the Raleigh Club (renamed the Geographical Club in 1854: Marshall-Cornwall, 1976),[4] and receiving its royal charter in February 1859 (Mill, 1930, 72—the Prince of Wales was a strong supporter in the early years: see Freeman, 1980a), the RGS was created to advance geographical knowledge and understanding. An account of its early history is provided by Markham (1881, who also in

[4] The Travellers' Club was founded almost contemporaneously.

the 1860s wrote some pioneering papers on deforestation in India). Freeman (1980a, 5) expresses the RGS's early goals as 'to know the world and to map it'. Much of its early activity was linked to the knowledge-creation role of geography through promoting expeditions and reporting of their findings to a catholic audience with a wide range of intellectual interests. This role has continued to the present day: in the early twentieth century, for example, the Society sponsored Antarctic expeditions—and one member of Scott's 'successful' South Pole expedition (Frank Debenham, an Australian) was the first holder of the chair of geography at Cambridge;[5] in the mid-century, it sponsored the expedition that first conquered Everest; and in the last decades it supported a number of scientific expeditions to, for example, the Orinoco Basin, Borneo, the Arabian peninsula, and northwest Australia (Cameron, 1980; Richards and Wrigley, 1996, 45, list recent projects), as well as providing advisory and other services for a range of scientific and other expeditions.[6]

Apart from their strong interest to the expanding London professional classes, these expeditions provided much valuable information for traders, investors, and settlers—and geography became associated with the provision of an information base regarding what is where (as in some of the early texts on the subject, notably Chisholm's *Commercial Geography*: Wise, 1975; Barnes, 2001). The discipline was thus implicitly, if not explicitly, tied to British trade and imperialism—not only its materialist goals but also the task of creating a feeling of 'national superiority'. In common with other European countries at the time, science and imperialism 'marched arm-in-arm' during the nineteenth century, as 'supremely ambitious, universalising projects concerned to know all, to understand all and, by implication, to control all' (Bell, Butlin, and Heffernan, 1995, 3), with geography better illustrating this 'imperial science' than any other discipline and the RGS (by far the largest and wealthiest of the nineteenth-century national scientific societies) central to the weaving of geography into 'the fabric of state imperial power' (Bell et al., 1995, 8; see also Driver, 2001a; for a discussion of how this was promoted through school texts, see Ploszajska, 1999).

[5] Another member of that expedition was an Australian, Griffith Taylor, who was a postgraduate in geology at Cambridge 1907–10 and later professor of geography at Sydney, Chicago and Toronto Universities (Taylor, 1958). Like Kenneth Mason, the first occupant of the chair at Oxford, Debenham had no formal training or background in geography. Taylor, Debenham, and a third polar explorer, Sir Raymond Priestly (later Vice-Chancellor of the University of Birmingham and President of the RGS 1961–3), were all related by marriage.

[6] As part of this role of expanding knowledge the Society also housed, at various times, the Public (now British) Schools Exploration Society, the Institute of Navigation, and the Hakluyt Society (which focuses on the history of exploration and discovery). The first edition of the RGS's (still published) *Hints to Travellers* appeared in 1854.

Freeman's (1961) survey of a century of British geography identified 'colonialism' as one of six trends that dominated the discipline until the 1950s, with the encyclopedic trend (exploring and the recording of observations) providing material with which geography's practical value was advanced. Indeed, many of the RGS's officers and active members were also leading players in the country's imperial agenda (Lord Curzon, for example, was Viceroy and Governor-General of India 1898–1905, and President of the RGS 1911–14: Gilmour, 1994) and the Society had close links with the Foreign, Colonial, and India Offices for many decades, with the Treasury providing an annual grant to maintain the RGS map collection until 1998. Very early in its existence the Society also established a library, which is now the largest private collection of geography books and journals in the world.

Alongside knowledge-creation and its dissemination to the commercial and intellectual classes, the RGS also adopted an important role in its propagation, especially within the country's expanding educational system. (Universal education to age 10 was introduced in 1872 in Scotland and 1880 in England. Under an 1870 Act, local authority (board) schools were first established in England in areas where voluntary provision by churches was insufficient, and the principle of elementary education for all children was legislated for in 1876.) The RGS promoted geography as a fundamental school discipline, providing a crucial grounding for citizenship. Key to this was a report commissioned by the RGS in 1884 from its newly appointed Inspector of Geographical Education, J. Scott Keltie (Jay, 1986; Wise, 1986). He surveyed geography teaching across Great Britain and several European countries, and his report unfavourably contrasted the British situation with that he had observed and learned of elsewhere (Keltie, 1886). He found most British geography teaching (in the public though not the board schools) to be of poor quality, a problem he linked to negative attitudes reflected in the discipline's absence from the country's universities. He called for the RGS to 'supply the necessary impulse to induce the bodies that rule or direct the course of British education to take up geography in an intelligent spirit' (cited in Wise, 1986, 372). This was the basis for a 'crusade' that involved not just the approaches to Cambridge and Oxford discussed below but also a series of other 'events', including a travelling exhibition of maps, atlases, books, apparatus, and pupils' work. Apart from the RGS officers (such as Freshfield, who was a major driving force: Middleton, 1991), others joined this crusade, including H. J. Mackinder (see below), and to assist in the promotion of the discipline in schools the RGS began to award prizes for essays in the 1880s—having established a range of annual medals and other prizes in the 1830s.[7] And then in 1895, when the International Geographical Congress met in London for the first time, the RGS President (Markham)

[7] There are now 18 of these, including two gold medals, seven other medals, and nine further awards.

reported that 'the authorities of the Universities of Great Britain are not even aware that geography is a distinct branch of knowledge' and called on the delegates to pass the following resolution, which was carried unanimously (the citations are taken from Wise, 1986):

> That the attention of the International Congress having been drawn by the British members to the educational efforts being made by the British geographical societies, the Congress desires to express its hearty sympathy with such efforts, and to place on record its opinion that in every country provision should be made for higher education in geography.

Geographical societies were also established later in the century in several of the main provincial cities—in four Scottish cities (The Scottish Geographical Society: Lochhead 1981, 1984) and Manchester in 1884,[8] in Newcastle-upon-Tyne in 1887, Liverpool in 1891, and Southampton in 1897. The rush to create them then, according to MacKenzie (1995), was related to imperial concerns: the RGS was portrayed as 'neither utilitarian nor imperial enough' (p. 95), and the 'new municipal imperialism' was exemplified in the third of the Manchester Society's objectives (MacKenzie, 1995, 95–6; see also Freeman, 1980b):

> To examine the possibility of opening new markets to commerce and to collect information as to the number, character, needs, natural products and resources of such populations as have not yet been brought into relation with British commerce and industry.

The contemporary contingent conditions related to the exploration, colonisation, and exploitation of Africa and Asia, and the societies' *raisons d'être* were consequently short-lived:[9] although a few continue to operate in some form more than a century later, outside London only the Scottish Geographical Society became firmly established as a major scientific society.[10]

Winning an important role for geography in the country's schools called for a supply of teachers trained in the discipline, and so the RGS sought to establish geography in the country's major universities—notably Cambridge and Oxford, both of which were approached (through letters to their respective vice-chancellors) in July 1871 (Stoddart, 1986, 83: see also Balchin, 1988).[11] There had been provision for the examining of physical

[8] The first lecture to both of these societies was delivered by the African explorer H. M. Stanley (Freeman, 1984c).

[9] The Manchester society's foundation was the result of the second of two attempts to promote what was initially called a 'Society of Commercial Geography' by a group of businessmen led by the then Bishop of Salford (later Cardinal Vaughan: Brown, 1971).

[10] The Manchester society met part of the costs of the first lectureship in geography at the university there (Freeman, 1980a).

[11] Scargill (1976, 439) notes that by 1885 there were 45 professorships of geography in European universities, but not one in the UK.

geography within both classical studies and natural science at Oxford by 1850, but no teaching was provided. The RGS overtures in the 1870s, alongside some internal pressure, did not yield developments until 1887, following the post-Keltie-Report approaches, when joint discussions finally resulted in the establishment of a readership, with half the costs borne by the RGS. The appointee was H. J. Mackinder, who lectured on both physical and historical geography, with the latter especially popular (notably with history students). He was soon arguing for more extensive provision, but in 1892 was appointed principal of the new extension college (later to become a university college) at nearby Reading, which post he held while continuing to occupy the Oxford readership (Scargill, 1976, 1999). In 1897, when he was also lecturing at the London School of Economics (LSE), Mackinder and the Society explored the possibility of a full honours school being established in London (Cantor, 1962), but when the negotiations failed discussions were re-opened with Oxford. A department was established in 1899: Mackinder was head, along with an assistant (A. J. Herbertson—like H. Y. Oldham, who was appointed at Cambridge, he had previously taught at Manchester) and two part-time teachers. Its mission was to provide courses for students in other honours schools, and also to prepare postgraduates for a teaching diploma.[12]

By the time negotiations to renew the RGS contribution were opened in 1903, Mackinder had become director of the LSE, where he had been teaching economic geography since 1895; he resigned from his Reading post then, and from Oxford in 1905. Herbertson replaced him at Oxford (becoming a professor in 1910), with the RGS contributing half of the total departmental cost. The RGS contribution was successively reduced, but only finally ended in 1924. An honours school was not established until 1930, however, and a chair was filled in 1932 with the appointment of Kenneth Mason (1932–53: Goudie, 1998; the RGS had a member of the appointment committee: Steel, 1987b).[13] There, as at Cambridge, the RGS involvement was more than financial, and the Society's officers (notably Sir Clements Markham, President 1863–1905) served on the relevant boards of study (see Mill, 1930, 179).

Cambridge did not respond to the RGS's original letter, or to a follow-up memorandum in 1884. A further letter did provoke a response,

[12] There was geography at King's College London in the mid-nineteenth century. C. G. Micelay was a lecturer in the subject from 1854 to 1858, when he was succeeded by William Hughes, who was made a professor in 1863 (Vaughan, 1985). He was succeeded in 1876 by H. G. Sealy who, although styled Professor of Geography, was a geologist-palaeontologist; he died in 1896.

[13] Mason was an explorer-surveyor with no geographical training or academic background. Goudie (1998) reports that in 1931 Mason met the then secretary of the RGS at a conference in Stockholm, was told about the chair at Oxford, and applied. He was elected without interview (presumably on the basis of strong references), and served until 1953.

however, and in 1887 the Society offered to meet 75 per cent of the costs of a lectureship for five years, together with an exhibition and prizes. With strong support from the professor of Zoology, a formal connection was established and a lecturer appointed a year later, but neither he nor his replacement the following year successfully stimulated interest in the subject. The RGS extended its contribution in 1893, however, and H. Y. Oldham was appointed from Manchester. But geography did not appear in the university examinations until 1903, when a Board of Geographical Studies with an examination leading to an ordinary BA degree was established: the first diploma was awarded in 1907. Dissatisfaction (largely with the slow progress at Cambridge compared to Oxford) led to a new agreement in 1908, with RGS money supporting two lectureships—with the understanding that when Oldham left a third would be established to replace his readership—and more money was pledged in 1911. Almost 100 students were being taught by 1913–14, but only for a (postgraduate) diploma, considered sufficient for intending schoolteachers. In October 1918, it was agreed to establish a geographical tripos (an honours degree programme), which was approved in January 1919, along with a readership. The RGS financial contribution was scaled down, finally ending in 1923.

Whereas the investment at Cambridge was slow to bear fruit, that at Oxford yielded a great deal from the outset in terms of prestige and visibility for the subject—through the activities of both Mackinder and Herbertson. As well as his work promoting higher education generally, Mackinder argued strongly for the advancement of geographical education and made significant contributions to the research literature through his ideas on geopolitics—some of the discipline's most seminal work of the early twentieth century, which still resonated in the discipline and beyond decades later. On geographical education, Mackinder's (1887) paper 'On the scope and methods of geography' promoted the discipline as occupying a bridging position between the natural sciences and the humanities, enabling it to appreciate the characteristics of individual places through the study of the interactions of physical and human phenomena (see Coones, 1987). His seminal paper in geopolitics (Mackinder, 1904) and subsequent publications (notably Mackinder, 1919) were immensely influential in the developing imagery of the geography of world power, underpinning such geopolitical constructions up to and including the Cold War following the Second World War. Mackinder himself became a politician (an elected Conservative MP, 1910–22, having been a Liberal Imperialist previously) and then a diplomat—and continued writing until 1943. (For his life and career, see Gilbert, 1972; Parker, 1982; Blouet, 1987; Toal, 1992; Mayhew, 2000.)

Herbertson, too, was a widely recognised scholar as well as a tireless worker for, and promoter of, geographical education—as exemplified by his contributions to the work of the Geographical Association, to whose committee he was elected in 1897 (see below). He died before he was 50—from overwork, several have suggested—but by then had made a lasting contribution to the establishment of geography both at Oxford specifically and in the British education system more generally (Gilbert, 1965; Jay, 1979.)

Although geography was established at Cambridge and Oxford universities earlier than elsewhere,[14] the first full honours degree school was opened at Liverpool in 1917, when P. M. Roxby—an Oxford graduate and a lecturer in regional geography at Liverpool since 1906—was appointed foundation holder of the John Rankin Chair of Geography (named after the donor, a prominent Liverpool businessman: Steel, 1967). It was followed in the next year by the University College of Wales, Aberystwyth, where H. J. Fleure was appointed to the Gregynog Chair of Geography and Anthropology (Steel, 1987a). Aberystwyth offered geography degrees in both arts and science faculties, which became common in most universities later in the century. Geography had been taught there intermittently since the college was founded in 1872; from 1892 it operated a day training centre for teachers, including geographers, that was restructured into a department for the training of teachers in secondary schools, with a lecturer in geography appointed with funds provided by a former teacher of geography at the college (Bowen, Carter, and Taylor, 1968, xix). Meanwhile, in 1902 Fleure was appointed to what he referred to as an 'odd-job post' as assistant lecturer in zoology, botany, and geology (Bowen, 1987, 27). A lectureship in geography (partly funded in the first place by the RGS—as were appointments at Edinburgh and Manchester: Mill, 1930, 251)[15]—was created in 1906, and after it had been held by a practising schoolteacher for the first year, Fleure asked that he be appointed so that he could develop his interests in geography and anthropology. He became a professor of zoology in 1910, however, but continued his promotional work for geography, and when a legacy of £20,000 was received 'for educational purposes. . .to be directed towards those aspects of education that dealt with international affairs and understanding' (Bowen, 1987, 28), Fleure was able to negotiate his transfer to the Gregynog Chair of Geography at the same time that the honours school (agreed before receipt of the legacy) opened.[16] The first chair in the discipline in this era was

[14] The total cost to the RGS of its support to Oxford during the period 1888–1923 was £11,000, to Cambridge £7,500, and to other universities (Edinburgh, Manchester and Wales) £2,500 (Balchin, 1993, 23).

[15] The RGS had circulated all universities and comparable institutions in 1901 pressing the case for geography in their curriculum (Withers, 2002).

[16] Fleure was replaced in 1930 by C Daryll Forde, a geography graduate from University College London, much of whose subsequent work was in archaeology, ethnography, and anthropology.

occupied at University College London by L. W. Lyde (his title was Professor of Economic Geography) in 1903,[17] although an honours school was not established there until 1919.[18]

Liverpool was one of the five redbrick universities that had emerged from a range of origins to attain that status at the end of the nineteenth century in England's major industrial cities—the others were Birmingham, Leeds, Manchester, and Sheffield. Manchester was the first to offer geography courses at Owens College, from the geology and history departments, with the first explicit appointment being that of H. Y. Oldham in 1892 (Freeman, 1954). At Birmingham, too, the first teaching of geography was in the geology department, whereas at Sheffield teaching of the subject started with a special appointment (R. N. Rudmose Brown, in 1908: Freeman, 1984b): at a dinner in the city to mark the granting of the University's Charter, the President of the RGS (Sir George Goldie) strongly pressed the discipline's case, and a few days later a prominent businessman gave money to support a lectureship for five years (Slater, 1988). As Slater points out, although with hindsight the 'need' for geography may seem obvious in a country where issues of empire, trade, and education were interwoven, nevertheless perception of and meeting that 'need' was not 'pre-ordained': 'ultimately the expansion of academic geography depended upon innovative and enthusiastic geographers who believed in what they were doing' (Slater, 1988, 179).

By the end of the 1930s, degree programmes had been established in the great majority of universities:[19] nevertheless, according to Wooldridge (1956, 26–7), this was largely because

> . . . reluctant Courts, Senates, Vice-chancellors and Deans have been forced to make provision for University Geography . . . [and] we remain an idea or a group of ideas which is slowly forcing its way to recognition but which remains unpopular, or at least unappreciated, and therefore unprivileged.

[17] On Lyde, see the obituary by Daryll Forde and Alice Garnett reprinted in Dickinson (1976), and also the discussion in Clout (2003).

[18] There was an earlier chair at UCL in the 1830s, occupied between 1833 and 1836 by Captain Alexander Maconochie, who was secretary of the Raleigh Club and founding assistant secretary of the RGS (1830–6)—before resigning to become a colonial administrator in Australia (Ward, 1960). See also footnote 12 above.

[19] In addition to Oxford and Cambridge, they included Aberdeen, Edinburgh, Glasgow, and St. Andrews in Scotland, Queen's University in Belfast, Aberystwyth and Swansea in the University of Wales, Bedford, Birkbeck, King's, Queen Mary, and University Colleges in the University of London, plus the London School of Economics (Beaver, 1987), the Universities of Birmingham, Bristol, Durham, Leeds, Liverpool, Manchester, and Sheffield, and the University Colleges (which taught London degrees) at Exeter, Hull, Leicester, Nottingham (Edwards, 1987), Reading, and Southampton. A full listing up to 1961 is given in Fleure (1961; reprinted in Dickinson, 1976, 16–21); see also Balchin (1993). The current University of Newcastle was until the 1950s a constituent college of the University of Durham. For a later list, see Kirk (1978, 48).

> Ignorant or perverse individuals lurking in the backwaters of the educational stream still suffice to retard and obstruct locally. . . . The dissatisfaction that we feel arises not from the lack of ostensible recognition but from the widespread belief among our colleagues and associates that we lack real academic status and intellectual respectability.

The departments were invariably small, with fewer than five staff in most cases and only a score or so of students: many lacked professorial status for their departmental head.[20] The department at Bristol, for example, founded in 1920, had only three staff until 1934, and first exceeded five in 1948 (Peel, 1975). At Liverpool, the first graduating class in 1920 contained 4 students; the average number graduating in the 1920s was 7 per annum, followed by 10, 9 and 18 in subsequent decades, and then 50 in the 1960s (to 1967: data in Steel, 1967). Bristol's department averaged 35 graduates in the 1960s, 58 in the 1970s and 1980s, and 68 in the 1990s (to 1995: Haggett, 1995). In London, the geographers in adjacent institutions (King's College and LSE) pooled their teaching resources in a Joint School in 1922—25 years before a separate geography department was created at King's (Stamp and Wooldridge, 1951; Balchin, 1997). But in general, the discipline was small and, compared to the situation in several European countries, relatively weak. French and German geographers, for example, supported a much wider range of journals well before the 1930s and produced many major monographs; some of their English counterparts were stimulated by such materials and the benefits of visits there (as recorded by Robert Dickinson in his autobiographical essay and field diaries: Johnston, 2001, 2002). In addition, a doctorate or equivalent was essential for a chair in France and Germany, so that the discipline's leaders there (and those aspiring to such a position) had much stronger research records than their UK contemporaries.[21]

The main task for these departments was to train individuals who would become teachers in the country's grammar schools. Many graduates took a diploma or certificate in education after obtaining their degree: indeed, it was their willingness to do this that obtained scholarships for them at a time when university education was not free. Student:staff ratios were not large in many of those departments, but teaching loads were invariably heavy because of the wide range of courses that had to be covered. Although most

[20] Estyn Evans (1978) provides insights to the problems of both managing these small departments and winning credibility for the discipline.

[21] Indeed, this situation continued well into the 1960s. I recall my first interview for an academic post when I was close to completing my MA, in which there was a discussion of whether it was worthwhile to do a PhD (I didn't get the job!). Nevertheless, some institutions pressed for this research qualification much earlier: for an appointee at the LSE in 1951 completing his PhD was a condition to be met before he could take up the post. At that time, however, a PhD was necessary for those aspiring to a post (whether temporary or permanent) at a US university, providing a stimulus to the enterprise!

staff had specialist interests, therefore, their teaching expertise had to be much wider (and may have hardly impinged on those interests). Research had a relatively low priority and the pressure to publish was not strong—though many did, and a few were prolific, notably Dudley Stamp who conducted the first major exercise in collaborative research by geographers, his 1930s Land Utilisation Survey (Stamp, 1946). (This was later used as the exemplar for an international project—Wise, 1997—and it undoubtedly led to his appointment as deputy chairman of the wartime Scott Committee on Land Utilisation in Rural Areas: Willetts, 1987.) Stamp also produced a large number of textbooks, widely employed in both universities and schools, throughout the British Empire—notably India, where geography has always been a strong discipline.

Research students pursuing higher degrees were also few, with many undertaking their studies part-time; initially most of these were MAs, the PhD only being introduced in the 1920s, and before that the only route to a doctorate was a DSc for published work. The Liverpool department, for example, averaged only 3 MAs per annum before the Second World War; its first PhDs were awarded in 1938 (both to overseas students), and only 11 were awarded in the next three decades (several to members of the department's own staff: the first 'home student' was awarded a PhD in 1950: Steel, 1967). The LSE department's first PhD went to P. W. Bryan in 1924 (who taught for some years at University College, Leicester), followed by C. J. Robertson in 1929, with others in 1931, 1937, 1943, 1945, 1946 and 1947, plus a DSc (Econ.) to one of its staff members (Stamp and Wooldridge, 1951):[22] between 1945 and 1994 the number of students registered for higher degrees there increased 12-fold, compared with a 250 per cent increase in undergraduates (Wise and Estall, 1995). The Bristol department did not award a master's degree until 1938, and the total conferred in the following six decades was just 47. Its first PhD was awarded in 1945; 15 in total were awarded prior to 1970, 55 in the following decade, a further 53 in the 1980s, and 35 in the next six years (Haggett, 1995). Taught masters' degrees were rare until the 1990s—the main exceptions were a general course available at the LSE and a number of specific courses (e.g. the Geography of North America) at Birkbeck College from the 1960s. The first of the 'new style' masters', designed to provide an introductory training in research methods prior to a PhD, was launched at Bristol in 1993, with three graduates then and 11 in the following year (Haggett, 1995); the Economic and Social Research Council (ESRC) now recognises 34 such courses for the award of studentships. In total, as noted above, UK departments offer over 170 such degree courses, many of them in specialised areas (such as GIS

[22] All PhDs from the University of London are granted by the federal university, not the individual college.

and environmental management) oriented at specific careers as well as more general research training.

Whereas in some institutions (notably Cambridge and Oxford) geography was introduced following external pressure and at others the major stimulus was a donation (as at Liverpool and Aberystwyth), elsewhere the creation of geography courses (and later degrees) and departments of geography reflected internal pressures and the demand for certain types of course. In some cases, as at King's College London, geography courses originated in geology departments; in others, the demand was for instruction in what was generally known as 'commercial geography' (Butlin, 1999a, reports that Ll. Rodwell Jones was appointed to an assistant lectureship in the Department of Economics and Commerce at Leeds in 1912, with one of his duties being to teach two courses in 'railway geography' for the North Eastern Railway Company); and yet others emerged from specialist courses reflecting the particular interests of a local institution—such as the Hartley Institution in Southampton (later Hartley College, then Hartley University College: Wagstaff, 1996), where there were courses in navigation, surveying, and fieldwork (taught for a time by Vaughan Cornish: Dickinson, 1976, 143; Freeman, 1980b, 210).

Sustaining the universities: geography in the schools

The successful institutionalisation of geography departments in UK universities in the first half of the twentieth century depended very much on their mutual interests with the emerging grammar and secondary schools: the schools employed the majority of the university graduates as teachers,[23] and in turn produced the students who wished to read for geography degrees at university. Thus sustaining geography in the universities, especially in the first half of the century, and providing a base for its later emergence as a research discipline, was very much tied up with its sustenance in the school system. (For an overview, see Walford, 2001.)

Although the initial work in the promotion of geography in schools was undertaken by the RGS, by the end of the nineteenth century it had been joined by a specialist organisation founded for just that purpose by Mackinder, Freshfield and ten teachers in 1893 (Fleure, 1943).[24] Working alongside the RGS and other bodies, the Geographical Association (GA)

[23] Little geography was taught at the major 'public schools', through much of the century: public school masters had been heavily involved in the GA's early days, but interest in the subject later declined in these schools, and was not revived until late in the century.

[24] The new GA admitted women from 1894 on, a very different situation from that in the RGS, where a 20-year protracted debate only ended with the decision to admit them in 1913 (Bell and McEwan, 1996).

rapidly established itself as a successful promoter of geographical education, and it was soon influencing important issues such as the content of school syllabuses and the setting of public examinations (Balchin, 1993). Its membership grew rapidly from 35 in its first year to 121 at the century's end and 1144 at the onset of the First World War. By 1921 there were over 4000 members, but a decline set in from 1924. Expansion resumed after the Second World War, and by 1992 a peak of 10,391 was reached—reflecting the popularity of geography in schools (shown also by the number of entrants for the main public examinations at ages 16 and 18: Graves, 1980; Lee, 1985).

Although the prime focus of the GA's work was geography in schools, it embraced the entire educational system and was strongly supported by many university-employed geographers, for whom its journal *Geography* (founded in 1901 as *The Geographical Teacher* to provide practical help to teachers: Fleure, 1963; Wise, 1993; it was retitled in 1925) provided an outlet for their research papers. Indeed, for much of the century the Association was served in its major administrative offices by a sequence of distinguished academics, whose commitment to its cause was central to geography's success in various debates over school curricula (as in the later 1980s and early 1990s over the national curriculum: Walford and Williams, 1985; Bailey, 1992; see also Gardner and Craig, 2001).[25] Herbertson was Honorary Secretary from 1900 to 1915, for example, followed by Fleure (1917–46), Alice Garnett (1947–67), Stan Gregory (1968–73), Malcolm Lewis (1973–9), and Bryan Coates (1979–84).[26] Their work, and that of their predecessors, has sustained geography as a major discipline in the country's schools, with very significant and substantial consequences for it in the universities (Wise, 1993)—a situation very different from that in the United States, where geography is largely absent from high-school curricula and recruitment to university courses is almost wholly reliant on attracting students once they have enrolled at the institution.[27]

Academic geographers were also involved in keeping schoolteachers up to date with disciplinary developments—activities now known as 'continuous professional development'. Some of this involved working through GA

[25] This promotion of geography included the risky, though ultimately successful, invitation to the then Secretary of State for Education (Sir Keith Joseph, who had insisted on changing the name of the Social Science Research Council) to address a joint meeting of the country's geographical societies, in the light of his previous statements on the history curriculum (Joseph, 1985): seven years later, a successor Secretary of State addressed the RGS (Clarke, 1992).

[26] Garnett, Gregory, Lewis, and Coates were all on the staff at the University of Sheffield, where the GA's headquarters were located from 1950, having been in Manchester previously. In addition, both Dudley Stamp and Michael Wise (later presidents of the RGS) acted as honorary treasurers to the Association.

[27] In this, the UK experience is much more similar to that of many continental European countries.

branches; others through university extension programmes. Some of the most notable of the latter included two organised at the University of Cambridge's Madingley Hall by Richard Chorley, Peter Haggett, and others, which introduced the 'new geography' of the time to schoolteachers, brought together some who became pioneers in developing curricula and materials, and led to the production of two major books (Chorley and Haggett, 1965, 1967), the second of which comprised major syntheses of contemporary work and was widely used as a textbook over the next decade.

As well as the university departments and the schools, geographers were also involved in promoting their discipline through institutions established to prepare teachers. For those intending to teach in the state grammar schools (for which a degree in geography was an essential prerequisite), most of the universities had institutes of education providing one-year postgraduate certificates in education (amongst which that at the University of London—initially the London Day Training College; later the University of London Institute of Education—was one of the leaders, with one of its early staff-members, James Fairgrieve, being very influential, not least through his many, widely used textbooks: Honeybone, 1984).[28] Most of the local authorities with educational responsibilities (basically the counties and county boroughs) had their own teacher-training colleges with small numbers of geographers on their staff, as did the many training colleges run by religious organisations to prepare teachers for their own schools: these provided the non-graduate qualifications for those wishing to teach in primary and secondary modern schools. (Most closed in the 1960s–1980s, being incorporated into the education faculties of local polytechnics and, in some cases, universities.)

The work of the RGS and the GA, and their officers in particular, was key to ensuring geography's role within the school curriculum throughout the twentieth century. This task was also sustained by a largely hidden body operating within government offices themselves—both central and local. Central government, for example, had an extensive inspectorate covering all important disciplines, and the chief inspectors for geography were important advice-givers to government ministers. Similarly, most local education authorities employed subject advisers for various disciplines, and many had one for geography. Their work was crucial to maintaining geography's strong presence in the schools.

Learned societies, research, and publication

The institutionalisation of an academic discipline involves not only the creation of teaching departments but also an infrastructure for disciplinary

[28] The first teaching of geography to teachers is believed to have been at St John's College, Battersea in 1840 (Vaughan, 1985; see also Baker, 1963).

promotion—both to itself and to the wider world. This usually has three main components: a formal organisation to 'represent' the discipline; journals in which practitioners' work can be published; and meetings at which those active in the discipline could meet and discuss common concerns, including their research work.

For geographers in the first third of the twentieth century the formal organisation was the Royal Geographical Society, which published a journal, *The Geographical Journal* (first published in 1893; previously it had produced both a *Journal*, from 1830 to 1880, and its *Proceedings*, 1855–92), and held regular meetings.[29] Regional and local societies also held meetings and some produced journals, although only a few—notably the *Scottish Geographical Magazine* (now the *Scottish Geographical Journal*: Freeman, 1984a, 1984c; McKendrick, 1997)—were successful on the national stage.

During the 1920s, many academic geographers gathered regularly at the annual conferences of the GA and of the British Association for the Advancement of Science. (The BAAS was founded in 1831, with geography and geology included in Section C; Section E, for geography and anthropology, was founded in 1914, with the RGS providing the organisation: Mill, 1930, 65.) But the agenda of these meetings were broader than that of exploring the findings of recent academic research. One of the first occasions when many of the country's academic geographers were involved in such a collective project came with the 1928 International Geographical Congress, held in Cambridge (Stoddart, 1983), for which a number of them collaborated to produce a book of essays on the country's geography (Ogilvie, 1928). In addition, heads of university departments began meeting annually in 1927 (Garnett, 1983).

By the 1930s, however, some academic geographers were becoming increasingly disillusioned with the attitudes of those in power within the RGS, in particular their views on some areas of research, notably though not only in human geography. At that time, the pages of *The Geographical Journal* were dominated, according to Freeman (1987, 19), by material on 'exploration, cartography and, occasionally, geomorphology'. In 1928, representations were made to the Society (led by C. B. Fawcett, recently appointed to the chair at University College London (see Clout, 2003), and P. M. Roxby, who held the chair at Liverpool) that insufficient space was given to human and regional geography (which dominated teaching in the discipline), with a claim being lodged for at least one-quarter of the journal to be devoted to such themes. The RGS established a committee

[29] The RGS property at 1 Savile Row from 1871–1913 was too small for meetings, which were held at King's College and elsewhere; after the move to its current house at 1 Kensington Gore the Society had adequate space in its large lecture hall, plus room for its library and map collection (substantially extended with a large lottery grant in 2001–2).

to frame a response, but its deliberations were unsatisfying to the plaintiffs, who proceeded to establish the Institute of British Geographers (IBG: see below); the committee was closed in February 1934 (Freeman, 1980a, 34–7). To some, including a founder member of the IBG, the RGS showed an '*intolerant attitude*. . . towards the newer spirit in geography' (Garnett, 1983, 33; emphasis in original), although to others the major issue was the difficulty in publishing work, and some of those involved (such as Garnett and Steel) stress that there was no great animosity between the IBG and the RGS, whose roles were presented as complementary (Steel, 1983).[30]

The stimulus for the creation of a rival body to the RGS originated among the increasingly large cohort of geographers employed in the various colleges of the University of London (Buchanan, 1954).[31] After meetings of all University of London professors and readers, followed by parallel meetings of the Conference of Heads of Department and of all academic geographers (both during the BA centenary meeting being held in London in September 1931), it was determined to establish an alternative organisation—the Institute of British Geographers (IBG)—which would both publish original research work in its *Transactions* (originally only monographs[32]) and hold annual conferences and field meetings. The document drawn up by a committee for circulation prior to the Institute's inception set out two objects:[33]

1 to hold meetings—one for the reading and discussion of papers in January of each year; short visits to university departments and for fieldwork; and an annual meeting; and
2 to arrange for the independent publication of research.

There were 73 foundation members in 1933 (Steel, 1983), who held an annual conference at a different university each year (usually at New Year:

[30] Freeman (1980b, 2), however, states that 'to academics the RGS has often seemed obscurantist, traditional, establishment-minded, imperialist'—although he then 'defends' it as it 'has endeavoured to remain politically [by which he clearly means party politically] neutral'.

[31] R. O. Buchanan (a New Zealander who did a PhD at the LSE and was on the staff at University College London before being appointed to a chair at the LSE) records that he was an officer of the IBG for all but one of its first 21 years existence. With Buchanan, the others involved in the initial discussions were S. W. Wooldridge (King's College) and H. A. Matthews (Birkbeck College).

[32] Beaver (1983) recalls that several early submissions were rejected, including one by him.

[33] A copy is in Steel (1983). The committee comprised members from Cambridge (J. A. Steers), Edinburgh (A. G. Ogilvie), King's College London (S. W. Wooldridge), Liverpool (P. M. Roxby), Manchester (H. J. Fleure), Oxford (J. N. L. Baker), Sheffield (A. Garnett), and University College London (C. B. Fawcett). Four of these—Fleure, Ogilvie, Steers, and Wooldridge—served on the Institute's inaugural committee which organised the first meeting (at York, in September 1932), along with P. W. Bryan (Leicester), C. D. Forde (University College London), Ll. Rodwell Jones (LSE), and H. A. Matthews (Birkbeck College, London).

Buchanan, 1954, reports up to 70 per cent of members attending these meetings, which clearly indicates the attraction of the venture), and field meetings in 1936–8 and 1947–57 (three of the last being in Ireland).

Elected officers were drawn from departments across the country and meetings were held every year (with the Council meeting quarterly). But membership remained below 100 throughout the 1930s and only a small number of monographs was published.[34] (Given the absence of journals—in stark contrast to the French and German situations— this suggests that the demand for publication outlets, and the amount of research then being undertaken, was small.) Membership increased rapidly after the Second World War, reaching 300 by 1950 and 600 a decade later; the peak came at the end of the 1980s, when there were nearly 2000 members (including many students, almost all of them postgraduates). By the 1960s, the *Transactions* had become a regular journal, and a decade later a second journal—*Area*—was launched. For some academics this all came too late, however, and in the mid-1950s they established a new journal—*Geographical Studies*—which they managed and published privately. The *raison d'être* of the members of the 'third generation' of academic geographers involved, according to the founding editor (Fisher, 1954), was to provide a forum for 'discussion and clarification of the major issues with which all geographers the world over are faced today'. But, as Steel (1983) suggests, one of their main concerns was to get their research published (something that the IBG was perceived as not doing) and avoid the editorial processes that were dominated by the discipline's established professors.[35] Nine issues appeared over the five years 1954–9, containing 41 substantive papers, but the journal then folded because of 'the increasing success of geographers in securing publication space in journals associated with cognate disciplines' (Steel, 1983, 37).

It was another decade before a large number of new journals was launched by commercial publishers as research productivity increased exponentially within the discipline, with most of those journals aimed at either specific groups within disciplines (such as the *Journal of Historical Geography*, *Political Geography*, and *Earth Surface Processes and Landforms*) or at interdisciplinary markets (such as those for the emerging spatial

[34] Several departments launched monograph and other series in the 1960s–1970s, using relatively cheap publication methods.

[35] Many of the 'third generation' had served in the Second World War before obtaining their degrees, and thus felt that time was not on their side: they wanted access to publishing media in order to promote their careers. It could well have been, therefore, that this was the first generation of British geographers for whom obtaining a PhD was both the norm and the foundation for a career in which research and publication played major roles. George Dury, then at Birkbeck College and later professor of geography at the universities of Sydney and Wisconsin, Madison, was a driving force behind the launch of *Geographical Studies*; others involved included Harold Brookfield, who also moved to Australia in the 1950s.

analysis area: *Urban Studies*—the first such journal, launched in 1964, *Regional Studies*, and *Environment and Planning*—both launched in 1969, the latter being expanded into four separate series in the 1970s and 1980s: all have been edited by geographers over much of the last 40 years and have attracted many papers from human geographers). Perhaps the most notable within geography was the decision by publishers Edward Arnold to publish a series of annual volumes on *Progress in Geography* in 1969 (edited by Christopher Board, Richard Chorley, Peter Haggett, and David Stoddart, who had been together in Cambridge at the time of the Madingley lectures), and then to replace these in 1977 by a pair of journals—*Progress in Human Geography* and *Progress in Physical Geography*—committed to publishing up-to-date reviews and summaries of disciplinary developments. Further specialist journals were founded in the 1990s, such as *Ecumene; Ethics, Place and Environment*; *Gender, Place and Culture; Health and Place*, and *Space and Polity*. Unlike the situation in continental Europe, however, where many university departments published their own serials, few local journals were launched: only the *East Midland Geographer* (produced at the University of Nottingham) flourished (from 1954 until it ceased publication in 1999) and attempts to revive others—such as the *Manchester Geographer*—were short-lived. Similarly, student journals have not achieved the prominence attained in other countries and disciplines (though on one, see Philo, 1998).[36] Postgraduate students were, however, involved in many of the discussion paper series launched by departments in the 1960s and 1970s as a way of circulating papers cheaply and quickly.

Geographers, like academics in all other academic disciplines, relied upon commercial publishers for the other staple of their publication activity—the textbook. Over the century, they were served by a number of firms and individual publishers, some of whom (such as John Davey—first of Arnold and then of Blackwell; Iain Stevenson—successively at Longman, Macmillan, Belhaven, and Wiley; and Vanessa Lawrence—at Longman) were strongly committed to the discipline, playing a major role in its promotion through a variety of projects. (John Davey, for example, was honoured for his contributions to geography by the RGS; he launched *Progress in Geography* and its successor journals, and conceived the idea for the *Dictionary of Human Geography* (Johnston et al., 2000). Vanessa Lawrence, a geography graduate from the University of Sheffield, launched GeoInformation as an offshoot of Longman, to promote publication in GIS; she is now Director-General of the Ordnance Survey.) At different times within the twentieth century, different publishers were instrumental in promoting geography: Methuen produced a major series of regional texts

[36] There was a short-lived northern universities student journal in the 1950s–1960s, and a web-based journal (*GeoPraxis*) was established by postgraduates in the 1990s.

in the 1950s and 1960s, for example, as well as their 'University Paperback' series; Longman had an extensive geography list from the 1960s to the 1980s; David and Charles undertook innovative geography publishing for a brief period in the 1970s, including reprinting classic texts; (Edward) Arnold played a leading, innovative role from the 1970s to the 1990s; both Allen & Unwin and Blackwell had distinguished geography lists in the 1970s to 1990s; and Routledge (which incorporated a number of other publishers, such as Croom Helm) provided wide textbook coverage in the last decades. Numerous academic geographers worked with them as advisers and editors, as did others for atlas publishers in the century's earlier decades (see Chapter 11). And from the 1980s on, these publishers launched a growing number of journals, edited by academics.

The post-1950s decades also saw an expansion of the meetings component of the IBG's mission—although the field meetings were ended in 1962, and fewer excursions were provided as part of the programme for the annual conference in early January. Increasingly, staff and students wanted to present their research papers for discussion at the annual conference and parallel sessions were organised. And then, in the 1960s, Study Groups were launched within the Institute's remit, beginning with the British Geomorphological Research Group (BGRG) in 1961 and followed soon after by Population Geography, Urban Geography, Statistical Techniques (later Quantitative Methods) and Historical Geography groups.[37] These organised their own sessions at the annual conference and, increasingly, separate specialist conferences (both one-day and residential) at other times of the year, soon becoming the central elements in the Institute's research role. From their meetings and other activities a range of publications (including journals) emerged; through them, and its later practice of establishing limited-life working parties to address particular issues and topics, the Institute played a major part in promoting research within the discipline in the last half of the century—although one consequence of this (probably unavoidable, given the increased diversity of research interests) was the fragmentation of the discipline into semi-isolated sub-disciplinary communities (Johnston, 1989).[38] The Institute also played a major role in the promotion of international cooperation through seminars, notably with geographers from East European countries. (A list of these, up to 1983, is given in Steel, 1983: the longest-established series involved Poland, with six held between 1959 and 1977, but there were also seminars with

[37] Relationships between the BGRG and the IBG were often difficult, with the former—which was also affiliated to the Geological Society—on several occasions threatening to 'break away' from the IBG.

[38] In general terms, the Institute has played a stronger role in human than in physical geography, especially since the 1970s; some physical geographers—such as climatologists and remote sensers—set up groups outwith the IBG remit and there were occasionally difficult relationships between the IBG and the BGRG.

geographers from Bulgaria, China, France, Germany, Hungary, India, Romania, and Soviet Russia.)

One of the Study Groups formed in the early 1980s was that on Women and Geography. Its creation was not achieved without both opposition and some ridicule within the Institute, yet it was based on a firm case both that the discipline was male-dominated (despite attracting many female undergraduate students) and that its scholarly preoccupations displayed a strong masculinist bias (as expressed by Rose, 1993). Up until then, the IBG had had only had one female president—Alice Garnett in 1966; and indeed it was not to have another before the merger with the RGS, after which Judith Rees was elected as chair of the Research Division (the equivalent of the previous presidency). There is no doubt that much has changed in the intervening two decades to counter the bias within the discipline (recorded in Johnston and Brack, 1983) and to bring feminist concerns to centre-stage within the discipline (as illustrated by Women and Geography Study Group, 1984, 1997: see also Chapter 19 below)—agenda items still far from completed.

In the early 1990s, strong moves were initiated to merge the RGS and the IBG. The two had cooperated for some decades—the IBG's administrative office was located in the RGS building, for example—and both were involved, along with the GA and other societies, in the creation of COBRIG (the Committee for British Geography), which eventually took over the international function formerly operated by the British National Committee for Geography.[39] The arguments for the merger were that a single organisation with a combined administration would be better placed to promote geography, through its resources (the RGS has some 15,000 fellows and a major building in London): a large, full-time staff could better develop contacts between the discipline and the public and private sectors (the IBG had only ever employed one administrator, and various moves to appoint an academic director had been defeated, the latest in 1978), and act as a 'voice for geography', including attracting major sponsorship for research and other activities. The merger was approved by substantial majorities of both organisations' memberships—though not without serious qualms within some areas of the former IBG regarding sponsorship and the likely attitudes of the new organisation to certain types of research.[40] The two were reconstituted as the Royal Geographical Society (with the Institute of British Geographers) in 1994, and in 1996 for

[39] The British National Committee for Geography was its formal contact with the International Council of Scientific Unions, and organised through the Royal Society until it was closed in the late 1980s (on which, see Wise's 1977 obituary of an executive secretary of the Royal Society). The IBG had one member on that committee from 1946 to 1952, when a second place was reserved for the president, ex officio (Buchanan, 1954).

[40] These qualms surfaced soon after the merger, with a debate at the 1996 annual conference (and thereafter) regarding sponsorship from Shell, in the context of the issues of environmental despoliation and civil rights in the Ogoni region of Nigeria (see Gilbert, 1999).

the first time an academic geographer—Rita Gardner—was appointed as the director. The former IBG, with its study groups,[41] now operates as the research arm of the merged society and continues to hold its annual conference each January. Meanwhile, the society has developed its promotional functions for geography as a teaching discipline, being contracted, for example, to produce the benchmarking document against which teaching in all university departments will be assessed by the country's Quality Assurance Authority (2000; Chalkley and Craig, 2001), and through the establishment of a nationally funded subject centre for geography (Gardner, 2000). It also functions as a source of geographical knowledge and expertise to be used beyond the academy—as exemplified by its popular journal *The Geographical Magazine*.

Constrained expansion: the 1960s on

As already noted, by 1939 there were departments of geography in most UK universities, and these began to grow rapidly from the 1950s on in response to growing student demand. This expansion was fuelled in part by a continued expansion in the demand for schoolteachers but also by the wider range of careers increasingly open to geographers, as in the new profession of town and country planning, following the 1947 legislation requiring all UK local authorities to develop plans for their areas (on which see Chapter 16). And when the first new university was established after the Second World War (at Keele, in 1950, initially as the University College of North Staffordshire), geography was one of the foundation subjects and departments (in an institution with a 'multidisciplinary mission').

Almost all of these departments grew substantially over the succeeding decades, as access to higher education was broadened and student demand for geography remained buoyant—increasingly, it seemed, because it offered both a general education and valuable transferable skills for a wide range of careers. Staff numbers increased—at Southampton, for example, from 4 in academic year 1950–1 to 12 in 1969–70, 16 in 1981–2, 19 in 1995–6 (Wagstaff, 1996) and 24 in 2001; the LSE department has very similar figures (Wise and Estall, 1995). At the century's end, the largest departments included those at Durham (41 staff), University College London (35), and the University of Cambridge (31, excluding a further 21 college posts: data from Royal Geographical Society, 2000). The volume of published research, if anything, increased at an even greater rate.

A survey of all geography departments entered for the 1989 Research Assessment Exercise (RAE, see below) recorded 603 academic staff

[41] There are now 22 such groups, which are listed on the RGS website: http://www.rgs.org.uk

working in the 36 departments (all of them in the 'old universities', excluding the polytechnics and other institutions that obtained university status in 1992), of whom over 40 per cent were born in the 1940s. (All the data reported here are taken from Johnston et al., 1990.) The departments ranged in size from 5 to 27 academic staff in 1988, with a mean just under 17 and a standard deviation of just over 5. In addition, the average department had 2.4 research staff and 2.7 research assistants, though the range in the former was from 0 to 24 and in the latter 0–12. The average student load per department comprised 171 undergraduates, plus 4 doing taught postgraduate degrees, and 16 doing research postgraduate degrees—though again there was a great deal of variation (undergraduate load ranging from 28 to 283, postgraduate taught load from 0 to 37, and research postgraduate load from 2 to 45). The average student:staff ratio was 11.3 (with a standard deviation of 2.0), but this increased massively over the next few years as student numbers grew and the amount of money available to universities per student fell substantially (Jenkins and Smith, 1993). Regarding publications, the survey recorded 962 articles published in refereed journals by the academic staff members in 1988, plus 429 chapters in edited collections and 158 individually or co-authored books—an average publication rate of 2.5 items per academic staff member per department in a single year.

The peopling of these departments involved substantial inter-university flows. These included British graduates returning from either or both of postgraduate training and university posts in North America and Commonwealth countries in Africa, Asia and Australasia. (In the 1950s and 1950s, when job opportunities in the UK were relatively few—at all levels—many graduates obtained positions overseas, being very influential in peopling the discipline in such countries and creating an 'Anglo-American' geography.) This was very different from the situation in many other European countries, where it is quite usual for departments to appoint their own graduates to their academic staffs (although a survey in the early 1980s showed that over the 50 years 1933–82 some 22 per cent of the 758 individuals who obtained a post in a British university department of geography did so in the university from which they obtained their first degree, and 26 per cent in the university from which they obtained their doctorate: Johnston and Brack, 1983.)[42] Among the various UK departments, a small number—notably those in Cambridge, Liverpool, London (all colleges), Oxford, and Wales—provided a substantial proportion of all of the staff recruited. Furthermore, of the 145 appointments to chairs during that period (of which 67 involved internal promotions), fully 63 involved individuals from just seven institutions (Bristol, Cambridge,

[42] This survey excluded the University of London, because data were not available on which college the individuals studied at: all degrees are awarded by the federal university.

Durham, Liverpool, LSE, Sheffield, and University College London).[43] Whether this concentration of appointments from certain departments strongly influenced the nature of geography as practised elsewhere is unclear—though it was suggested that the major changes in the discipline in the 1960s and 1970s were strongly focused on a 'Cambridge to Bristol axis' (Whitehand, 1970).[44]

In addition, up until the 1980s British universities provided many of the staff (including in many cases the foundation professors) for geography departments throughout the British Commonwealth (on which see Farmer, 1983: geography is much stronger as an academic discipline in the universities of the Commonwealth than most other parts of the world), as well as for numerous US universities (where a number of UK geographers received their postgraduate training in the 1950s to 1970s, before this became common at British universities, and others made sabbatical and other visits—notably to teach on the well-established North American summer schools). But contemporaneously with this, the contact between British geographers and their European counterparts declined—save with the Dutch, Swedish and, to a lesser extent, those in the other Scandinavian countries, who tended to publish in English. Within geography globally, an Anglo-American human geography emerged after the Second World War (Johnston, 1997; Sidaway and Samers, 2000), characterised not only by its particular paradigms but also its various organisational and institutional structures.

In contrast with the continued success and growth of geography departments in the established universities, however, the early 1960s saw the discipline fail to make its mark in the new universities founded then to cope with the growth in student numbers.[45] None of these planned for geography to be one of their foundation subjects, though it was rapidly introduced at Sussex (focused on a Geographical Laboratory whose staff were affiliated to a range of the inter-disciplinary schools), and only one other ever opened a fully fledged geography department (at Lancaster in the mid-1970s, acting on advice from assessors recommend

[43] Note that although Oxford produced a large number of staff for other universities, there were few professors among them.

[44] The bipolar pattern reflects the appointment of Peter Haggett (with Dick Chorley a major figure in the developments at Cambridge in the early 1960s) to a chair at Bristol in 1965 and the rapid growth there of a team working around him: he obtained the first large research grant awarded to a geographer by the SSRC. In previous decades, the influence of key individuals in placing their 'pupils' in posts around the country was very great: Prince (2000), for example, shows that 35 of Clifford Darby's former students held university posts in 1971, and 40 in 1991.

[45] They were at East Anglia (Norwich), Essex, Kent, Lancaster, Sussex (Brighton), Stirling, Warwick, Ulster (Coleraine), and York. Geography was also largely absent from the technological universities founded then, at Aston, Bath, Bradford, Cardiff, Heriot-Watt (Edinburgh), Salford, Surrey, and Strathclyde (Glasgow); of these, only Salford and Strathclyde had fully fledged geography departments.

by the University Grants Committee that establishing such a department could help overcome its student recruitment problems: McClintock, 1974).[46] Geography's absence from those universities was noted at the IBG's annual conference in 1960, and the issue was taken up by the RGS, which prepared a document promoting geography and offered to meet with the relevant officials at the new universities to discuss the issue; few availed themselves of the offer and the RGS did not repeat its success of the late nineteenth century in securing a foothold for geography in those new universities (Johnston, 2003). There was, however, a geographical presence in several of the new institutions that created Schools of Environmental Science (at East Anglia, Lancaster, Stirling, and Ulster—and in all cases, the founding professors were, or included, physical geographers) but not separate geography departments. And partly countering this geographical absence from many of the new institutions, there was a strong presence in the Open University, founded in 1967.

This absence of geography departments from such new institutions, many of which rapidly became major research centres, reflects one of geography's problems then. Although successful in attracting students in the established universities, it was clearly not seen as central to some of the burgeoning new areas of intellectual activity of the time—even though some of those universities (such as Essex) built much of their teaching in the humanities and social sciences around 'area studies', for which it would seem geography was well-qualified. Indeed, it was in comparison with the social sciences and the environmental sciences that geography apparently failed.[47] (Regarding environmental sciences, for example, many university geography departments were in the science as well as the arts—and later social science—faculties, offering BSc as well as BA degrees. But they were not as well funded as other physical science disciplines, and there was some antipathy and/or competition between geography and geology departments. Like geography, however, geology failed to gain a foothold as a separate discipline in the new universities of the 1960s and later.) With the exception of economics, most of the social sciences were relatively new disciplines, and were not as institutionalised in British academic life as geography. There were only two sociology departments in British universities in the mid-1950s (at Leicester and the LSE), and of the 54 academic posts listed at 16 universities in 1960, 15 (more than a quarter) were at one place—the LSE.[48] And the British Sociological Association was only

[46] The assessors included Keith Clayton, a Sheffield geography graduate who was on the staff at the LSE before being appointed a foundation Professor of Environmental Sciences at the University of East Anglia.

[47] Several of the pieces reprinted in Wooldridge (1956) refer to this negative attitude towards geography.

[48] I am grateful to Professor Jennifer Platt, who is writing a history of the British Sociological Association, for this information.

founded in 1951—one year after the Political Studies Association. But these disciplines prospered, taking root in all the new universities (by 1972, for example, there were 384 sociologists holding academic posts); they took little account of geography (as shown by Anderson's (1968), detailed analysis of British social science, which covered 16 separate disciplines—but included not a word on geography).[49] There had been, however, early links between geographers and sociologists through the Le Play Society, which organised 71 major field surveys between 1931 and 1960: its secretary for a period was a geographer, Stanley Beaver (1962), and its existence stimulated K. C. Edwards (founder of the geography department at the University of Nottingham) to establish a Geographical Field Group for graduate students and younger academics.

The reasons for geography's absence from the new universities—and hence association with some of what became the country's leading social science centres—are not readily discerned. What appears to be the case, however, is that geography had something of a 'dated' look about it at the time (having by the early 1960s not yet embraced the theoretical and quantitative revolutions that were sweeping the social sciences into favour, not least with policy-makers; geographers were still seen as the 'data gatherers': see Chisholm, 1971a)—a view that was not helped by a scornful critique of disciplinary practices (by a Manchester geography graduate: David, 1958) in a journal aimed at university administrators, to which geographers themselves provided no very convincing response (Fox and Lewthwaite, 1959; Parker, 1959). Furthermore, geography appeared to lack any promoters among the 'great and the good' who were very influential on the nature of the 'new universities'—each of which had an initial steering group to assist the lay and academic officers in determining the institution's profile and mission.[50] (At Warwick, for example, much of the groundwork leading to the university's establishment was done by a geography lecturer at the local technical college, Henry Rees, but he was eventually disappointed that his discipline was not included in the foundation subjects: Rees, 1989.[51]) And even though

[49] Several geographers switched to sociology in the 1960s and 1970s, and occupied chairs in the new departments, such as Ray Pahl at Kent and Duncan Timms at Stirling: the first geographer to become a professor of sociology was Bill Williams, trained at Aberystwyth and for a time on the staff at Keele before joining the sociology department at Swansea.

[50] By way of contrast, one of the country's leading anthropologists, Sir Raymond Firth, was involved in the establishment of the Research School of Pacific Studies at the Australian National University and strongly argued for geography's presence. Following his success, his then LSE colleague, Oskar Spate, was appointed to the foundation chair (Ward, 2001). Sir Raymond worked—'as a geographer' according to his *Times* (26 February 2002) obituarist—with Darby on the Naval Intelligence handbooks during the Second World War, and knew Dudley Stamp well at the LSE.

[51] Rees was an economic geographer with a PhD, supervised by Stanley Beaver, from the LSE.

some—such as Alfred Steers at Cambridge (Stoddart, 1987)—were very active in applied work, they do not appear to have promoted their discipline's cause at this time.[52]

This absence of geography from those new institutions was countered not only by its continued health and expansion in the older-established universities but also its presence in several of the polytechnics created in the late 1960s out of former technical colleges and associated institutions, including some teacher training colleges. The polytechnics became universities in 1992, and 16 of those institutions now have geography departments, with a geography presence in 13 others as well as in 14 degree-awarding Colleges of Higher Education (Royal Geographical Society, 2000; see also Wright and Jones, 1972).[53] These departments, whose degree provision was overseen by a central body (the Council for National Academic Awards) on which members of university departments were influential, provided alternative opportunities for development of the discipline in institutions whose mission was expressly applied and technological.[54] All of these institutions became universities after 1992, and some of their geography departments became substantial components of the discipline's teaching and research portfolio, with a few—such as Coventry University's undergraduate degree programme, which incorporated a 'sandwich' year of work experience—adding new elements to the provision.

That geography's absence from the 1960s 'new universities' was in some substantial way linked to its lack of recognition as a social science is also reflected in its experience with regard to the establishment of a Social Science Research Council during the same decade. In a research-driven academic community, a crucial element of infrastructural support comes in funding for projects and the training of postgraduates. Large projects were difficult to launch and sustain until the 1960s (see Dickinson, 1964), and the few that were successful—notably Dudley Stamp's Land Utilisation Survey of the 1930s and Alice Coleman's partial replication in the 1960s and 1970s—were very dependent on volunteer labour (in his case, schoolteachers and students of all ages: Stamp, 1946.) By the end of the 1960s, however, UK universities were funded by what was known as the dual-support system,

[52] In the early 1960s, the University of Reading appointed a geologist to its chair of geography, which drew concerned representations from the IBG! (A later attempt to appoint a geologist to another chair of geography there was not successful.) Two senior geographers, however (Charles Fisher and Michael Wise), argued with success that the University of Leeds should not close its School of Geography, following personality difficulties with its head, R. E. Dickinson (see Johnston, 2002).

[53] On one of those polytechnic departments, see Jenkins and Ward (2001). For many years, the GA had separate sections for teachers in public and preparatory schools, in primary schools, and in teacher training colleges/university departments of education.

[54] For geography, however, most of their degree schemes were very similar to those offered in the universities—probably as a response to the senior geographical adviser on the CNAA, Stanley Beaver.

whereby some of the government grant (an unspecified proportion) was for research support and research councils were established to provide grants for particular projects plus scholarships for postgraduates (previously, geography postgraduate funding had come directly from state scholarships in the arts and the sciences). Two research councils relevant to geographers' interests were established—the Natural Environment Research Council (NERC) and the Social Science Research Council (SSRC; later renamed the Economic and Social Research Council, ESRC). Geographical representation in the former was established relatively early—largely due to the efforts of a few individuals, such as David Linton (who was involved with its predecessor body, the Department of Scientific and Industrial Research, which allocated some research studentships for physical geographers) and Keith Clayton; Kenneth Hare, then Professor of Geography at King's College London (a climatologist) was one of the founding members.

When the SSRC was established, geography was excluded because of its 'bridging position' between the social and natural environmental sciences (Kirwan, 1965). At a time when the social sciences were booming, geography was seen not to be one of them (because of a perceived 'peculiar status', which it apparently shared with planning and social/economic history)—despite geographers being involved in the consultations undertaken at a range of universities prior to production of the Heyworth Report, which led to the SSRC's establishment (Heyworth, 1965: see also King, 1997, 1998). A campaign to change this was led by a small number of human geographers operating outside the learned societies, with the key documents being prepared by Michael Chisholm. A case (not supported by all university department heads) to include geography was made and accepted, though a full geography committee was never created; geography and planning were grouped together for the allocation of research grants, though each had its own allocation of research studentships (see Chisholm, 2001). Geography was thus late at the social sciences table in the country's research as well as its teaching infrastructure.

Until the restructuring of the SSRC/ESRC in the 1980s, human geography was separately recognised as a social science discipline, through its own committee for research studentship allocation and, with planning, access to research funds. From then on, there have been no separate disciplinary committees but many of the ESRC's operations reflect disciplinary divisions; geography receives an annual allocation of research studentships, for example, and at least one geographer served on each of the main awarding bodies in the 1980s and 1990s—the Research Grants Board and the Postgraduate Training Board (which Alan Hay chaired in the late 1980s). And geography departments have housed major ESRC centres, such as the Centre for Urban and Regional Development Studies at the University of Newcastle and a number of the regional centres set up under its GIS initiative in the late 1980s (see Chapters 12 and 13), some of them later linked to major computing centres—though geographers have not been as successful in establishing such centres as academics from a number of other social science disciplines.

This is a much more secure position than has been the case for physical geographers, who have never achieved separate disciplinary recognition in the NERC's structures; they have had to compete with other scientists for research grants and studentships within broad interdisciplinary remits,[55] and membership of NERC committees has been determined on the basis of individuals' scholarly standing, not as representatives of their disciplines. NERC has made substantial grants to geographers and geography departments, however, and supports a number of research centres and institutes—including the Environmental Systems Science Centre at the University of Reading, which was initially located within the geography department.[56]

This situation for physical geographers is better than that experienced until the end of the twentieth century by those human geographers whose main interests lie closer to the humanities than to the social sciences. No research council was established for the humanities, and responsibility for the allocation of the relatively small amount of available public funding (largely to support research studentships) was devolved to the British Academy—where there was no geography presence until 1967, when Sir Clifford Darby was elected, and only a relatively small one thereafter (Terry Coppock was elected in 1974, Jean Gottmann in 1977, and Sir Peter Hall in 1983: see below). Increasing pressure—allied with the changing nature of some research activities within the humanities—led to claims for a research council, and an Arts and Humanities Research Board (AHRB) was established in 1999, taking over some of the British Academy's former functions (Driver, 2001b). Geographers have been increasingly successful in obtaining British Academy awards, with 11 being awarded post-doctoral fellowships since 1985. The AHRB offers human geographers a further source of funding, which is now being realised, but no separate disciplinary presence.[57]

Despite these drawbacks, geography continued to prosper in most of the universities where it was well established and, with few exceptions, departments grew on the basis of buoyant undergraduate recruitment. Its

[55] There are currently nine divisions of its Science and Technology Board, seven of which have a peer review/expert group; three (earth sciences, freshwater sciences—which includes hydrology, and earth observation science) have members of geography departments, with that on freshwater sciences chaired in 2000 by a geographer—Angela Gurnell of the University of Birmingham. The other divisions are atmospheric science, marine science, polar science, science-based archaeology, terrestrial science, and interdisciplinary science.

[56] See Thrift and Walling (2000) for a listing of the many collaborative projects in which physical geographers are involved.

[57] Human geographers, of course, are able to tap other sources of research funds from a range of charities, including the largest—such as the Leverhulme, Nuffield, and Wellcome: Peter Haggett has been a governor of the Wellcome Trust.

status as a research discipline was high too, with an ESRC (1988, 2) report on the health of social science claiming that:

> British human geography is internationally of the first rank. The importance of British work was stressed by US informants, who stated that more innovative work was underway in Britain than in their own country. A large proportion of active US researchers were either of British origin or were inspired by UK work, particularly that of the younger British geographers of the last 10 years. British work was the inspiration for much US research on mobility, migration and the re-structuring of industry. UK informants also shared this view of UK human geography as intellectually thriving.

(For some of the UK evidence presented to this review, see Hall et al., 1987.)

Geographers reflecting on geography

One clear message from the preceding sections is that the last third of the twentieth century saw UK geography departments and geographers operating in very different contexts from those of earlier decades, contexts that have required positive responses if the discipline was to be sustained and enhanced. Some clear insights to these stimuli and responses are provided in a series of reviews commissioned since 1956 by the British National Committee for Geography (and its successors) for presentation to the quadrennial International Geographical Congresses (IGC): each has been written by two people comprising, after the 1972 report, one each from the fields of human and physical geography. Since 1960 the reviews have been published in *The Geographical Journal*, and copies presented to all IGC participants.[58]

Comparison of these documents identifies both geographers' changing intellectual concerns (issues that are the subject of other chapters in this book) and their reactions to the institutional setting. In the two early reviews, Edwards and Crone (1960) and Crone (1964) both reported on the discipline's growth within UK academia; in the first of the quadrennial series, Steel and Watson (1972) devoted three pages to geography's presence, and growth within, UK higher education, noting the growth in applications for degree places, the creation of geography programmes at the newly created polytechnics and other institutions, and the nature of the curricula offered.[59] Steel and Watson (1972, 148) also noted that individual geographers had been moving into senior positions in universities and government

[58] The 1956 report, by J. A. Steers, was not published. For 1964, when the IGC was held in London, there were three reviews: of geography (Crone, 1964), the RGS and exploration (Kirwan, 1964), and maps and charts (Harris et al., 1964). No review was commissioned for the 1968 IGC (at New Delhi).

[59] Interestingly, the absence of geography from the 1960s 'new universities'—see above—is covered by the somewhat misleading statement that 'new departments have sprung up in the majority (though not yet in all) of the new universities' (p. 151) and most of their discussion is based on returns to a questionnaire survey of departments offering geography degrees.

departments, so that 'geography has come into its own in its power to shape decisions at the highest level'. They were entirely positive and optimistic about geography's position in British universities, with bigger departments, more staff and a wide range of courses. This optimism is implicit in the 1976 review, in which Cooke and Robson (1976) devoted virtually all of their attention to intellectual concerns, with just two paragraphs on geography in schools and nothing on higher-education funding. Four years later, Doornkamp and Warren (1980) made passing comments in their first paragraph on 'straitened financial circumstances', difficulties in funding research students, and teaching skills, but then concentrated on academic inquiry. By 1984, however, there was a 'sea change'. Munton and Goudie (1984) devoted the first five pages of their review to institutional issues, noting that the subject's growth in the then universities had ended in 1978, that there had been a substantial drop in ESRC-funded postgraduates in human geography (from 90 to 39 over a four-year period), and that the discipline's popularity in the country's schools had started to decline.

This disquiet was repeated in later reviews. Bennett and Thornes (1988) also gave several pages to discussion of the changing context, emphasising the increased selectivity of government funding for research and the growing pressure for applied research. Gardner and Hay (1992) continued these themes of 'turbulent years' and 'major upheaval', noting the increased squeeze on public funding (notably less income per student leading to higher student:staff ratios) and the important role that the GA was playing in pressing geography's case in the new National Curriculum for schools. The following report (Richards and Wrigley, 1996) was the first to be produced after the massive growth in the universities following government decisions on expansion at the beginning of the 1990s. They noted that this growth (of 30 per cent in student numbers in the 'old' universities over a four-year period) was being undertaken with declining resources, so that student:staff ratios increased by over 20 per cent during the same period. Nevertheless, this expansion stimulated a reinvigoration of many university departments, with some experiencing 25 per cent increases in their staff complement; there was also growth in funding for research and postgraduates. But the expansion and resource squeeze were accompanied by more external review and audit of both research and teaching. And then Thrift and Walling (2000) ended the sequence by identifying five main phenomena in the discipline's intellectual environment:

1 insufficient money to sustain world-class research and teaching;
2 'audit fever';
3 labour-market uncertainty for graduates;
4 the growing influence of the RGS in disciplinary promotion; and
5 the increased attention to teaching, at all levels of the educational system.

Despite the reviewers' growing concerns about context and resources for geographers and geographical research, all present upbeat evaluations of the contributions being made to knowledge. Cooke and Robson (1976) decided that their period was best characterised by 'consolidation and reflection', but four years later Doornkamp and Warren (1980, 95) could refer to the discipline being 'extremely fruitful in new ideas, both from translated and from native stock'. Bennett and Thornes (1988) saw geography emerging from the turbulent years of the early to mid-1980s with 'increased vigour' and 'considerable enhancement of its reputation'. Although slightly more downbeat, Gardner and Hay (1992, 13) reported even more turbulence and pressures, but 'so far, the discipline appears to have withstood the variety of pressures and remains able to teach and research effectively'. Richards and Wrigley (1996) defined the mid-1990s as a 'maelstrom of change', during which physical geographers made 'important advances' and human geographers were experiencing 'a period of enormous theoretical vitality and excitement'. Four years later, Thrift and Walling (2000, 96) celebrated the century's end by proclaiming that:

> Geography is not just distinguished by its size, however. Increasingly it is also becoming a greater intellectual success. . .This report is optimistic but it is not triumphalist. British geography is clearly gaining greater respect from other disciplines in Britain and it is making important interventions in various kinds of policy. Moreover, it now has an important and growing international dimension. . . [but] there are still significant barriers to progress.

Out of adversity has come success, it would seem!

Geography at century's end

By the end of the twentieth century, geography was firmly established within the UK's universities and had a strong disciplinary infrastructure of its own. The discipline is taught to a large number of school students and attracts many applicants to universities, although the pressures of changing funding regimes and curricular changes in schools have negatively influenced some of the smaller university departments, notably though not only in the 'post-1992 universities', and there are growing concerns regarding the future role and strength of geography in the country's secondary schools.[60]

One major area of change in the discipline at century's end was in its research orientation. Until the 1980s, when resources were available most geography departments expanded their staff numbers by determining the areas of expertise of those appointed with particular regard to their teaching programmes—the dominant goal was disciplinary coverage. But by the end of that decade, the funding regime had been changed with the introduction

[60] Interestingly, geography is now strong in some of the country's largest 'public' schools.

of the regular Research Assessment Exercises (RAEs), which graded all of the departments for which submissions were made according to the perceived quality of their research,[61] and then funded their universities accordingly (on which see Johnston, 1995a, 1995b). Clayton (1985a)—a member of the University Grants Committee just before it introduced the RAEs—had already advised geographers that their departments would have 'to do less to do anything better', and increasingly this message was taken to heart.[62] Most departments developed research strategies that involved focusing on a few major clusters, which would encourage and sustain research productivity. In part, some were able to do this by competing successfully for 'new blood' appointments provided by the UGC (in part to counter some of the losses resulting from previous early retirement and redundancy packages offered to academic staff), many of which were in the more technical areas of the discipline. Both the RAE and the nature of the 'new blood' appointments were contested by some within the discipline (mainly in the pages of *Area*), but Clayton (1985b) defended them, hinting strongly at greater future central *dirigisme* in the direction of research and teaching priorities.

The external assessment of both research (through the RAEs) and teaching (through the Teaching Quality Assessment (TQA) programmes: Chalkley, 1996) has strongly influenced the practice of geography—as with most other UK university disciplines—in recent years, though the RAEs more so because large sums of money follow success in these (Johnston, 1995b), whereas the outcomes of the TQA audits do not influence university grants. Departments have thus become managed in much more detail as a consequence (Butlin, 1999b): geography remains a vibrant intellectual discipline, as the chapters in this volume amply demonstrate, but in order to be so it has had to become a successful business too. (Curran, 2000, 2001, for example, uses models of academic success to show that many university departments have the high levels of vitality that he associates with both innovation and substantial investment.)

Despite all this apparent success, geography has suffered from two related problems. The first is its fragmentation. All disciplines have become fragmented in recent decades, because of the great breadth and depth of knowledge: it is no longer possible for individual polymaths to

[61] Initially, they were funded on a four-fold scale ('outstanding', 'better than average', 'average', and 'below average'), which was replaced in 1989 by a numerical scale from 1 to 5 (with 5 the highest grade), but grade 3 was divided into two sub-grades for the 1992 RAE and in 1996 grade 5 was similarly subdivided.

[62] Peter Haggett was a member of the UGC at the time of the first two RAEs in 1986 and 1989 (see Johnston, 1995a). He was the only member of a geography department to serve on the UGC or its successors (UFC, HEFCE), although Keith Clayton preceded him as an environmental scientist and Ron Cooke served on HEFCE in the late 1990s and early 2000s, as a vice-chancellor.

gain more than a fleeting understanding of a wide range of subject matters, let alone synthesise and appreciate the massive research literatures. To advance in a discipline—even to teach in it—it is necessary to specialise; time pressures, if nothing else, allow no more. Thus geography has become a community of sub-communities, and individual departments—responding to the RAEs and the financial imperatives of success therein—have themselves broken into such separate (though far from insulated) compartments. The main divisions are between physical and human geography (which are not only substantive but also epistemological and methodological); in addition, there are major divisions within each—notably in human geography between those who adhere to the 'spatial analysis' approach, on the one hand (see Chapter 9 below), and those who favour the 'social theory' approach on the other.[63] Furthermore, scholars in many of these groups find much of their inspiration from, and identify with peer groups in, either other disciplines with which their interests overlap or newly emerging multi- and interdisciplinary groupings (as with environmental sciences in the 'new universities' of the 1960s). There are thus strong centrifugal forces within geography (as in many other disciplines), which in intellectual terms may outweigh the countervailing centripetal pressures, and which seek to sustain geography both politically (as a thriving discipline within the universities) and intellectually (as bringing a particular perspective on certain types of research problem and providing a locale within which they can be taught). The contemporary academy is characterised by tribes establishing and defending territories (Becher, 1989; Johnston, 1996).[64]

The second problem concerns geography's visibility—and to some extent its credibility. The discipline itself has a very strong foundation in its teaching activities linked to geography's strengths in the schools, though curriculum developments from the late 1990s onwards pose a threat to this continuing. Because of this, geography departments have established strong presences in many universities (including the country's most prestigious—another difference from the US situation) and, apart from some smaller departments, very few have been faced with either closure or 'forced merger' with another (cognate?) discipline. And UK geography research has a strong reputation internationally, having provided staff for many overseas universities and postgraduate training for many of their products.[65] The International Geographical Union held its Congress twice

[63] The terms are taken from Sheppard (1995); see also Johnston (1997).

[64] Although less so than in the United States (Johnston, 2000).

[65] In 1991, the Festival International de Géographie, held annually in September in Ste Die des Vosges, founded the Prix Vautrin Lud (which it promotes as geography's equivalent to a Nobel Prize). Peter Haggett was the first to be awarded it, followed by two geographers with strong British roots (Peter Gould and David Harvey) and then Doreen Massey, Ron Johnston, and Peter Hall.

in Britain during the twentieth century (at Cambridge in 1928 and London in 1964), and is scheduled to hold a third in Glasgow in 2004. The UK has also provided it with two of its presidents—Dudley Stamp (1960–4) and Michael Wise (1976–80), both professors at the LSE—as well as many of the leading figures in its many commissions and study groups.[66] And UK geography journals rank very highly internationally: *Transactions* (launched by the IBG and now produced by the RGS) is widely accepted as one of the top three general geographical journals in the English-speaking world, for example, and many of those launched in the 1960s to 1980s now have very significant international reputations.

But in the wider world of British academic life, geography and geographers have been much less visible. Although some individuals, such as Dudley Stamp and Peter Hall (both knighted for their services), have had high public profiles at certain stages of their careers and, along with others, have done much advisory work for governments and other bodies, the discipline's overall profile has been relatively low. (There was a great deal of specialist work done by geographers during the Second World War: Balchin, 1987; see also Chapter 7 below.) This was especially problematic in the 1960s, when geography failed to establish a foothold in either the 'new universities' or the initial stages of the SSRC. Too many people had an outdated view of the nature of geography ('What is the capital of Mali?'; 'Which is the longest river in Africa?', etc.), and when geographers were called upon it was usually to provide local information—as in the various BAAS handbooks produced in mid-century for those attending its conferences, some of which were edited by geographers and others had contributions from them.[67] Geography certainly came somewhat late to the approaches and concerns of the social sciences during that decade, and efforts were then made to illustrate what they had to offer in a number of books (such as Chisholm, 1971b, 1975; Chisholm and Manners, 1973; Chisholm and Rodgers, 1973). The century's last decades saw much less of this ignorance of geography's nature, but the discipline continues to labour under the misapprehensions and prejudices of earlier generations.

Outwith academia, some geographers have achieved high positions in both the public and the private sectors: Christie Willatts, for example, was a lecturer in geography at Birkbeck College after being organizing secretary of the Land Utilisation Survey, and then Principal Planner in the Department of the Environment, 1948–73; Brian Law, a lecturer in geography at University College London in the early 1950s (who introduced

[66] The current president is Anne Buttimer of the Republic of Ireland, who held a post at the University of Glasgow in the 1960s.

[67] Steel and Watson (1972) list those produced in the period 1964–71, all but one of which was edited by a geographer at the relevant local university.

Brian Berry to Lösch's work), rose to be Managing Director of Mars UK; and John Patten, a Cambridge graduate and then lecturer in geography at Oxford, was Secretary of State for Education and Science in the mid-1990s. But once in their new roles they were not identified as geographers—even though their self-identity may have included that. The same was true within the university sector, as with those geographers who joined environmental science departments. Geography in the UK lacks a presence other than for geographers teaching geography!

One of the basic themes of this essay is that although geography is made by geographers, within the constraints of local, regional, national, and even international contexts, its political advancement is very much dependent on pressure groups and individuals. Geography had a successful pressure group in the late nineteenth and early twentieth centuries in the RGS—assisted initially, and then very much overshadowed, by the GA with regard to geography in schools. But the RGS's influence waned with the creation of the IBG, and for nearly 50 years few senior geographers played major roles in its activities.[68] It was largely ineffective when geography needed a champion in the 1960s and the IBG offered no significant alternative: however successful it was as a learned society, it was somewhat introspective—and probably wary of the 'world outside'—and did not provide (or, with its limited resources, was unable to provide) the necessary leadership. The merger with the RGS in the 1990s is seen as a response to that, bringing together the discipline's intellectual vitality with a well-resourced and well-networked organisation to promote the discipline with renewed vigour and direction.

But leadership is provided by individuals, sometimes working alone within the existing institutional structures. And geography largely lacked this, whatever the limited successes of a few individuals. There were no popularisers to compete with, say, A.J.P. Taylor on television, or academics who gained prominence in the print media (even the more specialised journals, such as *New Society*, for which Peter Hall wrote extensively on planning issues). The IBG did, however, launch a successful initiative aimed at media coverage in the early 1980s. A freelance journalist was invited to an annual conference, and filed copy with the *Independent* newspaper. As a result of this, other papers became interested (including some 'down-market' tabloids as well as the broadsheets) and sent journalists to the conference. A press room was established as part of the annual organisation, with a conference press officer, and since then several papers have had one or two full pages of coverage of selected papers during

[68] Only three academic geographers have been presidents of the RGS, for example: Dudley Stamp (1960–4), Michael Wise (1980–2), and Ron Cooke (2000–). The Society started to draw on academic departments for its honorary secretaries in the early 1960s: Michael Wise was the first.

the three-day conference—a development that made the IBG (and now the RGS) envied by some other social science learned societies.[69]

In general, geography's senior figures operated within relatively limited spheres of influence—their own discipline and universities. Even with the latter, however, no geographer was appointed as a university vice-chancellor (or equivalent) between Mackinder at the LSE (1903–8) and Robert Steel's appointment as Principal of University College, Swansea in 1974.[70] It was only in the 1980s that geographers were appointed to the University Grants Committee and its successors (Keith Clayton and Peter Haggett, followed by Ron Cooke in the 1990s), and no geographer has led a research council.[71] Two of the highest-profile appointments have been Directors-General of the Ordnance Survey (David Rhind, 1990–7, and Vanessa Lawrence, 2000–), who have followed the tradition of geographers being involved in the country's cartographic enterprise (both R. A. Skelton and Helen Wallis, for example, were curators at the British Museum Map Room).[72] Furthermore, apart from the general association of geography with maps in the public imagination,[73] the discipline lacks a clear public constituency: unlike many others, there is no professional career path for geographers *sensu stricto*, apart from those who become teachers or researchers, so geography per se has no public profile—and is thus even

[69] One of the benefits for the discipline is that this conference takes place in the first week of January, when there is often relatively little news. There is the occasional downside to the coverage, if the press decide to poke fun at something discussed in the papers, and on one occasion a future president was accused (in a *Guardian* editorial: Johnston, 1991) of straying well beyond his discipline's limits, but the overall result has been very favourable to geography's public profile.

[70] This changed in the 1990s, with the appointments of Alan Wilson (Leeds), Ron Johnston (Essex), Ken Gregory (Goldsmith's), Peter Toyne (Liverpool John Moores), Ron Cooke (York), John Tarrant (Huddersfield), David Rhind (City), and Bill Ritchie (Lancaster); Ron Johnston and Ron Cooke were successively presidents of the IBG in the early 1990s. In addition, Kenneth Hare was Master of Birkbeck College in the University of London from 1966 to 1968.

[71] Peter Hall was appointed to chair the ESRC in 1987, but declined to take up the post following the decision to move its headquarters to Swindon.

[72] There are deep and long connections between geography and cartography, of course, and various commercial map-makers have been involved in disciplinary developments—the Edinburgh firm of Bartholomew, for example, played a part in the development of geography at the university there. But by the late twentieth century cartography had become a separate technical activity, linked to geography through research interests in remote sensing and GIS (see Chapter 12 below). Most geographical interest was in the history of cartography (see Chapter 11 below) with very little research—unlike the situation in the United States—on map construction, interpretation and use. An early director-general of the Ordanance Survey with strong geographical connections but no training was Arden-Close (1911–22): he was later president of both the RGS and the GA (Freeman, 1985).

[73] And even that association is muted. The British Library's map exhibition 'The Lie of the Land' (July 2001–April 2002) attracted three articles in *The Times* at the time of its opening, with none mentioning geography!

more reliant than some other professions on high-profile individuals to make its presence felt.[74] And geography lacks a popular audience for its books—unlike, say, history—and magazines such as the *National Geographic* and *Geographical Magazine* are much more adept at meeting the vernacular conception of geography than the academic—despite their good work in a variety of ways.[75]

Perhaps one of the clearest pointers to geography's lack of visibility (and status?) within the upper levels of the UK academic hierarchy, and thus of its power and influence, is the relative absence of its leading practitioners from the country's main academies. Only three geographers have been elected to the Royal Society: Wooldridge (as a geologist in 1959, a year before he was elected to the New York Academy of Sciences: Balchin, 1984); Fleure (in 1936: Garnett, 1970), again not explicitly as a geographer; and Rhind (in 2002).[76] And no geographer was elected to the fellowship of the British Academy (which presents itself as the academy for the humanities and the social sciences) until Clifford Darby in 1967. A small number was elected in subsequent years—and there was one per annum through the 1990s—but there were still only 18 geographers in the fellowship in 2001 (plus seven corresponding fellows overseas),[77] and geography is in a joint section with anthropology rather than having its own separate identity within the Academy. (Four geographers have served as the Academy Vice-President, however—Clifford Darby, Terry Coppock, Peter Haggett, and Roger Kain—and one President, Sir Tony Wrigley (1998–2001), was trained and initially practised as a geographer before switching his interests to economic and demographic history: Wrigley, 1965).[78] Three geographers were among the Foundation Academicians of the Academy of Learned Societies in the Social Sciences when it was launched in 1999, and others have since been elected (either as nominees of the RGS–IBG or by other societies affiliated to the Academy, reflecting their multidisciplinary activities). Outside the UK, several British

[74] In 2001, the RGS fellows voted to create a professional qualification of CGeog, but although this may help geographers in certain activities (such as the contest for consultancies), it is unlikely to enhance the public profile.

[75] The journal for schools, *Geography Review,* launched by academics at the University of Oxford in the mid-1980s, has done much to 'popularise' academic geography, however.

[76] David Rhind is a geographer who was a professor at Birkbeck College before being appointed as Director-General of the Ordnance Survey and then Vice-Chancellor of City University. David Walker, a biogeographer who worked in the Department of Geography at the Australian National University, was elected as a fellow in 1985.

[77] Only three of those elected have since died: Clifford Darby, Jean Gottmann, and Terry Coppock. In addition, two corresponding fellows died in 2000—Oskar Spate and Paul Wheatley.

[78] As did another Cambridge-trained historical geographer, Richard Smith FBA. Interestingly, the unit with which both Smith and Wrigley are now associated—the Cambridge Unit for the Study of Population—was integrated with the university's geography department in 2001.

geographers have received international recognition, either by other national geographical societies and academies, or by international awards, such as the Lauréat of the IGU and the Prix Vautrin Lud of the Festival International de Géographie. UK geography is well established and recognised internationally, but this success is not as obvious within the wider realms of British public and academic life.

And so geography enters the twenty-first century firmly established in the UK's university and school systems—though not without some pressing problems on its agenda—and with a substantial international reputation for the quality of its scholarship. And yet, despite the successes of those who launched, sustained, and promoted the discipline through the twentieth century, both individually and through its learned societies, geography does not have as high a profile nationally—either within or outwith academe—that might be the case. Perhaps that, while sustaining the scholarly reputation, is the main agenda item for the twenty-first century? Certainly the discipline seems well placed to do that: data on university arts, humanities, and social sciences departments showed that in 2000 some 60 per cent of all geography staff were aged 40 or under, a position only approached by psychology among other disciplines.

Acknowledgements

Many people have assisted me in the preparation of this chapter through providing information and reading/commenting upon drafts, among them Robin Butlin, Michael Chisholm, Keith Clayton, Ron Cooke, Felix Driver, Gary Dunbar, Peter Haggett, Peter Hall, Rita Johnston, Robert Mayhew, Bill Mead, Richard Munton, Charles Pattie, Jennifer Platt, Joe Powell, David Smith, Tim Unwin, Paul White, Jeremy Whitehand, Michael Williams, Charles Withers and, especially, Hugh Clout and Michael Wise.

References

Anderson, P. (1968) Components of the national culture. *New Left Review*, 50, 3–57.

Bailey, P. J. M. (1992) Geography and the National Curriculum: I A case hardly won: geography in the national curriculum of English and Welsh schools, 1991. *Geographical Journal*, 158, 65–74.

Baker, J. N. L. (1963) *The History of Geography*. Oxford: Blackwell.

Balchin, W. G. V. (1984) Sidney William Wooldridge 1900–1963. *Geographers: Biobibliographical Studies*, 8, 141–50.

Balchin, W. G. V. (1987) United Kingdom geographers in the Second World War. *Geographical Journal*, 153, 159–80.

Balchin, W. G. V. (1988) One hundred years of geography in Cambridge: a St. Catharine's view. *Cambridge*, 23, 39–53.

Balchin, W. G. V. (1993) *The Geographical Association: The First Hundred Years, 1893–1993*. Sheffield: The Geographical Association.
Balchin, W. G. V., ed. (1997) *The Joint School Story*. London: London School of Economics, The Joint School Society.
Barnes, T. J. (2001) 'In the beginning was economic geography': a science studies approach to disciplinary history. *Progress in Human Geography*, 25, 521–44.
Beaver, S. H. (1962) The Le Play Society and fieldwork. *Geography*, 40, 225–40.
Beaver, S. H. (1983) Reflections of a founder member. *Transactions, Institute of British Geographers*, NS8, 36–7.
Beaver, S. H. (1987) Geography in the Joint School (London School of Economics and King's College). In R. W. Steel, ed., *British Geography 1918–1945*. Cambridge: Cambridge University Press, 76–89.
Becher, T. (1989) *Academic Tribes and Territories*. Milton Keynes: Open University Press.
Bell, M., Butlin, R., and Heffernan, M. (1995) Introduction; geography and imperialism, 1820–1940. In M. Bell, R. Butlin, and M. Heffernan, eds, *Geography and Imperialism: 1820–1940*. Manchester: Manchester University Press, 1–12.
Bell, M. and McEwan, C. (1996) The admission of women Fellows to the Royal Geographical Society 1892–1914: the controversy and the outcome. *Geographical Journal*, 162, 295–312.
Bennett, R. J. and Thornes, J. B. (1988) Geography in the United Kingdom, 1984–1988. *Geographical Journal*, 154, 23–48.
Blouet, B. W. (1987) *Halford Mackinder: a Biography*. College Station, TX: Texas A & M University Press.
Bowen, E. G. (1987) Geography in the University of Wales, 1918–1948. In R. W. Steel, ed., *British Geography 1918–1945*. Cambridge: Cambridge University Press, 25–44.
Bowen, E. G., Carter, H., and Taylor, J. A. (1968) A retrospect. In E. G. Bowen, H. Carter, and J. A. Taylor, eds, *Geography at Aberystwyth: Essays Written on the Occasion of the Departmental Jubilee, 1917–18 – 1967–68*. Cardiff: University of Wales Press, xix–xxxvi.
Brown, T. N. L. (1971) *The History of the Manchester Geographical Society 1884–1950*. Manchester: Manchester University Press.
Buchanan, R. O. (1954) The I.B.G.: retrospect and prospect. *Transactions and Papers, Institute of British Geographers*, 20, 1–14.
Butlin, R. A. (1999a) Geography at Leeds: the early days. Typescript of lecture given at the 80th Birthday Conference of the School of Geography, University of Leeds.
Butlin, R. A. (1999b) Departmental strategies for balancing the demands of teaching and research. *Journal of Geography in Higher Education*, 23, 397–412.
Cameron, I. (1980) *To the Farthest Ends of the Earth: 150 Years of World Exploration*. London: Macdonald.
Cantor, L. M. (1962) The Royal Geographical Society and then projected London Institute of Geography 1892–1899. *Geographical Journal*, 128, 30–5.
Chalkley, B. (1996) Geography and Teaching Quality Assessment: how well did we do? *Journal of Geography in Higher Education*, 20, 149–58.
Chalkley, B. and Craig, L. (2000) Benchmark standards for higher education geography. *Journal of Geography in Higher Education*, 24, 395–422.
Chisholm, M. (1971a) Geography and the question of 'relevance'. *Area*, 3, 65–8.

Chisholm, M. (1971b) *Research in Human Geography*. London: Heinemann.
Chisholm, M. (1975) *Human Geography: Evolution or Revolution*. London: Penguin Books.
Chisholm, M. (2001) Human geography joins the Social Science Research Council: personal recollections. *Area*, 33, 428–30.
Chisholm, M. and Manners, G., eds (1973) *Spatial Policy Problems of the British Economy*. Cambridge: Cambridge University Press.
Chisholm, M. and Rodgers, B., eds (1973) *Studies in Human Geography*. London: Heinemann.
Chorley, R. J. and Haggett, P., eds (1965) *Frontiers in Geographical Teaching*. London: Methuen.
Chorley, R. J. and Haggett, P., eds (1967) *Models in Geography*. London: Methuen.
Clarke, K. (1992) Geography and the National Curriculum: II The Secretary of State's address to the Royal Geographical Society on geography in the national curriculum. 1991. *Geographical Journal*, 158, 75–8.
Clayton, K. M. (1985a) The state of geography. *Transactions, Institute of British Geographers*, NS10, 5–16.
Clayton, K. M. (1985b) New blood by (government) order. *Area*, 17, 321–2.
Clout, H. M. (2003) *Geography at University College London: A Brief History*. London: University College London, Department of Geography.
Cooke, R. U. and Robson, B. T. (1976) Geography in the United Kingdom, 1972–76. *Geographical Journal*, 142, 81–100.
Coones, P. (1987) *Mackinder's 'Scope and methods of geography' after a Hundred Years*. Oxford: School of Geography, University of Oxford.
Cormack, L. (1997) *Charting an Empire: Geography at the English Universities, 1580–1620*. Chicago, IL: University of Chicago Press.
Crone, G. R. (1964) British geography in the twentieth century. *Geographical Journal*, 130, 197–220.
Curran, P. J. (2000) Competition in UK higher education: competitive advantage and Porter's diamond model. *Higher Education Quarterly*, 54, 386–410.
Curran, P. J. (2001) Competition in UK higher education: applying Porter's diamond model to geography departments. *Studies in Higher Education*, 26, 223–51.
David, T. (1958) Against geography. *Universities Quarterly*, 12, 261–73.
Dickinson, R. E. (1964) *City and Region: A Geographical Interpretation*. London: Routledge & Kegan Paul.
Dickinson, R. E. (1976) *Regional Concept: The Anglo-American Leaders*. London: Routledge & Kegan Paul.
Doornkamp, J. C. and Warren, K. (1980) Geography in the United Kingdom, 1976–80. *Geographical Journal*, 146, 94–110.
Driver, F. (2001a) *Geography Militant: Cultures of Exploration in an Age of Empire*. Oxford: Blackwell.
Driver, F. (2001b) Human geography, social sciences and the humanities. *Area*, 33, 431–4.
Edwards, K. C. (1987) Geography in a University College (Nottingham). In R. W. Steel, ed., *British Geography 1918–1945*. Cambridge: Cambridge University Press, 90–9.

Edwards, K. C. and Crone, G. R. (1960) Geography in Great Britain 1956–60. *Geographical Journal*, 126, 426–41.

ESRC (1988) *Horizons and Opportunities in the Social Sciences*. Swindon: ESRC.

Evans, E. E. (1978) Beginnings. In J. A. Campbell, ed., *Jubilee 1928–1978: Geography at Queen's: An Historical Survey*. Belfast: Queen's University, 5–15.

Farmer, B. H. (1983) British geographers overseas, 1933–1983. *Transactions, Institute of British Geographers*, NS8, 70–9.

Fisher, C. A. (1954) Editorial: the third generation. *Geographical Studies*, 1, 1–3.

Fleure, H. J. (1943) The development of geography. *Geography*, 28(3), 69–77.

Fleure, H. J. (1961) Chairs of geography in British universities. *Geography*, 46, 349–53.

Fleure, H. J. (1963) Sixty years of geography and education. *Geography*, 48, 231–66.

Fox, J. W. and Lewthwaite, G. R. (1959) For geography: a review and commentary. *New Zealand Geographer*, 15, 84–93.

Freeman, T. W. (1954) Early developments in geography at Manchester University. *Geographical Journal*, 120, 118–19.

Freeman, T. W. (1961) *One Hundred Years of Geography*. London: George Duckworth.

Freeman, T. W. (1980a) The Royal Geographical Society and the development of geography. In E. H. Brown, ed., *Geography Yesterday and Tomorrow*. Oxford: Oxford University Press, 1–99.

Freeman, T. W. (1980b) *A History of Modern British Geography*. London: Longman.

Freeman, T. W. (1984a) The Manchester Geographical Society, 1884–1984. *The Manchester Geographer*, 5, 2–19.

Freeman, T. W. (1984b) Robert Neal Rudmose Brown 1879–1957. *Geographers: Biobibliographical Studies*, 8, 7–16.

Freeman, T. W. (1984c) The Manchester and Royal Scottish Geographical Societies. *Geographical Journal*, 150, 55–62.

Freeman, T. W. (1985) Charles Arden-Close 1865–1952. *Geographers: Biobibliographical Studies*, 9, 1–13.

Freeman, T. W. (1987) Geography during the inter-war years. In R. W. Steel, ed., *British Geography 1918–1945*. Cambridge: Cambridge University Press, 9–24.

Gardner, R. (2000) Establishment, role and relevance of a Subject Centre for UK Geography. *Journal of Geography in Higher Education*, 24, 157–62.

Gardner, R. and Craig, L. (2001) Is geography history? *Journal of Geography in Higher Education*, 25, 5–10.

Gardner, R. A. M. and Hay, A. M. (1992) Geography in the United Kingdom, 1988–92. *Geographical Journal*, 158, 13–30.

Garnett, A. (1970) Herbert John Fleure. *Biographical Memoirs of the Royal Society*, 16, 253–78.

Garnett, A. (1983) I.B.G.: the formative years—some reflections. *Transactions, Institute of British Geographers*, NS8, 27–35.

Geertz, C. (1983) *Local Knowledge: Further Essays in Interpretive Anthropology*. New York: Basic Books.

Gilbert, D. (1999) Sponsorship, academic independence and critical engagement: a forum on Shell, the Ogoni dispute, and the Royal Geographical Society (with the Institute of British Geographers). *Ethics, Place and Environment*, 2, 135–44.

Gilbert, E. W. (1965) Andrew John Herbertson: an appreciation of his life and work. *Geography*, 50, 313–31.

Gilbert, E. W. (1972) *British Pioneers in Geography*. Newton Abbott: David and Charles.

Gilmour, D. (1994) *Curzon*. London: John Murray.

Goudie, A. S. (1998) Kenneth J. Mason 1887–1976. *Geographers: Biobibliographical Studies*, 18, 67–72.

Granö, O. (1996) The institutional structure of science and the development of geography as professional practice. *Netherlands Geographical Studies*, 206, 17–29.

Graves, N. J. (1980) Geography in education. In E. H. Brown, ed., *Geography Yesterday and Tomorrow*. Oxford: Oxford University Press, 100–13.

Haggett, P. (1989) *The Geographer's Art*. Oxford: Blackwell Publishers.

Haggett, P. (1995) *Bristol Geography 1920–1995*. Bristol: University of Bristol, Department of Geography.

Hall, P., Jackson, P., Massey, D., Robson, B., Thrift, N., and Wilson, A. (1987) Horizons and opportunities in social science. *Area*, 19, 266–72.

Harris, L. J., Biddle, A., Thompson, E. H., and Irving, E. G. (1964) British maps and charts: a survey of development. *Geographical Journal*, 130, 226–40.

Heyworth, Lord (1965) *Report of the Committee on Social Studies*. London: HMSO (Cmnd 2660).

Honeybone, R. C. (1984) James Fairgrieve 1870–1953. *Geographers: Biobibliographical Studies*, 8, 27–34.

Jay, L. J. (1979) A. J. Herbertson. *Geographers: Biobibliographical Studies*, 3, 85–92.

Jay, L. J. (1986) John Scott Keltie, 1840–1927. *Geographers: Biobibliographical Studies*, 10, 93–98.

Jenkins, A. and Smith, P. (1993) Expansion, efficiency and teaching quality: the experience of British geography departments 1986–91. *Transactions, Institute of British Geographers*, NS18, 500–15.

Jenkins, A. and Ward, A. (2001) Moving with the times: an oral history of a geography department. *Journal of Geography in Higher Education*, 25, 191–208.

Johnston, R. J. (1989) The Institute, study groups, and a discipline without a core? *Area*, 21, 407–14.

Johnston, R. J. (1995a) The business of British geography. In A. D. Cliff, P. R. Gould, A. G. Hoare, and N. J. Thrift, eds, *Diffusing Geography: Essays for Peter Haggett*. Oxford: Blackwell, 317–41.

Johnston, R. J. (1995b) Geographical research, geography and geographers in the changing British university system. *Progress in Human Geography*, 19, 355–71.

Johnston, R. J. (1996) Academic tribes, disciplinary containers, and the realpolitik of opening up the social sciences. *Environment and Planning A*, 28, 1943–8.

Johnston, R. J. (1997) *Geography and Geographers: Anglo-American Human Geography since 1945*, 5th edn. London: Arnold.

Johnston, R. J. (2000) Intellectual respectability and disciplinary transformation? Radical geography and the institutionalisation of geography in the USA since 1945. *Environment and Planning A*, 32, 971–90.

Johnston, R. J. (2001) City regions and a federal Europe: Robert Dickinson and post-war reconstruction. *Geopolitics*, 6, 153–76.

Johnston, R. J. (2002) Robert E. Dickinson and the growth of urban geography: an evaluation. *Urban Geography*, 23, 135–44.

Johnston, R. J. (2003) Institutions and disciplinary health: two moments in the history of UK geography in the 1960s (forthcoming).
Johnston, R. J. and Brack, V. (1983) Appointment and promotion in the academic labour market: a preliminary survey of British university Departments of Geography, 1933–1982. *Transactions, Institute of British Geographers*, NS8, 100–11.
Johnston, R. J., Gregory, K. J., Haggett, P., and Sugden, D. E. (1990) *Research Activity in Departments of Geography in British Universities, 1984–1988: A Statistical Profile*. Privately published.
Johnston, R. J., Gregory, D., Pratt, G., and Watts, M. J., eds (2000) *Dictionary of Human Geography*, 4th edn. Oxford: Blackwell.
Joseph, K. (1985) Geography in the school curriculum. *Geography*, 70, 290–7.
Keltie, J. S. (1886) *Report of the Proceedings of the Royal Geographical Society in Reference to the Improvement of Geographical Education*. London: John Murray.
King, D. (1997) Creating a funding regime for social research in Britain: the Heyworth Committee on Social Studies and the founding of the Social Science Research Council. *Minerva*, 35, 1–26.
King, D. (1998) The politics of social research: institutionalising public funding regimes in the United States and Britain. *British Journal of Political Science*, 28, 415–44.
Kirk, W. (1978) Fifty years on. In J. A. Campbell, ed., *Jubilee 1928–1978: Geography at Queen's, An Historical Survey*. Belfast: Queen's University, 46–55.
Kirwan, L. P. (1964) The RGS and British exploration: a review of recent trends. *Geographical Journal*, 130, 221–4.
Kirwan, L. P. (1965) Geography as a social study. *Geographical Journal*, 131, 373–5.
Lee, R. (1985) Where have all the geographers gone? *Geography*, 70, 45–59.
Lochhead, E. (1981) Scotland as the cradle of modern academic geography in Britain. *Scottish Geographical Magazine*, 97, 98–109.
Lochhead, E. (1984) The Royal Scottish Geographical Society: the setting and sources of its success. *Scottish Geographical Magazine*, 100, 69–80.
McClintock, M. E. (1974) *University for Lancaster: Quest for Innovation*. Lancaster: University of Lancaster.
McKendrick, J. H. (1997) Regional journals in geography: a vision for the 21st century. *Journal of the Manchester Geographical Society*, 1, 2–17.
Mackenzie, J. M. (1995) The provincial geographical societies in Britain, 1884–1914. In M. Bell, R. Butlin, and M. Heffernan, eds, *Geography and Imperialism: 1820–1940*. Manchester: Manchester University Press, 93–124.
Mackinder, H. J. (1887) On the scope and methods of geography. *Proceedings of the Royal Geographical Society*, 9, 141–60.
Mackinder, H. J. (1904) The geographical pivot of history. *Geographical Journal*, 23, 421–37.
Mackinder, H. J. (1919) *Democratic Ideals and Reality: A Study in the Politics of Reconstruction*. London: Constable.
Markham, C. R. (1881) *The Fifty Years Work of the Royal Geographical Society*. London: Murray.
Marshall-Cornwall, Sir James (1976) *History of the Geographical Club, 1826–1975*. Privately published.
Mayhew, R. J. (1998a) Geography in eighteenth century British education. *Paedagogica Historica*, 34, 731–69.

Mayhew, R. J. (1998b) The character of English geography c.1660–1800: a textual approach. *Journal of Historical Geography*, 24, 385–412.

Mayhew, R. J. (2000) *Enlightenment Geography: The Political Languages of British Geography, 1650–1850*. London: Macmillan.

Middleton, D. (1991) Douglas W. Freshfield 1845–1934. *Geographers: Biobibliographical Studies*, 13, 24–31.

Mill, H. R. (1930) *The Record of the Royal Geographical Society*. London: Royal Geographical Society.

Munton, R. J. C. and Goudie, A. S. (1984) Geography in the United Kingdom, 1980–1984. *Geographical Journal*, 150, 27–47.

Ogilvie, A. G., ed. (1928) *Great Britain: Essays in Regional Geography*. Cambridge: Cambridge University Press.

Parker, W. H. (1959) Geography defended. *Universities Quarterly*, 13, 34–44.

Parker, W. H. (1982) *Mackinder: Geography as an Aid to Statecraft*. Oxford: Clarendon Press.

Peel, R. (1975) The Department of Geography, University of Bristol, 1925–1975. In R. Peel, M. Chisholm, and P. Haggett, eds, *Progress in Physical and Human Geography: Bristol Essays*. London: Heinemann, 411–17.

Philo, C. (1998) Reading *Drumlin*: academic geography and a student geographical magazine. *Progress in Human Geography*, 22, 344–67.

Ploszajska, T. (1999) *Geographical Education, Empire and Citizenship: Geographical Teaching and Learning in English Schools*. London: Historical Geography Research Group of the Royal Geographical Society.

Prince, H. C. (2000) *Geographers Engaged in Historical Geography in British Higher Education 1931–1991*. London: Historical Geography Research Series, 36.

Qualifications and Assessment Authority (2000) *Benchmarking Statement for Geography*. Gloucester: QAA.

Rees, H. (1989) *A University is Born: The Story of the Foundation of the University of Warwick*. Coventry: privately published.

Richards, K. S. and Wrigley, N. (1996) Geography in the United Kingdom, 1992–1996. *Geographical Journal*, 162, 41–62.

Rose, G. (1993) *Feminism and Geography*. Cambridge: Polity Press.

Royal Geographical Society (2000) *Directory of University Geography Courses 2001*. London: Royal Geographical Society (with the Institute of British Geographers).

Scargill, D. I. (1976) The RGS and the foundations of geography at Oxford. *Geographical Journal*, 142, 438–61.

Scargill, D. I. (1999) *The Oxford School of Geography, 1899–1999*. Oxford: University of Oxford, School of Geography.

Sheppard,. E. S. (1995) Dissenting from spatial analysis. *Urban Geography*, 16, 283–303.

Sidaway, J. D. and Samers, M. (2000) Exclusions, inclusions, and occlusions in 'Anglo-American geography': reflections on Minca's 'Venetian geographical praxis'. *Environment and Planning D: Society and Space*, 18, 663–6.

Slater, T.R. (1988) Redbrick academic geography. *Geographical Journal*, 154, 169–80.

Stamp, L. D. (1946) *The Land of Britain: Its Use and Misuse*. London: Longman.

Stamp, L. D. and Wooldridge, S. W., eds (1951) *London Essays in Geography: Rodwell Jones Memorial Volume*. London: Longman.

Steel, R. W. (1967) Geography at the University of Liverpool. In R. W. Steel and R. Lawton, eds, *Liverpool Essays in Geography: a Jubilee Collection*. London: Longman, 1–24.
Steel, R. W. (1983) *The Institute of British Geographers: The First Fifty Years*. London: Institute of British Geographers.
Steel, R. W. (1987a) The beginning and the end. In R. W. Steel, ed., *British Geography 1918–1945*. Cambridge: Cambridge University Press, 1–8.
Steel, R. W. (1987b) The Oxford School of Geography. In R. W. Steel, ed., *British Geography 1918–1945*. Cambridge: Cambridge University Press, 58–75.
Steel, R. W. and Watson, J. W. (1972) Geography in the United Kingdom 1968–72. *Geographical Journal*, 138, 139–53.
Stoddart, D. R. (1983) Progress in geography: the record of the I.B.G. *Transactions, Institute of British Geographers*, NS8, 1–13.
Stoddart, D. R. (1986) *On Geography and its History*. Oxford: Blackwell.
Stoddart, D. R. (1987) Alfred Steers: 1899–1987. A Personal and Departmental Memoir. Unpublished paper.
Taylor, G. (1958) *Journeyman Taylor: The Education of a Scientist*. London: Robert Hale.
Thrift, N. J. and Walling, D. E. (2000) Geography in the United Kingdom, 1996–2000. *Geographical Journal*, 166, 96–124.
Toal, G. (1992) Putting Mackinder in his place: material transformations and myth. *Political Geography*, 11, 100–18.
Vaughan, J. E. (1985) William Hughes 1818–1876. *Geographers: Biobibliographical Studies*, 9, 47–53.
Wagstaff, J. M. (1996) *Geography: the First Seventy-Five Years at the University of Southampton*. Southampton: University of Southampton, Department of Geography.
Walford, R. (2001) Geography's odyssey: the journey so far. *Geography*, 86, 305–17.
Walford, R. and Williams, M. (1985) Geography and the school curriculum: the recent role of the Geographical Association. *Area*, 17, 317–21.
Ward, R. G. (1960) Captain Alexander Maconochie, R.N., 1787–1860. *Geographical Journal*, 126, 459–68.
Ward, R. G. (2001) Oskar Hermann Khristian Spate (1911–2000). *Australian Geographical Studies*, 39, 253–5.
Whitehand, J. W. R. (1970) Innovation diffusion in an academic discipline: the case of the 'new' geography. *Area*, 16, 185–7.
Willetts, E. C. (1987) Geographers and their involvement in planning. In R. W. Steel, ed., *British Geography 1918–1945*. Cambridge: Cambridge University Press, 100–16.
Wise, M. J. (1975) A university teacher of geography. *Transactions, Institute of British Geographers*, 66, 1–16.
Wise, M. J. (1977) Sir David Martin. *Geographical Journal*, 143, 144–5.
Wise, M. J. (1986) The Scott Keltie report 1885 and the teaching of geography in Great Britain. *Geographical Journal*, 152, 367–82.
Wise, M. J. (1993) The campaign for geography in education: the work of the Geographical Association 1893–1993. *Geography*, 78, 101–9.
Wise, M. J. (1997) Sir Dudley Stamp and the World Land Use Survey. In A. L. Singh, ed., *Land Resource Management*. New Delhi: DK Publishers, 1–8.

Wise, M. J. and Estall, R. C. (1995) *A Century of Geography at LSE.* London: LSE Research Papers in Environmental & Spatial Analysis, 60.

Withers, C. W. J. (2002) A partial biography: the formalization and institutionalization of geography in Britain since 1887. In G. S. Dunbar, ed., *Geography: Discipline, Profession and Subject Since 1870.* Amsterdam: Kluwer, 79–119.

Withers, C. W. J. and Mayhew, R. J. (2002) Rethinking 'disciplinary' history: geography in British universities, c. 1580–1887. *Transactions, Institute of British Geographers,* NS27, 11–29.

Women and Geography Study Group (1984) *Geography and Gender: An Introduction to Feminist Geography.* London: Hutchinson.

Women and Geography Study Group (1997) *Feminist Geographies: Explorations in Diversity and Difference.* Harlow: Longman.

Wooldridge, S. W. (1956) *The Geographer as Scientist.* London: Thomas Nelson.

Wright, L. J. and Jones, J. H. (1972) Geography in the new polytechnics. *Geography,* 57, 120–6.

Wrigley, E. A. (1965) Changes in the philosophy of geography. In R. J. Chorley and P. Haggett, eds, *Frontiers in Geographical Teaching.* London: Methuen, 3–24.

II. Environment

3

Physical geography and geography as an environmental science

KEN GREGORY

The evolution of physical geography in Britain over the last 100 years cannot be divorced from developments elsewhere and by 2000 it had become increasingly difficult to distinguish physical geography from other disciplines. Some periods have shown a net gain, during which British physical geography assimilated and responded to trends developed elsewhere, whereas in others British trends provided a lead (especially reflecting the inspiration given by particular individuals) that has been perceived to be internationally influential. Whereas nineteenth-century geography was more holistic in character, it is ironic that for much of the twentieth century it became increasingly reductionist, with the development of many separate sub-fields succeeded by trends, very evident at the millennium, of a discipline seeking holism again—clearly linking with environmental science.

The evolution of physical geography can be traced over a century from its nineteenth-century foundations, through an evolutionary period that occurred in the twentieth century up to 1960, to the dramatic revolution in content and approach that metamorphosed physical geography to the end of the century, providing the basis for restructuring that was evident by 2000. This sequence allows some predication of the prospects that exist for the twenty-first century. My own undergraduate and postgraduate training occurred at the end of the second stage, so that I have subsequently been able to observe the sequential changes—from the time at the end of the 1950s when a few individuals were very dominant and pervasive, through the debates that arose and up to the present position, where there are pluralist approaches generating particular specialisations in individual institutions.

The foundations

The foundations of twentieth-century British physical geography undoubtedly included a major component derived from geology. In the nineteenth century the diluvialists were overcome by the establishment of stratigraphy and uniformitarianism, with the present as the key to the past prevailing as a guiding principle in the evolving approach to geomorphology. Other strands (Gregory, 2000) providing significant foundations were, first, the impact of the theory of evolution after Darwin's *The Origin of Species* (1859) and, secondly, exploration and mapping: the Ordnance Survey (founded in 1795) continued the production of large-scale survey maps with the Geological Survey (founded 1801), although a national soil survey was not inaugurated until 1949. Nineteenth-century geological surveyors provided information, often at the end of their memoirs, about the physical geography of the areas surveyed. This is illustrated by the work of De La Beche (1839) who, in his survey of the geology of Cornwall, Devon, and Somerset, provided a foundation for the physical geography of southwest England. Many of the first records of fragments of raised beaches and other major landscape features, often subsequently designated as Sites of Special Scientific Interest (e.g. in Gregory, 1997), were the result of painstaking field recording by the Geological Survey.

Other individuals, often without formal training, provided a foundation for the understanding of physical geography, an example being Colonel George Greenwood (Chorley, Dunn, and Beckinsale, 1964, 372), whose approach was clear from the title of *Rain and Rivers* (first published in 1857). Descriptive accounts of landscape and physical environment were instrumental in drawing attention to the characteristics of regions and to problems that should be addressed, in the way that the characteristics of southwest England were recorded by nineteenth-century writers (Shorter, Ravenhill, and Gregory, 1969, 23–8). Geologists such as Archibald Geikie (1859) used Scotland in his *Scenery of Scotland* to provide a basis to justify and establish the fluvial idea as the correct means for interpreting topography (Chorley et al., 1964, 407). At that time controversies were vehemently debated, including the extent to which drift deposits had been produced by a great inundation by water rather than by ice, and how an understanding of the impact of former glaciation was necessary for the understanding of scenery. Although Geikie (1882) contributed to the glacial theory on the structure of the Scottish glacial drifts and on glacial erosion, it was his younger brother James Geikie (1839–1915) whose book on *The Great Ice Age* (Geikie, 1874, 1877, 1894) established the Ice Age in the international literature for at least three decades after 1874 (Tinkler, 1985, 132) and exemplified the way in which British scientists had an influence well beyond Britain. Similarly influential were the ideas of Percy Fry Kendall, a geologist at the University of Leeds, who in a series of papers (Kendall, 1902, 1903) argued that anomalous glacial drainage channels, later described as overflow channels or meltwater channels, were the products of erosion by meltwater from vast proglacial lakes draining either around the ice margin or through a col and over a watershed away from the glacier (Sissons, 1958; Gregory, 1962, 1965).

Physical geography was not always dominated by landscape investigations. *Physiography* (Huxley, 1877), an approach characterised as a peculiarly Victorian science (Stoddart, 1975a, 1986) and which dominated education at all levels for a quarter of a century, began with the River Thames and proceeded to the less familiar springs, climate, denudation, glacial erosion, marine erosion, earth movements, and the formation of rocks and the earth as a planet. Although subsequently used as a term for geomorphology in the United States, physiography was used by Huxley in the sense of a description of nature. Stoddart (1975a) argues that Huxley's (1877) book marked the end of three centuries of the study of landforms in Britain, as espoused by Davies (1968), and the point at which the scientific study of landforms passed from professional geologists to geographers. In that passing, however, the import of the ideas of the American W. M. Davis (1850–1934) was to become extremely influential. His paper to the Royal Geographical Society introduced the geographical cycle in which he claimed that: 'All the varied forms of the lands are dependent upon—or, as the mathematician would say, are functions of—three variable quantities, which may be called structure, process and time' (Davis, 1899, 150). In addition to this trilogy Davis offered a cycle of erosion, the stages of which could be used to classify any landscape, and, to complement the normal cycle, he offered the arid and marine cycles and accidents glacial and volcanic. This imported foundation was to influence physical geography in Britain for the next 60 years, and its general importance is reflected in the fact that, of the first three substantial volumes on the history of the study of landforms, a single volume of 874 pages is devoted almost exclusively to Davis (Chorley, Beckinsale, and Dunn, 1973).

Evolution in the twentieth century

Gradual growth in the number of approaches to British physical geography from 1900 to 1960 came to be pervaded by the evolutionary ideas introduced by Davis. Then came the inevitable reactions inspired by them, accompanied by some suggested alternatives, followed by new strands—all paving the way for the revolution necessary to create a new physical geography after 1960.

A consequence of Davisian ideas (e.g. Davis, 1899) was that British physical geography came to be heavily influenced by denudation chronology. It was suggested (Day Kimball, 1948) that historical geology could be divided into two parts: stratigraphy, which is concerned with what is there, and denudation chronology, which is concerned with what isn't! Denudation chronology, in attempting to reconstruct landscapes that no longer existed, focused on southeast England, for which a monograph

(Wooldridge and Linton, 1939) provided a chronological model for Tertiary and Pleistocene landscape evolution embodying a number of key chronological stages. Although there were comparatively few researchers until the mid-twentieth century, this model prompted research in other areas, leading to debates, often between physical geographers and geologists. These are exemplified by a marine erosion explanation for surfaces on Exmoor (Balchin, 1952), a subaerial sequence for the whole of Wales (Brown, 1960), and controversy for the north of England and Scotland (Linton, 1933, 1951a). As several physical geographers, such as Wooldridge, had been trained as geologists (and it was his geological contributions that led to his election as Fellow of the Royal Society in 1959) it was perhaps inevitable that they brought a very strong landscape evolution approach to physical geography, an approach that culminated in studies of the sequence of river development in terms of terraces, for example along the Exe (Kidson, 1962), and in the identification of large numbers of erosion (later called planation) surfaces, some as little as 15 feet (c.5 m) apart (e.g. Brunsden, 1963; Orme, 1964).

With the benefit of hindsight it is easy to see why this approach focused attention on landscape evolution over some 50 million years, a period of time largely ignored by geologists in Britain, giving little if any attention to the last 1 million years, and effectively undertaking analyses which could be based upon investigation of as little as 5–10 per cent of the landscape (Young, 1964). Debates in the 1950s concerned the sequence of denudation chronology, geographical versus geological interpretations, and the origin of the remnants of former marine or subaerial surfaces. In a re-evaluation of the geomorphological system of Davis, and of the implications that had stemmed from it, Chorley (1965) outlined three major criticisms: that it led to a dogma of progressive, irreversible, and sequential change; that the emphasis was upon the historical sequence rather than upon functional associations; and that the approach was highly dialectical and semantic. The way in which physical geography came to be dominated by geomorphology (just one branch of the subject) is now difficult to comprehend, as is the acrimony that appeared in debates in the 1960s, when to some the Davisian method had become a stranglehold, or at least a sedative (Chorley et al., 1973, 753).

Although not explicitly recognised at the time, the evolutionary theme could also be detected in approaches to the atmosphere, to soil geography and to biogeography. Studies of the atmosphere, the most process-based part of physical geography at that time, were predominantly approached through climatic classification. *Climatology*, by Austin Miller (1931 and subsequent editions), who was trained in mathematics, proceeded from elements and factors of climate and air masses to climatic classification and a world treatment. Even the subsequent influence of the group at the Meteorological Institute at Bergen (Hare, 1951), leading to a classification of air masses and the foundations of dynamic climatology, was also one in which an evolutionary trend could be detected. Biogeography appeared in *Plant and Animal Geography*, in which Marion Newbigin (1936), trained as a biologist, presented taxonomies of plants, organisms and soils, focused upon distribution, and prepared the ground for biogeography to be based upon the ideas of succession—another strand of the evolutionary approach. This was the prelude to historical biogeography and the palaeo-ecological studies that came to dominate biogeography (Oldfield, 1963; Simmons, 1964). In *The Spirit and Purpose of Geography*, Wooldridge and East (1951) headed one chapter 'Physical Geography and Biogeography', so that it was not clear whether physical geography should include biogeography or not; they saw physical geography as resting 'upon specialist sciences like geology and meteorology'. Such views in this phase, when physical geography was effectively in the course of construction, were dominated by evolution, encouraging the notion that the subject was in some way different in character from specialist sciences—a notion that took time to dispel, although much later composite sciences such as geomorphology and ecology have been differentiated from basic sciences such as physics or biology (Osterkamp and Hupp, 1996, 436).

Some strands of physical geography, however, remained independent of the evolutionary emphasis, both within and without British physical geography. Investigations of glacial geomorphology achieved some independence from the main evolutionary theme: Vaughan Lewis, tragically killed in a car accident in 1961, was interested in processes and investigated Norwegian cirque glaciers (Lewis, 1960), and David Linton, maintaining his interest in glacial geomorphology, contributed papers on watershed breaching (Linton, 1951b) and also saw the potential of the more process-related implications of denudation chronology in the everlasting hills (Linton, 1957) by relating rates of erosion over geological periods to the timescales needed for the production of planation surfaces. Also independent of the evolutionary thrust in geomorphology was the study of coastlines. Arising from a request by the Ministry of Town and Country Planning in 1943, *The Coastline of England and Wales* (Steers, 1948) was important not only for the way in which it provided a survey of the coast, later followed by the *Coastline of Scotland* (Steers, 1973) and succeeding detailed studies of coastal segments such as Scolt Head Island (Steers, 1934), but also because it began to give attention to environmental processes, providing a survey of interest to government and the general public. Indeed this was the area of geomorphology most related to what was later to be thought of as applied physical geography.

Developments external to physical geography were also influential, including publications on *The Pleistocene Period* (Zeuner, 1945) and *Dating the Past* (Zeuner, 1946), which reflected the creation of an intellectual environment where emphasis was increasingly placed upon the range of evidence, including archaeological, that could be useful in interpreting the Pleistocene and establishing the basis for chronology. Unfortunately, physical geographers did not engage in those dating debates until later, perhaps because there were a few geologists who concentrated upon the Quaternary, including Wright (1914, 1937), author of *The Quaternary Ice Age*, and J. K. Charlesworth, whose encyclopaedic study *The Quaternary Era* (Charlesworth, 1957) stretched to two substantial volumes. Important for climatic classification, but also establishing the foundation for subsequent research interest by physical geographers in hydrology and water balance investigations, was Penman's (1948, 1950) work on evaporation and on vegetation and hydrology (Penman, 1963). A further external development, destined to have a considerable impact, was the introduction of the ecosystem concept by Tansley (1935), although it had not yet been absorbed by 1951 beyond a broad outline of plant ecology in *The Spirit and Purpose of Geography* (Wooldridge and East, 1951).

Partly catalysed by the influx into the profession of new physical geographers who had gained practical experience including meteorology and use of air photographs during the Second World War, a number of new strands began to emerge in the decade from 1950 to 1960. Atmospheric investigations began to focus upon dynamic climatology, inspiringly led by F. K. Hare, who had served with the Air Ministry (1941–5) and published *The Restless Atmosphere* (Hare, 1953) when he was in Canada before returning to King's College London in 1964. The great contribution made by this approach was its foundation upon the processes underlying weather, progressing to a regional treatment of climate on a continental scale. Local climate also began to attract process investigations in, for example, urban climatology, which ultimately culminated in *The Climate of London* (Chandler, 1965) but was discussed at the Royal Geographical Society in 1962 (Chandler, 1962).

New strands were also appearing in geomorphology: underfit streams, first recognised in the Warwickshire Itchen (Dury, 1952), challenged existing thinking (which was very stimulating for young physical geographers although unacceptable to some of their older colleagues), by demonstrating how rivers could be out of balance with their much larger meandering valleys, a demonstration that subsequently had many implications (see Dury, 1977). Although such examples were prompted by recent recruits to the profession, some new strands of research derived from existing doyens. David Linton had the ability to see where the major challenges lay: recognising that the problem of tors related to deep weathering and warmer climates (Linton, 1955); showing that mapping was an essential technique of data acquisition as, for example, in the delimitation of morphological regions (Linton, 1951c); extending the site concept (Wooldridge, 1932); and providing the forerunner for morphological mapping (Waters, 1958; Savigear, 1965). Among his later innovative contributions were the geography of energy, offering an energetic foundation for physical geography (Linton, 1965); the description of scenery as a resource (Linton, 1968), stimulating the scenic evaluation of landscape; and consideration of lunar landscapes (Linton, 1966). In a volume produced in his honour, it was suggested (Brown and Waters, 1974) that Linton was one of the first British geomorphologists to be born, academically speaking, into the subject, and that he saw its future potential, conceiving the idea for the British Geomorphological Research Group (BGRG), which first met in Sheffield in 1960. Also from an established geographer, *The Skin of the Earth* (Miller, 1953) introduced topographical anatomy and circulatory systems in the context of landscape described as 'compared to a symphony whose various elements, subtly interwoven, combine to assault the senses with a pleasure of fine sound'.

In approaching that symphony a focus was needed and it is arguable that the historical focus was a good one—one reminiscent of the dominance of the historical in human geography at the same time (see Darby, 1951). However, in the first 60 years of the century, when the foundations were being laid, the orchestra was not complete and the physical geographer not yet equipped to write a whole score. In some ways the breadth of the early decades of the century was not continued, in that animal geography was not developed and oceanography became established as the province of oceanographers. Three indices of the position of physical geography by 1960 can be seen. First, books available during the 1950s (Table 3.1) inevitably influenced the way in which new geographers were introduced to their subject. A growing number were authored by physical geographers, albeit some who had like Stamp, Miller, and Wooldridge trained in another discipline, and some directed towards the sixth-form market as well as towards the university audience. Examples of books available up to 1960 are more evenly distributed over the branches of physical geography than in the later twentieth century, but just 40 per cent were written by physical geographers, and the oceans were required reading at that time. Secondly, the papers given by British physical geographers at the 20th International Geographical Congress, which met in London in 1964 (Table 3.2), although not necessarily a representative sample, give an indication of how British physical geography appeared to the international audience in 1964.

Although by 1960 several strands were beginning to develop that reflected the dominance of the evolutionary approach, it was, thirdly, the power of particular individuals, especially in geomorphology, that was the key feature of those times. The flavour of this period is brilliantly captured by Stoddart (1997a). In their assessment of the development of geomorphology, Beckinsale and Chorley (1991, 285) concluded that Wooldridge was 'rightly regarded as a brilliant researcher, an unequalled field teacher and a jovial colleague', although he was hostile to alternative developments

Table 3.1 Chronological arrangement of examples of books available for physical geographers in Britain in the 1950s

Book	*First edn*	*Comment*
J. E. Marr, *The Scientific Study of Scenery*	1900	A geologically founded approach to scenery
J. Murray, *The Ocean*	1913	Oceans not subsequently researched by physical geographers
W. B. Wright, *The Quaternary Ice Age*	1914	A standard reference on the Quaternary
P. Lake, *Physical Geography*	1915	Standard text for many years
A. A. Miller, *Climatology*	1931	Included scheme of climatic classification and regional aspects of climate
G. W. Robinson, *Soils, their Origin, Constitution and Classification*	1932	Founded upon soil survey and classification methods employed by soil survey in Britain
S. W. Wooldridge and R. S. Morgan, *The Physical Basis of Geography*	1937	A geomorphology text, geological emphasis, ran to several subsequent editions
E. G. Bilham, *The Climate of the British Isles*	1938	Standard account for many years
M. I. Newbigin, *Plant and Animal Geography*	1936	Biological view of world vegetation and fauna, with a chapter on soils
W. G. Kendrew, *Climates of the Continents*	1937	Provided accounts with mean monthly values of temperature and precipitation
E. G. Bilham, *The Climate of the British Isles*	1938	Standard text on the British climate
A. G. Tansley, *The British Isles and their Vegetation*	1939	Extensive account of vegetation types with detailed physical background
S. W. Wooldridge and D. L. Linton, *Structure, Surface and Drainage in South East England*	1939	Provided an evolutionary model for the denudation chronology of southeast England
R. A. Bagnold, *The Physics of Blown Sand and Desert Dunes*	1941	A theoretical approach to desert processes, seen as more appropriate after 1960
A. Holmes, *Principles of Physical Geology*	1944	A very successful approach to scenery and landscape evolution by a geologist
F. E. Zeuner, *The Pleistocene Period: Its Climate, Chronology and Faunal Successions*	1945	Introduced the Pleistocene with reference to types of evidence that could be used for reconstruction of past environments

Table 3.1 (*Continued*)

Book	*First edn*	*Comment*
F. E. Zeuner, *Dating the Past: An Introduction to Geochronology*	1946	For many years provided the basic framework for dating methods
L. D. Stamp, *Britain's Structure and Scenery*	1946	The stratigraphy of Britain as a basis for understanding scenery
D. Brunt, *Weather Study*	1946	Textbook of meteorology especially for Air Force cadets
J. A. Steers, *The Coastline of England and Wales*	1946	Comprehensive survey with applied potential
F. D. Ommaney, *The Ocean*	1949	Oceans not subsequently researched by physical geographers
W. G. Kendrew, *Climatology*	1949	Well-illustrated account of elements of climate
A. E. Trueman, *Geology and Scenery in England and Wales*	1949	Popular account relating range of rock types to scenery
M. S. Anderson, *The Geography of Living Things*	1952	Succinct background for biogeography
G. Manley, *Climate and the British Scene*	1952	Successful account available to the general reader and introducing local climate
N. K. Horrocks, *Physical Geography and Climatology*	1953	A textbook, particularly for sixth forms, did not give any references
A. A. Miller, *The Skin of the Earth*	1953	Practical physical geography including analysis of maps and climatic data
G. V. Jacks, *Soil*	1954	Published in agricultural series
F. J. Monkhouse, *Principles of Physical Geography*	1954	Textbook for schools and HE, included oceans as well as other branches
Sir John E. Russell, *The World of the Soil*	1957	Intended for broad readership and included biological aspects of the soil
J. K. Charlesworth, *The Quaternary Era*	1957	Exhaustive two-volume summary
G. H. Dury, *The Face of the Earth*	1959	Stimulating and written to appeal to the general reader as well as the student with a new approach to processes in geomorphology

Notes: Books written by British physical geographers shown in bold. The books listed here are not included in the list of references unless cited in the text.

Table 3.2 Papers contributed by British physical geographers to the 20th International Geographical Congress, London, 1964

Area of physical geography	*British papers as a percentage of the total*	*Subjects of papers*
Biogeography	23.3	Pioneer of soil classification; Plant invasions in arid southwest of USA; Soil surveys in West Africa; Vegetation survey of Welsh uplands; Savanna woodland in Uganda; Migrant locusts; Man's impact on Barbados vegetation
Climatology	10.0	Rainfall patterns over Sierra Leone; Rainfall variability in tropical climates; Air pollution in Reading area
Hydrology, Oceanography and Glaciology	3.6	Pleistocene shorelines in Ireland
Geomorphology	9.7	Dry valleys in southeast England; Planation surfaces in south Atacama desert; Denudation chronology of Lanzarote; Playas and dunes in New Mexico; Aspect and landforms in northeast Yorkshire; Solution subsidence; Growth of Warren Farm spit Hampshire; Sources of beach shingles in England; Denudation chronology of South Wales; Early Tertiary planation surface in Britain; Pleistocene deposits in north Devon; Nivation cirques near Aberystwyth; Road-making materials in the tropics

Note: The subdivisions of physical geography are those used in the abstracts volume.

by what he styled 'periglacial extremists' and the 'morphometric squad', regarding geomorphology as primarily concerned with the interpretation of morphology, not the study of process—which could be left to physical geology (Wooldridge, 1958). If Wooldridge had not existed it would have been necessary to invent him (Beckinsale, 1997, 7) because he provided what was increasingly perceived to be an extreme view, one which required a reactionary trend. This view was sustained despite the more catholic approach of his students, David Linton and Eric Brown, and the contrasting approach of George Dury, whose experience of new process approaches in the United States was reflected in his stimulating paperback *The Face of the Earth* (Dury, 1959), which provided a breath of fresh air and the beginning of the physical geography revolution. Many forthright and public discussions occurred and Dury (1982) commented that Wooldridge 'could be, and often was, as forbidding in demeanour as he was impressive in presence, and it would have been a temerarious young researcher who would have dared to challenge him on his own ground'. Dury departed for the McCaughey Chair of Geography at Sydney in 1962, Wooldridge died suddenly in 1963, and the latter event accelerated the revolution that had already been initiated.

Revolution after 1960

The literature of available books (see Table 3.1) and papers conditioned the way in which students viewed physical geography around 1960, but as research numbers began to grow questions were asked and orthodox views were challenged, both inevitably influenced by awareness of ideas in other countries and facilitated by the formation of new societies and groups, such as the BGRG (founded 1960). Greater interaction of new academics and research students at more frequent meetings engendered healthy debates and a revolutionary atmosphere at a time when quantitative methods were increasingly being utilised (Gregory, 1985, 49–52). New books were required and one of the achievements of the 1960s and 1970s must be the wealth of books embodying new approaches. Of the many individuals who contributed to establishing the new order of physical geography, George Dury had already provoked outbursts and R. J. Chorley provided a pre-eminent influence through his papers (e.g. Chorley, 1962, 1965, 1971) and books, some produced in collaboration (e.g. Chorley and Haggett, 1965, 1967; Chorley, 1969, 1972, 1973; Chorley and Kennedy, 1971). These paved the way for a new physical geography, including influential textbooks on atmosphere (Barry and Chorley, 1976) and geomorphology (Chorley, Schumm, and Sugden, 1984). Dick Chorley's outstanding contribution is applauded in the volume compiled in his honour (Stoddart, 1997b).

A new approach to physical geography was assisted by the expansion in the numbers of staff and students in geography in higher education (Stoddart, 1967), by the burgeoning societies, and by the increasing numbers of journals (Table 3.3)—many initiated by British physical geographers. However, the revolution really depended upon the quest for explanation (e.g. Harvey, 1969) within new strands that centred upon environmental processes (Gregory, 1985), a more multidisciplinary approach to environmental change, human impacts, the utilisation of modelling and systems, and the intensification of applied as well as applicable research.

Table 3.3 Examples of growth in publication activity involving British physical geography after 1960

Indicator	*Examples (and dates)*
New journals established	*Journal of Biogeography* (1974) *Earth Surface Processes and Landforms* (1977) *Progress in Physical Geography* (1977) *Journal of Climatology* (1981) *Soil Survey and Land Evaluation* (1981) *Regulated Rivers* (1984) *Hydrological Processes* (1987) *Quaternary International* (1990) *The Holocene* (1991) *Global Ecology and Biogeography: Letters (1991)*
Increase in size of journals	*Earth Surface Processes and Landforms* 1976: 4 issues, 28 articles, 395 pages 2000: 12 issues, 98 articles, 1,489 pages
Geo Abstracts increase	Contained c. 2,000 abstracts per year in the 1960s; c.16,000 per year in the late 1990s. Contains abstracts of world literature
Edited volumes	Growth in physical geography has benefited from a great increase in edited volumes of research papers, especially arising from national and international conferences. Such volumes have been important elements in the growth of research activity and examples include Chorley and Haggett (1965, 1967); Chorley, (1972); Embleton, Brunsden, and Jones (1978); Goudie (1981a); Simmons and Tooley (1981); Brunsden and Prior (1984); Gregory (1987); Clark, Gregory, and Gurnell (1987); Hooke (1988); Viles (1988); Macmillan (1989); Kirkby (1994); Roberts (1994); Lewin, Macklin, and Woodward (1995); Petts and Amoros (1996); Stoddart (1997b); Thompson and Perry (1997). Many edited volumes have been published by Wiley, including the 23 cited in this chapter, and there were more than 30 relevant Wiley volumes on sale in 2002

Note: Other examples of new journals of interest to physical geographers are given in Gregory (2000, 107).

Environmental processes

Investigations of environmental processes already existed in meteorology and climatology and in studies of coastal environments (King, 1959; Kidson, 1964; Carr, 1965), but they were encouraged when existing models of denudation chronology invited comparison with more recent landscape change; when models of deglaciation demanded comparison with contemporary glaciers in the way that processes associated with contemporary proglacial lakes were compared with those in the British Quaternary (Sissons, 1958); and when it was inferred from knowledge of contemporary environments that subglacial drainage processes had operated during deglaciation in Britain (Sissons, 1960, 1961). Whereas field survey had previously been dominant, new stimuli were derived from the Columbia school of geomorphology in the United States (Strahler, 1992), which unleashed quantitative morphometric analysis, and from the great impact made by *Fluvial Processes in Geomorphology* (Leopold, Wolman, and Miller, 1964), which not only encouraged greater awareness of fluvial processes and hydrology but also influenced approaches to the investigation of slopes and to other branches of geomorphology. Greater attention to processes meant that new techniques had to be learned and developed and a fundamentally different approach was adopted. Morphometric analyses ran their course and were formative but not as durable (Werritty, 1997) as investigations of fluvial processes, which afforded a more fundamental approach.

Each branch of geomorphology needed to establish appropriate methods for process investigations, and to develop new methods of analysis (e.g. Whalley, 1978) in relation to clear conceptual experimental designs. Research on fluvial processes utilised systems of experimental catchments to gain data on rates of process operation and change (e.g. Gregory and Walling, 1973), adapting methods already successfully established in limestone basins (e.g. Smith and Newson, 1974), and on rates under controlled conditions such as urbanisation (Walling and Gregory, 1970; Walling, 1974). Enthusiasm for small catchment experiments (Ward, 1971) meant that some saw dangers in the approach being insufficiently founded upon clearly formulated objectives. In the two decades after 1960, when it was increasingly necessary to embrace the methods of hydrology, some commended the field of geographical hydrology (Ward, 1979), but some did not wish to be differentiated from earth and environmental sciences. It was from these foundations that results began to appear that were very significant internationally, not only leading to the greatly enhanced interpretation of sediment dynamics (Walling, 1983, 1996a), with sediment hydrograph analysis replacing the previously used rating curves, but also facilitating spatial analysis of process variations (Walling and Webb, 1980; Webb and Walling, 1980), the assimilation of additional parameters such as water temperature (Webb, 1995), the extension to a global scale (Walling, 1996b), and, perhaps most innovatively, the way in which radionuclides enabled process studies to link to investigations of environmental change (Walling and He, 1999). It is such progress that realised the potential to be unlocked from catchment experiments (Burt and Walling, 1984). Although research on small instrumented watersheds attracted criticism because the ultimate intellectual objectives were not always clear, subsequent investigations—which were multiscale, basin-based, related to ecology, and extended to recent environmental change, including the impact of human activity—productively engaged physical geography research.

Rapidly following and complementing investigations using catchment experiments, research on individual flood events (e.g. Anderson and Calver, 1977), on river channel dynamics and change (K. J. Gregory, 1978), and on gravel-bed rivers demonstrated how British physical geography could take a lead by initiating international conferences leading to very significant research volumes (Hey, Bathurst, and Thorne, 1982; Thorne, Bathurst, and Hey, 1987). It was in this way that geomorphologists began to engage with engineers, ultimately stimulating environmental river engineering (Hey, 1990) and becoming an integral part of flood studies (Newson, 1989). The dynamics of river channels invited investigations of channel planform changes (Lewin, 1977) and of the way in which these could be related to changes in the basin environment (Gregory, 1976). Placed in a more general context, river channel changes were being recognised as the result of a range of types of human activity (Gregory, 1977), subsequently stimulating a range of studies of flow regulation and reservoirs (Petts, 1984) and of river channelisation (Brookes, 1988).

Similar challenges confronted the investigation of hillslopes, perhaps even more difficult to instrument and monitor. Detailed field survey of slopes by morphological mapping methods, with analysis of the vast amounts of data (Gregory and Brown, 1966) or using best units (Young, 1971), were explored, but it was the investigation of slope processes utilising repeated field measurements at the Young pit (Young, 1960) or the investigation of mass movements (Prior and Stephens, 1972; Prior, 1978) from which process studies really developed. Aided by advances in slope modelling (e.g. Kirkby, 1971), this led to significant results being achieved from repeated surveys and measurements, such as those on the Dorset coast (Brunsden, 1984), engaged investigations alongside civil engineers (e.g. Brunsden and Prior, 1984), and provided the foundation for great advances in understanding hillslope processes (Anderson and Brooks, 1996).

Well-established investigations of processes in coastal environments were encouraged by the development of monitoring techniques such as radioactive tracers (Steers and Smith, 1956), accompanied by detailed historical analysis of coastal change, as exemplified by studies of changes at Spurn Point (de Boer, 1964), and also progressed towards modelling the context in the British seas and relations with oceanography (Hardisty, 1990). Process dynamics investigations became entrenched in other landscape processes, following the same sequence from monitoring small areas, through spatial pattern mapping, to modelling; but for some processes it was necessary to look overseas for field areas. In glacial processes British physical geographers looked to other countries for measurements, such as those undertaken in Norway (Lewis, 1960) and Canada (e.g. Andrews, 1963). As the interest of physical geographers in glacial mechanics grew, so did their contributions in papers published in the *Journal of Glaciology*, which had been established in Cambridge in 1947.

Investigations of arid region processes benefited from a seminal volume produced by Brigadier R. A. Bagnold (1941), which proved to be an essential foundation for investigations of the dynamics of processes in arid regions, and for subsequent impressive investigations, such as those of dust storms (Goudie, 1983; Goudie and Middleton, 1992). Early stimulus for investigations of periglacial processes derived from the identification of frozen-ground features in the east Midlands during geological survey (e.g. Kellaway and Taylor, 1953), but again it was overseas where further process investigations occurred (see Embleton and King, 1975). Studies of soils had progressed towards concern with dynamics, especially soil erosion (Kirkby and Morgan, 1980; Boardman, Foster, and Dearing, 1990), which provided a fruitful field for modelling (Kirkby, 1994). Less dramatically, in biogeography, process dynamics were assumed following the ecosystem approach but utilising energetics (e.g. Simmons, 1981, 1987).

Once the deficiency of understanding of the processes that had shaped the British physical environment was accepted, moves towards how to measure, analyse and interpret spatial and temporal patterns of process unleashed a profitable revolution in British physical geography, which gained from closer relations with other disciplines of hydrology, engineering, meteorology, and ecology. Greater research recognition of processes in physical geography was approached from both empirical and modelling viewpoints and for some time the field detail of the former was not easily reconciled with the assumptions that had to be made by the latter. The immediate requirement for textbooks gave comprehensive ones such as *Process in Geomorphology* (Embleton and Thornes, 1979), with 'form' and 'process' often appearing in the titles of books on slopes (Carson and Kirkby, 1972) and drainage basins (Gregory and Walling, 1973), and many others devoted to techniques (King, 1966; Goudie, 1981a, 1990) or to particular types of environment. This fruitful publishing period established such books in an international market, thus proselytising British physical geography, although the diffusion was itself aided by developments such as the *Geomorphological Abstracts* (1960–, later *Geographical Abstracts*), and the creation of new journals, many instigated by British physical geographers (Table 3.3), devoted to processes, such as *Journal of Biogeography* (1974–), *Earth Surface Processes and Landforms* (1977–), *Journal of Climatology* (1981–), *Regulated Rivers* (1984–) and *Hydrological Processes* (1987–). These serials were created at a time when other important new journals were established in Britain, such as *Progress in Physical Geography* (1977–), when other periodicals were being created for multi-disciplinary and international audiences, and when a dictionary of physical geography was edited by British physical geographers (Goudie et al., 1985). Physical geographers becoming involved in research council committees and their reviews (established to indicate where policy should lead) gave a further bond with other disciplines.

Many consequences followed from the focus on processes, including the historically based techniques necessary for the investigation of rates of environmental change (e.g. Hooke and Kain, 1982) and also for the theoretical underpinning of process investigations. In his chapter 'Bases for theory in geomorphology', Chorley (1978, 1) began with the aphorism 'whenever anyone mentions theory to a geomorphologist, he instinctively reaches for his soil auger', a statement subsequently often cited and one used at the beginning of two of the chapters in the volume published in Chorley's honour (Stoddart, 1997b). Although very apposite in 1978, when current problems and future prospects for geomorphology were collated (Embleton, Brunsden, and Jones, 1978), by the 1990s theory and modelling had become a much more integral part of physical geography, so that the wheel of theory may have come full circle from one teleology to another (Chorley, 1978, 11). Mathematical modelling was not always easily adopted and adapted because not all physical geographers had Mike Kirkby's asset of a first degree in mathematics. It is also paradoxical that at a time when processes were of increasing interest in physical geography, climatology research by physical geographers was declining relatively, influenced by the growth of university departments of meteorology and the inclusion of meteorology in physics and environmental sciences, a trend that has continued and one where we need to pay more attention to training the next generation of modellers (Unwin, 1989) with a distinctly geographic perspective (Henderson Sellers, 1989).

Environmental change

Studies of environmental change in Britain prior to 1960 had been dominated by studies in denudation chronology, with more attention given to Tertiary than to Quaternary landscape evolution, and with greater focus on the development of valley floor landscapes than on interpretation of drift deposits. Prior to the use of evidence from deep sea cores and ice cores, the sequence of drifts and the Quaternary chronology were not agreed from one area to another, let alone in Britain as a whole. An initial contrast in Quaternary investigations therefore tended to be between geologists, who focused more upon the deposits, regardless of the evidence in the scenery above them, and the physical geographers, who gave insufficient attention to what was below the surface. This dichotomy gradually reduced as physical geographers extended analysis of shoreline changes and beach platforms (e.g. Sissons, Smith and Cullingford, 1966), looked for modern analogues to explain patterns of deglaciation (e.g. Sissons, 1958; Price, 1973), explored techniques such as till fabric analysis (Andrews and King, 1968; Cullingford and Gregory, 1978), participated in new organisations such as the Quaternary Research Association (1963–) and provided books on glacial and periglacial geomorphology (Embleton and King, 1968). The shift in Quaternary investigations was supported by research in biogeography which, predominantly focusing upon historical investigations, utilised pollen analysis (Simmons, 1964) and other techniques (Oldfield, 1987) as the basis for the reconstruction of landscape change.

As in the case of studies of environmental processes, the growth and new direction of work on environmental change required new books to be written by physical geographers in Britain, many destined to achieve world recognition (e.g. Embleton and King, 1968; Sugden and John, 1976; Bowen, 1978—although his book was titled *Quaternary Geology*), participation in international as well as national multidisciplinary research groups, and gradual elucidation of the pattern of impact and Quaternary history of Britain. Contributions by physical geographers became very well embedded, as shown by recent books (e.g. N. Roberts, 1989; Bell and Walker, 1992; Lowe and Walker, 1997), by the presidency of the Quaternary Research Association (e.g. Professor Jim Rose), by collaboration with other disciplines (not only geology but also archaeology: Brown, 1997), by focus upon dating techniques (Lowe, 1991), and by involvement in some of the newly created journals (Table 3.3). Attention was not restricted to those areas glaciated at some time during the Quaternary; periglacial geomorphology, catalysed by studies in areas south of the maximum drift limits, allowed the use of ice wedges (Waters, 1961), comparison with areas of active periglaciation such as altiplanation terraces in Spitsbergen (Waters, 1962), and progressed to estimates of periglacial erosion in southern England (Williams, 1968) and to books on periglacial geomorphology (Embleton and King, 1975). The change in approach to environmental change was evident in the way in which a series of volumes on the geomorphology of the British Isles (King, 1976; Sissons, 1976; Davies and Stephens, 1978; Straw and Clayton, 1979) was able to devote considerable attention to the Quaternary stages.

Human impact

Human impact is an eminently geographical subject for investigation, linking as it does the human and physical environments, and it was addressed to remedy the paucity of investigations on the Holocene, to illuminate the disparity between process investigations affected by human activity and results of environmental change focused on periods before human activity was significant. Studies of Mediterranean valleys (Vita-Finzi, 1969) demonstrated the interaction between phases of human activity (often from the archaeological record) with valley sediments and morphology, proving to be a profitable theme that continued to engage British geomorphologists (Lewin, Macklin, and Woodward, 1995). Specific areas of the UK, such as the Norfolk Broads (Jennings, 1952), were appreciated as having been greatly affected by human activity (Lambert et al., 1970); much earlier, in *Man as a Geological Agent*, Sherlock (1922) had astutely noted that people are many times more powerful as an agent of denudation than all the atmospheric denuding forces combined. This might have stimulated earlier interest in human activity, as could have *Man and Nature* by George Perkins Marsh (1864), and *Man's Role in Changing the Face of the Earth* (Thomas, 1956), which arose from a Princeton symposium to which an historical geographer, H. C. Darby, contributed. Although there were a number of allusions to the theme (e.g. Wilkinson, 1963), investigations of processes directly affected by urbanisation (e.g. Walling and Gregory, 1970), and collations such as *Water, Earth and Man* (Chorley, 1969), it was not until 1970 that the subject was explicitly developed by Brown (1970). He characterised human beings as both a geomorphological process in relation to direct purposeful modifications of landforms, and also as indirectly effective through the human influence upon geomorphological processes—a theme reinforced by books on *Man and Environmental Processes* (Gregory and Walling, 1979, 1987) and *The Human Impact* (Goudie, 1981b), and by significant research by physical geographers in arid areas overseas (e.g. Goudie, 1990; Middleton and Thomas, 1992). Specific research in biogeography concentrated upon the evolution of landscape, land use and vegetation under the sequential impact of humans (Curtis and Simmons, 1976; Simmons and Tooley, 1981), showing the significance of particular stages, such as the Bronze and Iron Ages, in reducing forest cover of the British landscape. This also provided a theme for books on *The Ecology of Natural Resources* (Simmons, 1974), *Biogeography: Natural and Cultural* (Simmons, 1979), more recently *Environmental History* (Simmons, 1993a), and, for stressing the links to archaeology, *Alluvial Geoarchaeology* (Brown, 1997). Human impact was also a fruitful theme for soil geography, as it established the way in which soil profiles had evolved, influenced by vegetation and land use change (e.g. Bridges, 1978), and for climatology, particularly urban climatology, which reflected the same theme (e.g. Chandler, 1965). In all these sub-fields research was able to demonstrate the extent to which phases of human impact had changed the physical landscape and the dynamics of environmental processes (Gregory and Walling, 1987).

One extension of research on human impact led to studies of hazards and of particular environments, especially urban ones. Environmental hazards, especially the flood hazard, had been explored successfully in the United States, but there was considerable scope for research on hazards in Britain, illustrated by studies on the implications of the 1976 drought (Doornkamp, Gregory, and Burn, 1980), by the study of hazards in the British Isles (Perry, 1981) and more general statements about environmental hazards (Whittow, 1980). It was from work on hazards that applied research and consultancy could spring, well exemplified by the establishment of the Flood Hazard Research Centre at Middlesex Polytechnic, now the University of Middlesex (Penning Rowsell and Chatterton, 1977). Hazards continued to relate not only to floods (e.g. Smith and Ward, 1998) but also to risk and disaster reduction (Smith, 1992).

Urban environments attracted research discussions, and a symposium on the physical problems of the urban environment (Chandler, Cooke, and Douglas, 1976) concluded that 'physical geography can usefully contribute to the determination of public policies, with respect to the management and development of urbanised areas'. In addition to studies of urban climatology and recognition of urban heat islands and urban effects (e.g. Atkinson, 1987a), it was particularly in hydrology where the urban effects on discharge (Walling and Gregory, 1970; Hollis, 1975; Walling, 1979), on sediment yield (Walling, 1974), on water quality and pollutants (Ellis, 1979), as well as on stream channels (Gregory and Park, 1976; C. R. Roberts, 1989) were elucidated. Douglas (1981) argued that the city as a habitat or ecosystem should be linked to the city as a social system, subsequently elaborating the ecosystem view of the city (Douglas, 1983), and the *Urban Geomorphology of Drylands* (Cooke et al., 1982) arose from studies of applied geomorphology.

Theory and modelling

Reference has already been made to the increased utilisation of theory and modelling which, with systems analysis, was greatly facilitated by the quantitative revolution, the advent of models (Chorley and Haggett, 1967), and network analysis (Haggett and Chorley, 1969). Perhaps most significant was the adoption of general systems theory (Chorley, 1962). Subsequently introduced in discussing the role and relations of physical geography (Chorley, 1971), the analogy was drawn between the position of physical geography and that of a tightrope walker confronted by two ever-diverging tightropes: one representing research and teaching in earth and environmental science, the other playing a relevant role in an increasingly economically and socially oriented human geography. Of the three frameworks offered by Chorley (1971) at that time, model building, resource use, and systems analysis, the last appeared the most attractive and was promulgated in an extremely significant book—*Physical Geography: A Systems Approach* (Chorley and Kennedy, 1971). Although debated at the time, and questioned by some as providing an unnecessary and jargon-ridden statement of the obvious (Chisholm, 1967), or introducing confusion rather than clarification into empirical investigations (Smalley and Vita-Finzi, 1969), the systems approach gave a way of translating the jargon of the physical geographer into the lingua franca of much natural, physical, and engineering science, thus improving communication with those subjects (Cooke, 1971). Within two decades it was assimilated as a fundamental part of physical geography (Huggett, 1985), used as an approach for physical geography as a whole (e.g. King, 1980; White, Mottershead, and Harrison, 1984); for climatology, where it was argued that the application of systems theory had completely changed the subject of climatology (Lockwood, 1979); for biogeography in terms of soil and vegetation systems (Trudgill, 1977); and more broadly for resource systems and management and action frameworks (Simmons, 1991). However, although systems rapidly provided a useful conceptual approach, it took much longer to become integrated as the basis for a revised, and necessarily more complex, modelling strategy.

Embedding systems approaches in physical geography inevitably required a range of models, some of which were very complex, and since 1970 the efforts of Mike Kirkby have been outstanding in this field. The original propositions for modelling in the branches of physical geography (Chorley and Haggett, 1967) were succeeded more than two decades later by a volume on *Remodelling Geography* (Macmillan, 1989), in which it was suggested that the variety of modelling in physical geography was insufficiently accessible and that greater attention was required for the training of the next generation of modellers (Unwin, 1989). A consistently rigorous approach to theory, to models, or to experimental procedures was still lacking, so that it became increasingly important for physical geographers to have effective training in physical chemical and biological principles (Kirkby, 1989). Modelling had changed in style and subject matter in the 1980s as a result of the relaxation of subject boundaries within the earth sciences, the development of modelling at regional and global scales, the development of evolutionary models that revitalised flagging interest in long-term behaviour, and greater interest in areas such as vegetation, previously neglected in mega-models (Thornes, 1989).

As modelling became more elaborate it was necessary to identify a dominant theme, and energetics was explored (Gregory, 1987; 2000, 92–100), having long been employed in several scales of modelling of atmospheric process (Atkinson, 1987). It was suggested (Slaymaker and Spencer, 1998) that a sparsely inhabited niche that occurs between earth system science and Anglo-American physical geography could be occupied by a new revitalised physical geography. The considerable progress made in modelling in relation to specific fields is exemplified by geomorphological systems (Anderson, 1988) and *Process Models and Theoretical*

Geomorphology (Kirkby, 1994), and this has been extended by global models as indicated below.

Applied and applicable research

Foundations provided by the above themes established and invigorated applied and applicable research, although awareness had already arisen from work on coasts (e.g. Steers, 1948) and from hazards such as floods (Penning Rowsell and Chatterton, 1977; Smith, 1992). Ways in which geomorphology could be applied (e.g. Cooke and Doornkamp, 1974; Hails, 1977) were outlined, with land evaluation often a major theme (e.g. Dawson and Doornkamp, 1973; Dent and Young, 1981; Haines-Young, Green, and Cousins, 1993), applications of research becoming a popular subject in inaugural lectures (e.g. Chandler, 1970; Douglas, 1972) and presidential addresses (e.g. Cooke, 1992; Thornes, 1995). However, it was as a result of particular opportunities, or the work of specific individuals, that significant progress was made, although the advent of databases, GIS and remote sensing enhanced the feasibility of the application of physical geography research, and new journals such as *Applied Geography* (1981–) provided outlets for publication.

An example of coordinated applied research, established separately from universities, arose from the Bahrain surface materials resources survey (Brunsden, Doornkamp, and Jones, 1979, 1980), which involved a team of ten geologists, seven geomorphologists, two pedologists, two surveyors, and a cartographer producing an extremely detailed survey of this arid area and stimulating a number of research investigations such as that on salt weathering (e.g. Cooke et al., 1982). This subsequently led to the creation of Geomorphological Services (Brunsden, 1999) as a premier consultancy service independent of any single university but involving geomorphologists from a range of institutions and undertaking contract research investigations in the UK and overseas. An alternative model for focusing applied research was the establishment of a group or unit within a university institution: examples include the Geodata Institute at the University of Southampton and the Flood Hazard Research Centre, University of Middlesex. Since its establishment in 1970, the latter has undertaken research in the UK (e.g. Tunstall and Penning Rowsell, 1998) and overseas on a number of projects, including Venice (Penning Rowsell, Winchester, and Gardiner, 1998) and Malaysia (Parker and Chan, 1996), to achieve its mission: 'To help those who are vulnerable, by promoting the understanding of natural hazards, and providing insight, information, and policy advice whereby governments and other organisations provide sustainable assistance and protection for those at risk'. Its success in clearly establishing the way in which physical geography can and should be applied was recognised by the Queen's

Award for Industry in 2000, and influential research publications such as that on floods (Parker, 2000).

Moves towards a more applied physical geography with a broadening research agenda exposed new topics for application; multidisciplinary approaches catalysed and often facilitated applied investigations; and the progress made was illustrated in research papers, in books and in edited volumes (e.g. Hooke, 1988; McGregor and Thompson, 1995; Thompson and Perry, 1997). Many specific instances exist, such as the description of Sites of Special Scientific Interest (e.g. Gregory, 1997), and potential applications as outlined for example in relation to soils (e.g. Ellis and Mellor, 1995). However, a tension existed because for many years contract research was seen as inferior to research council sponsorship, which was also initially seen as inferior to unfinanced scholarship (Penning Rowsell, 1981, 11). This tension relaxed as it became necessary in physical geography, as in the earth and environmental sciences, to obtain research funding and to maintain ongoing programmes; but, despite an abortive attempt in the 1992 Research Assessment Exercise (RAE: see Chapter 2) to specify applied research, even by the time of the 1996 RAE the position of applied research was not explicitly clear. Furthermore, a lack of coherence in applied geomorphology required further interaction with scientists in other disciplines and marketing of potential expertise (Jones, 1995).

A further tension arose as environmental sciences and environmental studies had grown in the second part of the twentieth century in parallel with physical geography. The creation of a number of successful and distinctive departments of environmental science in British universities, such as those at East Anglia and Lancaster, was complemented in other universities by courses in environmental science constructed by consortia of departments, including geography departments. The position vis-à-vis environmental sciences was clouded by strong arguments made for greater familiarity with scientific and mathematical methods (e.g. S. Gregory, 1978), which could be achieved through closer alliance with other earth and environmental sciences (Clayton, 1980). Physical geography was observed in 1975 to be internally unbalanced, with geomorphology playing too dominant a role, with an integrated physical geography discovered by non-geographers as environmental science (Brown, 1975). The future of physical geography was seen as involving either disintegration or integration (Price, 1979) and geomorphology was seen to require a more geological approach (Worsley, 1979), although it was concluded that the most rapid change was most likely from fully integrated departments of environmental science (Clayton, 1980). By 1999 it was apparent that physical geographers had become increasingly specialised, 'perhaps working more as environmental scientists than as geographers', with the result that physical geography had fragmented into its component specialisms, so that the 'split by climatologists from the UK geographical community is threatened in other areas of the discipline' (Agnew and Spencer, 1999, 5). Despite the movement by climatology there is still an opportunity for the geographer climatologist who requires a thorough grounding in atmospheric science studies but is also conversant with the social sciences (Perry, 1995, 281). Collaboration of research groups is one way forward; the first collaboration (in 1994) between the BGRG and the more recently established Environmental Research Group produced a volume (Owens, Richards, and Spencer, 1997) in which environmental science rather than physical geography was seen as a foundation for environmental policy so that 'it may be necessary for policy making to parallel some of the nature and characteristics of that *(environmental)* science—namely the continuing dialogue between generalisation and specific context' (Trudgill and Richards, 1997, 11).

By the end of the twentieth century the shape and distribution of physical geography across university departments had altered dramatically as a result of the changes since the revolution initiated about 1960. This was a period in which research approaches and achievements, the textbooks produced, new journals initiated, and the influence of particular individuals such as Dick Chorley had all contributed to raising the standing of British physical geography on the international stage. Exploration of new fields of research, the complete metamorphosis of empirical investigations, and the development of modelling had characterised this development. However, as elsewhere in physical geography (Gregory, 2000), there had been a greater emphasis upon getting on with the research than upon the philosophy behind the approach, although some exceptions focused upon critical rationalist (Haines-Young and Petch, 1986) and realist (Richards et al., 1997) approaches. Although initially all university departments in Britain had endeavoured to keep pace with the expanding character of physical geography, and with its links to other disciplines, the technical demands for equipment and laboratory facilities, together with the need to attract specialist staff and funding, especially from the research councils, inevitably meant that not only did departments increasingly choose to specialise in the major branches of physical geography, but also to concentrate upon particular fields within those major branches. Thus initial concentration upon process studies in fluvial geomorphology and hydrology at the University of Exeter in the 1970s had become more intensified by the year 2000, as the requirements for laboratory analytical as well as field instrumentation requirements had become more exacting. In this regard physical geography specialisation followed the earlier trend of the physical sciences, and was reinforced by developments in remote sensing and GIS, as both fields contributed to advances in physical geography research and necessarily emphasised the distinctive character of specialised research centres. Research funding has become increasingly important and has been facilitated by increased interdisciplinary and international research collaboration (Thrift and Walling, 2000), both of which have intensified the specialisation that has emerged in specific university departments.

Restructuring physical geography

At the end of the twentieth century three alternatives were available for the future of physical geography (Gregory, 2000, 282). One, the status quo, could continue the reductionist trends of the twentieth century, perpetuate the increasingly weak links with human geography, and give insufficient attention to the needs and direction of future physical geography; this was seen as insufficient. A second alternative could be for physical geography and its sub-branches to disappear and to become independent parts of other independent disciplines. Although the growth of environmental science had apparently facilitated such a trend it would produce insufficient linkage with the social sciences (e.g. Perry, 1995) and, in Britain, would be at variance with geography; in 1999 geography was the fifth most popular subject at A-level (the examination for 17–18-year-olds) and in 2001 the seventh most popular subject at GCSE (16-year-olds), with 4.58% of papers taken. The third alternative (Gregory, 2000, 282) reaffirms the need for physical geography to reinvent the discipline to some extent (Gregory, 2000, 286).

A fourth phase of development, achieved by the time of the millennium, is therefore one of restructuring physical geography, seen as a culmination of the trends of the second half of the twentieth century, reinforced by new directions and focuses and the advocation of integration; the results of this restructuring are already manifest.

Culmination of trends

The culmination of trends in the twentieth century focuses attention on the character of physical geography by the millennium. Physical geography had become and should remain an effective natural science, with temporal as well as spatial models prominent and significant management consciousness (Clark, Gregory, and Gurnell, 1987, 384). Over the course of the second half of the twentieth century the subject of physical geography was greatly expanded, active engagement with other disciplines was legion, involvement with international research groups and organisations became commonplace throughout the branches of physical geography, and there was growing awareness of the need both to link study of the physical environment with society, and to present results to that society, as had been attempted two decades previously (Brunsden and Doornkamp, 1972).

By the end of the twentieth century a dramatic change in the techniques utilised by physical geographers had been achieved by the technical equipping of an increasingly professionalised physical geography. From realising that existing data collection sources were insufficient, it was appreciated that continuous monitoring at more numerous locations was required. Therefore the necessary techniques were absorbed, with monitoring networks eventually evolving to become multistage networks, such as that producing the admirable documentation of the Exe basin (Walling, 1996a), and sometimes providing continuous records of processes that it had not previously been possible to monitor so thoroughly, such as river bedload (Reid and Frostick, 1986) or stream channel bank erosion (Lawler, 1993). As dating techniques expanded, the accuracy and reliability of dating events and time slices was improved. They also opened up the possibility of sediment fingerprinting with ^{137}Cs, ^{210}Pb, and ^{7}Be, enabling soil erosion to be determined and associated with time changes greater than ever previously achieved (Quine et al., 1997) and the potential for the use of tracers was outlined (Foster, 2001). Major contributions were made with the availability of, and analysis from, remote sensing platforms, often using GIS, and this accompanied the use of increasingly sophisticated models (Huggett, 1991). Such developments meant that it became more usual for research in physical geography to employ a greater breadth of investigative techniques, so that empirical investigations could be validated with data from physical modelling, as in periglacial modelling of ice wedge casts, and from complex theoretical models.

Although there remains the danger that an increasingly sophisticated sledgehammer can be called upon to crack a relatively simple nut by obscuring the basic research objective, some of the most exciting physical geography research is now coming to fruition through the amalgamation of hitherto distinct approaches (Brown and Quine, 1999). One example of this is the way in which a cellular automaton model has been used to model the impacts of Holocene environmental change in an upland river catchment (Coulthard, Kirkby, and Macklin, 1999). This model has subsequently been employed to show how trace elements from mining sediments are progressively incorporated into the fluvial deposits moved downstream in Swaledale, enabling the position at 30-year periods to be projected; thus effectively combining several of the approaches that developed distinctively since 1960—processes, environmental change, human impact, and modelling. Ironically, more specialised techniques, associated with more reductionist approaches to the branches of physical geography, produced some integration because sub-branches of the subject used the same techniques: it became feasible, for example, to link process investigations of soil erosion (since isotopes were first released from weapons testing in 1954) with human impact on environmental change in a way not previously possible.

However, despite these trends it is clear that physical geography still remains unbalanced, with much more concentration upon the land surface and its processes than upon the atmosphere above and the soil immediately below. This is illustrated by the 1996 RAE: of the 1,119 individuals submitted to the geography panel, some 454 (41.9 per cent) could be recognised as researching on the physical environment. The subjects of more than 1,700 physical geography publications submitted shows how British physical geographers presented themselves and their key research achievements for the four-year period 1992–6: of the topics included, water-related research (19.7 per cent), Quaternary research (18.8 per cent), and biogeography (14.3 per cent) dominated, with relatively little research on climate or soils (5.6 per cent each). However, it was not easy to categorise publications into the traditional branches of geomorphology, climatology, and biogeography (including pedology). A further feature at the end of the twentieth century was holism, being stressed as an antidote to the reductionism inevitably arising from increasingly detailed research investigations. This complementing of reductionism had been achieved by building research groups that embraced expertise from several branches of physical geography and sometimes from several disciplines, and also by focusing upon problems at spatial scales greater than Britain. British physical geography research had been increasingly problem-focused upon overseas areas, so that it was possible to suggest a global physical geography being clearly discernible in 2000 (Gregory, 2000). Although such trends could emerge naturally from the expansion of process studies, such as the way in which Walling (1996b) was able to collate information upon erosion rate measurements to revise world maps of world sediment yield, there were also more explicit advocates of *Global Geomorphology* (Summerfield, 1991) emphasising global-scale processes and phenomena, and incorporating tectonics into the study of landforms.

Accompanying the increased awareness of global change prompted by world summit conferences, including Rio in 1992 and Kyoto in 1997, it was inevitable that a climatological approach should feature prominently in a global approach (e.g. Elsom, 1987; Kemp, 1990); several volumes subsequently directed attention to the global scale (e.g. Mannion, 1991; Roberts, 1994), some authored by a geographer in combination with other scientists, such as ecologists and geologists (Moore, Chaloner, and Stott, 1996). Such a shift in focus, although not as rapid as it might have been, to some extent redressed the tendency to focus upon the local environment (Clayton, 1991) in a parochial way (Kennedy, 1993), enabling researchers to seize the opportunity to consider the great flows of the planet (Simmons, 1991). It is in relation to climate change that global change made its first great impact, with knowledge of the world climate system taking a huge leap forward in the decades after 1970 (Huggett, 1991, ix). Not only in climatology but also in relation to river flow and fluvial erosion in England and Wales (Newson and Lewin, 1991), and to UK water resources (Arnell, 1996), have climate change scenarios and hydrological models been explored, and there continue to be many opportunities for physical geographers to explore potential impacts of spatial and temporal change in relation to specified environmental forcing (French, Spencer, and Reed, 1995). Not only have arguments been made in terms of the need for such impact studies, engaging in multidisciplinary and interdisciplinary investigation as necessary, but also the need to communicate the results; science on its own cannot solve global environmental change (O'Riordan, 1994, 10), and it is possible to conceive of civic science—negotiated science where future states are envisaged through open structures of learning, consensus-seeking, and bargaining (O'Riordan, 1994, 10). Opportunities remain for physical geographers to contribute to research on the impact of global change.

New focuses and directions

By the end of the twentieth century new focuses and directions had become increasingly evident (Table 3.4). Some have been expressed as new hybrid fields such as biogeomorphology (Viles, 1988), alluvial geoarchaeology (Brown, 1997), or the hydrosystem (Petts and Amoros, 1996); others as a concerted effort to concentrate upon certain sections of landscape, such as hillslopes (Anderson and Brooks, 1996), floodplains (Anderson, Walling, and Bates, 1996), and flood hazards (Parker, 2000); and yet others as the need for a more cultural physical geography (e.g. Gregory, 2000, 254). Cultural strands had emerged in many branches of physical geography (not just in those concerned with human impact), in relation to atmosphere (e.g. Thornes, 1999), and in biogeography, where the incorporation of culture has been a particularly significant force (e.g. Simmons, 1993b). It is from such a blend of physical geography and culture that consideration can be accorded to physical geographers' involvement in conservation (Gray, 1997, 2001) and to environmental design (Newson, 1995, 427; Gregory, 2000, 262), both areas of further potential. Physical geographers have focused on the impacts of human activity, and on the alternative courses of action available, but can also progress to evaluate the optimum strategy of environmental design rather than leaving this field exclusively to the landscape architect or to the ecologist. Within the more integrated holistic approach championed by a number of writers there is scope for what may appear to be a return to the ideals implicit in the nineteenth-century physiography of Huxley (Gregory, 2000). During discussion arising from a lecture on the content and relationships of physical geography (Brown, 1975) and consideration of the need for integration of its diverse trends, Stoddart (1975b) commented that it was a pity that the word physiography had fallen into disuse, for it 'expresses precisely what many of us feel about the scope and content of physical geography and its relationships both with other aspects of geography and with the natural sciences'.

Creation of new hybrid fields and more cultural physical geography are manifestations of change in British physical geography and have been

Table 3.4 Some integrating trends affecting British physical geography up to the millennium

Area	*Indication*	*Example*	*Example*
Field of research	New fields or branches	Hydrology	Wilby (1997)
		Quaternary environmental change	Lowe and Walker (1997)
		Landscape ecology	Haines-Young (2000)
	Type areas or events	Hazards: floods, droughts	Smith and Ward (1998) Parker (2000)
		Floodplains	Anderson et al. (1996)
		Urban areas	Douglas (1983)
	Themes	Techniques:	
		Research	Whalley (1978)
		Modelling	Thornes (1989)
		Systems	White et al. (1984)
		Environmental	Hooke (1988)
		planning and management	McGregor and Thompson (1995)
	Hybrid fields	Hydroclimatology	
		Hydrogeomorphology	Gregory (1979)
		Biogeomorphology	Viles (1988)
		Soil geomorphology	Gerrard (1993)
		Geoarchaeology	Brown (1997)
		Hydrosystems	Petts and Amoros (1996)
		Ecohydrology	Baird and Wilby (1999)
Research activity	Combination of research specialisms	Bahrain survey	Brunsden et al. (1980)
		Flood Hazard Research Centre	Parker (2000)

accompanied by the relaxation of its traditional threefold structure. This was illustrated by the publications submitted by physical geographers to the 1996 RAE: many academics no longer restrict their research to a single branch of physical geography and many (22 per cent) explicitly relate their research to applied topics. The tripartite division into geomorphology, climatology, and biogeography is still often quoted, but physical geography has been restructured (Gregory, Gurnell, and Petts, 2002) to embrace atmospheric sciences (including climatology and meteorology), geomorphology (including sedimentology and glaciology), hydrology together with oceanography, biogeography (including ecology and pedology), and Quaternary environmental change. Although the proportion of physical geography papers in the *Transactions of the Institute of British Geographers*

Table 3.4 (*Continued*)

Area	*Indication*	*Example*	*Example*
	International research organisations	International Association for Quaternary Research (INQUA) International Association of Geomorphologists (IAG)	11% of delegates at 1999 Durban conference were physical geographers (Werritty, 1995)
	Multidisciplinary research groups	Global change LOIS	Arnell (1996) Land Ocean Interaction Study; Quine et al. (1997)
Publication	Thematic volumes	Hillslope processes	Anderson and Brooks (1996)
	Edited volumes	Different branches of physical geography Different disciplines	Macmillan (1989) Thorne, Bathurst, and Hey (1987)
	Multidisciplinary journals	Editorial boards from different disciplines (e.g. *Regulated Rivers*)	In 2001 editors and editorial advisory board included 32% physical geographers, 32% from biological sciences, and the remainder from other disciplines

Note: The table gives examples of developments that have brought branches of physical geography together.

(19 per cent in 1988–97) is much smaller than the proportion of geographers who are broadly physical in their research orientation (slightly greater than 40 per cent in the 1996 RAE), the question of where all the physical geographers have gone (Agnew and Spencer, 1999) is answered by the fact that physical geographers have published increasingly in multidisciplinary journals and have indeed been very significant in the development of some of those journals (see Table 3.3).

These trends have all made it increasingly impossible for any one individual to cope with the advances of the discipline: research teams have increased in number; combined teams have facilitated the blending of different sub-disciplines, so that research problems can be approached from several directions; and sometimes the fusion of disciplines has benefited multidisciplinary investigations, as in the case of palaeohydrology (Gregory, Lewin, and Thornes, 1987). The inevitable consequence is that once again physical geography is unevenly spread over higher-education institutions in Britain as individual centres have developed reputations for research in particular fields. Departments in particular institutions of higher education, by adopting their particular emphasis in physical geography, have become increasingly distinctive (Gregory, 2001), and such diversity should be encouraged for the future because such pluralist approaches are healthy, economic, and pragmatic. Although the increasing specialisation of reductionist approaches and greater interaction with other disciplines in the later twentieth century increased the distance of physical from much human geography (e.g. Johnston, 1983), there were signs by 2000 of more holistic approaches, with an awareness of remedial and preventive approaches that overcome previous barriers to a better environment (Trudgill, 1990) and which could therefore come closer to human geography. Renewed conversations between physical and human geography (Massey, 1999), highlighting ways of looking at spatio-temporal representation for developing explanations (Raper and Livingstone, 2001) and, in physical geography, relating to closure employed to make things amenable to study (Lane, 2001), are probably a necessary characteristic of all kinds of research. Closure could be applied to physical and human geography so that geography could become increasingly pluralist in its distribution across British higher education.

Conclusion

Closure, in terms of space–time frameworks, as imposed on the discipline by British physical geographers in the twenty-first century, contrasts dramatically with that in the earlier phases of development, and the expertise in different university institutions in Britain presents an increasingly diverse picture. Reflecting on the four phases of development of British physical geography, which have inevitably emphasised the contribution of particular individuals and of books published, it is tempting to conclude that *description* prevailed up to 1900, succeeded by a *convergence* of approach in the branches of physical geography that adopted classification and an evolutionary focus characterising the second phase up to 1960, subsequently followed by *divergence* up to 2000, when the temporal and spatial scales for research investigations were greatly enlarged. Expansion, involving many more researchers, produced the reductionism of more detailed and finely delimited research enquiry in space and time dimensions, to such an extent that it was succeeded by a series of reactions urging *greater coherence* of holistic approaches (Table 3.4) in a fourth, restructuring phase. Closure in terms of the experience of an individual product of the reductionist phase inevitably qualifies presentation of British physical geography in terms of these four phases of development. However, we are now more aware of our predispositions and emphases, although one physical geographer has suggested (Stott, 1998, 2) that we may have to replace our northern-derived, historic metalanguage of equilibrium, sustainability, and balance with a different metalanguage that is more accepting of change, risk assessment, and non-equilibrium, so accepting new key signifiers.

The British physical landscape is now visualised very differently as a culmination of the twentieth-century development of physical geography, which has benefited from the internationalisation of its research. Reductionist trends in the second half of the century moved away from the landscape as a whole, but more multidisciplinary and holistic approaches may instigate a resurgence in awareness of the physical landscape. Environmental science, constructed in part from the existing sciences, has developed in the pluralist way that is increasingly appropriate for British physical geography, but a need remains for the interaction of environmental science with society; a restructured physical geography is appropriately configured to respond to that need. In twenty-first-century higher education, the basis for a seamless range of distinctive centres across the physical geography–environmental science spectrum is now established. British physical geography achieved considerable prominence in the late nineteenth century and, in its restructured form, reflecting the achievements of the twentieth century, could again take the lead to meet the exciting challenges that relate to the research topics potentially available. There is no doubt that the British physical geography academic scene presented when the 30th International Geographical Congress arrives in Britain in 2004 will be dramatically different, much more holistic, sensitive to environmental problems, and prepared to continue international leadership than that presented in 1964 (see Table 3.2), when the 20th Congress arrived.

References

Agnew, C. and Spencer, T. (1999) Editorial: Where have all the physical geographers gone? *Transactions, Institute of British Geographers*, NS24, 5–9.

Anderson, M. G., ed. (1988) *Modelling Geomorphological Systems*. Chichester: Wiley.

Anderson, M. G. and Brooks, S. M., eds (1996) *Advances in Hillslope Processes*. Chichester: Wiley (2 vols).

Anderson, M. G. and Calver, A. (1977) On the persistence of landscape features formed by a large flood. *Transactions, Institute of British Geographers*, NS2, 243–54.

Anderson, M. G., Walling, D. E., and Bates, P. W., eds (1996) *Floodplain Processes*. Chichester: Wiley.

Andrews, J. T. (1963) The cross-valley moraines of north-central Baffin Island, NWT: a descriptive analysis. *Geographical Bulletin*, 19, 49–77.

Andrews, J. T. and King, C. A. M. (1968) Comparative till fabrics and till fabric variability in a till sheet and a drumlin: a small scale study. *Proceedings Yorkshire Geological Society*, 36, 435–61.

Arnell, N. (1996) *Global Warming, River Flows and Water Resources*. Chichester: Wiley.

Atkinson, B. W. (1987a) Precipitation. In K. J. Gregory and D. E. Walling, eds, *Human Activity and Environmental Processes*. Chichester: Wiley, 31–50.

Atkinson, B. W. (1987b) Atmospheric processes—global and local. In M. J. Clark, K. J. Gregory, and A. M. Gurnell, eds, *Horizons in Physical Geography*. Basingstoke: Macmillan, 121–33.

Bagnold, R. A. (1941) *The Physics of Blown Sand and Desert Dunes*. London: Methuen (2nd edn 1954).

Baird, A. J. and Wilby, R. L., eds (1999) *Eco-Hydrology*. London: Routledge.

Balchin, W. G. V. (1952) The erosion surfaces of Exmoor and adjacent areas. *Geographical Journal*,108, 453–76.

Barry, R. G. and Chorley, R. J. (1976) *Atmosphere, Weather and Climate*. London: Methuen.

Beckinsale, R. P. (1997) Richard J. Chorley: a reformer with a cause. In D. R. Stoddart, ed., *Process and Form in Geomorphology*. London: Routledge, 3–12.

Beckinsale, R. P. and Chorley, R. J. (1991) *The History of the Study of Landforms or The Development of Geomorphology, Vol. 3: Historical and Regional Geomorphology 1890–1950*. London: Routledge.

Bell, M. G. and Walker, M. (1992) *Late Quaternary Environmental Change: Physical and Human Perspectives*. Harlow: Longman.

Boardman, J., Foster, I. D. L., and Dearing, J. A. (1990) *Soil Erosion on Agricultural Land*. Chichester: Wiley.

Bowen, D. Q. (1978) *Quaternary Geology*. Oxford: Pergamon.

Bridges, E. M. (1978) Soil, the vital skin of the earth. *Geography*, 63, 354–61.

Brookes, A. (1988) *Channelized Rivers: Perspectives for Environmental Management*. Chichester: Wiley.

Brown, A. G. (1997) *Alluvial Geoarchaeology: Floodplain Archaeology and Environmental Change*. Cambridge: Cambridge University Press Manuals in Archaeology.

Brown, A. G. and Quine, T., eds (1999) *Fluvial Processes and Environmental Change*. Chichester: Wiley.

Brown, E. H. (1960) *The Relief and Drainage of Wales*. Cardiff: University of Wales Press.

Brown, E. H. (1970) Man shapes the earth. *Geographical Journal*, 136, 74–85.

Brown, E. H. (1975) The content and relationships of physical geography. *Geographical Journal*, 141, 35–48.

Brown, E. H. and Waters, R. S. (1974) *Progress in Geomorphology: Papers in Honour of David L. Linton*. London: Institute of British Geographers, Special Publication No. 7.

Brunsden, D. (1963) The denudation chronology of the river Dart. *Transactions, Institute of British Geographers*, 32, 49–63.

Brunsden, D. (1984) Mudslides. In D. Brunsden and D. B. Prior, eds, *Slope Instability*. Chichester: Wiley, 363–418.

Brunsden, D. (1999) Geomorphology in environmental management: an appreciation. *East Midland Geographer*, 22, 63–77.

Brunsden, D. and Doornkamp, J. C. (1972) *The Unquiet Landscape*. Newton Abbot: David and Charles.
Brunsden, D. and Prior, D. B., eds (1984) *Slope Instability*. Chichester: Wiley.
Brunsden, D., Doornkamp, J. C., and Jones, D. K. C., eds (1979) The Bahrain surface materials resources survey and its application to regional planning. *Geographical Journal*, 145, 1–35.
Brunsden, D., Doornkamp, J. C., and Jones, D. J. C., eds (1980) *Geology, Geomorphology and Pedology of Bahrain*. Norwich: Geobooks.
Burt, T. P. and Walling, D. E., eds (1984) *Catchment Experiments in Fluvial Geomorphology*. Norwich: Geobooks.
Carr, A. P. (1965) Shingle spit and river mouth short term dynamics. *Transactions, Institute of British Geographers*, 36, 117–29.
Carson, M. A. and Kirkby, M. J. (1972) *Hillslope Form and Process*. Cambridge: Cambridge University Press.
Chandler, T. J. (1962) London's urban climate. *Geographical Journal*, 128, 279–302.
Chandler, T. J. (1965) *The Climate of London*. London: Hutchinson.
Chandler, T. J. (1970) *The Management of Climatic Resources*. Inaugural Lecture, University College London.
Chandler, T. J., Cooke, R. U., and Douglas, I. (1976) Physical problems of the urban environment. *Geographical Journal*, 142, 57–80.
Charlesworth, J. K. (1957) *The Quaternary Era*. London: Arnold (2 vols).
Chisholm, M. (1967) General systems theory and geography. *Transactions, Institute of British Geographers*, 42, 42–52.
Chorley, R. J. (1962) Geomorphology and general systems theory. *US Geological Survey Professional Paper*, 500-B, 1–10.
Chorley, R. J. (1965) A re-evaluation of the geomorphic system of W. M. Davis. In R. J. Chorley and P. Haggett, eds, *Frontiers in Geographical Teaching*. London: Methuen, 21–38.
Chorley, R. J., ed. (1969) *Water, Earth and Man*. London: Methuen.
Chorley, R. J. (1971) The role and relations of physical geography. *Progress in Geography*, 3, 87–109.
Chorley, R. J., ed. (1972) *Spatial Analysis in Geomorphology*. London: Methuen.
Chorley, R. J., ed. (1973) *Directions in Geography*. London: Methuen.
Chorley, R. J. (1978) Bases for theory in geomorphology. In C. Embleton, D. Brunsden, and D. K. C. Jones, eds, *Geomorphology: Present Problems and Future Prospects*. Oxford: Oxford University Press, 1–13.
Chorley, R. J. and Haggett, P., eds (1965) *Frontiers in Geographical Teaching*. London: Methuen.
Chorley, R. J. and Haggett, P., eds (1967) *Models in Geography*. London: Methuen.
Chorley, R. J. and Kennedy, B. A. (1971) *Physical Geography: A Systems Approach*. London: Prentice Hall.
Chorley, R. J., Beckinsale, R. P., and Dunn, A. J. (1973) *The History of the Study of Landforms, Vol. 2: The Life and Work of William Morris Davis*. London: Methuen.
Chorley, R. J., Dunn, A. J., and Beckinsale, R. P. (1964) *The History of the Study of Landforms, Vol. 1: Geomorphology before Davis*. London: Methuen.
Chorley, R. J., Schumm, S. A., and Sugden, D. E. (1984) *Geomorphology*. London: Methuen.

Clark, M. J., Gregory, K. J., and Gurnell, A. M. (1987) *Horizons in Physical Geography*. Basingstoke: Macmillan.
Clayton, K. M. (1980) Geomorphology. In E. H. Brown, ed., *Geography Yesterday and Tomorrow*. Oxford: Oxford University Press, 167–80.
Clayton, K. M. (1991) Scaling environmental problems. *Geography*, 76, 2–15.
Cooke, R. U. (1971) Systems and physical geography. *Area*, 3, 212–16.
Cooke, R. U. (1992) Common ground, shared inheritance: research imperatives for environmental geography. *Transactions, Institute of British Geographers*, NS17, 131–51.
Cooke, R. U. and Doornkamp, J. C. (1974) *Geomorphology in Environmental Management*. Oxford: Oxford University Press.
Cooke, R. U., Brunsden, D., Doornkamp, J. C., and Jones, D. K. C. (1982) *Urban Geomorphology in Drylands*. Oxford: Oxford University Press.
Coulthard, T. J., Kirkby, M. J., and Macklin, M. G. (1999) Modelling the impacts of Holocene environmental change in an upland river catchment, using a cellular automaton approach. In A. G. Brown and T. A. Quine, eds, *Fluvial Processes and Environmental Change*. Chichester: Wiley, 31–46.
Cullingford, R. A. and Gregory, K. J. (1978) Till ridges in Wensleydale, Yorkshire. *Proceedings, Geologists Association*, 89, 67–79.
Curtis, L. and Simmons, I. G. (1976) Man's impact on past environments. *Transactions, Institute of British Geographers*, NS1, 1–384.
Darby, H. C. (1951) The changing English landscape. *Geographical Journal*, 117, 377–98.
Darwin, C. (1859) *Origin of Species*. London: John Murray.
Davies, G. L. (1968) *The Earth in Decay: A History of British Geomorphology 1578–1878*. London: MacDonald Technical & Scientific.
Davies, G. L. and Stephens, N. (1978) *Ireland*. London: Methuen.
Davis, W. M. (1899) The geographical cycle. *Geographical Journal*, 14, 481–504.
Dawson, J. A. and Doornkamp, J. C., eds (1973) *Evaluating the Human Environment: Essays in Applied Geography*. London: Arnold.
De Boer, G. (1964) Spurn Head: its history and evolution. *Transactions, Institute of British Geographers*, 34, 71–90.
De La Beche, H. T. (1839) *Report on the Geology of Cornwall, Devon and West Somerset*. Geological Survey.
Dent, D. and Young, A. (1981) *Soils and Land Use Planning*. London: Allen & Unwin.
Doornkamp, J. C., Gregory, K. J., and Burn, A. S. (1980) *Atlas of Drought in Britain* 1975–76. London: Institute of British Geographers.
Douglas, I. (1972) *The Environment Game*. Inaugural lecture University of New England, Armidale, New South Wales.
Douglas, I. (1981) The city as an ecosystem. *Progress in Physical Geography*, 5, 315–67.
Douglas, I. (1983) *The Urban Environment*. London: Arnold.
Dury, G. H. (1952) The alluvial fill of the valley of the Warwickshire Itchen near Bishop's Itchington. *Proceedings Coventry District Natural History and Science Society* 2, 180–5.
Dury, G. H. (1959) *The Face of the Earth*. Harmondsworth: Penguin.
Dury, G. H. (1977) Underfit streams: retrospect, perspect and prospect. In K. J. Gregory, ed., *River Channel Changes*. Chichester: Wiley, 281– 93.

Dury, G. H. (1982) Review of D. K. C. Jones (ed.) (1980). *Progress in Physical Geography*, 6, 140–4.

Ellis, J. B. (1979) The nature and sources of urban sediments and their relation to water quality: a case study from north-west London. In G. E. Hollis, ed., *Man's Impact on the Hydrological Cycle in the United Kingdom*. Norwich: Geobooks, 199–216.

Ellis, S. and Mellor, A. (1995) *Soils and Environment*. London: Routledge.

Elsom, D. M. (1987) *Atmospheric Pollution: A Global Problem*. Oxford: Blackwell.

Embleton, C. E. and King, C. A. M. (1968) *Glacial and Periglacial Geomorphology*. London: Arnold.

Embleton, C. E. and King, C. A. M. (1975) *Periglacial Geomorphology*. London: Arnold.

Embleton, C. E., Brunsden, D., and Jones, D. K. C., eds (1978) *Geomorphology: Present Problems and Future Prospects*. Oxford: Oxford University Press.

Embleton, C. E. and Thornes, J. B. (1979) *Process in Geomorphology*. London: Arnold.

Foster, I. D. L., ed. (2001) *Tracers in Geomorphology*. Chichester: Wiley.

French, J. R., Spencer, T., and Reid, D. J. (1995) Editorial: geomorphic response to sea level rise—existing evidence and future impacts. *Earth Surface Processes and Landforms*, 20, 1–6.

Geikie, A. (1865) *Scenery of Scotland Viewed in Connexion with its Physical Geology*. London: Macmillan.

Geikie, A. (1882) *A Textbook of Geology*. London: Macmillan.

Geikie, J. (1874) *The Great Ice Age*. London: Dalby and Ibister (later editions 1877, 1894).

Gerrard, A. J. (1993) Soil geomorphology: present dilemmas and future challenges. *Geomorphology*, 7, 61–84.

Goudie, A. S., ed. (1981a) *Geomorphological Techniques*. London: George Allen & Unwin.

Goudie, A. S. (1981b) *The Human Impact: Man's Role in Environmental Change*. Oxford: Blackwell.

Goudie, A. S. (1983) Dust storms in space and time. *Progress in Physical Geography*, 7, 502–30.

Goudie, A. S. (1990) *Techniques for Desert Reclamation*. Chichester: Wiley.

Goudie, A. S. and Middleton, N. J. (1992) The changing frequency of dust storms through time. *Climatic Change*, 20, 197–225.

Goudie, A. S., Atkinson, B. W., Gregory, K. J., Simmons, I. G., Stoddart, D. R., and Sugden, D. E., eds (1985) *The Encyclopedic Dictionary of Physical Geography*. Oxford: Blackwell (2nd edn 1994).

Gray, J. M. (1997) Planning and landform: geomorphological authenticity or incongruity in the countryside? *Area*, 29, 312–24.

Gray, J. M. (2001) Geomorphological conservation and public policy in England: a geomorphological critique of English Nature's 'Natural Areas' approach. *Earth Surface Processes and Landforms*, 26, 1009–23.

Greenwood, G. (1857) *Rain and Rivers*. London.

Gregory, K. J. (1962) The deglaciation of eastern Eskdale, Yorkshire. *Proceedings Yorkshire Geological Society*, 33, 363–80.

Gregory, K. J. (1965) Proglacial lake Eskdale after sixty years. *Transactions, Institute of British Geographers*, 34, 149–62.

Gregory, K. J. (1976) Changing drainage basins. *Geographical Journal*, 142, 237–47.
Gregory, K. J., ed. (1977) *River Channel Changes*. Chichester: Wiley.
Gregory, K. J. (1978) Fluvial processes in British basins. In C. Embleton, D. Brunsden, and D. K. C. Jones, eds, *Geomorphology: Present Problems and Future Prospects*. Oxford: Oxford University Press, 40–72.
Gregory, K. J.(1979) Hydrogeomorphology: how applied should we become? *Progress in Physical Geography*, 3, 85–101.
Gregory, K. J. (1985) *The Nature of Physical Geography*. London: Arnold.
Gregory, K. J., ed. (1987) *Energetics of Physical Environment: Energetic Approaches to Physical Geography*. Chichester: Wiley.
Gregory, K. J., ed. (1997) *Fluvial Geomorphology of Great Britain*. London: Chapman and Hall (Geological Conservation Review Series, Joint Nature Conservation Committee).
Gregory, K. J. (2000) *The Changing Nature of Physical Geography*. London: Arnold.
Gregory, K. J. (2001) Changing the nature of physical geography. *Fennia*, 179, 9–19.
Gregory, K. J. and Brown, E. H. (1966) Data processing and the study of land form. *Zeitschrift für Geomorphologie*, 10, 237–63.
Gregory, K. J. and Park, C. C. (1976) Stream channel morphology in northwest Yorkshire. *Revue de Géomorphologie Dynamique*, 25, 63–72.
Gregory, K. J. and Walling, D. E. (1973) *Drainage Basin Form and Process*. London: Arnold.
Gregory, K. J. and Walling, D. E. (1979) *Man and Environmental Processes*. London: Dawson.
Gregory, K. J. and Walling, D. E. (1987) *Human Activity and Environmental Processes*. Chichester: Wiley.
Gregory, K. J., Lewin, J., and Thornes, J. B., eds (1987) *Palaeohydrology in Practice: A River Basin Analysis*. Chichester: Wiley.
Gregory, K. J., Gurnell, A. M., and Petts, G. E. (2002) Restructuring physical geography. *Transactions, Institute of British Geographers*, NS27, 136–54.
Gregory, S. (1978) The role of physical geography in the curriculum. *Geography*, 63, 251–64.
Haggett, P. and Chorley, R. J. (1969) *Network Analysis in Geography*. London: Arnold.
Hails, J. R., ed. (1977) *Applied Geomorphology*. Amsterdam: Elsevier.
Haines-Young, R. H. (2000) Sustainable development and sustainable landscapes: defining a new paradigm for landscape ecology. *Fennia*, 178, 7–14.
Haines-Young, R. H. and Petch, J. R. (1986) *Physical Geography: Its Nature and Methods*. London: Harper and Row.
Haines-Young, R. H., Green, D., and Cousins, S., eds (1993) *Landscape Ecology and Geographic Information Systems*. London: Taylor and Francis.
Hardisty, J. (1990) *British Seas: An Introduction to the Oceanography and Resources of the North-west European Continental Shelf*. London: Routledge.
Hare, F. K. (1951) Geographical aspects of meteorology. In G. Taylor, ed., *Geography in the Twentieth Century*. New York: Philosophical Library, 178–95.
Hare, F. K. (1953) *The Restless Atmosphere*. London: Hutchinson.
Harvey, D. W. (1969) *Explanation in Geography*. London: Arnold.
Henderson-Sellers, A. (1989) Atmospheric physiography and meteorologic modelling: the future role of geographers in understanding climate. *Australian Geographer*, 20, 1–25.

Hey, R. D. (1990) Environmental river engineering. *Journal of the Institution of Water and Environmental Management* 4, 335–40.
Hey, R. D., Bathurst, J. C., and Thorne, C. E., eds (1982) *Gravel Bed Rivers: Fluvial Processes, Engineering and Management*. Chichester: Wiley.
Hollis, G. E. (1975) The effect of urbanization on floods of different recurrence intervals. *Water Resources Research*, 11, 431–4.
Hooke, J. M., ed. (1988) *Geomorphology in Environmental Planning*. Chichester: Wiley.
Hooke, J. M. and Kain, R. J. P. (1982) *Historical Change in the Physical Environment: A Guide to Sources and Techniques*. London: Butterworth.
Huggett, R. J. (1985) *Earth Surface Systems*. Berlin: Springer Verlag.
Huggett, R. J. (1991) *Climate, Earth Processes and Earth History*. Berlin: Springer Verlag.
Huxley, T. H. (1877) *Physiography: An Introduction to the Study of Nature*. London: Macmillan.
Jennings, J. N. (1952) *The Origin of the Broads*. London: Royal Geographical Society, Research Series No. 2.
Johnston, R. J. (1983) Resource analysis, resource management, and the integration of physical and human geography. *Progress in Physical Geography*, 7, 127–46.
Jones, D. K. C. (1995) Environmental change, geomorphological change and sustainability. In D. F. M. McGregor and D. A. Thompson, eds, *Geomorphology and Land Management in a Changing Environment*. Chichester: Wiley, 11–34.
Kellaway, G. A. and Taylor, J. H. (1953) Early stages in the physiographic evolution of a portion of the East Midlands. *Quarterly Journal of the Geological Society*, 108, 343–76.
Kemp, D. D. (1990) *Global Environmental Issues: A Climatological Approach*. London and New York: Routledge.
Kendall, P. F. (1902) A system of glacier lakes in the Cleveland Hills. *Quarterly Journal of the Geological Society*, 58, 471–571.
Kendall, P. F. (1903) The glacier lakes of Cleveland. *Proceedings Yorkshire Geological Society*, 15, 1–39.
Kennedy, B. A. (1993) '. . . no prospect of an end'. *Geography*, 78, 124.
Kidson, C. (1962) The denudation chronology of the river Exe. *Transactions, Institute of British Geographers*, 31, 43–66.
Kidson, C. (1964) Dawlish Warren, Devon: late stages in sand spit evolution. *Proceedings Geologists Association*, 75, 167–84.
Kimball, D. (1948) *Denudation Chronology: The Dynamics of River Action*. London: University of London, Institute of Archaeology, Occasional Paper No. 8.
King, C. A. M. (1959) *Beaches and Coasts*. London: Arnold (2nd edn 1972).
King, C. A. M. (1966) *Techniques in Geomorphology*. London: Arnold.
King, C. A. M. (1976) *Northern England*. London: Methuen.
King, C. A. M. (1980) *Physical Geography*. Oxford: Blackwell.
Kirkby, M. J. (1971) Hillslope process–response models based on the continuity equation. In D. Brunsden, compiler, *Slope Form and Process*. London: Institute of British Geographers, Special Publication 3, 15–30.
Kirkby, M. J. (1989) The future of modelling in physical geography. In B. Macmillan, ed. *Remodelling Geography*. Oxford: Blackwell, 255–72.

Kirkby, M. J., ed. (1994) *Process Models and Theoretical Geomorphology*. Chichester: Wiley.

Kirkby, M. J. and Morgan, R. P. C., eds (1980) *Soil Erosion*. Chichester: Wiley.

Lambert, J. H., Jennings, J. N., Smith, C. T., Green, C., and Hutchinson, J. N. (1970) *The Making of the Broads: A Reconsideration of their Origin in the Light of New Evidence*. London: Royal Geographical Society, Research Series No. 3.

Lane, S. N. (2001) Constructive comments on D. Massey 'Space–time, "science" and the relationship between physical and human geography'. *Transactions, Institute of British Geographers*, NS26, 243–56.

Lawler, D. S. (1993) The measurement of river bank erosion and lateral channel change: a review. *Earth Surface Processes and Landforms*, 18S, 777–821.

Leopold, L. B., Wolman, M. G., and Miller, J. P. (1964) *Fluvial Processes in Geomorphology*. San Francisco, CA: Freeman.

Lewin, J. (1977) Channel pattern changes. In K. J. Gregory, ed., *River Channel Changes*. Chichester: Wiley, 167–84.

Lewin, J., Macklin, M. G., and Woodward, J. C., eds (1995) *Mediterranean Quaternary River Environments*. Rotterdam: Balkema.

Lewis, W. V., ed. (1960) *Investigations on Norwegian Cirque Glaciers*. London: Royal Geographical Society.

Linton, D. L. (1933) The origin of the Tweed drainage system. *Scottish Geographical Magazine*, 48, 146–66.

Linton, D. L. (1951a) Problems of Scottish scenery. *Scottish Geographical Magazine*, 67, 65–85.

Linton, D. L. (1951b) Watershed breaching by ice in Scotland. *Transactions, Institute of British Geographers*, 17, 1–16.

Linton, D. L. (1951c) The delimitation of morphological regions. In L. D. Stamp and S. W. Wooldridge, eds, *London Essays in Geography*. London: LSE, 199–218.

Linton, D. L. (1955) The problem of tors. *Geographical Journal*, 121, 470–87.

Linton, D. L. (1957) The everlasting hills. *Advancement of Science*, 14, 58–67.

Linton, D. L. (1965) The geography of energy. *Geography*, 50, 197–228.

Linton, D. L. (1966) Geomorphology by the light of the moon. *Tijdschrift K. ned. aardrijksk. Genoot.*, 83, 249–65.

Linton, D. L. (1968) The assessment of scenery as a natural resource. *Scottish Geographical Magazine*, 84, 219–38.

Lockwood, J. G. (1979) *Causes of Climate*. London: Arnold.

Lowe, J. J. (1991) *Radiocarbon Dating: Recent Applications and Future Potential.* Cambridge: Quaternary Research Association.

Lowe, J. J. and Walker, M. J. C. (1997) *Reconstructing Quaternary Environments*, 2nd edn. Harlow: Addison, Wesley, Longman (1st edn 1984).

Macmillan, B., ed. (1989) *Remodelling Geography*. Oxford: Blackwell.

Mannion, A. M. (1991) *Global Environmental Change: A Natural and Cultural History*. Harlow: Longman.

Marsh, G. P. (1864) *Man and Nature or Physical Geography as Modified by Human Action*. New York: Charles Scribner.

Massey, D. (1999) Space–time, 'science' and the relationship between physical and human geography. *Transactions, Institute of British Geographers*, NS24, 261–76.

McGregor, D. M. and Thompson, D. A., eds (1995) *Geomorphology and Land Management in a Changing Environment*. Chichester: Wiley.

Middleton, N. and Thomas, D. G. (1992) *World Atlas of Desertification*. London: Arnold.

Miller, A. A. (1931) *Climatology*. London: Methuen.

Miller, A. A. (1953) *The Skin of the Earth*. London: Methuen.

Moore, P. D., Chaloner, B., and Stott, P. (1996) *Global Environmental Change*. Oxford: Blackwell.

Newbigin, M. I. (1936) *Plant and Animal Geography*. London: Methuen.

Newson, M. D. (1989) Flood effectiveness in river basins: progress in Britain in a decade of drought. In K. Beven and P. Carling, eds, *Floods: Hydrological, Sedimentological and Geomorphological Implications*. Chichester: Wiley, 151–69.

Newson, M. D. (1995) Fluvial geomorphology and environmental design. In A. Gurnell and G. Petts, eds, *Changing River Channels*. Chichester: Wiley, 413–32.

Newson, M. D. and Lewin, J. (1991) Climatic change, riverflow extremes and fluvial erosion—scenarios for England and Wales. *Progress in Physical Geography*, 15, 1–17.

Oldfield, F. (1963) Pollen-analysis and man's role in the ecological history of the south-east Lake District. *Geografiska Annaler*, 45, 23–49.

Oldfield, F. (1987) The future of the past: a perspective on palaeoenvironmental study. In M. J. Clark, K. J. Gregory, and A. M. Gurnell, eds, *Horizons in Physical Geography*. Basingstoke: Macmillan, 10–26.

O'Riordan, T. (1994) Civic science and global environmental change. *Scottish Geographical Magazine*, 110, 4–12.

Orme, A. R. (1964) The geomorphology of southern Dartmoor. In I. G. Simmons, ed., *Dartmoor Essays*. Torquay: The Devonshire Association 31–72.

Osterkamp, W. R. and Hupp, C. R. (1996) The evolution of geomorphology, ecology and other composite sciences. In B. L. Rhoads and C. E. Thorn, eds, *The Scientific Nature of Geomorphology*. Chichester: Wiley, 415–41.

Owens, S., Richards, K., and Spencer, T. (1997) Managing the earth's surface: science and policy. *Transactions, Institute of British Geographers*, NS22, 3–5.

Parker, D. J., ed. (2000) *Floods*. London: Routledge (2 vols).

Parker, D. J. and Chan, N. W. (1996) Response to dynamic flood hazard factors in peninsular Malaysia. *Geographical Journal*, 162, 313–25.

Penman, H. C. (1948) Natural evaporation from open water, bare soil and grass. *Proceedings Royal Society London A*, 193, 120–45.

Penman, H. C. (1950) Evaporation over the British Isles. *Quarterly Journal of the Royal Meteorological Society*,76, 372–83.

Penman, H. C. (1963) *Vegetation and Hydrology*. Farnham Royal: Commonwealth Agricultural Bureau.

Penning Rowsell, E. C. (1981) Consultancy and contract research. *Area*, 13, 9–12.

Penning Rowsell, E. C. and Chatterton, J. B. (1977) *The Benefits of Flood Alleviation: A Manual of Assessment Techniques*. Farnborough, Saxon House.

Penning Rowsell, E. C., Winchester, P., and Gardiner, J. L. (1998) New approaches to sustainable hazard management for Venice. *Geographical Journal*, 164, 1–18.

Perry, A. H. (1981) *Environmental Hazards in the British Isles*. London: Allen & Unwin.

Perry, A. H. (1995) New climatologists for a new climatology. *Progress in Physical Geography*,19, 280–5.

Petts, G. E. (1984) *Impounded Rivers: Perspectives for Ecological Management*. Chichester: Wiley.

Petts, G. E. and Amoros, C., eds (1996) *Fluvial Hydrosystems*. London: Chapman and Hall.

Price, R. J. (1973) *Glacial and Fluvioglacial Landforms*. Edinburgh: Oliver and Boyd.

Price, R. J. (1979) The future of physical geography: disintegration or integration? *Scottish Geographical Magazine*, 94, 24–30.

Prior, D. B. (1978) Some recent progress and problems in the study of mass movement in Britain. In C. Embleton, D. Brunsden, and D. K. C. Jones, eds, *Geomorphology: Present Problems and Future Prospects*. Oxford: Oxford University Press, 84–106.

Prior, D. B. and Stephens, N. (1972) Some movement patterns of temperate mudflows: examples from north-eastern Ireland. *Bulletin Geological Society of America*, 33, 2533–44.

Quine, T. A., Govers, G., Walling, D. E., Zhang, X., Desmet, P. J. J., and Zhang, Y. (1997) Erosion processes and landform evolution in agricultural land: new perspectives from Caesium-137 measurements and topographic based erosion modelling. *Earth Surface Processes and Landforms*, 22, 799–816.

Raper, J. and Livingstone, D. (2001) Let's get real: spatio-temporal identity and geographic entities. *Transactions, Institute of British Geographers*, NS26, 237–42.

Reid, I. and Frostick, L. E. (1986) Dynamics of coarse bedload transport in Turkey Brook, a coarse-grained alluvial channel. *Earth Surface Processes and Landforms*, 11, 143–85.

Richards, K. S., Brooks, S., Clifford, N., Hams, T., and Lane, S. (1997) Theory, measurement and testing in 'real' geomorphology and physical geography. In D. Stoddart, ed., *Process and Form in Geomorphology*. London: Routledge, 265–92.

Roberts, C. R. (1989) Flood frequency and urban-induced channel change: some British examples. In K. Bevan and P. Carling, eds, *Floods: Hydrological, Sedimentological and Geomorphological Implications*. Chichester: Wiley, 57–82.

Roberts, N. (1989) *The Holocene*. Oxford: Blackwell.

Roberts, N., ed. (1994) *The Changing Global Environment*. Oxford: Blackwell.

Savigear, R. A. G. (1965) A technique of morphological mapping. *Annals of the Association of American Geographers*, 55, 514–38.

Sherlock, R. L. (1922) *Man as a Geological Agent*. London: Witherby.

Shorter, A. H., Ravenhill, W. L. D., and Gregory, K. J. (1969) *Southwest England*. London: Nelson.

Simmons, I. G., ed. (1964) *Dartmoor Essays*. Torquay: The Devonshire Association.

Simmons, I. G. (1974) *The Ecology of Natural Resources*. London: Arnold.

Simmons, I. G. (1979) *Biogeography: Natural and Cultural*. London: Arnold.

Simmons, I. G. (1981) *Biogeographical Processes*. London: Allen and Unwin.

Simmons, I. G. (1987) Order and truth: energetics in biogeography. In K. J. Gregory, ed., *Energetics of Physical Environment: Energetic Approaches to Physical Geography*. Chichester: Wiley, 145–60.

Simmons, I. G. (1991) *Earth, Air and Water Resources and Environment in the Late 20th Century*. London: Arnold.

Simmons, I. G. (1993a) *Environmental History: A Concise Introduction*. Oxford: Blackwell.

Simmons, I. G. (1993b) *Interpreting Nature: Cultural Constructions of the Environment*. London and New York: Routledge.

Simmons, I. G. and Tooley, M. J., eds (1981) *The Environment in British Prehistory*. London: Duckworth.

Sissons, J. B. (1958) Supposed ice-dammed lakes in Britain with particular reference to the Eddleston valley, southern Scotland. *Geografiska Annaler*, 40, 159–87.

Sissons, J. B. (1960) Some aspects of glacial drainage channels in Britain: Part 1. Scottish *Geographical Magazine*, 76, 131–46.

Sissons, J. B. (1961) Some aspects of glacial drainage channels in Britain: Part II. *Scottish Geographical Magazine*, 77, 15–36.

Sissons, J. B. (1976) *Scotland*. London: Methuen.

Sissons, J. B., Smith, D. E., and Cullingford, R. A. (1966) Late-glacial and post-glacial shorelines in southeast Scotland. *Transactions, Institute of British Geographers*, 39, 9–18.

Slaymaker, H. O. and Spencer, T. (1998) *Physical Geography and Global Environmental Change*. Harlow: Longman.

Smalley, I. J. and Vita-Finzi, C. (1969) The concept of 'system' in the earth sciences. *Bulletin Geological Society of America*, 80, 1591.

Smith, D. I. and Newson, M. D. (1974) The dynamics of solutional and mechanical erosion in limestone catchments on the Mendip Hills, Somerset. In K. J. Gregory and D. E. Walling, eds, *Fluvial Processes in Instrumented Watersheds*. London: Institute of British Geographers, Special Publication No. 6, 155–68.

Smith, K. (1992) *Environmental Hazards: Assessing Risk and Reducing Disaster*. London and New York: Routledge.

Smith, K. and Ward, R. C. (1998) *Floods: Physical Processes and Human Impacts*. Chichester: Wiley.

Steers, J. A., ed. (1934) *Scolt Head Island*. Cambridge: W. Heffer.

Steers, J. A. (1948) *The Coastline of England and Wales*. Cambridge: Cambridge University Press.

Steers, J. A. (1973) *The Coastline of Scotland*. Cambridge: Cambridge University Press.

Steers, J. A. and Smith, D. B. (1956) Detection of movement of pebbles on the sea floor by radioactive methods. *Geographical Journal*, 122, 343–5.

Stoddart, D. R. (1967) Growth and structure of geography. *Transactions, Institute of British Geographers*, 41, 1–20.

Stoddart, D. R. (1975a) 'That Victorian Science': Huxley's Physiography and its impact on geography. *Transactions, Institute of British Geographers*, 66, 17–40.

Stoddart, D. R. (1975b) Contribution to discussion. *Geographical Journal*, 141, 45–7.

Stoddart, D. R. (1986) *On Geography and its History*. Oxford: Blackwell.

Stoddart, D. R. (1997a) Richard J Chorley and modern geomorphology. In D. R. Stoddart, ed., *Process and Form in Geomorphology*. London: Routledge, 383–99.

Stoddart, D. R., ed. (1997b) *Process and Form in Geomorphology*. London: Routledge, 340–79.

Stott, P. (1998) Biogeography and ecology in crisis: the urgent need for a new metalanguage. *Journal of Biogeography*, 25, 1–2.

Strahler, A. N. (1992) Quantitative/dynamic geomorphology at Columbia (1945–60): a retrospective. *Progress in Physical Geography*, 16, 65–84.

Straw, A. and Clayton, K. M. (1979) *Eastern and Central England.* London: Methuen.

Sugden, D. E. and John, B. S. (1976) *Glaciers and Landscape: A Geomorphological Approach.* London: Arnold.

Summerfield, M. A. (1991) *Global Geomorphology.* Harlow: Longman.

Tansley, A. G. (1935) The use and abuse of vegetational concepts and terms. *Ecology*, 16, 284–307.

Thomas, W. L., ed. (1956) *Man's Role in Changing the Face of the Earth.* Chicago, IL: University of Chicago Press.

Thompson, R. D. and Perry, A., eds (1997) *Applied Climatology: Principles and Practice.* London: Routledge.

Thorne, C. R., Bathurst, J. C., and Hey, R. D., eds (1987) *Sediment Transport in Gravel-bed Rivers.* Chichester: Wiley.

Thornes, J. B. (1989) Geomorphology and grass roots models. In B. Macmillan, ed., *Remodelling Geography.* Oxford: Blackwell, 3–21.

Thornes, J. B. (1995) Global environmental change and regional response: the European Mediterranean. *Transactions, Institute of British Geographers*, NS20, 357–67.

Thornes, J. E. (1999) *John Constable's Skies.* Birmingham: University of Birmingham Press.

Thrift, N. and Walling, D. (2000) Geography in the United Kingdom 1996–2000. *Geographical Journal*, 166, 96–124.

Tinkler, K. J. (1985) *A Short History of Geomorphology.* London and Sydney: Croom Helm.

Trudgill, S. T. (1977) *Soil and Vegetation Systems.* London: Oxford University Press.

Trudgill, S. T. (1990) *Barriers to a Better Environment.* London: Belhaven Press.

Trudgill, S. T. and Richards, K. S. (1997) Environmental science and policy: generalizations and context sensitivity. *Transactions, Institute of British Geographers*, NS22, 5–12.

Tunstall, S. M. and Penning-Rowsell, E. C. (1998) The English beach: experiences and values. *Geographical Journal*, 164, 319–32.

Unwin, D. (1989) Three questions about modelling in physical geography. In B. Macmillan, ed., *Remodelling Geography.* Oxford: Blackwell, 53–7.

Viles, H., ed. (1988) *Biogeomorphology.* Chichester: Wiley.

Vita-Finzi, C. (1969) *The Mediterranean Valleys.* Cambridge: Cambridge University Press.

Walling, D. E. (1974) Suspended sediment and solute yields from a small catchment prior to urbanisation. In K. J. Gregory and D. E. Walling, eds, *Fluvial Processes in Instrumented Watersheds.* London: Institute of British Geographers, Special Publication No. 6, 169–91.

Walling, D. E. (1979) The hydrological impact of building activity: a study near Exeter. In G. E. Hollis, ed., *Man's Impact on the Hydrological Cycle in the United Kingdom.* Norwich: Geobooks, 135–52.

Walling, D. E. (1983) The sediment delivery problem. *Journal of Hydrology*, 65, 209–37.

Walling, D. E. (1996a) Suspended sediment transport by rivers: a geomorphological and hydrological perspective. In *Suspended Particulate Matter in Rivers and Estuaries. Archiv für Hydrobiologie Special Issues: Advances in Limnology*, 47, 1–27.

Walling, D. E. (1996b) Erosion and sediment yield in a changing environment. In J. Branson, A. G. Brown, and K. J. Gregory, eds, *Global Continental Changes: The Context of Palaeohydrology*. London: Geological Society, Special Publication 115, 43–56.

Walling, D. E. and Gregory, K. J. (1970) The measurement of the effects of building construction on drainage basin dynamics. *Journal of Hydrology* 11, 129–44.

Walling, D. E. and Webb, B. W. (1980) Patterns of sediment yield. In K. J. Gregory, ed., *Background to Palaeohydrology*. Chichester: Wiley, 69–100.

Walling, D. E. and He, Q. (1999) Changing rates of overbank sedimentation on the floodplains of British rivers during the past 100 years. In A. G. Brown and T. A. Quine, eds, *Fluvial Processes and Environmental Change*. Chichester: Wiley, 207–22.

Ward, R. C. (1971) *Small Watershed Experiments. An Appraisal of Concepts and Research Developments*. Hull: University of Hull, Occasional Papers in Geography No. 18, 254.

Ward, R. C. (1979) The changing scope of geographical hydrology in Great Britain. *Progress in Physical Geography*, 3, 392–412.

Waters, R. S. (1958) Morphological mapping. *Geography*, 43, 10–17.

Waters, R. S. (1961) Involutions and ice wedges in Devon. *Nature*,189, 389–90.

Waters, R. S. (1962) Altiplanation terraces and slope developments in VestSpitsbergen and south-west England. *Biuletyn Peryglacjalny*, 11, 89–101.

Webb, B. W. (1995) Regulation and thermal regime in a Devon river system. In I. Foster, A. Gurnell, and B. Webb, eds, *Sediment and Water Quality in River Catchments*. Chichester: Wiley, 65–94.

Webb, B. W. and Walling, D. E. (1980) Stream solute studies and geomorphological research: some examples from the Exe Basin, Devon. *Zeitschrift für Geomorphologie*, 36, 245–63.

Werritty, A. (1995) Geomorphology in the UK. In H. J. Walker and W. E. Grabau, eds, *The History of Geomorphology*. Chichester: Wiley, 457–68.

Werritty, A. (1997) Chance and necessity in geomorphology. In D. R. Stoddart, ed., *Process and Form in Geomorphology*. London: Routledge, 312–27.

Whalley, W. B., ed. (1978) *Scanning Electron Microscopy in the Study of Sediments*. Norwich: Geo Abstracts.

White, I. D., Mottershead, D. N., and Harrison, S. J. (1984) *Environmental Systems: An Introductory Text*. London: George Allen & Unwin (2nd edn 1992).

Whittow, J. B. (1980) *Disasters: The Anatomy of Environmental Hazards*. London: Allen Lane.

Wilby, R. A., ed. (1997) *Contemporary Hydrology*. Chichester: Wiley.

Wilkinson, H. (1963) *Man and the Natural Environment*. Hull: Department of Geography, University of Hull, Occasional Papers in Geography No. 1.

Williams, R. B. G. (1968) Some estimates of periglacial erosion in southern and eastern England. *Biuletyn Peryglacjalny*, 11, 89–101.

Wooldridge, S. W. (1932) The cycle of erosion and the representation of relief. *Scottish Geographical Magazine*, 48, 30–6.

Wooldridge, S. W. (1958) The trend of geomorphology. *Transactions, Institute of British Geographers*, 25, 29–35.

Wooldridge, S. W. and East, W. G. (1951) *The Spirit and Purpose of Geography*. London: Hutchinson.

Wooldridge, S. W. and Linton, D. L. (1939) *Structure, Surface and Drainage in South East England*. London: Philip.

Worsley, P. (1979) Whither geomorphology? *Area*, 11, 97–101.

Wright, W. B. (1914) *The Quaternary Ice Age*. London: Macmillan.

Wright, W. B. (1937) *The Quaternary Ice Age*. London: Macmillan.

Young, A. (1960) Soil movement by denudational processes on slopes. *Nature*, 188, 120–2.

Young, A. (1964) Discussion of slope profiles: a symposium. *Geographical Journal*, 130, 80–2.

Young, A. (1971) Slope profile analysis: the system of best units. In D. Brunsden, compiler, *Slope Form and Process*. London: Institute of British Geographers, Special Publication No. 3, 1–13.

Zeuner, F. E. (1945) *The Pleistocene Period: Its Climate, Chronology and Faunal Successions*. London: Hutchinson.

Zeuner, F. E. (1946) *Dating the Past: An Introduction to Geochronology*. London: Methuen.

4

The domestication of the earth

humans and environments in pre-industrial times

IAN SIMMONS

Weaving a text

The domestication of the earth entails the enfolding of 'nature' into human life and society, a topic discussed by many academic disciplines and practised by many millions more of our species. In this chapter, attention is particularly concentrated upon the millennia of the Holocene, when the human societies were food-collectors and agriculturalists who essentially lived off recently fixed solar energy. The period since about 1750 AD, when fossil fuels began to subsidise most of the relationships of humans to their surroundings, is dealt with in other chapters.

In the course of its last 100 years, geography has from time to time taken in, and focused its attention on, diverse approaches to its subject matter: these are discussed by the editors in the introduction. But as a ground bass to these variations the human–environment relation has persisted, though sometimes virtually at *sotto voce* level. In part, geography's attention has concentrated on landscapes as visible demonstrations, past and present, of these interrelations (as taken up by Michael Williams in Chapter 5), but it has also taken an approach based explicitly on late-twentieth-century ecological theory. Here, the dynamics of energy and matter flows that exhibit themselves as populations of living things (including humans), as well as inanimate matter in states of flux, form the epistemological basis of study. Hence, there was one unambiguous effort to attach the label 'Geography as Human Ecology' (Eyre and Jones, 1966). Historical geography has often looked to ecological models and living phenomena in studies of, for example, the rise of nature conservation (Sheail, 1976). Later approaches to the past have swung the emphasis towards, for example, the exercise of power within ecological systems and the way in which human societies control access to the resources of the natural world: 'Political Ecology' (Wilson and Bryant, 1997; Low and Gleeson, 1998; Stott and

Sullivan, 2000) has been one outcome. In the wider context of the social sciences there have been developments such as ecological anthropology and environmental economics. In the present context the emphasis must be on the sub-discipline of environmental history, pioneered in the United States by scholars such as William Cronon and Donald Worster (e.g. Cronon, 1983; Worster, 1994) but taken up by other nationalities in the European Society for Environmental History. The line taken here has usually reflected an emphasis on the local regulatory mechanisms and their political context: the minutiae of human agency and their environmental consequences. In any of these discourses, the use of ecological concepts has perforce had to move from ideas of balance and equilibrium to those of variability in space and time, and the effects of scale and temporal dynamics, all of which comprise change, chance, and contingency; in effect, the processes treated by complexity and chaos theories (Scoones, 1999).

The geographical approach has been especially characterised by the collection of immense bodies of data about the earth in both its natural (see Chapter 3) and its human forms—hence studies that draw upon the techniques of palaeoecology (e.g. pollen analysis, isotope ratios, radiometric dating, dendrochronology, and more recently, 'fossil' DNA). The investigation of human behaviour has probed from the spatial and aggregative (as in much 'regional' geography up until the 1950s) down to the finer tendrils of the human psyche and thus has much in common with the ethnographic methodology of, for example, anthropology and sociology. This move towards the individual has inevitably brought with it a greater role for the humanities in terms of the representation of the environment and its human inhabitants, and the analysis of the relationship in analytical modes such as philosophy and retrospective hermeneutics, and in the history of ideas. Such ways of thinking and writing have had their effect on the consideration of ecologically based ideas such as population growth and 'sustainability'.

As in many other discussions outside the natural sciences, the terms for the non-human, such as 'nature' and 'environment', are often between quotation marks because nobody is sure what such terms may mean. On the one hand, there is an inherited position of philosophical realism in which the non-human is ontologically autonomous—if humans were not present then it would continue to exist and evolve: the Holocene is simply a geological era dominated by humans (Crutzen and Stoemer, 2000). At the other extreme is the idealist position that holds that 'nature' is a totally cultural and indeed symbolic construction and therefore possesses a plasticity in the face of cultural demands and interpretations that is bounded only by laws like those of gravity and thermodynamics. An intermediate persuasion (adopted in this chapter) holds that there are indeed autonomous phenomena 'out there' but that our cognition of them (even with instrumentation to aid human senses) is always imperfect and may lead to false models of their essence. As J. B. S. Haldane put it, 'nature is not only queerer than we suppose, but it is queerer than we *can* suppose', thus implying the primacy of human language in any discussions of what the words might represent (quoted in Soper, 2000). He also indicates that neither nature nor its representation is a 'given' (Everden, 1992; Ingold, 2000). At a more realist level, the notion of a total metabolism of a society as a way of constructing a theory of society–nature interaction has been taken on from the functionalist tradition of cultural anthropology. The notion of 'colonisation of natural processes' is seen as an emergent property beyond the ideas of functional adaptation that were the focus of earlier studies (Fischer-Kowalski, 1998; Haberl, Batterbury and Moran, 2001). Not surprisingly, the concepts of human–machine hybrids, and the qualities of a human–machine–nature that disavows the existence of only two ontological categories and looks sharply at the creative presence of the non-human and the human–machine hybrid (Whatmore, 1999) is undergoing evaluation, though its extension to the early Holocene is as yet untested to any extent. Suggestions can, however, be made that pre-modern human groups live within hybrid worlds just as much as those of us who depend upon an 'advanced' technology (Strathern, 1996).

Any approach to even the incomplete investing of nature by humanity must therefore encompass two narratives. First, there has to be an agreed empirical chronicle of what happened, where and to what and whom. This is never easy because of the nature of historiography: research and its authors are likely to be positioned, for example. In any case, this story has to reach beyond the written records into the archaeological and palaeoecological spheres where the material evidence may well be sparse. In spite of difficulties, there has to be a continuing attempt to provide information about the interaction of hunter-gatherers with plants and animals, about the time and place when the actual genetic modification of biota by human groups becomes visible in the archaeological record, and about the way in which long-lasting agricultural systems such as *padi* rice production (to take just one example) have been adapted to climatic change and population growth, together with spiritual and technological innovation. In addition to any such account there must also be an appreciation of any cultural development of human societies that impinges on their use of the land and water of this planet. Metamorphoses of thought and outlook, the adoption of different technologies and the abandonment of certain levels of social control over land use can all be quoted as key factors in the human–environment relationship at some time in a particular location. There has to be an appreciation of domestication of the human mind as well as the cereals and the sheep. Writing, for example, is likely to force a different scheme of knowledge on a society, since making lists enforces a separation of one category from another not demanded by oral transmissions of information (Goody, 1977).

Beyond both of these intellectual endeavours there is the more difficult task of creatively mixing together the first and second strands of enquiry.

Without both, neither can be fully understood, but with both together the whole is immeasurably more than the sum of its parts. All the tasks of finding better languages, theories and data are still before us. As with early agriculture, there were grass seeds and there was water: somebody turned them into bread.

The worlds of the hunter-gatherers

Britain: nature and culture in metamorphosis

During the glacial phases of the Pleistocene, the British Isles were not a place to be seen. Though lower sea levels made a physical unity of the continent that was to become Europe, even the less vigorous advances of the ice left little habitat for the genus *Homo*. The intervals between ice-bound phases, both short interstadials and longer interglacial periods, were different. Sufficient time elapsed for the colonisation of temperate flora and fauna at the very least, and in some intervals a tropical set of ecosystems was present. More and better archaeological work has recovered human remains from these phases, as well as the stone tools that for many years were the markers for human presence, culture, and chronology, a picture now supplemented by many other types of evidence, most notably for absolute dating (Bridgland, 1998). There is little evidence to suggest anything other than the conventional picture: that the hominid groups were hunters and gatherers living at very low densities and taking the usufruct of land and water. There was a brief flurry of argument about a phase in the interglacial at Hoxne in Suffolk, where pollen analysis suggested the concatenation of worked flints, charcoal, and increased grass representation at the expense of trees. The biologists who commented thought that this denoted a phase of climatic cooling or aridity. Archaeologists were more inclined to the excitement of humans creating forest openings by controlling fire at the landscape scale: in this case *Homo erectus* (a species generally thought to have achieved that technique in Africa) was promoted as the first known creator of cultural landscapes in Britain. Later discussions of this and similar sites have, however, tended to homogenise around the climatic change explanation (Mullenders, 1993), and even the 'layers of grass pollen due to hippopotamus shit in the bottom of the lake' hypothesis seems to have been quietly sedimented down.

The quantity and quality of evidence improves markedly in the Devensian and early Holocene eras. Enough sites are found and the chronology of biotic change is so well established that firmer statements about the likelihood or otherwise of human manipulation of the environment can be made. Even so, ambiguity persists. In the late Devensian and earliest Holocene of the easternmost Vale of Pickering (Yorkshire), for example, the deposits reveal numerous episodes of high charcoal frequency, as do their equivalents in the Late Upper Palaeolithic of the Netherlands and Germany. Once definitely into the Early Mesolithic of the same area, then palaeoecology at Star Carr and eastwards in the Lake Flixton of the time shows phases of burning of the lake's marginal reedswamp during the early Holocene (Mellars and Dark, 1998). The dryland ecosystems of the time (mainly open woodland of birch and hazel) were little affected by fire. But were these deposits laid in lake deposits as a result of natural fires in a boreal landscape under a continental climate, or are they proxies for large-scale human activity of a resource-enhancing nature? Current levels of interpretation provide insufficient guidance for firm statements.

The Later Mesolithic is of course the last phase of hunter-gatherer occupation before the oncoming of agriculture; it also saw the insulation of these islands as sea level rose. In Great Britain the evidence for environmental relations tends to be best for the Highland Zone (*sensu* Sir Cyril Fox) because later land use has been less intensive and so destroyed less of the evidence, and also because downwash from hillslopes has buried the Mesolithic land surfaces of many valleys so that systematic plotting of sites is impossible. So the upland areas of England, Wales, and the Southern Uplands, together with the Scottish Highlands and Islands, provide us with many data about Mesolithic cultures and their linkages with nature. (Research was pioneered by G. W. Dimbleby as an Oxford forester and later a London environmental archaeologist, and put into a coherent theoretical framework by Cambridge archaeologist P. A. Mellars; current scholarly interactions have been summarised in Edwards and Sadler (1999) and by Mithen (1999).) An indication of the role of humans was provided for Dartmoor when it seemed that the growth of deciduous forest up the altitudinal gradient of the granite boss was being retarded by human activity, which centred on the use of fire. This suggestion has received ample confirmation from the Pennines, where work by the biologist J. H. Tallis has plotted the upper edge of deciduous woodland during the later Mesolithic and produced convincing evidence that grassy habitats plus hazel (*Corylus avellana*) scrub were preferred to closed deciduous woodland, a mix that could be achieved with the use of controlled fire (Tallis and Switsur, 1990).

The environmental relations of upland England and Wales have been explored in some detail. Here, fine-resolution pollen and stratigraphic analysis have allowed some reconstruction of the interaction of natural and cultural factors, as well as suggesting different ways of using the resources of the environment. It is reasonable to postulate a sequence of events leading to ecological change in which both natural and human-directed processes have important roles. A possible progression might be to start with mixed oak forest on the uplands, covering almost all the terrain, though broken where there was open water and mire accumulation in water-receiving sites. The upper edge of woody vegetation was marked by a hazel scrub. The forest underwent normal processes of death and renewal, which included gaps of various sizes caused by windthrow. The lower edge near streams underwent a sharp transition to an alder wood with other deciduous species, including elm. Openings in the forest with a ground cover of grasses and herbs attracted mammal herbivores differentially on a seasonal basis. This had two consequences: the concentration of animals made regeneration of the woodland less likely; and humans noticed the concentration of animals and wished to maintain and/or enhance it. Yet, none of the human-induced processes totally replaced natural events: natural openings continued to be formed by natural processes.

Increasing human populations (or more resource-hungry groups) may have tried to create extra openings in the woodlands or at their ecotones. The production of edges and openings away from natural margins was carried out by ring-barking rather than fire. This process was aided by the barking propensities of concentrations of red deer, especially in severe winters. They did this by opening the canopy by breaking off leafy branches, which were also useful in attracting animals to feed, especially in winter; their merits did not entirely depend upon human presence and they could be left behind as a general encouragement, for instance, rather than be immediate bait. This process resulted in some canopy lightening.

The management of clearings and forest edge was not directed solely at *Corylus* and its production of browse and nuts. Burning took place in the period of the autumn when conditions might be dry enough for a controlled ground fire. This might not occur every year. The incidence of fire might have interacted with gathering bracken rhizomes as a food. Visiting such areas for hunting might, however, have been a longer season: if *Bos* was being hunted then it might have coincided with the ground flora production season in the streamside woodlands; deer were also likely visitors since this is the season when they are most likely to feed off herbaceous material. At North Gill in the North York Moors, this type of woodland disappeared in the sixth millennium BP, to be replaced by encroaching blanket bog. More generally in the mid-Holocene, climatic change and soil maturation brought about the accumulation of *mor* humus over podsolic and gleyed soils. This predisposed scrub to turn to a heathy vegetation. Some of the openings underwent rises in water table and became invaded by rushes; thereafter peat accumulation began to get under way, even on ground that was water-shedding rather than water-collecting in a micro-topographic sense.

The presence of charcoal in late-glacial deposits in isolated places like the Isle of Rum, and good evidence for human-caused impact at around 10,320–9,910 BP, allow the interpretation that there was some very early occupation of those districts: there is even a tanged point of an Ahrensburgian type (Edwards, 1993). Likewise, the presence of fire often characterises discoveries in the Scottish Mesolithic. There are diachronous increases in charcoal at many sites and there is often charcoal in places without any other evidence of human presence in parts of the Western Isles and Shetland. Other pollen evidence of open vegetation has sometimes been interpreted as deriving from high densities of wild animals. The establishment of an overall occupation of Scotland by the later Mesolithic seems to have included environmental relationships such as the transport of deer meat (possibly already butchered) to islands such as Oronsay, year-round occupation at favourable places like the Solway Firth, and intensified use of land resources when part of the lower terrain was under a form of agricultural use (Finlayson and Edwards, 1997). The onset of cereal use seems to have been early by UK standards and its incorporation into a broad-band resource strategy suggests adoption of a new practice by an existing people rather than their replacement with an immigrant population.

The world context

Whereas many parts of Europe (and southern Scandinavia in particular) tell a similar story to that of Great Britain, there are events and processes elsewhere that seem to have no parallel within the confines of this little extension of the Eurasian continent. For example, the great extinctions that occurred at the end of the Pleistocene in North America, when many genera of large mammal herbivores were extirpated, have no apparent equivalent in any part of Europe, nor indeed in much of Asia. In Africa, the Upper Pliocene–Lower Pleistocene saw the extirpation of several species of large herbivores, sabre-toothed cats, giant cercopithecids, and the robust Australopithecines (Schuster and Schüle, 2000). These phenomena have naturally been the subject of controversy in terms of the balance between climatic factors that may have caused the diminution of such populations and the impact of the new predator whose social cooperativeness put even the wolf pack into a Second Division: more Swindon Town than Arsenal. A similar scale of extinctions occurs when humans arrive on formerly uninhabited islands (as in New Zealand, Madagascar, and Australia), and so the case for the dominance of the new hunters is strengthened. Mass killing has been detected in the High Plains of North America where Palaeoindian groups ran herds of buffalo over cliffs and into box canyons and dune crescents; fire was often used to help stampede the animals. There were so many of them that it took industrialisation to wipe them out, but nevertheless the killing was massive on a local scale; in southern Alberta there is a World Heritage Site commemorated as Head-Smashed-In Buffalo Jump (Krech, 1999). So far as is known, no West European mammal lived in herds of that scale once the tundra had been replaced by forests, so mass extermination was less likely. But for every place for which there is evidence of mass killing, there are many without any testimony for or against, and so the likelihood of it being common then is now always present. But the utility of fire to hunter-gatherers seems to have been virtually worldwide and detailed studies of near-recent groups in California, for example, show how many different environments and purposes were served by its use (Anderson, 1994); a parallel view of oak management in north-west Europe has also been published (Mason, 2000). Lewis (1994) has demonstrated for Australia the problems raised by viewing aboriginal use of fire through the lenses of our own time; a re-evaluation of it confirms that economic production and biodiversity maintenance go alongside social and religious needs (Yibarbuk et al., 2001).

Adaptations of materiality and mind

It was once common to label hunter-gatherers as 'children of nature', the implication being that they lived off the land but had no impact on it. This may have been true in a few environments: the north slope of North America for example, where animal hunting before European contact seems scarcely to have dented the populations of the migratory caribou (which remains undomesticated, unlike the reindeer) or the sea mammals. But in many places there is evidence that, like all children, the hunters were exploring the limits of what they could access and subjecting it to strains. These would have been especially manipulative if the ecosystems were under pressure from natural factors such as climatic change. One area of uncertainty is the antiquity of practices that could be called husbandry in the sense that they led to an increased level of attention from human groups without necessarily involving the conscious selection of genetic characteristics, which defines domestication in individual species. Regular watering of a stand of large-seeded wild grasses at one point in a seasonal round, for example, would leave little if any trace in the archaeological record. The example of the dog, however, ostensibly domesticated independently by separate hunter groups from more than one species of wolf, demonstrates the ability of humans to recognise potential and to organise another species to provide it.

What is normally read back into archaeological contexts is the assumption of a high level of 'traditional ecological knowledge'. This extends to plant and animal classification as well as to views of biological interrelationships. These forms of knowledge are analogous to western science but are much wider in their reach. They extend to social matters as well as spiritual relations and indeed cosmic ethics. So the principles upon which intervention in matters of resource and land management are based may well include not only knowledge of the living things themselves but also of the spirit world that illuminates and informs their entire worldview (Inglis, 1993; Fowler and Turner, 1999). It has to be acknowledged that the Palaeolithic finds that lead to inferences about the kind of religion practised might represent marginal and secondary features of it (Narr, 1987). It is a mistake, of course, to think of hunter-gatherer people as having a uniform cognition of the world: a division into shamanistic and totemistic societies may have led to different acceptances of the material world as well as expressions of it in art (Layton, 2000).

The domestication of plants and the dog raises questions of intentionality or purpose that cannot be ignored in any study of the human species and its cultures. In turn, this characteristic exists within a human world that is never totally material. Resources may not be used because their terrain is occupied by malign or evil spirits, for example, or giant shrews or titanic pike. But beyond such immediate prohibitions there lies an immense body of myth and religion which, as the anthropologist Tim Ingold (1986) has demonstrated, is replete with consequences for human relations with the natural world. Prehistoric art, for example, might be a way of engaging with the world rather than simply representing it. (So might Egyptian tomb paintings, and Renaissance oils, where the powerful line up with the Virgin; but somehow not Monet's later waterlilies.) There must have been a great deal of diversity in the past (direct evidence for Britain is lacking) and so generalisations are prone to exceptions. Nevertheless, common traits recur many times: in shamanistic groups there has been the notion that each class of animals has a spiritual master who may or may not present the animals to humans for use and who demands the proper treatment for them, for instance. Hunting secures the regeneration of those flows on which human life depends and killing animals is thus part of an act of world renewal, which unblocks the flow of life. The mediator between the levels of being—the wild animals, 'domestic' animals, humans, animal masters, and a Supreme being—is the shaman, who can travel between the different levels of the cosmos. For the circumboreal peoples, the bear was a special sort of shaman who was probably a person in disguise (the bear is a relatively anthropomorphic animal) who had no species master, only the Supreme being. To kill a bear or bear cub and then to treat the carcass to, for example, feasting was designed to attract more bears. Maritime people often had the same view of whales and walruses.

Comparative cultural anthropology contributes in another way to studies of human roles within environments. It is common for band-based societies to claim land based on the ownership of sites rather than areas, so their boundaries may be indefinite. Travel routes strongly influence the pattern of topographic features, which are not merely way-points but parts of a fundamental reference system in which an individual consciousness of the world and social identities are anchored. The surroundings of humans may also be aware, sensate, personified and can be offended. Overall, as the archaeologist Christopher Tilley (1994) has deduced from studies in, for example, North America and Australia, the flow of the land becomes a flow of the mind: moose are good to eat and moose are good to think. In essence, for hunter-gatherers there was no dichotomy between humans and nature but a single world embracing the people, the plants and animals, and the total landscape through which they all moved, and very likely the rest of the cosmos as well (Ingold, 2000). If we follow Mithen's (1998) arguments, then this came about because of the development of a cognitive fluidity during the period 60,000–40,000 years ago, when 'modern' humankind became the sole survivor of the genus. The notion of a change in consciousness can be carried through to the development of agriculture and might tentatively be placed in the arena of the history of ideas surrounding industrialism.

One theme that emerges from many accounts is the centrality of fire. In a probable continuity from hunting times, the reindeer herders of Russia 'fed' the blood of a sacrificed reindeer to the fire—the sacred centre of every dwelling The fireboard used to generate fire by drilling is a magical guardsman of the herd. The utility of fire at both landscape and domestic scales almost forces the acknowledgement that its utilisation is more mental than technical and that its mastery may well have brought about new social patterns in whichever hominid achieved that stage (Goudsblom, 1994). Fire must be seen also as an element in any nexus linking the material and non-material worlds (Pyne, 1994). Oelschlager (1991) asserts that religion does not develop in complete independence and isolation, but has a functional interaction with processes of an ecological and economic kind. (Marx said much the same kind of thing, with greater certainty of language.) Fire is one obvious expression of the intersection of all kinds of worlds: while off the mid-Atlantic coast of North America in 1632 the Dutch mariner David de Vries wrote of land that was smelt before it was seen.

Farming: evolution and revolution

Britain into domestication

A curiosity of an account like the present one is to treat Britain first when it was in fact on the periphery of developments. The actual evolution of *Homo* and its various species, and the learning of the control of fire, took place elsewhere and were only brought to Britain millennia later, as the ice loosened its grip from time to time. Likewise agriculture's major hearthlands were a considerable distance from a set of peri-Atlantic islands, and the products of the southwest Asian focus of domestication (notably cereals like wheat and barley, together with sheep, goats, cattle, and pigs) arrived as if transported by British railways rather than the TGV. Nor, by the time they arrived, was it possible to walk to and from the rest of non-insular Europe—an impossible feat until 1994.

So an economy that had its origins in the hill-lands of southwest Asia and that had become the foundation of the great riverine civilisations of the Old World starts to show up in archaeological and palaeoenvironmental evidence about 5,500 radiocarbon years before the present. The method of its transmission was once securely placed in the 'immigrant groups with new techniques' mode, but the alternative 'handing on of new ways to indigenous folk' has latterly gained ground, with evidence from DNA, for example, suggesting that in northwest Europe the diffusion lineages resulting from agricultural demes are quite small (Semino et al., 2000). British DNA does not support an extreme indigenist position but is inclined to support the notion of regional social interaction and intermarriage (Evison, 1999). The extent to which new ways encouraged the domestication of the environment have also been much debated, mostly in the archaeological community but with some contributions from geographers working in palaeoecological studies (Edwards and Sadler, 1999). In bald outline, the received picture of 'Neolithic' farmers has been that they engaged in land clearance, most often on the basis of shifting agriculture. The Danish *landnam* model of J. Iversen (1941) has been very influential, if indeed not foundational. The close analysis of Neolithic artefacts and remains of structures of all kinds, coupled with environmental information, suggested that more permanent settlement with fixed fields might have been possible, while a third approach places great store on the existence of local traditions and modes of subsistence, with regional variations likely in the amount of domestication of both species and landscapes. On the floodplains of Midland rivers, forest recession in the Neolithic was small scale, succeeded by larger-scale activity in the Bronze Age, and capped with almost total deforestation and apparent use for grazing in the Iron Age. Brown (1999) notes that the larger-scale clearance of the Bronze Age accompanies ritual use of the floodplains of the Soar and Nene and suggests that vegetational change may have been driven by the ritualisation of the landscape as much as its conversion to agricultural use. Changes in social order are held to have presaged movement onto different kinds of soils in southern England between the Late Neolithic and the Middle Bronze Age, not the other way around (Brück, 2000).

If, indeed, agriculture in the shape of domesticated plants and animals had been handed on rather than imposed, then this makes good sense provided either that the indigenous traditions were themselves sufficiently various to produce such a mosaic of practices or that the new ways were flexible enough to work in more than one fashion (Thomas, 1999). Certainly, the rather hesitant-looking spatters of cereal pollen in profiles whose radiocarbon dates could be either Mesolithic or Neolithic look like an existing community experimenting with something new, probably not always successfully (Simmons and Innes, 1996). This too fits in with a model of forest fragmentation that supposes that some recession of woodland came about from the inadvertent prevention of natural regeneration (Brown, 1997).

Out of the uncertainty of the earliest farming centuries comes a greater conviction that once into the ages of metal, the imprint of human society became greater and its traces more reliably detectable. Evidence from Dartmoor and the North York Moors, for example, shows large-scale land divisions, with the suggestion that communities divided up the landscape into 'estates' with access to a spectrum of natural resources in terms of land and water (Fleming, 1988; Spratt, 1989). Given that something of this order persisted through into the Iron Age, then the stage is set for historical geographers such as Glanville Jones (1990) to elaborate a 'multiple estate' model into the basis for much land-holding in early medieval times. Most reconstructions of the environment of Britain in the 'Dark Ages' and early medieval periods place considerable emphasis on the shrinkage of the truly wild, in the sense that many woodlands were now made into wood-pasture and coppice or cleared for agriculture. By 1200, all the woods in England were owned and often intensively managed; Rackham (1976) suggests that the last virgin (i.e. unmanaged) woodland was in the Forest of Dean in about 1150. These trees were felled to make timbers for the Dominican friary at Gloucester and were huge, being about 68.5 cm in diameter and 15 m in usable length. The Great Hall of the Bishop of Hereford also has some fine timbers of medieval date, emulating a stone structure. Created as open environments in prehistoric times, the uplands were certainly wild but their vegetation was in many places a result of grazing regimes as well as climatic changes. Some places were largely pristine ecologically: the Fens and the Somerset Levels, for example, were not devoid of human presence in say AD 600, but it was of the kind that was unlikely to have seriously affected bird and fish populations, and produced only low levels of successional deflection by summer grazing and reed harvesting. Later medieval peat cutting and marginal reclamation was another matter.

So, by the important (though not universally present) datum line of Domesday Book, the environment of lowland Britain was thoroughly domesticated, except where it was usually wet or very sandy: there were fens, bogs, saltmarshes, mudflats, beaches, and dunes that were little altered by human hands. Towns, farmlands, heaths, and woodlands depended for their form and function upon human societies. In the uplands, an illusion of wildness certainly prevailed, with mountains and moorlands being available resources (witness the number of *-shield* and *-sett* placenames in Norse-settled areas) but not areas that easily yielded their usufruct, and unlike today they were often occupied by domesticates only in summer. Where red deer populations survived, however, they were obvious locations for the pursuit of aristocratic pleasure. Though the written account of the chase of deer, fox, and wild boar chronicled in

Sir Gawain and the Green Knight probably dates from the 1370s, its emotional lineaments that treat of nature may well be older than the Norman conquest. The uplands too were often sources of tin, iron, and lead, with mining and smelting adding to the impress of resource use upon land and water.

The wider world of food production

If the establishment of agriculture in Great Britain still presents all kinds of problems of understanding, then it is not surprising that the very origins of the domestication of plants and animals is also shrouded in wreaths of differing interpretations. It is still open season on 'why', given that independent evolution in differing heartlands seems to be accepted. The major question that has emerged, especially from Barbara Bender's writings (e.g. Bender, 1978), is whether social development called forth agriculture as a response or whether it was agriculture that allowed the social and ideational changes of the kind mentioned below. In 1970, for example, Eric Isaac placed great store on the role of religion in the origins of agriculture, seeking ceremonial foundations for the bringing-in of animals to the human fold. On Fiji the population moved downslope in the face of soil erosion, and coping with the change from swidden to swamp taro seems to have confirmed the role of 'heroic chiefs': thus the economic may have depended upon the spiritual (Bayliss-Smith et al., 1988). The key role of a priesthood in controlling water distribution on Bali in recent times has rather confirmed long-extant ideas about the necessity for a system of central control in irrigation-based societies (Lansing, 1991; Lansing and Kremer, 1993; Anderson, 1996) in which 'co-adaptation' is so close-coupled that temporal priority seems an inapplicable concept.

Workers from many backgrounds have been concerned with the question 'why', rather more than the 'how'. Carl Sauer's (1952) provocative contribution to some extent merged the two. It provided an impetus for the collection of great quantities of archaeological, genetic, and folk-practice data, much designed to bolster or dethrone his argument that Mesolithic people living in woodlands with access to freshwater fish resources in southeast Asia were the cradle of true agronomy. His heir in terms of the systematising of the most reliable of the empirical evidence into a coherent set of processes that ought to have a wide applicability is D. R. Harris. The cultural–ecological nexus of settling down and breaking the ground in southwest Asia involved a phase of sedentary foraging. It then exploited the special opportunities of the region in the full incorporation of herd animals into the processes of agricultural production, which is seen as a progressive (if sometimes slow and intermittent) intensification of food production and processing (Harris, 1990; Harris and Hillman, 1989) (Fig. 4.1). Mithen (1998) argues that it was certainly not merely

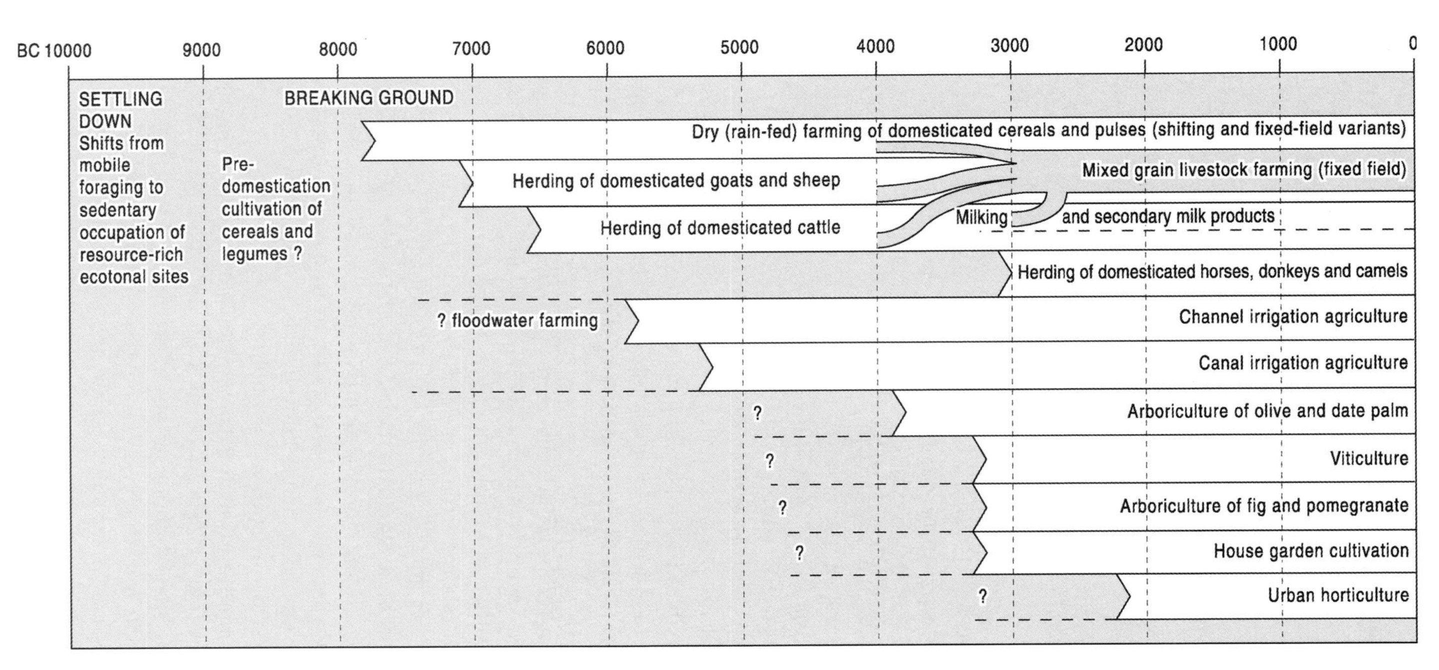

Fig. 4.1 A general model of the ecological relations of domestication of plants and animals in southwest Asia, as formulated by D. R. Harris. The key stage of 'settling down' must be emphasised, since this may have contained an important social component as well as a response to any relevant environmental factors.
Source: Harris (1990, fig. 4).

the accumulation of knowledge about plants and animals that led to agriculture. Using evidence from southwest Asia, he argues for the primacy of sedentarisation of some hunters when faced with short and sharp climatic changes and then for metamorphosis of the human ability to conceive of the world in new ways. Tool usage, power symbolism, and the ability to develop 'social relations' with plants and animals, allied to the lifting of any sanctions on manipulation of their populations, were the knot that tightened the new life (not always as well nourished as the old) and eventually, if sometimes indirectly, suffocated the old.

Once adopted as a successful mode of subsistence, the spread of agricultural practices proceeded in a number of ways. Vigorous groups engaged in territorial conquest; more reticent people engaged in trade or, occasionally, calm persuasion. Thus, the extension of domesticates into different environments necessitated adaptation. Probably the most widespread of these is the cultivation terrace, which in its high visibility on the land and spread across many cultures at one time attracted attention by geographers, who have mapped its regional and worldwide distributions (Spencer and Hale, 1961; Donkin, 1979) and, in the case of Paul Wheatley (1965), attempted a polymath excursion into origins that incorporated a dismissal of almost all existing work on the topic. The terrace proves to be one of a spectrum of techniques for maintaining the productivity of pre-industrial agroecosystems (and fisheries) which in total have the capacity to amaze by their thoroughness: the make-over of every cubic centimetre of ground seems to be complete once the soil has been broken (Klee, 1980). The work on the diffusion of techniques has been paralleled, indeed possibly exceeded, by accounts of the transfer of species across and between continents. Powerful states attracted many imports, as Schafer (1963) demonstrated for China during the T'ang dynasty (AD 618–906). Such processes have never been confined to the deliberate movement of immediately useful domesticates but have included accidental imports, the carriage and establishment of weeds and ruderals, and the conveyance of wild species and parasites, some large and some microscopic, together with other pathogens (Cohen, 1992). The influence of European-dominated trade and colonial expansion has been crucial in the transfer of species around the world from quite early times, as shown by the subtitle of A. W. Crosby's (1986) canonical book. The opposite tendency has been extinctions, which took place in pre-industrial agricultural times at both regional and local scales. In the case of small populations of endemic species (particularly those on islands and especially those of flightless birds) the extinction was in fact global. The dodo (*Raphus cucullatus*) can stand as an obvious if somewhat hackneyed example of a loss (in 1681) that needed no steam power to bring it about.

The involvement of geographers, often with co-workers, in combining palaeoecological techniques with knowledge of biogeography and land use has sometimes been deployed in parts of the world where early (i.e. pre-colonial and pre-industrial) land use history has only recently been recognised as an important factor in the evolution of domesticated landscapes and ecosystems. Such is the case in central Africa (Uganda, Rwanda, and Burundi), where D. Taylor and his co-workers have extracted sediment cores from swamps and lakes and demonstrated the changes in vegetation zones at the end of the Pleistocene (Taylor and Hamilton, 1997; Taylor, Marchant, and Robertshaw, 1999). The earliest farmers and metalworkers seem to have favoured the higher parts of the interlacustrine highlands, exhibiting forest clearance phenomena around 600 BC. A second phase of forest disturbance coincides with the shift of settlement from drier grassland to wetter and more forested regions during the fifteenth century AD. Here, as in Europe, an 'Iron Age' contributed massively to the creation of humanised landscapes. An analogous story is told for a very different kind of place: Easter Island. Here, there was deforestation at the hands of a growing human population on an isolated island. The results included human conflict as well as the demise of many species of plants and animals (Flenley et al., 1991). The plants included a species of palm which, it is speculated, was the means of rolling the famous statues into place. The vulnerability of isolated islands to initial human presence is shown especially well by the Hawaiian chain, where the first (Polynesian) presence in about 400 AD resulted in habitat destruction (coupled with predation by rats and dogs as well as being killed for food) below 1000 m above sea level. Forty species of birds were soon extinct and now only 25 per cent of the original taxa survive since a second wave of extinctions started in 1778 and continued in the twentieth century. Cattle, goats, sheep, and rats were the main agents of change in those years (Kirch, 1982).

The encounter with hitherto unknown species has always provoked a profound sense of confrontation with the Other; R. Grove (1995) records how, even before AD 1600, the transfer of species and environmental degradation were remarked upon in the written sources from ancient Mesopotamia onwards. In several ways, the ligature between physical, mental, and medical was the garden, which occupies a central place in most accounts of domestication of landscapes, though less so of species. Thus the garden and the zoo (still sometimes called 'the zoological garden') usually contain both the familiar and the exotic. Social scientists in general seem to have been less interested in delectations of the past than the grind of subsistence and so garden history, for example, features little in the work of geographers and anthropologists. Academic work on hunting for pleasure is a little more common since it can be seen as an exercise of power by one species over others, or as a cultural practice that sheds light on a society as a whole, for example (Allen and Smith, 1975).

There could certainly have been no pleasure in mining, yet metal ores are without doubt one of the great mediators between humans and environment when refined and fashioned into tools, replacing stone as the hard edge of utilitarianism. Though bronze had its uses, iron is harder and longer-lasting but at the same time less showy and thus more likely to be re-used than put in graves in a sort of post-mortem consumer splurge. Consider, nevertheless, the hectares of woodland either removed or converted to coppice to make charcoal for smelting, and the holes where the ore was extracted and the heaps of waste slag. As late as 1873, smelting operations at a mine near Eureka, Nevada, used wood to smelt 600 tonnes per day of gold and silver ores; they denuded the land of piñon pine for 80 km around (Clemner, 1985). The greatest waste of pre-industrial times might be thought to be that of warfare, and examples from early times of its destructive power (notably when fire was used as one weapon to e.g. raze crops or burn woodlands) are many (for Europe and its empires see Pyne, 1997). The effect was like swidden agriculture in the sense that the woods and crops mostly grew again eventually, even when, as at Carthage, the croplands were sown with salt when the city was pulverised by the Romans.

Many of these examples deal implicitly with the question of the rapidity of change. For many years this kind of question hovered in the background, to be attempted only in the rare cases where there was some kind of dependable dating evidence. Now that the range of reliable techniques has been extended (led by calibrated radiocarbon assay though by no means confined to it), firmer statements can be made. To some, the interchange between hunter and agriculturalist was a gradual process in which there were initial phases of contact with trial and error, followed at some chronological distance to a point of no return with the new economy (Zvelebil and Rowley-Conwy, 1984). Where agriculture was the first human economy in an environment not hitherto occupied by hunters, then it could bring about explosive change, as in the Pacific islands discussed above. Rapid ecological change in an area dominated by agriculture was also possible. Detailed examination of Mediterranean history, for example, points out that from very early recorded times social systems were largely those of 'proprietors of some economic muscle and . . . dependents of little or no autonomy' (Horden and Purcell, 2000, 275). Thus there were few restraints: the adoption of new practices, transfer of populations, dispossession of lower orders, piracy, and war—all might function as rapid changes of economic outlook and hence of production technologies and their ecological ramifications. Tenancies, for example, bound the servile to their masters with cords of seeds, water, or labour; colonisations ensured that the two need not be in close daily, appreciative, contact. One outcome was a set of micro-ecologies, encouraged by the fragmentation of the topography, with only the sea as the common element. Thus the idea of ecological change as a long cycle and gradual movement might have some validity as a time–space overview from the libraries of capital cities further north, but on the ground is punctuated by a series of rapid transformations that seem to define what should be regarded as the 'normal'. The notion of land use and land cover change as a punctuated non-equilibrium seems to be gaining ground in a more general spatial and temporal sense (Lambin and Geist, 2001).

Fire has a long history in the Mediterranean. Some of it was positive in the sense of creating habitat diversity which brought about genetic diversity in plants and animals, and without it the transition to agriculture seems impossible. Some was negative since it destroyed woodland and scrub that offered greater resource opportunities in an unburned—or less frequently burned—condition. Overall, however, fire was not a catastrophic influence in the Mediterranean, and here as elsewhere its role in transforming materials in a kind of proto-industrialism was profound. Ores became metals, seawater yielded salt, grape was transmuted to alcohol, and calcite rock to lime. Pyne (1997, 43) sees fire in early Europe as both a weapon of war and a metaphor that transformed thought as well as matter. Out of alchemy came chemistry, as well as a metaphor for assaying humans. So perhaps fire-hunting was imitated by fire-herding, in which pastures were modified and perpetuated by fire. In woodland environments, fire-fallow or swidden became a way of keeping fertility high and was transmissible out of the Mediterranean into more northerly latitudes; indeed, it was used as far north as Finland. Out of its use in agricultural economies came its most famous philosophical use in the image of Plato's cave.

The concepts contained in the word 'domestication' seem not to be confined to the rather narrow area of the modification of the genetics of living organisms but to a condition of bringing into the totality of the human fold any part of nature that is noticed by a society. Hence the centrality, perhaps, of the *domus* element in the expression, especially when sedentary settlement was the norm. It affirms and deepens Pierre Dansereau's (1957) largely forgotten proposition that biogeography was about the alteration of genetics *and* of ecosystems. In the widest of contexts, the role of land use changes associated even with the early domestication (in the broadest of meanings) of environments has taken on a new significance in the quest for the understanding of Holocene climatic change, especially where human hands may have contributed to modifications that have global applicability and social significance. What a shame that there were no satellites gathering land cover information in AD 1 (Simmons, 2000) to aid the work of the IGBP core project's unit PAGES, headed in the 1990s by geographer F. Oldfield (Alversen, Oldfield, and Bradley, 2000).

The human life-worlds of agriculture

If ideas about 'materiality and mind' can be put together for hunter-gatherers then what parallel notions can be erected for agriculture?

The task should be easier since the quantity of evidence is much greater, given that literacy was one outcome of successful large-scale food-producing societies. Two basic strands in the skein of understanding are those of continuity of worldview and its opposite, that of change. To give one instance of the continuity thread, Thomas thinks that the late Neolithic Dorset cursus in its setting brings 'a perceived unity of earth and sky, life and death, past and present, all being references to bring more and more emphasis on to particular spaces and places' (Thomas, 1999, 53). As the ecology of Neolithic agriculture is seen as a continuity with Mesolithic land management (Simmons and Innes, 1987), and as the genetic contiguity of the populations is realised, then such persistence of lifeworld is not surprising. In some degree of contrast, Oelschlager (1991, 29) maintains strongly that the ways in which the technologically possible transformation of economies was interactive with all kinds of social and ideological changes extended to no less an extent than that 'philosophy and theology sprang forth with a vengeance'. Out of a diverse set of possibilities of the metamorphoses of both humans and nature, the identification of the farmer with a settled home and a set of nearby fields that afford (or withhold) subsistence and then the wilderness 'outside', an Other, is likely very long-lasting. Its take-up into Greek thought ensured that culture, including agri-culture, was an achievement that separated humanity from the rest of the world. The non-humanised becomes a challenge to rational thought, with its emphasis on critical enquiry and reasoned argument; there may be a vestige of the hunter-gatherer in regarding natural environments as alive, but the main direction was in the imposition of human order. The seminal work of Clarence Glacken (1967) on the development of western ideas in pre-industrial agricultural times has not been followed, regrettably, by an equivalent for non-western cultures. The concentration on the heirs of Classical times (including Christianity) has perhaps obscured the elucidation of the likely variety in the extent to which a dichotomous separation of humans and nature was an outcome of the development of food production as distinct from food collection.

That endeavour was of course the seed-bed for many kinds of social development and also the incubator for industrialisation. Not every advanced agricultural society took off into a fossil-fuel-based economy but for those that did, fire once again proves a potent metaphor as well as a practical aid. In effect humans domesticated fire as hunter-gatherers, and once that had happened then constant tending was needed, like other domesticates. Harris (1994) has drawn attention to the correlation of fire with early sowing, planting, and cultivating, and with dependence on seed and nut production. Thereafter, the quest for fire became, as Pyne (1993) puts it, the pursuit of power. The ambiguity of 'power' in today's global order is one stage in a process in which agriculture was a necessary intermediate stage.

In a twist?

If we think of the data of all kinds that we have collected about the natural world, about humanity, and about their intermediates (of both mind and material), then we cannot deny the interactive processes that link them and without the knowledge of which no better understandings will be gained. Such processes can only be described in terms of metaphor and attempts to find an overarching image fail in the face of the complexity and dynamisms that accompanied the domestication of the globe. For example, different colours of wool strand could represent people, animals, and soils as they intertwine along their length—i.e. through time. Yet this image of a skein is not a happy one: the colours remain separate. The double helix of the DNA molecule might be better, although there are two separate strands here as well, even if they are connected by base pairs that might represent the transfer of ideas and materials between the two. We might be tempted to think instead of a Seurat-like *pointillisme* in which the colours are separate but when viewed at the necessary distance achieve emergent qualities of a new order. (No doubt computer graphics could produce a 3-D model, if anybody thought it worth the time.) None of these images convey a sense of time in which non-linear change is as common as gradual 'evolutionary' shifts. These are features as important historically as they may well be for our future; IGBP calls them 'low-probability/high catastrophe' events.

If we step back for a moment and look at the early millennia of pre-industrial economies, then certain broad-scale effects are clear. Today's research into global environmental change shows that land use change is the biggest driver of alterations on biodiversity (with climatic change as the next most important), and that relationship presumably applied often, *ceteris paribus*, to earlier times. But things were not always equal: the first-ever presence of humans cannot be repeated, and as discussed above, this was frequently a cause of species and genus extinctions in many different kinds of environment, both on continents and on islands. After these early inroads into biodiversity, most commentators focus on the post-1600-AD period, when the effects of increasing trade and exploration begin the accelerations of which the twenty-first century is the heir. Detailed investigation of the effects of biodiversity on early 'domestications' of environment are few: in Britain, one study (Simmons and Innes, 1988) showed an increase in the representation of pollen taxa during the time of maintenance of forest openings in the Mesolithic. Contractions in the range of many bird species in North America have been attributed to the impact of indigenous populations, though in the millennia between the time of 'Pleistocene overkill' and the arrival of the Europeans only two species of bird became extinct: a flightless duck of the Pacific Coast and a small southwestern turkey. By contrast, isolated islands show remarkable declines: at the time of Polynesian contact, Easter Island sustained 22–30 species of seabirds and six endemic species of land birds, a fauna soon greatly depleted (Steadman, 1996).

Could the period under discussion here have seen any human-caused effects on climate? Local-scale effects are not in doubt: soil temperatures and soil moisture content change when a forest clearing is made, for example. Larger spatial scales are difficult to document, and much of the work by climatologists refers to a rather unspecific 'pre-industrial period', usually taken to be the time between the mid-Holocene thermal optimum and 1600 AD. The long-term incidence of fire, especially if followed by soil erosion, is likely to have had some effect on the global carbon balance as well as regional climate. One scrutiny goes so far as to speculate that reduced fire incidence due to 'modern practices' reduced atmospheric carbon dioxide to the point where the Little Ice Age of the seventeenth to nineteenth centuries was a result. Human-led carbon dioxide emissions have been estimated at 0.3 GtC (gigatonnes of carbon) per year until about 1000 AD, rising to 0.7 GtC a year in 1700 AD. Such a level of emission (none of which is calculated to be from fossil sources) is in the order of the emissions from volcanic releases and three to four times smaller than the variation associated with an El Niño. In a Holocene perspective, the changes in carbon release from hunter-gatherers' environmental manipulation is usually reckoned to be minimal compared with even pre-industrial agriculture, but this evaluation may well need recalculation. Given the tendency of climates to 'flip', then the apparently marginal suggestion by Schüle (1992) that hunter-gatherers were able to trigger global climatic change may not be as unlikely as it sounds. More recent (since about 1600 AD) climatic history suggests that major fire years in the southwest United States and northern Patagonia follow the switch from El Niño to La Niña years, wherein the first produces fine fuels and desiccated conditions in the second favour large-scale fire.

The image of painting in dots of colour referred to above allows us to envisage two other global and long-term processes in both society and nature. Like Seurat and his followers (called 'divisionists' by some), the appearance is different from increasing distances, as satellite images have so effectively demonstrated. Nevertheless, it is possible to recognise that out of the matrices of forest and grassland that formed the habitat of most hunter-gatherers there has been created a much more fragmentary mosaic of fields, roads, copses, remnant wetlands, and urban–industrial phenomena. Yet what we now see was launched once agriculture became an irreversible economy: once, especially, agricultural surpluses allowed the stratification of societies so that there were those with time and energy to rule, count, write, and think. The creation of both physical and mental laagers brought about the differentiation of home and away, us and Other, and the latter often applied to nature, which could be exterminated, tortured, bred, and even kept in reserves and menageries. So even in the millennia before Christ, hunting, habitat change, conscious protection, and park and garden management were all present and all might require the erection of legal or actual fences around an ecosystem, preventing or altering the flows of energy and matter within it or with its surrounding areas. By contrast, there was also coalescence as trade and empires allowed or caused species to move into new environments along with new cultural ideas. Thus the Roman empire took many materials into its heartland and was simultaneously promoting the cultivation of wheat wherever possible, even unto North Yorkshire. Every shift of an army or a colonising movement must have carried with it the possibility of bacteria and viruses to which the recipient population had no immunity, as happened with the Spaniards in South America, and north Europeans in North America. The extermination of wild oxen in prehistoric Britain was compensated by the knowledge (and perhaps the germplasm itself, on the hoof) of how to breed domestic cattle. In spite of our knowledge of trade in, for example, Bronze Age Europe, and of the 'medieval' networks that presaged those of the nineteenth century (Abu-Lughod, 1989), the impression is given that coalescence was not as important as fragmentation in the phases under consideration here. There were still many independent systems within the cultural sphere and these were part of an ecological web where many of the systems were also independent, except to the extent that they were linked by global energy, carbon, nitrogen, and water systems, as indeed were (and are) the human orders. An attempt to visualise some of this was made by Yen (1989) for the whole of the Holocene (Fig. 4.2) in which he considered that societies became both socially and technologically more complex through time; this encouraged diffusion between systems and led to the loss of their independence. This loss can be seen as a proxy measure for the transfer of species, for example, as well as the diffusion of ideas about the 'proper' way to grow crops that sometimes led to the failure of colonial enterprises. Though pointillism may have more or less disappeared, in one historian's words it 'remade the world of art all the way from the look of a painting to the techniques that made it and the aesthetic theory behind it' (Everdell, 1997, 64).

The comment by Everdell can have a wider resonance for the history of the human environment. We can think—as outlined at the start—of the look of the land as described by millions of empirical studies, study the way in which (to take one example) agriculture took over from foraging, and place them in a number of different theoretical frameworks. For the latter, there is a choice all the way from monotheism to distributed cognition, with at any rate the more diffuse end being no doubt open to extension: from science to myth is only one possible axis (Thompson, 1989). That the social framework of all the domestications is important cannot be open to doubt, but that concept opens up the necessity for two other relativist

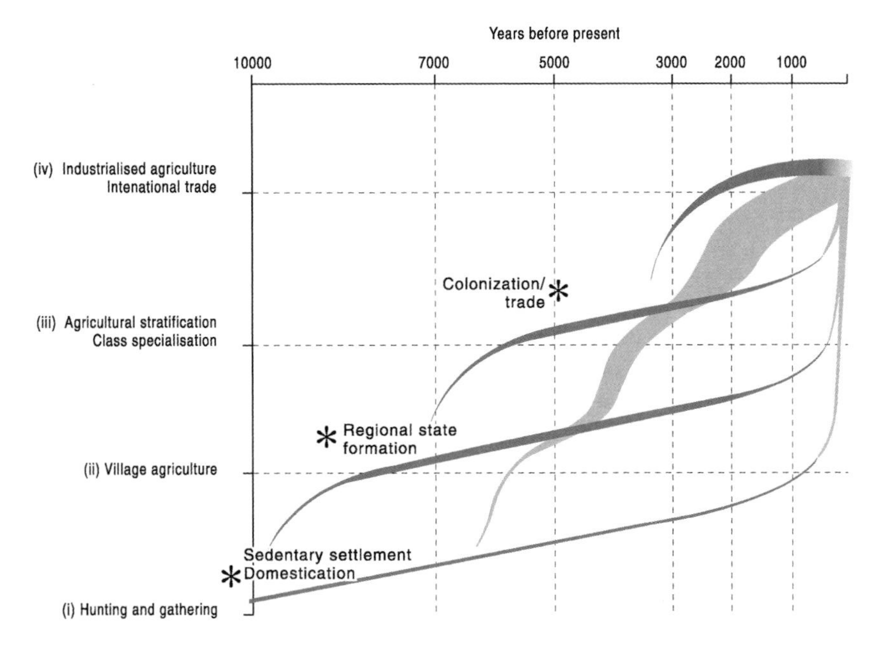

Fig. 4.2 The 'domestication of environment' as transformation from hunting-gathering (I) through agricultural forms of increasing complexity (II–IV). The inclination of the parallel lines of the systems through time represents the internal development of social and technological elaborations that affect environment, while the varying thickness of the lines indicates uneven amplitudes of such development. Diffusion between systems is portrayed by dotted lines which, in the seventeenth century and the beginnings of global colonisation, signalled the end to come of independence of subsistence systems I and II. Domestication, regional state formation, colonisation as single points, and system development as single lines, should be recognised as gross simplifications of history.
Source: Yen (1989, 70).

underpinnings to be inspected. First, the evidence that is collected by means other than modern science has been winnowed by all kinds of processes and its survival is to a large extent random. Yet when it comes in an approved package, its credence is so high that it may be used uncritically, as in the accounts of the soil erosion of Attica found in Herodotus. Nor are the natural sciences immune from the mundane: it was rumoured that most of the peats studied by the first investigators of Quaternary palaeoecology in the British Isles were within a quarter-mile of a public house. Secondly, however, much of today's environmental history beyond the local scale is written by people with an agenda, which is usually printed on green paper, though very likely of different shades. There is as well something of a dichotomy between writers who will only entertain the local and the particular, and those who—as in this contribution—are willing to try and make sense of what is otherwise a great pile of disparately derived statements, even if the only grand conclusion is that there is unlikely to be any grand conclusion because of the nature of the random and the contingent. Even so, there are intermediate stages, like the idea that independent transitions to agriculture occurred only a few times: perhaps once or twice in Eurasia and the Americas and once in Africa.

So the writing of accounts of the 'domestication' of the planet in pre-industrial times are full of tensions, some of which resolve from deep-rooted considerations of geographers, such as scale. The discussions of the relations between deforestation, agriculture, and soil change in the Himalayas exemplify that, as does analogous work on deforestation in Africa (see e.g. Kull, 2000) and the claims that population growth can lead to more 'rational' land use in East Africa (Tiffen, Gichuki, and Mortimore, 1994). Other tensions are less familiar: those relating to the biases of information sources are reasonably familiar, those relating to our own 'situated' knowledge have been thrust at us relatively recently—they would not have troubled us a century ago. Supposing we accept that disciplinary boundaries have little significance and even less virtue, we might agree that knowledge of the historic and cultural depths of change in humanity–nature relations needs compilation and interpretation for all kinds of reasons, not the least of which is some degree of liberation of both the human and the non-human (Anderson, 1997). At the intellectual scale there is the continuing problem of the West's desire to dichotomise and thus to erect dualities of almost everything (Ingold, 1993; Massey, 1995). At a more everyday level, we might want to see if there are ways of achieving stable population–environment relations that have hitherto been missed (Hoffman, 2001) and might be useful even in tomorrow's inevitably different world. If we conjure up the useful image first painted by Archilochus (*fl* 650 BC) and amplified by Isaiah Berlin, then geography has at various times adopted as totem the fox (who knows many things), then abandoned it for the hedgehog (who knows one big thing), and then taken up the fox again. In geographers' attempts to understand some of the early phases of the humanisation of the world, there seem to have been enough of us and enough habitats within the discipline's boundaries to maintain a flourishing population of both species, neither of which in Britain at least seem much diminished by either determined hunting or accidental death on the roads.

References

Abu-Lughod, J. L. (1989) *Before European Hegemony: The World System* AD *1250–1350.* New York and Oxford: Oxford University Press.

Allen, M. J. S. and Smith, R. (1975) Some notes on hunting techniques and practices in the Arabian peninsula. *Arabian Studies*, 2, 108–47.

Alversen, K., Oldfield, F., and Bradley, R. (2000) *Past Global Changes and their Significance for the Future*. Amsterdam: Elsevier.

Anderson, E. N. (1996) *Ecologies of the Heart: Emotion, Belief and the Environment*. New York and Oxford: Oxford University Press.

Anderson, K. (1997) A walk on the wild side: a critical geography of domestication. *Progress in Human Geography*, 21, 463–85.

Anderson, M. K. (1994) Prehistoric anthropogenic wildland burning by hunter-gather societies in the temperate regions: a net source, sink or neutral to the carbon budget? *Chemosphere*, 29, 913–34.

Bayliss-Smith, T. P., Bedford, R., Brookfield, H., and Latham, M. (1988) *Islands, Islanders and the World*. Cambridge: Cambridge University Press.

Bender, B. (1978) Gatherer-hunter to farmer: a social perspective. *World Archaeology*, 10, 204–22.

Bridgland, D. R. (1998) The Pleistocene history and early human occupation of the River Thames valley. In N. Ashton, F. Healy, and P. Petit, eds, *Stone Age Archaeology: Essays in Honour of John Wymer*. Oxford: Oxbow Books, 29–37.

Brown, A. G. (1997) Clearances and clearings: deforestation in Mesolithic/Neolithic Britain. *Oxford Journal of Archaeology*, 16, 133–46.

Brown, A. G. (1999) Characterising prehistoric lowland environments using local pollen assemblage. *Journal of Quaternary Science*, 14, 585–94.

Brück, J. (2000) Settlement, landscape and social identity: the Early–Middle Bronze Age transition in Wessex, Sussex and the Thames Valley. *Oxford Journal of Archaeology*, *19*, 273–300.

Clemner, R. O. (1985) The piñon-pine—old enemy or new pest? Western Shoshone Indians vs the Bureau of Land Management in Nevada. *Environmental Review*, 9, 131–49.

Cohen, M. N. (1992) The epidemiology of civilization. In J. Jacobsen and J. Firor, eds, *Human Impact on the Environment: Ancient Roots, Current Challenges*. Boulder, CO: Westview Press. 51–70.

Cronon, W. (1983) *Changes in the Land: Indians, Colonists and the Ecology of New England*. New York: Hill and Wang.

Crosby, A. W. (1986) *Ecological Imperialism: The Biological Expansion of Europe 900–1900*. Cambridge: Cambridge University Press.

Crutzen, P. J. and Stoemer, E. F. (2000) The 'anthropocene'. *IGBP Newsletter*, 41, 17–18.

Dansereau, P. (1957) *Biogeography: An Ecological Perspective*. New York: Ronald Press.

Donkin, R. A. (1979) *Agricultural Terracing in the Aboriginal New World*. Tucson, AZ: University of Arizona Press.

Edwards, K. J. (1993) Human impact on the prehistoric environment. In T. C. Smout, ed., *Scotland since Prehistory*. Aberdeen: Scottish Cultural Press, 17–27.

Edwards, K. J. and Sadler, J. P., eds (1999) Holocene environments of prehistoric Britain. *Journal of Quaternary Science*, 14, 477–635.

Everdell, W. R. (1997) *The First Moderns: Profiles in the Origins of Twentieth Century Thought*. Chicago, IL: University of Chicago Press.

Everden, N. (1992) *The Social Creation of Nature*. Baltimore and London: Johns Hopkins University Press.

Evison, M. P. (1999) Perspectives on the Holocene in Britain: human DNA. *Journal of Quaternary Science*, 14, 615–23.

Eyre, S. R. and Jones, G. R. J. (1966) *Geography as Human Ecology*. London: Edward Arnold.

Finlayson, B. and Edwards, K. J. (1997) The Mesolithic. In K. J. Edwards and I. M. B. Ralston, eds, *Scotland: Environment and Archaeology 8000* BC–AD *1000*. Chichester: Wiley, 109–25.

Fischer-Kowalski, M. and Haberl, H. (1998) Sustainable development: socio-economic metabolism and colonization of nature. *International Social Science Journal*, 50, 573–87.

Fleming, A. (1988) *The Dartmoor Reaves: Investigating Prehistoric Land Divisions*. London: Batsford.

Flenley, J. R., King, A. S. M., Jackson, J., Chew, C., Teller, J., and Prentice, M. E. (1991) The Late Quaternary vegetational and climatic history of Easter Island. *Journal of Quaternary Science*, 6, 85–115.

Fowler, C. S. and Turner, N. J. (1999) Ecological/cosmological knowledge and land management among hunter-gatherers. In R. B. Lee and R. Daly, eds, *The Cambridge Encyclopaedia of Hunters and Gatherers*. Cambridge: Cambridge University Press, 419–25.

Glacken, C. (1967) *Traces on the Rhodian Shore*. Berkeley and Los Angeles, CA: University of California Press.

Goody, J. (1977) *The Domestication of the Savage*. Cambridge: Cambridge University Press.

Goudsblom, J. (1994) *Fire and Civilization*. Harmondsworth: Penguin.

Grove, R. H. (1995) *Green Imperialism: Colonial Expansion, Tropical Island Edens and the Origins of Environmentalism 1600–1860*. Cambridge: Cambridge University Press.

Haberl, H., Batterbury, S., and Moran, E. (2001) Using and shaping the land: a long-term perspective. *Land Use Policy*, 18, 1–8.

Harris, D. R. (1990) *Settling Down and Breaking Ground: Rethinking the Neolithic Revolution*. Amsterdam: Nederlands Museum voor Anthropologie en Praehistorie.

Harris, D. R. (1994) Review of A. B. Gebauer and T. D. Price, eds, *Transitions to Agriculture in Prehistory*. *Antiquity*, 68, 873–7.

Harris, D. R. and Hillman, G. C., eds (1989) *Foraging and Farming: The Evolution of Plant Exploitation*. London: Unwin Hyman.

Hoffman, R. C. (2001) A longer view: is industrial metabolism really the problem? *Innovations*, 14, 143–55.

Horden, P. and Purcell, N. (2000) *The Corrupting Sea: A Study of Mediterranean History*. Oxford: Blackwell.

Inglis, J. T. (1993) *Traditional Ecological Knowledge: Concepts and Cases*. Ottawa: Canadian Museum of Nature.

Ingold, T. (1986) *The Appropriation of Nature: Essays on Human Ecology and Social Relations*. Manchester: Manchester University Press.

Ingold, T. (1993) An archaeology of symbolism. *Semiotica*, 96, 309–14.

Ingold, T. (2000) *The Perception of the Environment*. London and New York: Routledge.

Isaac, E. (1970) *Geography of Domestication*. Englewood Cliffs, NJ: Prentice-Hall.

Iversen, J. (1941) Landnam i Danmarks stenalder. *Danmarks Geologiske Undersøgelse* R II, 1–68.

Jones, G. R. J. (1990) Celts, Saxons and Scandinavians. In R. A. Dodgshon and R. A. Butlin, eds, *An Historical Geography of England and Wales*. London: Academic Press, 45–68.

Kirch, P. V. (1982) Impact of the prehistoric Polynesians on the Hawaiian ecosystem. *Pacific Science*, 31, 109–33.

Klee, G. A. (1980) *World Systems of Traditional Resource Management*. London: Edward Arnold.

Krech, S. (1999) *The Ecological Indian: Myth and History*. New York and London: W. W. Norton.

Kull, C. A. (2000) Deforestation, erosion and fire: degradation myths in the environmental history of Madagascar. *Environment and History*, 6, 423–50.

Lambin, E. F. and Geist, H. J. (2001) Global land use and cover change: what have we learned so far? *Global Change Newsletter*, no. 46, 27–30.

Lansing, J. S. (1991) *Priests and Programmers: Technologies of Power in the Engineered Landscape of Bali*. Princeton, NJ: Princeton University Press.

Lansing, J. S. and Kremer, J. N. (1993) Emergent properties of Balinese water temple networks' coadaptation on a rugged fitness landscape. *American Anthropologist*, 95, 97–114.

Layton, R. (2000) Shamanism, totemism and rock art: *Les Chamanes de la Pré'histoire* in the context of rock art research. *Cambridge Archaeological Journal*, 10, 161–86.

Lewis, H. T. (1994) Management fires vs corrective fires in Northern Australia: an analogue for environmental change. *Chemosphere*, 29, 949–63.

Low, N. and Gleeson, B. (1998) *Justice, Society and Nature: An Exploration of Political Ecology*. London and New York: Routledge.

Massey, D. (1995) Masculinity, dualisms and high technology. *Transactions, Institute of British Geographers*, NS20, 487–99.

Mason, S. L. R. (2000) Fire and Mesolithic subsistence: managing oaks for acorns in northwest Europe? *Palaeogeography, Palaeoclimatology, Palaeoecology*, 164, 139–50.

Mellars, P. A. and Dark, P. (1998) *Star Carr in Context*. Cambridge: McDonald Institute Monographs.

Mithen, S. (1998) *The Prehistory of the Mind: A Search for the Origins of Art, Religion and Science*. London: Phoenix.

Mithen, S. (1999) Mesolithic archaeology, environmental archaeology and human palaeoecology. *Journal of Quaternary Science*, 14, 477–83.

Mullenders, W. W. (1993) New palynological studies at Hoxne. In R. Singer, B. G. Gladfelter, and J. J. Wymer, eds, *The Lower Palaeolithic Site at Hoxne, England*. Chicago, IL and London: University of Chicago Press, 150–5.

Narr, K. J. (1987) Paleolithic religion. In M. Eliade, ed., *The Encyclopedia of Religion*. New York: Macmillan, 149–59.

Oelschlager, M. (1991) *The Idea of Wilderness: From Prehistory to the Age of Ecology*. New Haven, CT and London: Yale University Press.

Pyne, S. J. (1993) Keeper of the flame: a survey of anthropogenic fire. In P. Crutzen and J. G. Goldammer, eds, *Fire in the Environment: the Ecological, Atmospheric and Climatic Importance of Vegetation Fires*. Chichester: Wiley, 245–66.

Pyne, S. J. (1994) Maintaining focus: an introduction to anthropogenic fire. *Chemosphere*, 29, 889–911.
Pyne, S. J. (1997) *Vestal Fire*. Seattle and London: University of Washington Press.
Rackham, O. (1976) *Trees and Woodland in the British Landscape*. London: Dent.
Sauer, C. O. (1952) *Agricultural Origins and Dispersals*. New York: The American Geographical Society.
Schafer, E. H. (1963) *The Golden Peaches of Samarkand: A Study of T'ang Exotics*. Berkeley and Los Angeles: University of California Press.
Schüle, W. (1992) Anthropogenic trigger effects on Pleistocene climate? *Global Ecology and Biogeography Letters*, 2, 33–6.
Schuster, S. and Schüle, W. (2000) Anthropogenic causes, mechanisms and effects of Upper Pliocene and Quaternary extinctions of large vertebrates. *Oxford Journal of Archaeology*, *19*, 223–39.
Scoones, I. (1999) New ecology and the social sciences. What prospects for fruitful engagement? *Annual Review of Anthropology*, 28, 479–507.
Semino, O. and 15 others (2000) The genetic legacy of Paleolithic *Homo sapiens sapiens* in extant Europeans: a Y chromosome perspective. *Science*, 290, 1155–9.
Sheail, J. (1976) *Nature in Trust: The History of Nature Conservation in Britain*. Glasgow and London: Blackie.
Simmons, I. G. (2000) Making a mark: two thousand years of ecology, economy and worldview. *Journal of Biogeography*, 27, 3–5.
Simmons, I. G. and Innes, J. B. (1987) Mid-Holocene adaptations and later Mesolithic disturbance in northern England. *Journal of Archaeological Science*, 14, 1–20.
Simmons, I. G. and Innes, J. B. (1988) Disturbance and diversity: floristic changes associated with pre-elm decline woodland recession in north east Yorkshire. In M. Jones, ed., *Archaeology and the Flora of the British Isles: Human Influence on the Evolution of Plant Communities*. Oxford: Oxford University Committee for Archaeology Monographs, 7–20.
Simmons, I. G. and Innes, J. B. (1996) The ecology of an episode of prehistoric cereal cultivation on the North York Moors, England. *Journal of Archaeological Science*, 23, 613–18.
Soper, K. (2000) Future culture: realism, humanism and the politics of nature. *Radical Philosophy*, 102, 17–26.
Spencer, J. E. and Hale, G. A. (1961) The origin, nature and distribution of agricultural terracing. *Pacific Viewpoint*, 2, 1–60.
Spratt, D. A. (1989) *Linear Earthworks of the Tabular Hills of Northeast Yorkshire*. Sheffield: University of Sheffield Department of Archaeology and Prehistory.
Steadman, D. (1996) Human-caused extinction of birds. In M. L. Reaka-Kudla, D. E. Wilson, and E. O. Wilson, eds, *Biodiversity II*. Washington, DC: Joseph Henry Press, 139–61.
Stott, P. and Sullivan, S. (2000) *Political Ecology: Science, Myth and Power*. London: Arnold.
Strathern, M. (1996) Cutting the network. *Journal of the Royal Anthropological Institute*, NS2, 517–35.
Tallis, J. H. and Switsur, V. R. (1990) Forest and moorland in the south Pennine uplands during the mid-Flandrian period. II. The hillslope forests. *Journal of Ecology*, 78, 857–83.

Taylor, D. and Hamilton, A. (1997) Late Pleistocene and Holocene history at Mubwindi Swamp, southwest Uganda. *Quaternary Research*, 47, 316–28.

Taylor, D., Marchant, R. A., and Robertshaw, P. (1999) A sediment-based history of medium altitude forest in Central Africa: a record from Kabata swamp, Ndale volcanic field, Uganda. *Journal of Ecology*, 87, 302–15.

Thomas, J. (1999) *Understanding the Neolithic*. London and New York: Routledge.

Thompson, W. I. (1989) *Imaginary Landscape: Making Worlds of Myth and Science*. New York: St Martin's Press.

Tiffen, M., Gichuki, F., and Mortimore, M. (1994) *More People, Less Erosion: Environmental Recovery in Kenya*. Chichester: Wiley.

Tilley, C. (1994) *A Phenomenology of Landscape: Places, Paths and Monuments*. Oxford and Providence, RI: Berg.

Whatmore, S. (1999) Hybrid geographies: rethinking the 'human' in human geography. In D. Massey, J. Allen, and P. Sarre, eds, *Human Geography Today*. Cambridge: Polity Press, 22–39.

Wheatley, P. (1965) Discursive scholia on some recent papers on agricultural terracing and on related matters pertaining to northern Indochina and to neighbouring areas. *Pacific Viewpoint*, 6, 123–44.

Wilson, G. A. and Bryant, R. (1997) *Environmental Management: New Directions for the Twenty-first Century*. London: UCL Press.

Worster, D. (1994) *Nature's Economy: A History of Ecological Ideas*. Cambridge: Cambridge University Press.

Yen, D. E. (1989) The domestication of environment. In D. R. Harris and G. C. Hillman, eds, *Foraging and Farming*. London: Unwin Hyman, 55–75.

Yibarbuk, D., Whitehead, P. J., Russell-Smith, J., Jackson, D., Godjuwa, C., Fisher, A., Cooke, P., Choquenot, D., and Bowman, D. M. J. S. (2001) Fire ecology and Aboriginal land management in central Arnhem Land, northern Australia: a tradition of ecosystem management. *Journal of Biogeography*, 28, 325–43.

Zvelebil, M. and Rowley-Conwy, P. C. (1984) Transition to farming in northern Europe: a hunter-gatherer perspective. *Norwegian Archaeological Review*, 17, 104–28.

5

The creation of humanised landscapes

MICHAEL WILLIAMS

One of the prime focuses of geographical study has been, and still is, the study of the humanised landscape. Such study is impossible without combining in some way past human actions with description and analysis of present arrangements of space and scenery. Yet, the puzzle of how to traverse these debatable borderlands and write an 'incontestable geography' while including 'the compelling time sequence of related events which is the vital spark of history' (Whittlesey, 1945, 32) has proved difficult and has led to many experiments. Inevitably, time must be combined in some way with space, or put in disciplinary terms, history and geography must cohabit the same page (C. T. Smith, 1965; Harris, 1978; Bird, 1981; Renfrew, 1981; Lawton, 1983). The historical element and human action are implicit in the idea of the landscape. Such combinations, in various guises, often go under the name of Historical Geography. More latterly, the meaning of 'history', in its broadest sense, has been scrutinised closely because of the implicit subjective meaning embedded in any account of the past.

In the past writers had little problem with this cohabitation, seeing geography and history for what they really are—complementary and interdependent enterprises that are inseparable. In 1670, Richard Blome in his *A Geographical Description of the Four Parts of the World* was convinced that; 'without . . . Geography all History is a thing of little use, the affinity between them both being such, that they seem to centre both in one' (Blome, 1670, frontispiece), and he thought that geography was similarly enhanced by history.

But such common sense was lost sight of during the later nineteenth and early twentieth centuries. The 'new' academic disciplines like modern history, geography, English and modern languages fought for recognition in both the old and new universities against the opposition of the established schools, such as classics, ancient history, theology, philosophy and mathematics. Geography, though an ancient subject, was still battling for academic acceptance throughout the early twentieth century and was present in about two dozen departments only in England. 'It is difficult', wrote H. C. Darby (1983, 424), 'to realize how fragile and uncertain were the prospects for the subject'. On taking up his post in Birmingham in 1924 R. H. Kinvig was told by the dean of the faculty that 'geography was a subject for which the Faculty had not the slightest respect' and he was warned not to try and establish a department (Kinvig, 1955). In the late 1930s, Kenneth Mason in Oxford knew the continuation of the subject was on a knife-edge, despite having being established for over 50 years (Mason, 1958, 411). And, although it is difficult to appreciate today, modern history had a similarly protracted struggle, so that it was not until 1923 that the great pioneer of the Manchester School, T. F. Tout, could say: 'The battle for the recognition of the subject is as good as won' (Finberg, 1962, vii).

In the quest to establish academic credentials and distinctiveness Richard Hartshorne's *The Nature of Geography* (1939), based on the ideas of Immanuel Kant and Alfred Hettner, was influential, supplying geography with many of its disciplinary underpinnings. Its influence was immense in North America and Australia, but less so in Britain. Less it might be, but his conclusions about studies of the humanised landscape were not good for historical (and cultural) geographers. Hartshorne's damning statement that geographers were 'forced to distinguish between an historical and a geographical point of view' (Hartshorne, 1939, 188), and that much written in the name of geography was in fact history, stuck like a thorn in their flesh. Blome's 'centring both in one' of history and geography was replaced by firm academic pigeon-holes, and it is against this background that the first part of this essay is set.

Antecedents in history

In a purely chronological sense the puzzle of combining history and geography was first carried out almost entirely by historians.[1] Throughout the nineteenth century many historians were concerned to create the sense of place to accompany and 'ground' the changing histories of events. Jules Michelet's *Histoire de France* (1833) strove to include new elements of 'realism' in the standard fare of political relations and incidents by incorporating every aspect of social and economic endeavour. He complained that without a geographical base

> . . . the people, the makers of history, seem to be walking on air, as in those Chinese pictures where the ground is wanting. The soil too, must not be looked on only as the scene of action. Its influence appears in a hundred

[1] In writing this section I have benefited from reading and editing an unpublished ms of H. C. Darby, which is now in print (Darby, 2002), and the commentary by H. C. Prince on 'H. C. Darby and the historical geography of England' (Prince, 2002, 63–88).

ways, through food, climate, etc. As the nest, so is the bird. As the country, so are the men. (Michelet, 1864 edn, Preface)

He put his ideas into practice; Book Three (which comprises the first half of the second volume) is entitled *Tableau de la France* and describes each of the main provinces of the country, its climate, soils, and economy. French historians took this lesson to heart, and it is often forgotten that Vidal de la Blache's celebrated *Tableau de la géographie de la France* (1903), one of the first texts to describe the human impact on the landscape,[2] was the introductory volume of Lavisse's monumental *Histoire de France* (1903). By 1929 the formative figures of the historians Marc Bloch (1886–1944) and Lucien Febvre (1878–1956) had founded the *Annales d'historie économique et sociale*, which became one of the leading and most lively of history journals. No disciplinary boundaries were erected in its effort to elucidate the story of the past, and it drew on the work of climatologists, archaeologists, and geographers, as well as historians. Following a long tradition, then, geography students at French universities usually took courses in both geography and history and included historical narratives in their regional studies.

In Britain the divisions were more rigid. J. N. L. Baker (1936, 194; 1952, 406) thought that the beginnings of historical geography could be traced to the efforts of Thomas Arnold (1795–1842), Regius Professor of Modern History at Oxford, who was concerned to incorporate 'the geographical factor[s]' of climate and place in his *History of Rome* (1838, 157–63). Understandably it was the history of the classical and biblical world that inspired thought along these lines, and there followed many books in a similar vein. For example, in 1856, A. P. Stanley, who had heard Arnold lecture and read Karl Ritter, wrote his *Sinai and Palestine in Connection with their History*, setting out to trace the connection 'between the scenery, the features, the boundaries, the situations of Sinai and of Palestine on the one hand, and the history of the Israelites on the other' (Stanley, 1856, xiii). Similarly, Adam Smith's immensely popular *The Historical Geography of the Holy Land* (1884, and it went through 25 editions) followed much the same lines, but with more description that brought the 'contemporary' landscape alive.

The same approach inspired interpreters of the Classical world, and by the beginning of the twentieth century it was another Oxford scholar, J. N. L. Myres, the Wykeham Professor of Ancient History,[3] who campaigned for the founding of the Chair of Geography at Oxford and became

[2] The first chapter was published separately in 1918 as *The Personality of France*, translated by H. C. Brentnall (London: Christophers).

[3] The portrait of Myres hung in the common room of the School of Geography, Oxford, for many years, in some recognition of his seminal role in promoting geography. It seems now to have disappeared.

a part-time lecturer in the School (Koelsch, 1995, 57), and who by 1910 was giving lectures on 'The Geographical Study of Greek and Roman culture' and 'Greek Lands and the Greek People'. Later he said:

> All human history, then, is regional history, and loses value and meaning when its geographical aspect is overlooked: all geography, on the other hand, and (most obviously) all human geography, depends for its significance on the consideration that is contemplating not facts only, but events with causes and effects; processes, of which our map-distributions are momentary cross-sections. (Myres, 1928, 75)

He had no doubt that geography was essential to an understanding of archaeology and classical history.

For other historians 'historical geography' was concerned almost entirely with changes in political boundaries and the extent of nations and states, as in E. A. Freeman's *Historical Geography of Europe* (1881). This made geography a kind of reference aid to history, impregnating history with a territorial flavour. In time there was a shift from political boundaries to 'landscapes', their physical disposition, vegetation cover, and climate being involved in the explanation, for example, of the progress of Anglo-Saxon settlement. An early exponent of this was J. R. Green, who wrote: 'the ground itself, when we can read the information it affords, is, . . . the fullest and most certain of documents' (Green, 1881, vii). But the problem was that taken too far the geographical factor could tip over into a deterministic view that human action (history) was 'influenced' or, more strongly, was 'determined' by geography, and when practised *in extremis* by geographers it became a discredited philosophy and did the intellectual standing of geography much damage.

While some historians sought to make their studies more 'real' by including the 'geographical' factor, others were concerned to re-create the geography of past periods as a setting for their historical narrative. Perhaps most famous was Thomas Macaulay's comment in his *History of England* (1848, 1, 281): 'Could the landscape of England of 1685, be, by some magical process, set before our eyes, we should not know one landscape in a hundred, or one building in ten thousand, . . . Everything has changed'. Therefore it had to be re-created, as he did in the third chapter of the book. The approach became standard fare for successive generations of historians, from J. H. Clapham, G. M. Trevelyan, G. D. H. Cole, A. L. Rowse, and a whole host of successors who re-created the geography of regions at the national, regional, and local scale.

Likewise, prehistorians have tried to re-create the 'natural landscape' of vegetation, soils, and drainage as a background or stage on which to mount past human action, such as Cyril Fox's *Archaeology of the Cambridge Region* (1923), the Ordnance Survey's *Map of Roman Britain* (1931) which depicts woodland, and O. G. S. Crawford's *Map of Neolithic Wessex* (1932).

Of course, it is doubtful that such a pristine landscape ever existed, in the Neolithic, say, or any other age; so long as there have been humans on earth they have been manipulating the fauna and flora through fire, hunting, selection, and eventually cultivation. A landmark event in this re-creation of the past landscapes was the formation of the English Place Name Society in 1922. Its scholarly philological and etymological work, county by county, eliminated the guesswork about the light that place-names threw on past geographical conditions, and put their elucidation onto a sure scholarly footing (Darby, 1957).

For Americans, the division of history and geography posed far fewer intellectual problems because the evolution of their humanised landscape was integral to both disciplines. It could not be otherwise as most American history until the mid-twentieth century was the story of the conquering of space through time, and American geography the story of the change in spaces through time. What remained of a 'deeper' past in Indian culture was swept away and ignored in the great westward quest, its complexity, sophistication, and artefactual remains only being fully and widely appreciated during the last few decades (e.g. Denevan, 1992; Dolittle, 1992). History and geography were intermingled, time and space were blurred. Adam Hodgson, an early-nineteenth-century British traveller in America, commented perceptively that

> . . . in successive intervals of *space* I have traced society through the various stages which in most countries, are exhibited only in successive periods of *time*. I have seen the roving hunter acquiring the habits of the herdsman, the pastoral state merging with the agricultural; and the agricultural with the manufacturing and the commercial. (Hodgson, 1824, 318–19)

As places changed rapidly, the sensation of time being telescoped by distance was overwhelming.

Frederick Jackson Turner's 'The significance of the frontier in American history' (1894) was shot through with the geographical factor, and for him it was essential to engage in 'the study of the interactions between man and his environment' (Turner, 1908, 45), though later he did stress that he was not a student of a region but of a process, as if to emphasise where his disciplinary core lay. In another work well known to geographers, Walter Prescott Webb's *The Great Plains* (1927) was predicated on the thesis that the environment of the plains had help formed the economy and society of that part of the United States.

Geographical departures

The knowledge that all geographical accounts of place and landscape 'date' even as they are being written, and sooner or later became 'history', made the separation between history and geography philosophically impossible.

Simply, wrote Darby (1962a, 156): 'The present is but the past of some future'. This very impossibility gave heart to those geographers who wanted to introduce recognisably 'historical' elements into their humanised landscapes. It was an approach that appealed because, while employing historical data, its outcome was indisputably geographical in outlook and intent, and could not be confused with the old, discredited, determinism. Early lectures in geography at Oxford in 1906 by Sir Halford Mackinder adopted this approach. The geography of Britain was conceived of as a geography of Roman, Saxon, Norman, Medieval and Tudor, Stuart, and eighteenth- and nineteenth-century Britain. Other lecturers in the School of Geography, such as J. F. Unstead (1907, 25), adopted the idea of geography being the cutting of 'horizonal sections through time' interlaced with a study of interactions between humans and their environment. Many other geographers followed suit; for example, R. Jones (1926, 77), P. Roxby (1930, 289), and again Mackinder (1931, 268), who spoke of geography conceived of as being no more than 'the historical present'. The ferment of ideas was such that, in 1932, representatives of the Geographical Asociation and the Historical Association met to discuss 'What is historical geography'. While not agreeing entirely on one approach, the common view was that it was both a development of processes and a description of 'the present', whenever that 'present' happened to be (Anon, 1932; Gilbert, 1932).

Within geography, one of the earliest and most distinctive contributions to humanised landscapes came from the 'Aberystwyth School' of historically oriented human geography, which had an emphasis on anthropology, human ecology, and the western parts of Britain (Langton, 1988). In 1932, Emrys Bowen published the first of many works on the historical geography of settlement, the Celtic saints, and the spread of early Christianity in the British Isles (Bowen, 1932, 1954, 1969), and Peake and Fleure's collaborative ten-volume series *The Corridors of Time* (1927–56) and Daryll Forde's *Habitat, Economy and Society* (1934) were particularly influential. In addition, E. Estyn Evans started to produce a plethora of papers on Ireland, Irish archaeology, and what he called the 'Atlantic heritage' (Evans, 1996). But as the 1930s wore on two figures emerged who were to dominate the debate about history in geography—Carl O. Sauer in the United States and H. C. Darby in Britain.

Carl O. Sauer and the cultural landscape

The theme of reciprocal interactions between history and geography within academic history spilled over into American geography in the work of Albert Perry Brigham (1903), and particularly of Ellen Churchill Semple. These were written in language couched in determinism, with many references to geographical controls and influences. Semple's *Influences of the Geographic Environment* (1911) was an interpretation and re-statement of Friedrich Ratzel's *Anthropogeographie*, and was predicated on the simple assumption that 'man is a product of the earth's surface'.

Although Sauer had been taught by Semple and had an immense regard for her as a stimulating lecturer and person, he totally rejected the stultifying determinism of her physical environmentalism. Rather, he saw history (time) as an indispensable ingredient to a full understanding of cultural landscapes (Williams, 1983). In any region/area there was a *natural landscape* of physical qualities, such as vegetation, landforms, soils, minerals, and other resources significant to humans. There was also a *cultural landscape*, which was fashioned by human culture, manifest by human constructs such as fields, crops, houses, roads, plants, and animals, often the result of complex diffusion processes. 'Culture is the agent, the natural area is the medium, and the cultural landscape the result', he wrote (Sauer, 1925, 343). Consequently, historical geography was the series of changes that the cultural landscape underwent, and therefore it involved 'the reconstruction of past landscapes' (1925, 345). It was also an examination of some of the processes of human action.

Other than the creation of a widespread and influential school of 'Cultural Geography'—the Berkeley School—which produced literally hundreds of monographs and papers on landscapes, predominantly in the Americas, Sauer's other great contribution to the study of humanised landscapes was that he fully recognised that humans, through their culture, moulded, modified, and reshaped their environment, for good and for bad. Thus his landscape was a continuously evolving entity that depended on the analysis of processes, and became a tangible record of human adaptation and modification of an environment that was being continuously changed. Because 'retrospect and prospect' were 'different ends of the same sequence', the present was 'but a point on a line' that could be reconstructed from the beginning and projected into the future (Sauer, 1941, 360–1). Simply, time posed no problem in his geography; it was both immaterial but all-pervading, and in his view any geographer working in the purely 'short-time dimension of the contemporary scene is held by a peculiar obsession' (Sauer, 1941, 366).

Sauer's particular brand of geography had few direct adherents in Britain. Possibly the one exception was R. A. Donkin, who undertook deeply scholarly investigations into a wide variety of what were basically studies of the diffusion of practices, plants, and animals, including pearls, pearl-fishing, camphor, manna, Guinea fowl, the Muscovy Duck, the peccary, cochineal, and the extent of terracing in the New World (Donkin, 1979)—although he also made the definitive study of the origin, diffusion, and work of the Cistercian order in Europe (Donkin, 1978). One obvious explanation of the different approach in Britain was that its humanised landscape could only be explained by the long time spans of past cultures as revealed through archaeology, and that the visual and intellectual weight of history hung so heavily over the landscape that it could not be ignored, as perhaps it could be in North America. These elements of the past became incorporated into the dominant paradigm of the recoverable geography of the past as an explanation to the visual, humanised landscape. In that, Darby was the supreme exponent.

H. C. Darby and the systematisation of approaches

Despite the early labours of Bowen, Fleure, and others in Aberystwyth, H. C. Darby emerged as the main practitioner of historical geography and past landscapes during the later 1930s. He laboured and proselytised with a missionary zeal to establish historical geography as a '*self-conscious*' and distinctive subset of the discipline (Darby, 1983a, 423). To do that the 'new' historical geography had to be different from both contemporary human geography and from the powerful and long-established discipline of history. The reconstruction of the geographical past seemed to be the way to do this, and this was the basis of his immensely influential *An Historical Geography of England before AD 1800* (1936), which in fact contained not only reconstructions but also chapters on social, economic, and political events that produced changes. The authors were a roll-call of geographers working at the time—E. G. Bowen, E. W. Gilbert, S. W. Wooldridge, R. A. Pelham, E. G. R. Taylor, J. N. L. Baker, W. G. East, O. H. K. Spate, and Darby himself, accompanied by the historian D. T. Williams and the place-name scholar Eilert Ekwall.

Darby's PhD (the first in geography at Cambridge, awarded in 1931) was on the role of the Fenland in English history, and other than the fact that it was about a region, it was barely distinguishable from history. In a very amended form this was published in two volumes as an historical regional monograph of a process of landscape change: *The Draining of the Fens* (1940a) and *The Medieval Fenland* (1940b). It was while studying the Fenland that Darby had grasped the geographical significance of the Domesday record of 1086, and then developed his ideas in a number of papers and a chapter in *An Historical Geography of England before AD 1800* (Darby, 1936). He conceived of Domesday as *the* baseline cross-section in the humanised landscape of the country, and through painstaking analysis and ingenious cartography conveyed its contents, as in the maps of the distribution of slaves in England (Fig. 5.1), and the representation of the different measures of woodland in east and southeastern England in 1086 (Fig. 5.2). But it was a labour-consuming task, and for the next 30 years or more the geographical analysis of Domesday completely took him over.[4]

[4] The irony was that just when it was about complete it was possible to digitise the data and produce computer-generated maps in a fraction of the time.

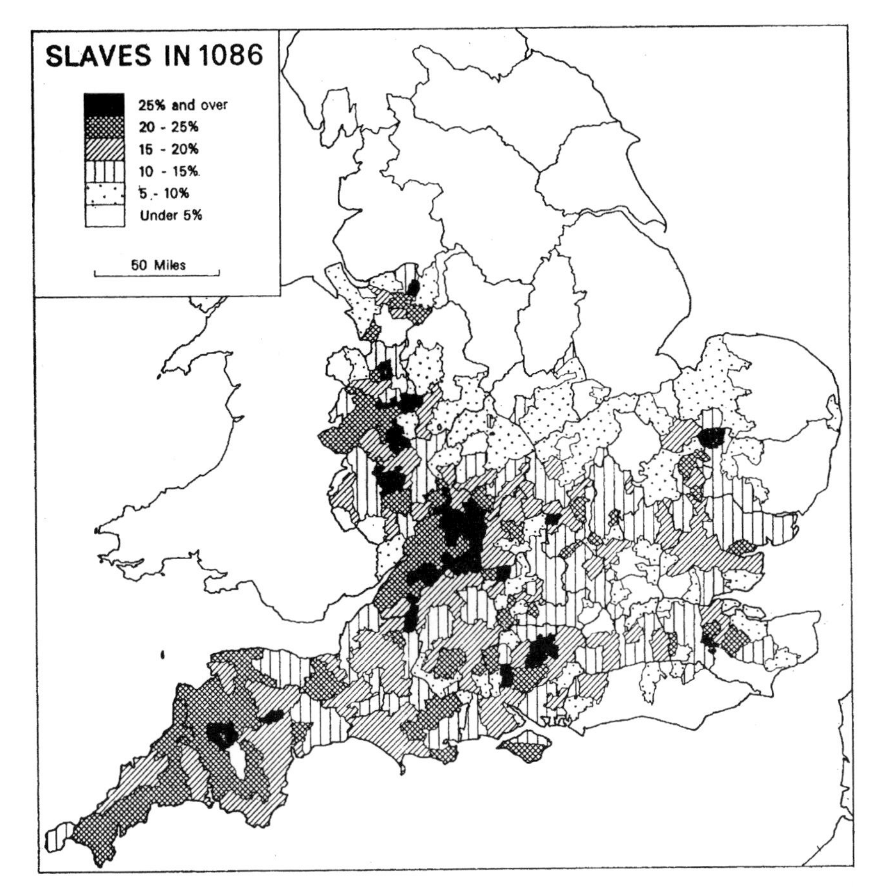

Fig. 5.1 The distribution of slaves in 1086.
Source: Darby (1977, 76, fig. 25). Reproduced by permission of Cambridge University Press.

With the outbreak of war Darby was appointed as editor-in-chief of the Cambridge centre for the production of Geographical Handbooks for the Naval Intelligence Division of the Admiralty, which gave him a unique opportunity to select other like-minded 'historical geographers' (see Chapter 7 in this volume) who saw regional studies as a largely historical enterprise in elucidating the evolution of humanised landscapes, liberally illustrated with maps. After the war, Darby was appointed to a chair at Liverpool, from where he went to University College London in 1949, and eventually back to Cambridge in 1966.

Darby was never the methodological theoretician, but was a pragmatist whose methodological investigations went as far as solving the problem of combining history with geography. The practical problem of 'how to do it'

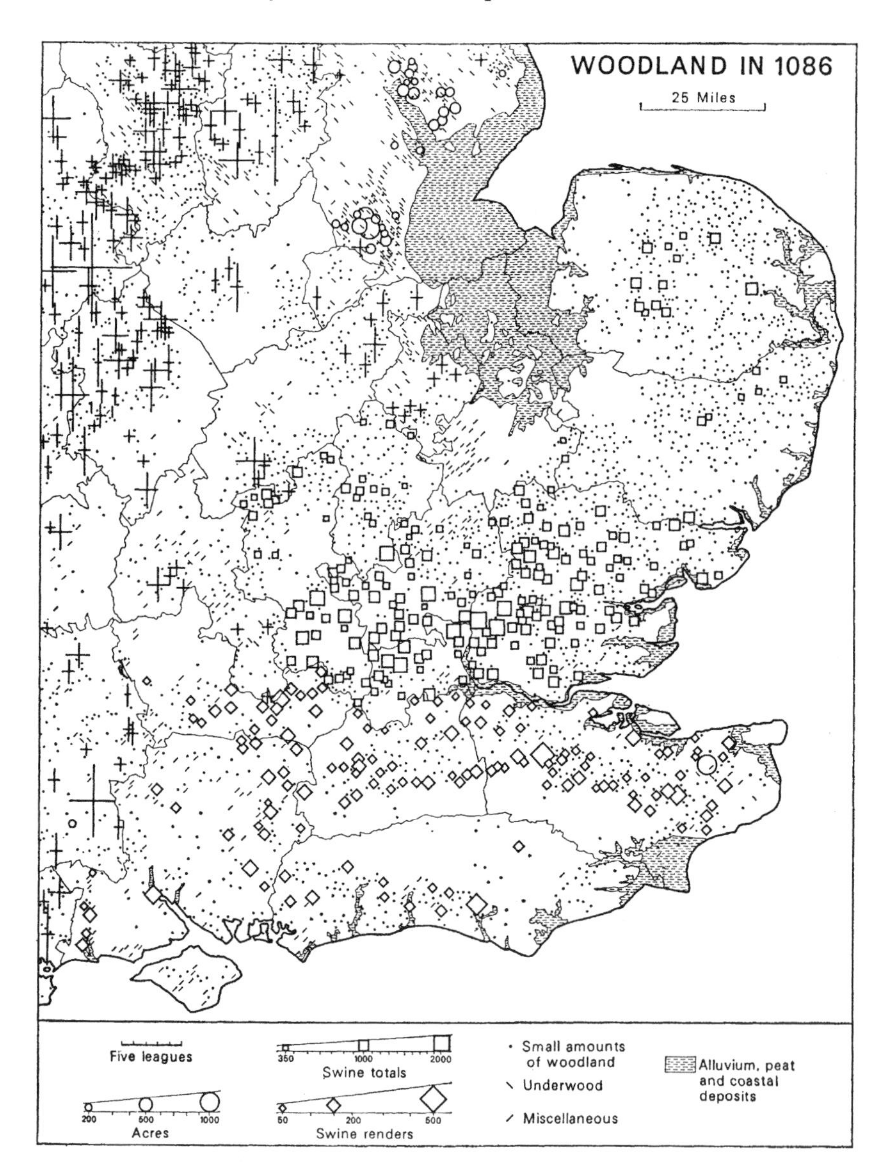

Fig. 5.2 The distribution of woodland in 1086 in east and southeast England. *Source*: Darby (1977, 183, fig. 62). Reproduced by permission of Cambridge University Press.

came to the fore in his 'On the relations of geography and history' (1953). He did not 'attempt to map with any precision the features of this debatable land' but recognised that 'here is a borderland with many trails and many different types of country', and he suggested four ways in which the two disciplines could be combined (Darby, 1953, 1). First, there was 'The geography behind history'. This was the approach of the early historians, who saw historical events as being explained by geographical factors. Secondly, there were 'Past geographies'—in effect re-creations of geographies of some past period and presented as a cross-section through time. Thirdly, came 'The history behind geography', what Darby later called 'vertical themes', which traced the development of some landscape feature or process through time to distinguish these from the horizonal cross-section. It is what today we would call 'the human impact on the landscape'. Finally, there was 'The historical element in geography', the time element present in any and all geographical work, but in this case concentrating on the vestiges in the landscape—what Whittlesey (1929) had earlier called 'sequent occupance'. These ideas were elaborated under slightly different titles and emphasis for the benefit of historians in his 'Historical geography' in H. P. R. Finberg's *Approaches to History* (Darby, 1962a), and then again in even greater detail for England, France, and the United States in the resurrected *The Relations of History and Geography* (Darby, 2002).

While the 1953 paper was about the intertwining of space and time, it was the humanised landscape, in some form or another, that figured largely in it. For Darby it was axiomatic that it was 'the purpose of geography to explain the landscape' and that 'an understanding of the landscape' formed an indisputable part of geographical study. The visual understanding of landscape became the touchstone of geographical relevance and purpose in the debatable history/geography frontier zone. In that paper the words *geography*, *-ies*, *-ical*, appear 58 times (other than in titles), variations of *history* 35 times, and *historical geography* 14 times, but *landscape* appears 24 times and *face of the country/countryside* 7 times (Williams, 1989).

A few years before, Darby had experimented with the third approach of 'The history behind geography' and published 'The changing English landscape' (Darby, 1951a). This explored changing landscape features, which he dubbed 'vertical' themes of change that included, for example, woodland clearing, marshland draining, urban expansion, and landscape garden development. To a certain extent this approach was the logical outcome of the pioneering work of the American polymath George Perkins Marsh (1864), whose *Man and Nature* showed that humans were agents of change (for good and bad) and landscape creators. Explaining the changing elements of the landscape became probably the most popular approach to humanised landscapes in later years.

The simple logic and common sense of Darby's arguments about cross-sections and vertical themes was influential in changing Hartshorne's views about the unbridgeable division between history and geography. In a revision of his earlier work he conceded (Hartshorne, 1959, 132) that there was no reason why a feature in all its changes through time was not geographical, and paid tribute to Darby's concept of the 'vertical theme' in geography. Secondly, he conceded that a continuous historical geography of an area could be formed from an unlimited number of cross-sections that were, in fact, changing areal variation through time. It was a vindication of the simple statement that Darby made later that geography and history 'refuse to be separated in practice, whatever theoretical distinctions we may draw between them.' (Darby, 2001, 171). Hartshorne's endorsement of Darby's ideas stripped away the last barriers between the two disciplines and many studies followed that strayed across the territory of each other. Historical geography and the study of landscape flourished and new posts were established in universities, many of them filled by Darby's students (Prince, 2000).

The 'swinging sixties'

But things do not stand still. Even as the new liberty freed historical geography and humanised landscape studies of its methodological confines, the 1960s brought new sets of wider intellectual influences to bear on them. Both history and geography changed as theoretical experiment, methodological approach, and focuses of study entered a period of radical rethinking, innovation and experiment. 'The "swinging sixties"', reflected Darby (1983a, 426) in later years, 'certainly left their mark upon our life and thinking', and the study of the humanised landscape could never be the same again.

First, what Darby had seen as a useful categorisation of approaches or working hypothesis for systematising ideas about space and time was seen by others as a rigid methodological statement that defined new working boundaries and approaches. Far from liberating geographers from the orthodoxies of the past, it seemed like a new constraint. In any case, Darby's schema could not easily capture the shift to seeing the landscape as a symbol expressing social, political, and economic ideologies, let alone class, race, and gender. In much the same way, though perhaps to a lesser extent and later, Sauer's more plastic cultural landscape seemed similarly dated and inflexible. While many kept their feet firmly on the ground by exploring places in the past, relict features, and changing landscapes, others pushed their heads into the clouds of theory in order to reinterpret historical data anew, whether through quantification or humanistic social theory, or Marxist theory. Both the reconstructed past and the cultural landscape became outmoded paradigms that were past their 'sell-by' dates. In North America, other geographers, like Andrew Clark, who were interested in the landscapes of the past, but rarely interested in quantification or social theory, rejected Darby's ideas and expressed 'little concern with making nice distinctions between geography and history or with dodging the appellation "cultural historians"' (Clark, 1954, 85). Such schema as

Darby's became 'procrustean operations' (Clark, 1972, 143) that the diversity and rapidity of change in the North American scene simply could not be made to fit. Here the overriding theme was 'change' and the 'rate of change' (Clark, 1959, 1962). The dominant influence was that of Sauer and his students, who concentrated on the cultural processes that produced geographies, for whom any distinction between past and present, said Clark, was 'illusory'.

Secondly, historians themselves began to experiment with new themes and new approaches, and moved over into geographical terrain. Paramount amongst these was W. G. Hoskins, whose *Making of the English Landscape* (1955) was the first detailed history of the man-made scene. 'The English landscape itself, to those who know how to read it aright, is the richest historical record we possess', declared Hoskins in the opening paragraph of the book, and then there followed a chronological sequence of general eras that created and affected the visual scene—the Celtic and Roman, the formative Anglo-Saxon, the era of medieval colonisation, the impact of the Black Death, the recovery and flowering of Tudor and Stuart England, the transformations of parliamentary enclosure, the landscapes of canals, roads, towns and railways, and so on. The avowed aim was to see 'the logic that lies behind the beautiful whole', which was a bold attempt to 'engage both reason and emotion, to combine explanation and evocation to a "many-sided pleasure" to the skilled historian and the sensitive traveller' (Meinig, 1979b, 197).

Local history and landscape history went hand in hand, and Hoskins was zealous in promoting both. In 1948, the Department of English Local History was founded at Leicester University, with Hoskins as reader-in-charge, and after a 14-year spell in Oxford (1951–65) he returned to Leicester as professor in 1965, only to relinquish the post three years later to devote himself entirely to writing and the promotion of local history. As part of this quest he had already launched a series with the publisher Hodder and Stoughton of well-illustrated histories of the 'making' of the landscapes of the counties of England. In all, 20 county volumes were published that were to form a rich repository in the visual and historical 'rediscovery' of the English landscape. Notably, two of the four published between 1954 and 1970 were written by geographers, W. G. V. Balchin's *Cornwall* (1954) and R. Millward's *Lancashire* (1955), the other two by historians—Finberg on *Gloucestershire* (1955) and Hoskins on *Leicestershire* (1957)—and then the series went dead. But it was resuscitated in 1970 with the general revival of interest in the landscape, and of the subsequent 17 volumes published between 1970 and 1985, nearly all were written by historians. In addition, Hoskins wrote some 28 books that involved landscape and local history, nine of them after his 'retirement' in 1965. Initially, the work of both Hoskins and his co-workers received only indifferent recognition from professional historians, who were not comfortable with either landscape or local history, and a wider public that was still to be convinced of the delights of unravelling the English scene. But after the spate of writing after his retirement he succeeded in breaking through the academic barrier to reach a wider audience in 1973 with a television film on the making of the English landscape, followed by a series of 12 shorter films entitled *One Man's England* (Hoskins, 1978), all of which were immensely popular (Meinig, 1979b, 200–1) and had a major impact on the public perception of landscapes of the past.

But that was not all: history blossomed throughout the post-war period and became a many-faceted enterprise with a strong local, landscape flavour that was educating the public and the profession to see landscape as history. During the early 1950s, new branches of history gained a firm footing and founded journals that became outlets that were eagerly patronised by geographers, who found that their interests did not fit easily into the strictly historical geographical schema devised in the past.

Agrarian history was one such departure. In 1951, the Museum of Rural Life opened in Reading, the Agricultural History Society followed in the next year, and the very successful journal *Agricultural History Review*, edited by H. P. R. Finberg and later by Joan Thirsk, in which many historical geographers found a ready outlet for their research on changing landscapes of farming, enclosure, settlement, and farming regions. The publication of the nine-volume *Agrarian History of England and Wales* between 1967 and 1991 was the culmination of this endeavour (Thirsk, 1967–91). A growing interest in the agrarian landscape was counterbalanced by an interest in the urban, arising out of the foundation of the Victorian Studies Centre at Leicester University by H. J. Dyos. His establishment of the *Urban History Yearbook* released a flood of research on every aspect of the urban (Dyos, 1968, 1973; Dyos and Wolff, 1973). In quick succession, so it seemed, the *Journal of Transport History* (again emanating from Leicester) was founded in 1953 by Michael Robbins and H. J. Dyos. Major reassessments of the development and competitive role of railways, canals, and roads in Britain's industrial change, and their impact on cities, followed (Simmons, 1961; Hadfield, 1968; Dyos and Aldcroft, 1969), which geographers like Appleton (1962), Pawson (1977), and Freeman (1977, 1980) took up with vigour.

In another development, the significance of the parish register, long the preserve of the local and amateur historian, became part of a national inventory of population growth and characteristics. The foundation of the Cambridge Group for the study of the History of Population and Social Structure (CAMPOP) in 1964 put population history onto a firm footing by conducting detailed investigations of households and families, reconstructing and back-projecting families from the early sixteenth century, and investigating long-term trends in population dynamics of fertility, mortality, migration, and nuptuality in a wide sample of parishes across the country. Along with historians like P. Laslett and R. Schofield, geographers such as E. A. Wrigley and R. Smith were leading lights, the monumental *The Population History of England: A Reconstruction* (Wrigley and Schofield, 1981) being a major product of the group research, as has been other work like Mary Dobson's *Contours of Death and Disease in Early Modern England* (1997).

Clearly, not all developments were within history; the founding in 1975 of the *Journal of Historical Geography* by J. Patten in Britain and A. H. Clark in the United States opened up a new outlet for traditional historical geographical studies of landscape, though in general it took a very eclectic view of what constituted historical geography.

So much for the disciplinary history and intellectual developments in both history and geography that preceded and influenced the study of humanised landscape, but what of the substantive study of the landscape itself? All landscapes are unique and consequently their stories beyond number and without end; hence not all landscapes can be examined. So attention here is focused on the British landscape, and in particular on two of the four approaches outlined by Darby that most relate to humanised landscapes: 'Past humanised landscapes' (past geographies) and 'Explaining humanised landscapes' (the history behind geography). The other two approaches have fallen into abeyance. The 'Geography behind history' is outmoded and discredited, while the 'Historical element in geography' is inevitable and self-evident, as is clear from other contributions to this volume. Many of the intellectual currents noted above surface in the study of these landscapes.

Past humanised landscapes

There are basically two approaches to understanding past humanised landscapes—the reconstruction of these landscapes from consistent and comprehensive sources, and the mapping of relict features. Increasingly both approaches rely on the interdisciplinary cooperation of historians, archaeologists, palaeobotanists, and other disciplines, relict landscapes perhaps more so than reconstructions.

Reconstructions of past landscapes

Darby's work on the Domesday record, which he thought of as the baseline for all further studies, was the quintessential study of a cross-section in the humanised landscape of the country. It culminated in seven major volumes. He wrote many of the county chapters and co-edited five of the regional volumes, and was the sole author of the first volume (Darby, 1952) and the final summary one for the country as a whole (Darby, 1977). The corpus of work was probably unrivalled as a sustained and single-minded effort by a single, individual geographical scholar, and it was

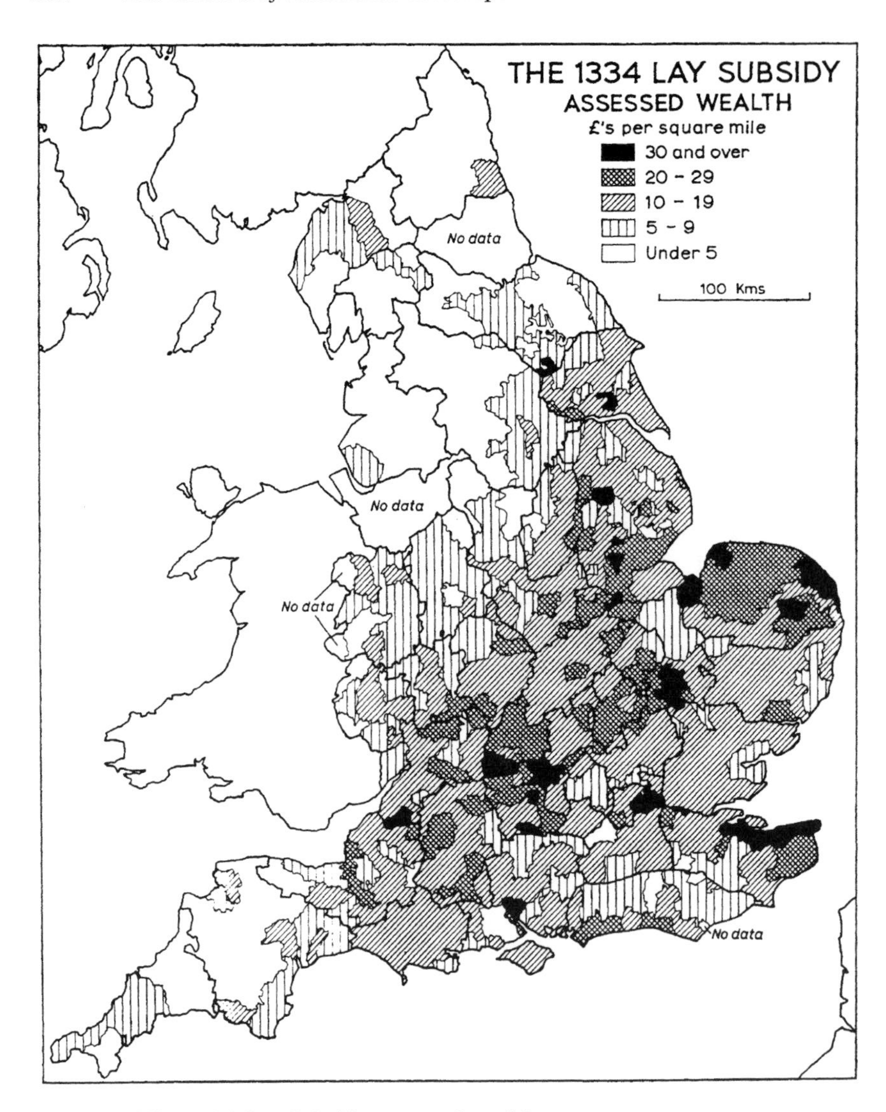

Fig. 5.3 The 1334 Lay Subsidy: assessed wealth.
Source: Glasscock (1973, 139, fig. 35). Reproduced by permission of Cambridge University Press.

for this contribution to learning that he was the first geographer elected to the British Academy, in 1967, his candidature supported by an anthropologists, a medievalist, and a modern historian.[5]

[5] He was elected to Section XII, Social and Political Studies, and the nomination papers were signed by Sir Raymond Firth (Chairman), M. M. Postan, Sir Denis Brogan, and Daryll Forde. I am indebted to Ms Susan Churchill of the British Academy for this information (13 August 2001).

With one cross-section established, others far too numerous to detail followed. Cohorts of colleagues and students analysed, for example, the Exchequer Lay Subsidy of 1334 (Fig. 5.3) (Glasscock, 1973, 1974), that of 1524 (Sheail, 1972), the Crop Returns of 1801 (e.g. Henderson, 1952; Thomas, 1963), the Tithe Returns (Kain and Prince, 1985), the 'General Views' of the agriculture of counties submitted to the Board of Agriculture between 1793 and 1815 (Darby, 1954), probate inventories (Overton, 1996), and the writing of individual travellers and observers such as Daniel Defoe (Andrews, 1956). A new source appeared with the lapsing of the 100-year confidentiality rule with each successive population census. Urban areas could now be analysed in great detail, an endeavour given new impetus by the urge to quantify during the 1960s, the rise of social area analysis in urban studies, the availability of computers, and the founding of the journal *Urban Studies*. This research first moved into high gear with the work of Richard Lawton (1955, 1968) and his students at Liverpool, and special publications of the Institute of British Geographers (Whitehand and Patten, 1977; Dennis 1979) showed how the new techniques could be used to provide views of the urban past. It was very much a *Geographical Interpretations of Historical Sources* (Baker, Hamshere, and Langton, 1970).

The repetition of one cross-section after another, while giving some sense of change in the landscape, was in fact very a-historical. The cross-sections, syntheses though they were, appeared like static slices of reality, lacking the preceding economic and social conditions that gave rise to them. Moreover, the cross-sections were only as good as their sources, which were frequently incomplete, biased, or flawed. Recognising this, Darby adopted for his *A New Historical Geography of England* (Darby, 1973) the structure devised by J. O. M. Broek (1932) in his study of the changing landscape of the Santa Clara Valley, California, in which chapters of historical narrative (explanation) separated chapters of geographical reconstructions (description), viz:

1. The Anglo-Saxon Foundations
2. Domesday England
3. Changes in the Early Middle Ages
4. England circa 1334
5. Changes in the Later Middle Ages
6. England circa 1600
7. The Age of the Improver, 1600–1800
8. England circa 1800
9. Changes in the Early Railway Age, 1800–1850
10. England circa 1850
11. The Changing Face of England, 1850–c.1900
12. England circa 1900

While being the epitome of the genre, *A New Historical Geography* was the last gasp of the source-led reconstructions, which, because of its long gestation, had sneaked into the 1970s, and it did not have a great influence or long life. Other more thematic and dynamic surveys of England and Wales appeared that emphasised the processes that helped to create spatial patterns rather than the reconstruction of past periods and the excavation of places. These incorporated ideas of quantification and contemporary social theory. At the regional level D. Gregory's study of the transformation of the Yorkshire woollen industry during the Industrial Revolution was path-breaking in using social theory in order to 'explicate the transformation of the woollen industry' and in attempting to put 'ordinary men and women' back into the picture of change (Gregory, 1982, 2, 1). But however laudable the aim, it was very difficult to capture ordinary lives in cartography; that is, of necessity, an exercise in generalisation. At the national level the volume of essays edited by R. A. Dodgshon and R. A. Butlin carried these ideas further in the hope that their text would 'stimulate the student's mind rather than just fill it' (Dodgshon and Butlin, 1978, viii). It was divided into two sections: one dealing with conventional 'past' geographies of the prehistoric, the Roman, and the Anglo-Scandinavian to the early and late medieval periods; the second dealing with themes of change in population, agriculture, industry, transport, and towns from 1500 to 1900. The survey by R. Lawton and C. Pooley (1992) was even more removed from the old model. It was divided into three periods, 1740s–1830s, 1830s–1890s, and 1890s–1940s, each with a repetitive quintet of themes as follows:

1 Political, social and economic context
2 Demographic change
3 The countryside
4 Industry and industrialisation
5 Urbanisation and urban life

No overt mention of 'past geographies' here, but processes and landscapes of change. The methodological niceties of what was geography and what was history in past humanised landscapes was a dead issue.

Relict landscapes

The visual vestiges of the humanised landscape of past occupation and use are tangible and evocative goads to investigation. As a cursory glance at any bookshop shelves will reveal, since the later 1970s, at least, there has been a near-unsatiable demand for this type of work from the general public, who readily understand its aims and methods, and whose interest has been stimulated by imaginative publishing and television. Undoubtedly, the pioneering television programmes by William Hoskins in 1973 were seminal in this awareness and interest, and the Cambridge University Collection of Air Photographs founded in 1949 provided a repository of landscape information that supplemented the geographer's traditional tool of the map with a graphic and understandable image (Baker and Harley, 1973; Glasscock, 1992).

But for the geographical world, the discovery by a geomorphologist, an historical geographer, and an archaeologist that the Norfolk Broads were in fact hollowed-out and flooded turf cuttings from the thirteenth and fourteenth centuries, from which 900 million cubic of peat had been cut, the term 'relict landscape' came to have new meaning (Lambert et al., 1960). Equally significant, though not so immediately dramatic because it happened over a longer period of time, was the work of the Medieval Village Research Group directed by the historian Maurice Beresford and the archaeologist John Hurst (Beresford, 1954; Beresford and Hurst, 1971). Eventually over 2,000 village sites were unearthed in England (Fig. 5.4), which not only added a completely new and tangible layer of a past humanised landscape but was frequently visible through the new medium of air photography. Other facets of landscape that were investigated were ridge and furrow (Mead, 1954, Harrison, Mead, and Pannett, 1965), strip lynchets (Wood, 1961), burgage plots and urban plans (Conzen, 1960; Slater, 1982), village forms and markets places (Roberts, 1977), landscape gardens and parks (Fig. 5.5) (Prince, 1967), pits and ponds (Prince, 1964), moated settlements in East Anglia (Fig. 5.6) (Emery, 1962), and many more.

The same approach suffused a new and vigorous investigation into traditional field patterns (Baker and Butlin, 1973; Dodgshon (1973, 1975) and parliamentary enclosure (Yelling, 1977; Turner, 1980), and even in neglected facets of urban areas, such as building cycles and the urban fringe (Whitehand, 1966, 1972, 1975), while Cherry (1972) probed the origins of urban form. Urban morphology fascinated historians and geographers alike, as shown by Beresford's survey of *Medieval England: An Aerial Survey* with J. K. S. St Joseph (1958), and his (1988) and Ward's (1962) investigations into the townscapes of Leeds. They showed conclusively that the landscape was a palimpsest, as the pre-urban cadastre of field boundaries influenced the patterns of housing development in the industrial cities of the north (Fig. 5.7). A similar approach of tracing the indelible influence of the pre-urban cadastre was employed by R. J. Johnston (1968) in his analysis of Melbourne's street patterns.

During the late 1950s, the study of the essentially rural-based humanised landscape was given an industrial twist when the study of the forgotten features of industrial archaeology blossomed under the aegis of Council for British Archaeology (Hudson, 1963). Now the relict features of the industrial revolution became a legitimate, respectable facet of study in the humanised landscape rather than something to be despised and disregarded (Buchanan, 1972).[6] The study of vernacular architecture, building materials, styles of construction, and planning added yet another new facet

[6] Several geographers contributed to this in the 1960s (see C. T. Smith, 1965).

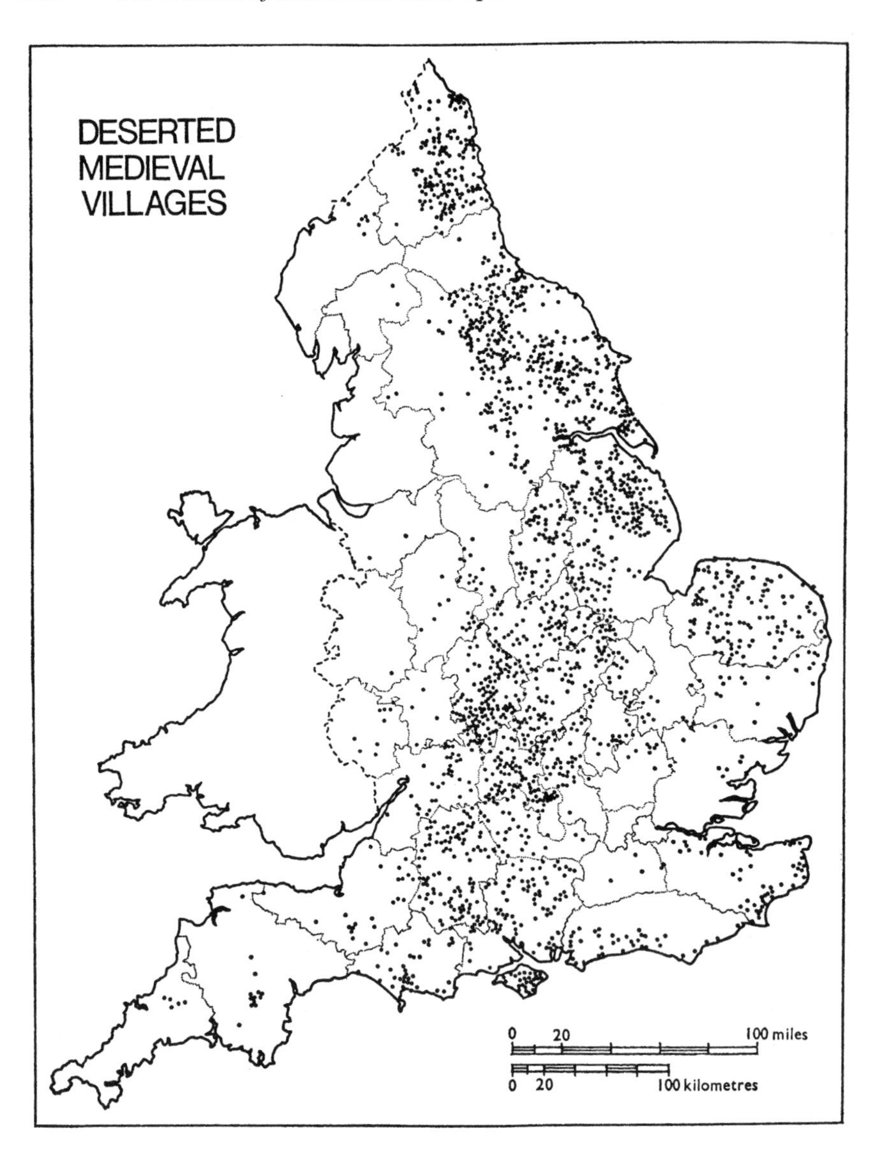

Fig. 5.4 The distribution of deserted medieval villages identified up to the end of 1968. *Source*: Beresford and Hurst (1971, 66, fig. 13). Reproduced by permission of James Clarke and Co. Ltd, Cambridge.

to both rural and urban scenes, with the work started by Nickolaus Pevsner's monumental survey of English architecture (1951 onwards), and continued by the Royal Commission on Historical Monuments for England, on which Darby served for 24 years (1953–77), and by individual scholars (e.g. Barley, 1961; Clifton-Taylor, 1962).

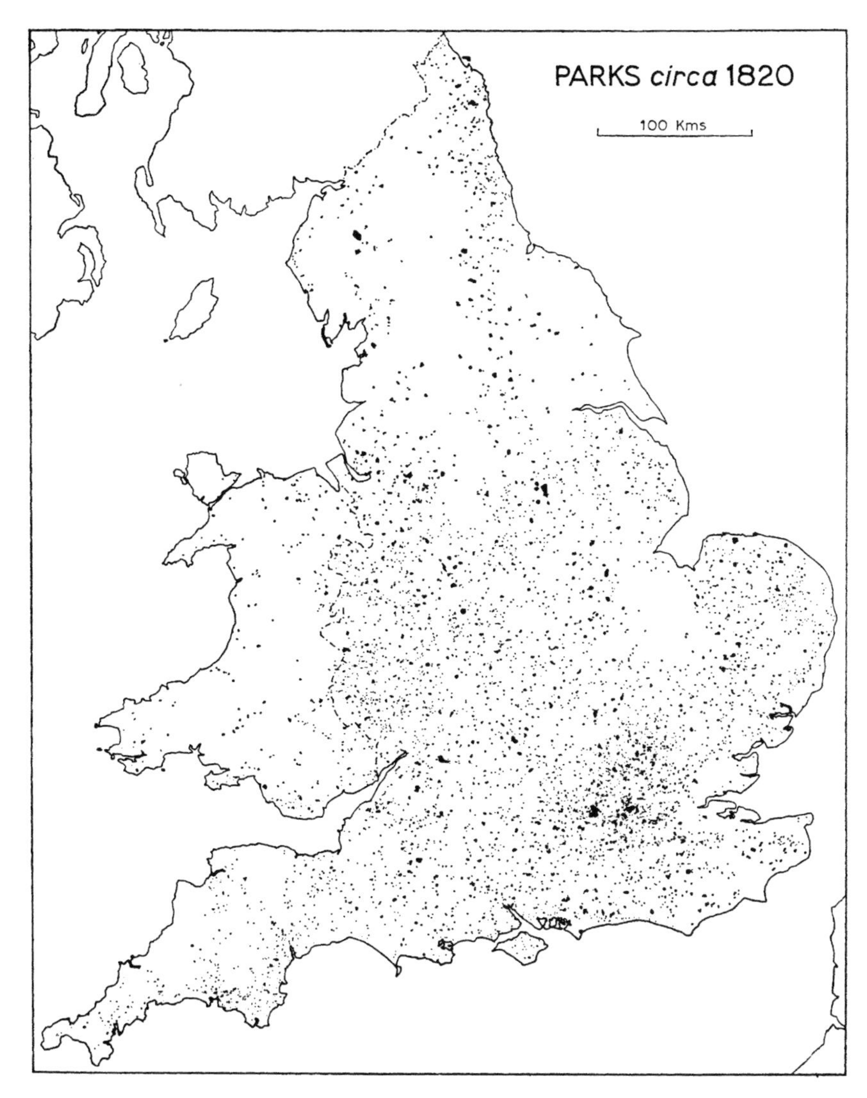

Fig. 5.5 Parks, c. 1820.
Source: Prince (1973, 42, fig. 90). Reproduced by permission of Cambridge University Press.

But this analysis of the minutiae of the humanised landscape could not withstand the pressures for generalisation and social interpretation, and while the study of the local landscape flourished during the 1960s and 1970s within British history, it was dying in geography by the 1970s and 1980s as new concerns were emerging.

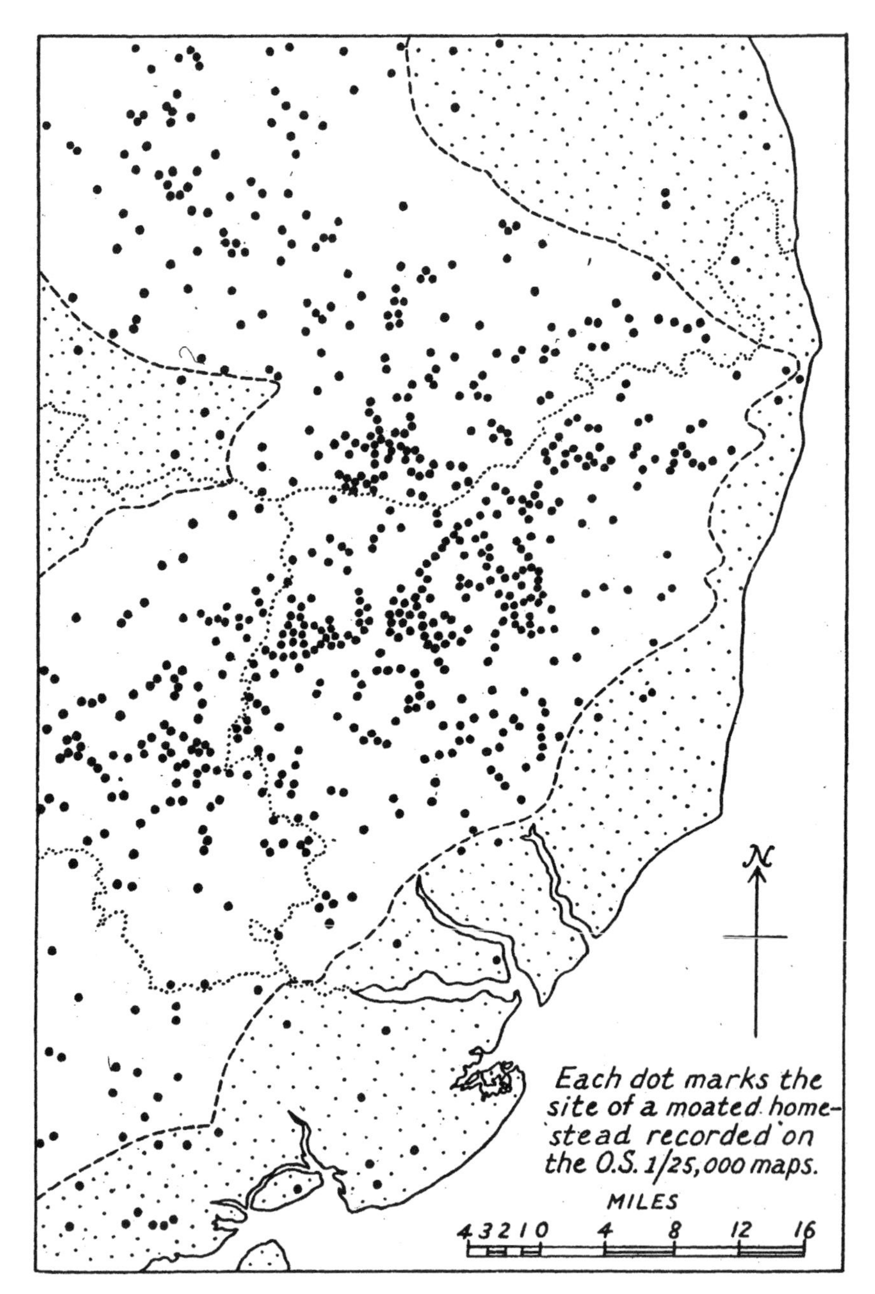

Fig. 5.6 The distribution of moated homesteads in East Anglia as marked on the Ordnance Survey maps at 1:25,000. The unshaded area denotes boulder clay and chalky loam plateaux.
Source: Emery (1962, 383, fig. 2). Reproduced by permission of the Geographical Association.

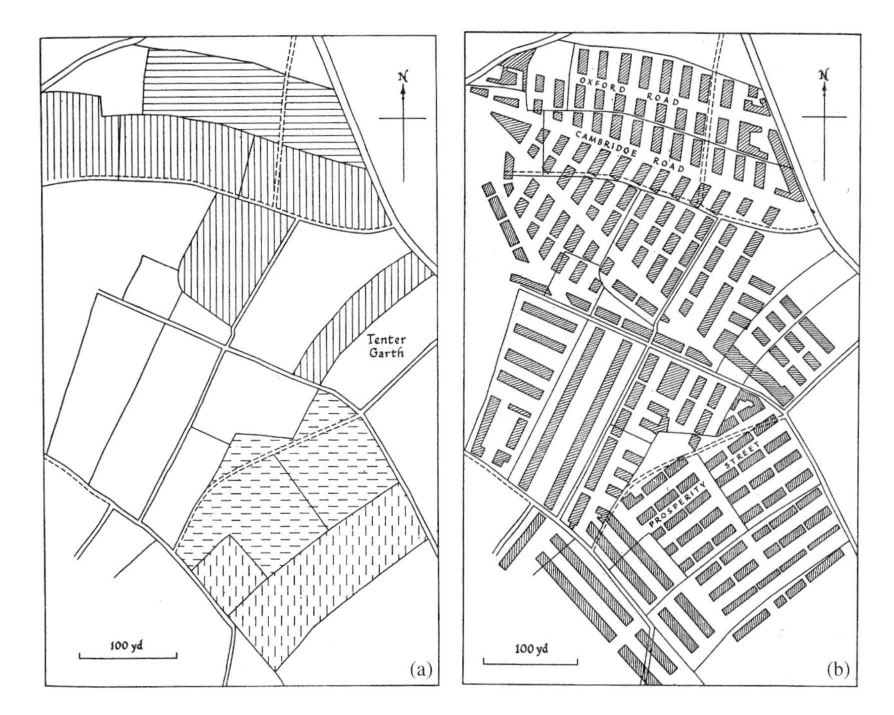

Fig. 5.7 (a) The pre-urban cadastre, southwest of Meanwood Road, Leeds, 1847. Different owners are shown by shading, a fifth owned the unshaded fields; (b) The same area finally filled by back-to-back houses.
Source: Beresford (1988, 9–10, figs 1 and 2). Reproduced by permission of Hambledon Press.

Explanation in the human landscape

While labouring away on his self-assigned task of the Domesday Book, Darby was, nevertheless, profoundly dissatisfied with the cross-section as a method of capturing the past of places. His Domesday work had been criticised for being a static entity that would have been more enlightening had it been supported by the interpretation of other medieval documents. Darby was aware that landscape elements did not arise just at one moment, however wide a band of 'time' was allotted to that moment, but were often the product of centuries, and were always in the process of becoming something else. Hence history (narrative) had to accompany geography (description) in order to provide the explanation and account of the process. Now any pretence, or even necessity, of being self-consciously 'geographical' was supplanted by the desire for a more total explanation of geographical change. In this light Darby's 'Changing English landscape' (1951a), sometimes dismissed as a minor piece, was quite revolutionary, and preceded by four years Hoskins's *Making of the English Landscape.*

The pity was that Domesday absorbed his energies, and he never pursued or developed this line of research to its logical conclusion, except in lecture courses. In any case, Darby's paper was 'hidden' in the *Geographical Journal*, while the work of Hoskins was highly visible as a Penguin book.

The *Landscape of England and Wales* by Coones and Patten (1986) was a hesitant attempt to further the idea of change in a fairly traditional account of the landscape that could best be described as a geographer's version of Hoskins's *Making of the English Landscape* of 30 years earlier. But it was too late; the Hoskins county volumes had successfully pre-empted the field, and its impact was limited if only because of the uniformly poor quality of the reproduction of illustrations, which was incomprehensible for any mid-1980s publisher of an essay on the visual.

Nevertheless, specific elements of landscape change in the 'natural' landscape became a favourite topic of geographers, particularly where the human activity was intimately and inextricably entwined with the 'natural' landscape. The draining of wetlands generated studies by geographers (Rollinson, 1964; Sheppard, 1966; Williams, 1970) and Darby himself (1940a, 1940b, 1983b), as well as by historians (Thirsk, 1953a, 1953b; Hallam, 1965) (see Fig. 5.8). Similarly, underdraining in clay land (Darby, 1964; Phillips, 1989), moorland reclamation (Parry, 1976), and woodland clearing (Darby, 1951b; Williams, 1989b, 2003) were other topics.

For those British-trained historical geographers working overseas who were influenced by the prevailing paradigms, of which there were many (Prince, 2000), the obvious making of new landscapes by European colonisation and settlement offered a ready set of themes. Within Australia, for example, Heathcote (1965), Powell (1970, 1977), and Williams (1974) all produced monographs on initial settlement and the creation of new geographies, which were detailed accounts of the explanation of the human landscape of large parts of the continent. Williams, for example, combined cross-sections to chart the rapid advance of settlement in South Australia, overlaying these with vertical themes of change in the constituent elements of the landscape, such as the rural survey system, the creation of towns, the clearing of woodland and forest, and the like (Fig. 5.9). One distinguishing feature of these Antipodean works was their awareness of the frequent mismatch in the environmental perception (or mis-perception) of this new land by ordinary settlers, on the one hand, and officials on the other. The capabilities of the environment were continually misread. Thus these authors partially anticipated the emphasis on cultural processes outlined below.

Almost by accident this brand of humanised landscape studies was soon to be at the very forefront of academic endeavour, being perfectly in tune with the rising interest in the human impact on the environment during the last quarter century. In reality it was an old theme, first adumbrated by George Perkins Marsh in his *Man and Nature, Or, Physical Geography as Modified by Human Action* as far back as 1864, but reinvigorated by the

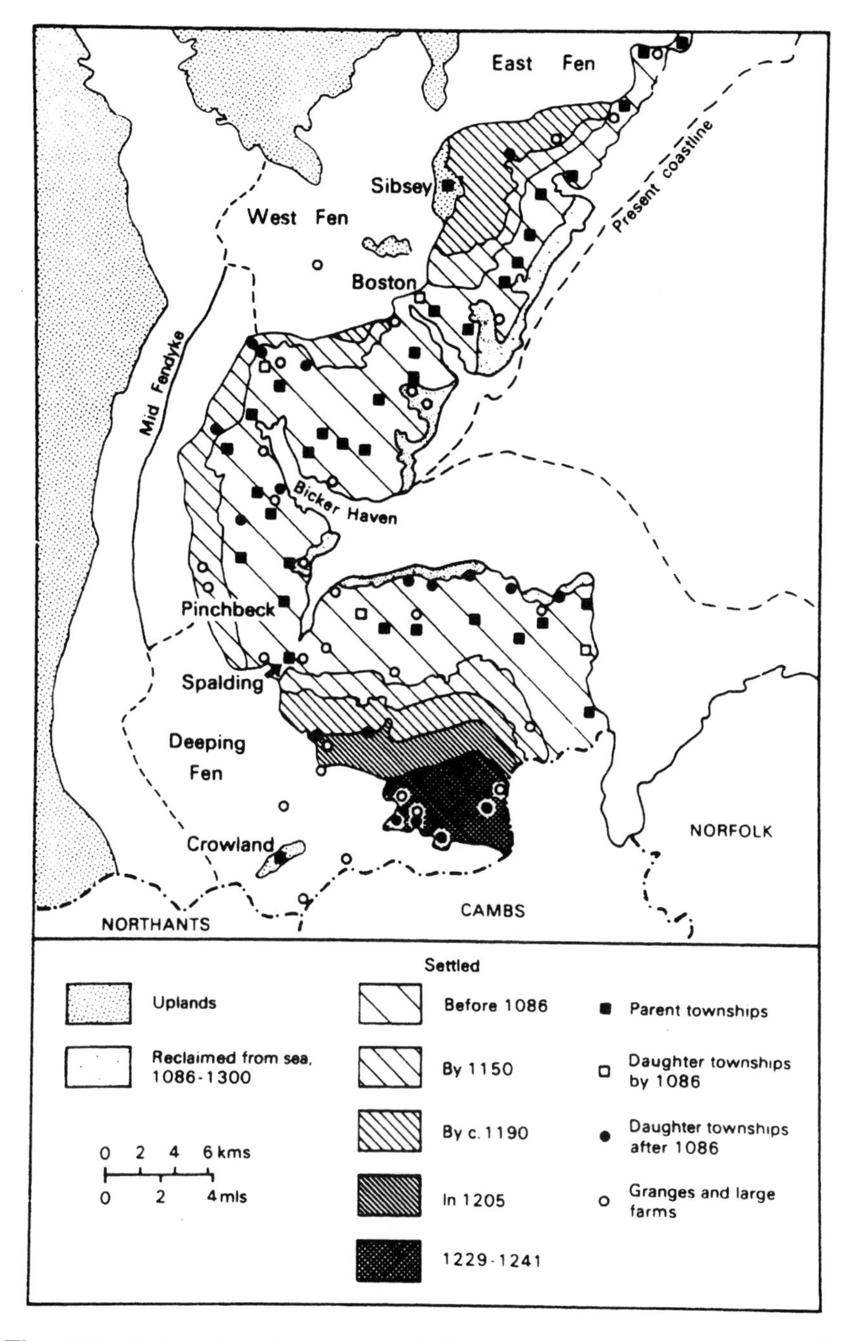

Fig. 5.8 Reclamation from sea and fen and new settlements in South Lincolnshire before 1307.
Source: Williams (1982, 98, fig. 4.4, based on Hallam, 1966, figs 1 and 2).

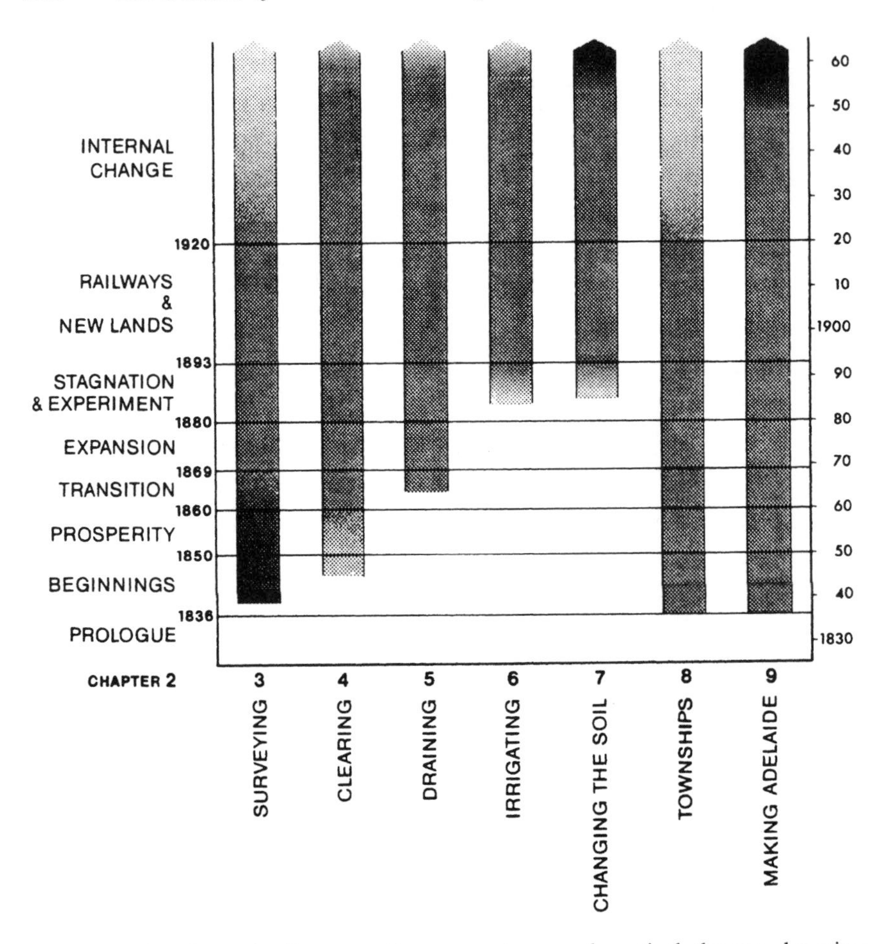

Fig. 5.9 A schematic diagram of cross-sections and vertical themes denoting chapters in the *Making of the South Australian Landscape*.
Source: Williams (1974, 3, fig. 1).

writing of physical geographers (e.g. Goudie, 1981; Simmons, 1989; and see also Chapters 4 and 14 in this volume) and some historical geographers (Powell, 1976; Williams, 1989b, 2003). It is alive and well in the work of many of the new environmental historians and the new journals such as *Environmental History* and *Environment and History*.

The humanised landscape: artefact or cultural process?

One can rarely divine a trend until it is well under way or even passed, and as far as studies of the humanised landscape are concerned, the onset of the 1970s was just one such beginning. The study of the landscape shifted imperceptibly from being principally either a physical entity or artefact or picture in words, as envisioned by Darby, Sauer, or even Hoskins, to become a cultural process. The plotting, mapping, and description of the past and its relations with geography, and the creation of landscapes through cultural modification, were replaced by a more symbolic interpretation of landscape and the interplay of that symbolism with the material (Williams, 1989a) that was coincident with the emergence of the 'new' cultural geography (Duncan and Duncan, 1988; Duncan, 1995). The construction of 'actual' landscapes gave way to the conceptualisation of 'ideal' landscapes, though the reciprocal influences between both were recognised.

Undoubtedly the rising interest and shift towards behaviouralist (Kirk, 1952), humanist, phenomenological, and perceptual approaches by geographers during the late 1960s and the 1970s (Lowenthal and Prince, 1965; Prince, 1971; Tuan, 1971, 1974; Lowenthal and Bowden, 1976), as they rejected positivism, generalisation, and number-crunching, was influential. They no longer attempted to understand the scenic or painterly sense of landscape (Tuan 1976), which is always the view of the detached observer—outside the frame gazing in. Such views were devoid of what Relph (1970) called the perspective of the 'existential insider'. By this reasoning, to see the world as a landscape would imply a particular and historically constructed way of seeing things that always 'reinforces a separation between society and nature, structures vision according to ownership and control, is highly gendered, tries to erase the facts of work from view and elevates selectivity to realism' (Mitchell, 1998, 14). In short, geographers became interested less in the facets and appearance of the landscape and more in the subjective meaning of places for people, focusing on symbolism, iconography, power, gender, aesthetics, and taste.

Beyond geography, three broadly Marxist works published during 1972 and 1973 were seminal. First, Raymond Williams's literary study, *The Country and the City* (1973), suggested that reading the landscape was essentially a creative act. Moreover, he asserted that the term 'landscape' was a privileged and inherently conservative way of seeing, and that 'a working country is rarely ever a landscape' (1973, 120). Secondly, the radical way of looking at art adumbrated in *Ways of Seeing* by the Marxist art critic John Berger (1973) was eagerly taken up by a widespread interdisciplinary group of landscape scholars. They could no longer look at landscape as an objective entity, but as an interpretation of economic, social and political forces. Thirdly, John Barrell's *The Idea of Landscape and the Sense of Place* (1972), soon to be followed by his *Dark Side of the Landscape* (1980), though founded in the poetry of John Clare and contemporary eighteenth- and early-nineteenth-century art, became a sustained critique of the 'Age of the Improver'. Landscape and its creation in parks, gardens, and productive farms was both a measure of the status of the new, rising middle and established upper classes, but was also redolent, symbolic, and even responsible for the far-reaching economic and social changes in the landscape perpetrated by them, as modernity and capitalism obliterated the past. Open fields were enclosed, common rights extinguished, manufacturing and urbanisation grew, and rural poverty and out-migration occurred. The rural workers were no longer integral to the rural scene, as in paintings in the past; they were no longer a part of some harmonious pastoral entity but were now relegated to the darker side of the landscape while the *nouveau riche* side was bathed in literal and metaphorical sunlight.

It is clear that these three works had a profound effect on geographical thinking as the new wave of historical/cultural geographers of the 1980s sought to break away from the realist interpretation and representation of the landscape. The Darbyesque landscape was already under attack as more statistically minded geographers thought the chances were missed of investigating telling relationships between the data (particularly Domesday) because of the absence of elementary correlation techniques (Gregory, 1976). Now cultural geographers castigated it as being 'bloodless', 'elitist', 'Whiggish', and 'bourgeois', only concerned with the big players and excluding the common people (Baker, 1982, 237; Baker and Gregory, 1984, 193, 186). The Sauerian cultural landscape was 'sterile', 'static', 'reactionary', 'conservative', characterised by an 'object fetishism', and evoking a superorganic autonomous culture that did not dig down into deeper meanings of landscape, or try to understand its symbolic qualities (Price and Lewis, 1993).[7] Both represented dead ends. Concepts like landscape, culture, nature—and even perception—seemed like black boxes or given structures that were 'inadequately theorized' (Mitchell, 1998, 16).

Thus, from now on all renderings of landscape were perceived as cultural, or iconic, or symbolic representations, 'be they earth, brick, verse, paint, ink or prose' (Seymour, 2000, 194), which could only be interpreted through the visual senses, which are conditioned by aesthetics, cultural conventions and the intellectual and power bias that resides in a particular class, gender, race, or economy (Barnes and Duncan, 1992). Now the work of artists (Prince, 1988), landscape designers (Daniels and Seymour, 1990), musicians, novelists (Daniels and Ryecroft, 1993), and poets became a major part of the repertoire of geographers. Even maps, the most traditional tool of the geographer, came under close scrutiny as being historically constructed cultural representations open to class, gender and ethnic biases (Harley, 1988; Alfrey and Daniels, 1990).

[7] This article is both a review of these criticisms and a spirited rebuttal.

An early geographical expression of this approach was in Cosgrove's *Social Formation and Symbolic Landscape* (1984). This adopted the approach of Berger's 'ways of seeing' and examined landscapes as varied as eighteenth-century English country estates, the Jeffersonian gridded American Mid West, L'Enfant's Washington, DC, Renaissance Italy, and the landscape of industrial capitalism as representations and reproductions of the particular class values of those who exerted power relations. The landscape was a 'visual ideology', a view taken up in Baker and Biger (1992). Elsewhere Cosgrove (1985) developed the theme that the bourgeois, western way of seeing was a product of Renaissance theories of space, the division of time and space, surveying (connected to arms and conquest), linear perspective, and territorial appropriation by a new urban class building up estates and reclaiming 'waste'. Some of these themes were developed in Cosgrove and Daniels's collection of essays on *The Iconography of Landscape* (1988), and in Cosgrove's (1993) extended study of their early manifestations in the fifteenth and sixteenth Venetian Republics, especially the mainland, Veneto side of the Lagoon with its extensive land-draining schemes, creation of 'new ' land, and the erection of the magnificent villas designed by Palladio.

The focus on class relations in the landscape and their material and literary manifestations by, for example, Barrell, Cosgrove, and even Raymond Williams, has in turn been criticised, both explicitly and implicitly, as being too narrow a view that excludes considerations of gender and sexuality. They are portrayed as western, patriarchal, and white ways of seeing (Nash, 1994, 1996; Hirsch and O'Hanlon, 1995), Rose (1993) going so far as to say that the Marxist emphasis on material artefacts and landscapes merely reinforces gender identity, particularly the literary *topos* of seeing nature/landscape as feminine, that should be 'improved', subjugated, and 'raped', a turn of imagery that few cultural geographers can resist. Peter Jackson (1989) sees landscapes not as an artefact but as manifestations of sexuality, Edward Said (1978) as the '"Othering" of the Orient', Mitchell (1994) as an 'imperial' way of seeing, and Matless (1995) as re-enforcing national identity. Moreover, recent work by anthropologists has criticised western ways of seeing as being privileged and specific views only. For example, the element of memory is eliminated from Australian aboriginal landscapes (Morphy, 1993), while non-western views of nature need greater recognition (Ellen and Fukui, 1996).

Taking the point further, Stephen Daniels has argued that landscape as represented through art and iconography was rarely what it seemed—it was duplicitous—(Daniels, 1989) and could be viewed in multiple ways (Daniels, 1993b; Daniels and Seymour, 1990), in each of which a complicated discourse of economic, political, social and cultural issues is encoded. Specific landscape features can also be interrogated. For example, trees and woodland can be symbolic of social and political structure, and provide

Fig. 5.10 'Mr and Mrs Robert Andrews', by Thomas Gainsborough (1748–50). Reproduced by permission of the National Gallery, London.

insights into a sense of national identity and/or authority (Daniels, 1988). In a sustained and dazzling treatment, *Landscape and Memory* (1995), the historian Simon Sharma explored the role of forests, mountains and rivers as repositories and residues of national and folk memory.

Nowhere have the multiple interpretations of a scene and its occupants come under such close scrutiny, and thereby exemplified the 'hidden meanings', as in Gainsborough's 1748 portrait of Mr and Mrs Robert Andrews at Bulmer, near Sudbury on the River Stour, Suffolk, probably painted on the occasion of their recent marriage (Fig. 5.10). In some ways it is a unique painting, not composed of a single individual or a family group, unlike so many of Gainsborough's portrait pieces, and consequently it has attracted much comment. At the risk of being at least the eighth academic to comment on them,[8] I will try to summarise these multiple views, and incidentally, in the hope that the couple, smug and self-satisfied as they undoubtedly appear, but who cannot answer back, may be left in peace. Berger (1973, 107–8) started the radical ball rolling by noting the 'proprietary attitude' in 'their stance and in their expression' as a result of their 'pleasure in seeing themselves depicted as landowners'. Daniels (1993a, 81–2), though not especially referring to them, uses their image to illustrate a more general theme of 'improvement', estate management, and the landscape garden in an essay on 'Humphrey Repton and the improvement of the estate'. In an altogether more analytical essay by Prince (1988, 102–5) on 'Art and agrarian change, 1710–1815' the actual location and background landscape is dissected more minutely. The background perspective has been rearranged by Gainsborough to enhance the view of their estate—'the Auberies'—and the status of its owners. But 'no harvesters, nor even a solitary shepherd intrude upon . . . [their] peace and privacy'.

> [The] means of production are taken for granted, the crop of wheat, the venerable oak, fat sheep, wild partridge and everything else are appropriated by the landowner who stands complacently in front of his accumulated assets . . . Mr and Mrs Andrews are too soft-skinned and unsuitably dressed to have toiled in the fields or even to have supervised others in doing so and none of the people employed in producing their good fortune appears in the picture.

Perceptively he notes that the view incorporates many features of the new, up-to-date farming—the wheat has been drilled, hawthorn hedges in the far distance neatly cut and laid, a new-style five-barred gate is displayed, and the sheep seem plump and well fed, probably on turnips and clovers. We also learn that Robert Andrews was an 'improving' farmer,

[8] It has been drawn to my attention that the portrait of *Mr and Mrs Andrews* has been used yet again recently in Holloway and Hubbard's *People and Place: The Extraordinary Geographies of Everyday Life* (2001).

and contributed an article on smut in wheat to the *Annals of Agriculture* 40 years later in 1786.

For Rose (1993, 91–3), the scene is more than a symbol of capitalist property relations; it is replete with gendered symbolism. She sets out to separate the couple, to 'prise them apart'. Robert Andrews is male and therefore mobile; Frances Andrews is female and therefore fixed firmly in her place. He stands 'gun on arm, ready to leave his pose and go shooting again; his hunting dog is at his feet, already urging him away'. She 'sits impassively, rooted to her seat with its wrought iron branches and tendrils, her upright stance echoing that of the old tree directly behind her. If Mr Andrews seems at any moment able to stride off into the vista, Mrs Andrews looks planted to the spot.' In addition, it is more than likely that Robert was the sole owner of the land while she was propertyless, and no more than a passive child-bearing machine: 'the shadow of the oak tree over her refers to the family tree she was expected to reproduce, and this role is itself naturalised by the references to trees and fields.'

Seymour (2000, 202–3) traces these nuances and even more, noting the pre-Berger and more innocent interpretations of the art historians and critics: Kenneth Clark, who saw them as 'a couple in nature', and Lawrence Gowing's idea that 'that they were involved in "the philosophic enjoyment of the natural world"'. What, one wonders, should one make of the fact that Frances Andrews had a small dog on her lap, which was subsequently painted out by Gainsborough?

The emphasis on the new cultural geography towards the humanised landscape should not blind us to the significant departures that had already been made in mainstream geography at the same time or even before the 'new' appeared. The break was not so abrupt or revolutionary as perhaps the work of Raymond Williams, Barrell, and Berger seemed to presage—they had already been taken on board, though not with a Marxist spin. In 1925 Sauer had urged geographers to 'go beyond science' in interpreting landscapes. *The Interpretation of Ordinary Landscapes* (Meinig, 1979), based on lectures given during 1975, had many innovations. Chapter headings give a hint of these: 'Axioms for reading the landscape' (Pierce Lewis), 'The beholding eye' (Meinig), 'The biography of landscape' (Samuels), 'Thought and landscape' (Yi-Fu Tuan), 'Age and artefact' (Lowenthal), 'The landscape of home' (Sopher), 'The order of a landscape' (J. B. Jackson), 'Symbolic landscapes' (Meinig), and 'Reading landscapes' (Meinig). Even earlier in Britain, there appeared the pioneering work of J. Appleton on *The Experience of Landscape* (1975), which explored the aesthetics and symbolism of scenes, both painterly and actual. It was simply before its time, and consequently overlooked. In addition, the British regional novel was a rich and obvious vein to mine. In 1948 Darby used Hardy's novels to reconstruct the regional geography of 'Wessex'. Later there came a string of pieces by Pocock on the North (1979), the regional novel generally (1981), and valued landscapes (1982), Birch (1981) on Wessex again, and Hudson (1982) on Arnold Bennett country. Darby grappled with the problem of reconciling the sequential nature of words with instantaneous apprehension of the visual scene (Darby, 1962b). Lowenthal explored perception, heritage, and attitudes to the past (e.g. 1966, 1997), and Wreford Watson used poetry to understand 'The soul of geography' (Watson, 1983; Johnston, 1993). Within history, there was Klingender's path-breaking but curiously neglected volume on *Art and the Industrial Revolution* (1947), and Hoskins's (1955, 19) plea to look at landscapes as a humane, historical art, which he likened to 'a symphony in which it is possible to enjoy as an architectural mass of sound, . . . without being able to analyse it in detail . . . [or] isolate the themes as they enter . . . [and] perceive the manifold subtle variations on a single theme'.

Conclusions

It seems inescapable that landscapes are inseparable from human beings, who are their creators, workers, representers, and interpreters through time. When all is said and done, landscapes are an almost totally human construct. They are a product of an amalgam of tangible objects, sight, and intellect, and they are open to many interpretations. Consequently, their meaning, use, and representation are highly controversial and contested. Looking back, nowhere have geographers been more 'prisoners of their own time and of their own cultural and intellectual world' (Darby, 1983a, 427) than in the study of the humanised landscape. The older school of studies battled to create disciplinary entities and withstand competition in the often false partitions of higher education. In that they were too 'earth-bound'. The new school of landscape geographers have rightly embraced the fluidity that the liberalisation of the intellectual structures has made available to them. But they may not be earth-bound enough as their studies spin off into cultural abstractions that give little sense of places as lived in by people.

It has been suggested that the 'old/new' duality smacks too much of a 'paradigm-trashing strategy that reinforces and reifies one component at the expense of the other' (Rowntree, 1988, 375), and the rather dismal dismissal of the good of the past with the bad in an effort to be revolutionary and radical, leading to a negativeness. The danger is already apparent that the concentration on the visual, symbolic content, important as it undoubtedly is in contributing to a full understanding of landscape, leads to a neglect of the tangible landscape that one encounters in such prosaic and mundane things as ditches, embankments, hedges, roads, buildings, and the whole panoply of vestiges of human production. Landscapes may be neither beautiful, ideal, nor socially just, but they are vital and tangible. The concept of landscape is far too important and central a topic for geographers to leave it fragmented and obscured in the jargon of the specialists in 'optics, psychology, epistemology and culture' (Meinig, 1979), or even art history and, one could add, race, gender, and ethnicity. There is a need for a *rapprochement* between the extremes so that the study of landscapes is seen as a humane art, at the very interface between the human and physical environment, and that landscape has both tangible and intangible nuances and realities. Landscapes *are* complex entities that demand the utilisation of every tool in the geographer's repertoire—from field work, mapping, GIS map abstracting, art, poetry, literature, history, and social critique—to make them understandable. The signs are there: the work of Heffernan (1995), for example, on the symbolic resonance of the macabre mass of imperial war graves on the Western

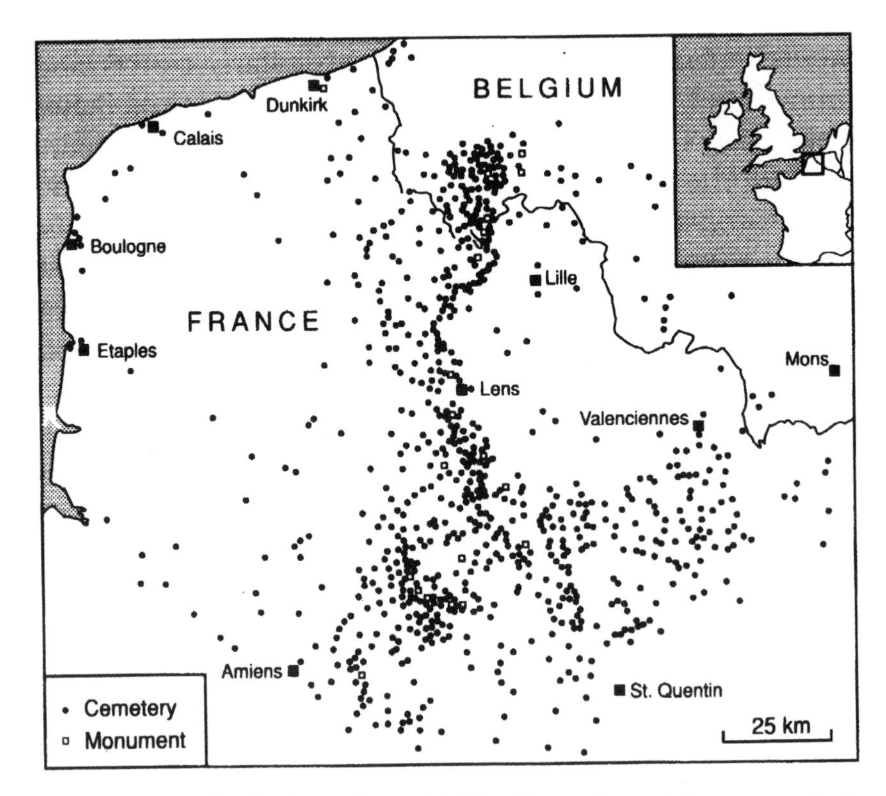

Fig. 5.11 The distribution of Imperial War Graves Commission cemeteries in France and Belgium.
Source: Heffernan (1995, 306, fig. 4). Reproduced by permission of *Cultural Geography*.

Front and the cult of remembrance in Britain show how view and distribution (Fig. 5.11) can enhance each other, and Daniels's extended study on *Humphrey Repton: Landscape Gardening and the Geography of Georgian England* (1998) is a satisfying blend of the 'old' with the 'new'. Another excellent and accessible account comes in a recent textbook by Atkins, Simmons, and Roberts (1998), based on a course they give at Durham University entitled 'Land, People and Time'. The work is an exciting amalgam of all types of evidence, related to actual places through time, and exemplifies the undoubted fascination that the complexity of landscape holds for all. When all is said and done, the art of good writing is to make that complexity intelligible and convey difficult ideas with apparent ease, not 'conveying simple ideas with conspicuous difficulty' (Billinge, 1983, 413).

Whatever the outcome, landscapes will remain central to the British geographical enterprise if for no other reason than that issues of environmental protection, heritage designation and preservation, regional planning and suburban 'spill', the ownership of the countryside and public access to it, and national identity (e.g. Daniels, 1993a, 80–111; Brace, 1999) will loom larger in the future, not less. As George Henderson (1998) commented in a recent review article 'Landscape is dead, long live landscape'. In addition, the past is always with us; as Michael Conzen has so aptly noted, 'like a weed, perhaps indomitable, for even if we as geographers may think we are finished with the past, the past is not finished with us' (Conzen, 1993, 89). A hundred years from now, on the occasion of the next Academy centenary, geographers (and others) will be looking back to the last century and still be discussing the structure, meaning, significance, and representation of what Meinig (1979b, 237) has called this 'grand earthly composition'—the ever-changing backcloth to the whole stage of human activity and the spirit of place.

References

Alfrey, N. and Daniels, S., eds (1990) *Mapping the Landscape: Essays in Art and Cartography*. Nottingham: University of Nottingham.

Appleton, J. H. (1962) *The Geography of Communications in Britain*. Oxford: Oxford University Press.

Appleton, J. H. (1975) *The Experience of Landscape*. London and New York: Wiley.

Andrews, J. H. (1956) Some statistical maps of Defoe's England. *Geographical Studies*, 3, 33–45.

Anon (1932) What is historical geography? *Geography*, 17, 39–45.

Arnold, T. (1838) *The History of Rome*. London: B. Fellowes, J. G. and Rivington, et al. (3 vols).

Atkins, P., Simmons, I., and Roberts, B. (1998) *People, Land and Time: An Historical Introduction to the Relations between Landscape, Culture and Environment*. London: Arnold.

Baker, A. H. R. (1982) On ideology and historical geography. In A. H. R. Baker, ed., *Period and Place: Research Methods in Historical Geography*. Cambridge: Cambridge University Press, 233–43.

Baker, A. H. R. and Biger, G., eds (1992) *Ideology and Landscape in Historical Perspective*. Cambridge: Cambridge University Press.

Baker, A. R. H. and Butlin, R. A., eds (1973) *Studies of Field Systems in the British Isles*. Cambridge: Cambridge University Press.

Baker, A. H. R. and Gregory, D. (1984) Some Terra Incognita in historical geography: an exploratory discussion. In D. Gregory and A. H. R. Baker, eds, *Explorations in Historical Geography*. Cambridge: Cambridge University Press, 180–94.

Baker, A. R. H. and Harley, J. B., eds (1973) *Man Made the Land: Essays in English Historical Geography*. Newton Abbot: David and Charles.

Baker, A. R. H., Hamshere, J. D., and Langton, J., eds (1970) *Geographical Interpretation of Historical Sources: Readings in Historical Geography*. Newton Abbot: David and Charles.

Baker, J. N. L. (1936) The last hundred years of historical geography. *History*, NS21, 193–207.

Baker, J. N. L. (1952) The development of historical geography in Britain during the last hundred years. *The Advancement of Science*, 8, 406–12.

Balchin, W. G. V. (1954) *Cornwall: An Illustrated Essay in the History of the Landscape*. London: Hodder and Stoughton.

Barley, M. W. (1961) *The English Farmhouse and Cottage*. London: Routledge and Kegan Paul.

Barnes, T. J. and Duncan, J. S. (1992) *Writing Worlds: Discourse, Text and Metaphor in the Representation of Landscape*. London: Routledge.

Barrell, J. (1972) *The Idea of Landscape and the Sense of Place: An Approach to the Poetry of John Clare*. Cambridge: Cambridge University Press.

Barrell, J. (1980) *The Dark Side of the Landscape: The Rural Poor in English Landscape Painting, 1730–1840*. Cambridge: Cambridge University Press.

Beresford, M. W. (1954) *The Lost Villages of England*. London: Lutterworth Press.

Beresford, M. W. (1984) Time and place: an inaugural lecture. In M. W. Beresford, ed., *Time and Place: Collected Essays*. London: Hambledon Press, 1–23.

Beresford, M. (1988) *East End, West End: The Face of Leeds during Urbanization, 1684–1842*. Leeds: Thoresby Society.

Beresford, M. W. and Hurst, J. G., eds (1971) *Deserted Medieval Villages*. London: Lutterworth.

Beresford, M. W. and St Joseph, J. K. S. (1958) *Medieval England: An Aerial Survey*. Cambridge: Cambridge University Press.

Berger, J. (1973) *Ways of Seeing*. Harmondsworth: BBC and Penguin Books.

Billinge, M. (1983) The Mandarin dialect: an essay in style in contemporary geographical writing. *Transactions, Institute of British Geographers*, NS8, 400–21.

Birch, B. P. (1981) Wessex, Hardy, and the nature of novelists. *Transactions, Institute of British Geographers*, NS6, 348–58.

Bird, J. (1981) The target of space and the arrow of time. *Transactions, Institute of British Geographers*, NS6, 129–51.

Blome, R. (1670) *A Geographical Description of the Four Parts of the World*. London: Printed by T. N. for R. Blome.
Bowen, E. G. (1932) Early Christianity in the British Isles: a study in historical geography. *Geography*, 17, 267–76.
Bowen, E. G. (1954) *The Settlement of the Celtic Saints in Wales*. Cardiff: University of Wales Press.
Bowen, E. G. (1969) *Saints, Seaways and Settlements in Celtic Lands*. Cardiff: University of Wales Press.
Brace, C. (1999) Looking back: the Cotswolds and English national identity, c 1890–1950. *Journal of Historical Geography*, 25, 502–16.
Brigham, A. P. (1903) *Geographical Influences in American History*. Boston, MA: Ginn and Co.
Broek, J. O. M. (1932) *The Santa Clara Valley, California: A Study in Landscape Change*. Utrecht: A. Oosthek's Uitgevers-Mij.
Buchanan, R. A. (1972) *Industrial Archaeology in Britain*. Harmondsworth: Penguin.
Cherry, G. (1972) *Urban Change and Planning: A History of Urban Development in Britain since 1750*. Henley-on-Thames: Foules.
Clark, A. H. (1954) Historical geography. In P. E. James and C. F. Jones, eds, *American Geography: Inventory and Prospect*. Syracuse, NY: Syracuse University Press for Association of American Geographers, 70–105.
Clark, A. H. (1959) *Three Centuries and the Island: A Historical Geography of Settlement and Agriculture in Prince Edward Island, Canada*. Toronto: Toronto University Press.
Clark, A. H. (1962) The sheep/swine ratio as a guide to a century's change in the livestock geography of Nova Scotia. *Economic Geography*, 38, 38–55.
Clark, A. H. (1972) Historical geography in North America. In A. R. H. Baker, ed., *Progress in Historical Geography*. Newton Abbot: David and Charles, 129–43
Clifton-Taylor, A. (1962) *The Pattern of English Building*. London: Batsford.
Coones, P. and Patten, J. (1986) *The Penguin Guide to the Landscape of England and Wales* Harmondsworth: Penguin.
Conzen, M. R. G. (1960) Alnwick, Northumberland: a study in town-plan analysis. *Transactions, Institute of British Geographers*, 27, 122 pp.
Conzen, M. P. (1993) The historical impulse in geographical writing about the United States, 1850–1990. In M. P. Conzen, T. A. Rumney, and G. Wynn, eds, *A Scholar's Guide to the Geographical Writing on the American and Canadian Past*. Chicago, IL: University of Chicago Press, University of Chicago Geography Research Paper, No. 235.
Cosgrove, D. (1984) *Social Formation and Symbolic Landscape*. Beckenham: Croom Helm.
Cosgrove, D. (1985) Prospect, perspective, and the evolution of the landscape idea. *Transactions, Institute of British Geographers*, NS10, 45–62.
Cosgrove, D. (1993) *The Palladian Landscape: Geographical Change and its Cultural Representations in Sixteenth Century Italy*. Leicester: Leicester University Press.
Cosgrove, D. and Daniels, S., eds (1988) *The Iconography of Landscape: Essays in Symbolic Design and the Use of Past Environments*. Cambridge: Cambridge University Press.

Crawford, O. G. S. (1932) Map of Neolithic Wessex. *Antiquity*, 6, 492–4.

Daniels, S. (1988) The political iconography of woodland. In D. Cosgrove and S. Daniels, eds, *The Iconography of Landscape: Essays on the Symbolic Design and Use of Past Environments*. Cambridge: Cambridge University Press, 43–82.

Daniels, S. (1989) Marxism, culture, and the duplicity of landscape. In R. Peet and N. Thrift, eds, *New Models in Geography: The Political-Economy Perspective*. London: Unwin Hyman, 196–220.

Daniels, S. (1993a) *Fields of Vision: Landscape Imagery and National Identity in England and the United States*. Cambridge: Polity Press.

Daniels, S. (1993b) Humphrey Repton and the improvement of the estate. In S. Daniels, *Fields of Vision: Landscape Imagery and National Identity in England and the United States*. Cambridge: Polity Press, 80–111.

Daniels, S. (1998) *Humphrey Repton: Landscape Gardening and the Geography of Georgian England*. Yale, CT: Yale University Press.

Daniels, S. and Ryecroft, S. (1993) Mapping the modern city: Allan Sillitoe's Nottingham novels. *Transactions, Institute of British Geographers*, NS18, 460–80.

Daniels, S. and Seymour, S. (1990) Landscape design and the idea of improvement, 1730–1900. In R. A. Dodgshon and R. A. Butlin, eds, *An Historical Geography of England and Wales*, 2nd edn. London: Academic Press, 487–520.

Darby, H. C., ed. (1936) *An Historical Geography of England before AD 1800*. Cambridge: Cambridge University Press.

Darby, H. C. (1940a) *The Medieval Fenland*. Cambridge: Cambridge University Press.

Darby, H. C. (1940b) *The Draining of the Fens*. Cambridge: Cambridge University Press.

Darby, H. C. (1948) The regional geography of Thomas Hardy's Wessex. *Geographical Review*, 38, 426–43.

Darby, H. C. (1951a) The changing English landscape. *Geographical Journal*, 117, 377–98.

Darby, H. C. (1951b) The clearing of the English woodlands. *Geography*, 36, 71–83.

Darby, H. C. (1952) *The Domesday Geography of Eastern England*. Cambridge: Cambridge University Press.

Darby, H. C. (1953) On the relations of geography and history. *Transactions, Institute of British Geographers*, 19, 1–11.

Darby, H. C. (1954) Some early ideas on the agricultural regions of England. *Agricultural History Review*, 2, 30–47.

Darby, H. C. (1957) Place-names and geography. *Geographical Journal*, 123, 387–92.

Darby, H. C. (1962a) Historical geography. In H. P. R. Finberg, ed., *Approaches to History: A Symposium*. London: Routledge and Kegan Paul, 127–56.

Darby, H. C. (1962b) The problem of geographical description. *Transactions, Institute of British Geographers*, 30, 1–14.

Darby, H. C. (1964) The draining of the English claylands. *Geographische Zeitschrift*, 52, 190–201.

Darby, H. C., ed. (1973) *A New Historical Geography of England*. Cambridge: Cambridge University Press.

Darby, H. C. (1977) *Domesday England*. Cambridge: Cambridge University Press.

Darby, H. C. (1983a) Historical geography in Britain, 1920–1980: continuity and change. *Transactions, Institute of British Geographers*, NS8, 421–8.
Darby, H. C. (1983b) *The Changing Fenland.* Cambridge: Cambridge University Press.
Darby, H. C. (2002) In M. Williams et al., eds, *The Relations of History and Geography: Studies in England, France and the United States*. Exeter: Exeter University Press.
Dennis, R. J., ed. (1979) *The Victorian City.* Special issue of *Transactions, Institute of British Geographers*, NS4, 125–319.
Denevan, W. M. (1992). The pristine myth: the landscape of the Americas in 1492. *Annals of the Association of American Geographers*, 82, 369–85.
Dobson, M. J. (1997) *Contours of Death and Disease in Early Modern England.* Cambridge: Cambridge University Press.
Dodgshon, R. A. (1973) The nature and development of infield–outfield in Scotland. *Transactions, Institute of British Geographers*, 59, 1–23.
Dodgshon, R. A. (1975) Infield–outfield and the territorial expansion of the English township. *Journal of Historical Geography*, 1, 327–45.
Dodgshon, R. A. and Butlin, R. A., eds (1978) *An Historical Geography of England and Wales*. London: Academic Press.
Dolittle, W. E. (1992) Agriculture in North America on the eve of conquest. *Annals of the Association of American Geographers*, 82, 386–401.
Donkin, R. A. (1978) *The Cistercians: Studies in the Medieval Geography of England and Wales.* Toronto: Pontifical Institute of Medieval Studies.
Donkin, R. A. (1979) *Agricultural Terracing in the Aboriginal New World.* Tucson, AZ: University of Arizona Press.
Duncan, J. (1995) Landscape geography, 1993–94. *Progress in Human Geography*, 19, 414–22.
Duncan, J. and Duncan, N. (1988) (Re)reading the landscape. *Environment and Planning D: Society and Space*, 6, 117–26.
Dyos, H. J., ed. (1968) *The Study of Urban History*. London: Edward Arnold.
Dyos, H. J. (1973) *Urbanity and Suburbanity*. Leicester: Leicester University Press.
Dyos, H. J. and Aldcroft, D. H. (1969). *British Transport: An Economic Survey from the Seventeenth Century to the Twentieth*. Leicester: Leicester University Press.
Dyos, H. J. and Wollf, M., eds (1973) *The Victorian City: Images and Realities*. London: Routledge and Kegan Paul.
Ellen, R. and Fukui, K. (1996) *Redefining Nature: Ecology, Culture and Domestication*. Oxford and Washington, DC: Berg.
Emery, F. V. (1962) Moated settlements in England. *Geography*, 47, 378–88.
Evans, E. E. (1996) *Ireland and the Atlantic Fringe: Selected Writings*, edited by J. Campbell. Dublin: The Lilliput Press.
Finberg, H. P. R. (1955) *Gloucestershire: An Illustrated Essay in the History of the Landscape*. London: Hodder and Stoughton.
Finberg, H. P. R., ed. (1962) *Approaches to History: A Symposium*. London: Routledge and Kegan Paul.
Forde, C. D. (1934) *Habitat, Economy and Society: A Geographical Introduction to Ethnology*. London: Methuen.
Fox, C. (1923) *The Archaeology of the Cambridge Region*. Cambridge: Cambridge University Press.

Freeman, E. A. (1881) *The Historical Geography of Europe*. London: Longmans, Green and Co. (2 vols).

Freeman, M. J. (1977) The carrier system in south Hampshire, 1775–1851. *Journal of Transport History*, 2nd series 4, 61–75.

Freeman, M. J. (1980) Transporting methods in the British cotton industry during the industrial revolution. *Journal of Transport History*, 3rd series 1, 59–74.

Gilbert, E. W. (1932) What is historical geography? *Scottish Geographical Magazine*, 48, 129–35.

Glasscock, R. E. (1973) England *circa* 1334. In H. C. Darby, ed., *A New Historical Geography of England*. Cambridge: Cambridge University Press, 136–85.

Glasscock, R. E. (1974) *The Lay Subsidy of 1334*. London: British Academy.

Glasscock, R. E., ed. (1992) *Historic Landscapes of Britain from the Air*. Cambridge: Cambridge University Press.

Goudie, A. S. (1981) *The Human Impact: Man's Role in Environmental Change*. Oxford: Blackwell.

Green, J. R. (1881) *The Making of England*. London: Macmillan and Co.

Gregory, D. (1982) *Regional Transformation and Industrial Revolution: A Geography of the Yorkshire Woollen Industry*. Basingstoke: Macmillan.

Gregory, S. (1976) Geographical myths and statistical fables. *Transactions, Institute of British Geographers*, NS1, 385–400.

Hadfield, C. (1968) *The Canal Age*. Newton Abbot: David and Charles.

Hallam, H. E. (1965) *Settlement and Society: A Study of the Early Agrarian History of South Lincolnshire*. Cambridge: Cambridge University Press.

Harley, B. (1988) Maps, knowledge and power. In D. Cosgrove and S. Daniels, eds, *The Iconography of Landscape: Essays in Symbolic Design and the Use of Past Environments*. Cambridge: Cambridge University Press, 277–312.

Harris, R. C. (1978). The historical mind and the practice of geography. In D. Ley and M. S. Samuels, eds, *Humanistic Geography: Prospects and Problems*. London: Edward Arnold, 123–37.

Harrison, M. J., Mead, W. R., and Pannett, D. J. (1965) A Midland ridge-and-furrow map. *Geographical Journal*, 131, 366–9.

Hartshorne, R. (1939) *The Nature of Geography: A Critical Survey of Current Thought in the Light of the Past*. Lancaster, PA: American Association of Geographers.

Hartshorne, R. (1959) *Perspective on the Nature of Geography*. Chicago, IL: Rand McNally for Association of American Geographers.

Heathcote, R. L. (1965) *Back of Bourke: A Study of Land Appraisal and Settlement in Semi-Arid Australia*. Melbourne: Melbourne University Press.

Heffernan, M. (1995) For ever England: the Western Front and the politics of remembrance in Britain. *Ecumene*, 2, 293–324.

Henderson, G. (1998) Landscape is dead, long live landscape: a handbook for sceptics. *Journal of Historical Geography*, 24, 94–100.

Henderson, H. C. K. (1952) Agriculture in England and Wales in 1801. *Geographical Journal*, 118, 338–45.

Hirsch, E. and O'Hanlon, M., eds (1995), *The Anthropology of Landscape: Perspectives on Place and Space*. Oxford: Clarendon Press.

Hodgson, A. (1824) *Letters from America Written During a Tour in the United States and Canada*. London: Hurst, Robinson (2 vols).

Hoskins, W. G. (1955) *The Making of the English Landscape*. London: Hodder and Stoughton.

Hoskins, W. G. (1957) *Leicestershire: An Illustrated Essay on the History of the Landscape*. London: Hodder and Stoughton.

Hoskins, W. G. (1978) *One Man's England*. London: BBC Publications.

Holloway, L. and Hubbard, P. (2001) *People and Place: The Extraordinary Geographies of Everyday Life*. Harlow: Prentice Hall.

Hudson, B. (1982) The geographical imagination of Arnold Bennett. *Transactions, Institute of British Geographers*, NS7, 365–79.

Hudson, K. (1963) *Industrial archaeology*. London: John Baker (5 vols). Reissued as *Industrial Archaeology: A New Introduction*. London: John Baker, 1976.

Jackson, P. (1989) *Maps of Meaning: An Introduction to Cultural Geography*. London: Unwin Hyman.

Johnston, R. J. (1968) An outline of the development of Melbourne's street pattern. *Australian Geographer*, 6, 433–65.

Johnston, R. J. (1993) The geographers' degree of freedom: Wreford Watson, postwar progress in human geography, and the future of scholarly work in UK geography. *Progress in Human Geography*, 17, 319–32.

Jones, L. R. (1926) Geography and the university. *Scottish Geographical Magazine*, 42, 65–79.

Kain, R. J. P. and Prince, H. C. (1985) *The Tithe Surveys of England and Wales*. Cambridge: Cambridge University Press.

Kinvig, R. H. (1955) *Newsletter*, Department of Geography Old Students' Association, University of Birmingham, 1(2), 5.

Kirk, W. (1952) Historical geography and the concept of the behavioural environment. In Indian Geographical Society, *Silver Jubilee Souvenir and N. Subrahmanyam Memorial Volume*. Madras: Indian Geographical Society, 152–60.

Klingender, F. D. (1947) *Art and the Industrial Revolution*. St Albans: Paladin (re-issued 1975).

Koelsch, W. A. (1995) John Linton Myres, 1869–1954. *Geographers: Biobibliographical Studies*, 16, 53–62.

Lambert, J. M., Jennings, J. N., Smith, C. T., Green, C., and Hutchinson, J. N. (1960) *The Making of the Broads*. London: Royal Geographical Society, Research Series No. 3.

Langton, J. (1988) Two traditions of geography, historical geography and the study of landscapes. *Geografiska Annaler*, 70B, 17–26.

Lavisse, E. (1903) *Histoire de France depuis les origines jusqu'à la Révolution*. Paris: no publisher.

Lawton, R. (1955) The population of Liverpool in the mid-nineteenth century. *Transactions, Historical Society of Lancashire and Cheshire*, 107, 89–120.

Lawton, R. (1968) Population changes in England and Wales in the later nineteenth century: an analysis of trends by registration district. *Transactions, Institute of British Geographers*, 44, 55–74.

Lawton, R. (1983) Space, place and time. *Geography*, 68, 193–207.

Lawton, R. and Pooley, C., eds (1992) *Britain, 1740–1950: An Historical Geography*. London: Edward Arnold.

Lowenthal, D. (1996) *The Past is a Foreign Country*. Cambridge: Cambridge University Press.

Lowenthal, D. (1997) *The Heritage Crusade, and the Spoils of History*. Cambridge: Cambridge University Press.
Lowenthal, D. and Bowden, M. J., eds (1976) *Geographies of the Mind*. New York: Oxford University Press.
Lowenthal, D. and Prince, H. C. (1965) English landscape tastes. *Geographical Review*, 55, 186–222.
Macaulay, T. B. (1848) *The History of England from the Accession of James II*. London: Longman, Brown, Green and Longmans (2 vols).
Mackinder, H. J. (1931) Discussion. In *Geographical Journal*, 78, 268.
Marsh, G. P. (1864) *Man and Nature: Or, Physical Geography as Modified by Human Action*. New York: Scribner.
Mason, K. (1958) Geography in Oxford. *The Oxford Magazine*, 56, 411–12.
Matless, D. (1995). The art of right living: landscape and citizenship, 1918–1939. In S. Pile and N. Thrift, eds, *Mapping the Subject: Geographies of Cultural Transformation*. London: Routledge, 93–123.
Mead, W. R. (1954) Ridge and furrow in Buckinghamshire. *Geographical Journal*, 120, 34–42.
Meinig, D. W., ed. (1979) *The Interpretation of Ordinary Landscapes: Geographical Essays*. London and New York: Oxford University Press.
Meinig, D. W. (1979b) Reading the landscape. In D. W. Meinig, ed., *The Interpretation of Ordinary Landscapes: Geographical Essays*. London and New York: Oxford University Press, 195–244.
Michelet, J. (1864) *Histoire de France*. Paris: Marpon and Flammarion.
Millward, R. (1955) *Lancashire: An Illustrated Essay on the History of the Landscape*. London: Hodder and Stoughton.
Mitchell, D. (1998) Writing western history: new western history's encounter with landscape. *Ecumene*, 5, 1–29.
Mitchell, W. J. T. (1994) Imperial landscape. In W. J. T. Mitchell, ed., *Landscape and Power*. Chicago, IL and London: University of Chicago Press, 5–34.
Morphy, H. (1992) Colonialism, history and the construction of place: the politics of landscape in Northern Australia. In B. Bender, ed., *Landscape, Politics and Perspectives*. Providence, RI and Oxford: Berg, 205–45.
Myres, J. L. (1928) Ancient geography in modern education. Presidential address to Geographical Section of British Association for the Advancement of Science. Reprinted in J. L. Myres (1958) *Geographical History in Greek Lands*. Oxford: Clarendon Press, 1–33.
Nash, C. (1994) Remapping the body/land: new cartographies of identity, gender and landscape in Ireland. In A. Blunt and B. Rose, eds, *Writing Women and Space: Colonial and Postcolonial Geographies*. New York: Guilford, 227–50.
Nash, C. (1996) Reclaiming vision: looking at the landscape and the body. *Gender, Place and Culture*, 3, 149–69.
Ordnance Survey (1931) *Map of Roman Britain*. London: HMSO.
Overton, M. (1996) *Agricultural Revolution in England: The Transformation of the Agrarian Economy*. Cambridge: Cambridge University Press.
Parry, M. L. (1975) Secular climatic change and marginal land. *Transactions, Institute of British Geographers*, 64, 1–13.

Pawson, E. (1977) *Transport and Economy: The Turnpike Roads of Eighteenth Century Britain*. London: Academic Press.
Peake, J. E. and Fleure, H. J. (1927–56) *The Corridors of Time*. Oxford and London: Clarendon Press (10 vols).
Pevsner, N. (1951–) *Buildings of England*. Harmondsworth: Penguin.
Phillips, A. D. M. (1989) *The Underdraining of Farmland during the Nineteenth Century*. Cambridge: Cambridge University Press.
Pocock, D. C. D. (1979) The novelist's image of the North. *Transactions, Institute of British Geographers*, NS4, 62–76.
Pocock, D. C. D. (1981) Place and the novelist. *Transactions, Institute of British Geographers*, NS6, 62–76.
Pocock, D. C. D. (1982) Valued landscapes and memory: the view from Prebend's Bridge. *Transactions, Institute of British Geographers*, NS7, 354–64.
Powell, J. M. (1970) *The Public Lands of Australia Felix*. Melbourne: Oxford University Press.
Powell, J. M. (1976) *Environmental Management in Australia, 1788–1914: Guardians, Improvers, Profit*. Melbourne: Oxford University Press.
Powell, J. M. (1977) *Mirrors of the New World: Images and Image-Makers in the Settlement Process*. Folkstone: Dawsons and Canberra: ANU Press (1978).
Price, M. and Lewis, M. (1993) The reinvention of cultural geography. *Annals of the Association of American Geographers*, 83, 1–17 (see also replies by Cosgrove, Duncan, and Jackson on 515–22.).
Prince, H. C. (1964) The origin of pits and depressions in Norfolk. *Geography*, 49, 15–32.
Prince, H. C. (1967) *Parks in England*. Shalfleet, Hants: Pinhorns.
Prince, H. C. (1971) The real, imagined and abstract worlds of the past. *Progress in Geography*, 3, 1–86.
Prince, H. C. (1973) England circa 1800. In H. C. Darby, ed., *A New Historical Geography of England*. Cambridge: Cambridge University Press, 389–464.
Prince, H. C. (1988) Art and agrarian change, 1710–1815. In D. Cosgrove and S. Daniels, eds, *The Iconography of Landscape: Essays in Symbolic Design and the Use of Past Environments*. Cambridge: Cambridge University Press, 98–118.
Prince, H. C. (2000) *Geographers Engaged in Historical Geography in British Higher Education, 1931–1991*. London: Historical Geography Research Group, Historical Geography Research Series No. 36.
Prince, H. C. (2002) H. C. Darby and the historical geography of England. In H. C. Darby, *The Relations of History and Geography: Studies in England, France and the United States*, with contributions by Michael Williams, Hugh Clout, Terry Coppock, and Hugh Prince. Exeter: Exeter University Press, 63–90.
Relph, E. (1970) An inquiry into the relations between phenomenology and geography. *Canadian Geographer*, 14, 193–201.
Renfrew, C. (1981) Space, time and Man. *Transactions, Institute of British Geographers*, NS6, 257–78.
Roberts, B. K. (1977) *Rural Settlement in Britain*. Folkstone: Dawson.
Rollinson, W. (1964) Schemes for reclamation of land from the sea in North Lancashire during the eighteenth century. *Transactions, Historical Society, Lancashire and Cheshire*, 115, 107–47.

Rose, G. (1993) *Feminism and Geography: The Limits of Geographical Knowledge.* Cambridge: Polity Press.

Rowntree, L. B. (1988) Orthodoxy and new directions: cultural/humanistic geography. *Progress in Human Geography*, 12, 575–86.

Roxby, P. M. (1930) The scope and aims of human geography. *Scottish Geographical Magazine*, 46, 276–90.

Said, E. (1978) *Orientalism.* New York: Columbia University Press.

Sauer, C. O. (1925) The morphology of landscape. *University of California Publications in Geography*, 2, 19–53. Reprinted in J. Leighly, ed. (1963) *Land and Life: A Selection from the Writings of Carl Ortwin Sauer.* Berkeley and Los Angeles, CA: University of California Press, 315–350.

Sauer, C. O. (1941) Foreword to historical geography. *Annals of the Association of American Geographers*, 31, 1–24.

Semple, E. C. (1911) *Influences of Geographical Environment.* Boston, MA: Houghton Mifflin.

Seymour, S. (2000) Historical geographies of landscape. In B. Graham and C. Nash, eds, *Modern Historical Geographies.* Harlow: Longman, 193–217.

Sharma, S. (1995) *Landscape and Memory.* London: HarperCollins.

Sheail, J. (1972) The distribution of population and wealth in England during the sixteenth century. *Transactions, Institute of British Geographers*, 55, 111–26.

Sheppard, J. (1966) *The Draining of the Hull Valley.* York: East Yorkshire Local History Society, Publication No. 20.

Simmons, I. G. (1989) *Changing the Face of the Earth.* Oxford: Blackwell.

Simmons, J. (1961) *The Railways of Britain.* London: Routledge and Kegan Paul.

Slater, T. R. (1982) Urban genesis and medieval town plans in Warwickshire and Worcestershire. In T. R. Slater and P. R. Jarvis, eds, *Field and Forest.* Norwich: Geo Books.

Smith, C. T. (1965) Historical geography: current trends and prospects. In R. Chorley and P. Haggett, eds, *Frontiers in Geographical Teaching.* London: Methuen, 118–46.

Smith, D. M. (1965) *The Industrial Archaeology of the East Midlands.* Newton Abbott: David and Charles.

Smith, G. A. (1894) *The Historical Geography of the Holy Land.* London: Hodder and Stoughton.

Stanley, A. P. (1856) *Sinai and Palestine in Connection with their History.* London: John Murray.

Thomas, D. (1963) *Agriculture in Wales during the Napoleonic Wars.* Cardiff: University of Wales Press.

Thirsk, J. (1953a) *English Peasant Farming: The Agrarian History of Lincolnshire from Tudor to Recent Times.* London: Routledge and Kegan Paul.

Thirsk, J. (1953b) *Fenland Farming in the Sixteenth Century.* Leicester: University of Leicester, Department of English Local History, Occasional Papers No. 3.

Thirsk, J., ed. (1967–91) *The Agrarian History of England and Wales.* Cambridge: Cambridge University Press (9 vols).

Tuan, Yi-Fu (1971) Geography, phenomenology and the study of human nature. *Canadian Geographer*, 15, 181–92.

Tuan, Yi-Fu (1974) *Topophilia: A Study of Environmental Perception, Attitudes and Values.* Englewood Cliffs, NJ: Prentice-Hall.

Tuan, Yi-Fu (1976) Humanistic geography. *Annals of the Association of American Geographers*, 66, 266–76.

Turner, F. J. (1894) The significance of the frontier in American history. *Proceedings, State Historical Society of Wisconsin*. 41, 79–112. Reprinted in E. E. Edwards, ed. (1938) *The Early Writings of Frederick Jackson Turner*. Madison, WI: University of Wisconsin Press, 87–181.

Turner, F. J. (1908) Report on the Conference on the Relations of Geography and History. In *Annual Report, American Historical Association*. Washington, DC: Government Printing Office, 1, 45–8.

Turner, M. E. (1980) *English Parliamentary Enclosure: Its Historical Geography and Economic History*. Folkstone: Dawson.

Unstead, J. F. (1907) The meaning of geography. *Geographical Teacher*, 59, 19–28.

Vidal de la Blache, P. (1905) *Tableau de la géographie de la France*. Paris: Hachette. The first chapter was translated and published separately as *The Personality of France*, London: Christophers, 1918.

Ward, D. (1962) The pre-urban cadaster and the urban pattern of Leeds. *Annals of the Association of American Geographers*, 52, 150–66.

Watson, W. (1983) The soul of geography. *Transactions, Institute of British Geographers*, NS8, 385–99.

Webb, W. P. (1927) *The Great Plains*. New York and Boston, MA: Ginn and Co.

Whitehand, J. W. R. (1966) Fringe belts: a neglected aspect of urban geography. *Transactions, Institute of British Geographers*, 41, 223–33.

Whitehand, J. W. R. (1972) Building cycles and spatial patterns of urban growth. *Transactions, Institute of British Geographers*, 56, 39–55.

Whitehand, J. W. R. (1975) Building activity and intensity of development at the urban fringe: a case of a London suburb in the nineteenth century. *Journal of Historical Geography*, 1, 211–24.

Whitehand, J. W. R. and Patten, J. H. C., eds (1977) *Change in the Town*. Special issue of *Transactions, Institute of British Geographers*, NS2, 257–416.

Whittlesey, D. (1929) Sequent occupance. *Annals of the Association of American Geographers*, 19, 162–5.

Whittlesey, D. (1945) The horizon of geography. *Annals of the Association of American Geographers*, 55, 1–36.

Williams, M. (1970) *The Draining of the Somerset Levels*. Cambridge: Cambridge University Press.

Williams, M. (1974) *The Making of the South Australian Landscape*. London: Academic Press.

Williams, M. (1982) Marshland and waste. In L. Cantor, ed., *The English Medieval Landscape*, London: Croom Helm.

Williams, M. (1983) 'The Apple of my Eye': Carl Sauer and historical geography. *Journal of Historical Geography*, 9, 1–28.

Williams, M. (1989a) Historical geography and the concept of landscape. *Journal of Historical Geography*,15, 92–104.

Williams, M. (1989b) *Americans and Their Forests: A Historical Geography*. New York and Cambridge: Cambridge University Press.

Williams, M. (2003) *Deforesting the Earth: From Prehistory to Global Crisis*. Chicago, IL: University of Chicago Press.

Williams, R. (1973) *The Country and the City*. London: Chatto and Windus.

Wood, P. D. (1961) Strip lynchets reconsidered. *Geographical Journal*, 127, 449–59.

Wrigley, E. A. and Schofield, R. S. (1981) *The Population History of England: A Reconstruction*. London: Edward Arnold.

Yelling, J. A. (1977) *Common Fields and Enclosure in England, 1450–1850*. London: Macmillan.

6

People and the contemporary environment

SALLY EDEN

Geographical approaches to human–environment relations have been diverse and dynamic over the last century. They have also been heavily influenced not only by academic disciplines outside geography but by popular and policy concerns outside academia. From an initial flurry of activity about how the environment influences society in the early part of the century, British geography then took a detour to other topics even as other disciplines discovered the environment as a topic of interest. This left geographers playing 'catch-up' in the late twentieth century, as the discipline sought to reoccupy the ground previously abandoned. This is not over yet: in the 1990s, research into 'the environment' and 'nature' was scattered across academia, as researchers from many disciplines began to explore not merely what these terms might mean but whether they have any meaning at all in a postmodern world.

Such a diversity of approaches, both disciplinary and interdisciplinary, makes the task of summarising in one chapter all the more difficult. To attempt to track these twists and turns of geographical attention, we need to look at not only geographical research but also the wider context in which geographers were working, the influences from other disciplines and from policy circles and public interest that both stimulated and fed on environmental research. Further, as we shall see, some would argue that all geography is *implicitly* about 'the environment' because it deals with physical or spatial processes, such as glacial retreat, river meandering, transport routes, urbanisation, and so on. But in this chapter, I shall focus only on work that *explicitly* illuminated our relationship with the environment, in an increasingly reflective way, and the vast majority of this, from geography at least, came in the last 30 years.

Early perspectives on the environment: 1900s to 1960s

Environmental influence

We should begin by noting how our twenty-first-century understanding of environmental issues differs from that in history. Today, we think of environmental issues such as global warming, water pollution, recycling, species loss, and deforestation, and particularly of how human activity is implicated in making these issues worse. Although many observers in the past did ponder the negative impacts of growing societies upon the environment and its resources, this was only a minor feature of environmental thinking before the late twentieth century (Glacken, 1956). More important were environmental influence and environmental resources. The bountifulness of the environment to provide resources of timber, meat, soil, water, workable ores, and other necessities or valuables was the key to the expansion of empires in the past. In Britain, the importance of geographical accuracy was tied to the successful navigation, and thus exploitation, of the contemporary environment; thus 'geography has frequently cast itself as the aide-de-camp to militarism and imperialism' (Livingstone, 1992, 352). Cartography was in the service of environmental use and the resources mapped were rarely conceived to be other than infinite, especially in the 'New World' of North America.

At the same time, geographical thinking was trying to explain the massive differences found across the globe, between climates, landscapes, peoples, and cultures. For centuries, the environment was used to explain geographical differentiation, especially between human groups—the original meaning of 'environmentalism' was precisely such environmental determinism (Livingstone, 1992). And such views were not merely a Victorian construction. In the early twentieth century, geographers were still debating how far 'environmentalism' applied, although they were beginning to reject the crudest forms of environmental determinism in favour of a complex of environmental factors (including human agency) in an interrelationship with social change (e.g. Fleure, 1947; see also Livingstone, 1992, 353). Indeed, Clark (1950, 15) noted even in 1949 that 'the environmentalist doctrine is by no means dead' in geographical teaching and research (as evident in Taylor, 1951).

And geographers were also starting to turn the cause–effect relationship around, to question how far society was changing the contemporary environment. Many were influenced by George Perkins Marsh's *Man*[1] *and Nature*,

[1] Because of the dominance of 'man' in the literature of this period, it would be tiring for the reader if all such usages were marked with scare quotes or *[sic]*. Therefore, let me simply make one comment stand for all: today, such usage is regarded as sexist and/or old-fashioned and/or inaccurate and will rarely be found in reputable academic work in the social sciences (the same is not true for the broadcast media, nor the natural sciences).

whether they acknowledged it or not. First published in America in 1864, Marsh's book was before its time in running against the prevailing current of environmental thinking.

> Appearing at the peak of American confidence in the inexhaustibility of resources, it was the first book to controvert the myth of superabundance and spell out the need for reform . . . Before Marsh wrote this book, few saw or fewer worried about how man affected his environment. Today, Marsh's insights are virtually taken for granted. (Lowenthal, 1965, ix)

Marsh set out not only to show how people were changing the environment but also explicitly to urge conservation and restoration of the natural harmonies disturbed by human actions, and he 'disputed not the desirability of exploiting nature but the bungling way it was being done' (Lowenthal, 2000, 406). Although his book helped to inspire US forest protection in the late nineteenth century, it then fell into neglect until rediscovered in the 1930s, when contemporary problems of agricultural erosion were gaining public and political attention.

As Lowenthal (1960) has argued, Marsh's approach would have suited geography well, in integrating physical and human influences, so why was it not picked up by the discipline for many decades? Even in the twentieth century, geographers still looked to the environment for the causes of human change, rather than vice versa. Exhortations to do otherwise were not taken up. For example, in addressing the British Association's Geography Section, Lucas (1914, 492) was inspired by Marsh to set what could have been an agenda for twentieth-century geography: 'Geographers have recorded what the world is according to Nature. I want them to note and teach others to note how under an all-wise Providence it is being subdued, replenished, recast, and contracted by man'. Similar sentiments were expressed by Sherlock (1923, 258), whose comment that 'the effect of Man on Nature seems to be almost entirely ignored' would remain relevant for another 40 years. From the beginning of the century, then, this trend was set: the discipline neglected environmental matters even as they rose in public prominence, and instead focused on 'vulgar environmentalism' even as other disciplines were turning away from it (Livingstone, 1992, 290).

Environmental conservation

In Marsh's book, we see the early origins of what we would term modern environmental management (Lowenthal 2000), especially with the aim of repairing damage or advocating more utilitarian and scientific approaches. Other roots for modern environmentalism come from less strictly scientific sources. In England, Gilbert White's 'arcadian' natural history was an early inspiration for greater attention to nature, mainly amongst the leisured classes (Worster, 1977). The Romantic movement, also in the late eighteenth and early nineteenth centuries, formed a backlash against industrialisation and urbanisation and an elegiac nostalgia for the old agricultural order in the countryside. Rejecting materialistic values in favour of aesthetic ones, especially as revealed through wild and rugged nature (Pepper, 1984), Wordsworth, Coleridge, Shelley, and others brought nature into the realm of the poetic, just as the rise in gardening and landscaping brought it in to the realm of the practical (Thomas, 1983). In the United States, in the later nineteenth century, the transcendentalism and biocentric philosophies of Thoreau, Emerson, and Muir headed a related backlash against the frontier ethic of conquering nature in the name of progress.[2] Instead, by revering nature, and especially wild, scenic nature, these ideas motivated the preservationist movement that produced the first national parks and forests by the late nineteenth century. For both UK and US writers, hence, nature was a source of inspiration and being in nature could uplift our senses and morals. And yet this also generated the seeds of later conflicts: Wordsworth's guide to picturesque viewpoints in the Lake District ironically encouraged the very touristic sight-seeing by the masses, untutored in the finer points of landscape appreciation, that he so disliked and that has later proved so difficult to manage.

Despite the earlier Romantic movement in England, landscape protection was institutionalised later here than in the United States, and was more obviously related to amenity than to strictly ecological integrity. After the First World War, with increasing leisure time and money, plus the vogue for outdoor activity, environmentally based recreation became much more common and so did calls for environmental space and access to be provided for it (Evans, 1997). The mass trespass of Kinder Scout in the Peak District in 1932, resulting in six arrests, was an intrinsically geographical and political act about the ordering of private-public space, but academic geography paid it little attention. Even as the new agenda was consolidated into the National Parks and Access to the Countryside Act 1949 as part of the post-war planning system, creating new geographical units and approaches, geographers were looking elsewhere for their research themes.

Hence, despite popular interest in environmental amenity, geographers did little to develop an environmental or conservation agenda of their own in the interwar period and even afterwards. One notable exception was J. A. Steers (1939, 1944, 1946a, 1946b), whose investigations into coastal amenity and degradation culminated in a survey of the entire coastline of England and Wales for the Ministry of Town and Country Planning. Steers organised symposia on the topic amongst geographers and showed

[2] There is a huge literature on all these literary and ecological figures, but useful introductions are found in Nash (1967), Worster (1977), and Palmer (2001).

the policy relevance of environmental research, involving John Dower (of the Dower Report on *National Parks in England and Wales*) in one symposium. Whilst Steers's coastal survey was comprehensive and well regarded, his aesthetic preservationism nevertheless betrayed a middle-class elitism (an accusation that environmentalism would face throughout the rest of the century), such as his comment that 'the cheap car and the week-end habit have played no small part in the deterioration of our coastal scenery' (Steers, 1944, 13). Even so, his was one of the few voices in academic geography before the 1960s to argue actively and persuasively for landscape protection: 'if we as a nation wish to preserve one of our finest heritages for the good of the people as a whole, we must act now and act vigorously on a national scale' (Steers, 1946a, 60). Vaughan Cornish (e.g. 1935) also sought to preserve the scenery of England through writings and representations to governmental and non-governmental bodies, although his approach has been criticised for its 'Wordsworthian mysticism' and 'rural reverence' (Livingstone, 1995, 367). Goudie (1971, 1) called Cornish 'the British equivalent of George Perkins Marsh' as a 'crusader' against environmental damage and for scenery preservation, but like Marsh his arguments were not pursued by other geographers, who were still looking the other way.

As well as such aesthetically driven conservation, more utilitarian conservation approaches developed that were geared towards 'productive use' and thus management that fostered 'multiple use' of the environment, rather than solely scenic appreciation (Hays, 1959). The practical bent and effective politicking of Gifford Pinchot made such approaches part of US Forestry Service culture by the beginning of the twentieth century (Williams, 1989). Conservation of forests in particular was institutionalised in Britain somewhat later, when the British Forestry Commission was established in 1919, after concern about the depletion of standing British timber (Evans, 1997). Both countries were therefore concerned to protect their environmental resources for the nation but specifically for future *use*. This embryonic appreciation of the finite limits to environmental resources led to some protection for environments but retained a close eye on continued production rather than preservation per se.[3]

Environmental management and ecological science

How we think about the environment has been inspired by ecological ideas, as well as by conservation practices, from Darwin onwards. In the twentieth

[3] Such utilitarian conservation lives on today in the Wise Use movement in the United States (e.g. McCarthy, 1998) and, to a diluted extent, in the sustainable development agenda, which emphasises economic growth alongside environmental protection, especially in Europe.

century, as ecology slowly rose to become a respectable academic science, a range of models was developed to describe and explain ecological entities and changes. In the United States, Frederic Clements argued that ecological change was dynamic, building through succession to a mature 'climax community' that could be seen as a single, complex organism (Worster, 1977). In Britain, Arthur Tansley preferred the concept of the ecosystem to that of the community, emphasising less its identity as an organism in favour of its identity as a mechanistic network. He also rejected the idea that all vegetation formations should lead to one 'natural' climax community, proposing instead a range of possible climax conditions, including human-created ones (Worster, 1977). This 'new ecology' led to the environmental approach that is internationally familiar today, based on thermodynamics, energy, and nutrient cycling (Bramwell, 1989; Golley, 1993), but Worster (1977) argued that this gradually excluded value judgements from ecology by relying upon increasingly 'economistic' models and dubbed it 'the dismal science'.

As well as academic developments, changes outside academia influenced thinking. The Dust Bowl period in the United States in the 1930s created massive public and policy concern about environmental damage and thus provided a key opportunity for ecological science to enhance its status, by demonstrating that the Dust Bowl problems were a product of ignoring natural processes and failing to manage the environment scientifically (Worster, 1977). In England, discussions about nature reserves and national parks also provided opportunities for input from ecological science (Bocking, 1993). Thus, ecological ideas and research provided a more scientific rationale for landscape preservation and management—helping the conservation movement—and also promoting ecology itself as the source of knowledge and expertise, thus helping the 'new' ecologists to gain status and power (Bocking, 1993; Golley, 1993).

This lesson was not lost on other disciplines and, as increasing evidence of environmental degradation accrued, work began to return to the themes raised by Marsh. In 1956, nearly a century after Marsh's book, the first major document on human damage to the contemporary environment was published, following a 1955 symposium at Princeton, New Jersey, entitled *Man's Role in Changing the Face of the Earth.* Initially proposed by the eminent American cultural geographer, Carl Sauer (see Chapter 5 in this volume), the symposium saw Marsh's *Man and Nature* as 'the first great work of synthesis in the modern period to examine in detail man's alteration of the face of the globe' (Thomas, 1956, xxix), yet Marsh's arguments took nearly 100 years to gain widespread attention. Typical of subsequent environmental research, the symposium and the publication that resulted from it involved many different disciplines, although geographers were the most numerous group, with 15 contributors, including Sauer, Clarence Glacken, and H. C. Darby (see Chapter 5 in this volume).

The need for multidisciplinary understanding was clearly stated then, as it has been continually since.

> Man, the ecological determinant on the planet, needs the insights of scholars in nearly all branches of learning to understand what has happened and is happening to the earth under man's impress. (Thomas, 1956, xxxvii)

However, the volume remained predominantly cautious and even optimistic about environmental futures. This attitude, Lowenthal (1990, 125; 2000) argued, showed most of the authors to be unrepresentative of their time, which was beginning to be greatly pessimistic about the consequences of new technologies (such as nuclear power, which was ignored in the volume) for environmental and human survival. To dismiss pessimism as the response of 'crackpots' (Lowenthal, 1990, 125) was typical of subsequent work in academia, despite wider public concern and pessimism in the 1960s and 1970s. The exceptions were names later to be associated with US conservation and environmentalism, including Kenneth Boulding, Lewis Mumford, and Frank Fraser Darling, as well as Carl Sauer himself.

By comparison with the United States, within British geography there were still few proponents of a more environmental or ecologically inspired view, even into the 1960s. Hare (1969) was similarly struck that, despite their early-twentieth-century interest in environmental matters, geographers had forgotten about the environment just as other disciplines had woken up to it. His comments in a very early edition of *Area* were prescient in suggesting that geography had missed out on a leading role in environmental research and has been slow on the catch-up ever since. This prompted several responses (e.g. Coppock, 1970), illustrating the discipline's re-awakening to concerns of the 1960s. Stoddart's (1970) response specifically justified geography having a greater role in environmental concerns because geography was better suited than conservation to provide management prescriptions, and that conservation, exemplified by Fraser Darling's 1969 Reith lectures, was too prone to doom-mongering. O'Riordan (1970b, 34) agreed with Hare: 'The geographer, who has long considered that he can play a "special role" as a synthesizer of academic endeavour in man–environment studies, now finds that other disciplines are interested in some at least of the same thing as he is.' O'Riordan's solution was to pursue interdisciplinary work, something that he has continued to advocate and practise, echoing Thomas above. Other geographers concentrated on bringing ideas from ecology into geography (e.g. Simmons, 1966; Harrison and Warren, 1970; Barkham, 1973), to provide new tools for addressing geographical problems.[4] This again reflected

[4] This tradition particularly continues through cultural ecology in the United States (Butzer, 1989; Zimmerer, 1994).

the rise in power of ecological science within academic and policy institutions and geography's need to play catch-up.

The new agenda: 1960s to 1990s

So it is clear that for some time geography failed to register the rising interest in ecology, conservation, and environmental protection. In the late 1960s and early 1970s, the environment experienced its second major phase of public and political attention through such high-profile initiatives as the UN Stockholm Conference on the Human Environment, the European Year of Conservation 1970 and Earth Day USA 1970. There had been previous waves of conservation interest in the late nineteenth century and between the world wars, but the phase that began in the 1960s was more global in both the problems and the solutions presented. It was also driven more by grassroots support for challenges to the status quo, not only through environmental movements but also anti-nuclear, anti-war, women's, and civil rights movements. Environmental matters became politicised and entered not only legislative debates but also electoral campaigns and governmental institutions.

Such public interest focused upon several key publications. Rachel Carson's *Silent Spring* (1962) is often credited as the springboard for environmental concern and as the book that made ecology a household word. Although written for a popular audience, the book reviewed scientific research into the effects of pesticides on ecological and human health in America. Echoing Marsh's reformism, Carson (1962, 12) was not against chemical pesticides, but against the bungling ways in which they were being used. Unlike Marsh, Carson's bestseller kicked off a decade of further investigation and campaigning, as well as retaliation from the chemical industry, which criticised her scientific rigour as well as her conclusions. In this new climate, Aldo Leopold's *A Sand County Almanac* (1949) was rediscovered, with its evocation of a community ethic for conservation, based on his work in ecology and restoration at the University of Wisconsin, as well as his hobbies of hunting, farming, and natural history. Somewhat later, *The Limits to Growth* was written for the Club of Rome by a team led from the Massachusetts Institute of Technology (Meadows et al., 1972) using computer modelling techniques, and *Blueprint for Survival* (Goldsmith et al., 1972) was written by the editors of the British journal, *The Ecologist*, who also drew on the Club of Rome's report. Both books set out apocalyptic visions of a future based on business-as-usual scenarios, and echoed neo-Malthusian arguments in emphasising population growth as a key problem. The MIT work particularly generated considerable academic critique of its assumptions, methods, projections, and agenda (e.g. Maddox, 1972; Harvey, 1974a; Sandbach, 1978), and thus marked a key point in the development of modern debates about environmentalism (Dobson, 1995, 35).

All three books were read as polemics and thus put vigour into the new environmental debate.

In such conditions of debate, therefore, geography could no longer ignore the environment and, from the 1970s, we see British geographers tackling environmental research in varied ways. Some focused on the local, some on the global, but the cumulative effect was to drag the discipline into the environmental arena and, after more than a half-century of false starts and neglect, begin to deal with environmental matters in the sense that their contemporary society understood them. When the environmental wave ebbed a little after the oil price crisis of the mid-1970s, geographical research continued sporadically into the 1980s. Then, in the late 1980s and early 1990s, the environment underwent another wave of public and political relevance, driven by the global profile of the 'Brundtland Report' (World Commission on Environment and Development, 1987) and the most famous environmental conference of the century, the UN Conference on Environment and Development (UNCED) in Rio de Janeiro in 1992. The ozone layer, the greenhouse effect, and tropical deforestation were becoming publicly recognised ideas, making headlines in newspapers and on television, and environmental NGOs saw their memberships soar. All this change provided both new topics for academic research and greater relevance to policy. Geography responded not only to these outward events but also to theoretical and methodological currents within social science and I shall use four focuses to illustrate this response.

Environmental policy and management

A diverse focus is around environmental management and policy. By the late 1960s, the interdisciplinary field of resource management, with its origins in utilitarian conservation and natural hazards, was well established in the United States (Emel and Peet, 1989). In Britain, the emphasis was more on environmental policy than natural conservation in academic circles. Often, research was stimulated by events outside academia—new legislation and policy, environmental accidents, and initiatives—and thus frequently either reflected these events or was relevant to them (e.g. O'Riordan, 1970a, 1971; Sheail, 1975; Adams, 1984, 1996; Munton, 1984; Harrison, 1996).[5] For example, after decades with no specifically 'environmental' coverage, the *Geographical Journal* grasped the new agenda from the top by reporting a symposium at the Royal Geographical Society (RGS) led by Maurice Strong of the United Nations in 1972, just prior to

[5] The early progress reports on the environment in *Progress in Human Geography* and its predecessor *Progress in Geography* principally review environmental policies and planning decisions rather than geographical research into them, of which there seems to have been little.

the UN Conference on the Human Environment in Stockholm, which Strong organised (he went on to organise UNCED in 1992).

Geographical work on environmental policy and management included two lines of intent. The first was to *influence* and *improve* environmental management and policy. Such work tended to be empirical and in physical geography. For example, in 1971, the British National Committee for Geography prepared a state-of-the-art report for the Royal Commission on Environmental Pollution on 'geographical studies of environmental pollution'. This highlighted the kind of geographical work that was felt to be policy-relevant under three headings: land pollution, especially through surveys of derelict land; air pollution, again through monitoring and surveys; and amenity and environmental quality, which was dominated by the United States, with little work yet completed in Britain except for land use and recreational use surveys.[6] Those supportive of geography's claim to environmental relevance urged geographers to become more useful to policy, for example through measuring, monitoring, and modelling environmental change; again, this is more common in physical geography. In practical terms, work in geography has fed directly into environmental management through commenting on and constructing conservation agendas (e.g. for the British Association of Nature Conservationists by Adams, 1996), through research and training (for example Warren and Goldsmith, 1974; Goldsmith and Warren, 1993) and through new journals like the *Journal of Environmental Management*, which was launched in 1973.[7]

But some were wary of politicised commentary or environmentalist rhetoric because they might endanger their reputation and policy influence. Thus, the politicisation and gathering momentum of environmental arguments outside academia were held to be rabble-rousing or doom-mongering and in deep need of some more cautious science. For example, the tone of the 1972 RGS conference and the *Geographical Journal* special issue mentioned above did not reflect the rising public concern—the speakers mainly rejected radical change and 'doom-sayers' within the 'cult of spaceship Earth' (e.g. Clayton, 1972) in favour of technological adaptation and rational management (e.g. Strong, 1972; Verney, 1972). So, the environment had made it into geographical print, but the approach taken to it was often sceptical and pragmatic rather than innovative or theoretical. This was to remain so, at least in physical geography, to the end of the century. For example, 20 years afterwards, an RGS conference on 'Earth Surface Resources Management in a Warmer Britain' (*Geographical Journal*, 1993), dominated by physical geographers, was praised for its

[6] It is interesting that water pollution, later to become a key environmental theme, is not itemised.

[7] Despite its name, *Environment and Planning* was originally launched to deal not explicitly with environmental matters but spatial change, modelling, and planning in the city.

cautious predictions, despite its title. Turner et al.'s (1990) compendium of environmental change over 300 years was intended as a quasi-sequel to Thomas et al. (1956, and therefore to Marsh 1864) and was similarly guarded in the implications of the changes it documented, pointing out the difficulties of prediction. Hare was perhaps more outspoken than most when he praised geographers for restraint in the face of the 'intoxication' of the environmental doom-mongers:

> We have not written any of the great books of the past two decades. But neither have we written much clap-trap. Avoiding error by silence is not very glorious; but it is better than ignorant loquacity. (Hare, 1980, 380)

Despite this, commentators felt that geography was failing to live up to its environmental promise, in terms of influencing policy and management (Hare, 1974, 1980; Newsom, 1991), and that other disciplines were taking the lead, despite geography's supposedly 'natural' fit with interdisciplinary environmental matters (e.g. Bryant and Wilson, 1998, 335). One argument in favour of this fit was that the discipline bridged the natural and social sciences through including physical and human geography, and thus was 'naturally' interdisciplinary and traditionally concerned with human–environment relationships. But there are two problems here. As I have shown, earlier geographers had *not* been very concerned with human effects on the environment, compared to the reverse, whereas other disciplines had. Any 'natural' lead had already been lost. More importantly, changes within post-war human geography towards Marxism, humanism, and social theory, and away from quantitative empiricism, fostered a disciplinary crisis that threatened the supposed 'unity' of physical and human geography. Reasserting this unity has been a key argument for those trying to make geography more environmentally relevant in the last 20 years. Coppock (1974, 9), in his presidential address to the IBG, not only warned that other disciplines were getting in first on the environmental turf but that the 'distinction between human and physical is increasingly meaningless' because both were needed to address environmental problems and to guarantee the geographer's policy-relevant role as 'general practitioner in a world of specialists'. Others have similarly argued for greater physical–human geography 'unity', again principally from the physical geography side (e.g. Jones, 1983; Douglas, 1986; Goudie, 1986; compare Johnston, 1983, 1986). By the 1990s, this was a difficult argument to sustain and it was mainly only physical geographers that attempted to do so. For example, Cooke (1992) considered that the division hampered both integrated environmental research, thus militating against its utility for policy and management and against the defence of geography from 'territorial incursions' by other disciplines, thus precisely reiterating the arguments put forward by Coppock (1974) 20 years earlier. By comparison, Johnston (1983, 142–3) considered that bridging human and physical geography would often result merely in 'academic tourism' and inferior scholarship, purely to address disciplinary neurosis. Indeed, Turner (2002) has recently resurrected the whole argument by claiming that only the human–environment tradition can save geography in its current moment of disciplinary crisis, through re-establishing its niche in a 'bridging role' between science and social science.

The second line of intent has been to *critique* environmental policy, rather than simply improve it. This tended to be in human geography and included empirical, conceptual, and political content. Academic critiques showed how the environmental policy-making process in general tends to be dominated by technocratic decision-making and low public participation and accountability, despite presenting itself as objective, non-political, rational, and fair. Work by Judith Rees (1973, 1985) and Timothy O'Riordan (1971, 1984, 1990; Garner and O'Riordan, 1982) has continually highlighted the problems of managing the environment and allocating resources in a modern society wherein powerful economic interests and the complexity of environmental change make simple decisions impossible. An example was the special issue of *Environment and Planning A*, which O'Riordan (1977, 1) edited in order 'to look at certain features of the ideological premises that underpin environmental decision making'. It included papers by Adams and Lowe: that by Adams was a trenchant critique of the use of models to (seemingly) make decisions in planning, so that planners become addicted to 'a comforting sensation of calm detachment' that is false; Lowe's was a precursor to Lowe and Goyder (1983, see below) and critiqued local decision-making and the elitism of conservation groups, underlining the inequalities in political efficacy according to wealth and education. Hence, work like this made clear that environmental decisions were—always—political decisions. Work in the 1990s continued to address the inadequacies of environmental management and policy and its technocentric, managerial approach (Adams, 1990; Westcoat, 1991; Owens, 1994; Bryant and Wilson, 1998).

One approach to this has been Marxist in origin. Early on, David Harvey (1974b) retaliated vehemently to Coppock's address (see p. 223) by urging geographers to resist the proto-fascist forces of the corporate state through challenging policy rather than prostituting the discipline in its service. He attacked the neo-Malthusian rhetoric of the *Limits to Growth* world models (Meadows et al., 1972) and similar publications from a Marxist standpoint (Harvey, 1974a, 1974b; also Sandbach, 1978).[8] Political economy has since been intermittently applied to environmental and resource allocation issues in order to critique environmental policies and how they reinforce inequalities of power and wealth (Sandbach, 1980;

[8] And also from outside geography from others (e.g. Maddox, 1972).

Schnaiberg, 1980; Redclift, 1984; Blaikie and Brookfield, 1987; Emel and Peet, 1989; Johnston, 1989). For example, Piers Blaikie (1985) cast the political-economic pressure upon small farmers to produce for distant markets as the cause of environmental degradation, thus linking local ecologies to international power. In common with the approach of 'cultural ecology' in North America, political economy was particularly applied to environmental questions in developing areas, seeking also mainstream recognition for its ability to bring nature and society together in politically aware research (e.g. Butzer, 1989; Turner, 1989).

Other sorts of critique were more pragmatic than theoretical. O'Riordan's continuing work criticised all manner of public policy decisions (e.g. O'Riordan, 1984 on Sizewell B nuclear power station proposals) but without adopting a theoretical platform. The appointment of Crispin Tickell as RGS President in 1990 brought a self-proclaimed environmental evangelist to lead the Society. Tickell's presidential addresses (1991, 1992) to the RGS repeatedly urged geographers to get involved with policy to make people aware of environmental damage and of how geography could contribute to identifying and remedying damage. Unlike his predecessors or successor, Tickell presented the environmentalist arguments simply, rejecting both scepticism and theoretical jargon.

By the 1990s, therefore, geographers were tackling environmental management and policy in diverse ways. The post-Brundtland era further promoted environmental concern and research, under the rubrics of sustainable development and sustainability (Owens, 1994, 2000; Sneddon, 2000). Much of it empirically surveyed the success (or otherwise) of sustainability initiatives, particularly in attempting to increase public participation in decision-making at national and local levels (e.g. Munton, 1997; O'Riordan and Voisey, 1997; Gibbs, Longhurst, and Braithwaite, 1998; Patterson and Theobald, 1996; Kitchen, Whitney, and Littlewood, 1997). Often, as we have seen before, such research has been reformist and under-theorised, practical and reactive rather than theoretical or focused, and further contributes to the diversity of environmental research in geography (Eden, 2000; Sneddon, 2000). Moreover, although intended to break environmental concerns out to connect with wider aspects of social justice, poverty, health, quality of life, and development, often sustainability has simply been elided with environmental agendas or come to dominate them. One obvious research need was to clarify and, where possible, operationalise the concept so as to avoid unwanted appropriation because 'its beguiling simplicity and apparently self-evident meaning have obscured its inherent ambiguity' (O'Riordan, 1989, 93). As with other forms of environmental policy, sustainability research has therefore experienced a perpetual tension between helping to improve policies by working with governmental or quasi-non-governmental groups and critiquing them from without.

Another inspiration in the 1990s came from the German sociologist Ulrich Beck's (1992) *Risk Society*. A public and academic success across Europe, his work was subsequently coupled with that of British sociologist Anthony Giddens (1991; Beck, Giddens, and Lash, 1994). Beck argued that we live today not with easily recognisable risks from starvation and disease, but invisible and unidentifiable risks from radiation and pollution that no one can escape, no matter how wealthy. His argument that the logic of risk distribution is now more powerful than the logic of wealth distribution in characterising the social condition in western society has been widely debated because it de-emphasised class, but for environmental research his work has suggested new ways to interrogate environmental risks and policy-making: by critiquing false rationality, uncertainty, indeterminacy, lack of public participation, and accountability (e.g. Shackley and Wynne, 1995; Irwin and Wynne, 1996; Lash, Szerszynski, and Wynne, 1996) and raising the profile of environmental science, studies, and policy. Although sociology is leading this work, geographers have also interrogated lay–expert relations in decision-making and the increasing scientisation of politics and politicisation of science, especially global climate science (e.g. Boehmer-Christiansen, 1995; Demeritt, 1996, 2001; Eden, 1996; O'Riordan and Jäger, 1997).

Environmental ideologies and groups

As well as considering practical matters of policy and management, geographers became interested in conceptual and ideological themes that underpinned them. The 'new' environmentalism of the late twentieth century was both complex and diverse, attracting critical analysis of its organisation, ideologies, development, and success. The shades of 'environmentalism' in this modern sense were examined in Timothy O'Riordan's (1976) classic *Environmentalism*, in which he divided environmental thinking into two modes: 'ecocentrism' and 'technocentrism'. He further subdivided the first into thinking based on bioethics and thinking based on self-reliant communities, and later revised this typology considerably in other publications (e.g. 1977, 1981, 1989), adding a subdivision also for technocentrism and summarising the modes in a diagram. For example, by 1989 (p. 85), he had renamed his subdivisions 'Gaianism', 'communalism', 'accommodation', and 'intervention'. O'Riordan was disinclined to provide a single agenda. He renounced theoretical meta-narratives and political prescriptions (although the work is struck through with political and ideological critique), and instead he reviewed a range of concepts that informed environmental ideologies from different social science disciplines, including economics, ecology, politics, and law. His ecocentric–technocentric typology has been widely used to more clearly analyse environmental approaches and brought home the fact that much 'environmental' thinking was largely driven by technological and economic considerations rather than environmental ones. However, the subdivisions he developed were less influential, mainly because modern environmentalism had become a much larger and more complex field of study by the 1990s, inspiring many works that classified and interpreted ideological material in different ways and under different labels (e.g. Eckersley, 1992; Merchant, 1992; Dobson, 1995; Dryzek, 1997). For Dobson (1995, 2) 'environmentalism is not an ideology at all' but merely the pragmatic, reformist, managerial approach to the environment common in western governments; instead the alternatives to evnvironmentalism, the more ideologically and politically developed notions that he considers under the heading of 'ecologism', have become the target of analysis in the 1990s.

Yet in many ways, O'Riordan's approach in *Environmentalism* is typical of what was to come from him and from other environmental researchers in the next 25 years. It is a synthesis of diverse disciplinary approaches, it is practically relevant but reformist rather than radical, it is dynamic and shifts with every revision, and it lacks a centre, thus serving, as one commentator put it, 'more as a panorama than as a perspective' (Westcoat, 1998, 590; see also Pepper, 1998). Environmental research in geography (and I might argue geography itself) has thus continually struggled with its lack of shape and its need to stay up to date, often at the expense of theoretical rigour.[9] Subsequent work in this stream was intermittent. David Pepper (1980, 1984) applied O'Riordan's typology to a critique of Hardin and what would later become known as NIMBYism, whilst developing his own socialist perspective on environmental ideologies (later set out in full in Pepper, 1996) and contributing to the literature on environmental history (see below). Other applications included Owens (1986; and see responses in Pepper, 1987; Owens, 1987), who critiqued the claimed 'greening' of mainstream British political parties as technocentric rather than radical. Simmons (1990; see also Bryant and Wilson, 1998) considered that geography itself tended to treat environmental issues in a reformist rather than a radical manner, because human geographers tended to leave environmental research to physical geographers, resulting in a more managerial and practical approach from the discipline as a whole. Marxist critiques of environmentalism were similarly concerned with its technocracy (especially following the *Limits to Growth* debates), its conservatism, and its reliance of scientific authority for validity (e.g. Lowe and Worboys, 1978; Sandbach, 1978).

In an empirically grounded monograph, Lowe and Goyder (1983) analysed local conservationist groups in England and showed the diversity

[9] Geography is not the only discipline to suffer such angst—environmental sociology offers a contrast in its clearer, more theoretically grounded attempts at self-demarcation and justification (e.g. Dunlap et al., 2002).

and the shifting coalitions (and conflicts) that exist under the banner of environmentalism. Consistent with Lowe's work elsewhere, this debate about the sociopolitical nature of environmentalism continues today, illustrating the healthy scepticism that geographers and others have applied to environmental ideologies as well as the diversity, membership, and location of environmental groups (e.g. Cowell and Jehlicka, 1995; Ward, 1996). Recently, political science and philosophy have taken the lead in analysing environmental ideologies and values, using political and philosophical traditions (e.g. Eckersley, 1992; Dobson, 1995). Again, geographers, especially human geographers, took their eyes off the ball and the discipline has not followed through on O'Riordan's early inspiration, despite the place-focus of many environmental protests in the 1990s (Routledge, 1997, is a notable exception). As Simmons noted (1990, 385), geographers always seemed to be looking elsewhere when environmental issues went up the agenda, despite the field seeming eminently suited to their talents and inclinations.

Environmental meanings

One area where geographers did run with the ball was in the related field of environmental perception. Initially inspired by psychology and behaviouralism (see Lowenthal's and Tuan's contributions to Lowenthal, 1967), this was later fed by the backlash against spatial science and spun out into analysis of environmental aesthetics (Appleton, 1975) and the individual experience of the environment (Tuan, 1974; Pocock, 1981, 1982, 1993). Forced to justify dealing with subjective notions in the face of 'objective' environmental damage (and adding a further twist to the problems of physical–human unity in geography identified above), such humanistic approaches, Lowenthal (1967, 3) argued, were necessary because 'without a prior understanding of the bases of perception and behaviour, environmental planning and improvement are mere academic exercises, doomed to failure because unrelated to the terms in which people think and the goals they select'. A generation on, his arguments still apply.[10] Perception research also fed into the wide, policy-oriented field of natural hazards research, led from the United States and focusing upon adaptive behaviour (Burton, Kates, and White, 1978; Aitken et al., 1989).

Humanism in particular inspired geographers to consider the symbolism of environments, especially those that are familiar or valued (Gold and Burgess, 1982). This field was supported also by landscape research and the rising star of social constructionism, from the late 1970s,

[10] The word 'prior' is important—subjective meanings are often tacked on to research for environmental policy purposes, sometimes only in terms of how well policy can be implemented post hoc.

in investigating the meanings and constructs of the 'environment' and 'nature' (e.g. Katz and Kirby, 1991; Demeritt, 1996, 1998). This work differs from the previous theme in that it considers taken-for-granted meanings rather than explicitly articulated policies and ideologies, and has been particularly important in sociology (e.g. Hannigan, 1995). In geography, directly addressing Lowenthal's argument, work has shown how the meanings of the environment affect the way that policy and planning is negotiated and changed (Whatmore and Boucher, 1993; Healey and Shaw, 1994; Hajer, 1995). Constructions of conservation as an environmental narrative and as a political movement have also shown how environmental thinking is shaped by specific places and contexts within Britain, from the Scottish Highlands to the 'heart' of England (Toogood, 1995; Cloke, Milbourne, and Thomas, 1996; Cosgrove, Roscoe, and Rycroft, 1996; MacDonald, 1998).

But most work has focused upon 'ordinary' people and the ways in which they process and create environmental meanings—in the abstract and in relation to specific 'real' places (e.g. Burgess, 1990; 1982; Eden, 1993). In the 1990s, this was also driven by policy commitments to public participation in sustainability programmes (which is easier said than done, as noted above) and by academic challenges to technocratic decision-making and to the dominance of 'expert' science in environmental debates (see Irwin, 1995; Irwin and Wynne, 1996; Owens, 2000; also Demeritt, 2001). This work has been led from University College London (Harrison and Burgess, 1994; Burgess, 1996; Harrison, Burgess, and Filius, 1996; Burgess, Clark, and Harrison, 1998) and Lancaster University (Macnaghten et al., 1995; Macnaghten and Urry, 1998; Myers and Macnaghten, 1998). Yet questions of scale have surprisingly been more implicit than explicit, although notable exceptions include Cosgrove (1994), Harrison and Burgess (1994), and Taylor and Buttel (1992; and see Buttel, Hawkins, and Power, 1990).

Environment and history

Supported by established work in historical geography and the study of landscapes in the discipline (see Chapters 4 and 5 in this volume), our fourth focus is the more clearly defined field of geographical work that has analysed environmental histories. By this, I mean not histories of particular environments or landscapes, although there are plenty of these, but histories of environmentalism, histories of ways of thinking about the environment and their implications in different societies. Environmental history has been more influential in the United States than Britain, perhaps due to the speed (and degree) of environmental change following European settlement in North America, which was highly accelerated and also well documented compared to change in Europe over millennia. A cornerstone of this endeavour, and nearly as weighty, was Clarence Glacken's *Traces on the Rhodian Shore* (1967), which analysed environmental thinking since 500BC. As Demeritt (1994b, 23) notes, this field has been led by historians rather than historical geographers, although Lowenthal's (1965, 2000) biographies of George Perkins Marsh combine not only cultural geography but also history and ecology. Further, this field has been led by North American historians, especially in accounts tracing the development of American notions of nature and wilderness (e.g. Nash, 1967; Worster, 1977, 1992; Oelschlager, 1991).

Let us contrast two cases. In *Man and the Natural World*, Keith Thomas (1983) presents an enjoyable review of the changing relationships between people and nature in England between 1500 and 1800, and is distinctively English in discussing the importance of animals as pets, the rise of gardening as a leisure pursuit, the re-imagination of landscape as countryside, and the distaste for eating many 'wild' species. His historical detail and breadth of coverage showed the development of contrary environmental attitudes and the origins of contentious debates that flowered in the nineteenth and twentieth centuries. In a distinctively American vein, William Cronon's *Nature's Metropolis* (1991) focuses on the rapid growth of Chicago in the nineteenth century, tracing the commodity and communication flows that bound it intimately with its hinterland to write a common history of city and countryside during the European settlement of the American West. This explicit intertwining of natural and social forces in historical development is typical of the best of environmental history and work like Cronon's (1983, 1991) has helped to inspire new interest in environmental history in geography, such as special issues on nature and the American West in *Antipode* and *Ecumene* in 1998. Both works reflect on the development of the dualisms between nature and society, between city and country, between tame and wild, in different cultural contexts.

What makes such works appealing for other disciplines is that they reinterpret human–environment relationships through analysing environmental meanings, often implicit or taken for granted but hugely powerful in driving environmental settlement, exploitation, damage, protection, and science (Cronon, 1983; Williams, 1989; Schama, 1995). In revisiting the origins of environmentalism, we see that how we think about the environment today is infused with ancient arguments about matter and science, about the nature of being, and about the human right to exploit nature's bounty. In explicitly focusing upon environmental ideas that challenged the prevalent tide, Thomas (1983) showed the diverse origins, both elite and lay, scientific and folkloric, of what today infuses environmental attitudes. In this sense, these works are not merely histories but often critiques of ideology, and often draw inspiration from Marxism and feminism as well as from historical approaches (e.g. Merchant, 1980; Pepper, 1984; Bramwell, 1989). Hence, an understanding of the origins of environmental thinking has become a starting point for students and researchers alike in getting to grips with contemporary environmental matters and, in particular, showing how our environmental ideas *change*. New journals were conceived to publish works in this new area—*Environment and History* (UK, first issue, 1995) and *Environmental History* (USA, first issue 1996) are the most recent—but work was also published in established journals in both history and geography.

Everywhere and nowhere in the 1990s

As we have seen, environmental work in geography has multiplied since the 1970s, drawing inspiration from initiatives within and outside the discipline. Because of these multiple inspirations, the last few years have been exciting but chaotic for environmental geographers, as the four focuses outlined above branched inwards and outwards. New journals, often interdisciplinary, were launched to cater for and promote environmental topics, such as: *Global Environmental Change* (first issue 1990), *Environmental Values* (first issue 1992), *Environmental Politics* (first issue 1992), *Ecumene*[11] (first issue 1994), *Ethics Place and the Environment* (first issue 1998), and *Journal of Environmental Policy and Planning* (first issue 1999). But the progress of 'environmental geography' within the British mainstream remained unclear. Regular disciplinary reports prepared for the *Geographical Journal* had trouble tracking it. Both Bennett and Thornes (1988) and Gardner and Hay (1992) thought that they could see human–environment issues re-emerging at geography's centre, but this was based more on optimism than evidence and tended to cite physical geography more than human geography.

Again, this brings us to the problems of 'unity'. First, the divisions between physical and human geography were growing in the 1990s—many human geographers were continuing to move away from quantitative methods and practical applications towards political economy, humanism, postmodernism, social theory, and cultural studies. Institutional rifts were also apparent as physical geographers built stronger links with the RGS than with the IBG, with obvious implications for environmental research after the merger. Environmental relevance found a home in the RGS's Environmental Forum, producing a series of conferences and special issues under such topics as 'Environmental transformations in developing countries' (*Geographical Journal*, 1996) and 'Coastal resilience and planning for an uncertain future' (*Geographical Journal*, 1998). But as human geographers leaned towards critiques of environmental science, modelling, and measure-

[11] To reflect changes in its content and purpose, *Ecumene* changed its name in 2002 to *Cultural Geographies*.

ment (e.g. Demeritt, 1996), the rifts became more contentious. *Transactions of the Institute of British Geographers* and *Area* were dominated by human geography, causing editors and others to appeal to physical geographers for contributions and to produce special issues from annual conference sessions on 'Managing the earth's surface: science and policy' and 'Altered and artificial landscapes' to redress this balance (both in *Transactions* in 1997). Secondly, human geography was also diversifying and fracturing within itself in theoretical and territorial wrangling between Marxists, feminists, cultural theorists, postmodernists, social constructionists, and others, as well as an outward turn to other disciplinary inspirations. Environmental research was caught up in the cacophony and was subjected to a whole range of different theoretical approaches, making it difficult to summarise or predict (though some have tried, e.g. Turner, 1997).

Both these disunities endangered the coherence of environmental research. Although many researchers have continued to do work that is interdisciplinary, policy relevant, and theoretically informed, they have come to environmental matters from different theoretical backgrounds, bringing diverse meta-narratives and methodological preoccupations to the party (e.g. Demeritt, 1994a). Thus, environmental research is more accurately described as multidisciplinary rather than interdisciplinary because of its 'multitude of mutually incommensurable methods and languages' (Richards and Wrigley, 1996, 53). Let me briefly illustrate these problems through the IBG's Environmental Research Group (ERG), set up in 1992. The ERG aimed to bridge human and physical geography research in environmental fields and thereby to grasp the opportunities that geography was missing because the discipline's environmental profile was too low and too diffuse to compete with that of other disciplines (echoing both Cooke and Coppock above). It also aimed specifically to raise the Institute's environmental profile in the face of competition from other 'scientific societies' and to address its 'asymmetry', skewed as it was towards the interests of, and members from, human geography (Lawler, 1992).[12] With plenty of members and good audiences at its annual conference sessions, the ERG initially thrived. But it soon became clear that its membership was a secondary identification—geographers were primarily 'historical', 'cultural', 'economic', or geomorphological' and only secondarily 'environmental'. The environment was becoming a second string to everyone's bow but nobody's particular speciality. Finding committee members and organisers to be active throughout the year became far more difficult than finding attendees for a day's conference; gradually the ERG saw its distinctiveness taken up by other research groups, finally merging with the Planning and Environment Research Group in 2000.

[12] Writing as a physical geographer, Lawler sourced most of his article from physical rather than human geography.

As well as institutional changes, environmental themes were influenced by new ideas. First, we saw a great interest in theorising 'nature' and challenging the nature/culture dualism, spurred by the cultural turn and by work in philosophy, cultural studies and theory, and history. In North America, environmental histories (see above) were generating debate about the nature of nature and of 'wilderness' in particular (e.g. Cronon, 1995; Proctor, 1998), and Donna Haraway's work (1992, 1996) championed the cyborg as the key metaphor in a world of nature–culture hybrids. In Europe, Bruno Latour's (1993) *We Have Never Been Modern* came out of work in the sociology of science but moved outward to tackle the interfusion of social and environmental matters through what became known as actor-network theory (for geographical perspectives, see *Environment and Planning A*, special issue 2000). Later linked with Haraway's work, this new focus on hybridisation of nature–culture–technology inspired geographers to re-theorise the connections between nature and society as well as the meaning of 'nature' itself (e.g. Braun and Castree, 1998; Whatmore and Thorne, 1998). Although diverse, these contributions to environmental geography share an interest in the power of nature as a concept and in how it is deployed to moralise, to motivate, and to exclude social groups. A recent outcropping is the geography of animals as symbols and as boundary markers of nature and wildness (Anderson, 1995; Wolch and Emel, 1998; Philo and Wilbert, 2000).

Second, interest in ethical concerns rose under an 'ethical turn' (see Chapter 20 in this volume), partly in opposition to postmodernism's lack of ethics (Gandy, 1996). This reinforced the application of political economy to environmental questions in order to tackle environmental and social injustice. Building on the earlier Marxist critiques and the application of political economy to environmental problems in developing areas (see above), recent work has sought to understand the transformation of 'original' or 'first' nature through labour and technology into socially produced or 'second' nature, entailing the commodification of nature and the alienation of labour (FitzSimmons, 1989; Dickens, 1996; Harvey, 1993, 1996; Smith, 1996; Castree, 1997). More specifically, under the rubric of political ecology, and drawing also from North American cultural ecology, work has analysed the environmental politics of production and its related inequalities of power and knowledge in the South, especially in agrarian systems (e.g. Redclift, 1984; Blaikie and Brookfield, 1987; Emel and Peet, 1989; Adams, 1990; Peet and Watts, 1996; Bryant, 1997). More generally, the ethical turn has challenged technocratic and exclusionary modes of decision-making, which tend to be presented as scientific and thus apolitical and amoral. This reflects the post-Rio agenda of participation and inclusion in environmental decisions (see above), but also the academic deconstruction of environmental problems as social and ideological constructs rather than as givens (Livingstone, 1995, 370). In Britain, such critiques have been associated with sustainability and sociology (e.g. Eden, 1996; Burgess et al., 1998; Owens, 2000), whereas in North America they have been associated with political ecology and philosophy (e.g. Peet and Watts, 1996; Proctor, 1998). Compared to the 1970s, when Hare and others (see above) were pleased that geographers were not politically involved, the 1990s saw increasing calls for the involvement of academics in popular struggles and the reflexive nature of the 'cultural turn' even made such involvement suitable material for academic conference papers and sessions. Within geography, the debate over Shell sponsorship of the RGS also illustrated this, as the initial Ogoni protests related to environmental management but spiralled out to encompass a range of ethical questions (see Chapter 20 in this volume).

Final word

Geography took its eye off the environmental ball for the first half of the twentieth century, and then got caught on the hop when environmental concern in the 1970s 'found geography fragmented, unprepared and perhaps unwilling to take a leadership role' (Turner, 1997, 204). Although geographers in Britain and elsewhere have explored a range of environmental themes since then, today 'the environment' is everywhere but nowhere in geographical research. Its interdisciplinarity, broad-ranging coverage, and popular interest makes its character as a specific sub-discipline difficult to define or pin down. In a way, that is a strength, because geographers are not rigidly tied to meta-narratives or grand theories. But it is also a danger for a discipline that wishes to grow, to be policy relevant, and to influence other disciplines. From this short review of the British experience, it is clear that environmental research has evaded pigeonholing and has freely drawn on a range of external inspirations for its continual reinvention. The resulting work has been lively and varied, but in the end somewhat shapeless. But if we compare it to our understanding of human–environment relationships a century ago, then clearly our vista has completely changed. From the dominance of environmental determinism, we have moved through phases of environmental policy contribution and critique to a wider appreciation of the complexity of our relationship with the environment that has enriched geographical work in diverse ways.

References

Adams, J. G. U. (1977) The national health. *Environment and Planning A*, 9, 23–33.

Adams, W. M. (1984) Sites of Special Scientific Interest and habitat protection: implications of the Wildlife and Countryside Act 1981. *Area*, 16(4), 273–80.

Adams, W. M. (1990) *Green Development*. London: Routledge.

Adams, W. M. (1996) *The Future of Nature*. London: Earthscan.

Aitken, Stuart C., Cutter, Susan L., Foote, Kenneth E., and Sell, James L. (1989) Environmental perception and behavioural geography. In Gary L. Gaile and Cort J. Willmott, eds, *Geography in America*. Columbus OH: Merrill, 218–38.

Anderson, Kay (1995) Culture and nature at the Adelaide Zoo: at the frontiers of 'human' geography. *Transactions, Institute of British Geographers*, NS20(3), 275–94.

Appleton, Jay (1975) *The Experience of Landscape*. Chichester: Wiley.

Barkham, J. P. (1973) Recreational carrying capacity. *Area*, 5(3), 218–22.

Beck, Ulrich (1992) *Risk Society*. London: Sage.

Beck, Ulrich, Giddens, Anthony, and Lash, Scott (1994) *Reflexive Modernization: Politics, Tradition and Aesthetics in the Modern Social Order*. Stanford, CA: Stanford University Press.

Bennett, R. J. and Thornes, J. B. (1988) Geography in the United Kingdom 1985–1988. *Geographical Journal*, 154(1), 23–48.

Blaikie, Piers M. (1985) *The Political Economy of Soil Erosion in Developing Countries*. Harlow: Longman.

Blaikie, Piers M. and Brookfield, Harold (1987) *Land Degradation and Society*. London: Methuen.

Bocking, Stephen (1993) Conserving nature and building a science: British ecologists and the origins of the Nature Conservancy. In Michael Shortland, ed., *Science and Nature*. Oxford: British Society for the History of Science, Monograph No. 8, 89–115.

Boehmer-Christiansen, Sonja (1995) Britain and the Intergovernmental Panel on Climate Change: the impacts of scientific advice on global warming. Part 1: integrated policy analysis and the global dimension. *Environmental Politics*, 4(1), 1–18.

Bramwell, Anna (1989) *Ecology in the 20th Century*. New Haven, CT: Yale University Press.

Braun, Bruce and Castree, Noel, eds (1998) *Remaking Reality*. London: Routledge.

British National Committee for Geography (1972) Geographical studies of environmental pollution. *Area*, 4, 114–21.

Bryant, Raymond L. (1997) Beyond the impasse: the power of political ecology in Third World environmental research. *Area* 29(1), 5–19.

Bryant, Raymond L. and Wilson, Geoff A. (1998) Rethinking environmental management. *Progress in Human Geography*, 22(3), 321–43.

Burgess, Jacquelin (1982) Filming the Fens: a visual interpretation of regional character. In John R. Gold and Jacquelin Burgess, eds, *Valued Environments*. London: Allen & Unwin, 35–54.

Burgess, Jacquelin (1990) The production and consumption of environmental meanings in the mass media: a research agenda for the 1990s. *Transactions, Institute of British Geographers*, NS15(2), 139–61.

Burgess, Jacquelin (1996) Focusing on fear: the use of focus groups in a project for the Community Forest Unit. *Area* 28(2), 130–5.

Burgess, Jacquelin, Clark, Judy, and Harrison, Carolyn M. (1998) Respondents' evaluation of a CV survey: a case study based on an economic valuation of the Wildlife Enhancement Scheme, Pevensey Levels in East Sussex. *Area*, 30(1), 19–27.

Burton, Ian, Kates, Robert W., and White, Gilbert F. (1978) *The Environment as Hazard*. New York: Oxford University Press.

Buttel, Frederick H., Hawkins, Ann P., and Power, Alison G. (1990) From limits to growth to global change: constraints and contradictions in the evolution of environmental science and ideology. *Global Environmental Change*, 1(1), 57–66.

Butzer, Karl W. (1989) Cultural ecology. In Gary L. Gaile and Cort J. Willmott, eds, *Geography in America*. Columbus, OH: Merrill, 192–208

Carson, Rachel (1962) *Silent Spring*. Boston, MA: Houghton Mifflin.

Castree, Noel (1997) Nature, economy and the cultural politics of theory: the 'war against the seals' in the Bering Sea, 1870–1911. *Geoforum*, 28(1), 1–20.

Clark, K. G. (1950) Certain underpinnings of our arguments in human geography. *Transactions and Papers of the Institute of British Geographers*, 16, 15–22.

Clayton, Keith (1972) Human environment: the impending crisis. *Geographical Journal*, 138(4), 424–5.

Cloke, Paul, Milbourne, Paul, and Thomas, Chris (1996) The English National Forest: local reactions to plans for renegotiated nature–society relations in the countryside. *Transactions, Institute of British Geographers*, NS21(3), 552–71.

Cooke, Ronald U. (1992) Common ground, shared inheritance: research imperatives for environmental geography. *Transactions, Institute of British Geographers*, NS17(2), 131–51.

Coppock, J. T. (1970) Geographers and conservation. *Area*, No. 2, 24–6.

Coppock, J. T. (1974) Geography and public policy: challenges, opportunities and implications. *Transactions, Institute of British Geographers*, 63, 1–16.

Cornish, Vaughan (1935) The cliff scenery of England and the preservation of its amenities. *Geographical Journal*, 86(6), 505–11.

Cosgrove, Denis (1994) Contested global visions: *One-World, Whole-Earth*, and the Apollo space photographs. *Annals of the Association of American Geographers*, 84(2), 270–94.

Cosgrove, Denis, Roscoe, Barbara, and Rycroft, Simon (1996) Landscape and identity at Ladybower Reservoir and Rutland Water. *Transactions, Institute of British Geographers*, NS21(3), 534–51.

Cowell, Richard and Jehlicka, Petr (1995) Backyard and biosphere: the spatial distribution of support for English and Welsh environmental organisations. *Area*, 27(2), 110–17.

Cronon, William (1983) *Changes in the Land: Indians, Colonists, and the Ecology of New England*. New York: Hill and Wang.

Cronon, William (1991) *Nature's Metropolis: Chicago and the Great West*. New York: W. W. Norton.

Cronon, William (1995) The trouble with wilderness; or, getting back to the wrong nature. In William Cronon, ed., *Uncommon Ground*. New York: W. W. Norton, 69–90.

Demeritt, David (1994a) The nature of metaphors in cultural geography and environmental history. *Progress in Human Geography*, 18(2), 163–85.

Demeritt, David (1994b) Ecology, objectivity and critique in writings on nature and human societies. *Journal of Historical Geography*, 20(1), 22–37.

Demeritt, David (1996) Social theory and the reconstruction of science and geography. *Transactions, Institute of British Geographers*, NS21(3), 484–503.

Demeritt, David (1998) Science, social constructivism and nature. In Bruce Braun and Noel Castree, eds, *Remaking Reality*. London: Routledge, 173–93.

Demeritt, David (2001) The construction of global warming and the politics of science. *Annals of the Association of American Geographers*, 91(2), 307–37.
Dickens, Peter (1996) *Reconstructing Nature: Alienation, Emancipation and the Division of Labour.* London: Routledge.
Dobson, Andrew (1995) *Green Political Thought*, 2nd edn. London: Routledge.
Douglas, Ian (1986) The unity of geography is obvious. . . . *Transactions, Institute of British Geographers*, NS11(4), 459–63.
Dryzek, John S. (1997) *The Politics of the Earth.* Oxford: Oxford University Press.
Dunlap, Riley E., Buttel, Frederick H., Dicken, Peter, and Gijswijt, August, eds (2002) *Sociological Theory and the Environment.* Lanham, MD: Rowman & Littlefield.
Eckersley, Robyn (1992) *Environmentalism and Political Theory.* London: UCL Press.
Eden, Sally (1993) Individual environmental responsibility and its role in public environmentalism. *Environment and Planning A*, 25, 1743–58.
Eden, Sally (1996) Public participation in environmental policy: considering scientific, counter-scientific and non-scientific contributions. *Public Understanding of Science*, 5, 183–204.
Eden, Sally (2000) Environmental issues: sustainable progress? *Progress in Human Geography*, 24(1), 111–18.
Emel, Jacque and Richard Peet (1989) Resource management and natural hazards. In Richard Peet and Nigel Thrift, eds, *New Models in Geography.* London: Unwin Hyman, 49–76
Evans, David (1997) *A History of Nature Conservation in Britain*, 2nd edn. London: Routledge.
FitzSimmons, Margaret (1989) The matter of nature. *Antipode*, 21(2), 106–20.
Fleure, H. J. (1947) Some problems of society and environment. *Transactions and Papers of the Institute of British Geographers*, 12.
Gandy, Matthew (1996) Crumbling land: the postmodernity debate and the analysis of environmental problems. *Progress in Human Geography*, 20(1), 23–40.
Gardner, R. A. M. and Hay, A. M. (1992) Geography in the United Kingdom 1988–92. *Geographical Journal*, 158(1), 13–30.
Garner, J. F. and O'Riordan, T. (1982) Environmental Impact Assessment in the context of economic recession. *Geographical Journal*, 148(3), 343–61.
Geographical Journal (1993) Special issue on Earth Surface Resources Management in a Warmer Britain. *Geographical Journal*, 159(2).
Geographical Journal (1996) Special issue on Environmental Transformations in Developing Countries. *Geographical Journal*, 163(2).
Geographical Journal (1998) Special issue on Coastal Resilience and Planning for an Uncertain Future. *Geographical Journal*, 164(3).
Gibbs, D., Longhurst, J., and Braithwaite, C. (1998) Struggling with sustainability: weak and strong interpretations of sustainable development within local authority policy. *Environment and Planning A*, 30, 1351–65.
Giddens, Anthony (1991) *Modernity and Self-identity.* Cambridge: Polity Press.
Glacken, Clarence (1956) Changing ideas of the habitable world. In William L. Thomas, Jr, ed., *Man's Role in Changing the Face of the Earth.* Chicago, IL: University of Chicago Press, 70–92
Glacken, Clarence (1967) *Traces on the Rhodian Shore.* Berkeley, CA: University of California Press.

Gold, John R. and Burgess, Jacquelin, eds (1982) *Valued Environments*. London: Allen & Unwin.

Goldsmith, F. B. and Warren, A., eds (1993) *Conservation in Progress*. Chichester: Wiley.

Goldsmith, Edward, Allen, Robert, Allaby, Michael, Davoll, John and Lawrence, Sam (1972) *Blueprint for Survival*. Boston, MA: Houghton Mifflin.

Golley, Frank Benjamin (1993) A *History of the Ecosystem Concept in Ecology*. New Haven, CT: Yale University Press.

Goudie, Andrew (1971) Vaughan Cornish: geographer. *Transactions, Institute of British Geographers*, 55, 1–16.

Goudie, Andrew (1986) The integration of human and physical geography. *Transactions, Institute of British Geographers*, NS11(4), 454–8.

Hajer, Maarten A. (1995) *The Politics of Environmental Discourse*. Oxford: Clarendon Press.

Hannigan, John A. (1995) *Environmental Sociology*. London: Routledge.

Haraway, Donna (1992) The promises of monsters: a regenerative politics for inappropriate/ed others. In Lawrence Grossberg, Cary Nelson, and Paula Treichler, eds, *Cultural Studies*. New York: Routledge, 295–338.

Haraway, Donna (1996) *Modest_Witness@Second_Millennium*. London:Routledge.

Hare, F. K. (1969) Environment: resuscitation of an idea. *Area*, 4, 52–5.

Hare, F. K. (1974) Geography and public policy: a Canadian view. *Transactions, Institute of British Geographers*, 63, 25–8.

Hare, F. K. (1980) The planetary environment: fragile or sturdy? *Geographical Journal*, 146(3), 379–95.

Harrison, Carolyn M. (1996) 'The Best Available Place'—reconciling leisure uses of the countryside and their environmental impact. *Area*, 28(3), 339–46.

Harrison, Carolyn and Burgess, Jacquelin (1994) Social constructions of nature: a case study of conflicts over the development of Rainham Marshes. *Transactions, Institute of British Geographers*, NS19(3), 291–310.

Harrison, Carolyn M. and Warren, Andrew (1970) Conservation, stability and management. *Area*, 2, 26–32.

Harrison, Carolyn, Burgess, Jacquelin, and Filius, Petra (1996) Rationalizing environmental responsibilities: a comparison of lay publics in the UK and the Netherlands. *Global Environmental Change*, 6(3), 215–34.

Harvey, David (1974a) Population, resources, and the ideology of science. *Economic Geography*, 50(3), 256–77.

Harvey, David (1974b) What kind of geography for what kind of public policy? *Transactions, Institute of British Geographers*, 63, 18–24.

Harvey, David (1993) The nature of environment. *The Socialist Register*, 1–51.

Harvey, David (1996) *Justice, Nature and the Geography of Difference*. Oxford: Blackwell.

Hays, Samuel P. (1959) *Conservation and the Gospel of Efficiency*. Cambridge, MA: Harvard University Press.

Healey, Patsy and Shaw, Tim (1994) Changing meanings of 'environment' in the British planning system. *Transactions, Institute of British Geographers*, NS19(4), 425–38.

Irwin, Alan (1995) *Citizen Science*. London: Routledge.

Irwin, Alan and Wynne, Brian, eds (1996) *Misunderstanding Science?* Cambridge: Cambridge University Press.

Johnston, R. J. (1983) Resource analysis, resource management and the integration of physical and human geography. *Progress in Physical Geography*, 7(1), 127–46.

Johnston, R. J. (1986) Four fixations and the quest for unity in geography. *Transactions, Institute of British Geographers*, NS11(4), 449–53.

Johnston, R. J. (1989) *Environmental Problems: Nature, Economy, and State.* London: Belhaven Press.

Jones, David K. C. (1983) Environments of concern. *Transactions, Institute of British Geographers*, NS8(4), 429–57.

Katz, Cindi and Kirby, Andrew (1991) In the nature of things: the environment and everyday life. *Transactions, Institute of British Geographers*, NS16(3), 259–71.

Kitchen, T., Whitney, D., and Littlewood, S. (1997) Local authority/academic collaboration and Local Agenda 21 policy processes. *Journal of Environmental Planning and Management*, 40, 645–59.

Lash, Scott, Szerszynski, Bronislaw, and Wynne, Brian, eds (1996) *Risk, Environment and Modernity.* London: Sage.

Latour, Bruno (1993) *We Have Never Been Modern.* Hemel Hempstead: Harvester Wheatsheaf.

Lawler, Damian (1992) Environmental geography in the IBG: the new Environmental Research Group. *Area*, 24(3), 309–16.

Leopold, Aldo (1949) A *Sand County Almanac, and Sketches Here and There.* New York: Oxford University Press.

Livingstone, David N. (1992) *The Geographical Tradition.* Oxford: Blackwell.

Livingstone, David N. (1995) The polity of nature: representation, virtue, strategy. *Ecumene*, 2(4), 353–77.

Lowe, P. D. (1977) Amenity and equity: a review of local environmental pressure groups in Britain. *Environment and Planning A*, 9, 35–58.

Lowe, Philip and Goyder, Jane (1983) *Environmental Groups in Politics.* London: Allen & Unwin.

Lowe, Philip and Worboys, Michael (1978) Ecology and the end of ideology. *Antipode*, 10(2), 12–21.

Lowenthal, David (1960) George Perkins Marsh on the nature and purpose of geography. *Geographical Journal*,126(4), 413–17.

Lowenthal, David (1965) Introduction to reprint of George Perkins Marsh, *Man and Nature.* Cambridge MA: Belknap Press of Harvard University Press, ix–xxix.

Lowenthal, David, ed. (1967) *Environmental Perception and Behaviour.* Chicago, IL: Department of Geography, University of Chicago, Working Paper No. 109.

Lowenthal, David (1990) An awareness of human impacts: changing attitudes and emphases. In B. L. Turner II, William C. Clark, Robert W. Kates, John F. Richards, Jessica T. Mathews, and William B. Meyer, eds, *The Earth as Transformed by Human Action.* Cambridge: Cambridge University Press with Clark University, 121–35.

Lowenthal, David (2000) *George Perkins Marsh: Prophet of Conservation.* Seattle: University of Washington Press.

Lucas, Charles P. (1914) Man as a geographical agency. *Geographical Journal*, 44(5), 477–92.

Maddox, John (1972) *The Doomsday Syndrome.* Basingstoke: Macmillan.

McCarthy, James (1998) Environmentalism, Wise Use and the nature of accumulation in the rural West. In Bruce Braun and Noel Castree, eds, *Remaking Reality*. London: Routledge, 126–49.

MacDonald, Fraser (1998) Viewing Highland Scotland: ideology, representation and the 'natural heritage'. *Area*, 30(3), 237–44.

Macnaghten, P., Grove-White, R., Jacobs, M., and Wynne, B. (1995) *Public Perceptions and Sustainability in Lancashire*. Lancaster: Lancaster University.

Macnaghten, Phil and Urry, John (1998) *Contested Natures*. London: Sage.

Marsh, George Perkins (1864) *Man and Nature*. New York: Scribner.

Meadows, Donella H., Meadows, Dennis L., Randers, Jørgen, and Behrens, William W. III (1972) *The Limits to Growth*. New York: Signet.

Merchant, Carolyn (1980) *The Death of Nature*. San Francisco, CA: Harper & Row.

Merchant, Carolyn (1992) *Radical Ecology*. New York: Routledge.

Munton, Richard (1984) Resource management and conservation: the UK response to the World Conservation Strategy. *Progress in Human Geography*, 8(1), 120–6.

Munton, Richard (1997) Engaging sustainable development: some observations on progress in the UK. *Progress in Human Geography*, 21, 147–63.

Myers, G. and Macnaghten, P. (1998) Rhetorics of environmental sustainability: commonplaces and places. *Environment and Planning A*, 30, 333–53.

Nash, Roderick (1967) *Wilderness and the American Mind*. New Haven, CT: Yale University Press.

Newsom, Malcolm (1991) Space, time and pollution control: geographical principles in UK public policy. *Area*, 23(1), 5–10.

Oelschlager, Max (1991) *The Idea of Wilderness*. New Haven, CT: Yale University Press.

O'Riordan, Timothy (1970a) Spray irrigation and the Water Resources Act 1963. *Transactions, Institute of British Geographers*, 49, 33–47.

O'Riordan, Timothy (1970b) New conservation and geography. *Area*, No. 4, 33–6.

O'Riordan, Timothy (1971) *Perspectives on Resource Management*. London: Pion.

O'Riordan, Timothy (1976) *Environmentalism*. London: Pion.

O'Riordan, Timothy (1977) Guest editorial. *Environment and Planning A*, 9, 1–2.

O'Riordan, Timothy (1981) *Environmentalism*, 2nd edn. London: Pion.

O'Riordan, Timothy (1984) The Sizewell B inquiry and a national energy strategy. *Geographical Journal*, 150(2), 171–82.

O'Riordan, Timothy (1989) The challenge for environmentalism. In Richard Peet and Nigel Thrift, eds, *New Models in Geography*. London: Unwin Hyman, 77–102.

O'Riordan, Timothy (1990) On the greening of major projects. *Geographical Journal*, 156(2), 141–8.

O'Riordan, Timothy and Jäger, Jill, eds (1997) *Politics of Climate Change*. London: Routledge.

O'Riordan, Timothy and Voisey, Heather, eds (1997) *Sustainable Development in Western Europe*. London: Frank Cass.

Owens, Susan (1986) Environmental politics in Britain: new paradigm or placebo? *Area*, 18(3), 195–201.

Owens, Susan (1987) A rejoinder to David Pepper. *Area*, 19(1), 77–9.

Owens, Susan (1994) Negotiated environments? Needs, demands, and values in the age of sustainability. *Environment and Planning A*, 29, 571–80.
Owens, Susan (2000) 'Engaging the public': information and deliberation in environmental policy. *Environment and Planning A*, 32, 1141–8.
Palmer, Joy, ed. (2001) *Fifty Key Thinkers on the Environment*. London: Routledge.
Patterson, A. and Theobald, K. S. (1996) Local Agenda 21, compulsory competitive tendering and local environmental practices. *Local Environment*, 1, 7–19.
Peet, Richard and Watts, Michael, eds (1996) *Liberation Ecologies: Environment, Development, Social Movements*. London: Routledge.
Pepper, David (1980) Environmentalism, the 'life-boat ethic' and anti-airport protest. *Area*, 12(3), 177–82.
Pepper, David (1984) *The Roots of Modern Environmentalism*. Beckenham: Croom Helm.
Pepper, David (1987) Environmental politics: who are the real radicals? *Area*, 19(1), 75–7.
Pepper, David (1996) *Eco-socialism*. London: Routledge.
Pepper, David (1998) Classics in human geography revisited: O'Riordan, T., 1976, *Environmentalism*, London, Pion. Commentary 2. *Progress in Human Geography*, 22(3), 591–2.
Philo, Chris and Wilbert, Chris, eds (2000) *Animal Spaces, Beastly Places*. London: Routledge.
Pocock, Douglas (1981) Sight and knowledge. *Transactions, Institute of British Geographers*, NS6, 385–93.
Pocock, Douglas (1982) Valued landscape in memory: the view from Prebends' Bridge. *Transactions, Institute of British Geographers*, NS7, 354–64.
Pocock, Douglas (1993) The senses in focus. *Area*, 25(1), 11–16.
Proctor, James D. (1998) Geography, paradox and environmental ethics. *Progress in Human Geography*, 22(2), 234–55.
Redclift, Michael (1984) *Development and the Environmental Crisis*. London: Methuen.
Rees, Judith (1973) The demand for water in South-East England. *Geographical Journal*, 139(1), 20–36.
Rees, Judith (1985) *Natural Resources*. London: Methuen.
Richards, Keith and Wrigley, Neil (1996) Geography in the United Kingdom 1992–1996. *Geographical Journal*, 162(1), 41–62.
Routledge, Paul (1997) The imagineering of resistance: Pollok Free State and the practice of postmodern politics. *Transactions, Institute of British Geographers*, NS22(3), 359–76.
Sandbach, Francis (1978) Ecology and the 'limits to growth' debate. *Antipode*, 10(2), 22–32.
Sandbach, Francis (1980) *Environment, Ideology, and Policy*. Montclair, NJ: Allanheld, Osmun & Co.
Schama, Simon (1995) *Landscape and Memory*. New York: Random House.
Schnaiberg, Allan (1980) *The Environment, from Surplus to Scarcity*. New York: Oxford University Press.
Shackley, Simon and Wynne, Brian (1995) Global climate change: the mutual construction of an emergent science-policy domain. *Science and Public Policy*, 22, 218–30.

Sheail, John (1975) The concept of National Parks in Great Britain 1900–1950. *Transactions, Institute of British Geographers*, 66, 41–56.
Sheail, John (1976) *Nature in Trust*. Glasgow: Blaikie.
Sherlock, R. L. (1923) The influence of man as an agent in geographical change. *Geographical Journal*, 61(4), 258–68.
Simmons, I. G. (1966) Ecology and land use. *Transactions, Institute of British Geographers*, 38, 59–72.
Simmons, I. G. (1990) No rush to grow green? *Area*, 22(4), 384–7.
Smith, Neil (1996) The production of nature. In George Robertson, Melinda Mash, Lisa Tickner, Jon Bird, Barry Curtis, and Tim Putnam, eds, *FutureNatural*. London: Routledge, 35–54
Sneddon, Christopher S. (2000) 'Sustainability' in ecological economics, ecology and livelihoods: a review. *Progress in Human Geography*, 24(4), 521–49.
Steers, J. A. (1939) Recent coastal changes in south-eastern England: 1 The nature of coastal changes. *Geographical Journal*, 93(5), 399–408.
Steers, J. A. (1944) Coastal preservation and planning. *Geographical Journal*, 104(1–2), 7–27.
Steers, J. A. (1946a) Coastal preservation and planning [part 2]. *Geographical Journal*, 107(1–2), 57–60.
Steers, J. A. (1946b) *The Coastline of England and Wales*. Cambridge: Cambridge University Press.
Stoddart, D. R. (1970) Our environment. *Area*, No. 1, 1–4.
Strong, Maurice (1972) Human environment: the impending crisis. *Geographical Journal*, 138(4), 411–17.
Taylor, Griffith, ed. (1951) *Geography in the Twentieth Century*. New York: The Philosophical Library.
Taylor, J. A. (1974) The ecological basis of resource management. *Area*, 6(2), 101–6.
Taylor, Peter J. and Buttel, Frederick H. (1992) How do we know we have global environmental problems? Science and the globalization of environmental discourse. *Geoforum*, 23(3), 405–16.
Thomas, Keith (1983) *Man and the Natural World*. New York: Pantheon Books.
Thomas, William L. Jr, ed. (1956) *Man's Role in Changing the Face of the Earth*. Chicago, IL: University of Chicago Press.
Tickell, Crispin (1991) Presidential address. *Geographical Journal*, 157(3), 326–9.
Tickell, Crispin (1992) Presidential address. *Geographical Journal*, 158(3), 322–5.
Toogood, Mark (1995) Representing ecology and Highland tradition. *Area*, 27(2), 102–8.
Tuan, Yi-Fu (1974) *Topophilia*. New Jersey: Prentice Hall.
Turner, B. L. II (1989) The specialist-synthesis approach to the revival of geography: the case of cultural ecology. *Annals of the Association of American Geographers*, 79(1), 88–100.
Turner, B. L. II (1997) Spirals, bridges and tunnels: engaging human–environment perspectives in geography. *Ecumene*, 4(2), 196–217.
Turner, B. L. II (2002) Contested identities: human–environment geography and disciplinary implications in a restructuring academy. *Annals of the Association of American Geographers*, 92(1), 52–74.

Turner, B. L. II, Clark, William C., Kates, Robert W., Richards, John F., Mathews, Jessica T., and Meyer, William B., eds (1990) *The Earth as Transformed by Human Action.* Cambridge: Cambridge University Press with Clark University.

Verney, R. P. (1972) The management of natural resources. *Geographical Journal,* 138(4), 417–20.

Ward, Neil (1996) Surfers, sewage and the new politics of pollution. *Area,* 28(3), 331–8.

Warren, A. and Goldsmith, F. B., eds (1974) *Conservation in Practice.* Chichester: Wiley.

Westcoat, James L. (1991) Resource management: the long-term global trend. *Progress in Human Geography,* 15(1), 81–93.

Westcoat, James L. (1998) Classics in human geography revisited: O'Riordan, T., 1976, *Environmentalism,* London, Pion. Commentary 1. *Progress in Human Geography,* 22(3), 589–90.

Whatmore, Sarah (1997) Dissecting the autonomous self: hybrid cartographies for a relational ethics. *Environment and Planning D: Society and Space,* 15, 37–53.

Whatmore, Sarah and Boucher, Susan (1993) Bargaining with nature: the discourse and practice of 'environmental planning gain'. *Transactions, Institute of British Geographers,* NS18(2), 166–78.

Whatmore, Sarah and Thorne, Lorraine (1998) Wild(er)ness: reconfiguring the geographies of wildlife. *Transactions, Institute of British Geographers,* NS23(4), 435–54.

Williams, Michael (1989) *Americans and their Forests.* Cambridge: Cambridge University Press.

Wolch, Jennifer and Emel, Jody (1998) *Animal Geographies.* London: Verso.

World Commission on Environment and Development (1987) *Our Common Future.* Oxford: Oxford University Press.

Worster, Donald (1977) *Nature's Economy.* San Francisco CA: Sierra Club.

Worster, Donald (1992) *Under Western Skies: Nature and History in the American West.* New York: Oxford University Press.

Zimmerer, Karl S. (1994) Human geography and the 'new ecology': the prospect and promise of integration. *Annals of the Association of American Geographers,* 84(1), 108–25.

III. Place

7

Place description, regional geography and area studies

the chorographic inheritance

HUGH CLOUT

This essay examines a form of study that for many generations was an accepted focus of geography but that has largely disappeared from current academic practice in many British universities. Fewer and fewer professional geographers would claim that spatial knowledge of one or several parts of the world, with all the linguistic expertise and cultural skills required, is the central part of their activity (Wooldridge, 1950; Mead, 1963, 1980; Farmer, 1973). Instead, most now focus on systematic analyses which, arguably, may be applicable anywhere on the globe but are usually rooted, exclusively and almost blindly, in the Anglo-American academic sphere and in literatures articulated in the English/American language. This retreat from area studies is all the more bewildering, since colleagues in other academic disciplines, media professionals, and members of the general public expect geographers to be informed about places and regions within and beyond their own home country. As television interviews, newspaper reports, and book reviews reveal only too explicitly, the regional experts of the twenty-first century are to be found in departments of anthropology, archaeology, history, and international affairs, rather than among academic geographers. Novelists, essayists, and journalists delight in writing about places but such fascination has diminished in geography departments.

The uncertain relationship between studying regions (or places) and focusing on themes (or systematic processes) has characterised geography since the ancient Greeks, with 'chorography' being the term employed to define the art of 'describing the parts of the Earth', to echo Strabo (Livingstone, 1992). The specific words may have differed and the precise arguments reconfigured across the centuries, but this ambiguous duality has appeared repeatedly over the past 2000 years (Darby, 1962). On the eve of the Second World War, Richard Hartshorne (1939) delivered his powerful defence of geography as 'areal differentiation', and Robert Dickinson (1976, 3) insisted that 'geography is a chorological science—that is, it is the regional science. Chorography implies description'. A few years later John Fraser Hart (1982, 1) asserted regional study to be 'the highest form of the geographer's art'. When surrounded by all the cultural and linguistic skills that convincing practice of area studies demands, a strong case may be made for agreeing with Fraser Hart's claim, but when the writing of regional geography is reduced to encyclopedic presentations of descriptive facts and figures, with little or no insight, direction, or passion, then the result is banal. In this guise, regional geography is, indeed, a trivial pursuit. Such mindless chorography is best abandoned to the academic archive or the computerised database, for retrieval whenever factual support is required to be added to an argument. Against this background, this essay seeks to trace the manifestation of regional studies in British geography during the twentieth century, paying attention to the institutional structures and some of the individuals involved, as well as the publications that appeared. A handful of these works were brilliant and memorable but the majority were obsessed with the encyclopedic and failed to excite the mind of even the most tolerant student.

The Oxford model

Geography had been taught in British universities since at least the sixteenth century, when students at Oxford were instructed about maps, globes, and discoveries (Gilbert, 1972; Scargill, 1999, 12). In the 1830s, Professor Alexander Maconochie delivered lectures on parts of Asia, Africa, and South America at the University of London (University College London—UCL: Ward, 1960, 464). Across the Channel, French defeat in the Franco-Prussian War of 1870 was attributed partly to superiority of Prussian numbers and military technique, but also to accurate knowledge of the terrain that would be traversed. This information was imparted at military academies in Berlin and other cities where Ratzel's geography was professed. Under the French Third Republic, geography was recognised as a useful subject to instruct pupils about nationhood and the fabled unity of France, and also to prepare military men for war. Schoolteachers needed to be trained to deliver such messages and institutes of geography were duly created in French universities, with staff drawing initially on German publications, maps, and atlases. In due course, geography was to appear as a university discipline in Britain, with the Royal Geographical Society (RGS) endowing a readership at Oxford in 1887 (and later at Cambridge: Coones, 1989; Stoddart, 1989; Johnston, this volume, Chapter 2). The post at Oxford was held by Halford Mackinder, who found his first audiences among historians; not until 1899 was a diploma in geography established.

The early post-holders at Cambridge were not directly influential, but F. H. H. Guillemard, who lectured only in 1888, went on to edit the slender 'Cambridge County Geographies' (published by Cambridge University Press) that were widely used in schools. In Scotland, George G. Chisholm pioneered the teaching of geography in Edinburgh, a quarter of a century after the Royal Scottish Geographical Society had been founded in 1884.

Not surprisingly, Mackinder and his colleagues based much of their teaching on work that appeared in Germany and especially in France, since French publications, authored by Paul Vidal de la Blache and his disciples or published in the *Annales de Géographie*, were more accessible linguistically. As in earlier times, geography at Oxford was conceived as being concerned with thematic issues, such as climatology and commerce, but also with knowledge of the continents and 'natural regions'. Particular emphasis was placed on area studies in textbooks written for schoolteachers by Mackinder, by his successor Andrew John Herbertson, and by Lionel Lyde, who was appointed to a chair of geography at UCL in 1903. Mackinder's *Britain and the British Seas*, which introduced the 'Regions of the World' series published by Heinemann, appeared in 1902, one year before the influential *Tableau de la Géographie de la France* (Vidal de la Blache, 1903). Most of its chapters were thematic, although metropolitan England, industrial England, Scotland, and Ireland were described individually.

In the years up to 1914, over 850 schoolteachers attended five summer schools in geography at Oxford (over 250 in 1910 alone) and large numbers of London teachers went to evening classes at UCL and other colleges of London University. By the time Herbertson died in 1915, 77 students had received the Oxford diploma in geography and a further 80 had obtained a certificate. Eva G. R. Taylor, Blanche Hosgood, and Charles Fawcett attended courses at Oxford and would duly take the messages of Mackinder and Herbertson to the London colleges. Although trained as historians at Oxford, Percy M. Roxby and Alan G. Ogilvie were enthused by Mackinder's lectures and would teach geography in Liverpool and Edinburgh universities (Freeman, 1967, 156–86). Herbert John Fleure was a natural scientist by training but he too attended Mackinder's classes and would take the regional tradition to Aberystwyth and Manchester (Dickinson, 1976, 44). Textbooks and wall maps played an essential role in propagating regional geography, with the publisher George Philip selling half a million copies of Mackinder's texts and the Clarendon Press in Oxford printing 1,400,000 copies of the textbooks written by Herbertson and his wife (Gilbert, 1972, 194). According to Balchin (1993, 13) it was, however, Lyde who 'dominated the textbook market at the beginning of the twentieth century', selling over 4 million copies of his works (Balchin, 1993, 100; Clout, 2003a).

The regional approach became firmly established in classrooms throughout Britain. However, there was one important difference between regional writing on the two sides of the Channel. The early regional monographs, written under the direction of Vidal de la Blache, were substantial doctoral theses, involving large quantities of fieldwork and rigorous archival research. Each had specific objectives and focus, as well as contributing to a mosaic of geographical knowledge that might eventually cover the whole of France. By contrast, British geography in the early twentieth century, even as it existed in universities, was a subject taught by schoolteachers in order to train the next generation of schoolteachers. British academic geographers sought to model their regional teaching on the *Tableau* (1903), but this was not a research monograph but rather a geographical introduction to a series dealing with the history of France before 1789. Copies of the *Tableau* would be used reverentially by staff and students in some British universities until mid-century; Professor H. C. Darby found a teaching set when he arrived at Liverpool in 1945. S. J. K. Baker, a pupil of Roxby's in the 1920s, read the book 'during his first term at Liverpool. Always afterwards the volume was handily placed upon his bookshelves, although he could cite it spontaneously' (McMaster, 1993, 263). He was not alone in this enthusiasm.

The concern of British geographers was with preparing textbooks for students and for pupils rather than undertaking research. Doctorates were rare at this stage, the few early holders, such as Charles Fawcett, Marion Newbigin, and Hilda Ormsby, achieving this title by submitting published works rather than by assembling a cohesive thesis. In 1927, the Scottish geographer Arthur Geddes, who was strongly influenced by the writings of Frédéric Le Play and the philosophy of Rabindranath Tagore, obtained a *doctorat d'état* from Montpellier for a regional study of part of south India supervised by Jules Sion (Farmer, 1983, 72). Not until 1931 was the first geography doctorate awarded at Cambridge, to Henry Clifford Darby (Steel, 1987, 119).

Place description in war and peace

During and immediately after the First World War, the Admiralty War Staff Intelligence Division prepared a series of regional handbooks dealing with selected parts of the world. The project was coordinated by Henry Dickson (professor of geography at Reading, and formerly Mackinder's assistant), with about 70 writers and over a dozen draughtsmen (Freeman, 1980, 103; Darby, 1983, 14). The authorship of most volumes was not indicated, though a handful of academic geographers were involved, notably Rudmose Brown, Howarth, Macfarlane, and Rishbeth. Each handbook was organised thematically but the range of topics and the emphasis placed upon each varied. By virtue of their strategic importance, much space was allocated to topography, roads, and railways. For example, the volume on *Macedonia and Surrounding Territories* (1916) devoted 360 of its 524 pages to roads, tracks, and railways, while that on *Bulgaria* (1917) allocated 364 of its 536 pages to these matters. By contrast, some handbooks were more rounded in their approach, such as those on *German East Africa* (1916), *Netherlands India (Dutch East Indies)* (1918), and *Alsace-Lorraine* (1919), which contained important discussions of physical, historical, and human geography.

The level of illustration varied considerably from volume to volume, presumably according to what was readily available at the Admiralty and in the RGS library and map room. Thus, *Macedonia and Surrounding Territories* contained detailed city maps (e.g. Salonica and Sofia) that identified ethnic quarters and important buildings, whereas *The Uganda Protectorate* (1920) had only a single fold-out map showing administrative areas, rivers, and towns. The handbook on *Finland* (1919) contained some fine maps and paid tribute to the work of the Geographical Society of Finland, its periodical *Fennia* (with many articles or summaries in German, French or English), and the second edition of the national atlas (1910), with its accompanying text in French. The volume on *Syria (including Palestine)* (1919) was not only large (723 pages) but also distinctive in its composition, with half its chapters devoted to specific regions of Palestine and an array of photographs being included in a rear pocket. Geographers, civil servants, and colonial administrators had collaborated to realise a scholarly kind of strategic regional geography in time of war, producing a total of 50 handbooks and 130 short reports. The conclusion of peace enabled some, particularly Ogilvie, to help draw up the new political map of the Balkans and other parts of Europe (imparting knowledge that would be of great value when another set of handbooks were compiled during the next world war: Miller and Watson, 1959; Steel, 1987, 4). The new challenge was to present geography as a viable academic subject when the imperatives of war were not pressing.

A clear opportunity was offered since some universities had already created geography departments (e.g. the London School of Economics (LSE) in 1895, UCL in 1903, Birkbeck College London in 1909, Liverpool in 1909, and Aberystwyth in 1917) and others were ready to do so. By 1928, two dozen departments would be in existence. The London Board of Studies in Geography was particularly powerful since it specified a common syllabus for teaching in the London colleges and also controlled programmes delivered in five provincial university colleges (Exeter, Hull, Leicester, Nottingham, Southampton) that followed the London external system (Garnett, 1983, 29; Steel, 1987, 91). After much debate, the Board chose to emphasise 'general regional geography', with five of the nine honours papers requiring a synthetic study of natural regions and their subdivisions. This formula was hardly surprising since many lecturers had been trained at Oxford, while others, such as Lyde

(UCL) and J. F. Unstead (Birkbeck), had been in close contact with Herbertson.

The Oxford–London connection had been strengthened in 1905 when Mackinder moved to the LSE, where he held a lectureship (and eventually a chair) in economic geography, and served as director during 1905–8 (Coones, 1989). After studying under Mackinder at the LSE and lecturing at Leeds, Rodwell Jones moved back to London in 1919 and would succeed Mackinder as professor in 1925 (Wise, 1980; Butlin, 1999). His elder sister, Hilda Ormsby (neé Jones), also lectured at the LSE, having assisted Mackinder at Oxford. The early world of academic geography was remarkably small and most of its members were closely interlinked. Alliances could be strong but, on the other hand, personal dislikes could be intense and intellectual conflicts devastating. By the 1920s, the driving force behind regional writing had moved from Oxford to the capital, where staff at the LSE and other colleges would play a notable role. As the number of students taking geography in schools and universities was set to grow, so publishers sought to commission authors who could cover components of the syllabuses. Schools and colleges around the Empire formed significant markets for English-language textbooks that could be modified for local use. Royalties from textbooks supplemented the meagre pay of university teachers.

In 1924 Rodwell Jones initiated the 'Advanced Geographies' series from the Methuen publishing house (located five minutes' walk from the LSE) with a volume on *North America* written with P. W. Bryan. (The book would run to ten editions and was still in use by sixth-formers in 1960.) The next volume was *South America* (1927) by Edward Shanahan, a lecturer in commerce at the LSE. In 1926, Dr L. Dudley Stamp, a young geologist from Kings College London and previously professor of geology and geography in Rangoon, was appointed reader in economic geography at the LSE. His Methuen textbook on *Asia* appeared in 1929 and would run to 12 editions. These pioneer volumes were followed by other regional texts by lecturers at London University or by graduates of the London system. Hilda Ormsby's *France* was published in 1931 (but the manuscript of her companion volume on Germany would later be destroyed in an air raid: Harrison Church, 1981, 96). Then followed *Southern Europe* (1932) by Marion Newbigin, the doyenne of Edinburgh geographers who held a London DSc, and *Africa* (1934) by Walter Fitzgerald from Manchester, who had imbibed the Oxford diet of regional geography from Roxby. Could Mackinder have been the original link with Methuen?

In the early 1930s Stamp collaborated with his young LSE colleague Stanley H. Beaver to produce *The British Isles* (1933) in the 'Geographies for Advanced Study' for Longman. Like the Methuen series, this collection contained thematic as well as regional texts and once again the 'London network' was fully activated. Wartime mobilisation and severe paper shortages would bring both these projects to a temporary halt in the early 1940s, but both would flourish after the Second World War, with Stamp, and increasingly Beaver, acting as academic brokers. The Methuen and Longman volumes had no pretensions to be other than textbooks. While comparable in bulk with the great regional monographs by French doctoral candidates, they were by no means comparable in terms of research. These books were, in effect, digests of material to assist lecturers and teachers to prepare classes, and to inform a minority of their students who managed to absorb their detailed factual contents. Ideas and concepts were certainly few and far between.

The meeting of the International Geographical Congress (IGC) at Cambridge in 1928 provided an opportunity for British geographers to present the regions of Britain in a single volume, both to foreign academics and to lecturers and undergraduates in the UK. Ironically, this had already been done by Albert Demangeon, who had travelled extensively in Britain and Ireland before and after the war, and published *Les Iles Britanniques* (1927a). He had read widely in English, German, and French; many of his own photographs appeared in the book. The text would be translated into English by E. D. Laborde but would not be published until 1939; Demangeon's 'home rule' rhetoric on Ireland was toned down and partly replaced by facts and figures. The book would go through many editions and, my colleagues tell me, was still in use in British universities in the early 1960s. *Great Britain: Essays in Regional Geography* (1928) was published by Cambridge University Press and edited by Ogilvie (reader at Edinburgh), whose background had been at Oxford and Berlin. After a general introduction and an essay on climate, 13 chapters discussed the component parts of Britain but excluded Ireland. These units differed from the *Provinces of England* (1919) proposed by Fawcett at the end of the war. The authors were academics from various parts of the country, and their writing reflected the methodological uncertainty of the fledgling discipline. Many had qualified in other subjects and were growing into geographers. Some favoured geology or history as guiding themes in their chapters, rather than displaying a convincing brand of regional geography. Darby (1983, 6) noted that the authors were 'often working under far from ideal conditions and frequently [were] lacking the full confidence of their colleagues in other subjects'.

A few chapters were based on their authors' original research but most were drawn from reading, experience, and local observation. Two essays are of particular interest, both being written by former Oxford students: Roxby's study of East Anglia expanded on his earlier research into agricultural geography and used historical sources to define the *pays* of the eastern counties; and Ogilvie's chapter on central Scotland emphasised exploitation of resources and how the cultural landscape was shaped, before examining component areas. Other notable essays included Fleure's discussion of Wales, which considered mounting problems of unemployment in the southern valleys. The chapter on Cumbria by F. J. Campbell emphasised rural decline, accurately anticipating that much land would be allocated to water gathering, forestry, and a national park. The short essay on the Fenlands by Frank Debenham (reader at Cambridge) is of contextual interest, since it was written just before his student H. C. Darby was to embark on a doctoral thesis to be titled 'The role of the Fenland in English history' (1931) (Darby, 2002, 2; Clout, 2004).

The challenges of the economically depressed 1930s, with rural decline and industrial collapse in much of Britain contrasting with concentration of population growth and conversion of farmland to urban uses in the southeast, opened the way for writing a new kind of area study that would be seen through the prism of a comprehensive Land Utilisation Survey. This gargantuan enquiry was orchestrated at the LSE by Dudley Stamp and his protégé E. Christie Willatts. The project received an initial grant from the Rockefeller Research Fund of the LSE in 1930 but would subsequently involve Stamp's own money (including royalties from his school textbooks) as well as a truly enormous amount of his energy (Wise, 1983, 45). It embraced two distinct parts: first, land use was mapped (and checked) on a field-by-field basis by schoolchildren, teachers, students, and lecturers in local universities, as well as by students and staff (past and present) of the LSE; then, 92 county reports were prepared to classify the evidence of the land use maps and sometimes to place them in their historical context. It was at this second phase that the LSE (and Joint School of Geography, with Kings College) network truly came into play. For example, Andrew C. O'Dell, an LSE graduate, former lecturer at Birkbeck College London and subsequently at Aberdeen, wrote the county reports for Roxburghshire, Fife, Orkney, and Shetland. This latter report developed from his previous research on Shetland, which had been encouraged by Eva Taylor at Birkbeck and by Stamp (O'Dell, 1939; Turnock, 1987, 112). Many other LSE graduates contributed to the completion of the project by drafting reports. Stamp himself wrote the reports for Surrey (which had to be surveyed and written up as a condition of the Rockefeller grant) and his own home county of Kent, as well as five others. Some reports were entirely the work of local geographers, such as K. M. Buchanan (Worcestershire), Mary Marshall (Oxfordshire), and K. C. Edwards (Nottinghamshire)—who was known to Stamp through the Le Play Society. Two dozen reports earned their authors masters' degrees and three were original enough to merit doctoral awards: Renfrewshire (Morag Moyes), Middlesex and London (E. C. Willatts), and Norfolk (John Mosby). The latter two incorporated substantial historical discussions and set the current land use into a wider time frame. Stamp would assemble the results of this vast enterprise in *The Land of Britain: Its Use and Misuse* (1948), presenting messages of great relevance to post-war planning.

The Land Utilisation Survey reports complemented several urban planning studies with notable contributions by geographers (Glass, 1948; West Midland Group, 1948; Daysh, 1949).

Geographers from the LSE, and especially Stamp and Beaver, were heavily involved in the Le Play Society, which had emerged from the Sociological Society following the inspiration of Patrick Geddes. It provided opportunities for small numbers of lecturers, teachers, and students to undertake fieldwork that would explore the interrelationship of place, folk, and work (Wise, 1975, 21). In the spirit of Le Play's research in nineteenth-century Europe, field parties investigated community life in parts of the continent during the 1930s, affording participants a range of experiences very different from industrial Britain in economic depression. Surveys were made in rural locations ranging from southwest France to Albania, Bulgaria, Romania, and Finland. Detailed studies were made of land use, house types, and household budgets, and in some instances short monographs were published. These were not substantial scholarly contributions but they had the merit of reporting original research and exemplify how small, well-focused field classes may be run on limited funds. With unavoidable interruptions, the surveys continued to be undertaken under the watchful eye of K. C. Edwards after the war (Steel, 1983, 115); Dudley Stamp managed to find time to lead seven study groups (Dickinson, 1976, 71).

In a totally different vein, Darby completed his regional study of the Fens in 1940, the two volumes building upon, but very different from, his doctorate. With their rigorous historical analysis and sharp focus on the cultural landscape they evoke the great French monographs, notably Roger Dion's *Le Val de Loire* (1934). The Fenland studies may also be compared with Demangeon's discussions of land reclamation and the resultant landscapes in the damp valleys and coastlands of *Picardie* (1905), and in the Dutch polderlands and Belgian Kempenland that he described in *Belgique, Pays Bas, Luxembourg* (1927b) in the *Géographie Universelle* initiated by Vidal de la Blache and Lucien Gallois (Clout, 2003b, c).

Sustained research overseas was rare among British geographers during the depressed interwar years and very few undertook enquiries in colonial territories. This dearth of overseas regional writing reflected the relatively weak position of geography in domestic and colonial universities and the absence of funds for lecturers to travel or postgraduates to be supported. W. R. Mead (1998, 131) remarked that had a generous uncle not bequeathed him money he might not have started graduate work on northern Europe. However, some courageous academics left for posts in the Empire; for example, in 1938 Dr Oskar Spate moved to the University of Rangoon and began his long involvement with South Asia. He was not to remain there long, having to flee to India when the Japanese overran Burma (Farmer, 1983, 73). His commitment to India would be reinforced by wartime service and by post-war opportunities.

Area studies serving the nation in wartime

The outbreak of the Second World War offered British geographers new opportunities to produce regional studies. The Inter-Services Topographical Department had been established in the School of Geography at Oxford, and subsequently a vast set of Naval Intelligence Handbooks was commissioned (Wilson, 1946; Freeman, 1980, 143; Balchin, 1987, 170). Unlike their predecessors these were to be written to a relatively standardised format. They were intended to provide a background for more intensive operational reports and to be of longer-term value for educating naval officers in peacetime (Darby, 1945). A total of 58 volumes would be printed under 31 titles, each covering one or several countries. Further manuscripts were written but never published, while many others were being prepared or planned when the war ended. The handbooks represent the largest volume of regional writing ever accomplished, greater than both the *Géographie Universelle* conceived by Vidal de la Blache and Gallois and its successor collection coordinated by Roger Brunet in the 1990s (Gosme, 1997; Clout and Gosme, 2003).

Two production centres were established, one in the School of Geography at Oxford (under the direction of Professor Kenneth Mason) and the other at the Scott Polar Institute at Cambridge (initially managed by James M. Wordie but effectively run by H. C. Darby, recalled from the Intelligence Corps). The volumes were to be scholarly in tone and illustrated with a large number of photographs and high-quality maps; Oxford and Cambridge had fine libraries and map collections to assist their writers. Operations were organised differently in the two cities. The Oxford centre called on the department's own staff working part time (especially R. P. Beckinsale, E. W. Gilbert, C. F. W. R. Gullick, and A. F. Martin, as well as Professor Mason), together with recent geography graduates (e.g. Robert and Eileen Steel, Mary Marshall), and various scientific experts, colonial administrators and Fellows of the Royal Geographical Society. For example, the four-volume handbook on Italy was compiled by Beckinsale, Gullick, and Martin (with assistance from non-geographers) and the two volumes on Turkey were the work of Mason and Marshall. The inexhaustible Brigadier H. St J. L. Winterbotham, formerly Director-General of the Ordnance Survey, wrote sections of the handbooks on Morocco, Belgian Congo, French Equatorial Africa, French West Africa, and Albania.

Academic geographers from the Cambridge department could not be made available in this way and so Darby needed to build up a core team of full-time writers. These included long-term associates (e.g. Elwyn Davies from Manchester) and recent Cambridge graduates (e.g. F. J. Monkhouse, F. W. Morgan). The distinguished artist Gwen Raverat, who prepared many block diagrams and line drawings, reinforced a team of cartographers. Darby insisted that cartography should be of the highest standard; his management of the drawing office would be repeated later at UCL when thousands of maps for his Domesday geographies and for his colleagues would be prepared. Other geographers worked on Cambridge projects during their university vacations (e.g. Alice Garnett on Yugoslavia, T. W. Freeman on China, to enhance the slow delivery of P. M. Roxby, who had conducted fieldwork in the Far East), or were commissioned to write specific sections of text (e.g. G. R. Crone, Margaret Davies, A. A. Miller, S. W. Wooldridge: see Table 7.1). Norman Pounds, who compiled material on the Greek islands, told me of the challenge of working fast and accurately since lives could depend on what was written. The work of the geographers was supplemented by contributions from historians and other specialists. In addition, Cambridge received academics evacuated from the LSE whose services could be called upon, notably the versatile geographer S. H. Beaver, who wrote on many parts of the world, and the distinguished anthropologist Raymond Firth, who contributed to four volumes on the Pacific Islands and to the handbook on Indo-China. Sir Raymond made it clear to me that members of the core team took responsibility for the final product, although contributors advised on maps and other illustrations as well as supplying text. He noted that his manuscripts were accepted without modification, since the core team had no experience of the Pacific.

The division of labour between the two centres was determined by the expertise of the personnel available, or thought likely to become available. Oxford tended to concentrate on the Mediterranean, Africa, and the Middle East, while Cambridge focused on Western Europe, the Balkans,

Table 7.1 Geographers involved in the production of four titles from the Cambridge centre

France (4 vols)	*Germany (4 vols)*	*Greece (3 vols)*	*Yugoslavia (3 vols)*
H. C. Darby (4)	E. Davies (1)	J. R. James (3)	I. L. Foster (3)
S. H. Beaver	F. J. Monkhouse (1)	S. H. Beaver	S. H. Beaver
A. A. Miller	E. J. Passant (1)	H. C. Darby	H. C. Darby
F. J. Monkhouse	H. C. Darby	E. Davies	A. Garnett
F. W. Morgan	F. W. Morgan	J. B. Mitchell	L. Latham
A. C. O'Dell	A. F. A. Mutton	F. J. Monkhouse	M. Mann
H. Ormsby	A. C. O'Dell	A. G. Ogilvie	F. J. Monkhouse
S. W. Wooldridge	H. Ormsby	N. J. G. Pounds	A. E. Moodie

Note: Authors are listed alphabetically, not according to the quantity of their contribution. Editors are indicated at the top of each list, with figures in brackets indicating the number of volumes edited. Non-geographers, and geographers making only minor contributions, are not shown.

East Asia, and the Pacific. For practical reasons some work was reallocated; for example, the handbook on Albania was redirected to Oxford where volumes on Italy had been prepared, rather than remaining at Cambridge where Greece and Yugoslavia were compiled. Working under enormous pressure and endeavouring to achieve the highest standards, the Oxford centre produced 17 titles (28 volumes) and the Cambridge centre 14 titles (30 volumes) by the time the series was halted. The handbooks included over 5,000 maps and diagrams and almost 6,500 photographs. The respective university presses published them, with print runs averaging 3,000 but ranging from 1,750 for Corsica to 4,125 for France (Gosme, 1997). Manuscripts on Thailand (Stamp), Argentina (Rodwell Jones), Brazil (R. H. Kinvig), and other sections of the world were not put into production by the end of the war. Rather than being archived, the manuscript on Czechoslovakia was returned to its author, Harriet Wanklyn, who would produce a monograph on the country a decade later. Further volumes on Finland (with which the name of the young researcher W. R. Mead had been associated), Russia, Japan, and many other countries were contemplated.

By virtue of their nature as Naval Intelligence documents, these vast, scholarly compendia, embracing detailed discussions of physical geography, ports, railways, trade, and socio-economic history, remained restricted for use by members of the armed forces until ten years after the war. Then they were remaindered for acquisition by universities, local libraries, academics, and members of the public. Undoubtedly they assisted geography lecturers who had to prepare regional courses in the 1950s and were sources of inspiration for textbook writers. Books by Monkhouse (1959) on Western Europe, Alice Mutton (1961) on Central Europe, A. E. F. Moodie (1945) on the Italo-Yugoslav boundary, F. W. Morgan (1952) on European ports and harbours, and O'Dell (1956) on railways doubtless reflected the reading that these scholars had done to contribute to the handbooks. Frank Monkhouse's monograph on *The Belgian Kempenland* (1949), in a short series edited by Darby, may well have originated in reading for the handbook on Belgium. It is not inconceivable that *Maps and Diagrams*, co-authored by Monkhouse and H. R. Wilkinson (1952) when they were in Darby's department in Liverpool, reflected lessons learned in the Cambridge drawing office. The textbook on *Southern Europe* (1975) by Robert Beckinsale and his wife and Edwards's various papers on Luxembourg (1961, 1969) are lineal descendants of the Naval Intelligence project. However, some research that was commissioned for the handbooks was not subsequently developed by their authors. For example, T. W. Freeman did not return to Chinese themes (although his master's degree at Leeds had dealt with migration in Asia), nor did Beaver write about the East Indies once the volumes were complete (Freeman, 1983, 92). These remarkable handbooks were suited for study in the library or for preparation for invasion; they were not intended for use in battle, but rather for strategic planning.

The Naval Intelligence project focused the minds of some British geographers on compiling desk studies of sections of the world. Others acquired unexpected and involuntary experience of distant parts of the globe through military service, sometimes enduring captivity as prisoners of war, and subsequently would turn their knowledge to regional writing. The activities of British geographers in wartime have been coordinated in a valuable report by Balchin (1987). In this fashion, B. H. Farmer went as a Royal Engineer surveyor to India, Ceylon, Singapore, Thailand, and the East Indies (Farmer, 1983, 73). Charles A. Fisher was also posted as a surveyor to the Far East, where he was to spend three-and-a-half years in a Japanese prisoner-of-war camp, serving on the notorious Burma–Siam railway. After fleeing from Burma to India, Spate worked with the Inter-Services Topographical Department in Ceylon, while W. B. Fisher used wartime experience in the Middle East to build on regional work started in the late 1930s under the direction of Demangeon and Jacques Weulersse at the Sorbonne. Research by R. J. Harrison Church (1943) on railways and development in West Africa, which had been initiated at the Sorbonne and later presented for a London (LSE) doctorate, was used to help compile the handbook on French West Africa (1944). Subsequently, it would prove invaluable in writing several textbooks on the region (Harrison Church, 1951, 1957, 1964).

Regional textbooks: the ossification of a paradigm?

With peace restored, publishing houses sought to expand and revise their lists of regional titles once paper shortages had come to an end. Whilst physical geography and cartography had developed their own thematic identities, human geography remained largely synonymous with regional studies in most universities. The main exception was Cambridge, where thematic aspects of human geography were already well identified. Only one of the six papers in Part I of the tripos was devoted to regional geography, and none of the following six papers in Part II (Darby, 1983, 24). The emphasis on teaching was systematic and would, in due course, be exemplified in a series of introductory textbooks published at mid-century by the English Universities Press. In addition to Darby's Fenland studies (1940a, 1940b), Harriet Wanklyn's books on *The Eastern Marchlands of Europe* (1941, with a preface by Ogilvie) and *Czechoslovakia* (1954) were the main contributions to regional writing at Cambridge. B. H. Farmer's important work on South Asia would appear later. By contrast, area studies remained very prominent in London, at the provincial university colleges (which taught the London external programme), at Oxford, and at colleges and universities in the Empire, as well as in the final years of pre-university training.

Geography editors, notably E. W. Parker and J. R. C. Yglesias at Longman and their counterparts at Methuen, Nelson, and other publishing houses, were keen to sign up new authors. As before, Stamp and Beaver used the London network to full advantage, calling upon at least four types of regional expertise. First, there were authors who had worked on the Naval Intelligence handbooks and had relevant material to hand, notably Monkhouse and Mutton. Secondly, geographers with military service overseas could contribute first-hand experience in regional writing, such as W. B. Fisher on *The Middle East* (1950) and C. A. Fisher on *South-East Asia* (1964). Charles Fisher would write passionately of his arrival in Malaya in 1941, which was a case of 'love at first sight' (p. vii). He declared:

> not even the ensuing three and a half years of prison camp life succeeded in appeasing my appetite for foreign travel ... And, indeed, while the somewhat unorthodox form of enforced nomadism, which between 1942 and 1945 took me into ... other parts of the region, imposed obvious constraints upon the pursuit of geographical research—though it is surprising how much field-work can be accomplished in the absence even of a stout pair of boots—it nevertheless afforded certain valuable compensations. (p. vii)

Spate's wartime experiences in South Asia were reinforced by intensive reading in India House, just across the Aldwych from the LSE, where he held a readership until moving to Canberra in 1954. In his text on *India and Pakistan* (1954) he expressed his gratitude to the Indian students whom he supervised in London for what they taught him about the realities of the sub-continent (p. xi). After having published on Spain, Ernest H. G. Dobby had gone to Raffles College, Singapore, just before the outbreak of war. Subsequently he ran intelligence units in India and the East Indies to monitor Japanese broadcasts and drew on that experience as he wrote his text on *Southeast Asia* (1950). In 1952 Paul Wheatley moved to Malaya from UCL and would research an erudite monograph entitled *The Golden Khersonese* (1961) on the historical geography of the peninsula.

James Wreford Watson from Edinburgh was based in Canada from 1939 to 1954 and used that long experience to write *North America* (1963). Several volumes coordinated by the Africanist Robert Steel, who had been trained at Oxford, worked on three handbooks (Algeria, Morocco, French West Africa) and then researched in the Gold Coast, provide a sample of this work (Steel and Fisher, 1956; Steel and Prothero, 1964). As well as providing the essential link between British geographers and academics in English-speaking African countries, Steel nurtured an important focus of work on the developing world at the University of Liverpool that would be carried forward by R. Mansell Prothero (1969) and others.

John H. Wellington, who had graduated from Cambridge in 1931, held the chair of geography at Cape Town and wrote a two-volume study of *Southern Africa* (1955). Monica Cole, who studied in London and then lectured at Cape Town and Witwatersrand (1947–51), incorporated her extensive fieldwork in *South Africa* (1961). K. M. Buchanan and J. C. Pugh (1953), and W. B. Morgan with J. C. Pugh (1969), wrote texts based on their research while at Nigerian universities. Complementary work was undertaken in the Sudan (Barbour, 1961) and in East Africa (W. T. W. Morgan, 1973). From his base in Hong Kong, T. R. Tregear (1965) wrote about the neighbouring giant of Mainland China. Many other regional geographies, of varying weight, rigour, and inspiration, were compiled by British academics on the basis of their teaching overseas, either while in post or after their return to the UK. Serving as a member of the Ceylon Land Commission from 1955 to 1958, B. H. Farmer (1957, 1974) produced important monographs on peasant colonisation on the island and in parts of India.

In addition to incorporating translations of regional texts by French geographers in order to expand their lists, the major British publishing houses turned to geographers with regional expertise in the British Isles or in Europe. Wilfred Smith's long-awaited *Economic Geography of Great Britain* (1949) explored recent conditions and their antecedents. T. W. Freeman prepared a regional text on *Ireland* (1950), based on over a decade teaching at Trinity College Dublin and extensive fieldwork (often by bicycle) that enabled him to report on sub-regions and local areas in considerable detail. This text was followed by a remarkable research monograph on *Pre-Famine Ireland* (1957). Emrys Bowen and colleagues at Aberystwyth wrote a text on *Wales* (1957, 1959) that combined systematic and regional approaches and built on his earlier (1941) book. O'Dell brought together his wide travelling experience in northern Europe to write *The Scandinavian World* (1958). James M. Houston, who had been Ogilvie's student in Edinburgh, drew on the Oxford tradition of work in southern Europe as well as on his doctoral research on Spain, to produce a vast tome on *The Western Mediterranean World* (1964). Closer to home, geographers wrote about *The Weald* (Wooldridge and Goldring, 1953) and *The Peak District* (Edwards, 1962) for the 'New Naturalist' series, which included geographical texts on Britain's scenery (Stamp, 1946), climate (Manley, 1952), and coasts (Steers, 1953).

Robert E. Dickinson, who had taught at UCL from 1928 to 1947, made abundant use of his detailed knowledge of Europe to compile several texts on Germany. These originated in the 1930s when he learned German to complement his French, travelled widely in Germany (including spending time with Walther Christaller), and researched in libraries in both Germany and France (Johnston, 2000, 2001). Early in the war he continued to lecture to UCL students evacuated to Aberystwyth, but was then brought into RAF Intelligence work at Oxford, with secondment as a special consultant to RAF Bomber Command and the Ministry of Home Security. He did not, however, contribute to the Naval Intelligence volumes on Germany that were compiled at Cambridge. *The German Lebensraum* (1943) was his first book-length statement on Germany and was followed by *The Regions of Germany* (1945), which was of value to the allies in defining the post-war *länder*. These books served as prototypes for his massive *Germany* (1953), which replaced the text drafted for Methuen by Hilda Ormsby that was lost in an air raid. He focused on interwar and earlier conditions, making only brief statements about bombing and reconstruction. *City Region and Regionalism* (1947) and *The West European City* (1951) reflected his close familiarity with the work of continental geographers and his extensive travels. They demonstrated two themes that he had explored in his doctoral thesis on East Anglia (1933), namely the interpretation of urban morphology and the recognition of urban spheres of influence. With their detailed knowledge, encyclopedic and repetitive tendencies, and lack of concluding message, Dickinson's enormous books exemplify the weaknesses as well as the strengths of traditional regional writing. Facts were marshalled and left to speak for themselves; many readers could not hear the message amidst the noise of information.

By contrast with these authors of standard texts, W. R. Mead and E. Estyn Evans produced decidedly different types of regional writing. After a master's degree on the geographical background to Finland's foreign trade in 1937 (supervised by Dr Ormsby at the LSE), Bill Mead started work on a doctorate dealing with northern Europe (Mead, 1983). Like others of his generation, his work was interrupted by the outbreak of war. He was stationed in Canada and was involved in setting up an RAF flying school. Incidentally, his nearest geographical neighbour was Wreford Watson. (He would turn all this experience to good advantage through subsequent teaching at UCL and a highly original textbook, which stressed interpretation of large-scale maps: Mead and Brown, 1962.) On the basis of extensive reading, his doctorate on 'The geographical background to community of interest among the north European peoples' was defended at the LSE in 1947 and provided a springboard for a lifetime that would be dedicated to studying northern Europe (Mead, 1947). A Rockefeller Scholarship enabled him to spend 1949 living among Karelian refugee farming families who had been resettled on 'cold farms' in Finland (Mead, 1953). During the 1950s he researched numerous Scandinavian themes on the basis of extensive fieldwork and contacts in universities throughout northern Europe (Mead, 1953). His *Economic Geography of the Scandinavian States and Finland* (1958) did not aspire to be comprehensive (or to compete with O'Dell's text) but, rather, focused imaginatively on past and present aspects of the economies of northern Europe. It was essential support for students at UCL in the 1960s who opted for his course on northern Europe. In 1963 Mead wrote about 'the adoption of other lands' by geographers. His corpus of work over six decades confirms his fascination with the people and places of northern Europe and the fact that he has been academically 'adopted' by several generations of Scandinavian colleagues (Mead, 1963, 1981, 1993).

Working in a different tradition, with close affinity to anthropology and archaeology, and drawing on much experience in the field, the Welsh geographer E. Estyn Evans produced a cluster of regional writings that examined rural life throughout Ireland (1947, 1957, 1973), in the Atlantic ends of other parts of Europe, and also in one much-loved area, the *Mourne Country* (1951) of South Down in Ulster. Here, too, was a passionate expression of regional study by a scholar who not only had made Northern Ireland his home but through his dedication and commitment was truly adopted by the Irish (Steel, 1987, 40). A fine volume, which includes a long essay by his widow Gwyneth, sets his major books in context and reprints some of his lesser-known essays (Graham, 1994; Evans, 1996).

Despite the very real intellectual limitations of the encyclopedic regional approach, during the 1950s academic publishers continued to launch new series that privileged area studies. The economic geography collection of the Bell publishing house (based just across the road from the LSE) was edited attentively by R. O. Buchanan, with volumes on New England (Estall, 1966), London (Martin, 1966), Yugoslavia (Hamilton, 1968), and East Africa (O'Connor, 1966) using the expertise of London geographers. The scope of this series was spatially wide but concentrated thematically on economic approaches. By contrast, a 12-volume collection from the Nelson publishing house and edited by W. Gordon East of Birkbeck College sought to cover all parts of the British Isles in a traditional way. In fact, the published monographs would vary according to the specialism of individual authors and the characteristics of the region under discussion. The first list of authors incorporated colleagues from London University as well as local experts, of whom some had London connections. The early volumes on *North England* (1960) by A. E. Smailes (of Queen Mary College and formerly UCL) and on *The East Midlands and the Peak* (1963) by G. H. Dury (formerly of Birkbeck College) were lavishly illustrated with maps and photographs. By contrast, *The Bristol Region* (1972) by Frank Walker (a former student of Roxby) was curiously ill balanced, with only one chapter on current conditions and just a handful of maps to illustrate a 410-page monograph. Only six volumes in the series materialised, since some contributors died and others failed to deliver. In any case, the market for lengthy regional texts on the British Isles was far from buoyant and Nelson decided to terminate the series. Many years later, an updated version of J. H. Johnson's contribution on Ireland (1994) would appear from a different publisher as a well-illustrated text concentrating solely on human geography.

National and international scientific events continued to provide geographers with opportunities to write about component parts of Britain. The British Association for the Advancement of Science published special handbooks on the occasion of their annual meetings in provincial cities. Very often these volumes were edited by local geographers who collaborated with colleagues to produce balanced and well-illustrated accounts. These did much to propagate geography among other scientists. Local draughtsmen were able to display their talents to good effect. The tradition was long established, with *A Scientific Survey of the Cambridge District* (1938), edited by Darby, being an impressive early example. Geographical writing occupied varying proportions of these handbooks, but four are particularly memorable: *Birmingham and its Regional Setting* (Wise, 1950), compiled in the immediate aftermath of the war; *Merseyside* (Smith, 1953); *Swansea and its Region* (Balchin, 1971); and *The Potteries* (Phillips, 1993). Each is dominated by the research and writing work of local geographers, and contains many fine maps.

The months preceding the 20th International Geographical Congress in London in 1964 gave rise to three major books, as well as specialist texts on London (Clayton, 1964; Coppock and Prince, 1964) and journal articles presenting the British Isles to the global community of geographers. *The British Isles* (Watson and Sissons, 1964) comprised 30 systematic chapters, while *Field Studies in the British Isles* (Steers, 1964) contained supporting essays for official excursions. *Great Britain: Regional Essays* (Mitchell, 1962) was the key text that involved a network of Cambridge geographers and local experts. Three introductory chapters dealt with relief, climate, and population, and the remaining 27 worked their way from the southwest peninsula to the Orkneys and Shetlands, exactly like the British Isles course that I attended at UCL in 1963–4. The book was published by Cambridge University Press, with a dozen essays being written by Cambridge staff or graduates. Apart from containing recent socio-economic information and being authored by geographers who had trained as such, the volume bore an uncanny resemblance to its predecessor of 1928 (Edwards, 1974, 3). Indeed, one author, R. H. Kinvig, contributed chapters to both volumes. The essays were worthy, straightforward and well illustrated, but they were not particularly memorable. Yet another opportunity to give area studies some appeal had been allowed to slip by.

Regrets: I have a few

And then, it all started to dry up. Within months of the 20th IGC, members of another generation of Cambridge geographers were preaching a new geography that had no place for traditional area studies. *Frontiers in*

Geographical Teaching (1965) and *Models in Geography* (1967), both edited by Richard Chorley and Peter Haggett, were launched by Methuen, whose directors decided to back a totally different paradigm than the regional approach exemplified in their existing geography list. Another publishing house, Edward Arnold, published *Locational Analysis in Human Geography* (Haggett, 1967) and *Network Analysis in Geography* (Haggett and Chorley, 1969), further reinforcing the message. Haggett's *Geography: a Modern Synthesis* (1972) was designed for undergraduates and was followed by a flurry of school textbooks by Cambridge disciples that conveyed this new geography of models, theories, nodes, and surfaces. The revolution was well and truly under way and would flow with varying speed through the British educational system. Regional analysis, regional science, and statistical recognition of regions and sub-regions figured strongly in the new paradigm, but these studies were far removed from forms of regional description practised earlier in the century.

In the 1970s, members of a whole generation of area studies experts retired from British universities, or were encouraged to take early retirement, but rarely were they replaced by young geographers with any overseas expertise or the kind of immersion in language, culture, and a foreign education system that credible work in area studies demands (C. A. Fisher, 1970). Charles Fisher's harsh experience as a prisoner of war offered an extreme version of an 'outsider' becoming an 'insider', a British scholar adopting another land, and an academic learning geography in the hardest of ways. To quote his telling words:

> In living for several years at a level little different from that of millions of Asian peasants, and making the best one could of an environment that often seemed to pose more problems than it afforded opportunities, I learned to some degree to look at South-east Asia from within rather than ... from without, and this is surely the beginning of geographical understanding. (Fisher, 1964, vii)

Of course, the regional paradigm did not disappear overnight. John House's volume entitled *France: An Applied Geography* appeared in 1978 and Tom Elkins produced a handsome monograph on *Berlin* (1988), but Janice Price, then geography editor at Methuen, told me that harsh commercial pressures meant that no other regional volumes would come from that publishing house. Nonetheless, British historical geographers continued (and indeed continue) to write high-quality research monographs set in discrete spatial settings. With their systematic emphasis and well-defined objectives, these area studies are totally different from the encyclopedic textbooks of the past. Academic Press published a range of important volumes on the historical geography of the Netherlands (Lambert, 1971), Yugoslavia (Carter, 1972), Trinidad (Newson, 1976), the Western Mediterranean (Delano-Smith, 1979), and Scandinavia

(Mead, 1981). Cambridge University Press has published an impressive suite of regional studies in its historical geography series, with monographs dealing with the West Indies (Watts, 1987), Poland (Carter, 1994), Siberia (Bassin, 1999), and many other regions. The monumental study by Alan Baker (1999) of peasant associations in part of the Loire valley evokes the best regional monographs written by French geographers. British historical geographers also made notable contributions to the attractive 'Making of the English Landscape' series, from Hodder and Stoughton, edited initially by the historian W. G. Hoskins and then by the geographer Roy Millward (for example, Brandon, 1974; Emery, 1974; Balchin, 1983). Adopting a similar approach, Robert and Monica Beckinsale drew on decades of personal research in their monograph of the countryside and market towns surrounding Oxford entitled *The English Heartland* (1980). The requirement of large publishing houses to publish only 'profitable' titles means that regional monographs have become difficult to place since large sales cannot be guaranteed (Clout, 1996). This problem, of course, affects academic historians as well as historical geographers.

Notwithstanding the flow of high-quality regional writing from British historical geographers, programmes of study in universities and schools were revised in the 1970s to reduce and, in some cases, to eliminate a regional dimension. As a result, recent cohorts of British undergraduates have come to appreciate their world as a manifestation of globalising systems articulated through writings in the English/American language, but they know little about the human dimension of survival strategies in different parts of the world that only geographers with lived experience of such places can convey. I remain to be convinced that the multiple brands of cultural geography that are flourishing at present can satisfy that requirement.

It would, of course, be untruthful to suggest that the vast regional catalogues compiled in the past were anything other than tedious, lacking in passion, and unable to hold the attention span of even the most hard-working undergraduate. The passing of what Bill Mead (1980, 295) has called 'the dinosauric age' of regional textbooks is surely a blessed relief that no one could deny. The large regional texts of the 1950s and early 1960s, published by Methuen, Longman, Bell, Nelson, the University of London Press, and others, which slumber dusty, unloved, and unwanted on the shelves of second-hand bookshops, remind us that a profound revolution has occurred. However, one of the outcomes has been that more than the 'bathwater' has been thrown away (Fisher, 1970): for example, work by Harry Wilkinson (1951) on the ethnographic mapping of Macedonia that would be forgotten for a quarter of a century now has a terrible relevance. Most professional geographers in Britain now seem unable to cope with credible area studies and appear largely unconcerned that a portion of their academic territory has been colonised by members of other disciplines, who are interviewed on television, write news briefings, and produce best-sellers. British geography undergraduates are ignorant of writings other than those in English and fail to appreciate that some geographers in other parts of the world perceive, represent, and analyse issues in differing ways from those in current favour in the UK. Whatever the cause, and regardless of whether it was academic suicide or academic murder, professional geography in Britain has virtually abandoned the practice of area studies and, in so doing, has rejected part of its birthright.

Not all has yet been lost; insightful regional volumes continue to be written (notably Beynon, Hudson, and Sadler, 1984, on Teesside, and Graham and colleagues, 1997, on Ireland). However, most contributions to regional studies at home and overseas by British geographers at the dawn of the new millennium are far from impressive. Often they are confined to short contributions in 'scissors-and-paste' edited textbooks. The role of geographers in area studies centres set up at UK universities following the Hayter Report (1961) and the Parry Report (1965) is sharply in decline (Farmer, 1973; Harrison Church, 1976). Regional expertise is being conveyed at these centres, and in other places, by anthropologists, linguists, 'historians of the present time', and a galaxy of others who recognise that spatial knowledge ('place') genuinely does matter in our globalising world. At UCL, Professor Emeritus Bill Mead, now beyond his mid-eighties, writes about the cultures, landscapes, and economic activities of Finland, Norway, and the other Scandinavian states with the same zeal that he demonstrated in much earlier years (Mead, 1993, 2002). But is there anyone to succeed him in that role? The answer is negative. Clearing the bookshelves and box files of my colleague Professor Frank Carter, who died in 2001, was depressing, since his amazing personal library could have been a gold mine for a young geographer with wide-ranging East European expertise. There is, however, no one in British geography with appropriate linguistic skills and interests that span historical geography, political studies, migration, environmental management, and conservation.

Some geography departments have retained a handful of experts on parts of the globe that are perceived to be distant and possibly even exotic in terms of traditional, but not internet, spatial relations. Their courses covering South Asia, the Pacific Rim, tropical Africa, southern Africa, Latin America, the post-Soviet states—and even the European Union and the British Isles—are generally well received by students, who appreciate grounding their knowledge in specific regions and places (Clout, 1989). Such modules are not encyclopedic but focus discussion around an appropriate theme, be it urbanisation, identity, poverty, exclusion, or whatever, within a specific spatial context, in full recognition of the ever-increasing manifestation of globalisation. However, my annual rounds as visiting examiner convince me that area studies experts, even in their thematic guise, are an endangered species. Their confidence has not been enhanced by the complete omission of an area studies dimension from Research Assessment Exercises. By contrast, research on Latin America still flourishes in US geography departments, and French geographers continue to undertake important regional work, with young researchers using their fluent English to work in many parts of Africa and South Asia. As I conclude this essay, I cannot escape the feeling that I am raking through the dying embers of what was once an essential part of our discipline and one that is now in desperate need of revitalisation. Globalisation, the standardisation of information through the medium of English on the internet and on satellite television, and the tedious uniformity of shopping experiences in many parts of the world have, of course, triggered a counter-reaction that values, privileges, and even invents 'authenticity'. If something of this mindset can be allowed to permeate into educational circles, then area studies may still have a chance in British academic geography. One can still hope that a new flame may burn brightly in the years ahead so that future generations of geographers may appreciate that they are citizens of the world and not only of the UK.

Acknowledgements

I wish to thank Bill Mead, Hugh Prince, and Michael Wise for their advice. Lady Eva Darby kindly supplied me with an unpublished report by Sir Clifford Darby. While at UCL, Cyril Gosme investigated the Naval Intelligence Handbooks and prompted me to research the social networks involved in their production. My thanks go to Dr Iain Stevenson for advice on how publishers shaped the geographical agenda, and to former members of UCL for recollections of decades past when area studies held sway. I dedicate this essay to the memory of Professor Frank Carter (1938–2001), whose geographical knowledge of East Central Europe was unsurpassed.

References

Balchin, W. G. V., ed. (1971) *Swansea and its Region*. Swansea: University College Swansea.

Balchin, W. G. V. (1983) *The Cornish Landscape*. London: Hodder and Stoughton.

Balchin, W. G. V. (1987) United Kingdom geographers in the Second World War. *Geographical Journal*, 153, 159–80.

Balchin, W. G. V. (1993) *The Geographical Association: The First Hundred Years*. Sheffield: Geographical Association.

Baker, A. R. H. (1999) *Fraternity among the French Peasantry: Sociability and Voluntary Association in the Loire Valley, 1815–1914*. Cambridge: Cambridge University Press.

Barbour, M. B. (1961) *The Republic of the Sudan*. London: University of London Press.
Bassin, M. (1999) *Imperial Visions: Nationalist Imagination and Geographical Expansion in the Russian Far East, 1840–1865*. Cambridge: Cambridge University Press.
Beckinsale, M. and R. (1975) *Southern Europe: The Mediterranean and Alpine Lands*. London: University of London Press.
Beckinsale, R. and M. (1980) *The English Heartland*. London: Duckworth.
Beynon, H., Hudson, R., and Sadler, D. (1994) *A Place called Teesside: A Locality in a Global Economy*. Edinburgh: University of Edinburgh Press.
Bowen, E. G. (1941) *Wales: A Study in Geography and History*. Cardiff: University of Wales Press.
Bowen, E. G., ed. (1957) *Wales*. London: Methuen.
Bowen, E. G. (1959) Le Pays de Galles. *Transactions, Institute of British Geographers*, 26, 1–23.
Brandon, P. (1974) *The Sussex Landscape*. London: Hodder and Stoughton.
Buchanan, K. M. and Pugh, J. C. (1953) *Land and People in Nigeria*. London: University of London Press.
Butlin, R. A. (1999) Geography at Leeds: the early days. Unpublished manuscript.
Carter, F. W. (1972) *Dubrovnik (Ragusa): A Classic City State*. London: Academic Press.
Carter, F. W. (1994) *Trade and Urban Development in Poland: An Economic Geography of Cracow, From its Origins to 1795*. Cambridge: Cambridge University Press.
Chorley, R. J. and Haggett, P., eds (1965) *Frontiers in Geographical Teaching*. London: Methuen.
Chorley, R. J. and Haggett, P., eds (1967) *Models in Geography*. London: Methuen.
Clayton, R., ed. (1964) *The Geography of Greater London*. London: George Philip.
Clout, H. (1989) Regional geography in the UK: a trend report. In L. J. Paul, ed., *Post-War Development of Regional Geography*. Utrecht: Netherlands Geographical Studies, 86, 25–41.
Clout, H. (1996) *After the Ruins: Restoring the Countryside of Northern France After the Great War*. Exeter: University of Exeter Press.
Clout, H. (2003a) *Geography at UCL: A Brief History*. London: University College London.
Clout, H. (2003b) In the shadow of Vidal de la Blache: Albert Demangeon and the social dynamics of French geography in the early twentieth century. *Journal of Historical Geography*, 29 (in press).
Clout, H. (2003c) Albert Demangeon, 1872–1940: a life in geography. *Scottish Geographical Journal*, 119 (in press).
Clout, H. (2004) French influences on British historical geography: the teaching and writing of H. C. Darby. In P. Boulanger and J.-R. Trocher, eds, *Où en est la Géographie Historique?* Paris: L'Harmattan (in press).
Clout, H. and Gosme, C. (2003) The Naval Intelligence Handbooks: a monument in geographical writing. *Progress in Human Geography*, 27, 153–73.
Cole, M. (1961) *South Africa*. London: Methuen.
Coones, P. (1989) The centenary of the Mackinder Readership at Oxford. *Geographical Journal*, 155, 13–22.
Coppock, J. T. and Prince, H. C., eds (1964) *Greater London*. London: Faber.

Darby, H. C. (1931) The role of the Fenland in English history. Unpublished PhD thesis, University of Cambridge.

Darby, H. C., ed. (1938) *A Scientific Survey of the Cambridge District*. London: British Association for the Advancement of Science.

Darby, H. C. (1940a) *The Medieval Fenland*. Cambridge: Cambridge University Press.

Darby, H. C. (1940b) *The Draining of the Fens*. Cambridge: Cambridge University Press.

Darby, H. C. (1945) Outline History of the Work of NID5, Cambridge. 1940–1945. Unpublished manuscript.

Darby, H. C. (1962) The problem of geographical description. *Transactions, Institute of British Geographers*, 30, 1–14.

Darby, H. C. (1983) Academic geography in Britain, 1918–1946. *Transactions, Institute of British Geographers*, NS8, 14–26.

Darby, H. C. (2002) *The Relations of History and Geography*. Exeter: University of Exeter Press.

Daysh, G. H. J. (1949) *Studies in Regional Planning*. London: George Philip.

Delano-Smith, C. (1979) *Western Mediterranean Europe: An Historical Geography of Italy, Spain and Southern France since the Neolithic*. London: Academic Press.

Demangeon, A. (1905) *La Picardie et les régions voisines*. Paris: Armand Colin.

Demangeon, A. (1927a) *Les Iles Britanniques*. Paris: Armand Colin.

Demangeon, A. (1927b) *Belgique, Pays-Bas, Luxembourg*. Paris: Armand Colin.

Demangeon, A. (1939) *The British Isles*, trans. E. D. Laborde. London: Heinemann.

Dickinson, R. E. (1933) The distribution of urban settlements in East Anglia. Unpublished PhD thesis, University of London.

Dickinson, R. E. (1943) *The German Lebensraum*. Harmondsworth: Penguin.

Dickinson, R. E. (1945) *The Regions of Germany*. London: Kegan Paul.

Dickinson, R. E. (1947) *City, Region and Regionalism*. London: Routledge and Kegan Paul.

Dickinson, R. E. (1951) *The West European City*. London: Routledge and Kegan Paul.

Dickinson, R. E. (1953) *Germany*. London: Methuen.

Dickinson, R. E. (1976) *Regional Concept: The Anglo-American Leaders*. London: Routledge and Kegan Paul.

Dion, R. (1934) *Le Val de Loire*. Tours: Arrault.

Dobby, E. H. G. (1950) *Southeast Asia*. London: University of London Press.

Dury, G. H. (1963) *The East Midlands and the Peak*. London: Nelson.

Edwards, K. C. (1961) The historical geography of the Luxembourg iron and steel industry. *Transactions, Institute of British Geographers*, 29, 1–16.

Edwards, K. C. (1962) *The Peak District*. London: Collins.

Edwards, K. C. (1969) Luxembourg: how small can a nation be? In J. W. House, ed., *Northern Geographical Essays in Honour of G. H. J. Daysh*. Newcastle: University of Newcastle-upon-Tyne, 256–67.

Edwards, K. C. (1974) Sixty years after Herbertson: the advance of geography as a spatial science. *Geography*, 59, 1–9.

Elkins, T. H. (1988) *Berlin: The Spatial Structure of a Divided City*. London: Methuen.

Emery, F. V. (1974) *The Oxfordshire Landscape*. London: Hodder and Stoughton.

Estall, R. C. (1966) *New England: A Study in Industrial Adjustment*. London: Bell.
Evans, E. E. (1947) *Irish Heritage: The Landscape, the People and their Work*. Dundalk: Dundalgan Press.
Evans, E. E. (1951) *Mourne Country: Landscape and Life in South Down*. Dundalk: Dundalgan Press.
Evans, E. E. (1957) *Irish Folk Ways*. London: Routledge and Kegan Paul.
Evans, E. E. (1973) *The Personality of Ireland: Habitat, Heritage and History*. Cambridge: Cambridge University Press.
Evans, E. E. (1996) *Ireland and the Atlantic Heritage: Selected Writings*. Dublin: Lilliput Press.
Farmer, B. H. (1957) *Pioneer Peasant Colonisation in Ceylon*. Oxford: Oxford University Press.
Farmer, B. H. (1973) Geography, area studies and the study of area. *Transactions, Institute of British Geographers*, 60, 1–15.
Farmer, B. H. (1974) *Agricultural Colonisation in India since Independence*. Oxford: Oxford University Press.
Farmer, B. H. (1983) British geographers overseas, 1933–1983. *Transactions, Institute of British Geographers*, NS8, 70–9.
Fawcett, C. B. (1919) *The Provinces of England*. London: Williams and Norgate.
Fisher, C. A. (1964) *South-East Asia*. London: Methuen.
Fisher, C. A. (1970) Whither regional geography? *Geography*, 55, 373–89.
Fisher, W. B. (1950) *The Middle East*. London: Methuen.
Fitzgerald, W. (1934) *Africa*. London: Methuen.
Freeman, T. W. (1950) *Ireland*. London: Methuen.
Freeman, T. W. (1957) *Pre-Famine Ireland*. Manchester: Manchester University Press.
Freeman, T. W. (1967) *The Geographer's Craft*. Manchester: Manchester University Press.
Freeman, T. W. (1980) *A History of Modern British Geography*. Harlow: Longman.
Freeman, T. W. (1983) A geographer's way. In A. Buttimer, ed., *The Practice of Geography*. Harlow: Longman, 90–102.
Garnett, A. (1983) IBG: the formative years: some reflections. *Transactions, Institute of British Geographers*, NS8, 27–35.
Geddes, A. (1927) *Au Pays de Tagore*. Paris: Armand Colin.
Gilbert, E. W. (1972) *British Pioneers in Geography*. Newton Abbott: David and Charles.
Glass, R. (1948) *The Social Background of a Plan: A Study of Middlesbrough*. London: Routledge and Kegan Paul.
Gosme, C. (1997) Les Geographical Handbooks de la NID5. Unpublished *mémoire de maîtrise en géographie*, Université de Paris IV.
Graham, B. (1994) The search for the common ground: Estyn Evans' Ireland. *Transactions, Institute of British Geographers*, NS19, 183–201.
Graham, B., ed. (1997) *In Search of Ireland: A Cultural Geography*. London: Routledge.
Haggett, P. (1967) *Locational Analysis in Human Geography*. London: Edward Arnold.
Haggett, P. (1972) *Geography: a Modern Synthesis*. New York: Harper and Row.
Hamilton, F. E. I. (1968) *Yugoslavia: Problems of Economic Activity*. London: Bell.
Harrison Church, R. J. (1943) The railways of West Africa: a geographical and historical analysis. Unpublished PhD thesis, University of London.

Harrison Church, R. J. (1951) *Modern Colonisation*. London: Hutchinson.
Harrison Church, R. J. (1957) *West Africa*. London: Methuen.
Harrison Church, R. J. (1964) *Africa and the Islands*. Harlow: Longman.
Harrison Church, R. J. (1976) British geographical research overseas. In J. I. Clarke and P. Pinchemel, eds, *Human Geography in Britain and France*. London: SSRC and IBG, 54–8.
Harrison Church, R. J. (1981) Hilda Ormsby 1877–1973. *Geographers: Biobibliographical Studies*, 5, 95–7.
Hart, J. F. (1982) The highest form of the geographer's art. *Annals of the Association of American Geographers*, 72, 1–29.
Hartshorne, R. (1939) *The Nature of Geography*. Lancaster, PA: Association of American Geographers.
House, J. W. (1978) *France: An Applied Geography*. London: Methuen.
Houston, J. M. (1964) *The Western Mediterranean World*. London: Longman.
Johnson, J. H. (1994) *The Human Geography of Ireland*. Chichester: Wiley.
Johnston, R. J. (2000) City-regions and a federal Europe: Robert Dickinson and post-World War II reconstruction. *Geopolitics*, 5, 153–76.
Johnston, R. J. (2001) Robert E. Dickinson and the growth of urban geography: a re-evaluation. *Urban Geography*, 22, 702–36.
Jones, L. R. and Bryan, P. W. (1924) *North America*. London: Methuen.
Lambert, A. M. (1971) *The Making of the Dutch Landscape: An Historical Geography of the Netherlands*. London: Academic Press.
Livingstone, D. N. (1992) *The Geographical Tradition*. Oxford: Blackwell.
Mackinder, H. J. (1902) *Britain and the British Seas*. London: Heinemann.
Manley, G. (1952) *Climate and the British Scene*. London: Collins.
Martin, J. E. (1966) *Greater London: An Industrial Geography*. London: Bell.
McMaster, D. N. (1993) Samuel John Kenneth Baker, 1907–1992. *Transactions, Institute of British Geographers*, NS18, 263–6.
Mead, W. R. (1947) The geographical background to community of interest among the North European peoples. Unpublished PhD thesis, University of London.
Mead, W. R. (1953) *Farming in Finland*. London: Athlone Press.
Mead, W. R. (1958) *An Economic Geography of the Scandinavian States and Finland*. London: University of London Press.
Mead, W. R. (1963) The adoption of other lands: experiences in a Finnish context. *Geography*, 48, 241–54.
Mead, W. R. (1980) Regional geography. In E. H. Brown, ed., *Geography: Yesterday and Tomorrow*. Oxford: Oxford University Press, 292–302.
Mead, W. R. (1981) *An Historical Geography of Scandinavia*. London: Academic Press.
Mead, W. R. (1983) Autobiographical reflections in a geographical context. In A. Buttimer, ed., *The Practice of Geography*. Harlow: Longman, 44–61.
Mead, W. R. (1993) *An Experience of Finland*. London: Hurst.
Mead, W. R. (1998) All chance, direction. *Acta Universitatis Upsaliensis*, C63, 131–40.
Mead, W. R. (2002) *A Celebration of Norway*. London: Hurst.
Mead, W. R. and Brown, E. H. (1962) *The United States and Canada: A Geographical Study of Regional Problems*. London: Hutchinson.
Miller, R. and Watson, J. W., eds (1959) *Geographical Essays in Memory of Alan G. Ogilvie*. London: Nelson.

Mitchell, J. B., ed. (1962) *Great Britain: Geographical Essays*. Cambridge: Cambridge University Press.
Monkhouse, F. J. (1949) *The Belgian Kempenland*. Liverpool: University of Liverpool Press.
Monkhouse, F. J. (1959) *A Regional Geography of Western Europe*. Harlow: Longman.
Monkhouse, F. J. and Wilkinson, H. R. (1952) *Maps and Diagrams*. London: Methuen.
Moodie, A. E. F. (1945) *The Italo-Yugoslav Border Region*. London: George Philip.
Morgan, F. W. (1952) *Ports and Harbours*. London: Hutchinson.
Morgan, W. B. and Pugh, J. C. (1969) *West Africa*. London: Methuen.
Morgan, W. T. W. (1973) *East Africa*. Harlow: Longman.
Mutton, A. F. A. (1961) *Central Europe*. Harlow: Longman.
Newbigin, M. I. (1932) *Southern Europe*. London: Methuen.
Newson, L. A. (1976) *Aboriginal and Spanish Colonial Trinidad*. London: Academic Press.
O'Connor, A. M. (1966) *An Economic Geography of East Africa*. London: Bell.
O'Dell, A. C. (1939) *The Historical Geography of the Shetland Islands*. Lerwick: Manson.
O'Dell, A. C. (1956) *Railways and Geography*. London: Hutchinson.
O'Dell, A. C. (1958) *The Scandinavian World*. Harlow: Longman.
Ogilvie, A. G., ed. (1928) *Great Britain: Essays in Regional Geography*. Cambridge: Cambridge University Press.
Ormsby, H. (1931) *France*. London: Methuen.
Phillips, A. D. M., ed. (1993) *The Potteries*. Stroud: Sutton.
Prothero, R. M., ed. (1969) *A Geography of Africa*. London: Routledge and Kegan Paul.
Scargill, D. I. (1999) Geography's centenary. *Oxford Magazine*, Hilary Term, 8th Week, 12–13.
Shanahan, E. W. (1927) *South America*. Methuen: London.
Smailes, A. E. (1960) *North England*. London: Nelson.
Smith, W. (1949) *An Economic Geography of Great Britain*. London: Methuen.
Smith, W., ed. (1953) *Merseyside: A Scientific Survey*. Liverpool: University of Liverpool Press.
Spate, O. K. (1954) *India and Pakistan*. London: Methuen.
Stamp, L. D. (1929) *Asia*. London: Methuen.
Stamp, L. D. (1946) *Britain's Structure and Scenery*. London: Collins.
Stamp, L. D. (1948) *The Land of Britain: Its Use and Misuse*. Harlow: Longman.
Stamp, L. D. and Beaver, S. H. (1933) *The British Isles*. Harlow: Longman.
Steel, R. W. (1983) Kenneth Charles Edwards, 1904–1982. *Transactions, Institute of British Geographers*, NS8, 115–19.
Steel, R. W., ed. (1987) *British Geography, 1918–1945*. Cambridge: Cambridge University Press.
Steel, R W. and Fisher, C. A., eds (1956) *Geographical Essays on British Tropical Lands*. London: George Philip.
Steel, R. W. and Prothero, R. M., eds (1964) *Geographers and the Tropics: Liverpool Essays*. Harlow: Longman.
Steers, J. A. (1953) *The Sea Coast*. London: Collins.
Steers, J. A., ed. (1964) *Field Studies in the British Isles*. London: Nelson.

Stoddart, D. R. (1989) A hundred years of geography at Cambridge. *Geographical Journal*, 155, 24–32.

Tregear, T. R. (1965) *A Geography of China*. London: University of London Press.

Turnock, D. (1987) Andrew Charles O'Dell, 1909–1966. *Geographers: Bio-bibliographical Studies*, 11, 111–22.

Vidal de la Blache, P. (1903) *Le Tableau de la Géographie de la France*. Paris: Hachette.

Walker, F. W. (1972) *The Bristol Region*. London: Nelson.

Wanklyn, H. G. (1941) *The Eastern Marchlands of Europe*. London: George Philip.

Wanklyn, H. G. (1954) *Czechoslovakia*. London: George Philip.

Ward, R. G. (1960) Captain Alexander Maconochie, 1787–1860. *Geographical Journal*, 126, 459–68.

Watson, J. W. (1963) *North America*. Harlow: Longman.

Watson, J. W. and Sissons, J. B., eds (1964) *The British Isles: A Systematic Geography*. London: Nelson.

Watts, D. (1987) *The West Indies: Patterns of Development, Culture and Environmental Change Since 1492*. Cambridge: Cambridge University Press.

Wellington, J. H. (1955) *Southern Africa*. Cambridge: Cambridge University Press (2 vols).

West Midland Group (1948) *Conurbation: A Planning Scheme for Birmingham and the Black Country*. London: Architectural Press.

Wheatley, P. (1961) *The Golden Khersonese*. Kuala Lumpur: University of Malaya Press.

Wilkinson, H. R. (1951) *Maps and Politics: A Review of the Ethnographic Cartography of Macedonia*. Liverpool: University of Liverpool Press.

Wilson, L. S. (1946) Some observations on wartime geography in England. *Geographical Review*, 36, 597–612.

Wise, M. J., ed. (1950) *Birmingham and its Regional Setting: A Scientific Survey*. Birmingham: British Association.

Wise, M. J. (1975) S. H. Beaver: an appreciation. In A. D. M. Phillips and B. J. Turton, eds, *Environment, Man and Economic Change*. Harlow: Longman, 9–27.

Wise, M. J. (1980) Llewellyn Rodwell Jones 1881–1947. *Geographers: Bio-bibliographical Studies*, 4, 49–51.

Wise, M. J. (1983) Three founder members of the IBG: R. Ogilvie Buchanan, Sir Dudley Stamp, S. W. Wooldridge: a personal tribute. *Transactions, Institute of British Geographers*, NS8, 41–54.

Wooldridge, S. W. (1950) Reflections on regional geography in teaching and research. *Transactions, Institute of British Geographers*, 16, 1–11.

Wooldridge, S. W. and Goldring, F. (1953) *The Weald*. London: Collins.

8

The passion of place

DOREEN MASSEY
NIGEL THRIFT

Introduction

We all know that it exists. That feeling of being in a place. That feeling of something happening there above and beyond the mere rush of existence. But, just like consciousness, that feeling proves a remarkably elusive thing to describe. In some ways, this is an embarrassing admission for two geographers to have to make. After all, it could be argued that place is their stock in trade. But, whether we think of blow-by blow descriptions of place laden with statistics or humanist expositions of place based on a Thoreauian sublime, there is no doubt that geographers have struggled to know what they are meant to know.

No wonder that place has long been a key element in geographical thought and writing. Along with 'region' it has been a core conceptual focus of what geography, or certainly human geography, has been thought to be about. In some ways, indeed, it is hard to separate region from place or place from region. The concerns of German regional geographers, of the French Vidalian School, and of the cultural geography pioneered by Carl Sauer in the United States, were both 'regional' and concerned issues that are central to a focus on 'place'. Indeed, in principle, 'place' might be thought of as the generic term for a spatial unit, location or territory. And yet, in geographical discourse, there is a set of understandings, often implicit, that are evoked when the term 'place' is mobilised. Moreover, these understandings have given rise to, or been the lightning-conductors for, a number of long-running debates that have been important in the definition and nature of geography as a discipline. Implicitly 'places' are often imagined in various ways: as at a smaller spatial scale than regions; as being loci of spatial specificity; as being humanly meaningful loci of spatial specificity; as, in their character and in their study, entailing a synthesis of elements through from the economic to the geological and geomorphological, to the cultural and imaginative.

Imagined in this way, 'places' as objects of conceptualisation and of research raise some crucial issues that have long been the concern of geographers; indeed at times it has been these concerns that have marked off the specificity of geography as a discipline. Thus:

- there is the issue of spatial variation, the conceptualisation of space, and the passivity or influence of the spatial realm;
- there is the thorny 'problem' of specificity and uniqueness, of the significance of these and of how (indeed whether) they can be 'scientifically' analysed;
- there are issues around the conceptualisation of 'identity'; and
- there are the problems and possibilities of geography's supposed character as a synthesising discipline.

All these issues not only go to the roots of geography as a discipline but also relate to debates within a wider academic field. In addressing conundrums such as these geography is entangled with other debates—in the natural sciences, in the social sciences, in the humanities and in philosophy. And within the discipline of geography, debate over place has shifted with the rhythms of these wider debates, sometimes reflecting them, at other times forcing the pace.

Certainly in recent years these debates have become more intense. The stakes have been upped. There is more to play for because that something called place has become so central to modern life even as its meaning and practice have become more and more diverse. In one way, this is because (in the 'First World' at least) the most traditional sense of place—as a location for natural communion—is being threatened as the bulldozers move in. In another way, it is because a sense of place framed as something 'local' (whether this means as the supposed force of face-to-face communication or the range of everyday dealings) has been disrupted by a range of different kinds of movements that link what were often regarded as discrete and contained places in unexpected combinations. In one other way it is because place has been quite literally refigured. Every place comes to us bound up in a matrix of cultural descriptions laid down over many centuries and yet always available to be reworked again—an almost endless stream of associations able to be activated by the moving and still images of photography, film, and television, by the textual incarnations to be found in books and magazines, and even by our own body movements, which expect to encounter particular kinds of places and are pre-programmed to move in quite different ways in certain places as opposed to others. For all of us, as never before on quite the same scale, place now comes pre-edited in myriad ways. And then there is one more way in which place has become pivotal. For whatever reason, place has become one of the key means by which the social sciences and humanities are attempting to lever open old ways of proceeding and tell new stories about the world, whether as a vast space of flows in which place is gradually being erased or as the sensuous rediscovery of the pleasures of the specific in which place is being rediscovered as something in-between (see e.g. Thrift, 1996, 1999). So we live in exciting times and what we most want to convey in this chapter is some flavour of these times and why place has now become so central to the concerns of so many. But though place may no longer be the exclusive disciplinary badge of what it is to be a geographer, that hardly makes it less central to the discipline. Indeed, we would argue that it makes it more central.

The first part of the chapter therefore recounts a history of the place of place in British geography. In this section, we want to show how the notion of place was progressively narrowed as a result of a belief in technical solutions but was then progressively widened out again as a concern with process has questioned how place, so often conceived as static and unchanging, can be thought of as in a continual state of becoming. The second part of the chapter then consists of our own account of place, an account that builds on this widening history (not least because we have been a part of it) in order to show why place will keep on coming back to haunt the discipline's concerns.

The history of place

Until the 1960s, place was a key element of geographical writings. It may be that place was sometimes treated simplistically—though nowhere near as simplistically as some subsequent writers have claimed—but geographers recognised the need to fill out any account of naked locational information in order to gain some sense of place. For example, one often derided aspect of field classes was the drawing of sketch maps, a skill that many nowadays think might be well worth reviving! But, in the 1960s, geography turned to number in an attempt to simulate a generally misread notion of 'science'. So places became bounded entities that interacted in ways that could be captured as indexes and flows, and were therefore modellable. Of course, this quantitative impulse had been evident in the discipline since the days of summarising Empire but now it took on a new form, not just as a compendium of trade figures but as a full-blown attempt to enumerate the world and then derive general principles. Each place would be summed up by cleanly drawn, intersecting planes of numbers. Each place could then be dynamised by general equations that espoused the modernist hope that science could intervene everywhere in everything.

In retrospect, it may seem to have been an odd ambition. But perhaps not. It was very much within the spirit of the times, often stimulated by 'progressive' political impulses (for example, a modernist commitment to a wider social good). And it also demonstrated an ambition to keep the discipline of geography up to date with the supposedly scientific march of other disciplines. However, if various technological revolutions and a war on poverty and social injustice provided an enabling backdrop to the project, as geographers soon became only too aware, there was another story, of the machine logics that underpinned the Vietnam War, preoccupied various negative analyses of the capitalism of the time, and seemed to be leading to the thinking computer HAL going mad in the movie *2001*.[1]

At this moment, it is fair to say, geography as a discipline was tailing other areas of academe. Looking back now, though with an acute awareness that it is always so very easy to do this—and that in years to come others may similarly look back on us—it appears as though this might have been a moment when geography lost its nerve. In return for supposed scientific acceptability in the wider, and hegemonic, ranks of academe, it traded in some of its own curious characteristics, abandoned some of its older preoccupations. Ironically, as we shall argue, it is those characteristics and preoccupations that have re-emerged in recent years to propel 'geography', or at least a kind of spatial consciousness, into the centre of a range of intellectual movements and debates.

But that is to run ahead. What geography lost, in the period when 'place' was viewed primarily through the lens of spatial science, included the following:

- a recognition of specificity—the whole impulse was to derive general principles;
- any notion of 'identity' that was more than an aggregation of indexable characteristics;
- any notion of 'synthetic' complexity in the fuller sense; and
- any notion of place as experienced.

One of the responses to this consequent sense of loss was to simply invert these characteristics by turning to a kind of humanism which, in geography at least, had its roots in the tradition of those who understood place and its study as a slow and steady conversation between human and environment, each imprinting the other, often in a Romantic reaction to the excesses of an industrial age that seemed likely to silence that conversation once and for all by destroying this gradual process of the accretion of landscape. Though always more of a characteristic of North American than British geographical thought (indeed most of its key British exponents have ended up in North America), 'humanist geography'

[1] There is an irony here, of course. The attempt by those in GIS and other technical disciplines to put a number to every place has now produced the preconditions in which such a vision can be made to come into existence. The world is made to fit fields of numbers, rather than vice versa.

(Ley and Samuels, 1978) had its effects through a general emphasis on the meaning of place and consequently on the qualitative and the experiential registers. But it would be fair to say that by the end of the 1970s the history of attempts to understand place seemed to have settled around two desires, both of which came in for criticism from an emerging feminist geography: on the one hand a desire to try to capture the meaningfulness of it all; on the other, nonetheless and in the end, to capture it, pin it down, to *know*.

Different territories

The next moves in geography's history of the consideration of place were attempts to evade this binary opposition and to begin to sketch out another, rather different, territory of possibility. These moves were succoured by a more general theoretical turn in geography in Britain, which began with readings of Marx in particular, but then rapidly broadened out. In contrast to North America, British geographers' readings of Marx in any case tended to be rather eclectic, the result in particular of the influences of an Althusserian anti-humanism that was strongly focused on thinking the world as process, an historically inclined political economy of the kind found in the work of Maurice Dobb and Eric Hobsbawm, and the kind of 'culturalism' to be found in the work of Raymond Williams, Stuart Hall, E. P. Thompson, Pierre Bourdieu, and others—a culturalism that came imbued with an uncommonly strong sense of place because of its emphasis on the importance of 'experience'. The main thrust of the Marxist position was, of course, not to take capitalism for granted, to treat it as natural and inevitable, but to make its existence and its workings explicit, in particular with regard to the economy and the state, and with a particular focus on class. It was in this form that it was wielded in geography in the analysis of specificity and difference.

It can be argued, however—indeed we would argue—that this set of propositions had a bigger and more structured impact in the discipline of geography than it did elsewhere. One of the inheritances of the previous period, and in particular of spatial science and of much of the empirical (indeed empiricist) work that accompanied it, was a tendency to explain geographical distributions *by* geographical distributions. One geographical pattern was taken to 'explain' another. The spatial was explained by the spatial. In that sense, geography was a self-contained discipline. Its object of study and its proposed explanatory mechanisms (the spatial, and spatial mechanisms) were held to be defining of and internal to the discipline. The argument of Marxism was to disrupt all that (as in Harvey 1973, 1982). And indeed in this guise Marxism was one element in a wider movement sweeping the discipline whose argument was that, as the slogan of the time had it, 'the spatial was socially constructed'. In other words, geographers must look again—as they once had—to wider mechanisms to explain the geographical variations, the characteristics of place, which it was their aim to analyse. The argument of the Marxist geographers, then, lay within this wider proposition but crucially added to it the argument that these wider mechanisms should be analysed in terms of capitalism and class. This was a significant intervention not only in terms of the discipline's self-understanding but also in terms of the engagement of geographers in policy and political debate. This was a period when uneven development—geographical variation—in many guises was at the heart of policy discussion. Internationally, there was a strong Third-World movement; within the UK there was heated contest over the form and efficacy of regional policy, while this in turn later came to be overshadowed by the increasingly desperate plight of the inner cities.

The response of the political classes to these 'problems of place' was to interpret them in terms of the characteristics of place. So 'problem regions' were lacking in entrepreneurship or adaptability. Inner cities were home only to the unemployable or the feckless. This was a political discourse that mirrored the intellectual strategy of explaining the spatial by the spatial. In terms of political strategy, it was the specifically geographical version of the more general manoeuvre of blaming the victim for its own problems. It is worth pausing for a moment here to consider what this response implied about the implicit conceptualisation of place that was being deployed. Effectively it implied, as did spatial science, a notion of places as bounded, and as having characteristics (identities) that were somehow a product of them themselves. What was not part of this conceptualisation was the possibility that the identity of place might significantly arise from the place of that place within a wider and more general nexus of (social) relations.

In general terms, the response of Marxist analysts of place was to reverse the terms. It was not, for instance, the inner cities that were responsible for their industrial decline but industrial decline that was responsible for the plight of the inner cities, and that decline itself was a product of the workings of the capitalist economic system (Massey and Meegan, 1982). Even in the very early days, however, there was not one 'Marxist approach' but many variants. They ranged from heavy-duty readings that attempted to mobilise the labour theory of value in an understanding of the specificity of place, with much deployment of the terminology of surplus value, to more relaxed approaches that drew on Marx far more for an understanding of social relations and the relational construction of identity and for a recognition of the significance of class. These early differences would subsequently emerge into sharper focuses of disagreement. But the overall message was consistent and powerful: one must look 'beyond' a place, in some sense, for an explanation of its identity and, crucially, one must look to capitalist economic forms and to the social relations of class. The geographical organisation of society was a part of its wider capitalist character.

There were already issues here that were immediately ripe for debate and/or further development. There was the early paucity of attention to socio-cultural and economic structures other than those of class. There was the tendency for all explanation to be 'top-down'—the global explained the local, in today's parlance; but there was not much influence, early on, the other way. And there was the unfortunate result that, through its tendency always to find the cause (the *real* character of an area) in the one same thing—capitalism—some Marxist analyses were in the end quite bad at respecting and at paying attention to that specificity of place that had historically been one of the primary passions of geography.

At this point, however, this stream of argument once again joined up with another more general move within geography. The broader proposition in the context of which the Marxist approach first launched its case was, as we have seen, that 'space is socially constructed'; in other words, that space is a product of society and its constitutive social relations (Gregory and Urry, 1985). This argument was crucially important to the discipline and, it is probably fair to say, has continued ever since to be widely acknowledged. It was, however, soon to be pushed a step further. For what thoughtful analyses increasingly showed was that 'space' (and thus places and the characteristics of places) was not only a resultant of social processes; rather it was an active participant in them too. Not only was space to be conceptualised as socially constructed, but society was inextricably spatially constituted too, and that makes a difference. In relation to the specific issue of an understanding of place, this move marked the beginning of a questioning of a simply 'top-down' analysis (from global to local) and a beginning too of a re-recognition of the significance both of geographical variation in general and of the uniqueness of place in particular.

The development of a 'spatial divisions of labour' approach (Massey, 1984) to the analysis of place took off within this context. As with some other originally Marxist-inspired approaches the emphasis here was on developing a relational understanding of space and place. That is to say that, rather than space being viewed as a container 'within which' the world proceeded, social space was itself understood as being a product of social relations. This also challenged that view of space as somehow 'naturally' divided up into segments (territories, regions, places), which had characterised many earlier approaches. Rather, an imagination of space as constructed out of nets of interconnections began to take shape. This had major implications for the conceptualisation of place. Place could no longer be taken as a given, as an already bounded entity whose characteristics were somehow produced within it (out of the soil, as it were). Rather a 'place' was now, in this view, understood as being a moment in a wider relational space—a particular point in the wider intersections of social relations. Place was thus not closed; its very specificity was constructed out of relations with elsewhere; and for that reason in turn there was no eternal authenticity to be hailed or claimed; the character of a place was always in construction. This was an understanding that would be mobilised to challenge the range of exclusivist claims to place that were entering the political arena: for instance, in the late 1980s, the 'new' nationalisms, parochialisms, and regionalisms. And it was a view, again, captured in a number of aphorisms: that we were moving from a space of places to a network society (Castells, 1996), that a 'progressive' sense of place would be a global sense of place, and so forth.

Once again, too, these theoretical and conceptual moves were related to shifts in the wider social sciences and humanities. Rethinking the nature of place in this way represented geography's participation in a much more general move, precipitated in particular by feminist and postcolonial studies, to reconceptualise what was meant by 'identity'. While feminism, postcolonialism, and other strands of thought within the social sciences and humanities were challenging the universals that modernism in its various guises, including Marxism, had proposed, they were at the same time not only insisting on a fuller recognition of 'difference' but also reformulating how difference and identity could be thought. This is far too complex a debate to render adequately here but what does seem to be true is that the reformulation of place in this period was geography's contribution to this more general formulation of the nature of identity as relational. As in the wider debate, moreover, the character of this relationality varied between proponents. Some (e.g. Sibley, 1995) stressed the discursive and material production of geographical difference through attempts at the purification of space through 'othering'. Others (e.g. Massey, 1991b) stressed the relational nature of identity through the positive heterogeneity of the constellation of relations whose intersection produced the uniqueness of place.

Differing territories

By the mid-1980s, taking their initial cue from these kinds of debates, geographical reckonings of space and place were focusing on the task of making determinist approaches like Marxism more open to contingency by taking greater note of diverse processes of social and cultural reproduction. In its essence, this meant that three different forms of variegation were injected into geographical work on place with the aim of producing accounts that could reach out to the diversity of places without losing the theoretical thread. One was a much greater attention to processes of subjectification. Drawing on explicitly geographical work, such as the intricate time-geographic diagrams of Torsten Hägerstrand (Thrift and Pred, 1982), the remarkable depth of the works of Michel Foucault, and Anthony Giddens's structuration theory (which not only itself called on Foucault but also paid considerable attention to time-geography: see Giddens, 1979), more room was opened up for a geographical consideration of the processes by which flesh is individuated and gathered as social groups, a process which—it became increasingly clear—involved not just the machinery of spatial and temporal layout but also social and cultural definitions of what counted as space and time itself which were continually being written in to the body (Thrift, 1983).

Another was a much greater interest in matters of culture. On one level this had the simple effect of multiplying the kinds of spaces that were thought to be legitimate matters for geographers' concern. So, for example, much greater attention was paid to both transient and fixed spaces of sexuality, and especially various gay and lesbian spaces (Valentine, 1989). On another level, however, it was part of a more far-reaching struggle to politicise spaces. So, following the same example, those spaces previously simply assumed to conform to hetero-normative rules and procedures were disrupted by writers on new gay and lesbian queer spatialities, which could form the building blocks of new kinds of politics. Then, a final note of contingency was injected by the sharp debate over the study of 'localities'. The locality studies were a large and ambitious ESRC (SSRC)-funded attempt to take the spatial-divisions-of-labour approach further by studying different 'places' within the UK; the aim was both to investigate the very different ways in which what were thought of as macro social and economic processes were being played out in different parts of the country (in other words, the nature and the mechanisms of the continued reproduction of the uniqueness of place) and to ponder the implications and effects of that variability (in other words, how geography was mattering: Cooke, 1989).

The emphasis on a recognition of the unique had already been attacked by some in relation to the spatial-divisions-of-labour approach. But now the debate became broader and more heated, and the lines of divide more clearly defined. For those who were opposed to a focus on place, the very terms 'local' and 'uniqueness' reverberated with negative connotations. There were a number of aspects to this. Most of the antagonists of place were of the Marxist school and committed to pursuing a Marxism of the type described earlier. A main aim of this line of thought had been to point to the international capitalist relations underlying all local specificity, but this valuable insight often ran the risk of overlooking specificity altogether. Specificity was seen, therefore, as a side-track to what should be the main point. This position was very understandable, given geography's long history of a purely descriptive relation to place. But it was also a deeply problematic reaction since it could be argued that part of the 'curious' character of geography is precisely its explicit concern with variation—with specificity.

Moreover, the arguments ran deeper than this. The very term 'local' came in for denigration: things were 'only' local issues, 'only' local struggles, and so forth, rather than the global and universal issues which, it was argued, were more worthy of study. As was frequently pointed out, there was a strong element of 'social-scientific masculinity' in play here (see Rose, 1993): we should be tackling the big issues, like class struggle, rather than local struggles (amongst which feminism typically ranked). We should not be fragmenting struggles (through raising feminism or issues of local areas) but gather together under the universal banner. In reply, it was pointed out that the 'universal' banners of the working class and of globalism had themselves only been maintained by smothering difference. It was vital to *address* difference rather than to ignore it. Further, it was also pointed out (Sayer, 1984; Massey, 1991a) that there was a conflation going on here between universal/global/theoretical on the one hand and specific/local/descriptive on the other. In fact, it was replied, the global is just as 'specific' as the local; the 'local' can demand as much theoretical sophistication as the global in its analysis.

This debate was happening at the time when the wider groves of academe were embattled between modernism and postmodernism, and when poststructuralism was becoming more widespread and more assured within British social science (Gregory, 1994; Jacobs, 1996). In some ways the debate over locality studies both reflected some of those concerns and expressed geography's particular take on it. It is a debate that continues into the twenty-first century. What was really at issue, however, was the manner in which place should be conceptualised (Barnes and Duncan, 1991; Keith and Pile, 1993). If places are indeed understood as bounded territories with internally generated authenticities, then localisms can indeed be problematical for the generation of wider understandings and wider movements (the problem of militant particularisms). But if the specificity of place is always understood as generated relationally then there is no simple divide between inside and outside, between local and global, between local struggles and wider movements. The question of the significance of 'place' within geography was thus intimately related to the manner of its conceptualisation.

In the 1990s, perhaps the single greatest theoretical influence on the theorisation of place was poststructuralism. It would be simply inaccurate to imply that the various forms of poststructuralism arrived as some kind of theoretical shockwave (Doel, 1999)—British geographers had already had a long involvement with Foucault, for example—but it would be fair to say that the exact implications of poststructuralist principles were predominantly worked through in that decade. In particular, it is important to point to five main consequences. One was to cement the kind of anti-humanism that had already been a feature of so much British theoretical work (going back to at least Althusser). In fact, British geographers were actively involved in the derivation of 'transhuman' approaches, which began to understand social processes as shifting ethologies of unlike things. Thus, work on nature moved away from the cosy ethical precepts that were possible when the world was divided into nature and society to precepts in which human flesh became simply a part of a realm of always hybrid quasi-objects, demanding new ethical stances that are only now being worked through (Whatmore, 1998).

Another consequence was to pay much more attention to the sheer work of building spaces and keeping them together. This kind of emphasis paid much more attention to objects as important actors in their own right, able to act not just as passive intermediaries but as active 'mediaries'. In turn, this kind of work on the actual labour of reproduction led to the study of the profusion of practical epistemologies to be found in a whole set of new spaces of knowledge that now hoved into view, each with their own characteristic means of transforming time and space (May and Thrift, 2001; Livingstone, 2003). So the actual empirical construction of space into place, via a vast range of texts and instruments, began to be seen as a material concern in its own right. What had often been relegated to the sidelines came to be seen as central.

A third consequence was to provide different understandings of power (Sibley, 1995; Philo, Paddison, and Sharp, 2001; Hudson, 2001; Allen, 2003). Power was understood as a set of dynamic procedures and discourses, open to change, which were focused on human conduct rather than as fixed social templates from which different degrees of influence could be read off. In particular, power was seen as written in to the body, leading to a much greater attention to embodiment as a kind of fleshy script; each movement of the body in space and time had its own half-forgotten genealogy of numerous attempts at inscribing power. And the body itself existed in many registers—not just vision but also sound, touch, smell, taste, and movement itself—which were all susceptible to manipulation.

Fourthly, geographers took up poststructuralist authors' concern with process. And since the world was seen as in the making, rather than as a succession of fixes in space, much more attention had to be focused on matters of representation. The fact that objects of all kinds (including subjects) were seen as in process rather than as fixes meant that they had their own compounding time signatures, which required means of representation that could take in instability and change as normal states of affairs, rather than stable depictions. This structure was made particularly clear by the kind of identity/non-identity politics found in many feminisms and postcolonialisms, where repetition vied with the potential of each new interval. And it applied even more strongly to place: places were increasingly studied as transient and multipurpose phenomena, within which numerous discursive frames jockeyed for position and from which the new could issue, as well as the reassertion of tradition.

Then, finally, geographers became interested in methodologies that could capture at least some of the liveliness of place. That meant turning to a host of different approaches that could work both with and against the grain of places to produce revelatory effects, on however a modest a scale. That might mean paying more attention to the 'simple' act of walking, as a means of showing up the various dimensions of bodily effort that go to make up places; or it might mean turning to the extensive archive based around performance in order to make everyday places incongruous, even strange; or it might mean using modern technologies to link up places around the world in order to show how social relations are continually working not so much across space as to construct points at which different spaces can cross (Thrift, 2000). Whatever the case, the clear intention was to make places show up differently, so that they might be worked with differently.

In turn, these kinds of allegiances had consequences. One was to send British geographers scurrying off on a quest for process-based approaches of all kinds that could provide possible illuminations of the power of place, but a place now conceived as constantly in motion, changing its nature almost as often as the weather (Pile and Thrift, 1995; Pile and Keith, 1997). No doubt, this can be seen as a moment of theoretical promiscuity as geographers headed off in all kinds of directions.[2] But it might better be seen as a period of catching up with process-based writings hitherto ignored, many of which, in any case, shared significant bloodlines. The second consequence was to produce new allegiances, most particularly with developments in continental Europe. For example, geographers have become a part of the project of 'actor–network theory and after', taking a role as full partners in the development of this approach. Perhaps this European orientation helps to explain an increasing divergence of theoretical style with the United States, the traditional partner of British geographical endeavour.

Certainly, it is interesting to reflect on the British theoretical trajectory in comparison with that of US geography over the same period in order to see just how distinctively different the trajectory of British geography often seems to have become as it has struggled to find new ways to think place (Anderson et al., 2003). The US geographic tradition of studying place exhibits a strong allegiance to a Thoreauesque moral certainty and centredness that keeps recurring in the practice of US geography. It is there in the predominance of a declarative Marxism that is keen to police transgression and to constantly pronounce its pain on behalf of the oppressed (at least in part as an index of its extreme distance from political power compared with Europe: see Duell, 2000). It is there in a conception of the global and the local as essentially different spheres of action, notwithstanding attempts to

[2] From a rediscovery of certain kinds of phenomenology through to the headiest realms of practice-based theory, from the most abstruse forms of Deleuzian pragmatics through to the constant iterations of Butlerian feminist theory, from the emphasis on a technologised mimesis found in the writings of Benjamin through to meditations on the place of self as found in the later writings of Foucault and Irigaray (see Foucault, 1975/2001; 2001).

link them through a heavy emphasis on notions of scale and re-scaling (but see Gibson-Graham, 1996). It is there in US geography's allegiance to a humanism concerned with revealing a universal 'human' experience of place which, though tempered by a contextualism drawn from poststructuralism, still seems to want to find a home in place (e.g. Adams, Hoelscher, and Till, 2001). And it is there in a postcolonialism that in the United States is often more concerned to uncover and assert the rights of subjected peoples than to deconstruct the whole notion of peoples (Gilroy, 1992).

This pocket history of British geography's recent engagements with place should perhaps begin to give the lie to Hugh Clout's elegy for another way of doing geography (see Chapter 7 in this volume). Clearly, modern British human geography has very different political and theoretical engagements but it also shares some of the same allegiances—to detail as not incidental, to the specificities of the event as having their own power, to the need to touch and tarry in places—that can and do connect past and present. Indeed, as contemporary historical geographers have dug deeper into this past, so they have found strange, even exotic, resonances in some of the less-well-known regional thinking of the early twentieth century, which have provided new attachments for the thinking of the present. For example, Cameron and Matless's (2003) work on Marrietta Pallas shows the way in which elements of current human geographical thinking on place were prefigured and can now themselves act as resources for thinking place anew.

Looking Ahead?

A long journey that can only continue, then. This section attempts to provide a synthetic account of place as it might now be understood. It is important to underline that word 'synthetic', of course. This is an account that pulls together the writings of a number of geographers and is, as a result, faithful to no particular account. Whilst this strategy has its disadvantages, it at least allows us the thoroughly artificial but still very useful luxury of the kind of coherence that could never be found in an area that is still criss-crossed by all manner of fierce debates.

To understand why this new kind of approach to place has grown up we need to briefly recap by returning to the 1960s and to the growth of a tradition of understanding the world that showed up first in the widespread use of quantitative methods but then broadened out significantly. Roughly speaking, we might interpret this as a move away from a regional tradition dependent upon the compilation of surveys, lists, classifications, and the like to an analytic tradition based upon principles of reduction in order to produce general law-like explanations, thus paralleling the first stage of 'scientific' thought (Foucault 1975/2001). In geography, as one might expect, this move tended to manifest itself as a concern with spatial pattern as a kind of analytical diagram. Some of the shortcomings of this kind of approach rapidly become clear and stimulated other ways of proceeding, most notably certain kinds of Marxism and humanism. But, in their early manifestations at least, these approaches often still tended to cleave to some of the explanatory values that typified quantitative work—most notably a search for general principles that could apply everywhere and a tendency to want to define themselves either as scientific or the exact binary opposite. However, at one and the same time, these approaches also began to prepare the ground for a different kind of approach, one that was willing to think about space as more than just a series of templates on which various behaviours took place, as more than just a series of different economic tectonic plates, as more than just a passive participant in the game of life.

Underlying this nascent approach to places were some quite specific ambitions. The first of these was to somehow capture processes. Ideas like the 'spatial division of labour' and 'locality' were meant to capture something more than a slavish succession of blocks of space, one block replacing another in almost military order. There was also the intention to challenge the role and content of space and time themselves. The nature of the dimensions came to be seen as part of the problem of process, rather than a neutral frame, so that investigations of place were increasingly couched as investigations of how space and time had been empirically constructed as givens rather than universals.

The second ambition was to make more room for spatial complexity. The kind of patterns that had been conjured up by spatial analysis sought clarity above all things, usually (though not always) within the framework of a Euclidean spatial and temporal framework. But it rapidly became clear that this clarity was won at the expense of what might be called 'enactment' (Law and Hetherington, 2000), the sheer material effort of making spaces, which meant that spaces very rarely mapped neatly and passively on to one another. Indeed this very messiness was a stimulus to producing new spaces. Similarly, the patterns conjured up by spatial analysts were usually couched in terms of distance, speed, and other Euclidean primitives that hid as much as they revealed because they bought into a particular culturally given version of space.

The third ambition followed on. It was to move away from spaces viewed as if from on high right into the action, and especially into the press of embodiment, thereby making the subject position of the observer crucial. Gender, ethnicity, sexuality, class ... all had a bearing on how and what kinds of spaces could be enacted. It followed that if more than one of these spatial stories was going on, then all kinds of spatial stories could be taking place at once, and continuous trade could take place between them. So social groups and identities could no longer be understood as occupying fixed territories but had continually to revise their positions—even if a lot of effort was actually being invested in staying still.

Then there was one more ambition. And that was to influence political agendas by producing open and creative responses, often outside what would conventionally be regarded as 'the political'. In retrospect, the kind of responses that were being sought were rather like Deleuze's 'becomings': the constant emergence of the new (Deleuze and Parnet, 1987). In other words, search out openings not closures, find the conditions under which something new can be produced, value the multiple, and beware of fixed templates and finished theories.

Very gradually these kinds of ambitions have started to produce a body of work that has widened and deepened as it has gone on, a developing vocabulary of 'good words' that attempt to redefine the world—new descriptions of place if you like. Without a doubt, they constitute a different way of tracing and constructing a field, but not necessarily one totally at odds with what came before (so that when Hugh Clout laments the death of regional geography in Chapter 7, we have a certain sympathy, but would also point to its regeneration in this different guise). In particular, we would argue for the following continuities. First, there is a commitment to the specificity of places, and to (often minute) observation. The second continuity is a commitment to the curious, not as an orientalist obsession with othering the other, but as a kind of wonder about the strange and charmed nature of the world, and its capacity to add new contributions, produce new arrangements, force new hybrids (Fisher, 1998)—a passion for the wonder of difference. Thirdly, there is a corresponding commitment to exploration. Again, the word conjures up some heavy colonial freight (Driver, 2000), but we think that it needs to be reclaimed from the oppressive colonial past, as a positive, reciprocal, and passionate activity of encountering.

Four Steps

So how can we understand places in this way? We want to make a four-step argument, which, we hope, nails up some of the broad elements of an evolving doctrine.

The first step is materialism. Materialism is a sticky word, often nowadays associated with certain forms of Marxism. Yet in recent years the whole question of materialism has reappeared on the intellectual agenda as the question of what counts as matter, material substance, the 'thisness' of bodies and other objects. What, we think, most evidently characterises this new-found interest is an interest in 'matter': its instability, its ability to constantly mix and transmute, coupled with the means by which such an active state of affairs can be represented. In order to illustrate this new-found interest, there is no better move to make than that made by a number of commentators recently—back to Marx's early work on

Democritus and Epicurus (e.g. Tiffany, 2000; Bennett, 2001). Marx's turn to these early Greek atomist philosophers was part of his need for a philosophy that gave priority to the sensuous natural world but did not have to conceive of the world as a mechanistic realm of necessity. Following Epicurus, for Marx matter was not inert; 'it was too recalcitrant or unruly to be part of an order of natural necessity' (Bennett, 2001, 19). However, Marx then went on to identify this wilfulness at the base of the natural order as providing a characteristic only of human action, thereby 'losing touch with the remarkable appreciation of agency within nature that Epicurus actively allows' (Bennett, 2001, 121). So the non-human is downgraded, the animated nature of Epicurus becomes the inanimate and alienated nature of Marx. Recent work by geographers has precisely attempted to re-animate nature, to bring it back within the commonwealth as a full participant by emphasising the multiple qualities of things, rather than treating them as unities or totalities. Moving away from the kind of philosophical anthropology represented by movements like humanism, this 'transhuman' work tends to depict human/nature as a set of more or less temporary ethologies in which all kinds of unlike elements may set up house together as periodic recurrences (Whatmore, 2002), what Deleuze and Guattari (1988) call 'refrains' or 'vibrations'.

Second step: if there is a characteristic of materiality that is crucial, it is what we might call relationality. But what actually is a relation? To begin with, a relation always signifies a relation to an other (Gasché, 2000). Then it always signifies a between, not a unified object but a means of connection. So objects like humans are made up of the encounter of many relations, they are multiple—'a bloc of variable sensations' as Deleuze would have it; and they are continuously interactive, eclectic, and syncretic. This view of relations has consequences. The first is that it destabilises fondly held descriptions of the world, and does so decisively. As an example, take the division between the human and the environment. Recent work shows that any such division is not just hard to draw but fundamentally mistaken. For example, the work of Ingold (2001) shows how the human mind and the environment resonate together in such a way that knower and known cannot be separated in any meaningful way. It is worth quoting Ingold at some length:

> From well before birth, the infant is immersed in a world of sound in which the characteristic patterns of speech mingle with all the other sounds of everyday life, and right from birth it is surrounded by already competent speakers who provide support in the form of contextually grounded interpretations of its own vocal gestures. This environment, then, is not a source of variable input for a preconstructed device, but rather furnishes the variable conditions for the growth or self-assembly, in the course of early development, of the neurophysiological structures underwriting the child's capacity to speak. As the conditions vary, so these structures will take manifold forms,

> each differentially tuned both to specific sound patterns and to other features of local contexts of utterance. These variably attuned structures, and the competencies they establish, correspond of course to what appear to observers as the diverse languages of the world. In short, language—in the sense of the child's capacity to speak in the manner of his or her community—is not acquired. Rather it is continually being generated and regenerated in the developmental contexts of children's involvement in worlds of speech. What applies specifically in the case of language and speech also applies, more generally, to other aspects of cultural competence. Learning to walk in a particular way, or to play a certain musical instrument, or to practice a sport like cricket or tennis, is a matter not of acquiring *from* an environment representations that satisfy the input conditions of pre-constituted cognitive devices, but of the formation, *within* an environment, of the necessary neurological connections, along with attendant features of musculature and anatomy, that underwrite the various skills involved. The notion that culture is transmissible from one generation to the next as a corpus of knowledge, independently of its application in the world, is untenable for the simple reason that it rests on the impossible precondition of a ready-made cognitive architecture. In fact, I maintain, nothing is really transmitted at all. The growth of knowledge in the life history of a person is a result not of information transmission but of guided rediscovery, where what each generation contributes to the next are not rules and representations for the production of appropriate behaviour but the specific conditions of development under which successors, growing up in a social world, can build up their own aptitudes and dispositions. (Ingold, 2001, 271–3)

Thus we arrive at what we might call a logic of sense, consisting of the constant fine tuning of 'knower and known' as they are caught up in and have to sort out particular events, as being more typical of human goings-on than an analytic logic. In turn, we come to a notion of place as caught up in this constant activity of fine tuning of knower and known, as part of the way that what we call 'human' awareness (or, perhaps we should say 'a-where-ness') takes shape.

The second consequence of the relational approach follows naturally on and is the stress it puts on the *work* of mediation between knower and known, especially through a vast range of texts, images, and instruments. Bruno Latour puts this well in surveying the changes that have recently taken place in the history and philosophy of science as that subject has moved to a remarkably spatialised view of the world (see Livingstone, 2003):

> The active locus of science, portrayed in the past by stressing its two extremities, the Mind and the World, has shifted to the middle, to the humble instruments, tools, visualisation skills, writing practices, focussing techniques, and what has been called 're-representation'. Through all these efforts, the mediation has eaten up the two extremities; the representing Mind and the represented World. (Latour, 1998, 422)

We can only know places through frames and practices inhabited by texts, images, and instruments, which are themselves active 'mediaries', rather than passive 'intermediaries'.

The third consequence of the relational approach is that it gives us a quite different view of space. Space is thought of as process, as constantly shifting ground as different relational geometries agitate it. So space is no longer understood as a container, but as a convoluted surface in which some things can be near though they are far, and some things can be far though they are near, according to the nature of the relations at issue. 'As on a pure surface, certain points of one figure in a series refer to the points of another figure: an entire galaxy of problems with their corresponding dice-throws, stories and places, a complex place, a "convoluted story"' (Deleuze, 1990, xiv). Certain relational geometries can 'freeze' space—indeed this may be one definition of power—but even they must struggle to stay still. They suffer constant encroachment and mutation as they co-evolve with other geometries, creating new 'assemblages'. And this constant encroachment and mutation is often now thought of in terms of 'events', emergent moments when something new is produced, most especially though a shift in spatio-temporal register (which might be something as 'little' as a gesture or as 'large' as an act of war), which produces a new intensity, a new 'individuation'.[3] Events do not, of course, have to be seen in purely human terms; they can be 'transhuman'. 'What we're interested in, you see, are modes of individuations beyond those of things, persons or objects: the individuation, say, of a time of day, of a region, of a climate, a river or a wind, of an event. And maybe it's a mistake to believe in the existence of things, persons, subjects' (Deleuze, 1992, 26).

Third step: to see the world again as strange and full of wonder, rather than as just a commodified wasteland. All kinds of impulses have fuelled a growing interest in the idea that contemporary spaces might be seen as having wondrous dimensions. One is the increasing scepticism that theory can act like a know-it-all in the face of the world's fullness and multiplicity. Increasingly, theory is being seen as a modest aid to practice, a set of 'ontological sights'; not a sure set of coordinates precisely mapping out the whys and wherefores of the world (Pleasants, 1999). And this view of theory goes hand in hand with another—that we really cannot know a lot of what is going on. Unintended consequences are ubiquitous. There is a whole shadow realm of events and connections that are hard to relate to any intent (Shotter, 1993; Oyama, 2000). But that does not mean trying to set theory aside; rather it emphasises what we might call theory's 'lyric'

[3] A lot of contemporary art is involved primarily in trying to show up these events. 'Following Bergson, we might say that as beings in the world we are caught in a certain spatio-temporal register: we see only what we have already seen (we see only what we are interested in). At stake with art, then, might be an alterity, a switching, of this register' (O'Sullivan, 2001, 127).

quality in line with the idea that the world itself has lyric qualities—a lyric substance, even (Tiffany, 2000)—and that we need all kinds of new and inventive images, analogies, and tropes if we are ever to group and conjugate these qualities of secret ministry.

Perhaps this is one of the reasons that so much attention is now being given to human behaviours like play, which upset single-minded seriousness by posing all kinds of 'what-ifs', and to the associated quality of the 'virtuality' of an event, the creative resolution of the problematic complex involved in an event that is produced in concert with the circumstances encountered (which is never opposed to the 'real' because it is how 'reality' comes about). Again, perhaps this is one of the reasons why so many are now looking at the commodity anew, not just as a cipher of capitalist production built to amuse until death (though it is certainly that) but also—sometimes, just sometimes—as a means of enchantment, a set of pleasures whose force, it might be added, may be susceptible to ethical appeal. And it is certainly one of the reasons why there is so much interest in affective complication—the passions, moods, emotions, feelings, desires—which a certain version of theory wants to banish to the sidelines, both because it is so threatening to the constitution of a disembodied and epistemic agency and because it is indifferent to means–end difference. 'It is enjoyable to enjoy. It is exciting to be excited. It is terrorising to be terrorised, and angering to be angered. Affect is self-validating with or without any further referent' (Tomkins, cited in Sedgwick, 1995, 7). Yet it is entirely possible to argue that affect is proper and essential to the exercise of reason, and that we need to expand our notions of reason accordingly (Fiumara, 2001).[4]

Seeing the lyric qualities of the world in turn requires new senses of place within which things like selves develop, places that are often transitory and riverine but no less effective. Take the case of music, and its ability to call worlds into being with only the slightest stimulus.

> Music feels very near to us also because we experience it so manifestly *within* ourselves. When we look at a painting or a play or a movie, the act of seeing is one in which we are inevitably required to deal with the distance or space that lies between us and the object we behold. We must map our visual field and establish relations (nearer, farther, smaller, larger) of the objects within it because these are not given immediately by sense impressions. Even reading requires translation from letter to word to image. Music, by contrast, is experienced as immediate. We might know perfectly well that it originates at a distance—across the room, perhaps—but we still experience it as

[4] Even academics are, for example, often gripped by a problem. And they know the sheer power of attention to the text, that sustained and intense engagement that is partly theatrical in its trance-like qualities. What affect produces is a certain definition of what counts as an experience and, at the time, a certain awareness of how that experience might (and only might) shift registers.

> contiguous with ourselves, as pressing upon us; it has already entered the canals of our ears and penetrated to our brain. No void or empty space lies between us and it. Music is in us, its rhythms an echo or objectification of rhythmic qualities of our embodied life; of the heart beating, of the lungs expanding and contracting, of the legs walking.
>
> But music not only annihilates the space that lies between us and it, it also establishes a feeling of space *inside* us, and when it opens up this space, it enlarges us. We feel that only a force intimate with and even coextensive with our self could accomplish this enlargement, and when we behold the existence of space within ourselves, our relationship to spatiality itself changes. Space is no longer an absence, or void, between us and the objects around us. Space is our familiar. Space is what we ourselves enclose. Our self, we see, is not a dense, inner kernel of compacted identity but more like a membrane, the filmiest of screens between the space within and the space without. (Bromell, 2000, 73)

We come then to the last step. So far we have tried to outline a more contingent and a more animated geography of place and place for geography, one that is trying to re-imagine the material world as vibrant, quirky, overflowing. But this is not being done for purely intellectual reasons. It is a fundamentally political move. It arises, to begin with, out of a deep-felt sense that current notions of democratic politics lack a crucial dimension, which is difficult to name but crucial to get right. This dimension lies somewhere between the capacity to give and receive surprises (a capacity Deleuze described as 'becoming-otherwise'), the expansion of what is understood as knowing to include 'aesthetic' experiences (such as music), and the consideration of how different kinds of bodily dispositions (and especially subintentional dispositions) are inculcated (Thrift, 2002). This dimension is, in turn, conceived as a means of cultivating new sensibilities that presume generosity to the world, 'rendering oneself more open to the surprise of other selves and bodies and more willing and able to enter into productive assemblages with them' (Bennett, 2001, 130), by means of new modes of perception and new forms of comportment.

This is not, however, a Romantic project. It is, rather, determinedly ordinary (Dunn, 2000), based in forms of politics (often drawn from performance and other knowledges of surprise) that can break through habit. The arts of commotion that this kind of politics tries to foster should not be seen as 'small' just because they are located in the ordinary. And this links to another point. For this kind of politics also tries to re-imagine place as a constantly shifting set of connections that may stretch many thousands of miles and yet be trivial, or occupy only a few square metres and be crucial. In other words, this is a politics that does not take scale as a given condition of the world. And this leads to a final aspect of this politics: its relative catholicity as to what counts as political material. Because the approach is non-humanist, all kinds of actors insist on their place at the political table. Intersubjective relations may beckon political space, but so might human–animal and organism–machine relations. Often politics must intervene in 'fuzzy aggregates' of 'forces, densities, intensities' that may not be thinkable in themselves but can still be operated on and with.

None of this, of course, is to set aside what are conventionally regarded as political projects. Rather, it is precisely meant to strengthen them, for without the kinds of 'enchantment' this added political dimension brings we will have less 'energy and inspiration to enact ecological projects, or to contest unjust modes of commercialisation, or to respond generously to humans and non-humans that challenge our settled identities' (Bennett, 2001, 174). The places we inhabit will be greyer, the futures we want for them far less likely.

Recapping place

Given this hard-won theoretical background, how might we conceive of place? One thing to say is that there can be no fixed definition: places are constantly adding qualities that give them new potentials for action, thus making them into new 'wheres' and new 'a-where-nesses'. But we can now provide at least a brief (and so necessarily caricatured) reprise.

To begin with, places are gatherings of habitual practices that are necessarily attached to particular locations. So we drive past the same large tree on the corner each day and vaguely notice how it changes with the seasons. Gradually, we build up more and more detail (Henry James's 'solidity of specification'). In turn, this sedimentation of detail feeds into our general expectation of how the world will turn up. We have produced a kind of staging that directs our senses and from which we can make all manner of routine encounters that require skilled improvisation to maintain them.

But, secondly, these stagings are not free choices. In large part, they are produced by the imperatives of networks of organisations that issue the commands and edicts that course through our flesh and which we have become used to enacting and come to regard as normal elements of how places turn up. We drive past the same large tree on the corner each day because we are going to work or to shop. We drive in a car that itself is the product of these large networks of organisations. And we are able to drive only because of a web of intermediaries (for example, all the minutiae of roads and parking), which thread together so many networks of organisations that they themselves have force in the world as dumb but potent actors with their own inhuman agendas.

But, though places are gatherings that are diagrammed by the various networks of organisations, they are also confluences of all kinds of planes of affect that exceed these networks, even as they are manipulated by them. There are all kinds of ways in which affect can be felt in places: as discontents and general longings that do not have to be grand to have effects ('this place is boring', 'this house is too small', 'the neighbours get on our nerves', and so on), as practices of memory and witness that act as cues to both call up and lay down memories (as simple as tags like 'I did this there' or as complex as some ritual spaces that are machines intended to produce certain memory effects), as an impetus to fantasy about other circumstances, and so on. In other words, places are massively important elements of the social imaginary and one of the key means by which other circumstances can be imagined.

A considerable amount of care needs to be taken with such a skeletal account, and geographers are already filling it out in numerous ways. For one, it is very easy to make this account seem as though it is concerned only with human 'a-where-ness'. But, as we have seen, places figure in transhuman ways and other kinds of actor have their own circulating ethologies of encounter that give primacy to place. So, for example, much attention is now being given to the ways in which animals figure in such circulations in ways that can give animals prospects of their own, set off from human imperatives (e.g. Whatmore, 2002). Then, it is also very easy to make this account seem as though it is concerned only with single sites. But, as we have stressed again and again, in a process-based account places cannot be closed off. This has always been the case: like nations, places are as much made up of crossings and hybridities as they are fixed origins and tightly drawn unities. And the historical development of agencies like the media, mass travel, and tourism has only underlined this point. Places have more and more resonances with events that occurred outside their boundaries and these events are constitutive, not incidental (Urry, 2000). That ornament from the holiday in Bali that we think of as part of what we call home; that travel programme on television documenting the place we used to live in that spurs us to look again at the photograph album of relatives strung out across the globe—more and more, we look outside a place to confirm its existence. Which brings us to another point: places are increasingly doubled by the multiplicity of means of representation available to document them (for example, texts, photographs, videos, and the new machinery of location like GPS, GIS, and RFID tags) and the resultant weight of cultural machinery now available to confirm (and contest) their existence as places. From local guides and histories, which now seem to document almost every occupied square metre of the globe, to the multiple means of keeping places frozen in time that are now available (whether maps or photographs or websites), the potential for places to become part of a vast and sprawling cultural archive continues to grow.

This brings us to one more point. The constant untying and re-assembling of place that is a feature of many parts of the modem world has forged new kinds of subject who have begun to have what one of us once called a global sense of place (Massey, 1991b). For many, sense of place is no longer concentrated in just a few places as it once was. Rather, it is strung out across the world, either as a particular kind of way of life (for example the Punjabi-Mexican-American struggling to forge a new kind of cultural heritage) or as a commitment to places that are 'far away' and that require the construction of modes of hospitality and political sympathy that are becoming much easier to make now that so much place knowledge circulates. We should not overdo this tendency, but it does suggest all manner of political possibilities that were not available before (or were much more difficult to realise).

Conclusions

We will end on an ironic note. The original spatial analysis wanted, above all, to make geography into something approximating a 'science', replete with certainties and predictions and able to provide common cause between physical and human geography. But the process-based view of place that we have expounded here is not so very different in some of its fundamentals from recent developments in certain areas of science, not just those areas where scientific protocols have always been more fluid (like biology) but also even in some areas of chemistry and physics where the boundary between the 'imaginary' and the 'real' is now rarely drawn. J. B. S. Haldane once argued that 'the universe is not only queerer than we suppose, but queerer than we can suppose'. It is to produce a human *and* physical geography (Massey, 1999) that can sense the sense of this statement that we now turn.

References

Adams, P., Hoelscher, S., and Till, K. E., eds (2001) *Textures of Place: Exploring Human Geographies.* Minneapolis, MN: University of Minnesota Press.

Allen, J. (2003) *Lost Geographies of Power.* Oxford: Blackwell.

Anderson, K., Donosh, M., Pile, S., and Thrift, N. J., eds (2003) *The Handbook of Cultural Geography.* London: Sage.

Barnes, T. and Duncan, J. S., eds (1991) *Writing Worlds.* London, Routledge.

Bennett, J. (2001) *The Enchantment of Modern Life: Attachments, Crossings and Ethics.* Princeton, NJ: Princeton University Press.

Bromell, N. (2000) *Tomorrow Never Knows: Rock and Psychedelia in the 1960s.* Chicago, IL: University of Chicago Press.

Cameron, L. and Matless, D. (2003) Benign ecology: Marrietta Pallas and the floating fen of the delta of the Danube. *Cultural Geographies* (forthcoming).

Cooke, P., ed. (1989) *Localities.* Basingstoke: Macmillan

Deleuze, G. (1990) *The Logic of Sense.* New York: Columbia University Press.

Deleuze, G. (1992) *Negotiations, 1972–1990.* New York: Columbia University Press.

Deleuze, G. and Guattari, F. (1988) *A Thousand Plateaus.* Minneapolis, MN: University of Minnesota Press.

Deleuze, G. and Parnet, C. (1987) *Dialogues.* London: Athlone Press.

Driver, F. (2000) *Geography Militant.* Oxford: Blackwell.

Doel, M. (1999) *Poststructuralist Geographies.* Edinburgh: Edinburgh University Press.

Duell, J. (2000) Assessing the literary: intellectual boundaries in French and American literary studies. In M. Lamont and L. Thevenot, eds, *Rethinking Comparative Cultural Sociology: Repertoires of Evaluation in France and the United States.* Cambridge: Cambridge University Press, 94–124.

Dunn, T. L. (2000) Wild things. In J. Dean, ed., *Cultural Studies and Political Theory.* Ithaca, NY: Cornell University Press, 258–68.

Fisher, P. (1998) *Wonder: The Rainbow and the Aesthetics of Rare Experiences.* Cambridge, MA: Harvard University Press.

Fiumara G. C. (2001) *The Mind's Affective Life: A Psychoanalytic and Philosophical Inquiry.* London: Brunner-Routledge.

Foucault, M. (1975/2001) Events and eventuality. In *The Essential Works of Michael Foucault: Volume 3. Power.* Harmondsworth: Penguin.

Foucault, M. (2001) *The Order of Things.* London: Routledge.

Gasché, R. (2000) *Of Minimal Things.* Stanford, CA: Stanford University Press.

Gibson-Graham, J. K. (1996) *The End of Capitalism—As We Knew It.* Oxford: Blackwell.

Giddens, A. (1979) *Central Problems in Social Theory.* Basingstoke: Macmillan.

Gilroy, P. (1992) *The Black Atlantic.* London: Verso.

Gregory, D. and Urry, J., eds (1985) *Social Relations and Spatial Structures.* Basingstoke: Macmillan.

Hudson, R. (2001) *Producing Places.* New York: Guilford Press.

Ingold, T. (2001) From complementarity to obviation: on dissolving the boundaries between social and biological anthropology, archaeology and psychology. In S. Oyama, P. Griffiths, and D. Gray, eds, *Cycles of Contingency: Developmental Systems and Evolution.* Cambridge, MA: MIT Press, 255–80.

Jacobs, J. (1996) *Edge of Empire: Postcolonialism and the City.* London: Routledge.

Keith, M. and Pile, S., eds (1993) *Place and the Politics of Identity.* London: Routledge.

Latour, B. (1998) How to be iconophilic in Art, Science and Religion? In C. A. Jones and P. Galison, eds, *Picturing Science: Producing Art.* New York: Routledge, 412–30.

Law, J. and Hetherington, K. (2000) Materialities, spatialities, globalities. Department of Sociology, University of Lancaster at http://www.comp.lancs.as.uk/sociology/soc029jl.html

Ley, D. and Samuels, M., eds (1978) *Humanistic Geography.* London: Croom Helm.

Livingstone, D. (2003) *Putting Science in its Place.* Chicago, IL: University of Chicago Press.

Massey, D. (1984) *Spatial Divisions of Labour.* Basingstoke: Macmillan.

Massey, D. (1991a) The political place of locality studies. *Environment and Planning A*, 23, 267–81.

Massey, D. (1991b) A global sense of place. *Marxism Today*, 24–9.

Massey, D. (1999) Space–time, science and the relationship between physical geography and human geography. *Transactions, Institute of British Geographers*, NS24, 261–76.

Massey, D. and Meegan, R. (1982) *The Anatomy of Job Loss.* London: Methuen.
Massey, D., Allen, J., and Sarre, P., eds (1999) *Human Geography Today.* Cambridge: Polity Press.
May, J. and Thrift, N. J., eds (2001) *Time–Space.* London: Routledge.
O'Sullivan, S. (2001) The aesthetics of affect: thinking art beyond representation. *Angelaki,* 6, 125–35.
Oyama, S. (2000) *Evolution's Eye: A System's View of the Biology–Culture Divide.* Durham, NC: Duke University Press.
Pile, S. and Keith, M., eds (1997) *Geographies of Resistance.* London: Routledge.
Pleasants, N. (1999) *Wittgenstein and the Idea of a Critical Social Theory.* London: Routledge.
Rose, G. (1993) *Feminism and Geography.* Cambridge: Polity Press.
Sayer, A. (1984) *Method in Social Science: A Realist Approach.* London: Hutchinson.
Sedgwick, T. K., ed. (1995) *Shame and its Sisters.* Durham, NC: Duke University Press.
Shotter, J. (1993) *Cultural Politics of Everyday Life: Social Constructionism, Rhetoric and Knowledge of the Third Kind.* Milton Keynes: Open University Press.
Sibley, D. (1995) *Geographies of Exclusion.* London: Routledge.
Thrift, N. J. (1983) On the determination of social action in space and time. *Environment and Planning D: Society and Space,* 1, 23–57.
Thrift, N. J. (1996) *Spatial Formations.* London: Sage.
Thrift, N. J. (1999) Steps to an ecology of place. In D. Massey, J. Allen, and P. Sarre, eds, *Human Geography Today.* Cambridge: Polity Press, 295–321.
Thrift, N. J. (2000) Afterwords. *Environment and Planning D: Society and Space,* 18, 213–55.
Thrift, N. J. (2002) Summoning life. In P. Cloke, P. Crang, and M. Goodwin, eds, *Envisioning Geography.* London: Arnold.
Thrift, N. J. and Pred, A. (1982) Time-geography: a new beginning. *Progress in Human Geography,* 5, 277–86.
Tiffany, D. (2000) *Toy Medium: Materialism and Modern Lyric.* Berkeley, CA: University of California Press.
Urry, J. (2000) *Sociology Without Societies.* London: Routledge.
Valentine, G. (1989) A geography of women's fear. *Area,* 21, 34–41.
Whatmore, S. (1998) Dissecting the autonomous self: hybrid cartographies for a relational ethics. *Environment and Planning D: Society and Space,* 15, 37–53.
Whatmore, S. (2002) *Hybrid Geographies.* London: Sage.

IV. Space

9

Order in space

geography as a discipline in distance

RON JOHNSTON

The 1960s saw a series of major changes in geographical practice in the UK, which interacted with similar changes in North America, where they started in the mid-1950s. To some, these constituted a 'conceptual revolution' (Davies, 1972a), creating a 'new geography'. Others, such as Chisholm (1975), argued that evolution better described the changes. Whether revolution or evolution, however, the changes were substantial.

The stimulus for change included increasing dissatisfaction with much contemporary geographical practice, especially that associated with the 'regional approach', with criticisms from both established scholars (e.g. Kimble, 1951; Freeman, 1961), which included a bitingly satirical piece lampooning regional geography (Buchanan, 1968; Johnston, 1999),[1] and younger members of the profession (e.g. Wrigley, 1965)—though there were some spirited defences, not least an early one from Wooldridge (1950). Alongside this internal discontent was a growing awareness of new research agendas in other disciplines—especially the burgeoning social sciences—with which geographers previously had little contact. For many, the first clear evidence that change was in the air came from the two collections of essays derived from courses on the 'new geography' offered by the University of Cambridge's extramural department (Chorley and Haggett, 1965a, 1967).[2] In addition, there were texts that presented new syntheses of geographical material (Haggett, 1965), new methods for analysing

[1] Keith Buchanan's paper was published under a pseudonym, but it was widely known that he was the author: I heard him give an illustrated seminar based on it!

[2] By then, the change had more than 'been in the air' for over a decade in the United States, but there is little evidence of it in the UK literature prior to 1965. Freeman (1961), for example, does not refer to Schaefer's (1953) seminal piece, and while the Schaefer–Hartshorne debate may have been discussed in some UK department common rooms and seminars (e.g. at the LSE) there is very little evidence that it had a wide impact.

geographical data (Gregory, 1962; Cole and King, 1968; see also Cole, 1969, on the removal of vagueness from geographical descriptions), and new philosophical underpinnings to geographical inquiry (Harvey, 1969a). The departments of geography at the universities of Cambridge and Bristol were placed at the core of the 'revolution' (Whitehand, 1970),[3] though these two were rapidly joined by a number of others—notably Leeds, where Alan Wilson's appointment to a chair in 1969 was a catalyst for major developments, with one of the few examples in UK geography departments of focused development on a particular line of work.[4] Many of those involved in the initial stages had recently visited North American universities and been strongly influenced by changes there (Johnston, 1997a).

The 'revolution' comprised several inter-related components:

1 *A concern for scientific rigour.* Much current geographical practice was portrayed as theoretically weak—'vague description' was a frequent pejorative comment on regional geography—and lacking the perceived objectively neutral approach associated with both the 'natural sciences' and, increasingly too, the 'social sciences'. Harvey's (1969a) *Explanation in Geography* was a clarion call for geographers to adopt the hypothetico-deductive 'scientific method', and concluded with the statement 'by our theories you will know us' (p. 486): explanation and prediction were to be the goals of human geographic research.
2 *An argument that quantitative methods formed a necessary component of this more rigorous approach* to the portrayal and analysis of geographic information, although not all of the early proponents of this cause tied it directly to the philosophical claims regarding 'scientific method'; some saw the adoption of standard statistical procedures as simply the proper way to use data (as in Stan Gregory's pioneering *Statistical Methods and the Geographer in* 1962).
3 *A claim that human geographers should focus on searching for spatial order in the patterning of human activities*, rather than on the definition of regions characterised by their uniqueness. Haggett's (1965) *Locational Analysis in Human Geography* was the clearest early statement of this, with many of the chapters in *Models in Geography* (Chorley and Haggett, 1967) providing fuller syntheses of relevant material and

[3] Chorley and Haggett were both at Cambridge in the early 1960s: Haggett moved to Bristol, in 1965.

[4] In most departments, as discussed in Chapter 2, teaching coverage was the main determinant of appointment policies, and although expansion stimulated departmental growth, few used the additional resources to concentrate on particular specialisms—and few built major research centres on outside funds until the 1980s.

approaches.[5] Geographers should switch their attention to 'horizontal order' from the 'vertical (land–society) inter-relationships' that lay at the core of the regional approach.

4 *A desire that human geographers' work should be applied to a wide range of 'real-world' problems.* Many geographers were concerned that their discipline lacked its deserved status among decision-makers (Wooldridge, 1956, 33–4; the issue still rankled in the mid-1970s: Coppock, 1974; Steel, 1974). Other social science disciplines—notably economics but increasingly also sociology, psychology, and political science—were much more influential, in part because they took the perceived more rigorous approach to problem-solving associated with the 'scientific method' and 'quantification'. Geographers should promote their expertise in the analysis and creation of spatial order—increasingly needed with the growth in town and country planning.

These changes were substantive, epistemological and methodological, therefore, and marked a clear break in the conception of geography's aims and methods. Some harbingers were produced in earlier decades, however. Fawcett (1919), for example, proposed a regionalisation of Great Britain as the basis for devolved government, using functional regions, the tributary areas of major cities. His ideas were taken up by one of his students, Dickinson, who identified hierarchies of such regions in his PhD work on East Anglia (and thereby predated the publication of Christaller's classic work on central places) and applied them in proposals for devolved government in post-Second-World-War Germany (Dickinson, 1938, 1945). From his wide reading in urban studies, and his travels during the 1930s across both Europe and the United States, he introduced UK geographers to a great range of material on city regions and their internal structure (Dickinson, 1947), which should be the basis for major geographical contributions to the emerging practice of town and country planning (see Dickinson, 1942). His only contributions to debates about the philosophical bases of geographical study related to the centrality of the functional region as a geographical concept (Dickinson, 1939); he did not explore either 'scientific method' or quantitative applications (Johnston, 2001a, 2002).

Others similarly pre-dated the main changes in economic geography, notably though not only some of those who read for the joint degrees in geography and economics on offer at the London School of Economics (LSE) and at University College London (UCL). R. O. Buchanan

[5] Note, however, Wise's comment (1968, xii) that in Wilfred Smith's economic geography 'order was there to be discovered, given the tools'; these were the geographer's 'special tools, those of field survey and of the map'.

(a New Zealander who had studied at the LSE in the late 1920s, and who was then on the staff at UCL for two decades before moving to the LSE in 1949), for example, was influenced by work on the localisation of industries by economists such as Alfred Marshall and Weber, applying 'well developed economic theories and principles, of the division of labour, of comparative advantage, of economies of scale, of transport costs' (Wise and Rawstron, 1973, 6); similarly, Stanley Beaver used the works of Weber, Hoover, and others in his teaching at the LSE,[6] and Brian Law, who obtained BSc and MSc degrees in geography and economics at UCL, introduced location theories to his courses there between 1951 and 1955 (before leaving for a business career that included being managing director of Mars). But although material from economics may have underpinned their approach to economic geography, little evidence of its impact appeared in their published works. There is, for example, no mention of it in Beaver's major text (with Stamp until the 1960s) on the British Isles (Stamp and Beaver, 1971); the third edition of Estall and Buchanan's (1973) *Industrial Activity and Economic Geography* has a two-page section on 'location theory and models' in which they explicitly eschew their construction (p. 24)—noting, for example, that one of their major disadvantages is 'their divorcement, in smaller or greater degree, from real world conditions', and in the fourth edition (1980) they argued, briefly, that Weberian models were no longer very appropriate in a changing world. Brian Law did, however, introduce Brian Berry to Lösch; Berry (1993) reports devouring the book en route to the United States to begin graduate school in 1953. An exception to the general absence of location theory in published work was Wilfred Smith (who was a student and then on the staff at the University of Liverpool, 1921–55). Weber's work influenced his studies of the location of industry (Smith, 1949, 1952, 1955; Wise, 1968), but there were few following his lead for more than a decade and his writings were not very influential later. A similar fate also befell Watson's (1955) essays on 'Geography as a discipline in distance' and the social geography of cities (Watson, 1953; see Johnston, 1993; on geographers and economists before the 1950s, see Lawton, 2001; Rawstron, 2002).

Applied geography was a major focus for other pre-1960s scholars, notably Stamp, whose Land Utilisation Survey of Great Britain was used in land use planning before and after the Second World War (Stamp, 1946); he also wrote the first book entitled *Applied Geography* (Stamp, 1960). Stamp was very much an empiricist, however, and far from enamoured by the 'theoretical and quantitative revolutions'; indeed, he

[6] Michael Wise (1956) reviewed Lösch's (1954) book in *The Geographical Journal*, emphasising its importance as a stimulus to geographers interested in 'the location problem'—on which Hartshorne's arguments regarding the nature of geography 'may be disappointing'.

suggested that Haggett's (1964) paper on quantitative analysis of land use change in Brazil could be interpreted as using 'an enormous steam hammer . . . in the cracking of nuts'.[7] Quantitative methods were being applied by a small number of geographers before the 1960s, too—notably by climatologists (see Crowe, 1936).[8] The big change was therefore epistemological, but the substantive and methodological concerns of a few became the focus of very many more.

Geography, functional regions, and spatial order

Dickinson's criticisms of regional geography focused on the concept of the natural region generally employed in the 1930s, founded on characteristics of the physical environment. He argued that regions are human creations, made by people structuring the world spatially. Much of that structuring involved the flows of people and commodities, in patterns focused on urban centres organised hierarchically. Such functional (or nodal) regions were the basic geographical units—a point also noted by Kimble (1951, 159) in his characterisation of traditional methods of regional delimitation as 'drawing boundaries that don't exist around areas that don't matter . . . from the air it is the links in the landscape that impress the observer, not the boundaries'.

Functional regions were the organising template in Haggett's (1965) seminal text. Their existence was largely taken as given, however, with little discussion of their nature or definition. The first edition drew on Fawcett and Dickinson to establish their salience, and the second additionally claimed that:

> The arguments for adopting the nodal region as the basic spatial unit are persuasive . . . A growing proportion of the world's population is

[7] Stamp's comments are in the discussion of Haggett's paper, published immediately after it in *The Geographical Journal* (Haggett, 1964, 380). The comments end with an exclamation mark, with the implication that these may not have been Stamp's own views, though his call for those using quantitative techniques to 'tear up some of their results' suggests a negative attitude to at least some of the 'new lines of thought'. Haggett had defended his approach—prior to Stamp's summing-up—by noting that 'While method has loomed large in this paper I hope that it has not blotted out the fact that . . . traditional problems of forest depletion are reaching critical levels. Statistical analysis can help to clarify some of the problems and provide some checks on intuitive solutions' (p. 380). It is also recorded that a few days after the paper was given, Haggett was told by his Head of Department at Cambridge that he was 'bringing the subject of geography into disrepute by applying such mathematical methods' (Chorley, 1995, 360).

[8] Stan Gregory was a student at King's College London when Crowe was at Queen Mary College and teaching climatology on inter-collegiate courses. He began to teach statistics to geography students at Liverpool in the 1950s. (When I was applying to go to University in 1958 I was 'warned off' Liverpool by my geography master because the course there was 'very mathematical'; I went to Manchester, where I was taught by Crowe!).

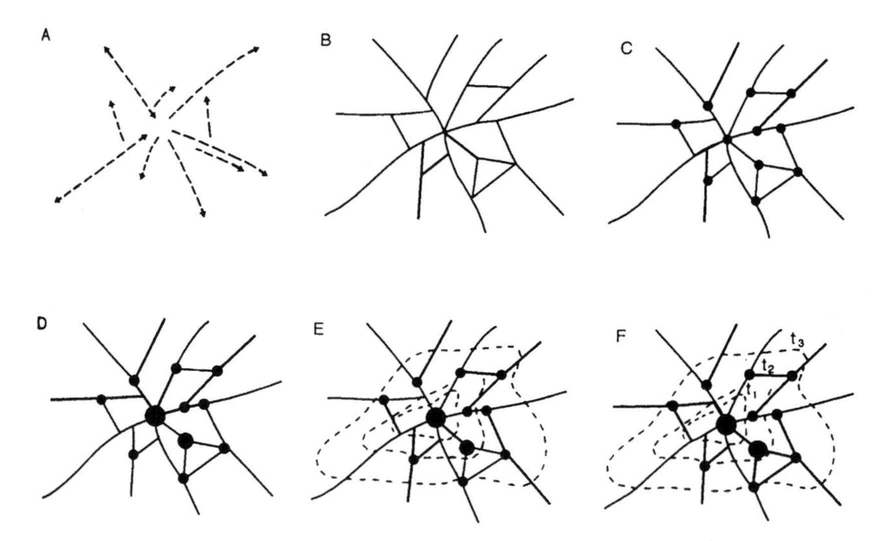

Fig. 9.1 The sixfold structure of a nodal regional system (after Haggett et al., 1977). The six components are: A Interaction; B Networks; C Nodes; D Hierarchies; E Surfaces; and F Diffusion.

> concentrated in cities, and consequently man's organization of the globe in increasingly centred about its major population nodes. Cities form easily identifiable and mappable nodal regional units . . . Second, city regions stress the comparability of different parts of the world, and thus encourage the search for general theories of human spatial organization. (Haggett, Cliff, and Frey, 1977, 8)

Both editions focused on the region's main components, as did the core of Haggett's introductory text, *Geography: a Modern Synthesis* (retitled in the latest edition as *Geography: A Global Synthesis*: Haggett, 2001). He initially identified five spatial components to nodal regions—flows, routes, nodes, hierarchies, and surfaces—adding a sixth (diffusion) in a second edition (Haggett et al., 1977) (Fig. 9.1). Haggett found many stimuli in work outside geography (such as the location theories of von Thünen, Hoover, and Lösch) as well as within (notably Christaller and Hägerstrand),[9] and used these—together with a wide range of illustrative examples, many from non-geographical sources—in the first part of the book, on 'Models of locational structure'. The second part dealt with 'Methods in locational analysis', outlining both the bases of the hypothetico-deductive method (later much expanded by Harvey, 1969a) and various standard statistical

[9] Christaller gets 18 entries in the index to the first edition, as does Isard, Lösch gets 24 and von Thünen 12. Apart from himself, the most-referenced geographers in the index are Chorley (18 mentions), Berry (17), Bunge (16), Garrison (15), and Hägerstrand (13).

procedures for the analysis of point, line, flow, surface, and area patterns. Harvey's (1969a) book, in particular, was immensely influential, presenting a detailed exposition of the contours of the positivist route to explanation based on a wide reading of philosophical materials, and introductions to related issues such as probability, geometry (as the language of geography), and systems: intriguingly, Harvey himself rejected that route soon after the book was published (Gale, 1972; Harvey, 1972, 1973; for a further influential critique, see Gregory, 1978).

Locational Analysis promoted this 'new geography' by example rather than by rhetoric; there was little proselytising. Haggett's focus—as was made clear in a later, autobiographical, volume (Haggett, 1989)—was the search for spatial order. This was placed at the heart of geography while recognising that order and chaos were very much individual interpretations: Haggett saw the distinguishing characteristic of scholarly work as its focus on 'universals . . . [on] pattern, structure and meaning' (p. xvi).[10]

There was a much wider range of exemplars in the second edition of *Locational Analysis*. Other books expanded the material on several of the schema's components: on line patterns (Haggett and Chorley, 1969: in this, Haggett, unlike most of his contemporaries, sought to sustain close links between human and physical geography); on regionalisation procedures using flow and other data (Cliff et al., 1975); in work on surfaces, which used the technique of trend surface analysis to analyse three-dimensional map patterns using an adaptation of general linear models (Chorley and Haggett, 1965b; Wrigley, 1977); and, most notably, a series of books on diffusion, focusing on disease epidemics (e.g. Cliff, Haggett, and Ord, 1967; Cliff, Haggett, and Smallman-Raynor, 1983; Haggett, 2000). The methods employed in disease diffusion studies were also applied to a range of other data—notably unemployment rates—but attempts to model and predict changes over space and time (as lead–lag sequences) were not particularly successful. (This was exemplified in Chisholm, Frey, and Haggett's 1970 volume on regional forecasting, as in the chapter by Bassett and Haggett on spatial variations in unemployment over time; see also Haggett, 1971.) Others with strong links to Haggett contributed parallel works on point patterns (Getis and Boots, 1978), and by the early 1970s a large corpus of work had been reported on these various topics. Meanwhile others explored some of the underpinning theories in considerable detail—Chisholm (1962) and Hall (1966) on von Thünen,[11] and Smith (1971) on industrial location theories. This work was used to 'sell' human geography

[10] On Haggett's career, see Chorley (1995) and Thrift (1995).

[11] Chisholm and Hall were both students with Haggett at St Catharine's College, Cambridge, in the early 1950s, where they were students of A. A. L. Caesar, who supervised a large number of students who became leaders of the discipline from the 1960s on (Chisholm and Manners, 1971). His lectures were strongly influenced by economic approaches to location theory—e.g. von Thünen—as well as Christaller's central place theory.

to the social sciences (as in Chisholm, 1971a; see also Chapter 2 above)—in parallel to a similar exercise in the United States (Taaffe, 1970)—and became the core material in contemporary undergraduate texts (e.g. Lloyd and Dicken, 1972).

Haggett's sixfold decomposition of a nodal region did not fully cover all of its aspects (Johnston, 1997b), paying little attention, for example, to the division of the surfaces into bounded 'regions' (though see Haggett, 1981, on the impact of what he terms 'categorical spaces' on the study of spatial patterns, and also Taylor, 1971b). Many urban geographers were attracted by sociological studies of social areas in cities (Jones, 1960; Williams and Herbert, 1962; Clark and Gleave, 1973; Rees, 1978), producing conclusions regarding typical British forms (Robson, 1969; Timms, 1971). Statistical procedures for data reduction (e.g. identifying the main patterns in a series of maps through principal components analysis: Johnston, 1978; in studies of urban social areas these became known as factorial ecologies: Rees, 1971, 1978; Davies, 1984) and the classification of areas attracted much attention, with regionalisation presented as a special case of general scientific procedures of classification (Grigg, 1967). But formally bounded spaces in spatial organisation at various scales were largely ignored, so that political geography, which focuses on such spaces—states and their subdivisions—was largely excluded from the 'new geography project' (Johnston, 2001b). Later work on territoriality (notably two essays by Taylor, 1994, 1995) also lay outside the spatial analysis theme.

The value of statistical procedures is precision and accuracy, for both summarising large datasets (descriptive statistics) and testing hypotheses (inferential statistics: Hay, 1985a); they allow confidence in making generalisations. Successful tests of hypotheses provide not only explanations (as a basis for theory-building) but also tools for predictions: Haggett (1994, 6) maintained that 'prediction plays an essential part in the building of models of geographical systems', a claim illustrated with reference only to epidemiology.[12] The availability of computers facilitated quantitative work, much of it in the early years done in urban geography, with a twin focus on systems of cities (the hierarchical arrangement of urban places in a country or region) and cities as systems (the internal structuring of land uses within cities).

The study of flows attracted considerable attention, in part because of their relevance to contemporary planning concerns with urban growth

[12] Gregory (1994, 59) argues that Haggett's 'main concerns have always been descriptive-predictive', and that he has little interest in 'theoretical questions' and explanation. Haggett himself agreed with this in a published interview (Browning, 1982, 54). In response to a statement that 'It seems to me that you are naturally inclined towards models but not necessarily theory', he replied: 'Yes . . . I think that the broader theories of the subject are not something I can cope with easily'.

and traffic patterns. Very popular was the gravity model, which predicted flows as a function of characteristics of the origin and destination areas plus the distance between them. A major breakthrough in its application was made by Alan Wilson (1970, 1974), who brought mathematical insights to bear on this modelling strategy.[13] His adaptation of entropy-maximising procedures to the modelling of flows was integrated with other theories of location to produce general location-allocation models that could be used both to analyse contemporary urban forms and to design new ones—of what is done where, and of how much movement there is between places (see Wilson, 1970; Batty, 1976, 1978).

Wilson's approach differed from the statistical methods preferred by many; rather than test hypotheses about parts of the whole (the relationship of the volume of traffic to distance, for example), he sought to model the whole. This was also the case in work on spatial demography, which introduced spatial variations to traditional models of population change that focused almost exclusively on fertility and mortality (e.g. Rees and Wilson, 1977): Woods and Rees (1986, i) introduced their edited volume as 'a book with a message. It argues the case for viewing demographic patterns, structures and systems from a spatial perspective'. Wilson's methods were later adapted for other purposes, such as the resolution of the ecological inference problem discussed below for estimating spatial variations in various aspects of electoral behaviour (Johnston and Hay, 1982; Johnston, Pattie, and Allsopp, 1988; Johnston and Pattie, 2000, 2001). Wider uses of entropy, as a measure of information that can be used to generalise map patterns, were explored by Chapman (1977).

Theory—what theory?

The proposed epistemological shift posed some of the biggest problems in this putative revolution, because there was virtually no theory within the discipline on which geographers could draw, and little appreciation of what might be available elsewhere. British geography prior to the 1960s was very strongly empiricist. The work of Dickinson (1947) and Smailes (1953) on urban hierarchies, for example, was almost entirely descriptive and, although it was referenced, Christaller's (1933) original formulation of central place theory had no apparent impact on how they approached their subject.[14] Similarly, only some of the economic geographers interested in the location of industry paid much attention to theoretical work on this subject by economists. Most of the generalisations derived by geographers

[13] Gould (1972) referred to Wilson's (1970) book as one of the hardest, but also one of the most important, he had read in geography.

[14] Dickinson was certainly well aware of Christaller's work, and spent time with him in 1935 when on his extended stay in Germany (Johnston, 2001c).

before the 1970s—as in Conzen's (1969) work on townscape development—were inductive, however; the general approach (demonstrated in Dickinson's 1951 book, *The West European City*) involved case studies leading to classifications of general types, though in ignorance of sociologists' work on the role of ideal-types in theory development.[15]

The search for a theoretical base for geographical studies thus involved a major rupture in disciplinary practice. Since travel costs are almost invariably related to distance travelled, then minimising distance should be a key variable in decision-making—and since geography is about space, then distance should be the key variable that geographers bring to social science. Alongside American contemporaries, therefore, British geographers (a number of whom had obtained higher degrees in the United States) explored the literature of other social sciences, notably economics, sociology, and the newly-created regional science, for stimuli. The work on central places initiated by Christaller and Lösch (1954) attracted much attention, as did the industrial location theories of Hoover (1948),[16] Weber (1929), and others, the agricultural location theories of von Thünen and its application by other economists (e.g. Clark, 1951),[17] Perroux's (1955) ideas on growth poles, and, a little later, the diffusion theories of Torsten Hägerstrand (1968). Other than Christaller, Hägerstrand—a Swede—was the sole trained and practising geographer among this group of pioneering locational theorists, and his work only gained widespread appreciation after his visit to one of the early centres of quantitative work in the United States (Duncan, 1974). From these, geographers derived a general argument that, if those making locational decisions are driven by the desire to maximise profits, they should seek the least-cost locations with regard to transport costs. Similarly, those who purchase goods and services should seek to minimise their travel costs to the point of purchase, thereby maximising their available amount to spend on the goods and services themselves.

'Geography as a discipline in distance' was the title of Watson's (1955) inaugural professorial lecture at the University of Edinburgh, but it made little impact on the wider community (Johnston, 1993).[18] In part, perhaps,

[15] There was little valuable cross-fertilisation with physical geography, either; that too was largely inductive in its approach until the 1960s, and relied very strongly on ideal types (such as Davis's model of landscape development: Chorley, Beckinsale, and Dunn, 1973) with no hypothesis testing of either the fundamental assumptions or the assumed consequences.

[16] Hoover's theories were taught by Stanley Beaver at the LSE just after the Second World War.

[17] Michael Chisholm, who later published one of the first geographical critiques of von Thünen's work (Chisholm, 1962), was at the Oxford Institute for Research in Agricultural Economics with Clark in the 1950s.

[18] It got one brief mention, in the chapter by Garner on urban geography, in *Models in Geography* (Chorley and Haggett, 1967, 304), for example, an even briefer mention in *Locational Analysis* (Haggett, 1965, 13), and just a little more in *Explanation in Geography* (Harvey, 1969a, 210).

this was because he took a rather idiosyncratic definition of distance as 'the extent to which objects adapt themselves to, or dominate, their environment' (p. 2), although he later wrote of 'cost' as well as 'geographical' distances, which 'came to exert a preponderant influence in human geography' (p. 3), and he expanded the discussion to incorporate what he called time distance and social distance as well. But it was only in the 1960s and 1970s that the theme was developed by others, however, and extended to a wide range of subject matter. Within urban geography, for example, distance was seen as a key variable underlying decisions such as which shopping centre to patronise and where to seek a new home (Johnston, 1971, 1973). The 'frictions of distance' became a key phrase.

The focus on distance underpinned geographers' modelling and hypothesis-testing regarding both location patterns and flows between points and areas: central place theory (which posited a nested arrangement of settlements on a hexagonal grid) was the foundation for much work on point patterns and the gravity model played a similar role for studies of flows. Some of the models were deductive, deriving optimum location and transportation strategies from 'first principles': they suggested the 'best' locations for activities and the most efficient pattern of flows through a transport network. Many of the 'predictions' were relatively imprecise, however. The gravity model, adapted by American economists and sociologists as a 'law of least effort', predicted that the greater the distance between two places of similar size, the smaller the amount of movement (e.g. of migration or of telephone traffic) between them, holding their relative size constant. This could be tested by regressing the volume of movement against distance, expecting a negative slope coefficient—but with no suggestion as to its size (and thus the steepness of the distance-decay pattern: Taylor, 1971a). Such tests allowed general statements to be made about spatial relationships and patterns, but no more. Although the regression and correlation coefficients gave apparently precise statements of the nature of the relationship for the dataset analysed, they were only determined empirically: predictive ability was not advanced substantially, since little was known about why the coefficients varied between case studies. (An early presentation of this case was Sayer, 1978.) Suggestions for alternatives to the gravity model included an 'intervening opportunities' formulation, which postulated that the amount of movement between two places was not simply a function of the distance between them, but rather of the number of intervening opportunities: more people will travel ten miles to work between *A* and *B* if there are no closer alternatives than will be the case if there are jobs at *C* only five miles away. A further formulation, a competing destinations model, argued that people first select a cluster of potential destination choices and then choose a particular destination from within that, spatially more restricted, set (Fotheringham, Charlton, and Brunsdon, 2000, ch. 9).

The gravity model was built on a relatively weak theoretical base, therefore. Wilson placed the empirical relationship on a much sounder mathematical foundation. He showed that in its usual formulation the model could readily produce 'nonsense' forecasts of the amount of movement between two places. For example, if the model is being used to forecast commuter flows between areas, then the maximum flow between any pair cannot exceed either the number of commuters from the origin area or the number of jobs in the destination area. His constrained formulation—fitted using entropy-maximisation procedures which identify the most-likely pattern of flows, given the characteristics of both origins and destinations plus the capacity of the links between them—limited the forecasts to these maxima, and also to a cost function reflecting the total amount expended in the system. (However, the constraints, especially the total amount of energy/money expended on movement, were usually empirically based.) This model was extended in a variety of ways, providing a mathematical theory on which to base studies of a wide range of flow phenomena. (See also Johnston's 1976 work applying the gravity model in studies of international trade, and the debate on the interpretation of regression coefficients in such constrained situations: e.g. Cliff, Martin, and Ord, 1974; Johnston, 1975; Sheppard, 1984.)

The mathematical modelling introduced by Wilson involved a more holistic approach to the analysis of spatial structures (at various scales: Johnston, 1973) than that adopted by those who focused on testing hypotheses about one of their components (such as a flow pattern). But all elements in a system influence most others, either directly or indirectly, through a variety of feed-forward and feedback effects; analysing the relationships between some components only may thus be difficult because of the problems of 'holding everything else constant'. To counter this, some delved into systems theory and analysis, with Stoddart (1967) arguing that the latter provided geographers with a 'unifying methodology'. Wilson (1974, 1981) introduced a range of models that were applied in particular contexts (Wilson, Rees, and Leigh, 1977); Bennett and Chorley (1978) extended the range of material covered across both human and physical geography;[19] and others explored the nature of systems relationships in geographical contexts (e.g. Langton, 1972; Huggett, 1980). Some were attracted by the notion of a general systems theory that could place human geographical endeavours within a broader interdisciplinary framework—although Chisholm (1967) argued this was little more than an 'irrelevant distraction'. Various forms of system thinking continue to provide stimuli, however, as in Batty and Longley's (1994) use of fractal

[19] Although a physical geographer, Chorley was highly influential on many human geographers in the 1960s and 1970s, through his work with Haggett and as a tutor at Cambridge. His 1962 paper on systems analysis was seminal, for example.

geometry (relating to objects 'whose irregularity as a non-smooth form is repeated across many scales': p. 59) to identify order in the physical form of cities (and using modern visualisation techniques—see below—to display those forms).

Simple models related behaviour to distance, as in the classic gravity model. But what is the exact relationship with distance—linear, exponential, power, or what? The time and cost involved in movement of goods have two components: the terminal costs and the movement costs. Since the latter may be fixed, whatever the length of the journey, then the total cost will only increase as a function of the relationship between movement costs and distance, thereby introducing a non-linear element. Rather than exploring these aspects of measuring the impact of distance, however, most geographers either took an inductive, empiricist approach, finding the best-fit relationship within a dataset (as explored in detail in Taylor, 1971a)—and then often did not explore why a particular relationship provided the best fit, or assumed a particular functional form without much theoretical foundation for that decision.

Recognition that although distance is related to many aspects of behaviour the exact form of that relationship can vary has been incorporated into some analyses. Distance as a barrier/constraint to movement can be manipulated by human ingenuity, with regard to the speed of movement along different routes, for example: it is not a fixed constraint, but a plastic one (which ultimately may disappear, some argue, with information flows that are almost instantaneous). Explorations of the plasticity of space have investigated the 'distortion' of spatial structures, as with bringing places with more frequent, faster public transport services between them 'closer' in space as well as time, relative to those with less provision (on such map transformation methods see Forer, 1978, and Gatrell, 1983; Cliff and Haggett, 1998, produced examples from their work on disease diffusion involving 'forced spaces'—the maps of 'nearness' derived from flow patterns).

Two difficulties were identified with the deductive approach to modelling and testing. The first, as just discussed, was that many of the hypothesised expectations were overly general—as were the conclusions drawn from the empirical tests. The second was that the models themselves were over-simple, based on restricted assumptions regarding decision-making—particularly linking profit-maximisation to distance-minimisation. (Interestingly, these models have recently been 'rediscovered' by economists who are promoting a 'new economic geography' based on the same simplifying assumptions with regard to the role of distance: see Martin, 1999.) The concept of satisfying decision-making attracted attention, therefore, notably among those interested in industrial location: decision-makers clearly need to make profits, but may not wish to maximise them, with other factors influencing their final choices. Rawstron (1958) developed three principles of industrial location—physical restriction, economic restriction, and technical restriction—which allowed him to constrain locational choices, and these ideas were developed further by Smith (1971). Once the spatial margins within which profitable decisions could be made were identified, then empirical work could explore why particular sites were chosen.

Spatial pattern and spatial behaviour

In general, therefore, geographers in the early years of spatial analysis concentrated on the outcomes of locational decision-making (which they displayed in aggregate form) rather than on the processes producing them. Simple (or naive) assumptions of human behaviour were used to derive models of how the world should look, and these were the ideal types against which mapped patterns were evaluated. Any 'deviations'—i.e. the world didn't look as it was supposed to (Lösch, 1954, argued that 'the laws revealed through theory are the sole economic guide to what *should* take place . . . the question of the best location is far more dignified than the actual one')—were 'explained away' (e.g. by deviations from the uniform plain on which regular hexagonal arrangements of central places were supposed to develop) rather than taken as reasons for re-assessing the assumptions and models. Eventually, however, the need for a re-assessment was appreciated and new approaches explored: the search for order was not dropped, but there was much less emphasis on the goals of deriving law-like generalisations with wide applicability and comparing 'reality' against idealised spatial arrangements based on simplistic theories.

In this extended search for the principles underpinning spatial decision-making geographers began to investigate individual motivations and behaviour patterns in more detail. (One of the first papers to explore this was Kirk, 1963; see Boal and Livingstone, 1989.) They argued that information is key to decision-making, and that this is not universally available: it has to be sought out, which takes time and thereby imposes a cost, so that it may be better to make an 'imperfect', but 'satisfying', decision rather than spending more time and money on gaining further information that might enable a slightly 'better' outcome. Furthermore, the flow of information is itself spatially structured. Much of it passes through interpersonal networks, many of which are spatially concentrated; the closer you are to the source of some information (and in many cases a particular interpretation of it), the more likely you are to receive it. And the search for information is also often spatially structured: people looking for a new home, for example, may concentrate on particular areas, perhaps close to their current home or to an important point in their life, such as the school they wish their children to attend. (For a British review of this work, much of which was North American in provenance, see Gold, 1980; one major area of investigation involved uncovering 'mental maps'—see Pocock and Hudson, 1978.) In this context, some simply explored 'behaviour in space', whereas others sought to generalise from their unique studies by deriving rules of 'spatial behaviour' (much of which is reviewed in Golledge and Stimson, 1997).

Work on information flows reintroduced the concept of place to geography, albeit in a new guise—and not the only one adopted, as later chapters in this book illustrate. Places—or locales, or communities, or regions, or . . .—are contexts within which decisions are made. Understanding what people do needs an appreciation not only of what they are (what Thrift, 1983, called a compositional approach) but also where they are (or have been: a contextual approach). Hägerstrand's notion of 'time geography' provided a major impetus to this form of thinking (Thrift, 1977). Since all behaviour occurs at spatial and temporal coordinates,[20] then these two fundamental variables combine as major constraints—you can only be in one place at one time. Hägerstrand identified three separate constraints to such interaction: capability constraints (such as the ability to cover distances); coupling constraints (the need for those involved in an activity to be in the same place at the same time); and authority constraints (the bars to some people being in some places at some times). These provided a framework for behavioural analyses (Carlstein, Parkes, and Thrift, 1978; Parkes and Thrift, 1980).

One set of ideas in which place is crucial comprises the concept of a neighbourhood effect, of considerable importance in electoral geography. Pioneering work by Cox (1969) on spatially patterned flows of information relevant to electoral decision-making suggested how local politically biased information flows generate spatial clustering of partisan support patterns. Subsequent studies produced evidence of spatially varying voting patterns consistent with this model (see Johnston, 1986) but provided only circumstantial evidence that similar people act differently in different milieux. Identifying the characteristics of the local areas around the homes of individual respondents to large social/electoral surveys—their bespoke neighbourhoods—has produced more convincing evidence (e.g. Johnston et al., 2001; MacAllister et al., 2001), including studies of the evaluation of local economic conditions as part of the voters' calculus (Johnston et al., 2000a). Work on patterns of conversation has directly addressed the processes underpinning such geographies (Pattie and Johnston, 1999, 2000).

This behavioural work involved a subtle shift in the focus of geographical work—with place joining space as a core geographical concept (Taylor, 1999). The role of distance was downplayed somewhat: instead of studying

[20] An argument that some might now challenge with regard to cyberspace and 'the end of geography' (though see Dodge and Kitchin, 2000).

how information flowed over space, for example, geographers turned to how the informational content of different places was created, how it influenced the behaviour of individuals there, and how it was subsequently changed through that behaviour. Giddens (1984), drawing on the spatial constraints identified in Hägerstrand's time-geography, referred to these as structuration processes operating in locales. The world comprises a myriad places—some bounded, some overlapping, some nested into others—which act as the contexts for learning (Johnston, 1991). In studying them, geographers have re-emphasised a further basic spatial variable—scale: places as contexts operate at a variety of scales, all of which may significantly influence behaviour (Johnston et al., 2000b). Bringing space and place together involves, in Hudson's words:

> Space can be thought of as a product of stretched-out social relationships, place as the condensation of intersecting social relationships in a specific time–space context. (Hudson, 2001, 282)

Although it is possible to develop deductive models of behaviour employing a wider range of factors than just distance, and then to derive testable hypotheses, most of those attracted to this 'behavioural geography' adopted inductive approaches (Harvey, 1969b). Individual decision-making was explored through social surveys. Design of questionnaires and sampling frameworks for their administration became important geographical exercises, and statistical procedures for analysing the collected data were adopted—such as modifications to regression analysis in order to handle binary and multinomial (categorical) data (Fingleton, 1984; Wrigley, 1985; O'Brien, 1992).

A combination of deductive and inductive approaches characterised much of the empirical work undertaken within this geographical paradigm over the last third of the twentieth century. To some, this was a retreat from theory that severely constrained geographers' ability to predict future spatial forms: the traditional theories were thus still used as the basis for suggesting optimum decision-making strategies—in, for example, the location of new service facilities, as in the work of a consulting company (GMAP) founded in the School of Geography at the University of Leeds. (The general direction of their work was outlined in Clarke and Wilson, 1987: for examples, see Longley and Clarke, 1995, and Birkin et al., 1996.) Others argued that the inductive route offered greater potential returns in applied areas (Openshaw, 1994), as in work on geodemographics, which took existing methods of classifying small areas using census data, added data available from studies of consumer behaviour, and developed profiles of area types that could be used by commercial firms in their direct marketing strategies (Longley and Harris, 1999).

Against this more empiricist approach, others argued that inspiration should be sought in areas of the social sciences other than neo-classical economics and the associated discipline of regional science (Chisholm, 1966, 1971b). Variants of Marxian economics—not only the work of Marx himself but also twentieth-century theorists such as Wallerstein—attracted considerable attention, with geographers introducing spatial components to their models of patterns of economic development. Both groups argue that space is key to the structuring and restructuring of economies (Harvey, 1982; Taylor, 1982), but they are unable to present simple spatial representations of the geography of various development levels: the detailed contours at any time reflect the contingent conditions then, within the context of inherited patterns. In this approach to social-spatial science, the general rules cannot predict what the particular outcomes may be—as argued in Sayer's (1984) presentation of realist approaches to science, which can be applied in both the human and the natural worlds. There are general tendencies, but these are implemented in particular contingent circumstances, which may not be repeated. Thus explanation can unravel the cause-and-effect relationships for a particular case, but not generalise from them because the circumstances may never reappear. In the positivist approach to science, explanation is based on replicability—if *A*, then *B*—in closed systems: most of the subject matter of human geography involves open systems, where *A* never recurs in the same context, and so the same outcome cannot be predicted. (Indeed, because decision-makers are continually learning from their experiences, the same context may never reappear, a point made by a human geographer even before the 'quantitative and theoretical revolutions' were launched: Clark, 1950.)

Geography, statistics, and spatial statistics

The first edition of *Locational Analysis in Human Geography* (Haggett, 1965) presented standard statistical procedures (based on the general linear model) as the means for testing many hypotheses quantitatively. These were widely adopted, especially regression and related techniques (analysis of variance, principal components analysis etc.)—increasingly so as reliable statistical packages became available (Wrigley and Bennett, 1981, have a comprehensive set of reviews of that work). But these methods were excluded from the second edition 12 years later, because of a very substantial problem associated with their application to geographic data, identified and explored by one of Haggett's colleagues, working in collaboration with a statistician—the *spatial autocorrelation* problem (Cliff and Ord, 1973, 1981).[21]

[21] This was one of the issues arising from 'the peculiar nature of geographical data' introduced to statisticians in the mid-1970s (Taylor and Goddard, 1974).

Haggett et al. introduced this problem in the following way:

> Since the first edition of this book appeared, the use of conventional statistical models and tests of inference . . . has become common place in human geography. It is not the purpose of this chapter to outline the procedures for carrying out such analyses. Standard accounts of the methods exist . . . rather, the aim of this chapter is to examine those properties of geographical data which make their analysis using statistical methods of the sort cited more difficult than might at first appear. (Haggett et al., 1977, 329)

The issue was the problem of spatial dependence. Economists had identified temporal autocorrelation as presenting problems with applying the general linear model to time series. Because one value in a series is likely to be influenced by previous values of that variable at the same place (house values in city x at time t are likely to be closely related to their values there at time $t{-}1$), use of regression with such a dataset violates one of the method's key assumptions—the independence of the observations; as a consequence, the estimated regression parameters will be inefficient. This problem is more severe in geographical applications, because whereas temporal autocorrelation is unidirectional (history always moves forwards!), spatial autocorrelation is multidirectional—the value of house prices at one point in a city can be related to their value at other places in all directions from that point, at both the same time and at previous times—though there may be directional variations in the strength of the relationship. Indeed, Haggett et al. (1977, 330) argued that 'the assumption of independence cannot be sustained for spatial data' so that: 'If geographers continue to apply the usual forms of many of the basic statistical models to spatially autocorrelated data, then a very severe risk is run of reaching misleading conclusions' (p. 336).

There was something of a paradox in the spatial autocorrelation issue, therefore. A considerable proportion of geographic theory suggested that like events cluster together, in part at least because they spread out from core areas. (Waldo Tobler (1970, 236) once expressed this as 'the first law of geography: everything is related to everything else, but near things are more related than distant things'.) But because of this, standard statistical methods for identifying such patterns and their relationships to others, as shown through quantitative map comparison, cannot be used for substantial technical reasons. Cliff and Ord (1973, 1981) explored the issues raised by spatial, temporal, and space–time autocorrelation, and identified means of circumventing them. This involved specific procedures, which they and Haggett applied in their continuing studies of diffusion patterns—especially of infectious disease. Others (e.g. Hepple, 1974; Bennett, 1978; Haining, 1990) took the ideas forward, developing families of procedures for tackling the problems—procedures that have probably been adopted less than should have been the case.

The importance of the spatial autocorrelation problem looms large in a late-1970s survey of quantitative geography (Wrigley and Bennett, 1981) and was one of the bases for contacts between geographers and statisticians. These included joint conferences organised through the IBG's Quantitative Methods Study Group (see Chapter 2 in this volume), a special issue of *The Statistician* aimed at introducing geographers' work to a wider audience (*The Statistician*, 1974),[22] and two presentations at meetings of the Royal Statistical Society (Cliff and Ord, 1975; Bennett and Haining, 1985; Bennett, Haining, and Wilson, 1985).

Identification of the spatial autocorrelation issue stimulated a major bifurcation in geographers' quantitative work. Many continued to rely on the basic statistical procedures—either declining to take account of the warnings or, in a few cases, countering by arguing that the issue was not relevant in all circumstances (e.g. Johnston, 1984). Many applications were facilitated by the increasing availability of standard statistical packages, and textbooks aimed at this market. Few books on statistical methods for geographers focused on such issues, or indeed on the analysis of spatial data: they concentrated on standard statistical procedures, with geographical examples but little or no recognition of problems involved in their application to spatial data. Indeed, only a few texts have focused on spatial (or map) analysis. Unwin (1981), for example, dealt with point, line, area, and surface patterns, along with map comparison, providing a clear emphasis on spatial statistical analysis; Lewis (1977) dealt with point and line patterns for a more advanced audience; Upton (a statistician) combined with Fingleton (a geographer) in an advanced two-volume treatment of aspects of spatial data analysis (Upton and Fingleton, 1985, 1989); Gatrell, in collaboration with a statistician, published a text with worked examples on a CD-ROM (Bailey and Gatrell, 1995); and Fotheringham et al. (2000) brought out a book as 'a statement on the vitality of modern quantitative geography . . . where the [recent] emphasis has been on the development of techniques explicitly for *spatial* data analysis' (pp. xi–xii). Similarly, relatively few followed the paths set out in two editions of a book on mathematical models (Wilson and Kirkby, 1975; Wilson and Bennett, 1985). To some extent, spatial statistical analysis—as against statistical analysis in geography—was a minority activity.

Developments in spatial analysis

An important technical issue identified and worked upon by geographers related to a wider issue in data analysis for social scientists, who frequently want to draw conclusions about individual behaviour but only have

[22] This included papers on probability, Markov models, spatial series analysis, spatial demography, multivariate analysis, and factor analysis.

information on population aggregates relating to everybody in a particular area. For example, they may want to show that wealthier people live in bigger homes, but their data only allow them to conclude that areas with high proportions of wealthy people also have high proportions of big homes, from which they may infer that wealthy people live in big homes. That is an *ecological inference* only; all the big homes could be occupied by relatively poor people, and vice versa. This is an example of an *ecological fallacy*, of concluding that a relationship that holds at one scale—the aggregate—also holds at another—the individual. In striving to avoid it, one can only indicate that the aggregate relationship implies the individual relationship, but that the evidence is not conclusive.

This fallacy presents a potentially difficult problem for many geographical analyses using aggregate data, such as those derived from censuses. Wilson's development of entropy-maximising methods has been adapted to infer individual characteristics and relationships from aggregate data with some degree of certainty, as in electoral geography to produce estimates of spatial variations in relationships such as the number of voters in a given area who voted for the same party at two consecutive elections (Johnston and Pattie, 2000, 2001). Alternative approaches to the same issue include a modelling strategy developed by Steel and Holt (Wrigley et al., 1996) and the use of micro-simulation to estimate local geographies (e.g. of where 'yuppies' live within a city) from various aggregate datasets collated at larger scales (see Voas and Williamson, 2000).

Geographers extended appreciation of the ecological fallacy through their work on the *modifiable areal unit problem* (*'maup'*), first brought to geographers' notice in a substantive way by Openshaw and Taylor (1979: the term comes from Yule and Kendall's (1950) work on 'modifiable units' with regard to the correlation of wheat and potato yields in the UK). They argue that there are no 'natural areal units' for geographical analysis of statistical data, and most studies use sets of areas constructed as amalgams of contiguous smaller areas—often for administrative convenience rather than statistical analysis, such as the counties employed by Yule and Kendall. Openshaw and Taylor showed that using different amalgamations of the same set of small areas as observation units may (indeed, is very likely to) produce different statistical results in ecological analyses. The correlation between the percentage of the population aged over 60 in Iowa and the percentage of the population who voted Republican at the 1968 Congressional election varied from .2651 to .8624 across five different aggregations of the State's counties into six large regions, and was .3466 when the observation units were the 99 counties. The 'maup' is produced by two separate components: an aggregation issue—different aggregations of small areas into larger ones produce different results, with the extent of the difference a function of the degree of spatial autocorrelation in the data; and a scale issue—larger aggregations tend to produce different results from smaller ones, and usually the larger the areas (and thus the smaller the number of areas) the larger the correlations.[23] At the extreme, their findings suggest that it may be possible to generate any result you want—i.e. seek the correlation (either negative or positive) that best fits your theory (Openshaw, 1977, 1996)! More sedately, Openshaw and Taylor recommend caution, recognising that different results are possible depending on the construction of the dataset used (over which the geographer may have no control; many ecological analyses use areal data provided by census and like authorities).[24]

This work on 'maup' was linked to other studies of spatial aggregation—such as the delimitation of parliamentary constituencies. The geography of support for different political parties could result in different outcomes when the votes cast are translated into seats, depending on the particular spatial aggregation chosen. Gudgin and Taylor (1979) developed a statistical account for this. If a set of constituencies was defined for partisan gain, a bias in the election results favouring the party involved (i.e. gerrymandering) should occur: they showed that a non-partisan approach—such as that employed by the independent Boundary Commissions in the UK—would also favour one party over another (what they termed non-partisan gerrymandering). Further evaluation of this has extended appreciation of such non-partisan cartography (Rossiter, Johnston, and Pattie, 1999) and shown that the amount of bias in the UK electoral system has increased over recent decades (Johnston et al., 2001).

Other work has adapted methods developed elsewhere for geographical applications. One of the most fruitful has been multi-level modelling, introduced by educational statisticians to unravel the relative importance of individual and school factors in pupils' performance. Jones (1991) generalised this method to examine the impact of local context on a range of behaviours and conditions, inquiring whether similar people act differently in different contexts. (For a full description, see Jones and Duncan, 1996, 2002.) In the original formulation, children's' performance is modelled as a function of their individual and home characteristics, plus those of the school they attend (and perhaps also their form within that school and the school district in which it is located). Some of the data refer

[23] This issue was recognised by American sociologists in the 1960s (Duncan, Cuzzort, and Duncan, 1961), and applied to a popular area within geography—urban ethnic segregation—more than a decade later (Woods, 1976); for more recent work on that issue, see Johnston, Forrest, and Poulsen (2001).

[24] Although in another piece they argued that three reasons are advanced to ignore the 'maup': 'it is insoluble; . . . it is of trivial importance; . . . [and] that to acknowledge its existence would cast doubt on nearly all applications of quantitative techniques to zonal data. . . . the first two of these are incorrect, the third is quite correct' (Openshaw and Taylor, 1981, 67). For one example of the impact of zoning and aggregation issues—on gravity model parameters—see Fotheringham and Wong (1991).

to the individual, and some to groups of individuals within the analysed population, and multi-level modelling provides an efficient way of distinguishing the various influences (i.e. 'nature' and 'nurture'). Geographical applications have included studies of voting (in which the relationship between the class structure of a constituency in the UK and the percentage of votes won by the Labour party was shown to vary between different parts of the country: Jones, Johnston, and Pattie, 1992; Jones, Gould, and Watt, 1998) and a number of other geographies, such as those of unemployment and various aspects of health and health care (as illustrated in a special issue of *Environment and Planning A*: Bullen, Jones, and Duncan, 1997).

A further form of geographical data analysis concerns spatially varying relationships. It is commonly, if implicitly, assumed in spatial analyses that relationships do not vary across the selected study area. Geographically weighted regression (GWR: Fotheringham, Brunsden, and Charlton, 1998) addresses whether this is so by extending a procedure for spatially decomposing regressions (the expansion method: Jones and Casetti, 1991) to continuous relationships. Over an entire region there may be an observed relationship between, say, school performance and the social class of its catchment area, but GWR—by fitting the same regression to different sub-areas within that region—may identify variations in the relationship's strength, extent, and even direction, showing geographical variations in behaviour that are concealed by the global regression (Fotheringham, Charlton, and Brunsden, 2001). GWR is one example of a class of methods known as 'local statistics', of which there are three main types (Unwin, 1996; Unwin and Unwin, 1998): statistics for non-overlapping areas; statistics for sets of adjacent, overlapping areas (as with moving averages in two dimensions); and statistics where local values are standardised against global values.

Some aspects of spatial analysis have received greater attention than others. Point patterns, for example, have attracted more interest than line patterns, as in a series of publications by Openshaw. In studies of the nuclear industry, for example, he evaluated the potential impact of bombing different targets and of different locations for nuclear power stations and hazardous waste disposal (Openshaw, Steadman, and Green, 1983; Openshaw, 1986; Openshaw, Carver, and Fernie, 1989). The latter work linked with his later studies of the clustering of rare events round certain points—such as particular cancers linked to possible radiation sources, inductively suggesting cause-and-effect relationships: if significant clusters are identified, this provides strong circumstantial evidence that there is a local cause of the disease, worthy of closer investigation. He began this work by building a 'geographical analysis machine', which inductively sought all possible clusters in a point pattern, such as one for cases of childhood leukaemia (Openshaw et al., 1987), and was extended to identify clusters of points with similar characteristics (such as poor- and high-performing schools: Openshaw and Turton, 2001). Such 'geographical analysis and explanations machines' relied on massive computer power (by the standards of the time) to analyse very large datasets, and were further developed by the proposed use of artificial intelligence procedures (Openshaw and Openshaw, 1998).

Many advances in spatial analysis were greatly facilitated by the development of Geographical Information Systems (GIS), combined hardware and software for the capture, storage, checking, integration, manipulation, display, and analysis of spatially referenced data (Longley et al., 1999; see also Chapter 12 below). While much of the development has been undertaken by commercial companies, realising the great potential of GIS in a wide range of applications, geographers have played a central role in developing and applying analytical software for use within these systems (Haining, Wise, and Ma, 1996, 1998, 2000). This was initially stimulated by a major ESRC initiative—the Regional Research Laboratories (Masser, 1987)—from which a number of research and training centres evolved (such as that concerned with epidemiology and emergency service planning at the University of Lancaster; for examples of GIS in spatial analysis see Longley and Batty, 1996). The investment in these major centres, along with a small number of 'new blood' appointments in the early 1980s, was a major stimulus to the continued strength of quantitative work in UK human geography.

In addition, the massive computing power now readily available—including desktop PCs—has enabled substantial advances in visualisation, not only the production of a wide range of maps and cartograms (e.g. Dorling, 1995) but also the depiction of statistical relationships in the exploratory stages of analysis and modelling. Seeing, it is argued, is a good way to understanding (Hearnshaw and Unwin, 1994); advances in computing power allowed graphical presentation of data as part of 'exploratory data analysis', whereby patterns in datasets are evaluated descriptively as a prior stage to hypothesis development and testing (Sibley, 1987; Haining et al., 1998).

The role of technology

Very few of these developments could have occurred without cumulative rapid changes in technology, which allowed the massive volume of required computation to be undertaken increasingly easily and quickly, at decreasing relative prices (see Longley et al., 2001, 445–6). Computers became generally available for research in universities in the 1960s, and geographers were soon using them—as illustrated by an early catalogue of programs (Tarrant, 1968) and texts on computer use by and for geographers (Dawson and Unwin, 1976; Unwin and Dawson, 1985). There were two major areas of use from the outset—multivariate statistical analyses and computer mapping (Haggett, 1969). With the former, most geographers relied on packages developed for wide use—such as SPSS, MICROTAB and MINITAB—although developments in specialised fields of spatial analysis (such as spatial autocorrelation) required custom-built programs.

It was with regard to computer graphics and mapping that work specifically by and for geographers became important, though the procedures had wider applications. One of the two early centres involved was the Department of Geography at the University of Edinburgh (the other was Harvard University), but this was joined by others—such as that for census research at the University of Durham—where mapping spatial patterns was a key component of their research (Census Research Unit, 1980). As computer power was increased—making map creation easier and quicker—so visualisation as a geographical tool was enhanced.

A major breakthrough in this area was stimulated by the 900th anniversary of the production of the Domesday Book of 1086. The British Broadcasting Commission (BBC) took the initiative in developing a hardware–software package that replicated the original in a variety of ways on video discs. Geographers played major roles in this project (Goddard and Armstrong, 1986): mapping systems for 33 different sets of areas and over 20,000 different data files were produced (Openshaw, Wymer, and Charlton, 1986b), which allowed analysis of the land and people of Britain in geographic detail (Rhind and Mounsey, 1986). Openshaw, Rhind, and Goddard (1986a) described it as a revolutionary project, involving an investment of more than £2.5 million (much of it by the BBC[25]) on innovations in data storage (on optical disks) and their manipulation (including graphical and cartographical form) using micro-computers.

The Domesday project was a prototype in the development of GIS, discussed in greater detail in Chapter 12. More generally, it marked the growing use of computers in geographical research for the intensive manipulation and inductive exploration of large and complex datasets, unconstrained by the assumptions of classical statistics.[26] Termed 'geocomputation', this incorporates such topics as data visualisation and transformation, representing and modelling spatial interactions, and generalisation regarding spatial processes (Longley, 1998). Furthermore, according to Macmillan, such is the computing power now available that it allows more than just a step change in the geographer's ability to manipulate information to answer traditional questions:

[25] Which hoped to recoup its investment through sales of its computers, to schools and elsewhere.

[26] The use of computers is not restricted to quantitative research. Increasingly, those employing textual analyses employ packages such as NUD*IST (Crang et al., 1997).

> Technical advances have been so advanced that the problems we can think about have changed. Moreover, there has been a transformation in the kinds of thoughts we can have . . . [by analogy with computing power expansion and mathematics] As it became possible to find solutions to mathematical problems computationally, so it became possible to think about new classes of problems that could be solved. (Macmillan, 1998, 257)

These same improvements in the speed of handling large datasets have enabled improvements in other areas of analysis. Previously, because of the time and cost involved in a single analysis, few geographers explored their data in any detail—investigating whether they met the criteria for application of the general linear model, for example. Too often, a single regression model was fitted, and the results reported. Techniques of exploratory data analysis have enabled much more detailed, and yet quick, appreciation of the nature of datasets and the relationships within them—frequently using interactive graphics (including maps) to identify 'rogue' observations, non-linearities, and so forth (Wrigley, 1984). These facilitate much more rigorous hypothesis-testing and greater confidence in the final research findings. In 1965, Haggett suggested that mapping residuals from regression was the best way forward when seeking to improve a model's level of explanation: improvements in computer methods for visualisation in the 1990s made that much more feasible—as did the ability to transform cartograms to highlight different aspects of the geography.

The path to knowledge: the epistemology of spatial analysis

Whereas many geographers followed the lead of the 1960s pioneers in the adoption of statistical approaches to the analysis of geographical data, relatively few addressed the epistemological and ontological issues in any detail. By far the most detailed was Harvey's (1969a) exploration of the philosophy of the positivist approach to science, and the role of theories, models, and laws in the hypothetico-deductive schema. Davies (1972b, 9) referred to the adoption of these procedures as a 'more rational scientific methodology' stimulating a 'more coherent approach to knowledge' (p. 11). Nevertheless, to some of those involved—such as Wilson (1972)—theoretical developments were more important than the technical ones, though he opposed the hypothetico-deductive approach because it usually involves micro-scale studies from which macro-scale, comprehensive theories are difficult to construct. His goal, as expressed consistently over some 35 years (see Wilson, 2000), was developing theories of large-scale systems directly, via model-building.

The approach codified by Harvey, and generally accepted by geographers, took the verification route to scientific generalisation, whereby 'experiments' were conducted to validate hypotheses. A few suggested that geographers should employ the falsification route proposed by Popper: all knowledge is provisional, and the goal of any experiment should be to falsify a hypothesis, not to verify it. Because knowledge is tentative, then its nature does not change if a further experiment sustains the hypothesis, since a general law is presumably only valid for so long as it works—further confirmatory evidence is provided, but no more, and scientific advancement is slow. Critical rationalists, on the other hand, design experiments to try and falsify a hypothesis: once a general belief has been proved false, its value has gone, and an alternative must be sought (Haines-Young and Petch, 1985; Hay, 1985b; Bird, 1989). But most have preferred the slow accumulation of positive findings as the way of advancing knowledge, through understanding, explanation, and prediction, than the more 'destructive'—if potentially more valuable—approach involving a critical experiment followed by a search for alternatives.

The general—often little more than implicit—adoption of the verification approach meant that much geographical work was associated not just with the positivist approach to science—based on empiricism and measurement—but also with the logical positivist school. This claimed that the positivist procedure provided the only valid route to knowledge; anything else was mere cosmology, generating beliefs that were unsustainable by recourse to systematic evaluation of evidence. The 'logical positivist' label was used by those who favoured alternative epistemologies to describe work done by spatial analysts in a pejorative way, leading to retorts that much quantitative work was not that constrained (see Bennett, 1981); rather it represented a rigorous way of sifting large masses of data and finding patterns within them that called for closer examination. We live in a mass society, in which there is much mensuration using common metrics (such as money), and to understand how that society operates involves dealing with it on its own terms; spatial analyses of various types are central to many decision-making processes (Johnston, 2000). Indeed, Openshaw (1989) argued against the hypothetico-deductive approach to studying spatial patterns (in part because he claimed that geography is 'theory-poor'), maintaining that 'mathematical modelling' should be replaced by 'computer modelling' (now known as geocomputation). In an increasingly data-rich society with massive computing power, geographers should return to inductive ways of thinking, to 'modelling data without any great depth of understanding'. In this way they would identify patterns in data, and through their spatial analysis 'add geographical value' to them and so develop commercial markets for geographical expertise in data provision.

Although the 'scientific method' promoted by geographers in the 1960s and 1970s now plays a much smaller role within the discipline, which is characterised by epistemological and methodological pluralism, nevertheless terms such as theory, model, and hypothesis are not obsolete—though often used differently to that employed by Harvey (1969a).[27] Much empirical work is set within the general context of realism—introduced to geographers by Sayer (1984, 1992)—which argues that explanation can only be achieved through intensive study, usually of just a small number of cases, using a range of qualitative methodologies (Limb and Dwyer, 2001). Nevertheless, this approach also recognises the need in some circumstances for a preliminary stage of extensive research—one that asks questions such as 'What are the regularities, common patterns, distinguishing features of a population? How widely are certain characteristics or processes distributed or represented?' (Sayer, 1992, 243). Answers to these lack 'explanatory penetration' because they concern 'similarity, dissimilarity, correlation and the like, rather than causal, structural and substantial' explanation (p. 246), and so they need to be followed up by intensive investigations. The two approaches are complementary, therefore (and so qualitative methodologies are now included alongside quantitative in several recent 'techniques' textbooks; e.g. Flowerdew and Martin, 1997; Kitchin and Tate, 2000). The approaches and methods introduced and honed by geographers over the last 40 years as means of teasing out regularities remain important to the search for understanding contemporary society—much of which is numerically based and large scale in its structure (Johnston, 2000).

A continuing theme

Burton's (1963) claim that by the mid-1960s geography had undergone a successful quantitative and theoretical revolution was undoubtedly at least premature, if not a mis-statement—especially when read in hindsight. But the forms of thinking and analysis introduced to geography at that time undoubtedly substantially changed the ways in which geographers were trained and undertook research. Very soon all undergraduates in the discipline were required to take introductory courses in statistical methods, and many took advanced courses too. Increasingly, as well, students were introduced to computer use and programming and took (compulsory in many departments) courses in 'the philosophy and history of geography'. Despite subsequent changes in research interests and orientations, many of these courses remain obligatory, and have been joined by others in GIS and its applications (increasingly seen as a 'core transferable skill' for geographers). A new 'attitude of mind' had been engendered, with 'an organized point of view, theoretical structure and a problem orientation' replacing 'a morass of regional description at a subjective level' (Davies, 1966).

[27] On theory as currently understood, see Barnes (2001).

But the discipline has not stood still for 40 years. Critiques of its 'new' approach were soon launched (as reviewed in Johnston, 1997a), some arguing that it be abandoned because, in particular, of the mechanistic way it treated human decision-making, with its emphases on explanation and prediction. Whether there is order in the world as studied by human geographers, which can be investigated within the protocols of positivist science and using the generalising methods of statistics and mathematics, became a focus of contention—with some spirited defences offered by those still committed to that position. Haggett (1989, 2000), for example, has continued to argue the case for forecasting as part of the modelling procedure, though very largely in the particular case of epidemic disease diffusion, where human agency is relatively unimportant and prediction reasonably successful (or at least post-diction, which suggests that predictive models should be feasible: Haggett, 2001, 499); and Wilson (2000) has continued to promote the value of classical least-cost location models as a basis for planning optimal (or commercially viable) futures, claiming that 'most of the equilibrium models developed in the 1960s and 70s have stood the test of time in offering at least reproductive power—and particularly in applications where relatively short time-scales are involved' (Wilson, 2001, 21).

This approach came under substantial attack from some quarters in a volume conceived to mark the twentieth anniversary of the publication of *Models in Geography* (Chorley and Haggett, 1967; Macmillan, 1989). Harvey, for example, argued that:

> Those who have stuck with modelling . . . have largely been able to do so, I suspect, by restricting the nature of the questions they ask. I accept that we can now model spatial behaviours like journey-to-work, retail activity, the spread of measles epidemics, the atmospheric dispersion of pollutants, and the like with much greater security and precision than was once the case. (Harvey, 1989, 212–13)

But this does not enable geographers to address many of the contemporary major issues facing society (third world debt, geopolitical tensions etc.), so that:

> There must be thousands of hypotheses proven correct at some appropriate level of significance in the geographical literature by now, and I am left with the impression that *in toto* this adds up to little more than the proverbial hill of beans. (Harvey, 1989, 213)

Somewhat similarly, Cosgrove argued that the

> . . . faith in a rational and unilinear social and environmental progress, in technology as a means of achieving it, in the efficacy of abstract numerical relations to gain a purchase on 'objective truth', and the belief in planning as a way of overcoming the contingency and flux of unpredictable human life

> and its environmental relations are all characteristically modernist. (Cosgrove, 1989, 243)

This was in contrast to contemporary, postmodern, distrust 'for a privileged path to truth or to accurate representation of a single reality . . . the relationship between words and things has always been a contingent, socially constructed and unstable one, and this is equally true of non-verbal representations of the world, including mathematics' (Cosgrove, 1989, 243). Thus, mathematical modelling is but one approach to describing (and understanding) the world, and has no privileged position within the academic division of labour.

Others more pragmatically argued that in a world dominated by quantities there is a need to understand their use and abuse (Johnston, 2000), and that generalised theoretical claims—such as those that bring together thinking in economic and cultural geography (Lee and Wills, 1998; see also Amin and Thrift, 2000)—need to be formally scrutinised, perhaps leading to testable hypotheses (Martin and Sunley, 2001; Plummer and Sheppard, 2001; Plummer and Taylor, 2001). Models still play a role, according to Flowerdew (1989, 252), in simplifying and communicating about the world, as steps on the path to understanding and changing it (though the contents of Peet and Thrift's (1989) *New Models in Geography*, subtitled *The Political-Economy Perspective*, are very different from those of the predecessor volume, with very few diagrams, little statistics and mathematics, and only slight attention to the search for the type of spatial order that characterised the 1960s). Generalisation is not the only scientific goal pursued by geographers, and models (simplified descriptions of reality, in whatever format—mathematical or other) may play some part in identifying significant patterns, relationships, and trends: they are not 'useless, evil or counter-evolutionary', but simply means to ends, to understanding causes and effects.

The location theories that stimulated human geographers in the 1960s are now rarely referred to, therefore (though Wilson, 2000, has an appendix on 'geography's classical theorists').[28] Indeed, the definition of theory has changed with the decades. To those who first started to use it in the 1960s and 1970s, it meant a set of connected statements (laws and constraining conditions) used as the basis for explanation and for deducing testable hypotheses. This was entirely consistent with the positivist epistemology embraced then (albeit only implicitly in some cases). But in other epistemologies later adopted by geographers—such as realism—it was a more abstract means of constructing mental frameworks for apprehending reality, although in some cases the abstract theorising has taken precedence over the empirical apprehension. Theory reigns—but not the same type of theory.

[28] Von Thünen, Weber, Palander, Hoover, Hotelling, Burgess, Hoyt, and Harris and Ullman: only the last two were trained and/or practising geographers!

Nevertheless, the search for pattern and order in datasets (both spatially organised and otherwise) remains a central component of what many geographers do, often as initial siftings of information in the exploration of what is important and worthy of closer study; it is also important for work in archaeology, a discipline that has adopted and adapted many of the models and techniques employed by geographers in the 1960s (Clark, 1972, 1977; Hodder and Orton, 1976). Much work is still being done in that vein, with increasing technical sophistication. As other approaches to geography have moved towards centre stage, relatively little work in the spatial analysis mould is published in the discipline's mainline journals (the *Transactions of the Institute of British Geographers*, for example), and less is appearing in the multidisciplinary journals founded in the 1960s to sustain the 'new geography' and its close relations (such as *Environment and Planning A*). Instead, it is being published in the specialist journals serving the spatial analysis community—such as *Geographical Analysis*, *Geographical and Environmental Modelling*, *International Journal of Geographical Information Science*, and so on. Geography at the century's end was a pluralistic discipline, with the various sub-communities being somewhat semi-independent and self-sustaining. Most who continue to work in the spatial analysis tradition recognise the difference between an extensive approach to research, which addresses the world and poses important questions about it, and intensive approaches, which seek explanation and understanding through more detailed, frequently qualitative, research methods. To some, geography is no longer a discipline in distance and spatial analysis is passé: to others, such an orientation still has much to offer in an overall strategy aimed at understanding and then improving the world we have made—and continue to re-make.

Acknowledgements

I am grateful to Hugh Clout, Peter Dicken, Danny Dorling, Stewart Fotheringham, Tony Gatrell, Rita Johnston, Paul Longley, Charles Pattie, Ian Simmons, David Smith, Alan Wilson, and Michael Wise for comments on drafts of this chapter, and much useful help and advice.

References

Amin, A. and Thrift, N. J. (2000) What kind of economic theory for what kind of economic geography? *Antipode*, 32, 4–9.

Bailey, T. C. and Gatrell, A. C. (1995) *Interactive Spatial Data Analysis*. Harlow: Longman.

Barnes, T. J. (2001) Retheorizing economic geography: from the quantitative revolution to the 'cultural turn'. *Annals of the Association of American Geographers*, 91, 546–65.

Bassett, K. and Haggett, P. (1970) Towards short-term forecasting for cyclic behaviour in a regional system of cities. In M. Chisholm, A. E. Frey, and P. Haggett, eds, *Spatial Forecasting*. London: Butterworth, 389–414.

Batty, M. (1976) *Urban Modelling: Algorithms, Calibrations, Predictions*. Cambridge: Cambridge University Press.

Batty, M. (1978) Urban models in the planning process. In D. T. Herbert and R. J. Johnston, eds, *Geography and the Urban Environment: Volume 1*. Chichester: Wiley, 63–134.

Batty, M. and Longley, P. A. (1994) *Fractal Cities: A Geometry of Form and Function*. London: Academic Press.

Bennett, R. J. (1978) *Spatial Time Series: Analysis, Forecasting and Control*. London: Pion.

Bennett, R. J. (1981) Quantitative and theoretical geography in Western Europe. In R. J. Bennett, ed., *European Progress in Spatial Analysis*. London: Pion, 1–34.

Bennett, R. J. and Chorley, R. J. (1978) *Environmental Systems: Philosophy, Analysis and Control*. London: Methuen.

Bennett, R. J. and Haining, R. P. (1985) Spatial structure and spatial interaction: modelling approaches to the statistical analysis of spatial data. *Journal of the Royal Statistical Society, A*, 148, 1–36.

Bennett, R. J., Haining, R. P., and Wilson, A. G. (1985) Spatial structure, spatial interaction and their integration: a review of alternate models. *Environment and Planning A*, 17, 625–46.

Berry, B. J. L. (1993) Geography's quantitative revolution: initial conditions, 1954–1960; a personal memoir. *Urban Geography*, 14, 434–41.

Bird, J. H. (1989) *The Changing Worlds of Geography: A Critical Guide to Concepts and Methods*. Oxford: Clarendon Press.

Birkin, M., Clarke, G., Clarke, M., and Wilson, A. G. (1996) *Intelligent GIS: Location Decisions and Strategic Planning*. Cambridge: GeoInformation International.

Boal, F. W. and Livingstone, D. N., eds (1989) *The Behavioural Environment*. London: Routledge.

Browning, C. E. (1982) *Conversations with Geographers: Career Pathways and Research Style*. Chapel Hill, NC: University of North Carolina Press.

Buchanan, K. M. (1968) A preliminary contribution to the geographical analysis of a Poohscape. *IBG Newsletter*, 6, 54–63. Reprinted in *Progress in Human Geography*, 23 (1999), 253–66.

Bullen, N., Jones, K., and Duncan, C. (1997) Modelling complexity: analysing between-individual and between-place variation—a multilevel tutorial. *Environment and Planning A*, 29, 585–609.

Burton, I. (1963) The quantitative revolution and theoretical geography. *The Canadian Geographer*, 7, 151–62.

Carlstein, T., Parkes, D. N., and Thrift, N. J., eds (1978) *Timing Space and Spacing Time*. London: Edward Arnold (3 vols).

Census Research Unit (1980) *People in Britain: A Census Atlas*. London: HMSO.

Chapman, G. P. (1977) *Human and Environmental Systems: A Geographer's Appraisal*. London: Academic Press.

Chisholm, M. (1962) *Rural Settlement and Land Use*. London: Hutchinson.

Chisholm, M. (1966) *Geography and Economics*. London: Bell.

Chisholm, M. (1967) General systems theory and geography. *Transactions, Institute of British Geographers*, 42, 45–52.
Chisholm, M. (1971a) *Research in Human Geography*. London: Heinemann.
Chisholm, M. (1971b) In search of a basis for location theory: micro-economics or welfare economics? In C. Board et al., eds, *Progress in Geography 3*. London: Edward Arnold, 111–34.
Chisholm, M. (1975) *Human Geography: Evolution or Revolution?* Harmondsworth: Penguin.
Chisholm, M. and Manners, G., eds (1971) *Spatial Policy Problems of the British Economy*. Cambridge: Cambridge University Press.
Chisholm, M., Frey, A. E., and Haggett, P., eds (1970) *Regional Forecasting*. London: Butterworth.
Chorley, R. J. (1962) *Geomorphology and General Systems Theory*. Washington, DC: US Geological Survey, Professional Paper 500-B.
Chorley, R. J. (1995) Haggett's Cambridge: 1957–1966. In A. D. Cliff et al., eds, *Diffusing Geography: Essays for Peter Haggett*. Oxford: Blackwell, 355–74.
Chorley, R. J. and Haggett, P., eds (1965a) *Frontiers in Geographical Teaching*. London: Methuen.
Chorley, R. J. and Haggett, P. (1965b) Trend-surface mapping in geographical research. *Transactions, Institute of British Geographers*, 37, 47–67.
Chorley, R. J. and Haggett, P., eds (1967) *Models in Geography*. London: Methuen.
Chorley, R. J., Beckinsale, R. P., and Dunn, A. J. (1973) *The History of the Study of Landforms, Vol. 2: The Life and Work of William Morris Davis*. London: Methuen.
Christaller, W. (1933) *Central Places in Southern Germany*. Englewood Cliffs, NJ: Prentice-Hall (this English translation was published in 1966).
Clark, B. D. and Gleave, M. B., eds (1973) *Social Patterns in Cities*. London: Institute of British Geographers, Special Publication No. 5.
Clark, C. (1951) Urban population densities. *Journal of the Royal Statistical Society A*, 114, 490–6.
Clark, D. L. (1972) *Models in Archaeology*. London: Methuen.
Clark, D. L. (1977) *Spatial Archaeology*. London: Methuen.
Clark, K. G. T. (1950) Certain underpinnings of our arguments in human geography. *Transactions and Papers, Institute of British Geographers*, 16, 13–22.
Clarke, M. and Wilson, A. G. (1987) Towards an applicable human geography: some developments and observations. *Environment and Planning A*, 19, 1525–42.
Cliff, A. D. and Haggett, P. (1998) On complex geographical space: computing frameworks for spatial diffusion processes. In P. Longley, S. M. Brooks, R. MacDonnell, and B. Macmillan, eds, *Geocomputation: A Primer*. Chichester: Wiley, 231–56.
Cliff, A. D. and Ord, J. K. (1973) *Spatial Autocorrelation*. London: Pion.
Cliff, A. D. and Ord, J. K. (1975) Model building and the analysis of spatial pattern in human geography. *Journal of the Royal Statistical Society B*, 37, 297–384.
Cliff, A. D. and Ord, J. K. (1981) *Spatial Processes: Modelling and Applications*. London: Pion.
Cliff, A. D., Haggett, P., and Ord, J. K. (1967) *Spatial Aspects of Influenza Epidemics*. London: Pion.

Cliff, A. D., Martin, R. L., and Ord, J. K. (1974) Evaluating the friction of distance parameter in gravity models. *Regional Studies*, 8, 281–6.

Cliff, A. D. et al. (1975) *Elements of Spatial Structure: A Quantitative Approach.* Cambridge: Cambridge University Press.

Cliff, A. D., Haggett, P., and Smallman-Raynor, M. (1983) *Measles: An Historical Geography of a Major Human Viral Disease from Global Expansion to Local Retreat, 1840–1990.* Cambridge: Cambridge University Press.

Cole, J. P. (1969) Mathematics and geography. *Geography*, 54, 152–63.

Cole, J. P. and King, C. A. M. (1968) *Quantitative Geography*. London: Wiley.

Conzen, M. R. G. (1969) *Alnwick, Northumberland: A Study in Townplan Analysis.* London: Institute of British Geographers, Publication No. 27.

Coppock, J. T. (1974) Geography and public policy: challenges, opportunities and implications. *Transactions, Institute of British Geographers*, 63, 1–16.

Cosgrove, D. E. (1989) Models, descriptions and imagination in geography. In B. Macmillan, ed., *Remodelling Geography*. Oxford: Blackwell, 245–54.

Cox, K. R. (1969) The voter decision in spatial context. In C. Board et al., eds, *Progress in Geography 1*. London: Edward Arnold, 81–118.

Crang, M., Hudson, A., Reimer, S., and Hinchcliffe, S. (1997) Software for qualitative research: I. Prospectus and overview. *Environment and Planning A*, 29, 771–87.

Crowe, P. R. (1936) The rainfall régime of the Western Plains. *The Geographical Review*, 26, 463–84.

Davies, W. K. D. (1966) Theory, science and geography. *Tijdschrift voor Economische en Sociale Geografie*, 57, 125–30.

Davies, W. K. D., ed. (1972a) *The Conceptual Revolution in Geography*. London: University of London Press.

Davies, W. K. D. (1972b) The conceptual revolution in geography. In W. K. D. Davies, ed. *The Conceptual Revolution in Geography*. London: University of London Press, 9–17.

Davies, W. K. D. (1984) *Factorial Ecology*. London: Gower.

Dawson, J. A. and Unwin, D. J. (1976) *Computing for Geographers*. Newton Abbott: David and Charles.

Dickinson, R. E. (1938) The regions of Germany. *The Geographical Review*, 29, 609–26.

Dickinson, R. E. (1939) Landscape and society. *Scottish Geographical Magazine*, 55, 1–4.

Dickinson, R. E. (1942) The social basis of physical planning. *Sociological Review*, 34, 51–67 and 165–82

Dickinson, R. E. (1945) *The Regions of Germany*. London: Kegan Paul, Trench, Trubner and Co.

Dickinson, R. E. (1947) *City Region and Regionalism*. London: Routledge & Kegan Paul.

Dickinson, R. E. (1951) *The West European City*. London: Routledge & Kegan Paul.

Dodge, M. and Kitchin, R. (2000) *Mapping Cyberspace*. London: Routledge.

Dorling, D. (1995) *A Social Atlas of Britain*. Chichester: Wiley.

Duncan, O. D., Cuzzort, R. P., and Duncan, B. (1961) *Statistical Geography*. New York: The Free Press.

Duncan, S. S. (1974) The isolation of scientific discovery: indifference and resistance to a new idea. *Science Studies*, 4, 109–34.
Estall, R. C. and Buchanan, R. O. (1973) *Industrial Activity and Economic Geography*, 3rd edn. London: Hutchinson.
Estall, R. C. and Buchanan, R. O. (1980) *Industrial Activity and Economic Geography*, 4th edn. London: Hutchinson.
Fawcett, C. B. (1919) *The Provinces of England*. London: Hutchinson (1960 re-issue).
Fingleton, B. (1984) *Models of Category Counts*. Cambridge: Cambridge University Press.
Flowerdew, R. (1989) Some critical views of modelling in geography. In B. Macmillan, ed., *Remodelling Geography*. Oxford: Blackwell, 245–54.
Flowerdew, R. and Martin, D., eds (1997) *Methods in Human Geography*. Harlow: Longman.
Forer, P. C. (1978) A place for plastic space? *Progress in Human Geography*, 2, 230–67.
Fotheringham, A. S. and Wong, D. W. F. (1991) The modifiable areal unit problem in multivariate statistical analysis. *Environment and Planning A*, 23, 1025–44.
Fotheringham, A. S., Brunsdon, C., and Charlton, M. E. (1998) Geographically weighted regression: a natural evolution of the expansion method for spatial data analysis. *Environment and Planning A*, 30, 1905–27.
Fotheringham, A. S., Charlton, M. E., and Brunsdon, C. (2000) *Quantitative Geography: Perspectives on Spatial Data Analysis*. London: Sage.
Fotheringham, A. S., Charlton, M. E., and Brunsdon, C. (2001) Spatial variations in school performance: a local analysis using geographically weighted regression. *Geographical & Environmental Modelling*, 5, 43–66.
Freeman, T. W. (1961) *A Hundred Years of Geography*. London: Duckworth.
Gale, S. (1972) On the heterodoxy of explanation: a review of David Harvey's *Explanation in Geography*. *Geographical Analysis*, 4, 285–322.
Gatrell, A. (1983) *Distance and Space: A Geographical Perspective*. Oxford: Oxford University Press.
Getis, A. and Boots, B. N. (1978) *Models of Spatial Processes*. Cambridge: Cambridge University Press.
Giddens, A. (1984) *The Constitution of Society*. Cambridge: Polity Press.
Goddard, J. B. and Armstrong, P. (1986) The 1986 Domesday project. *Transactions, Institute of British Geographers*, NS11, 290–5.
Gold, J. R. (1980) *An Introduction to Behavioural Geography*. Oxford: Oxford University Press.
Golledge, R. G. and Stimson, R. J. (1997) *Spatial Behavior: A Geographic Perspective*. New York: Guilford.
Gould, P. R. (1972) Pedagogic review. *Annals of the Association of American Geographers*, 62, 689–700.
Gregory, D. (1978) *Ideology, Science and Human Geography*. London: Hutchinson.
Gregory, D. (1994) *Geographical Imaginations*. Oxford: Blackwell.
Gregory, S. (1962) *Statistical Methods and the Geographer*. London: Methuen.
Grigg, D. B. (1967) Regions, models and classes. In R. J. Chorley and P. Haggett, eds, *Models in Geography*. London: Methuen, 461–510.

Gudgin, G. and Taylor, P. J. (1979) *Seats, Votes and the Spatial Organisation of Elections*. London: Pion.

Hägerstrand, T. (1968) *Innovation Diffusion as a Spatial Process*, trans. Allan Pred. Chicago, IL: University of Chicago Press.

Haggett, P. (1964) Regional and local components in the distribution of forested areas in southeast Brazil: a multivariate approach. *Geographical Journal*, 130, 365–77.

Haggett, P. (1965) *Locational Analysis in Human Geography*. London: Edward Arnold.

Haggett, P. (1969) On geographical research in a computer environment. *Geographical Journal*, 135, 497–507.

Haggett, P. (1971) Leads and lags in inter-regional systems: a study of cyclic fluctuations in the South West economy. In M. Chisholm and G. Manners, eds, *Spatial Policy Problems of the British Economy*. Cambridge: Cambridge University Press, 69–95.

Haggett, P. (1981) The edges of space. In R. J. Bennett, ed., *European Progress in Spatial Analysis*. London: Pion, 51–70.

Haggett, P. (1989) *The Geographer's Art*. Oxford: Blackwell Publishers.

Haggett, P. (1994) Prediction and predictability in geographical systems. *Transactions, Institute of British Geographers*, NS19, 6–20.

Haggett, P. (2000) *The Geographical Structure of Epidemics*. Oxford: Clarendon Press.

Haggett, P. (2001) *Geography: A Global Synthesis*. Harlow: Pearson Education.

Haggett, P. and Chorley, R. J. (1969) *Network Models in Geography*. London: Edward Arnold.

Haggett, P., Cliff, A. D., and Frey, A. E. (1977) *Locational Analysis in Human Geography*, 2nd edn. London: Edward Arnold.

Haines-Young, R. and Petch, J. R. (1985) *Physical Geography: Its Nature and Methods*. London: Harper & Row.

Haining, R. P. (1990) *Spatial Data Analysis in the Social and Environmental Sciences*. Cambridge: Cambridge University Press.

Haining, R. P., Wise, S., and Ma, J. (1996) The design of a software system for interactive spatial statistical analysis linked to a GIS. *Computational Statistics*, 11, 449–66.

Haining, R. P., Wise, S., and Ma, J. (1998) Exploratory spatial data analysis in a geographic information system environment. *The Statistician*, 47, 457–69.

Haining, R. P., Wise, S., and Ma, J. (2000) Design and implementing software for spatial statistical analysis in a GIS environment. *Journal of Geographical Systems*, 2, 257–86.

Hall, P. (1966) *The Isolated State* (being a translation, by Carla Wartenberg, of J. H. von Thünen's *Die isolieert Staat*, originally published in 1826). Oxford: Pergamon Press.

Harvey, D. (1969a) *Explanation in Geography*. London: Edward Arnold.

Harvey, D. (1969b) Conceptual and measurement problems in the cognitive-behavioral approach to location theory. In K. R. Cox and R. G. Golledge, eds, *Behavioral Problems in Geography: a Symposium*. Evanston, IL: Northwestern University Studies in Geography, 17, 35–68.

Harvey, D. (1972) On obfuscation in geography: a comment on Gale's heterodoxy. *Geographical Analysis*, 4, 323–30.

Harvey, D. (1973) *Social Justice and the City*. London: Edward Arnold.
Harvey, D. (1982) *The Limits to Capital*. Oxford: Blackwell.
Harvey, D. (1989) From models to Marx: notes on the project to remodel geography. In B. Macmillan, ed., *Remodelling Geography*. Oxford: Blackwell, 211–16.
Hay, A. M. (1985a) Statistical tests in the absence of samples: a comment. *Professional Geographer*, 37, 334–48.
Hay, A. M. (1985b) Scientific method in geography. In R. J. Johnston, ed., *The Future of Geography*. London: Methuen, 129–42.
Hearnshaw, H. A. and Unwin, D. J., eds (1994) *Visualisation in GIS*. Chichester: Wiley.
Hepple, L. W. (1974) The impact of stochastic process theory upon spatial analysis in human geography. In C. Board et al., eds, *Progress in Geography 6*. London: Edward Arnold, 89–142.
Hodder, I. and Orton, C. (1976) *Spatial Analysis in Archaeology*. Cambridge: Cambridge University Press.
Hoover, E. M. (1948) *The Location of Economic Activity*. New York: McGraw Hill.
Hudson, R. (2001) *Producing Places*. New York: Guilford Press.
Huggett, R. J. (1980) *Systems Analysis in Geography*. Oxford: Oxford University Press.
Johnston, R. J. (1971) *Urban Residential Patterns: An Introductory Review*. London: George Bell.
Johnston, R. J. (1973) *Spatial Structures: Introducing the Study of Spatial Systems in Human Geography*. London: Methuen.
Johnston, R. J. (1975) Map pattern and friction of distance parameters: a comment. *Regional Studies*, 9, 281–3.
Johnston, R. J. (1976) *The World Trade System: Some Enquiries into its Spatial Structure*. London: George Bell.
Johnston, R. J. (1978) *Multivariate Statistical Analysis in Geography: A Primer on the General Linear Model*. Harlow: Longman.
Johnston, R. J. (1984) Quantitative ecological analysis in human geography: an evaluation of four problem areas. In G. Bahrenberg, M. M. Fischer, and P. Nijkamp, eds, *Recent Developments in Spatial Data Analysis*. Aldershot: Gower, 131–44.
Johnston, R. J. (1986) The neighbourhood effect revisited: spatial science or political regionalism? *Environment and Planning D: Society and Space*, 4, 41–55.
Johnston, R. J. (1991) *A Question of Place: Exploring the Practice of Human Geography*. Oxford: Blackwell.
Johnston, R. J. (1993) The geographer's degrees of freedom: Wreford Watson, postwar progress in human geography and the future of scholarship in UK geography. *Progress in Human Geography*, 17, 319–32.
Johnston, R. J. (1997a) *Geography and Geographers: Anglo-American Human Geography since 1945*, 5th edn. London: Arnold.
Johnston, R. J. (1997b) W(h)ither spatial science and spatial analysis. *Futures*, 29, 323–35.
Johnston, R. J. (1999) Classics in human geography revisited: Keith Buchanan 'A preliminary contribution to the geographical analysis of a Poohscape'. *Progress in Human Geography*, 23, 253–66.

Johnston, R. J. (2000) On disciplinary history and textbooks: or where has spatial analysis gone? *Australian Geographical Studies*, 38, 125–37.

Johnston, R. J. (2001a) City-regions and a federal Europe: Robert Dickinson and post-World War II reconstruction. *Geopolitics*, 5, 153–76.

Johnston, R. J. (2001b) Out of the 'moribund backwater': territory and territoriality in political geography. *Political Geography*, 20, 677–94.

Johnston, R. J. (2001c) Robert E. Dickinson and the growth of urban geography: an evaluation. *Urban Geography*, 22, 702–36.

Johnston, R. J. and Hay, A. M. (1982) On the parameters of uniform swing in single-member constituency electoral systems. *Environment and Planning A*, 14, 61–74.

Johnston, R. J. and Pattie, C. J. (2000) Ecological inference and entropy-maximizing: an alternative estimation procedure for split-ticket voting. *Political Analysis*, 8, 333–45.

Johnston, R. J. and Pattie, C. J. (2001) On geographers and ecological inference. *Annals of the Association of American Geographers*, 91, 281–2.

Johnston, R. J., Pattie, C. J., and Allsopp, J. G. (1988) *A Nation Dividing? Britain's Changing Electoral Map, 1979–1987*. Harlow: Longman.

Johnston, R. J., Forrest, J., and Poulsen, M. F. (2001) Sydney's ethnic geography: new approaches to analysing patterns of residential concentration. *Australian Geographer*, 32, 149–62.

Johnston, R. J. et al. (2000a) Local context, retrospective economic evaluations, and voting: the 1997 general election in England and Wales. *Political Behavior*, 22, 121–43.

Johnston, R. J. et al. (2000b) The neighbourhood effect and voting in England and Wales: real or imagined? In P. J. Cowley, D. T. Denver, A. T. Russell, and L. Harrison, eds, *British Elections and Parties Review, Vol. 10*. London: Frank Cass, 47–63.

Johnston, R. J. et al. (2001) Housing tenure, local context, scale and voting in England and Wales, 1997. *Electoral Studies*, 20, 195–216.

Jones, E. (1960) *A Social Geography of Belfast*. Oxford: Oxford University Press.

Jones, J. P. and Casetti, E., eds (1991) *Applications of the Expansion Method*. London: Routledge.

Jones, K. (1991) Specifying and estimating multilevel models for geographical research. *Transactions, Institute of British Geographers*, NS16, 148–59.

Jones, K. and Duncan, C. (1996) People and places: the multilevel model as a general framework for the quantitative analysis of geographical data. In P. A. Longley and M. Batty, eds, *Spatial Analysis: Geographical Modelling in a GIS Environment*. New York: Wiley, 1–16.

Jones, K. and Duncan, C. (2002) Modelling context and heterogeneity: applying multilevel models. In E. Scarbrough and E. Tanenbaum, eds, *Research Strategies in the Social Sciences: A Guide to New Approaches*. Oxford: Oxford University Press, 94–123.

Jones, K., Gould, M., and Watt, R. (1998) Multiple contexts as cross-classified models: the Labour vote in the British general election of 1992. *Geographical Analysis*, 30, 65–93.

Jones, K., Johnston, R. J., and Pattie, C. J. (1992) People, places and regions: exploring the use of multi-level modelling in the analysis of electoral data. *British Journal of Political Science*, 22, 343–80.

Kimble, G. H. T. (1951) The inadequacy of the regional concept. In L. D. Stamp and S. W. Wooldridge, eds, *London Essays in Geography*. London: Longman.
Kirk, W. (1963) Problems of geography. *Geography*, 48, 357–71.
Kitchin, R. and Tate, N. (2000) *Conducting Research into Human Geography: Theory, Methodology and Practice*. London: Prentice-Hall.
Langton, J. (1972) Potentialities and problems of adapting a systems approach to the study of change in human geography. In C. Board et al., eds, *Progress in Geography 4*. London: Edward Arnold, 125–79.
Lawton, R. (2001) Two economic geography texts. *Progress in Human Geography*, 25, 303–9.
Lee, R. and Wills, J., eds (1998) *Geographies of Economies*. London: Arnold.
Lewis, P. W. (1977) *Maps and Statistics*. London: Methuen.
Limb, M. and Dwyer, C., eds (2001) *Qualitative Methodologies for Geographers*. London: Arnold.
Lloyd, P. E. and Dicken, P. (1972) *Location in Space: A Theoretical Approach to Economic Geography*. New York: Harper & Row.
Longley, P. A. (1998) Foundations. In P. A. Longley, S. M. Brooks, R. MacDonnell, and B. Macmillan, eds, *Geocomputation: A Primer*. Chichester: Wiley, 3–16.
Longley, P. A. and Batty, M., eds (1996) *Spatial Analysis: Modelling in a GIS Environment*. Cambridge: GeoInformation International.
Longley, P. A. and Clarke, G. P., eds (1995) *GIS for Business and Service Planning*. Cambridge: GeoInformation International.
Longley, P. A. and Harris, R. (1999) Towards a new digital infrastructure for urban analysis and modelling. *Environment and Planning B*, 26, 855–78.
Longley, P. A., Brooks, S. M., MacDonnell, R., and Macmillan, B., eds (1998) *Geocomputation: A Primer*. Chichester: Wiley.
Longley, P. A., Goodchild, M. F., Maguire, D. J., and Rhind, D. W., eds (1999) *Geographical Information Systems*. New York: Wiley (2 vols).
Longley, P. A., Goodchild, M. E., Maguire, D. J., and Rhind, D. W. (2001) *Geographic Information Systems and Science*. Chichester: Wiley.
Lösch, A. (1954) *The Economics of Location*. New Haven, CT: Yale University Press.
MacAllister, I. et al. (2001) Class dealignment and the neighbourhood effect: Miller revisited. *British Journal of Political Science*, 31, 41–60.
Macmillan, B. (1998) Epilogue. In P. Longley, S. M. Brooks, R. MacDonnell, and B. Macmillan, eds, *Geocomputation: A Primer*. Chichester: Wiley, 257–64.
Macmillan, B. ed. (1989) *Remodelling Geography*. Oxford: Blackwell.
Martin, R. L. (1999) The new 'geographical turn' in economics: some critical reflections. *Cambridge Journal of Economics*, 23, 65–91.
Martin, R. L. and Sunley, P. J. (2001) Rethinking the 'economic' in economic geography: broadening our vision or losing our focus. *Antipode*, 33, 148–61.
Masser, I. (1987) ESRC's regional research laboratory initiative. *Area*, 19, 274–5.
O'Brien, L. (1992) *Introducing Quantitative Geography: Measurement, Methods and Generalized Linear Models*. London: Routledge.
Openshaw, S. (1977) A geographical solution to scale and aggregation problems in region-building, partitioning and spatial modelling. *Transactions, Institute of British Geographers*, NS4, 459–72.

Openshaw, S. (1986) *Nuclear Power: Siting and Safety*. London: Routledge.
Openshaw, S. (1989) Computer modelling in human geography. In B. Macmillan, ed., *Remodelling Geography*. Oxford: Blackwell, 70–88.
Openshaw, S. (1994) Computational human geography: towards a research agenda. *Environment and Planning A*, 26, 499–505.
Openshaw, S. (1996) Developing GIS-relevant zone-based spatial analysis methods. In P. A. Longley and M. Batty, eds, *Spatial Analysis: Geographical Modelling in a GIS Environment*. New York: Wiley, 55–73.
Openshaw, S. and Openshaw, C. (1998) *Artificial Intelligence in Geography*. Chichester: Wiley.
Openshaw, S. and Taylor, P. J. (1979) A million or so correlation coefficients: three experiments on the modifiable areal unit problem. In N. Wrigley, ed., *Statistical Applications in the Spatial Sciences*. London: Pion, 127–44.
Openshaw, S. and Taylor, P. J. (1981) The modifiable areal unit problem. In N. Wrigley and R. J. Bennett, eds, *Quantitative Geography: A British View*. London: Routledge, 60–70.
Openshaw, S. and Turton, I. (2001) Using a Geographical Explanations Machine to explore spatial factors relating to primary school performance. *Geographical & Environmental Modelling*, 5, 85–101.
Openshaw, S., Steadman, P., and Greene, O. (1983) *Doomsday: Britain after Nuclear Attack*. Oxford: Basil Blackwell.
Openshaw, S., Rhind, D. W., and Goddard, J. B. (1986a) Geography, geographers and the BBC Domesday project. *Area*, 18, 9–13.
Openshaw, S., Wymer, C., and Charlton, M. (1986b) A geographical information and mapping system for the BBC Domesday optical discs. *Transactions, Institute of British Geographers*, NS11, 296–304.
Openshaw, S., Charlton, M., Wymer, C., and Craft, A. (1987) A Mark I Geographical Analysis Machine for the automated analysis of point data sets. *International Journal of Geographical Information Systems*, 1, 335–58.
Openshaw, S., Carver, S., and Fernie, J. (1989) *Britain's Nuclear Waste: Siting and Safety*. London: Belhaven Press.
Parkes, D. N. and Thrift, N. J. (1980) *Times, Spaces and Places*. Chichester: Wiley.
Pattie, C. J. and Johnston, R. J. (1999) Context, conversion and conviction: social networks and voting at the 1992 British general election. *Political Studies*, 47, 877–99.
Pattie, C. J. and Johnston, R. J. (2000) 'People who talk together vote together': an exploration of contextual effects in Great Britain. *Annals of the Associations of American Geographers*, 90, 41–66.
Peet, R. and Thrift, N. J. (1989) *New Models in Geography: The Political-Economy Perspective*. Boston: Unwin Hyman (2 vols).
Perroux, F. (1955) Note sur la notion de la 'pôle de croissance'. *Economie Appliquée*, 8, 307–20.
Plummer, P. and Sheppard, E. S. (2001) Must emancipatory economic geography be qualitative? *Antipode*, 33, 194–9.
Plummer, P. and Taylor, M. (2001) Theories of local economic growth. Part 1. Concepts, models and measurement, and Part 2. Model specification and empirical validation. *Environment and Planning A*, 33, 219–36 and 385–98.

Pocock, D. C. D. and Hudson, R. (1978) *Images of the Urban Environment*. London: Macmillan.
Rawstron, E. M. (1958) Three principles of industrial location. *Transactions, Institute of British Geographers*, 25, 135–42.
Rawstron, E. M. (2002) Do-it-yourself economic geography. *Progress in Human Geography*, 26, 831–6.
Rees, P. H. (1971) Factorial ecology: an extended definition, survey and critique. *Economic Geography*, 47, 220–33.
Rees, P. H. (1978) *Residential Patterns in American Cities 1960*. Chicago, IL: University of Chicago, Department of Geography, Research Paper 189.
Rees, P. H. and Wilson, A. G. (1977) *Spatial Population Analysis*. London: Edward Arnold.
Rhind, D. W. and Mounsey, H. (1986) The land and people of Britain: a Domesday record, 1986. *Transactions, Institute of British Geographers*, NS11, 315–25.
Robson, B. T. (1969) *Urban Analysis*. Cambridge: Cambridge University Press.
Rossiter, D. J., Johnston, R. J., and Pattie, C. J. (1999) *The Boundary Commissions: Redrawing the UK's Map of Parliamentary Constituencies*. Manchester: Manchester University Press.
Sayer, A. (1978) The incompatibility of dynamic modelling and conventional regional science. In R. L. Martin, N. J. Thrift, and R. J. Bennett, eds, *Towards the Dynamic Analysis of Spatial Systems*. London: Pion, 65–76.
Sayer, A. (1984) *Method in Social Science: A Realist Approach*. London: Hutchinson.
Sayer, A. (1992) *Method in Social Science: A Realist Approach*, 2nd edn. London: Routledge.
Schaefer, F. K. (1953) Exceptionalism in geography: a methodological examination. *Annals of the Association of American Geographers*, 43, 226–49.
Sheppard, E. S. (1984) The distance-decay gravity model debate. In G. L. Gaile and C. J. Willmott, eds, *Spatial Statistics and Models*. Dordrecht: Reidel, 367–88.
Sibley, D. (1987) *Spatial Applications of Exploratory Data Analysis*. Norwich: GeoBooks, CATMOG 37.
Smailes, A. E. (1953) *The Geography of Towns*. London: Hutchinson.
Smith, D. M. (1971) *Industrial Location: An Economic Geographical Analysis*. New York: Wiley.
Smith, W. (1949) *An Economic Geography of Great Britain*. London: Methuen.
Smith, W. (1952) *Geography and the Location of Industry*. Liverpool: University of Liverpool Press.
Smith, W. (1955) The location of industry. *Transactions, Institute of British Geographers*, 21, 1–18.
Stamp, L. D. (1946) *The Land of Britain*. London: Longman.
Stamp, L. D. (1960) *Applied Geography*. London: Penguin Books.
Stamp, L. D. and Beaver, S. H. (1971) *The British Isles: A Geographic and Economic Survey*. London: Longman.
The Statistician (1974) Special Issue on Geography and Statistics. *The Statistician*, 33(3/4), 149–308.
Steel, R. W. (1974) The Third World: geography in practice. *Geography*, 59, 189–207.

Stoddart, D. R. (1967) Organism and ecosystem as geographical models. In R. J. Chorley and P. Haggett, eds, *Models in Geography*. London: Methuen, 511–48.

Taaffe, E. J. (1970) *Geography*. Englewood Cliffs, NJ: Prentice-Hall.

Tarrant, J. R. (1968) Computers in geography. *Institute of British Geographers Newsletter*, 6, 11–25.

Taylor, P. J. (1971a) Distance decay curves and distance transformations. *Geographical Analysis*, 3, 221–38.

Taylor, P. J. (1971b) Distances within shapes: an introduction to a family of finite frequency distributions. *Geografiska Annaler* B, 53, 40–54.

Taylor, P. J. (1982) A materialist framework for political geography. *Transactions, Institute of British Geographers*, NS7, 15–34.

Taylor, P. J. (1994) The state as container: territoriality in the modern world-system. *Progress in Human Geography*, 18, 151–62.

Taylor, P. J. (1995) Beyond containers: internationality, interstateness and interterritoriality. *Progress in Human Geography*, 19, 1–15.

Taylor, P. J. (1999) Places, spaces and Macy's: place–space tensions in the political geography of modernities. *Progress in Human Geography*, 23, 7–26.

Taylor, P. J. and Goddard, J. B. (1974) Geography and statistics: an introduction. *The Statistician*, 23, 149–55.

Thrift, N. J. (1977) *An Introduction to Time Geography*. Norwich: GeoBooks, CATMOG 13.

Thrift, N. J. (1983) On the determination of social action in space and time. *Environment and Planning D: Society and Space*, 1, 23–57.

Thrift, N. J. (1995) Peter Haggett's life in geography. In A. D. Cliff et al., eds, *Diffusing Geography: Essays for Peter Haggett*. Oxford: Blackwell Publishers, 375–96.

Timms, D. W. G. (1971) *The Urban Mosaic: Towards a Theory of Residential Differentiation*. Cambridge: Cambridge University Press.

Tobler, W. R. (1970) A computer movie simulating urban growth in the Detroit region. *Economic Geography*, 46, 234–40.

Unwin, A. (1996) Exploratory spatial analysis and local statistics. *Computing and Statistics*, 11, 387–400.

Unwin, A. and Unwin, D. J. (1998) Exploratory spatial data analysis with local statistics. *The Statistician*, 47, 415–21.

Unwin, D. J. (1981) *Introductory Spatial Analysis*. London: Methuen.

Unwin, D. J. and Dawson, J. A. (1985) *Computer Programming for Geographers*. Harlow: Longman.

Upton, G. J. G. and Fingleton, B. (1985) *Spatial Data Analysis by Example: 1. Point Pattern and Quantitative Data*. Chichester: Wiley.

Upton, G. J. G. and Fingleton, B. (1989) *Spatial Data Analysis by Example: 2. Categorical and Directional Data*. Chichester: Wiley.

Voas, D. and Williamson, P. (2000) An evaluation of the combinatorial optimisation approach to the creation of synthetic microdata. *International Journal of Population Geography*, 6, 349–66.

Watson, J. W. (1953) The sociological aspects of geography. In G. Taylor, ed., *Geography in the Twentieth Century*. London: Methuen, 453–99.

Watson, J. W. (1955) Geography: a discipline in distance. *Scottish Geographical Magazine*, 71, 1–13.
Weber, A. (1929) *Alfred Weber's Theory of the Location of Industries* (being a translation, by C. J. Friedrich, of the German version published in 1909). Chicago, IL: University of Chicago Press.
Whitehand, J. W. R. (1970) Innovation diffusion in an academic discipline: the case of the 'new' geography. *Area*, 17, 277–83.
Williams, W. M. and Herbert, D. T. (1962) The social geography of Newcastle-under-Lyme. *North Staffordshire Journal of Field Studies*, 2, 108–26.
Wilson, A. G. (1970) *Entropy in Urban and Regional Modelling*. London: Pion.
Wilson, A. G. (1972) Theoretical geography: some speculations. *Transactions, Institute of British Geographers*, 57, 31–44.
Wilson, A. G. (1974) *Urban and Regional Models in Geography and Planning*. Chichester: Wiley.
Wilson, A. G. (1981) *Geography and the Environment: Systems Analytical Models*. Chichester: Wiley.
Wilson, A. G. (2000) *Complex Spatial Systems: The Modelling Foundations of Urban and Regional Analysis*. Chichester: Wiley.
Wilson, A. G. (2001) Land-use/transport interaction models: past and future. *Journal of Transport Economics and Policy*, 35, 3–26.
Wilson, A. G. and Bennett, R. J. (1985) *Mathematical Models in Human Geography and Planning*. Chichester: Wiley.
Wilson, A. G. and Kirkby, M. J. (1975) *Mathematics for Geographers and Planners*. Oxford: Oxford University Press.
Wilson, A. G., Rees, P. H., and Leigh, C. M. (1977) *Models of Cities and Regions*. Chichester: Wiley.
Wise, M. J. (1956) Economic geography and the location problem. *Geographical Journal*, 122, 98–100.
Wise, M. J. (1968) Wilfred Smith: an appreciation. In M. J. Wise, *Wilfred Smith: An Historical Introduction to the Economic Geography of Great Britain*. London: George Bell, ix–xxxiii.
Wise, M. J. and Rawstron, E. M. (1973) *R. O. Buchanan and Economic Geography*. London: Bell.
Woods, R. I. (1976) Aspects of the scale problem in the calculation of segregation indices: London and Birmingham 1961 and 1971. *Tijdschrift voor Economische en Sociale Geografie*, 67, 169–74.
Woods, R. I. and Rees, P. H., eds (1986) *Population Structures and Models: Developments in Spatial Demography*. London: Allen & Unwin.
Wooldridge, S. W. (1950) Reflections on regional geography in teaching and research. *Transactions and Papers, Institute of British Geographers*, 16, 1–11.
Wooldridge, S. W. (1956) *The Geographer as Scientist*. London: Thomas Nelson
Wrigley, E. A. (1965) Changes in the philosophy of geography. In R. J. Chorley and P. Haggett, eds, *Frontiers in Geographical Teaching*. London: Methuen, 3–20.
Wrigley, N. (1977) Probability surface mapping: a new approach to trend surface mapping. *Transactions, Institute of British Geographers*, NS4, 129–40.
Wrigley, N. (1984) Quantitative methods: diagnostics revisited. *Progress in Human Geography*, 8, 525–35.

Wrigley, N. (1985) *Categorical Data Analysis for Geographers and Environmental Scientists*. London: Longman.
Wrigley, N. and Bennett, R. J., eds (1981) *Quantitative Geography: A British View*. London: Routledge & Kegan Paul.
Wrigley, N., Holt, T., Steel, D., and Tranmer, M. (1996) Analysing, modelling, and resolving the ecological fallacy. In P. A. Longley and M. Batty, eds, *Spatial Analysis: Geographical Modelling in a GIS Environment*. New York: Wiley, 24–40.
Yule, G. U. and Kendall, M. G. (1950) *An Introduction to the Theory of Statistics*. London: Charles Griffin.

10

Global, national and local

PETER J. TAYLOR

> One of the characteristic features of geographical research is its concern with a particular scale of reality. (Haggett, 1965b, 164)

This chapter describes how geographical scale has been handled in British geography over the last century. With geography meaning, literally, 'writing about the Earth' it would seem that the matter of scale should be largely settled: geographers are the 'global scientists'. In practice this has not been the case and, in fact, the global has been a relatively neglected scale of geographical analysis. Most geographers have spent most of their time 'writing about their country' and those from Britain have been no exception. In this chapter I explore how and why this privileging of the national scale has operated in a century of British geography.

As a topic in its own right, scale has attracted a minute amount of the total effort of British geographers but, and it is a 'big but', choice of scale is implicit in everything they teach and research. Geographical scale is something you cannot 'get out of'; all studies take place at a given scale. Thus, far from having too little to review, the potential literature for reference in this chapter is massive, literally all of British geography. In fact, although remaining broad in scope, I do place limitations on the subject matter considered here. I am concerned with scale only in human geography and focus largely on selected works through the twentieth century that have had a disproportionate effect on British geography. The result is a story of economic, political, and urban geographies weaving their way through global, national and local scales.

I take a traditional chronological approach to my storytelling. Put simply, British geographers began with a focus upon global and national scales, recovered a local scale which led on to a focus on the national and the local, and currently the emphasis is on the global and the local. In order to understand these alternate combinations of scale preferences it is necessary to begin with a brief discussion of geographical scale as a social construct.

The natures of geographical scale

At first glance, geographical scale is not a particularly complex concept. I was able to use it in the preamble above without defining it. This implies a straightforward concept, one that readers will easily understand: different geographical scales mean different sizes in the areal extent of study. But of course there are dangers in this simplicity. Certainly I do not want to imply that there is anything 'natural' about the geographical scales that geographers employ in their researches. Two questions are raised here as a way of problematising the concept. First, why are there three scales specified by the title of this chapter—why not four or more? Secondly, and in any case, have these scales kept their meaning over the last century: for instance, is the idea of 'global' in 2000 the same in it was in 1900?

Social construction: state as pivot

Geographical scale is a social construction. It may seem to 'just appear' in our studies as a 'given' but behind every choice of scale there is a bundle of social processes that have created it. In our modern world scale seems so simple because its construction is, to a large degree, the outcome of a single institution, the state. The rise of the nation-state and its dominance over our collective geographical imaginations is directly reflected in the three scales of our title. The state defines the pivotal scale ('national'), with a scale above (previously called 'international', today 'global' is preferred) and a scale below ('intra-national', commonly referred to today as 'local'). The social sciences have been built upon this largely unexamined spatial order of scale: the core concerns of political science (national government), sociology (national society) and economics (national economy) are at the pivot scale. (The use of the adjective 'national' is not necessary; replace it simply with 'the' and the scale meaning remains in all three cases.) International (political) relations and local politics, international comparative sociology and community sociology, and international economics and regional economics, are all poor relations within their respective disciplines. Elsewhere I have referred to this specific ordering of knowledge as the result of 'embedded statism' (Taylor, 1996).

Geographical scale is, therefore, a largely hidden dimension within contemporary social knowledge, a testimony of the power of the state to define our modern world. Human geography is a social science, of course, but we would expect geographical scale to be rather less taken for granted in this discipline. And this is indeed the case. The state has been an important player in geographical discourses but it has not always been treated as the pivotal scale because of geographers' traditional use of the regional concept to frame their studies. Regions as areal units of study are immensely flexible. The study region may indeed be defined by a state's boundaries but this is not necessarily so. Regions can be large (continental scale) or small (neighbourhood scale) and can cover all areal extents in between. Other social sciences use regions, of course, but here they are less flexible and are employed to indicate studies above and below the state, such as regional power blocs (of states) in international (political) relations and regional divisions of states in regional economics. Defining regions, with or without the use of states, has been a major preoccupation of human geographers and this has led to a rich tradition of researches at a variety of scales. However, over the last century, human geographers have come to conform more and more to social science norms, resulting in a severe erosion of this distinctive tradition.

Social meaning: inconstancy and nexuses

Today it is the global scale that dominates most social thinking. 'Globalisation' is a buzzword of our times and all the social sciences are adapting their concepts in efforts to come to terms with this 'new' scale of operations. Human geography is no exception. The irony is, however, that geographers have a strong global tradition that has had little or no impact on current thinking. For instance, a century ago Mackinder (1904, 22) was describing processes with a 'worldwide scope' in which 'every explosion of social forces . . . will be sharply re-echoed from the far side of the globe'. In effect, human geography has studied 'two globals', first as imperial/political geography then, and second as corporate/economic geography now. The key point is that geographical scales do not 'stand still,' as it were, while political, social, and economic processes continue to operate unabated. This follows from the identification of geographical scale as a social construct, an outcome of those very processes that are forever changing. In other words, the meaning of geographical scales is not a constant.

These changes in meaning are perhaps most obvious for the global scale but it is equally the case for other scales. When the local was imported into British geography from the French regional school, it was very much a rural idea of place; today localities are usually defined in urban terms. As for the 'national' scale, states change their boundaries, states divide, states coalesce, new states form, the nature of states changes—clearly the meaning of a scale based on this institution has perforce been dynamic. And all this, like the global, is reflected in changing human geographies and the scales they study.

Finally, it is important not to see different geographical scales as somehow separate from one another. Scales may be treated separately for analytical purposes but in reality scales exist together—when we are local we are simultaneously national and global. There may be processes we study that have more salience at one scale, say the global, but it is impossible for 'global processes' to operate only globally; they will be 'grounded' at particular local places and be subject to many specific mixes of local cultures and national jurisdictions. The processes that make up distinctive geographical scales operate together. This is where the simplicity of scales gives way to complexities. Below I use the idea of a causal nexus to indicate interrelations between different scales whose processes are so intertwined that analytical separation is quite counter-productive. Focusing on three scales produces three possible nexuses—global–national, national–local and global–local—and I will argue that each nexus has had a period when it dominated British geography in the last century.

A global–national nexus: imperial geographies

For most of the modern era geography has been closely intertwined with exploration: it was largely the reporting of the practices of explorers. Since the exploration was by Europeans of non-European lands, Stoddart (1982) has dubbed geography 'the European science'. But this exploration leading to 'discovering' new lands is not a simple encounter whereby Europeans brought 'new worlds' into contact with Europe (Taylor, 1993). There were very clear power relations in these encounters that led to European domination of the world, consolidated in the late nineteenth century by the 'new imperialism'.

After 1900 exploration and geography separated. The former had always been a very practical pursuit but, with no new 'exploitable' lands to discover, exploration was reduced to its 'heroic' dimension: new 'adventures' to the poles and to the tops of high mountains. In contrast, geography became consolidated as an academic discipline taught in schools and universities. Basically, geography took over the practical dimension from its exploration forebear and this meant, rather than being a European science, it was to be, more specifically, an imperial science. Thus this first geography discipline was perforce global in its scale of enquiry. The global and the practical came together explicitly in Herbertson's (1905, 1910) work on 'natural regions': he devised environmental land divisions of the world as regions that he subsequently suggested as areal units for imperial planning by European states.

The new geography of the universities included two particular 'sub-disciplines' that encapsulated the global–national nexus at the heart of the discipline: commercial geography (Chisholm, 1922) and political geography (Mackinder, 1904). Both were concerned with the power and wealth of states within their territories and with how worldwide relations between states, both economic and political, affected levels of power and wealth. Here I focus briefly on the work of two leading British geographers in the early twentieth century—Halford Mackinder and C. B. Fawcett.

Halford Mackinder (1904, 1919) is well known for his political geography and in particular for his 'heartland theory'. Basically this provided a framework for British foreign policy as it moved from imperial rivalries outside Europe to concern for those rivalries being expressed within Europe. Mackinder warned of the dangers on Britain's doorstep of continental land alliances that could overthrow its world leadership, based as it was on long-standing naval supremacy. This was a global–national nexus expressed directly as statecraft. But Mackinder was more than a political geographer. Much less well known are his writings on commercial geography, where he again took a global view. In this work his concern was for the 'great trade routes' that defined Britain's economic position within the international economy (Mackinder, 1900). Of course, Mackinder did not see these as two distinct global–national nexuses. Trade policy (tariff reform) was central to the politics of his era and, indeed, his ideas on this led to him changing political parties. These two topics were brought together when he explicitly highlighted the national scale in his text *Britain and the British Seas* (Mackinder, 1907). Most contemporary geographers would find the reference to the 'British seas' a little odd but for Mackinder this was crucial to his argument: the seas were the conduit of trade and the basis of defence. After a physical and regional description of Britain, the book concludes with chapters on 'strategic geography', 'economic geography' and 'imperial Britain'—this is a regional geography that does not neglect relations with the rest of the world.

C. B. Fawcett has been less celebrated in geography but we find a very similar global–national nexus in his work. His *Provinces of England* (Fawcett, 1960) has the most longevity among his writings and in this he offered a geographer's regional division of the country as a new basis for government administration—part of the 'home rule all round' reaction to Irish nationalism. This 'domestic' statecraft is complemented by his magisterial geography of the British Empire (Fawcett, 1933). As well as physical and economic surveys of the constituent territorial components of the Empire, he emphasises the worldwide connections with chapters on 'Seaways and airways' and 'World relations of the British Empire'. However, it is in the organisation of the text that the global–national nexus is most clearly shown: there are two chapters each on the major territories (Britain, Canada, Australia, South Africa, India), one considering 'internal' matters, the other 'external'. Topics such as Britain's position in the 'Ocean gate of Europe' and Canada's domination by the United States take this text way beyond a territorial 'survey' of the empire; it is a truly global geography text for a British audience.

Recovering the local: a national–local nexus

Fieldwork is one of the few vestiges of exploration to be found within the discipline of geography. Merging with a parallel influence promoting 'social surveys' from Patrick Geddes (1915) and the civics and planning movement, fieldwork and field courses have long since been raised to the level of a fetish in geographical teaching compared to other social studies in schools and universities (see Wooldridge and East, 1951, ch. 9). Fieldwork is inevitably local in scale and therefore the potential for local study to become an important element in the development of British geography was there from the beginning. Initially overshadowed by the global, as imperial–national concerns gradually gave way to national and local planning concerns in the middle of the twentieth century, a new scale nexus emerged linking the national with the local. This was expressed in three different developments in British geography: regional geographies of the country, contributions to national and local planning, and the development of the new sub-discipline of urban geography.

Regional texts on Britain: geographical contributions to the theory and practice of Britain

The mid-twentieth century was the heyday of regional geography, now most definitely defined at the national and sub-national scales. The key text for this new focus was Ogilvie's (1928) *Great Britain: Essays in Regional Geography*, which was published for the International Geographical Union (IGU) conference in Britain in 1928. Going through many editions, this set the style for a particular synthetic ideal of regionalism in which the social was described as the product of the physical. Chapters are devoted to a total of 23 regions, with each chapter beginning with a region's physical geography before describing its human geography. The physical basis of this exercise in regionalisation is illustrated by the identification of regions such as 'The London Basin' and 'The Fenlands'. This formula for geographical study was applied to undergraduate dissertations for several decades as students were obliged to 'pick their region' and then proceed 'from geology to industry via agriculture' in a hopeless search for 'regional synthesis'. The strength of this form of regional geography is shown by the imitation of Ogilvie's book some 34 years later in *Great Britain: Geographical Essays* (Mitchell, 1962). This text has chapters devoted to 27 regions, all organised with the physical geography first, and although 'London Basin' has now become 'London Region', the Fenlands keeps its status as a region with its own chapter.

The idea of a synthesis between land and people is, of course, at the heart of all nationalist projects. Nations need to be integrated with their homelands. Thus, the scale implications of this regional geography are plain to see: Britain consists of a collection of physical units that together constitute the land of Britain. The nexus between the local units and the nation-state is to be found in the presentation of the sovereign territory as a coterie of countrysides. Geographical scale is being used to promote a British national image that privileges the rural over the urban in a new theory of Britishness. This is the discipline of geography's contribution to the broader intellectual and political twentieth-century project that was eliminating Britain's industrial heritage from 'real Britain' and, in particular, from 'real England' (Colls and Dodds, 1986; Taylor, 2000).

Synthetic regional geography did not go unchallenged in this period. By the time the IGU returned to Britain in 1964, the text prepared for the event overtly proclaimed a different type of geography: *The British Isles: A Systematic Geography* (Watson and Sissons, 1964). This followed an alternative 'one-scale approach' to the geography of Britain, pioneered by Stamp and Beaver (1933), in which different themes were studied across the whole country. This approach proved to be of much more practical utility than the synthetic studies.

Geography for national and local planning

The 1930s depression followed by the Second World War stimulated programmes of national reconstruction in which British geographers were much to the fore. Planning was conceived at three nested scales: national economic (demand) planning (Keynesianism), regional economic (supply) planning, and local government (physical) planning. While the former was the domain of economists in the Treasury, geographers made important contributions to the latter two levels of planning and, in particular, showed how they fitted into the larger whole (see also Chapter 16 below).

The Land Utilisation Survey initiated by Dudley Stamp in 1930 remains the largest national exercise in applied geography in Britain. Harnessing a huge team of schoolchildren, the land use of every field across the whole of Britain was surveyed in the early 1930s. The idea was to produce a basic database for British agriculture at the time of its nadir. From the scale of the individual field, information was aggregated to three higher scales and published as: local map sheets (1 inch to a mile) covering all of England and Wales and Scotland except for the less populated Highlands; 92 county reports; and a national volume *The Land of Britain: Its Use and Misuse* (Stamp, 1950). This research became very important for wartime food planning during the German blockade and especially for post-war planning. In 1941 Stamp was made Vice-Chairman of the Committee on Land Utilisation in Rural Areas, whose 1942 report contributed to the Town and Country Planning Act 1947, the Agriculture Act 1947, and the National Parks Act 1949. Note that this most urbanised of countries did not organise city or urban planning, rather it is 'town and country' that is to be planned. In reality this was in essence what Peter Hall and his colleagues (1973) have termed, in the title of their classic critique, *The Containment of Urban England.* Town and country planning was a scale-strategy to save all of the British countryside from urban people through the employment of 'green belts' around cities and towns. In this way the power of rural Britain continued to impact directly on urban dwellers, resulting in large numbers of them being herded into high-rise flats in the 1960s. This is a particularly British national–local nexus creating a great British planning disaster.

Geographers, of course, did not ignore industrial and urban Britain. In 1938 the Royal Geographical Society provided evidence to the Royal Commission on the Geographical Location of the Industrial Population in the form of numerous maps prepared by E. G. R. Taylor (1938). The idea of industry being concentrated in an 'axial belt' from Lancashire through the Midlands to London was included in the Commission's report and excited much controversy in British geography. Following Mackinder, Baker and Gilbert (1944) suggested there was no single 'belt' but rather two separate regions, Metropolitan England and Industrial England. Clearly the latter critique is more about thinking in terms of processes rather than simply reading patterns off maps, but it is also a matter of scale. Any spatial scale can be divided into component parts; informed choice requires understanding the salience of different scales for a given purpose. Methods to facilitate such decision-making involve much more than traditional mapping and did not become available in geographical research until after the quantitative revolution, as described in Chapter 9.

Urban geography: new contents, new scales

In the mid-century classical statement of the nature of geography by two leading British geographers of the time (Wooldridge and East, 1951), there are chapters on physical geography, historical geography, economic geography, political geography, and regional geography—but not on urban geography. In fact this sub-discipline not only fails to have a chapter devoted to it, it does not even feature in the book's index. As we have seen, what geographical interest there was in the local was largely rural, the coterie of landscapes that required protection from the people. For this land-obsessed discipline, any democratic expression of urban interests was, as one geographer put it, 'happily . . . quite unthinkable' (Freeman, 1958, 56).

Despite this infertile disciplinary landscape, to use a perhaps apt metaphor, an urban geography did emerge as a strong tradition of British geographical research that had its roots in the mid-century. Not surprisingly, the stimulus to this development came from the discussion around the industrial population distribution map, which was soon realised to actually be a map of towns and cities. Thus to interpret the distribution map required understanding cities and how they relate one to another. The person who appreciated this was Arthur Smailes, who produced two classic papers at the time describing the urban hierarchy of England and Wales. This led him to describe the 'axial belt' as simply an 'urban mesh' (Smailes, 1944. 1946). With Dickinson (1947), who brought German studies of urban regions to the notice of British geographers, Smailes set the framework of a new 'local' form of study involving two scales: internal (urban land use regions) and external (urban service fields). This organisation carried through into quantitative geography in which a vibrant urban geography was able to find models from other social science to fit the internal/external division of urban knowledge.

The new quantitative geography and its scale problem

Regional geography severely declined in the decades after mid-century. Quite simply, treating the description of regions as an end in itself gradually lost its intellectual appeal as the core of the discipline. Led by economic geographers who had long used regions as a tool of explanation rather than as the goal of their researches, and abetted by urban geographers whose emphasis on functional regions had a similar motive, regional geography as a synthesis of multiple factors in homogeneous areas was attacked as an intellectually sterile pursuit. Synthetic regional geography largely disappeared in the 1960s, to be replaced by a 'new geography' based upon an altogether different paradigm (Haggett and Chorley, 1967). Thus this was not just a victory of a more systematic geography; additionally geography portrayed itself as self-consciously 'scientific' in contrast to the 'art' of regional geography. Generally going under the name of the 'quantitative revolution', this new geography can be interpreted as a 'modernisation' of the discipline to bring it into line with the other social sciences. For many years the odd one out in social studies, human geography was transformed from an old-fashioned synthesising discipline into a specialised social science using the techniques and methods of the latter to inform spatial patterns and mechanisms. In the process, a new rigour was brought to bear on the question of geographical scale.

This transformation of geography was largely accomplished by geographers based in the United States and came to the notice of most British human geographers only after the 'revolution' was deemed to have occurred (Burton, 1963). Nevertheless, British geographers were instrumental in consolidating the new geography in the mid-1960s with six influential texts: *Statistical Methods for Geographers* (Gregory, 1963), *Frontiers in Geographical Teaching* (Chorley and Haggett, 1965), *Locational Analysis in Human Geography* (Haggett, 1965a), *Models in Geography* (Chorley and Haggett, 1967), *Quantitative Geography* (Cole and King, 1968), and *Explanation in Geography* (Harvey, 1969). Not only do these books mark a watershed in British geography, they also show the beginnings of distinctive British offerings moving from reviews of largely American studies to fresh research contributions. Perhaps the clearest case of the latter is the way Peter Haggett brought questions of geographical scale to the heart of the new geography. With Haggett (1965a, 1965b; see also Haggett and Chorley, 1967), geographical scale became an explicit concern of quantitative researchers as 'scale problems' (Haggett, 1965b).

The statistics of geographical scale

The basic problem of using statistical techniques in spatial analyses is that there is no 'geographical individual' to act as the basic object of study (Chapman, 1977). Measurements are collected for a given set of areal units, analysis is executed, and results obtained, but interpretation cannot simply end there. This is because the results of all such spatial analyses are scale-dependent. For instance, correlations between two variables for small census units will be different from the correlation for the same variables calculated using city regions as areal units. In other social sciences this 'aggregation problem' is somewhat less severe because there are individuals—decision-making social agents—who constitute the key level of analysis, whereas in spatial analysis the key level of analysis—geographical scale—varies with the purpose of the research. Thus transferring rigorous quantitative analyses from mainstream social sciences to human geography provided interesting intellectual puzzles for the new geographers.

Haggett (1965b) identified three scale problems. First, the problem of information available at different scales required a means of standardisation. Haggett and his colleagues suggested a new measure of areal extent, the G-scale, based upon logarithmic units of the earth's surface (Haggett, Chorley, and Stoddart, 1965). This seemingly 'natural' measure allowed direct comparison between the scales of any sets of geographical information. Secondly, the areal coverage problem was about how geographers could adequately cover their domain, the land surface of the earth. The solution to this was sampling, which led directly to the third problem, the scale linkage problem. The latter was the crucial issue: how to relate results at one scale of analysis to other scales. For instance, interpreting local fieldwork as a sample of a larger spatial unit (a region or country) is problematic if we expect different results to obtain at different scales. The solution involved one of the most innovative elements of quantitative geography whereby scale was incorporated into the analysis as the crucial variable. In such 'scale component analysis' the operation of different processes can be identified at different geographical scales (Haggett, 1965b, 265–9).

Notice that the very empirical nature of quantitative geography tended to reinforce the local scale of analysis. The emphasis on data collection strengthened the fieldwork tradition in geography (Board, 1965; Collins, 1965). Further, quantitative methods were inserted into the geography curriculum in the 'practical' slot, thus supplementing or even superseding mapwork. However, the whole matter of the scales of study in quantitative geography reveals more about the new geography than first meets the eye.

Scale revealing the critical limitations of the new geography

The analytic treatment of scale in the new geography had severe critical limitations. In becoming a social science, human geographers took on more than simply a new bundle of techniques and models, they engrossed a privileged scale of study—the national. The social sciences as creations and creatures of the state divided up all human activities into three policy spheres—economic, social, and political—and therefore human geography was reorganised into the necessary trilogy: economic geography, social geography, and political geography. However, along with the new models available from economics, sociology, and political science there came the unexamined geographical assumption in the form of the embedded statism of the social sciences (Taylor, 1996). As we have seen, human geography was certainly very familiar with operating at this scale, as in statecraft and national planning, but this was different: regions and their scale flexibility were replaced by an inflexible and critically unquestioned national scale. This may have produced a better social science but it impoverished geography as a distinctive discipline.

Table 10.1 classifies by scale all the human geography information contained in tables and diagrams for four of the key quantitative geography texts mentioned above. (Gregory, 1963, and Harvey, 1969, are omitted from the table because of the paucity of such data in these texts, which are essentially about statistics and methodologies.) The results are as expected.

Table 10.1 Geographical scales in British quantitative geography

Text	*International* (%)	*National* (%)	*Sub-national* (%)	*Local* (%)
Chorley and Haggett (1965) (n = 24)	8.3	41.7	12.5	37.5
Haggett (1965) (n = 147)	7.5	29.9	39.5	23.1
Chorley and Haggett (1967) (n = 43)	0.4	46.5	23.3	30.2
Cole and King (1968) (n = 76)	7.9	51.3	27.6	13.2

Note: n indicates the number of items of evidence in tables and diagrams for each book. Only human geography evidence is counted.

The new need for data creates two outcomes. First, the local scale is well represented as a result of data collection in the field. Secondly, where such collection is not undertaken, there is a reliance on the state to provide for data needs, so that the national and sub-national scales are well represented. The neglected scale in this process is the international scale and, in particular, the global. In fact the latter rarely appears in these texts: nearly all the evidence at this scale concerns international comparisons at far less than a global scale.

The irony in all of this is that for all its claims to revolutionary credentials and the enhanced concern for scale, the new quantitative geography focused upon the same local and national scales as did the geography it was replacing. However, the treatment of these two scales differed somewhat. Whereas we have described a national–local nexus for the previous geography, this would be an inappropriate characterisation for the new geography. The analytic bent of the latter led to a separation of scales, as in the scale component analysis that measures distinguishing scale effects. The key point is that the linkage between scales being sought in the new geography was not a connection but rather a division of variance.

Reinforcing the national scale: development without a national–global nexus

The transplant of embedded statism into human geography as social science is best exemplified in studies of economic development. Since this is supposed to be about lessening material inequalities at a global scale we might expect that here we could find evidence for new post-imperial studies at this scale. In fact this is not the case. Importing the social science approach to development led to a development geography that, far from reintroducing the global, actually reinforced the national scale of analysis in geography.

In the social sciences the process of development, like its social twin modernisation, was conceptualised as a national project. Each country was considered to be a separate laboratory for development policies that would project the country forward. The forward march was assumed to be the same for all countries, who were thus all marching along parallel paths to development (Taylor, 1989). The only difference between countries is that they are located at different points along the same journey. Therefore any differences in material wealth are merely a matter of timing; eventually, given the correct policies, all will reach the goal of, as the most famous model described it, 'high mass consumption'. There are many critical points that can be made about this model but what is noteworthy here is the elimination of the global scale. This just does not exist for development social science; there are only countries to develop.

It was Keeble (1967), in his chapter in *Models in Geography*, who provided the main introduction of this social science into British geography.

His argument is set up by the usual dismissal of previous geographical studies of development in favour of model-building (pp. 243–6). Given the new geography's interest in scale, Keeble does introduce geographical scale into his consideration of economic growth models. Initially he provides a table with three scales, which he describes thus: national (areas are state units), sub-national (areas are regions of state units), and supra-national (areas are continents, or even whole world). The use of 'even' implies the rarity with which this scale is used; in fact in his table Keeble provides examples of models for the two lesser scales but not for the supra-national, whether global or not. After discussing Rostow's (1960) famous 'non-spatial' stages-of-growth model, with its promise of high mass consumption for all countries, Keeble organises his text using the three scales listed above but has by far the least to say about the supra-national, using just one main example, the pattern of economic differences in Western Europe. Symbolically, the map of Europe is portrayed as 18 separate country maps (Keeble, 1967, 272). In the same volume, Hamilton (1967, 399–400) draws upon the most famous 'spatialisation' of Rostow's (1960) model—Taaffe, Morrill, and Gould's (1963) transport development model—and converts it into a national industrial development model. Whether general development or just industrialisation, the point is that geographers are studying one country at a time without recourse to anything resembling a global scale. Thus instead of a new development geography heralding a new global–national nexus, we have only more national analyses.

Challenging the national: new local and new global scales

In the aftermath of the quantitative revolution, if there were to be a challenge to the dominance of the national scale of study in human geography the most likely challengers would seem to be urban geographers. This sub-discipline strongly consolidated its position within geography in the 1960s and 1970s and clearly offered the possibility of enhancing the local scale of analysis. This did not happen: urban geography research increasingly took on a national agenda. This trend had its origins in late 1960s US geography, where 'urban problems' were seen as indicative of the ills of society as a whole. Whether studying national urban systems for national planning or intra-urban patterns and processes for national social policies (such as in housing), rather than this marking a return to a national–local nexus there was a generalised absorption of one scale (the local) into another (the national), clearly marked by the national scale being commonly designated 'an urban society'. This was exacerbated by the radical challenges to public policy heralded by David Harvey's (1973) highly influential *Social Justice and the City*, in which he turned from searching for policies to make cities more just places to raising questions about the nature of society that prevented cities becoming just places. From this point onwards the thrust of researches increasingly turned to the internal side of urban geography to study social processes in a capitalist state. The overall outcome is reflected in a comprehensive review of urban geography at the end of this period in which the external models so popular during the quantitative period (e.g. central place theory) are conspicuous by their total absence (Bassett and Short, 1989). Cities had become objects of study as places where society's problems were accentuated, and in this sense urban geography was no longer local.

But the local did re-emerge as a powerful theme within British geography. The new 'rise of the local' had a very specific origin in a new neo-Marxist regionalism. At about the same time there was a new 'rise of the global', which had a much more multifarious origin. However, both re-engagements with non-national scales emerged from British geographers trying to make sense of the worldwide economic downturn that was particularly severe in its effects on Britain in the 1970s and 1980s.

The localities project

The downturn in the world economy had different impacts on different parts of Britain and thus rekindled the long-running 'north–south debate' in national economic policy. This crude political characterisation of the geography of recession was interpreted in a more geographically sensitive manner in a new neo-Marxist theory of regional development. In the classic statement of this approach, Doreen Massey's (1984) *Spatial Divisions of Labour*, there is a return to the idea, banished by the quantitative revolution, that all places are unique (p. 117). However, unlike the physically unique sub-regions of traditional regional geography, Massey defined regional uniqueness in terms of the social inheritance of places. Every location has had a different experience of past periods of influential social change so that it brings to the contemporary situation its own particular historical mix of social and economic attributes and relations. In the famous geology metaphor of this theory, every region has different layers of investment reflecting its past successes and failures within the national and world economy. Thus the national economy is made up of a coterie of local regional economies, each differentially affected by the world economic downturn.

In 1984 the Economic and Social Research Council (ESRC) set up a research programme based upon this new regional theory. Entitled the Changing Urban and Regional System (CURS), the research focused upon seven study areas, three from the south and four from the north and midlands. The aim was to understand how national and global economic restructuring was impacting in different ways in these different places. The results are reported in Cooke's (1989) *Localities*, in which the policy issues are discussed, particularly the social and political contexts, both local and national. The basic question asked was in what circumstances and how could local economic initiatives be helpful? With the local firmly back on the geographer's research agenda, this new regionalism has become a standard approach for relating different scales in British human geography. Integrating the localities research with other similar work on places that focuses more on cultural and political processes, Johnston (1991) has been able to encompass a wide range of topics in this highly scale-sensitive manner.

Putting the 'geo' back into geography

The economic downturn in the world economy put the global back on the agenda in British geography. Self-evidently not a local or national problem, to understand social change clearly required a return to a consideration of supra-national processes. Hence there were soon calls for putting the 'geo' back into geography in which British geographers were prominent (Johnston, 1985; Thrift, 1985, Taylor, 1989). In practice there were two main avenues through which the global made its reappearance: economic geography and political geography.

As we have seen, the economic restructuring in the localities research included an important global dimension but there were more explicit treatments of the global in 1980s economic geography. In particular three books stand out. First, Taylor and Thrift (1982) brought together a series of essays on multinational corporations. These large firms were rewriting the rulebook on industrial location theory by investing in manufacturing beyond their home countries and, more importantly, beyond the 'developed world' itself. This creation of a 'new international division of labour' was obviously just as significant for policy in the 'developed' countries as in the 'developing world'. Secondly, Grigg (1985) produced an extensive overview of the world food situation, its production and maldistribution. Thirdly, the first edition of Peter Dicken's (1986) *Global Shift* appeared, which was to become the standard reference for the economic geography of the world economy, going through new editions to the present day. Quite simply, by the end of the 1980s it was not possible for students to study economic geography without incorporating a major global dimension.

In contrast, and as we have seen above, political geography had a strong global tradition going back to the imperial influence on geography. Although this interest waned somewhat after the Second World War due to its association with German geopolitics, Mackinder and his heartland theory remained a staple of the fare provided to political geography students. Thus it features in Muir's (1975) textbook, which began the tradition of organising political geography into the three scales—local, national, and international. Explicitly followed by Short (1982), political geography became the most scale conscious sub-discipline of geography. The weakness of this work was to be found in the deficit of linkages between the three scales, but a solution was provided by bringing a world-systems approach to bear on a materialist political geography (Taylor, 1982, 1985). Political geography proved to be the main vehicle into which this most global of social theories was transferred in geography. By the end of the 1980s the global dimension was once again integrated into political geography studies in both teaching and research.

It was not just in economic and political geography that the global scale became important again. One set of essays published at this time (Johnston and Taylor, 1986) brought the global scale to human geography in general with a title that proclaimed *A World in Crisis?*, though the question mark indicated uncertainty over whether a perceived crisis (by some) was in fact 'real'. This included essays on demographic and cultural issues as well as economic and political subjects. But the key point was that, with hindsight, we can see that the roots had been established for British human geographers to make important contributions to the study of globalisation that came to dominate much social science research in the 1990s.

Today's global–local nexus

Globalisation is a remarkable concept. Although the processes it encompasses have been variously traced to decades before 1990, the idea only really took off as a way of seeing the world in the 1990s. By the end of the decade social scientists were publishing on globalisation at the rate of 32 items per week (Taylor, Watts, and Johnston, 2002)! Why this massive interest compared with what was still a minority social science concern in the 1980s? Two changes in world geography seem pertinent. First, the end of the Cold War in 1989–91 meant the division into First, Second, and Third Worlds called out for rethinking and therefore revision. Secondly, the rise of Japan and consequent economic success of Pacific Asia meant that the old assumption that the core of the world economy was a West European/North American monopoly also needed rethinking and therefore revision. Both rethinkings could and did lead to revisions suggesting a more integrated 'one world'. Globalisation is the term that expresses both political and economic 'new worlds' and became the heavily favoured term to describe the bundle of many processes that promote transnational outcomes. The term is, of course, not neutral in the way it is employed and its contentiousness is part of its fascination. I am not concerned with the politics of the concept here but rather with how it has been researched in British human geography relative to other geographical scales.

It seems to me that the exploration of linkages between geographical scales is one of the prime contributions that geographers can make to globalisation debates. This is why I have used the phrase 'global–local nexus' to describe this section. As a nexus this implies that geographers are doing more than recognising processes at different scales, they are investigating the causal mechanisms through which processes at different scales reinforce each other. The fact that it is the local that is commonly paired with the global in such thinking indicates the way globalisation implies an erosion of the power of the state: key processes are moving to above and below the national scale. It is, of course, the only scale nexus identified in this story that does not include the national. In practice national-level processes are not as neglected as this formulation might suggest—they are sometimes included in the 'local' category in opposition to the global, as we will see below.

Glocalisation: connections across scales

Since geographers were not 'late' discoverers of the global, it follows that they had transcended elementary contemplation/description of this new focus in social science by the time globalisation had become such a popular topic. Relations between the global and other scales were widely discussed in the literature but mostly the detailed research itself did not go beyond a chosen scale. This was largely the case with the localities project, but in a follow-up study, *Towards Global Localization*, Cooke et al. (1991) do address a series of key questions at different scales. Analysis of the communication and computing industries in different countries shows how innovation in firms create enabling technologies that impact on corporate strategies, producing industrial shifts. It is this single bundle of processes that is simultaneously producing new local and global scales of activity. Subsequently, one of Cooke's co-authors, Swyngedouw (1992), coined the term 'glocalization' to describe a rescaling of economic activity as part of corporate strategies of flexible accumulation. This term emphasises that the contemporary politics of scale is not a singular scale effect, as globalisation implies, but operates precisely through constructing *different* scales of activity.

This multi-scalar approach to understanding globalisation is best illustrated in the popular Open University texts *Geographical Worlds* (Allen and Massey, 1995) and *A Shrinking World?* (Allen and Hamnett, 1995). In the former the chapter on 'local worlds' is explicitly about how various places are linked together by 'global connections' that are continually reconstituting local worlds (Meegan, 1995). More generally, the books describe processes producing a very uneven globalisation, one made up of many different local regions continually being remade by processes that are simultaneously local (e.g. inheritance), national (e.g. regulation), and global (e.g. corporate). The very same inter-scalar arguments can be found in another popular text that appeared at the same time, *Geographies of*

Global Change (Johnston, Taylor, and Watts, 1995). In this collection of geographical essays it is the range of processes in which the global scale is implicated that is emphasised: geoeconomic, geopolitical, geosocial, geocultural, and geoenvironmental changes are all charted. Global change may be the focus but in every case it is multi-scalar processes that are identified.

The most famous attempt to provide a framework for linking geographical scales is Amin and Thrift's (1992) conceptualisation of 'nodes' of economic activity operating through 'global networks' of flows. They were referring generally to concentrations of firms within 'neo-Marshallian' districts. Although such districts are traditionally 'industrial' in nature, one of the most explicit examples of concentrations of firms operating through global networks is to be found in the service sector. The activities of advanced producer service firms located within world cities constitute the classic example of a global–local nexus within contemporary globalisation.

World cities as a local–global nexus

In contemporary research world cities are usually defined as major urban concentrations of economic activities that have worldwide connections. In the words of Peter Hall (1966, 7) in his pioneering book on the subject, world cities are where 'a disproportionate share of the world's most important business is conducted'. From this starting point, a world city literature has grown that is both very international and very interdisciplinary. Nevertheless, British geographers have continued to make prominent contributions, not least Hall himself (e.g. Hall, 2000). With London commonly identified as a premier world city, British geographers, notably Nigel Thrift (1987, 1994), have had an important role in developing ideas about how world cities work.

The key point made in research on world cities is that no city can be studied in isolation. This reverses the 'internalist turn' in urban geography reported above. Cities operate within networks of cities and therefore understanding any city is impossible without knowledge of its geographical connections. World cities, themselves, are a special case in which those links are intensely global. But this renewed interest in the external relations of cities has not meant a consequent neglect of relations internal to the city. World cities are global–local nexuses in which the local intensive concentration of economic activity creates global economic connections and the global reach provides for more local economic opportunities in a causal spiral of mutually reinforcing scale effects. Again, we can use Open University texts to illustrate how contemporary British geography is developing these new ideas on geographical scale. In Massey, Allen, and Pile's *City Worlds* (1999), the two key essays are on 'worlds within cities' (Allen, 1999) and 'cities in the world' (Massey, 1999), which between them provide us with an image of networks within cities within a network of cities. This argument is comprehensively backed up by the essays in a companion volume, *Unsettling Cities* (Allen, Massey, and Pryke, 1999), which cover a wide range of economic, political, social, and cultural processes that link the local with the global. In a somewhat narrower vein, Taylor (2001) has specified a world city network as an interlocking network where global service firms, through their worldwide office locations, 'interlock' cities in the servicing of global capital. This work is part of a global internet project, the *Globalization and World Cities (GaWC) Study Group and Network*, an international and interdisciplinary centre for world city research but very much a product of British geography and thus with a direct interest in globalisation as rescaling.

Globalisation with its reliance on worldwide instantaneous communication has sometimes been interpreted as heralding the 'end of geography'. The diminishing importance of distance in some social and economic activities does not foreclose geography, it is an integral part of the creation of new geographies. And at the heart of these new geographies there comes an enhanced importance for geographical scale. For much of the last century questions of scale have been hidden by the embedded statism of the social sciences. However, from a position where the question had been settled at the national scale, with globalisation, the concept of scale has been freed to imbue the whole of the social sciences. Whereas it was mainly human geographers who incorporated scale relations in their social analyses, today relations between the global, national, and local are crucial issues in all social sciences. Within contemporary British geography 'global, national, and local' compose a critical agenda for making sense of the twenty-first century.

References

Allen, J. (1999) Worlds within cities. In D. Massey, J. Allen and S. Pile, eds, *City Worlds*. London: Routledge, 53–98.

Allen, J. and Hamnett, C., eds (1995) *A Shrinking World?* Oxford: Oxford University Press.

Allen, J. and Massey, D., eds (1995) *Geographical Worlds*. Oxford: Oxford University Press.

Allen, J., Massey, D., and Pryke, M. (1999) *Unsettling Cities*. London: Routledge.

Amin, A. and Thrift, N. (1992) Neo-Marshallian nodes in global networks. *International Journal of Urban and Regional Research*, 16, 571–87.

Baker, J. N. L. and Gilbert, E. W. (1944) The doctrine of an axial belt of industry in England. *Geographical Journal*, 103–4, 49–73.

Bassett, K. and Short, J. (1989) Development and diversity in urban geography. In D. Gregory and R. Walford, eds, *Horizons in Human Geography*. Basingstoke: Macmillan, 175–93.

Board, C. (1965) Fieldwork in geography, with particular emphasis on the role of land-use survey. In R. J. Chorley and P. Haggett, eds, *Frontiers in Geographical Teaching*. London: Methuen, 186–214.

Burton, I. (1963) The quantitative revolution and theoretical geography. *Canadian Geographer*, 7, 151–62.

Chapman, G. P. (1977) *Human and Environmental Systems: A Geographer's Appraisal*. London: Academic Press.

Chisholm, G. G. (1922 [1889]) *Handbook of Commercial Geography*. London: Longman, Green and Co.

Chorley, R. J. and Haggett, P., eds (1965) *Frontiers in Geographical Teaching*. London: Methuen.

Chorley, R. J. and Haggett, P., eds (1967) *Models in Geography*. London: Methuen.

Cole, J. P. and King, C. A. M. (1968) *Quantitative Geography*. Chichester: Wiley.

Collins, M. P. (1965) Field work in urban areas. In R. J. Chorley and P. Haggett, eds, *Frontiers in Geographical Teaching*. London: Methuen, 215–38.

Colls, R. and Dodds, P., eds (1986) *Englishness: Politics and Culture 1880–1920*. London: Croom Helm.

Cooke, P., ed. (1989) *Localities*. London: Unwin Hyman.

Cooke, P., Moulaert, F., Swyngedouw, E., Weinstein, O., and Wells, P. (1991) *Towards Global Localization*. London: UCL Press.

Dicken, P. (1986) *Global Shift: Industrial Change in a Turbulent World*. London: Paul Chapman.

Dickinson, R. E. (1947) *City Region and Regionalism*. London: Routledge and Kegan Paul.

Fawcett, C. B. (1933) *A Political Geography of the British Empire*. London: University of London Press.

Fawcett, C. B. (1960 [1919]) *The Provinces of England*. London: Hutchinson.

Freeman, T. W. (1958) *Geography and Planning*. London: Hutchinson.

Geddes, P. (1915) *Cities in Evolution*. London: Williams and Norgate.

Gregory, S. (1963) *Statistical Methods and the Geographer*. London: Longman.

Grigg, D. (1985) *The World Food Problem*. Oxford: Blackwell.

Haggett, P. (1965a) *Locational Analysis in Human Geography*. London: Arnold.

Haggett, P. (1965b) Scale components in geographical problems. In R. J. Chorley and P. Haggett, eds, *Frontiers in Geographical Teaching*. London: Methuen, 164–85.

Haggett, P. and Chorley, R. J. (1967) Models, paradigms and the new geography. In R. J. Chorley and P. Haggett, eds, *Models in Geography*. London: Methuen, 19–41.

Haggett, P., Chorley, R. G., and Stoddart, D. R. (1965) Scale standards in geographical research: a new measure of areal magnitude. *Nature*, 205, 844–7.

Hall, P. (1966) *The World Cities*. London: Heinemann.

Hall, P. (2000) Global city-regions in the twenty first century. In A. J. Scott, ed., *Global City-Regions*. Oxford: Oxford University Press.

Hall, P., Gracey, H., Drewett, R., and Thomas, R. (1973) *The Containment of Urban England*. London: Allen & Unwin.

Hamilton, F. E. I. (1967) Models of industrial location. In R. J. Chorley and P. Haggett, eds, *Models in Geography*. London: Methuen, 362–424.

Harvey, D. (1969) *Explanation in Geography*. London: Arnold.
Harvey, D. (1973) *Social Justice and the City*. London: Arnold.
Herbertson, A. J. (1905) The major natural regions of the world. *Geographical Journal*, 25, 300–12.
Herbertson, A. J. (1910) Geography and some of its present trends. *Geographical Journal*, 36, 468–79.
Johnston, R. J. (1985) To the ends of the Earth. In R. J. Johnston, ed., *The Future of Geography*. London: Methuen, 326–38.
Johnston, R. J. (1991) *A Question of Place*. Oxford: Blackwell.
Johnston, R. J. and Taylor, P. J., eds (1986) *A World in Crisis? Geographical Perspectives*. Oxford: Blackwell.
Johnston, R. J., Taylor, P. J., and Watts, M. J., eds (1995) *Geographies of Global Change*. Oxford: Blackwell.
Keeble, D. E. (1967) Models of economic development. In R. J. Chorley and P. Haggett, eds, *Models in Geography*. London: Methuen, 243–302
Mackinder, H. J. (1900) The great trade routes: the connection with the organization of industry, commerce and finance. *Journal of the Institute of Bankers*, 21, 1–6, 137–55, 266–73.
Mackinder, H. J. (1904) The geographical pivot of history. *Geographical Journal*, 23, 421–2.
Mackinder, H. J. (1907 [1902]) *Britain and the British Seas*. Oxford: Clarendon Press.
Mackinder, H. J. (1919) *Democratic Ideals and Reality*. London: Constable.
Massey, D. (1984) *Spatial Divisions of Labour: Social Structures and the Geography of Production*. Basingstoke: Macmillan.
Massey, D. (1999) Cities in the world. In D. Massey, J. Allen, and S. Pile, eds, *City Worlds*. London: Routledge, 99–156.
Massey, D., Allen, J., and Pile, S., eds (1999) *City Worlds*. London: Routledge.
Meegan, R. (1995) Local worlds. In J. Allen and D. Massey, eds, *Geographical Worlds*. Oxford: Oxford University Press, 53–104.
Mitchell, J. (1962) *Great Britain: Geographical Essays*. Cambridge: Cambridge University Press.
Muir, R. (1975) *Modern Political Geography*. Basingstoke: Macmillan.
Ogilvie, A. G. (1928) *Great Britain: Essays in Regional Geography*. Cambridge: Cambridge University Press.
Rostow, W. W. (1960) *Stages of Economic Growth*. Cambridge: Cambridge University Press.
Short, R. J. (1982) *An Introduction to Political Geography*. London: Routledge and Kegan Paul.
Smailes, A. E. (1944) The urban hierarchy of England and Wales. *Geography*, 29, 41–51.
Smailes, A. E. (1946) The urban mesh of England and Wales. *Transactions, Institute of British Geographers*, 11, 85–101.
Smailes, A. E. (1953) *The Geography of Towns*. London: Hutchinson.
Stamp. L. D. (1950) *The Land of Britain: Its Use and Misuse*. London: Longman.
Stamp, L. D. and Beaver, S. H. (1933) *The British Isles: A Geographic and Economic Survey*. London: Longman.
Stoddart, D. R. (1982) Geography: a European science. *Geography*, 67, 289–96.

Swyngedouw, E. (1992) The Mammon quest: 'glocalisation, interspatial competition and the monetary order: the construction of new scales'. In M. Dunford and K. Kafkalas, eds, *Cities and Regions in the New Europe*. London: Belhaven Press, 39–68.

Taaffe, E. J., Morrill, R., and Gould, P. R. (1963) Transport expansion in underdeveloped countries: a comparative analysis. *Geographical Review*, 53, 503–29.

Taylor, E. G. R. (1938) Discussion on the geographical distribution of industry. *Geographical Journal*, 92, 22–39.

Taylor, M. J. and Thrift, N., eds (1982) *The Geography of Multinationals*. London: Croom Helm.

Taylor, P. J. (1982) A materialist framework for political geography. *Transactions, Institute of British Geographers*, NS7, 15–34.

Taylor, P. J. (1985) *Political Geography: World-Economy, Nation-State, Locality*. Harlow: Longman.

Taylor, P. J. (1989) The error of developmentalism in human geography. In D. Gregory and R. Walford, eds, *Horizons in Human Geography*. Basingstoke: Macmillan, 303–19.

Taylor, P. J. (1993) Full circle, or new meaning for the global? In R. J. Johnston, ed., *The Challenge for Geography. A Changing World: A Changing Discipline*. Oxford: Blackwell, 181–97.

Taylor, P. J. (1996) Embedded statism and the social sciences: opening up to new spaces. *Environment and Planning A*, 28, 1917–28.

Taylor, P. J. (2000) Which Britain? Which England? Which North? In D. Morley and K. Robins, eds, *British Cultural Studies*. Oxford: Oxford University Press, 127–44.

Taylor, P. J. (2001) Specification of the world city network. *Geographical Analysis*, 33, 181–94.

Taylor, P. J., Watts, M. J., and Johnston, R. J. (2002) Geography/globalization. In R. J. Johnston, P. J. Taylor, and M. J. Watts, eds, *Geographies of Global Change*, 2nd edn. Oxford: Blackwell, 1–14.

Thrift, N. (1985) Taking the rest of the world seriously? The state of British urban and regional research in a time of economic crisis. *Environment and Planning A*, 17, 7–24.

Thrift, N. (1987) The fixers? The urban geography of international commercial capital. In J. Henderson and M. Castells, eds, *Global Restructuring and Territorial Development*. Beverly Hills, CA: Sage, 203–33.

Thrift, N. (1994) On the social and cultural determinants of international financial centres: the case of the City of London. In S. Corbridge, R. Martin, and N. Thrift, eds, *Money, Power and Space*. Oxford: Blackwell, 327–55.

Watson, J. W. and Sissons, J. B. (1964) *The British Isles: A Systematic Geography*. London: Arnold.

Wooldridge, S. W. and East, W. G. (1951) *The Spirit and Purpose of Geography*. London: Hutchinson.

V. Geography in action

11

Geography displayed

maps and mapping

ROGER KAIN
CATHERINE DELANO-SMITH

Maps enjoy today an unprecedented ubiquity. As commodities, they decorate ornaments, clothes, domestic rooms, and streets alike. There are numberless maps just a mouse-click away on the internet. They are presented daily on televised weather forecasts. Scarcely a single issue of a major western newspaper lacks at least one small-scale map, usually pinpointing the location of a newsworthy event or clarifying a political issue. Larger-scale maps—often in colour—are spread over the newssheet as an aid in the reconstruction of a major disaster. On 12 September 2001, no serious newspaper omitted a map of Lower Manhattan to show the location of the World Trade Center, and many reporters sharpened their analyses of the implications of the attack with a map showing the worldwide distribution of terrorist groups.[1] It is also as hard to imagine an entirely map-less household as it is to conceive of a map-less mass media. Relatively few households have no need of a road atlas, while city dwellers and visitors alike have occasion to turn with relief to a street map. Pocket diaries contain Henry Beck's map of the London Underground network, still the essential way-finding aid in London (Fig. 11.1). Anyone who walks for pleasure in England knows and uses large-scale Ordnance Survey maps, which portray footpaths. Those who sail in English estuaries and inshore waters should have access to a hydrographic chart, if only for their safety (Fig. 11.2). We might draw a sketch map to help the first-time visitor find our home.

[1] Hastily produced newspaper maps made for instant communication can be notoriously unreliable. *Private Eye* (19 October 2001) drew attention to the *Daily Mirror*'s map of 2 October 2001, which purported to show where terrorist attacks had taken place around the globe since 11 September. *Private Eye* retorted 'should someone tell the Tamil Tigers that they seem to have blown up Madagascar twice, while Jerusalem appears to have moved to Cyprus?'

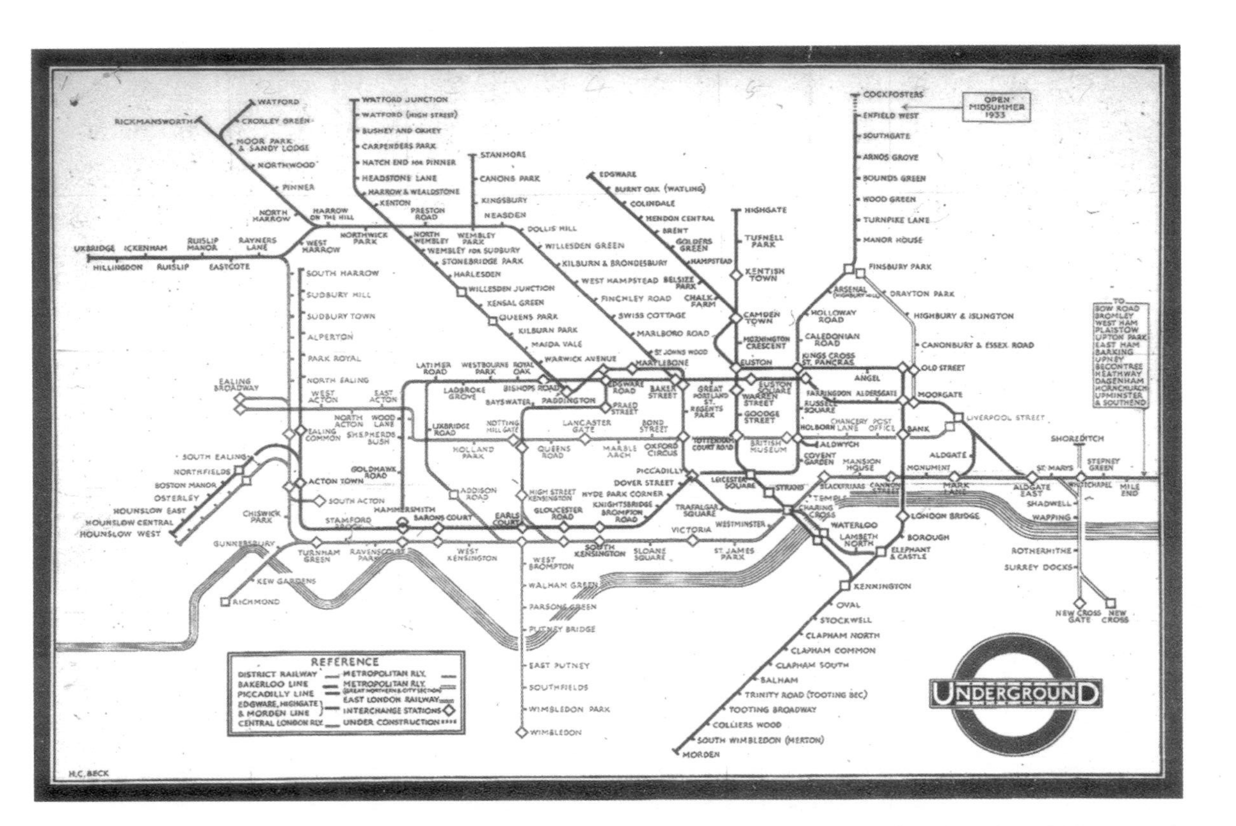

Fig. 11.1 The first card folder edition of Henry Beck's topological *First Diagram* of the London Underground.
Source: Victoria and Albert Museum, E.815-1979. Reproduced by permission of the Victoria and Albert Museum, London.

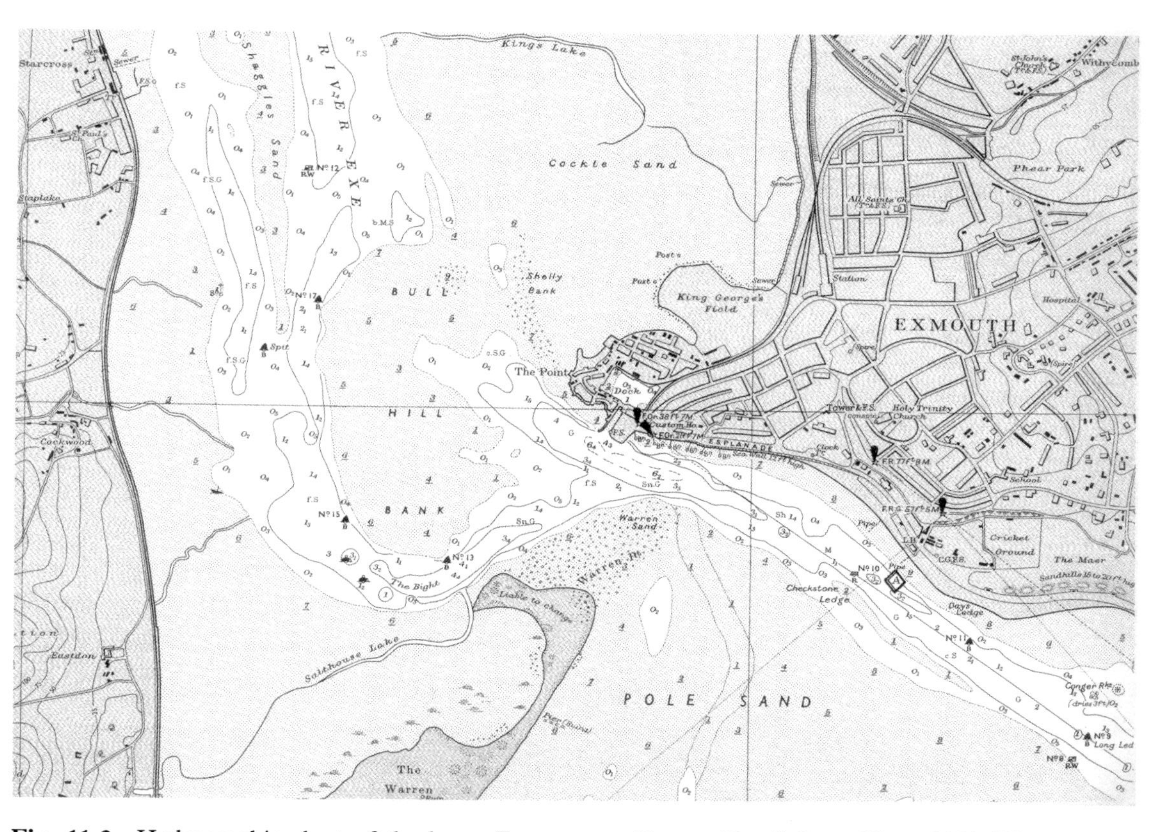

Fig. 11.2 Hydrographic chart of the lower Exe estuary, Devon. Detail from Chart 2290 'Plans on the South Coast of Devon, Exmouth Harbour', scale 1:1,250, published at the Admiralty, London, 1960. Reproduced by permission of the Department of Geography, University of Exeter.

The unremarkable use of maps in everyday life is mirrored in the world of scientific inquiry. Professional geographers use maps, as they have always done, to reveal spatial patterns in their data and to seek correlations in those patterns, although they no longer always do so by the time-honoured method of simple visual comparison of distributions on a series of paper maps. Geographical Information Systems (GIS) are founded on maps stored electronically. Dorling and Fairbairn (1993, 125) may be correct when they assert that 'just as the introduction of the printing press allowed an explosion in the quantity of books in the world, the geographical information system has easily resulted in more maps being "drawn" in the last decade than were created in all previous human history'. A prescient observation made in 1977 foresaw, correctly as it turns out, that many maps at the end of the century would be temporary maps, existing for short times only on a display screen (Robinson, Morrison, and Muerhcke, 1977). Even the conventional printed Ordnance Survey topographical map can now be delivered digitally for printing in a high-street shop at a customer-specified scale, and with the print centred on a site of the customer's choosing. Satellite photographs of the earth taken from space and remotely sensed images, at different scales and at different resolutions, are in a sense maps inasmuch as they 'have effectively replaced conventionally surveyed maps as the most practical way of accurately representing the earth's surface and its physical geography' (Cosgrove, 1999a). A variety of techniques—orthophotography in which the photographic geometry has been rectified to that of a map (Petrie, 1977), photogrammetry, differential GPS, and remote-sensing—have all been developed over the last third of the century to render traditional techniques of land survey effectively obsolete for both large- and small-scale mapmaking.

In one form or another, maps have been around for a long time. They predated writing in prehistoric societies, they have been widely used in non-literate societies, and they were studied by the ancients (Delano-Smith, 1987; Lewis, 1998b; Jacob, 1992). Modern maps may use an increasingly universal language of conventional signs to communicate across the barriers of language and culture, but the key semiotic concepts remain the same as millennia ago. Despite the high profile maps enjoy in modern society at large, a certain amount of unease has been expressed for a number of years over what is perceived to be a decline in the use of maps by geographers (Balchin, 1976; Muerhcke, 1981; Board, 1984). In most British geography departments, cartography has been distanced from geography. According to recent reviews, the teaching of traditional cartography and mapwork to geography students has all but disappeared, its place in the curriculum being filled by GIS (Green, 1995; Board, 1999; and see Rhind and Adams, 1980). In 1999, the editor of *The Geographical Journal* commented on what he saw as a decline in research, teaching, and resourcing of cartography in British geography departments (Millington, 1999). Even so, it was somewhat challenging, to say the least, to read in the following year an editorial in the *Transactions of the Institute of British Geographers* headed 'In memory of maps' (Martin, 2000).

While not disputing the basic premise that the role of maps in geography had indeed changed over the century, we would contend that an obituary for maps in geography is premature. Professional geographers may not be using printed paper topographical maps as they once did, and as society in general still does. They may well be removing well-built map chests from their offices to replace them with computer workstations.[2] They may no longer teach surveying or map projections to geography students or, arguing that technological developments have rendered such techniques obsolete, conduct measured land surveys in their own fieldwork. And they may no longer *study* maps. In all those senses the divorce between cartographers and geographers is indeed virtually complete. But geographers do indeed still *use* maps. What has happened in the period discussed in this book is that maps have been radically shifted from their former central position in geographical study to the margins of the subject. The traditional prerogative of the geographer has been largely abandoned to scholars from other disciplines.

Maps and society at the dawn of the century

In 1900, geography as an autonomous university discipline scarcely existed in Britain. University geography began in 1887, when the Royal Geographical Society (RGS) offered to help the universities of Oxford and Cambridge finance lecturers in the subject, and gained independence in May 1897 when the School of Geography at Oxford was founded as the first separate department for the discipline (Scargill, 1976, 446–9; see also Chapter 2, above). Hence it was not university-based academic geography that set the pace of the general public's interest in maps at the turn of the century but a long-established tradition of general and professional map use in society at large. Figure 11.3 serves as reminder, especially to those for whom the word 'map' tends to evoke nothing more exotic than an Ordnance Survey *Landranger* sheet or a page from a road atlas, that for centuries all sorts of maps had been made for and were used by individuals from different levels in society and from a broad spectrum of professions and activities. It reminds us too that the remarkable diversity of map form and function was no less impressive at the start of the twentieth century than at the end.[3]

[2] Another casualty of the current lack of interest in traditional maps is the insidious destruction of departmental archives, swept away in ignorance of their intellectual (and sometimes also monetary) value.

[3] The Ordnance Survey describes the 1:50,000 *Landranger* series on the cover as 'the all-purpose map with public rights of way and tourist information'. For the range of pre-modern English maps see Delano-Smith and Kain (1999).

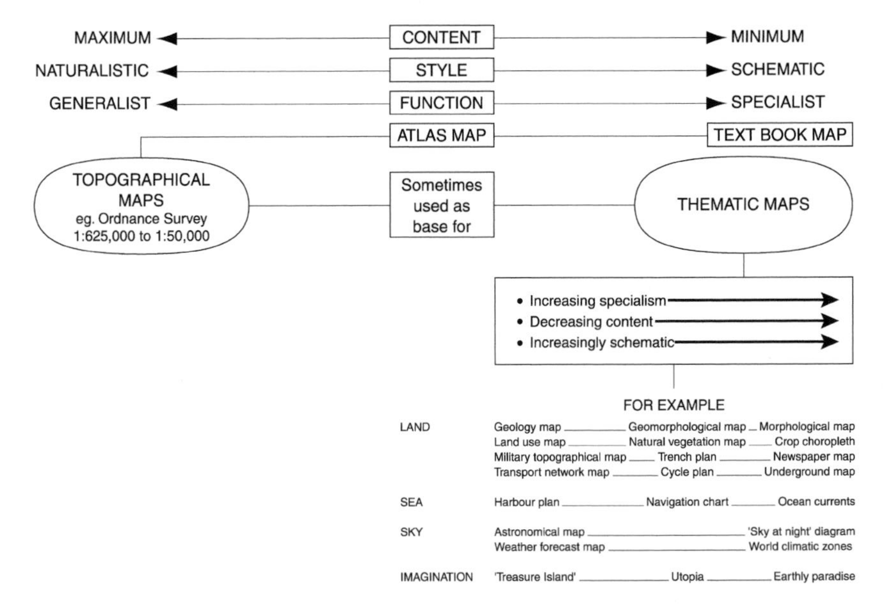

Fig. 11.3 The world of maps. Diagram summarising the relationship between map content, style, and function, with the generalist topographical map and the specialist thematic map as examples.

From well before the emergence of departments of geography in British universities, Fellows of the RGS had been exploring, mapping, and taking possession of parts of the unknown world, helping to extend the Empire to its furthest reaches. The close of the nineteenth century marked the high point of this activity. For the imperialists the map was the principal geographic tool: 'by representing the huge complexity of a particular physical and human landscape cartographically in a single image, geographers provided the European imperial project with arguably its most potent device'(Bell, Butlin, and Heffernan, 1995, 4).[4] As the nineteenth century turned into the twentieth, both geography and the role of maps within the subject were on the point of change, not least as a result of Halford Mackinder's paper 'On the scope and methods of geography', which set out an agenda for the development of geography (Mackinder, 1887; Bell et al., 1995, 4).[5]

Before we start to trace the changes in the relationships between British academic geography and maps throughout the twentieth century, note

[4] For the new worlds generally see Kain and Baigent (1992); for Africa see Stone (1995).

[5] Mackinder's comment was that 'We are now near the end of the roll of great discoveries'; see Coones (1987).

should be taken of some of the ways in which maps were embedded in the daily life of the nation in the last decades of the nineteenth century and first decades of the twentieth century. There was, for example, a myriad specialist maps for specialist use in urban contexts. Some of the most intriguing were the maps produced for insurance companies, for whom urban fires were presenting an increasingly expensive problem. Charles Goad had emerged as the major English company producing accurate large-scale town plans designed expressly as aids to risk assessment (Rowley, 1984, 1985). Goad's fire insurance plans portray urban layout in incredible detail: not only are individual buildings depicted in plan, and their commercial uses indicated, but details such as the type of building materials, the interior arrangement of staircases, party walls, doors (with their direction of opening) and windows are all specified (Plate 1).[6]

Maps also had a place in an era of burgeoning social reform in both town and country. By 1903, Charles Booth's survey of poverty in London, *Life and Labour of the People in London* (1902–3), had run to seventeen volumes and included his *Descriptive Map of London Poverty*. Booth's objective was to make a practical contribution to understanding London's social problems by making reliable information available and thus to 'lift the curtain' to provide a more informed basis for discussion of contemporary problems than the usual sensationalised 'people of the abyss' depictions (Reeder, 1984; Bales, 1991). Booth obtained his data from School Board visitors who were responsible for carrying out house-by-house surveys in their districts to locate and enrol children for schooling. From their information, he classified streets on a seven-point scale, ranging from poverty to comfort, to reveal the greatly contrasting conditions of life in London. His *Poverty Map* emerged as an eloquent expression of the geographical components of the class system imposed by nineteenth-century *laissez-faire* capitalism. It confirmed the already well-known distinction between east and west London, but it also revealed a socially far more heterogeneous London than had been suspected, a London composed of distinct residential areas (Hyde, 1975).

At the turn of the century, political groupings such as the Land Reform Movement sought to institute a free trade in land to break up Britain's large landed estates and the political system they fostered, which was exacerbating the polarisation of wealth in British society. The solution tried by Prime Minister Lloyd George's new Liberal government to redistribute landed wealth, if not the ownership of land itself, was to tax increases in property values over time. The Finance Act of 29 April 1910 made provision for the valuation of all land in the United Kingdom, as a basis from which subsequent increases in value could be taxed at the time of a property transfer or sale. The valuation was conducted by identifying

[6] Plates are located between pp. 398 and 399.

and numbering each piece of separately occupied landed property, valuing each of these 'hereditaments', and then recording their boundaries on large-scale Ordnance Survey maps (Kain and Baigent, 1992; Short, 1997).

Smaller-scale topographical maps published by commercial firms, such as George Bartholomew, as well as by the Ordnance Survey, were in demand from ordinary citizens in pursuit of leisure. Increasing numbers of individuals and their families came to see maps as their charter to the 'countryside and its innermost recesses' and as 'an unending revelation of its minutest features', as one of the major popularisers of landscape writing was later to put it (Batsford, 1940, 63 ; see Brace, 2001). The Ordnance Survey was primarily a military concern but by 1919 the one-inch-to-one-mile map certainly merited being dubbed the 'Popular Edition', a map designed and promoted explicitly to serve the growing mass market for a general topographical map (Hodson, 1999). The early years of the century saw the development of country walking, bicycling, and motoring as leisure pursuits, activities celebrated in the illustrations on the covers of the Popular Edition Ordnance Survey maps by artists such as Arthur Palmer, J. C. T. Willis and, above all, Ellis Martin (Nicholson, 1983; Browne, 1991; Matless, 1995) (Plate 2).

The notion that the countryside could be 'unfolded' as the map was unfolded (Matless, 1997, 146) extended into literature and the arts. The imperialistic theme of the blank spaces on European maps of Africa and the Far East played a catalytic role in Joseph Conrad's stories (most famously perhaps in his *Heart of Darkness*, 1898–9, 1902). An avowed map enthusiast all his life, Conrad was still reflecting in his last years how 'map gazing brings the problems of the great spaces of the earth into stimulating and directing contact with sane curiosity and gives an honest precision to one's imaginative faculty' (*Last Essays*, 1928, 19). In parallel vein C. E. Montague saw map-reading as analogous to reading music. In an essay called 'When a map is in tune', he remarked that as 'a musician's mental ear, can hear directly, when he reads a succession of notes on ruled paper, the rise and fall of an air . . . So the reader of maps is freed, before long, from the need to go through a conscious act of interpretation when gazing at the mapped contours of a mountain that has never been seen' (Montague, 1924, 40.)

Maps as geography: the regional paradigm

When E. G. R. Taylor published in 1921 a slim school textbook, *Sketch Map Geography: A Text Book of World and Regional Geography for the Middle and Upper School*, few regional geographers can have questioned the inextricable association of maps and geography. Taylor's book, with its preponderance of map to text, was reprinted 15 times before being revised with

Eila M. J. Campbell's collaboration in 1950 (final edition 1957). Maps had a crucial role to play in a branch of geography in which the central concern was the delimitation and description of unique regions as the end products of the study of areal differentiation. As explained by P. W. Bryan, Senior Examiner in Geography for the London Matriculation Examination, the map is the best medium for showing distributions and identifying patterns: the map, he wrote, is 'peculiarly the vehicle of the geographer in his effort to bring out clearly the relationship between man's activity and the physical setting' (Bryan, 1933, 65). Like Taylor, Bryan also recognised that in the writing of regional geography it was sometimes reasonable for verbal text to be considered supplementary to the graphic image, and to allow maps to express 'the essence of the relationships between man and the environment in the region studied' (p. 83). As the classicist J. L. Myres noted on the occasion of the 50th anniversary of the Geographical Association, 'geographers express themselves in maps . . . from this dependence on maps, a geographer derives an essential part of his training . . . in his maps he expresses geographical ideas, as well as geographical facts' (Myres, 1933, 72–3). Later in the 1930s, Richard Hartshorne held a similar view. Geography, he wrote, 'depends first and fundamentally on the comparison of maps depicting the areal expression of individual phenomena' (Hartshorne, 1939, 463; Johnston, 1997). So fundamental was the place of maps in geography in Hartshorne's view that he considered that if a problem was not conducive to solution by map interpretation and comparison, its legitimacy as a field of study in geography was open to doubt. From the perspective of a historical cultural geographer, Carl Sauer (1941, 6) also thought that 'the ideal formal geographic description is the map'.

Where maps with data that could assist in defining regions did not exist, they were to be made. Within limits, the more detailed such maps could be, the better. Fieldwork was preceded by the study of existing maps and it was also the principal source of data for an interim mapping. The next step was the analysis of the draft map, usually by visual comparison, so that the required regional boundaries could be decided upon and then represented on yet another map or maps. The iterative process of data encoding and decoding onto and from maps revealed a potential disjuncture between a 'cartographer's map' and a 'geographer's map'. The objective in a cartographer's map was to record as much detail as possible, while the geographer's map was to present a customised selection of salient facts. The emphasis was on matters of maximum importance, even to the extent of suppressing minor details, on the basis that underlying patterns are often revealed only when some detail is filtered by the mapping and re-mapping process.

In their post-Second-World-War review of the 'spirit and purpose' of geography, Wooldridge and East (1951, 70) proclaimed that 'no one claiming the title of geographer, however humbly, is entitled to be ignorant of how maps are made'. Nor, they continued, 'can it seem other than incongruous if the geographer cannot make at least a simple map, or cannot determine his latitude'. Many regional geographers working away from Britain had to be able to supply reconnaissance maps where local topographical maps were still unavailable or unsuitable, or to be able to assess critically those maps that were available. They needed at least some grasp of how to plot data quarried from maps and/or fieldwork. Above all, geographers needed to be able to *use* maps, to interpret their messages, and to read the geography contained within them. This meant, given the technology of the interwar and immediate post-war years, that geographers also needed to know about simple methods of land survey and how to represent statistical and observational data. Textbooks such as F. J. Monkhouse and H. R. Wilkinson's *Maps and Diagrams* (first edition 1952) were considered part of the regional geographer's essential basic education, even though the geographer's sense of responsibility towards maps was not exclusive: the book addressed not only geographers but also the 'historian, the economist, the geologist, the ethnographer, in fact all who are concerned with the depiction of spatial distribution' (Monkhouse and Wilkinson, 1964 edition: half-title).

A classic example of the mid-century geographer's unquestioning acceptance that geography and maps could not be disassociated is H. C. Darby's Domesday Geography of England project (1937–77). Darby's reconstruction of Domesday geography was rooted in mapping. The set of codexes known as Domesday Book contains only text. It is a description of the resources of the 13,000 or so settlements in as much of England as King William's clerks managed to cover in 1086. The original survey was carried out geographically, village by village, but the data were compiled tenurially, according to the feudal rank of the land-holders within each county. To be able to reconstruct the geography of early medieval England, Darby found it necessary that the work of the clerks 'be undone, and the survey set out once more upon a geographical basis' (Darby, 1952, ix). This entailed tabulating all the data given for each place. From that, it was 'only one step forward to the construction of maps' (Darby, 1952, 15). No justification was felt to be required for the use of a mapping methodology.

Domesday scholars before Darby, some from disciplines other than geography, also saw mapping as the way forward. The historian F. W. Maitland was speculating on the future of Domesday studies when he remarked that 'those villages and hundreds which the Norman clerks tore into shreds will have been reconstituted and pictured in maps' (Maitland, 1897, 520). For Darby, the 1930s was 'an exciting time', when 'the picture was changing from week to week' as completed maps could be fitted together to reveal 'how the distributions carried over from one county to another' (Darby, 1977, 375). The outcome of this experimentation was Darby's idea, formulated in 1937, of a collaborative, regionally based, multi-volume 'Domesday geography' of England. Research was interrupted by war but eventually seven volumes were published between 1952 and 1977. Five volumes (four involving collaborators) contained the regional studies, one provided a gazetteer of Domesday place-names, and the last offered a synthesis of the regional geography of Domesday England as a whole.[7] Throughout the entire project, and despite some changes between its inception in 1937 and publication of the first volume (Eastern England) in 1952, the founding principle of a map analysis of the documentary evidence was never in question. The six substantive volumes contain some 800 maps in total, setting out the geography of late-eleventh-century England 'as objectively as possible', together with tables of statistical data. The maps in the volume that deals with England as a whole provided generalised distributions against which regional variations could be measured. The general map of Domesday population density in England, for example, highlights the fundamental regional contrast between a northern England (characterised by low densities) and a southern England (characterised by high densities), divided by a line between the Humber and Severn estuaries (Fig. 11.4).

The regional ideal was not sought only by human geographers. In physical geography, the regional approach underpinned most research and publication in the first 60 or so years of the twentieth century and in many instances was explicitly based on mapping. In climatology, the work of classificatory scientists such as W. Köppen (in Germany) and C. W. Thornthwaite (in the United States) led to the mapping of climatic regions by British geographers such as A. Austin Miller (1951; Barry and Chorley, 1968: Appendix 1) (Fig. 11.5). In geomorphology, the regional paradigm encouraged the study of particular areas and the evolution of their local landscapes (see especially, Wooldridge and Linton, 1939). It was regionalism, too, that stimulated the thinking behind the division of the world into hierarchies of morphological or 'natural' regions, a genre in which British geographers were influenced by the work of their American colleagues, who had earlier compiled a map of the physiographic divisions of the United States (Fenneman, 1916). In Britain, David Linton adopted the same methodology for a map of 'the physiographic divisions of western Europe and their component physiographic provinces' (Linton, 1951, 206) (Fig. 11.6). According to Linton's hierarchical system of morphological regions, whole continents could be coherently described from the smallest, indivisible component—the

[7] The seven volumes, all published by Cambridge University Press are *Eastern England,* by H. C. Darby, 1952; *Midland England,* edited by Darby and I. B. Terrett, 1954; *South-East England,* edited by Darby and E. M. J. Campbell, 1962; *Northern England,* edited by Darby and I. C. Maxwell, 1962; *South-West England,* edited by Darby and R. Welldon-Finn, 1967; *Domesday Gazetteer,* by Darby and G. R. Versey, 1975; and *Domesday England,* by Darby, 1977.

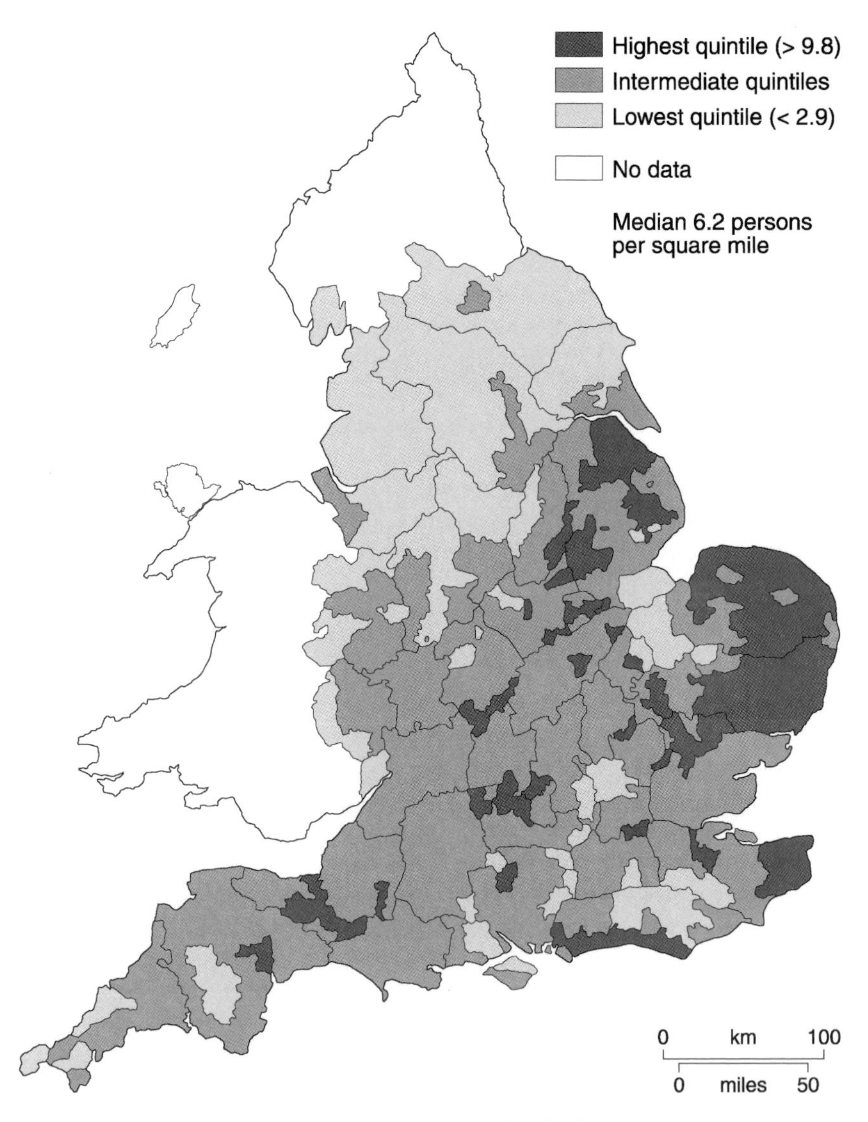

Fig. 11.4 Domesday population density in England, after Darby (1977).

flats and slopes of which all landscapes are composed—right up with increasing diversity to the top of the regional hierarchy (Waters, 1958). Morphological mapping was developed as a technique by which detailed micro-relief could be represented on maps by plotting field-derived information on breaks of slope, their form (convex or concave), and gradient (Whittow, 1984) (Fig. 11.7). Indeed, the British Geomorphological

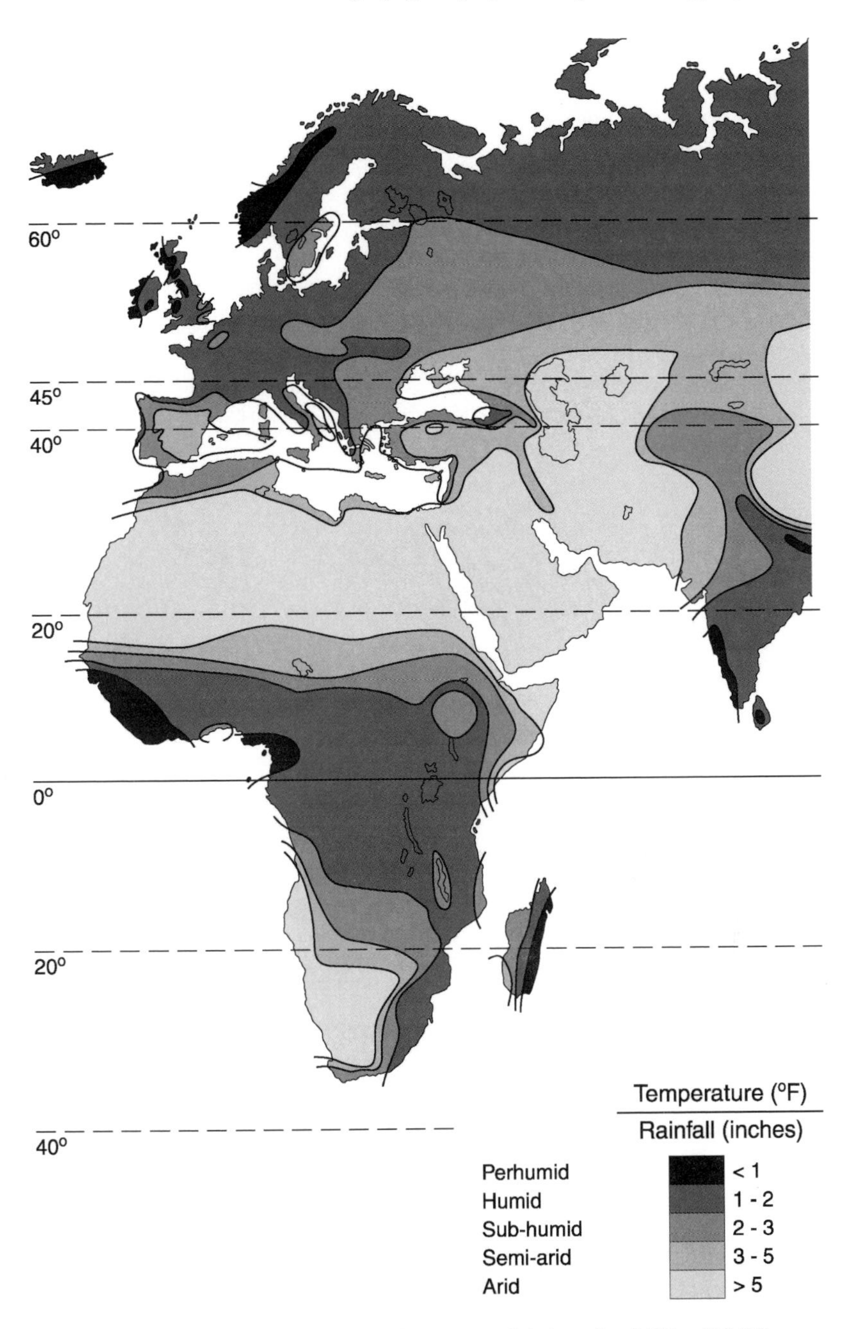

Fig. 11.5 Climate regions of Africa, Europe and Asia, after Miller (1951).

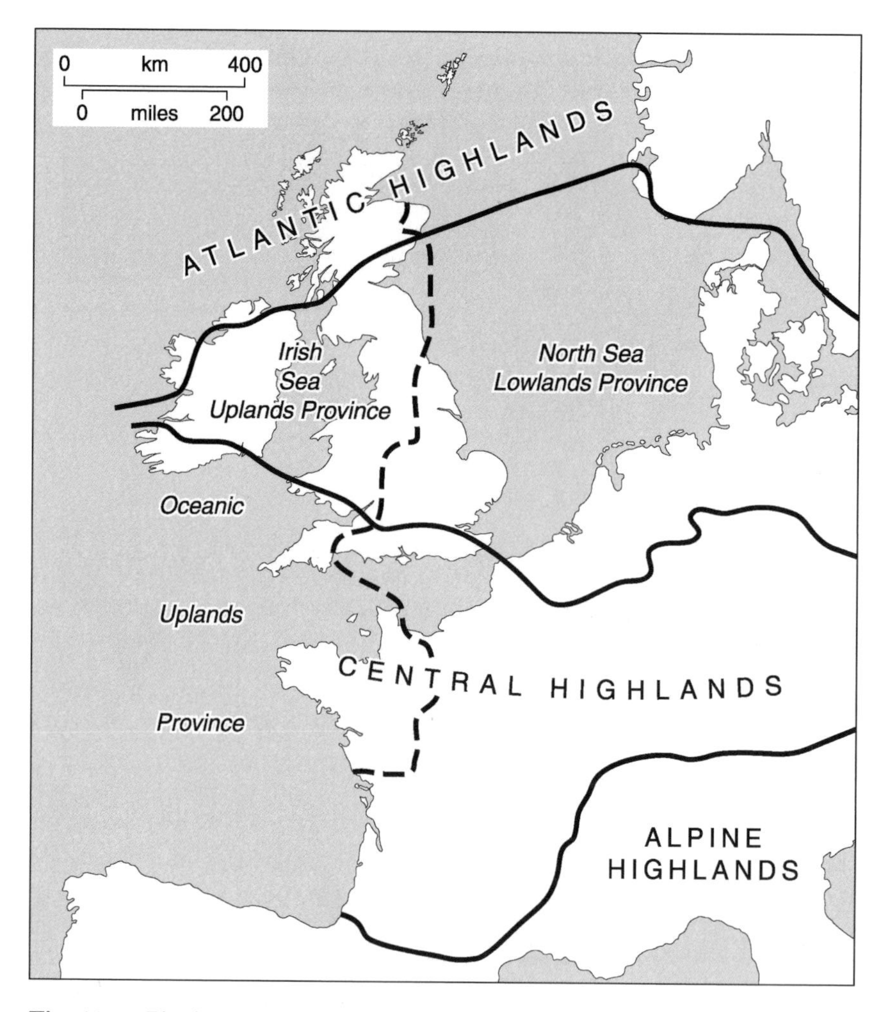

Fig. 11.6 Physiographic divisions of Europe, after Linton (1951).

Research Group began life as a group of geomorphologists who organised themselves nationally to produce a series of morphological maps.

Maps, quantification, and spatial science

As discussed by Hugh Clout in this book (Chapter 7), the tenets of regional geography came under sustained attack after about 1950. The so-called 'quantitative revolution' was a major discontinuity in geography, albeit a more drawn-out process than the word revolution usually implies. The impact on the position of maps in geography in the new era of spatial

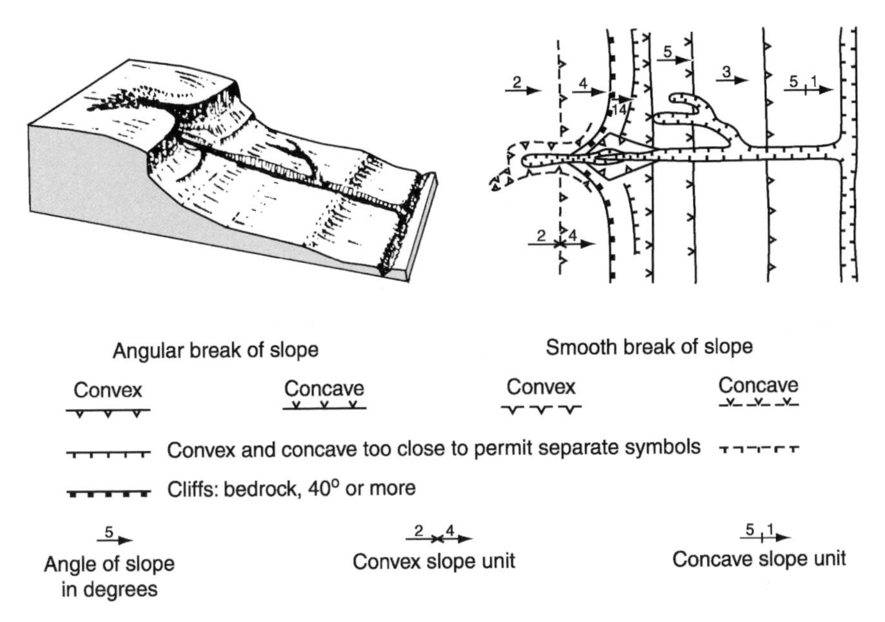

Fig. 11.7 The technique of morphological mapping, after Cooke and Doornkamp (1974).

science and logical positivism has been thought to have been profound. The boast from America was that 'the mathematics of space-geometry' could (or rather, should) be utilised 'with an efficiency never achieved by other sciences' (Bunge, 1962, 201; Johnston, 1997). Correlations of spatial distributions were to be considered 'statistically and dynamically'. Maps, geographers were solemnly informed, were 'old hat' (Chisholm, 1971).

In fact, maps were by no means old hat, even if many of those wedded to spatial theory and the notion of geography as a science failed to recognise the semantic sleight of hand, the substitution of 'model' for 'map'. Quantitative geographers did not see themselves as mappers but rather as modellers. In contrast, David Harvey pointed to the relatively commonplace use of the mapping concept by physical scientists when seeking to explain scientific theories (Harvey, 1969). Scientists, he said, adhered to a belief in the map as something that 'allows you to find your way around in reality, . . . allows you to say things about places you have never been to, and so on [just as] a theory helps you to find your way around and to say things about phenomena not yet observed' (Harvey, 1969, 370–1). Christopher Board (1967), in his contribution to the second Madingley Lectures in Geography, subsequently published as *Models in Geography*, had already argued that maps might be considered as formal models of reality. Board contended that instead of using maps as tools for region building and for communicating regional boundaries, they might be regarded as conceptual models or frameworks for representing and communicating the nature of the real world. He went on to summarise the ways in which factors such as those relating to the map-maker (which account for the human or subjective element in mapping) and those concerning the map user (such as the purpose of the map, data selected, scale used, degree of generalisation, use of signs and colour) act collectively, so that 'only a fraction of the information in the real world eventually finds its way to the map reader from the map or representational model of the real world' (Board, 1967, 704). In short, 'the less a map is like the real world, the more abstract it is, the more it is a model of that real world' (Board, 1967, 707) (Fig. 11.8).

Quantification in geography introduced the application of mathematical techniques to the interpretation of mapped spatial data. Studies such as that by Clarke (1959) indicated that visual interpretation of maps of statistical data varied very much from one individual to another. Clarke had asked a group of undergraduate students to interpret different map symbols and was startled to discover that fewer than 10 per cent of the students' estimates were accurate—a sobering lesson for the map-maker and a caution to all who interpret maps visually. More rigorous procedures for pattern analysis were brought in to geography in the 1960s. Trend-surface mapping, for example, was developed to improve the interpretation of the patterns in mapped spatial data by separating the intrinsic message (trend) from unexplained, random variations. Illustrating how the technique might be applied to commonly used isarithmic maps of continuous surfaces,

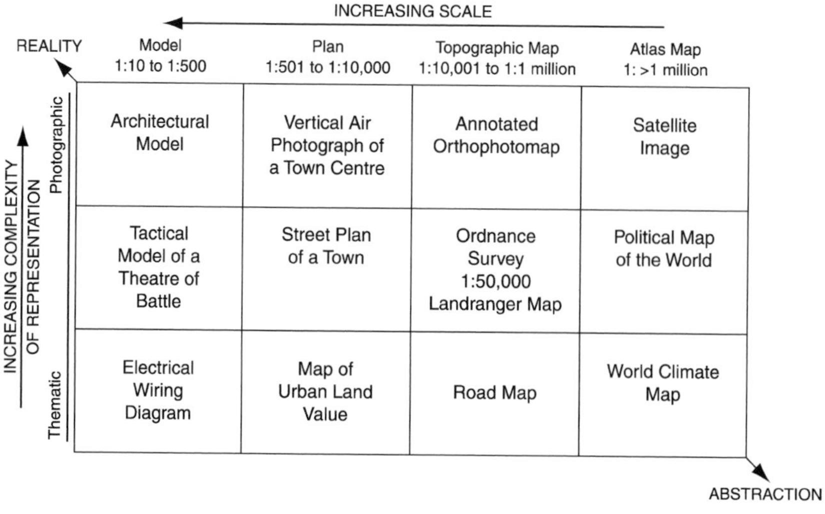

INCREASING SCALE ←

REALITY

INCREASING COMPLEXITY OF REPRESENTATION ↑	Model 1:10 to 1:500	Plan 1:501 to 1:10,000	Topographic Map 1:10,001 to 1:1 million	Atlas Map 1: >1 million
Photographic	Architectural Model	Vertical Air Photograph of a Town Centre	Annotated Orthophotomap	Satellite Image
	Tactical Model of a Theatre of Battle	Street Plan of a Town	Ordnance Survey 1:50,000 Landranger Map	Political Map of the World
Thematic	Electrical Wiring Diagram	Map of Urban Land Value	Road Map	World Climate Map

ABSTRACTION

Fig. 11.8 The map as a model, after Board (1967).

Chorley and Haggett (1965, 65) concluded that 'there seems no reason, other than convention and lethargy, why [this type of map] should not be very widely adapted for use in all branches of geography, both physical and human, in the immediate future'. Allied to computerised data processing, quantification also made possible new types of maps. Multidimensional scaling, for example, enabled maps based on other than linear distances to be constructed, the most commonly employed dimensions being time and cost (Haggett, 1990).

Paralleling the quantitative revolution in geography was the emergence of cartography as an academic discipline in its own right. Christopher Board recalls how a number of geographers interested in map communication, design, production, and interpretation were tempted to migrate to this new discipline in the 1960s, himself included.[8] Key events in the institutionalisation of cartography in Britain were the foundation of the British Cartographic Society in 1963 (the first issue of the *Cartographic Journal* appeared in 1964) and the publication of the Society of University Cartographers' first *Bulletin* (in 1966). The International Cartographic Association was founded in 1960 and established a number of Commissions, which carried forward work in various cartographic specialisms.

Maps, then, were still present in geography in the era of spatial science, but they increasingly involved mathematical rather than visual interpretation. The role of maps in the research process was not all that different from that stated by Wooldridge and East from their perspective within the regional paradigm, as 'pre-eminently the geographer's tool both in investigation of his problems and the presentation of his results' (Wooldridge and East, 1958, 64)—except for the word 'pre-eminently'. The spatial scientists of the 1960s did not rely solely on maps but by then had developed other (mathematical and statistical) tools for describing and analysing spatial patterns and relationships. For all the severity of the disjuncture of the paradigms of regional geography and spatial science, maps still had a place in the scientific method of hypothesis-testing and theory-building. What had changed was that maps were no longer central, as they had been in regional geography.

Maps in use

Geography, maps, and public policy

At the start of the twentieth century, it was not only geographical teaching that was held to have been greatly improved by the 'modern' use of maps (Holdich, 1902, 417). In his presidential address to the British Association

[8] Christopher Board, personal communication.

for the Advancement of Science at Belfast in 1902, T. H. Holdich highlighted a number of contexts in which maps were wanted by administrators: for general administration at home and in the colonies; in the political department for defining limits and boundaries; in the planning of economic development, especially for 'colonial and frontier progress'; and not least for the military officers responsible for 'preserving peace and good order' (Holdich, 1902, 419; Stoddart, 1992). 'In short', he went on, 'the cheapest, the quickest, the surest, indeed the only satisfactory method' of regulating the military control of a large and growing colony, or of a long stretch of military frontier, 'is to be armed with a perfect summary of what that country contains in the shape of a geographical map'. When such topographical maps had been used in the South African (Boer) War of 1899–1902, Holdich noted, 'the apathy shown by many of our foremost generals and leaders on the subject of maps' arose from a 'well-founded doubt of their own ability to make use of them' (p. 427). The outbreak of the First World War immediately put an end to any enduring apathy, and by the Second World War (1939–45), the American geographer J. K. Wright (1942, 527) was sufficiently confident to comment that 'maps are indispensable instruments of war'. Maps provided information for strategic decisions, help form public opinion, and would be essential afterwards to assist in the reconstruction of war-ruined landscapes.

A fair bit is known about the activities of British geographers and the use of maps in military and civil operations during the Second World War. Accordingly, attention here is directed mainly at the First World War, in which the number of academic geographers available for any kind of military or government service was small in comparison with those available to the government in 1939. The first department of geography (Oxford) was scarcely 15 years old when war broke out in the summer of 1914 (Scargill, 1976). Academic geographers who were involved worked mostly on intelligence and on mapping, often through the RGS, whose relations with both school and university geography were close. The RGS had offered its premises the day before war was declared to the Geographical Section of the General Staff of the War Office. The building was immediately used to house a 'hastily recruited' team of civilians, which Arthur R. Hinks (formerly a lecturer at Cambridge, but then assistant secretary at the RGS and, from early 1915, secretary) set to work. The two most urgent tasks were the compilation of lists of the place-names on the French topographical map of 1:80,000 and the creation of a wall-map of Britain for use in planning home defence strategies (Mill, 1930; Heffernan, 1996, 2002). Hinks's enduring preoccupation, though, was the international 1:1,000,000 map, a former peacetime project whose strategic importance was slowly, and not without a good deal of dispute, being recognised (Heffernan, 1996, 2002). Also in the RGS premises was the Admiralty's Naval Intelligence Division. Under the direction of Henry Dickson (once a lecturer at Oxford but by that time at Reading), Rudmose Brown (a botanist who had attended A. J. Herbertson's summer school in Oxford), John McFarlane (a historian who in 1903 became a lecturer in geography at Manchester), and others started work on the regional handbooks, thematic manuals, and reports that were needed to inform discussion of naval, military, and political problems, both during the war but especially at its conclusion. By 1920, some 157 publications had been produced. At the War Trade Intelligence Division, Arthur White, one-time secretary of the Scottish Geographical Society, was working there from 1915 to 1919.[9] Other academic geographers contributed much-needed cartographical expertise to the war effort. Charles Fawcett lectured at Southampton while also working for the Ordnance Survey, where some 33 million maps were produced for military needs. A. E. Young compiled map tables, which were later published, and textbooks on surveying were written. Advice on projections and of the need for an international grid, instruction on surveying for artillery and surveying in the field, and relief models were made available. The printing of maps was also carried out behind the front line in France, where the Ordnance Survey, fearing that cross-Channel communications would be blockaded by the enemy, had set up an overseas department in 1917 (Chasseaud, 1991, 1999).

At the end of the war British geographers, unlike their French and, especially, American colleagues, were singularly absent from the 1919 Peace Conference at Versailles. The notable exceptions were Alan Ogilvie, who before the war was an assistant in the School of Geography at Oxford, and William Stanford, who at the beginning of the century was also a member of staff at Oxford (see Scargill, 1976, Plate IV). Besides working for the Geographical Section of the General Staff, Ogilvie had seen war service in the Balkans and his first-hand experience of the problems of boundaries and national minorities would have served him well at Versailles (Heffernan, 1996). Ogilvie never said much about his wartime experiences. Apart from a mention in his inaugural lecture at Edinburgh in 1923 and indirectly in an allusion to the way 'members of other Delegations' appreciated Isaiah Bowman's 'gifts of judgement and organisation' and his responsibility for 'the admirable American presentations' at the Peace Conference Territorial Commissions—'where patience was often sorely tried', Ogilvie published only a short booklet relating to his experiences on the boundary commission (Ogilvie, 1922; 1950, 227; see Baker, 1959). Why Halford Mackinder was not invited as an expert geographer to the

[9] The War Trade Intelligence Division was set up to investigate reports that neutral countries were supplying the enemy with goods of a potential military application (such as Swiss gravel from the Rhine valley, which could be used for making concrete pill-boxes): personal communication from Peter Chasseaud.

Peace Conference is an intriguing issue and may have something to do with his geopolitical views (Mackinder, 1904; Kearns, 1985). Heffernan (1966, 521) is surely right in suggesting that the general absence of British geographers from Versailles related to 'the largely amateur nature of the discipline in the U.K. at the time', itself a function of the subject's status in British universities. Finally, note should be taken of those geographers whose academic discussions, at the RGS and in the universities, and especially on the redefinition of European frontiers and national self-expression, contributed to contemporary geographical opinion (e.g. Lyde, 1915). What maps were used by the British at the Peace Conference and in all these discussions has not been documented.[10]

When in 1985 the RGS initiated the collection of material relating to service in the Second World War of as many geographers as could be traced (some 325 individuals), the door was opened for explorations into the much larger contribution made by British geographers to military and civil mapping in that war than had been the case a quarter of a century earlier. The result of the Society's efforts, coordinated and classified by W. G. V. Balchin, is what is known as the 'Royal Geographical Society World War 2 Archive' (Balchin, 1987). Much more is thus also known about the use of maps in the Second World War. Geographers were now involved not only in the making of maps but also—and critically—in their interpretation. The geographer's trained eye for terrain analysis and expertise in regional description were invaluable in assessing the nature of a piece of country that had to be crossed, the potential for defensive sites, and the existence of natural hazards. The unprecedented demand for maps that the war created led directly to a widespread interest in maps as huge numbers of people learnt to use maps in all sorts of situations. The need to have accurate topographical and thematic maps and relief models became an accepted fact (Robinson et al., 1977; Pearson, 2002).

At the commencement of hostilities in 1939, the Inter-Services Topographical Department highlighted the paucity of cartographical information. Balchin (1987, 169) recalls how the only maps available to Bomber Command pilots engaged over southern Norway following the German invasion in 1940 were those in the 1912 edition of Baedeker's *Scandinavia*. Even more dismayingly, old British maps were being used by German Military Intelligence as the basis for aerial attacks on London in the early 1940s (Board, 1995). By early 1941, the Naval Intelligence Division had established two geographical handbook centres, one at

[10] Harold Nicolson refers frequently to 'my maps' in his memoirs, as well as to the Americans' 'vast relief map in sections depicting the Adriatic, very beautiful', and describes Lloyd George confusing a relief map with an ethnographic map (Nicolson, 1943, 223 and 333). In contrast, see Palsky (2002) for an exposition of the activity of French geographers and their maps at the Peace Conference.

Oxford, directed by Kenneth Mason, and the other at Cambridge, directed by H. C. Darby. The two centres produced 58 descriptive handbooks covering all the main theatres of war. Geographers were the principal contributors and by the end of the war, 'there were few academic geographers who had not been involved in some way with the NID geographical handbook project' (Balchin, 1987, 171).[11]

The experience of both world wars underlined the critical role of maps in decision-making. Other projects in the early part of the century also underscored the usefulness of maps in the public policy arena, notably the movement for regional surveying inspired by Patrick Geddes. From the 1890s, Geddes had argued for regional survey and planning but his influence within academic geography was probably most significant after 1918 'when geographers played an increasingly active role in the growing movements for regional survey and town and country planning' (Matless, 1992, 465; 1999). The premise was that a survey of the present state of affairs was an essential preliminary to improvement through planning. The centrality of maps was emphasised by C. C. Fagg and Geoffrey Hutchings (1930) in what became the standard text on the conduct of regional surveys. Base maps, they pointed out, were needed for plotting survey results since 'wherever possible the most satisfactory method of recording regional observations is on a map [and] the number of maps that may be prepared is almost unlimited' (p. 93). They put maps into three classes. First were those that could be compiled from existing data, such as relief and population maps. Second were those that required fieldwork but no special or technical expertise, such as maps of woodland or public open space. The majority of maps needed in regional surveying, Fagg and Hutchings felt, would fall into one or other of these two categories. The third category comprised maps for which specialist technical knowledge was essential, as in the case of mapping differences in soil types. Whatever the category, the necessary maps were then to be assembled in a 'regional survey atlas', which would provide details of the physical background of the region in question, its history, and its economic and social characteristics.

Particularly supportive of the educational value of regional surveying was the schoolteacher membership of the Geographical Association. The Association had a committee for regional surveys, of which the honorary secretary in 1928 was Dudley Stamp (Board, 1968). Stamp's own goal was 'a uniform survey of the whole country' which would be secured by ensuring schools employed the same base map, the six-inch Ordnance Survey sheets that covered their respective parishes, and by adhering to the plotting of

[11] The Royal Geographical Society/Institute of British Geographers holds a complete set of the handbooks and an archive detailing the activities of the Oxford and Cambridge centres. See also Hugh Clout's chapter in this volume (Chapter 7).

simple surface utilisations, 'woodland, meadow, arable, etc.' (Stamp, 1928, 348; 1962). What Stamp in fact proposed was simply the substitution of a national coverage of land use for the kind of detailed, multi-thematic mapping advocated by the regional survey movement (Stamp, 1931).

Fieldwork for what became the First Land Utilisation Survey of Great Britain was conducted by school pupils and their teachers. Seven categories of land use were plotted, field by field, on to 1:10,560 Ordnance Survey maps. The drafts were then reduced and the sheets published at the scale of one inch to one mile (1:63,360). The first maps appeared on New Year's Day 1933 (Plate 3). By 1935 the bulk of the fieldwork had been accomplished, and in 1946 the last of the county reports accompanying the 170 map sheets of the Survey also saw the light of day. The results of the Survey constitute a remarkable record of rural land use in the decade before the great changes induced by the Second World War and the ploughing up of grassland for crops. Although the Survey cannot be regarded as a strictly synchronous cross-section, the maps of the First Land Utilisation Survey approached the ideal of a total cross-section more closely than any other survey before remotely sensed digital data became available from the LANDSAT satellite thematic mapper.

In the 1960s Alice Coleman attempted to replicate the 1930s survey (Coleman, 1961, 1964). Her methodology differed in detail, but the objective was broadly similar: the collection of data through fieldwork and plotting these data on to maps for printing and publishing. Both the data collection and the printing components of land use surveying were, however, rendered technically anachronistic before a great deal had been achieved. Ground survey was being replaced by remote-sensing and the capture of data on multispectral images. GIS saw to the manipulation and merging of data, as well as the plotting and representation (printing) of the output. Notwithstanding these developments, Coleman still defended her field mapping approach (Coleman, 1980).

As the end of the Second World War came into sight, planning for the post-war period became a matter of increasing urgency on the government's agenda (Willatts and Newson, 1953). E. C. Willatts was appointed as maps officer in the new Ministry for Town and Country Planning, where the information recorded on the First Land Utilisation Survey land use maps was employed in a number of ways by the nascent town and country planning movement (Willatts, 1971). It was used to regulate land use in general and also, more specifically, to promote planning policies, such as the establishment of a green belt around each major conurbation; to decide on the best sites for new towns; and to establish priorities in the reclamation of derelict land (Rycroft and Cosgrove, 1995). Planners such as Patrick Abercrombie (1938) had shown an interest in the results of land use surveying from an early stage. In 1937, the geographer Eva Taylor submitted a portfolio of maps to the Royal Commission on the Geographical

Location of the Industrial Population (the Barlow Commission) and, in collaboration with Dudley Stamp, continued to argue throughout the Second World War for the compilation of a 'national atlas' for use by proposed bodies such as the Central Planning Authority (Taylor, 1940; Willatts, 1971). By 1942, the Ordnance Survey was printing the first ten sheets of just such an atlas. Today maps are accepted without question as key instruments in the UK planning system. Development plans are published as maps and written statements, and land scheduled for specific uses is demarcated on the maps. In 1960, the geographer's position in relation to the use of maps in planning was virtually unassailable. In his pioneering *Applied Geography*, Dudley Stamp (1960, 9–10) argued that the geographer's unique contribution to the team investigations that preceded the formulation of a plan was in representing data from surveys on maps and then undertaking the interpretation and analysis of the map. As practitioners from other disciplinary backgrounds became increasingly familiar with maps, however, geographers needed to equip themselves with additional (or different) skills if they were to continue to contribute in the planning field (Chisholm, 1971).

Mapped land use data have also been used in environmental management, usually in parallel with town and country planning. Although national agencies print their own geological, soil, and topographical maps, there are benefits to be obtained by developing the data further for administrative purposes. Thus physical geographers have created the geomorphological map which, in addition to describing the general morphology of a locality, also provides highly detailed information about slopes and drainage, rock type, soil characteristics, and so on, all of which can be obtained from separate printed geological, soil, and topographical maps but which, when collated on to one map, gives the planner a formidable reference tool. 'In practice geomorphological maps rarely show the distribution of processes, only the distribution of landforms resulting from defined processes' (Cooke and Doornkamp, 1974, 371). Land systems mapping, while not concerning itself with locating various kinds of landforms, sets out to delimit the boundaries of areas containing similar landform properties. For example, areas liable to land slippage might be distinguished from more stable areas based on a synthesis of relief, materials, and processes. Land systems maps have many practical uses, such as assessing the suitability of land for different types of agricultural exploitation. Yet again mapping procedures have been revolutionised by the introduction of remote sensing for terrain analysis, although the basic principle of studying landform distributions remains a continuing objective (Gregory, 1985).

The map as source

A map is a repository of information relating to the time at which it was compiled. Topographical maps in particular have been long and widely used in geography as a quarry for data to employ in reconstructing aspects of both the physical and the human environment of a specific date or period. Recent discussions have broadened the consideration of maps from being seen as windows revealing a world of observable 'facts' to an appreciation of the way they also contain 'a social construction of the world expressed through the medium of cartography' (Harley, 1990a, 4). Harley's warning that maps are not simple mirror-like reflections of the 'real' world but need decoding as social documents was heeded by Robin Butlin (1993, 90), who noted that maps are 'more than sources of hard evidence and means of representation of information of a factual kind'. By and large, though, the majority of geographers who have used maps as a source in a wider research project have been concerned with extracting information from the face of the map, decoding its conventional signs, and importing the resultant data into a geographical study. Such was the approach applied to the tithe maps of the mid-nineteenth century, from which land use data were extracted for comparison with the patterns revealed by the First Land Utilisation Survey of Great Britain (Willatts, 1933; Henderson, 1936; Kain and Prince, 1985, 2000) (Fig. 11.9). Similar approaches have made maps an invaluable source in studies of field systems (Baker and Butlin, 1973), urban morphology (Conzen, 1960), and rural settlement morphology (Roberts, 1977).

For some studies in physical geography, maps are a prime if not a unique source of data. Techniques of morphometric analysis have been developed as an aid to the study of map-derived data on landforms at the reconnaissance stage and as a preparation for detailed work in the field in studies of drainage basin dynamics and landscape evolution (Gardiner, 1974). Almost all such studies are based on data derived from published topographical maps or special 'pulls' that show only the 'blue line' stream networks. Techniques of stream ordering decompose the network within each drainage basin into a number of discrete segments to facilitate the measurement from the map of a range of parameters, such as total channel length, basin area, and basin relief (Gregory and Walling, 1973).

Retrospective evidence can be obtained from maps by referring to a sequence of non-current maps. This type of historical approach is increasingly used by physical geographers to provide long-term perspectives for studies of contemporary physical processes, for understanding the nature and causes of environmental change in the historical period, and above all for measuring the magnitude of the impact of human activities on the physical environment (Hooke and Kain, 1982). Modern process studies often involve too short a timescale to reveal the full extent of the impact of known activities, and the evidence derived from early maps can be used for calibrating predictive models. Process studies have also shown that many environmental changes are far more rapid, and extensive, than had formerly been thought; a full span of extant maps allows such changes to be captured and recorded. In a classic study, J. A. Steers (1926) used maps

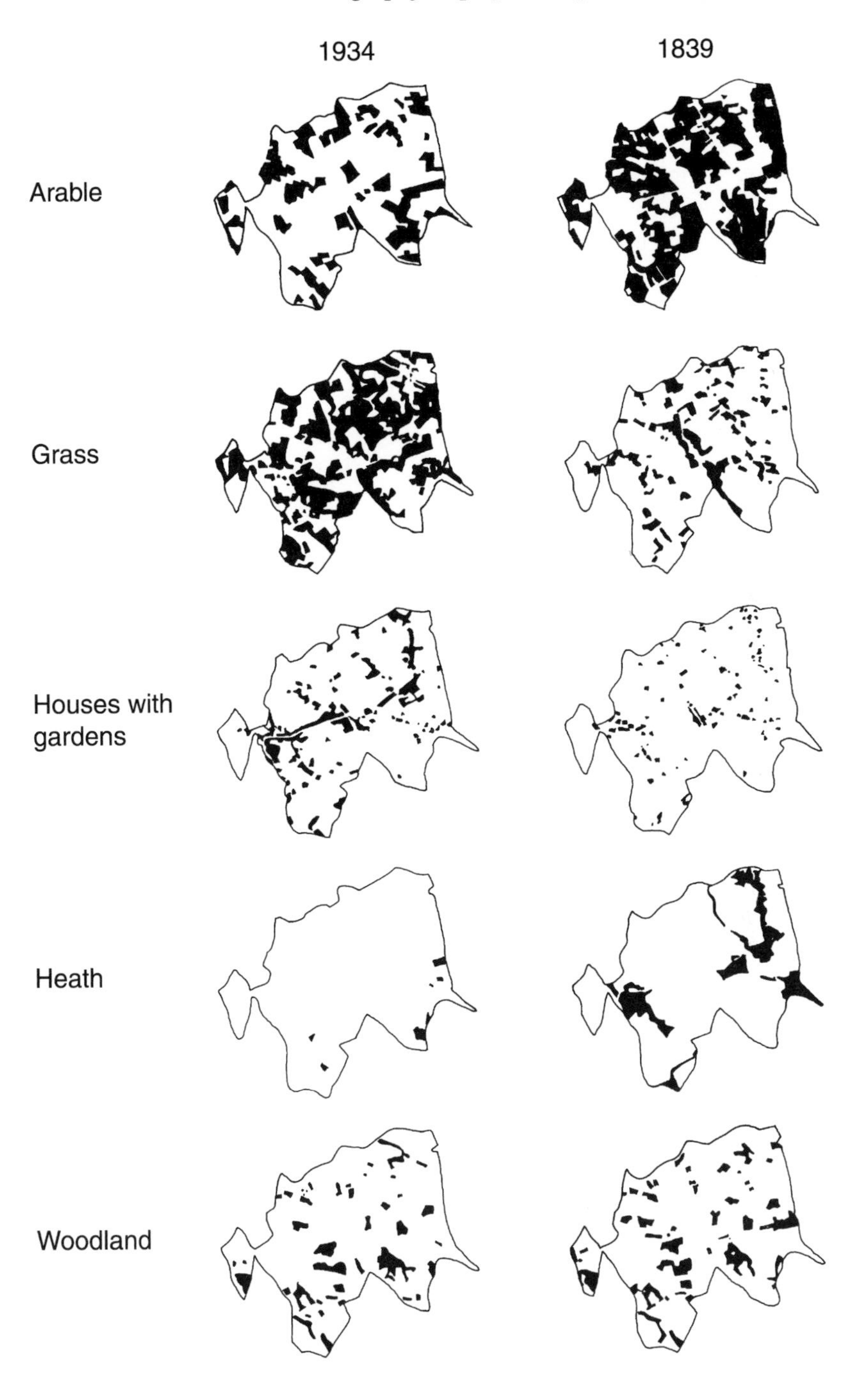

Fig. 11.9 Land use in a Chiltern parish, Great Missenden, Buckinghamshire in 1934 and 1839, after Willatts (1937).

from the sixteenth to the nineteenth centuries to trace the growth of coastal spits at Orford Ness, Suffolk. Likewise, George de Boer (1964) traced the evolution of Spurn Point, Humberside, from 1684 to 1852, and Catherine Delano-Smith (1986) reconstructed the silting of the Magra estuary in Tuscany, using sedimentary evidence supplemented with a map drawn by Ercole Spina in 1592 (Fig. 11.10). In many studies of river channel changes, maps have also proved to be a most important source (Hooke, 1995). Map analyses of some Devon rivers by Janet Hooke (1977) over the period 1843–1962 revealed rates of channel movement of up to 3 m per year and of a strikingly consistent nature: bends tend to become less sinuous, the

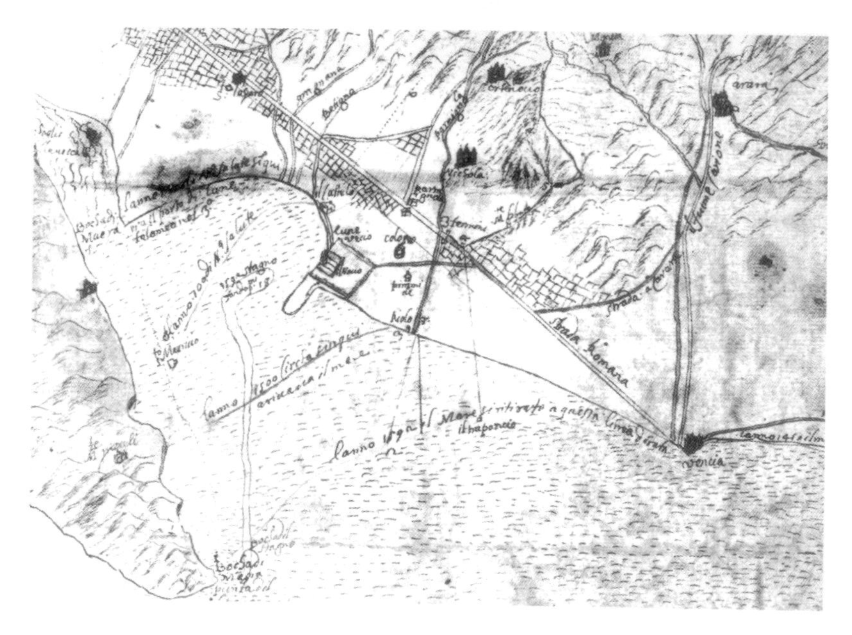

Fig. 11.10 Detail from Ercole Spina's manuscript map of the Gulf of La Spezia and the mouth of the River Magra (1592) on which Spina dated the succession of shorelines marking the advance of post-Roman siltation up to the end of the sixteenth century. The Roman colony of Luni was established on the eastern side of the estuary early in the second century BC at a point where a large sand bar (shown on the map) had already started to form. Spina's suggested Roman coastline is indicated by the solid line (*L'anno --- de Nostra Salute sin qui era il porto di Lune;* the date is illegible). His second coastline is dated *L'anno 700 di Nostra Salute*; the third as *L'anno 1500 circha sin qui arrivava il mare*; and the fourth, that of his own day, as *L'anno 1592 il mare si ritirato a questa linea dorata*.
Source: Spina's map is in Ippolito Landinelli, 'Origine dell'antichissima città di Luni e sua distruzione', MS 1610, Biblioteca Comunale di Sarzana. Reproduced with permission from the copy in the Biblioteca Civica di Genova, ASGe, Manoscritto numero 433.

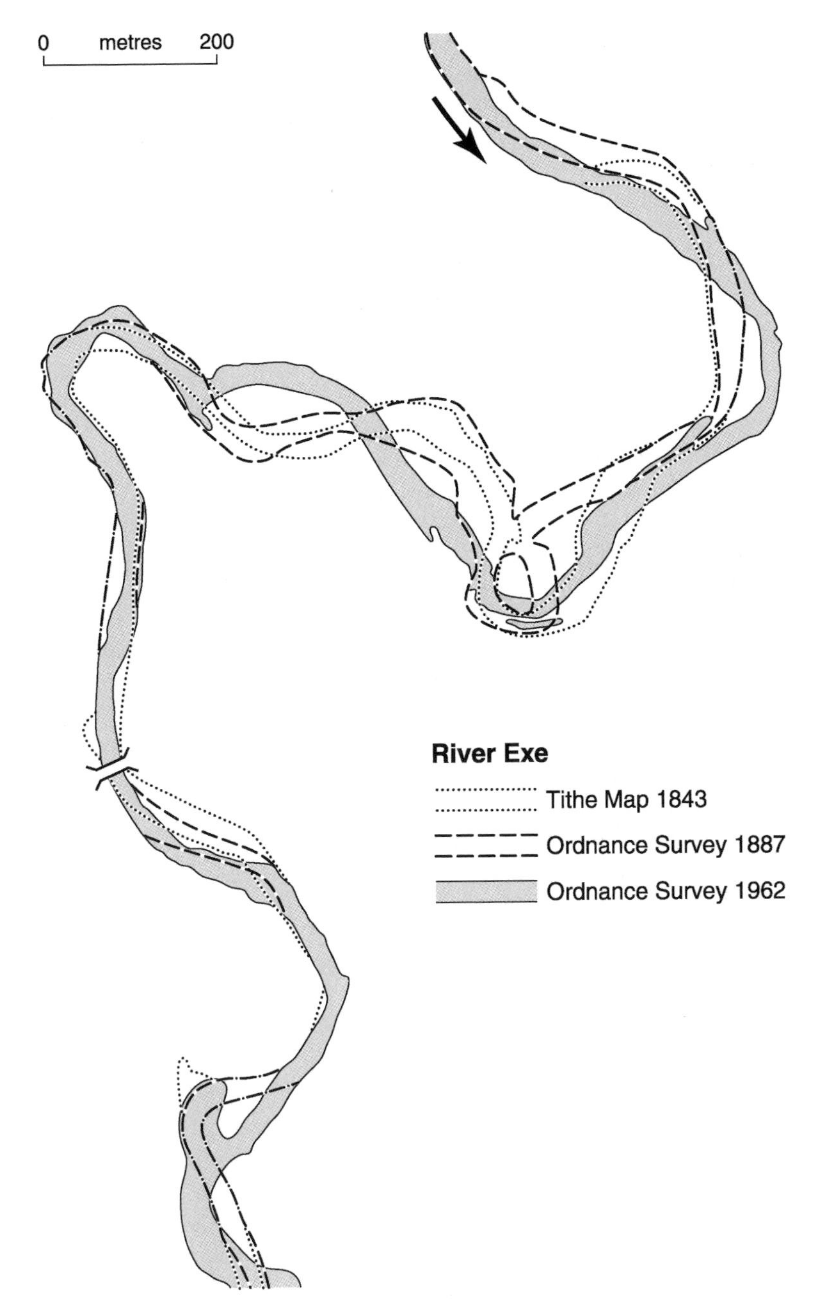

Fig. 11.11 Channel changes of the river Exe, Devon, 1840–1962, after J. M. Hooke in Hooke and Kain (1982).

amplitude of meanders increases, and the meanders shift downstream (Fig. 11.11). The use of historical maps to provide a longer-term perspective on physical changes is often a required component of environmental investigations today: for example, their examination and analysis is specified in the government guidelines for Shoreline Management Plans; and river conservation schemes usually involve restoring rivers to their 'natural' courses, which means that analysis of early maps is essential for reconstructing the previous courses of now straightened rivers.

Thematic maps

All maps are selective and, in the sense that no single map can present and communicate all the potentially mappable information for the area concerned, all maps are in theory thematic. In practice though, the degree of specificity on maps varies from, at one end of the spectrum, a maximum portrayal of content (as on a topographical map) to the most highly specific maps in terms of content at the other (see Figs 11.3 and 11.8). Heuristic motives explain why many thematic maps are compiled; the map is the instrument by which spatial structures are revealed, enabling the correlation of the 'geography' of one mapped distribution with that of other variables in the search for causal relationships.

Although maps representing a single class of information date back to at least early modern times, it was the increasing use of thematic maps in the natural sciences and the emerging social sciences in the nineteenth century that demonstrated to a wider public than previously the powerful didactic as well as heuristic role of maps. Thematic mapping came to be associated with the wave of reform that swept Western Europe after the French Revolution of 1789. In general, thematic maps helped establish the idea amongst human geographers that the spatial characteristics and spatial behaviour of human beings ought to be as predictable and as subject to laws as were the physical aspects of the environment. With hindsight, we can see that this concern for physical explanations eventually led human geography into the cul-de-sac of environmental determinism. Another hazard is to see significance in correlations where none exists. In the specialist field of modern medical mapping, the authors of the seminal *Atlas of Disease Distributions* warned that inappropriate mapping techniques may 'suggest false concentrations and . . . start false trails' (Cliff and Haggett, 1988, x). The bonus is that the thematic map can still be as revealing as it was for John Snow in 1854 (Snow, 1855) in the epidemiology of many diseases, particularly infectious diseases: 'although the prospect is still a distant one, it may be possible to forecast maps and so be in a position to provide early warning of disease events and their public-health consequences' (Cliff and Haggett, 1988, xii).

The thematic mapping of socio-economic data has also been used to help throw the spotlight on the spatial inequalities of human life chances. In 1977,

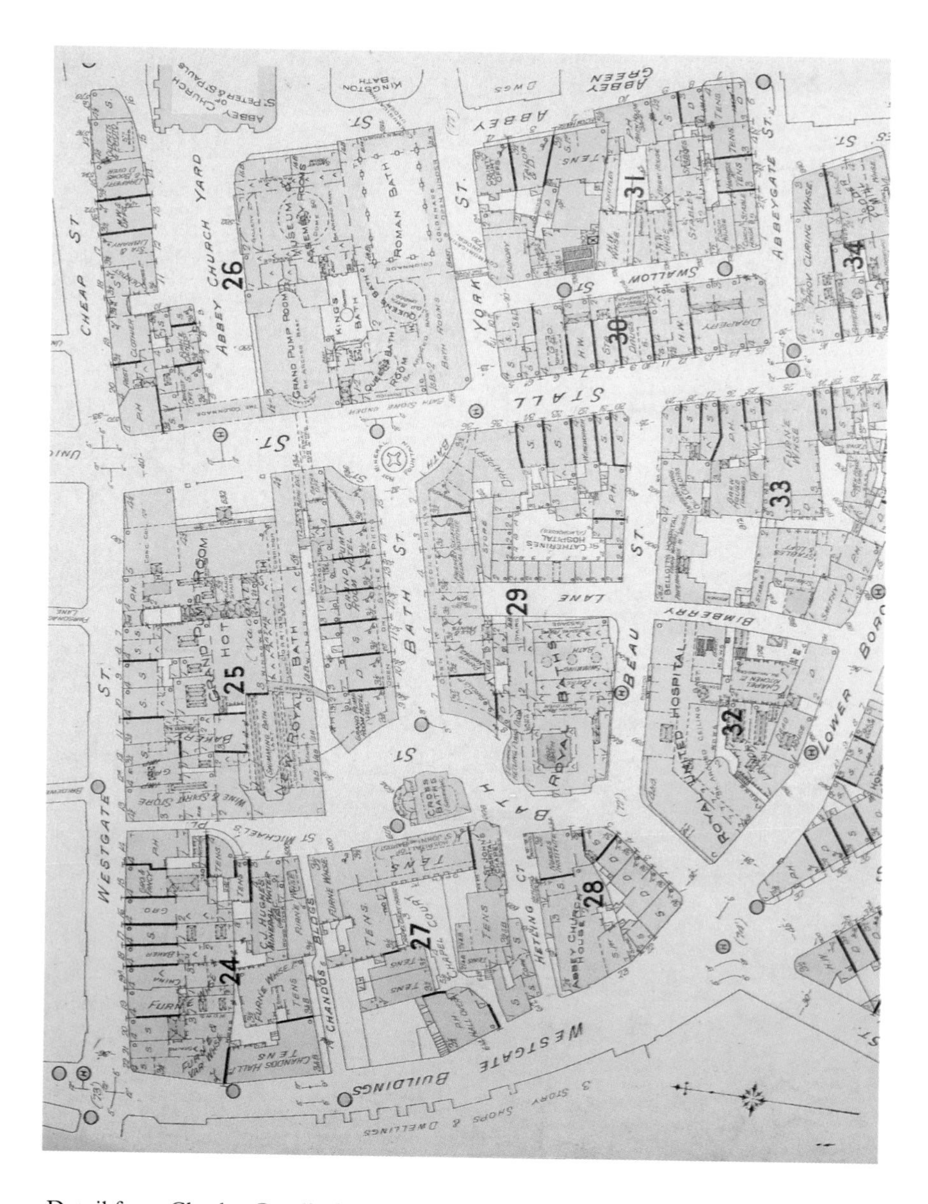

Detail from Charles Goad's fire insurance map of Bath, 1902. The so-called 'Goad maps' were produced to satisfy the demands of insurance underwriters who needed large-scale town plans in order to identify particular buildings and to assess insurance risks.

Source: British Library, Maps 145.b.9.(1). Reproduced by permission of the British Library Board.

Ellis Martin's illustration for the cover of the Ordnance Survey Popular Edition one-inch-to-one-mile map of the Chilterns, 1932.
Source: Private collection.

First Land Utilisation Survey of Great Britain: sheet 114, parts of Middlesex and Surrey, 1931–2.
Reproduced by permission of Department of Geography, University of Exeter.

PLATE 4

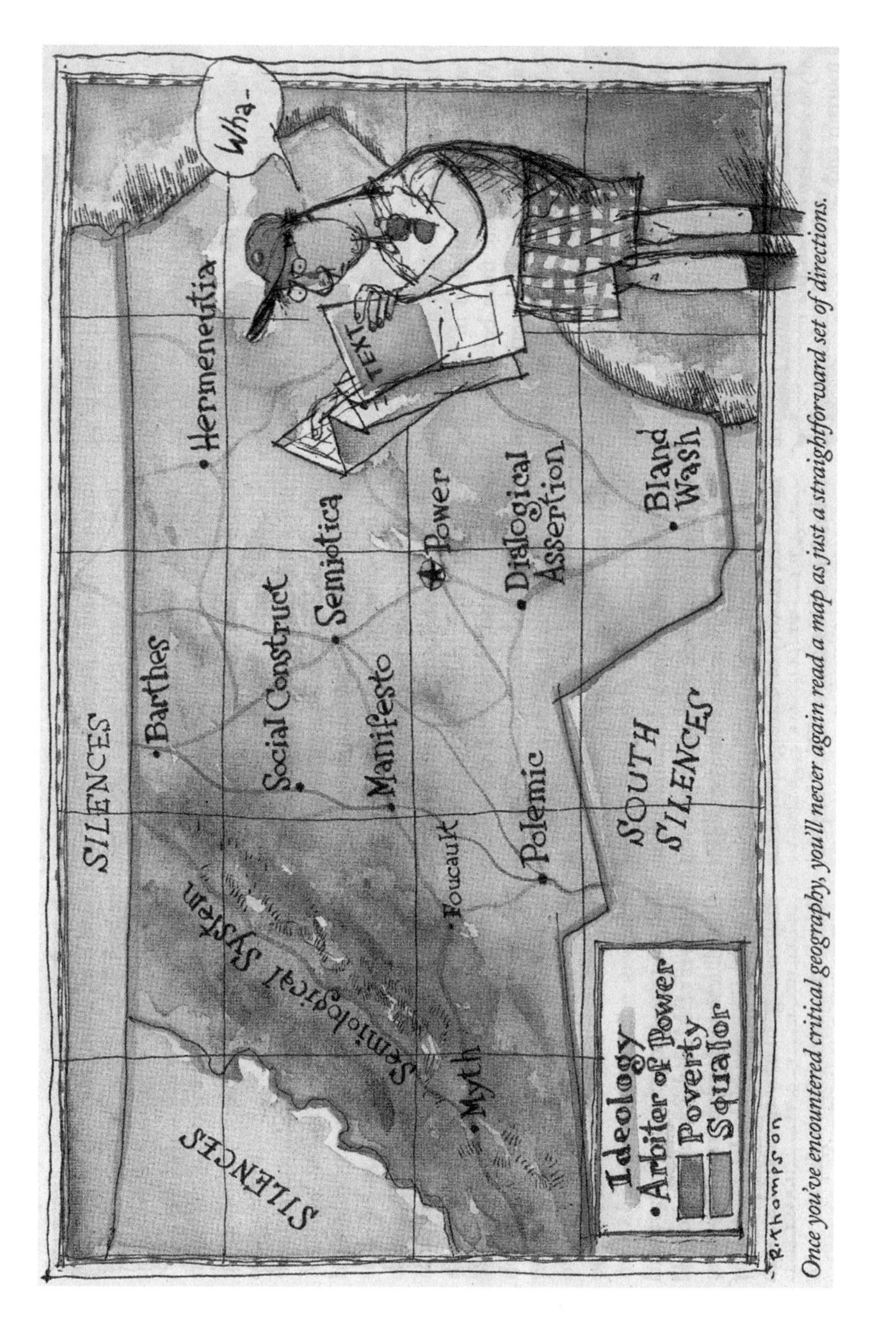

Once you've encountered critical geography, you'll never again read a map as just a straightforward set of directions.

Source: From N. Lemann (2001) Atlas shrugs: the new geography argues that maps have shaped the world. *The New Yorker*, 9 April, 131–4. Reproduced by permission of *The New Yorker*.

A map as street furniture: Chicago.

PLATE 6

(a) The entirety of the 'geographical framework of Britain' in the 1960s. Information was stored as 40 million copies of maps in the Ordnance Survey's West Building (tall building in centre foreground).
Source: Photograph kindly supplied by Ordnance Survey.
(b) The equivalent information can now be held in Compact Disc format in one hand and copied in a few minutes.
Source: Photograph courtesy of Landmark Information Group. Copyright © Prodat Systems Plc 2002. © Crown copyright 2002. All rights reserved. Licence number 1000240449.

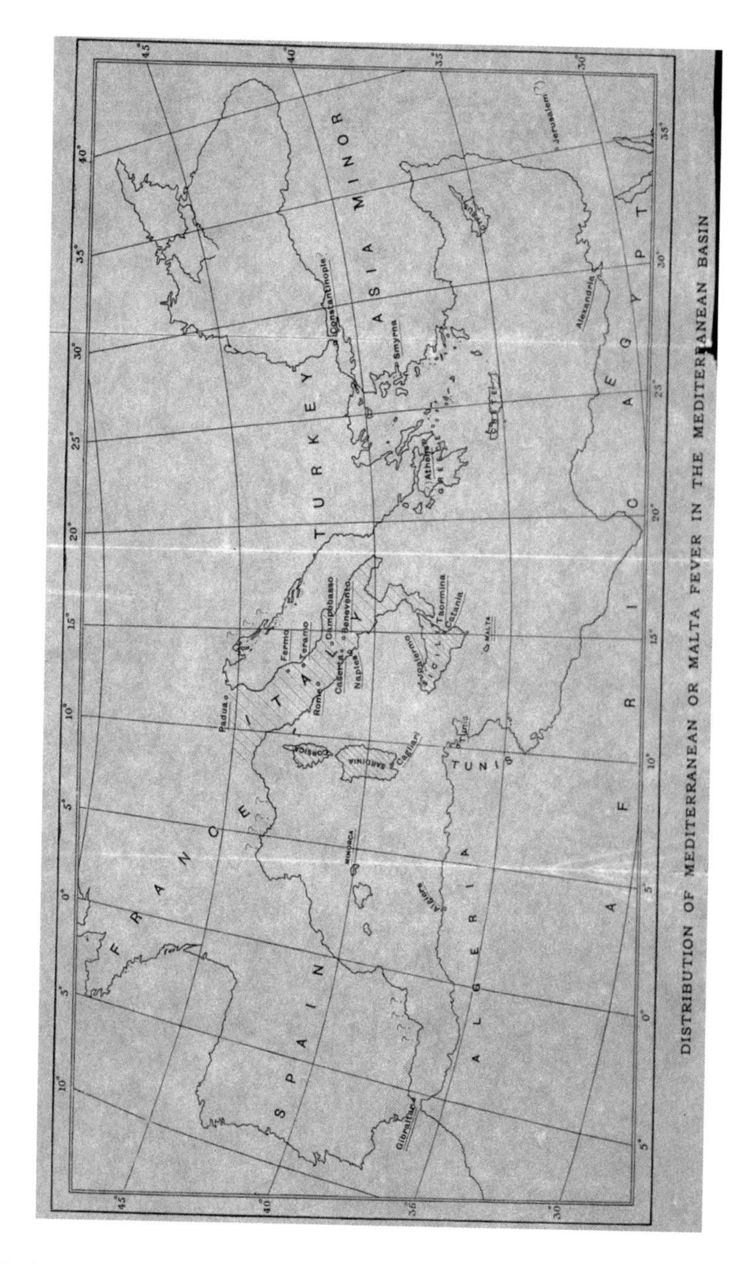

Medical mapping at the start of the century. Clemow's map of the distribution of Malta or Mediterranean fever from one of the earliest medical geography texts of the century.

Source: From F.G. Clemow (1903) *The Geography of Disease*. Cambridge: Cambridge University Press, facing p. 284.

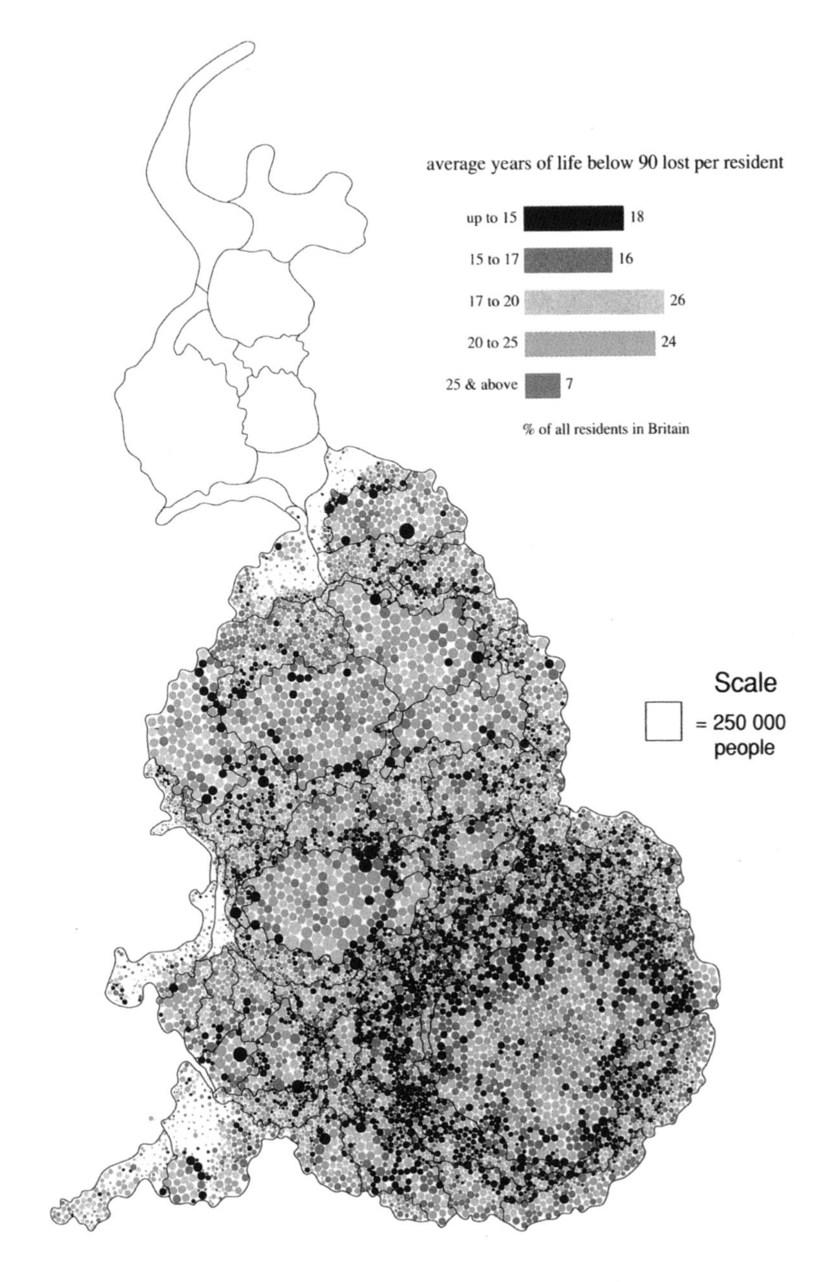

Dorling's maps of mortality in England and Wales using years of life lost 1981–1989. Wards are plotted as circles proportionate in size to the population.
Source: From D. Dorling (1995) *A New Social Atlas of Britain*. Chichester: Wiley, 163.

geographers were invited by David Smith to 'address the question of what *should* be, along with the traditional concern with what *is*' (Smith, 1977, xi). Mapping social indicators would reveal the inequitable distribution of quality-of-life factors and the uneven distribution of 'levels-of-living' (Lewis, 1968). It would also raise questions of whether or not these inequalities are just (Knox, 1975; and see Pacione, 1995, for a modern map-based study). In university geography, mapping social indicators provided the foundations for a 'relevant' human geography in the 1970s. An early study was Coates and Rawstron's mapping of data relating to personal income, employment, immigrant populations, health care, mortality, and education in the belief that 'variations from place to place in selected aspects of human geography are a cause for concern and are worth measuring' (Coates and Rawstron, 1971, 290). They saw their study as analogous to the way maps function in meteorology, where the prime objective of compiling a series of weather maps separated by small intervals of time is to forecast the weather. According to Coates and Rawstron, economic and social 'storms' should also be mapped and tracked so that their impact could be predicted and remedial measures put in hand. Through their maps, Coates and Rawstron confirmed the unequal rise of relative prosperity, for example, and its marked concentration in southeast and south-midland England (Fig. 11.12). The methodology has worn well but, 30 years on, the Great Britain Historical

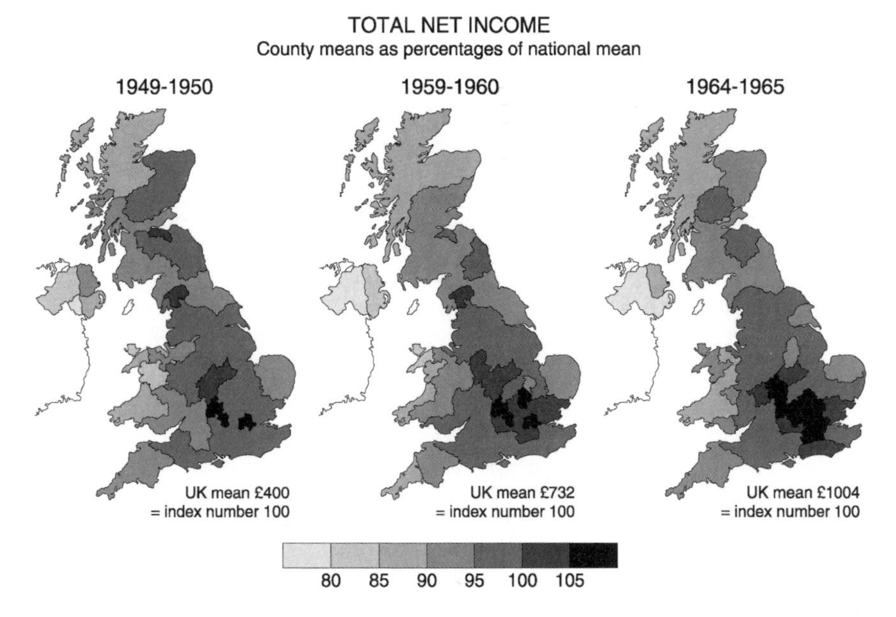

Fig. 11.12 Personal income in the United Kingdom, 1939–65, after Coates and Rawstron (1971).

GIS project (Gregory and Southall, 1998) has enabled the analysis of geographical variations in social indicators at a level of detail of the census ward instead of the county (Gregory, Dorling, and Southall, 2001). Interestingly, although the analyses employ the full power of GIS, the results are communicated by a series of printed maps.

Compiling atlases

In *The Use of Geography*, Frank Debenham (1950, 54) claimed that there was likely to be an atlas of sorts in every British home. He never clarified the evidence behind his assertion, and he went on to dismiss most of these family atlases as little more than dictionaries, used as a means of locating a place listed in the index on one of the maps. As he was well aware, though, an atlas can be far more than a dictionary. We should see it rather as a kind of graphic encyclopedia, not merely a place-finding system and still less just a bound book of maps: simply encasing a bundle of unsorted maps is not sufficient to create an atlas (Akerman, 1995). An atlas needs an editor to impose a structure with an identifiable logic to the selection, design, and arrangement of the maps. The twentieth century saw a flourishing of the atlas genre.

In the early part of the century, British world atlases seem not to have displayed anything especially 'geographical' in either organisation or content. The *Times Survey Atlas of the World* (1922), big though it is, and despite being prepared at the 'Edinburgh Geographical Institute' under the direction of J. G. Bartholomew, holder of the RGS Victoria medal, is no more than a straightforward collection of topographical maps with a place-name gazetteer. Tellingly perhaps, Bartholomew styled himself 'Cartographer to the King', and the *Times Survey Atlas* is indeed a cartographical rather than a geographical compilation. Later atlases incorporated a thematic content with the traditional topographical element, adding maps giving information on climate, geology, natural resources, population, economic activity, and culture. Geographers were often involved as editors to ensure the current geographical viewpoint was reflected in the content of the atlas. H. C. Darby, for example, edited the *Library Atlas* through many editions. In 1978 its entire content, thematic as well as topographical, was redesigned and updated (Fullard and Darby, 1978). Michael Wise and Robin Butlin contributed substantial chapters on 'Modern Britain' and 'The historical geography of Britain' respectively to the *Ordnance Survey Atlas of Great Britain* (Ordnance Survey, 1982). British geographers also acted as consultants or facilitators to overseas governments wishing to develop national atlases, as K. C. Edwards did for the Luxembourg Ministère de l'Education Nationale. Edwards explained:

> Neighbouring countries each possess an atlas and it is fitting that the Grand-Duché now completes the mosaic of West European national atlases. The individual maps contained in this atlas reflect changes and developments occurring in Luxembourg since the Second World War, together with some of the consequences of the broader economic groupings during that period. (Ministère de l'Education Nationale, 1971: preface)

Edwards had a long-standing interest in the geography of the Duchy, having studied its iron and steel industry for his doctorate, and had extensive contacts there. Luxembourg lacked its own university and Edwards undertook to have the atlas compiled in his department at the University of Nottingham, using statistical and other material sent over from Luxembourg. The maps were all drawn at Nottingham and the atlas was printed and bound in Luxembourg.

From providing general atlases with thematic maps, it was but a step to the production of purely thematic atlases. The involvement of British geographers in the production of economic atlases seems to have been well established by the early years of the twentieth century. The first of the Oxford University Press's series of economic atlases was *An Atlas of Economic Geography* (1914) by J. G. Bartholomew and with an introduction by L. W. Lyde of University College London. By mid-century, Oxford geographers had a substantial involvement. For example, C. F. W. R. Gullick was geographical adviser for the *Oxford Economic Atlas of the World* (Oxford University Press, 1954) and C. Gordon Smith similarly led the production of the *Oxford Regional Economic Atlas: The Middle East and North Africa* (Oxford University Press, 1960).

A historical atlas constructs 'images of the past' (Black, 1997a). As a genre, historical atlases go back not only to the nineteenth century but to early modern atlases like Abraham Ortelius's *Parergon* (1579). The modern heyday of the historical atlas started in the 1970s, when H. C. Darby was one of the editors of the atlas published as the final volume of the *New Cambridge Modern History* (Darby and Fullard, 1970). Although many twentieth-century historical atlases were intended to provide some of the 'geography behind history', especially through the plotting of changing political and administrative boundaries, historians rather than geographers figured more prominently amongst their editors (Darby being a notable exception). Ramsay Muir (1911), for example, was professor of history at Liverpool University, the maps in Reginald Lane Poole's *Historical Atlas of Modern Europe* (1896–1902) were compiled by historians, and Jeremy Black, the modern writer on historical atlases, is himself a historian.

Leaving historical atlases mainly to the historians, geographers later in the century compiled a whole range of other thematic atlases. J. N. L. Baker's *Atlas of the War* (1939) included maps in which the economic dimensions of the conflict were emphasised. Michael Freeman's *Atlas of Nazi Germany* provides a thematic analysis 'intended for 6th formers, university students and the serious general reader' (1987, preface). There were also more specialist atlases, aimed less at the general public than the academic researcher. For his *Atlas of Agriculture*, J. T. Coppock (1964) constructed maps of Great Britain using the Ministry of Agriculture's annual parish-based statistics. Melvyn Howe compiled a *National Atlas of Disease Mortality in the United Kingdom* which contains two principal sets of maps, one for 1954–8 and a second for 1959–63 (Howe, 1970; see also Learmonth, 1972). This theme has been further developed by Robert Woods and Nicola Shelton in *An Atlas of Victorian Mortality* (1997). Roger Kain (1986) reconstructed an earlier agricultural geography in a series of county and national maps in his *Atlas and Index of the Tithe Files of Mid-Nineteenth-Century England and Wales* (Fig. 11.13). The incidence and effects of the drought of 1976 were analysed in an atlas sponsored by the Institute of British Geographers (Doornkamp and Gregory, 1980).

The arrival of computerised data-processing and mapping greatly speeded the production process for the maps of an atlas well before the introduction of GIS. The cartographical analysis of otherwise almost unmanageable datasets, such as the decennial censuses of population, could now not only be undertaken but also achieved on a timescale that would be useful to policy-makers.[12] A prime objective of the *Census Atlas of South Yorkshire* (one of the earliest of the computer-based atlases) was to have it ready for publication by 1 April 1974, to mark the coming into being of the new metropolitan county of South Yorkshire (Coates, 1974). It was prepared from 1971 census data, some of which was not available until the end of 1973, yet it was published on time—a remarkable achievement considering that, although its compilers had access to computers, this was still the age of punch cards for data input! By 1991, computers were of a different generation, and the increasingly widespread application of computer cartography to the massive database generated by the 1991 population census marched alongside a growing social concern in human geography. Analysis of the spatial components of society resulted in a number of atlas studies (e.g. Fielding, 1993; Forrest and Gordon, 1993). An innovative presentation is found in Daniel Dorling's *New Social Atlas of Britain* (1995), in which topological geometry, in the form of cartograms, was employed to map units of equal population rather than the actual boundaries of census wards, focusing attention on people rather than the administrative unit in which they reside (Dorling, 1995). This technique has been developed further in a study of life chances and health in Britain (Mitchell, Dorling, and Shaw, 2000).

The millennial year 2000 provided a special stimulus for another spate of publications in the by then well-established genre of county historical atlases: for example, East Yorkshire, Suffolk, Sussex, and Lincolnshire

[12] A. J. Hunt's collaborative project to map 1961 census data was not published until 1968.

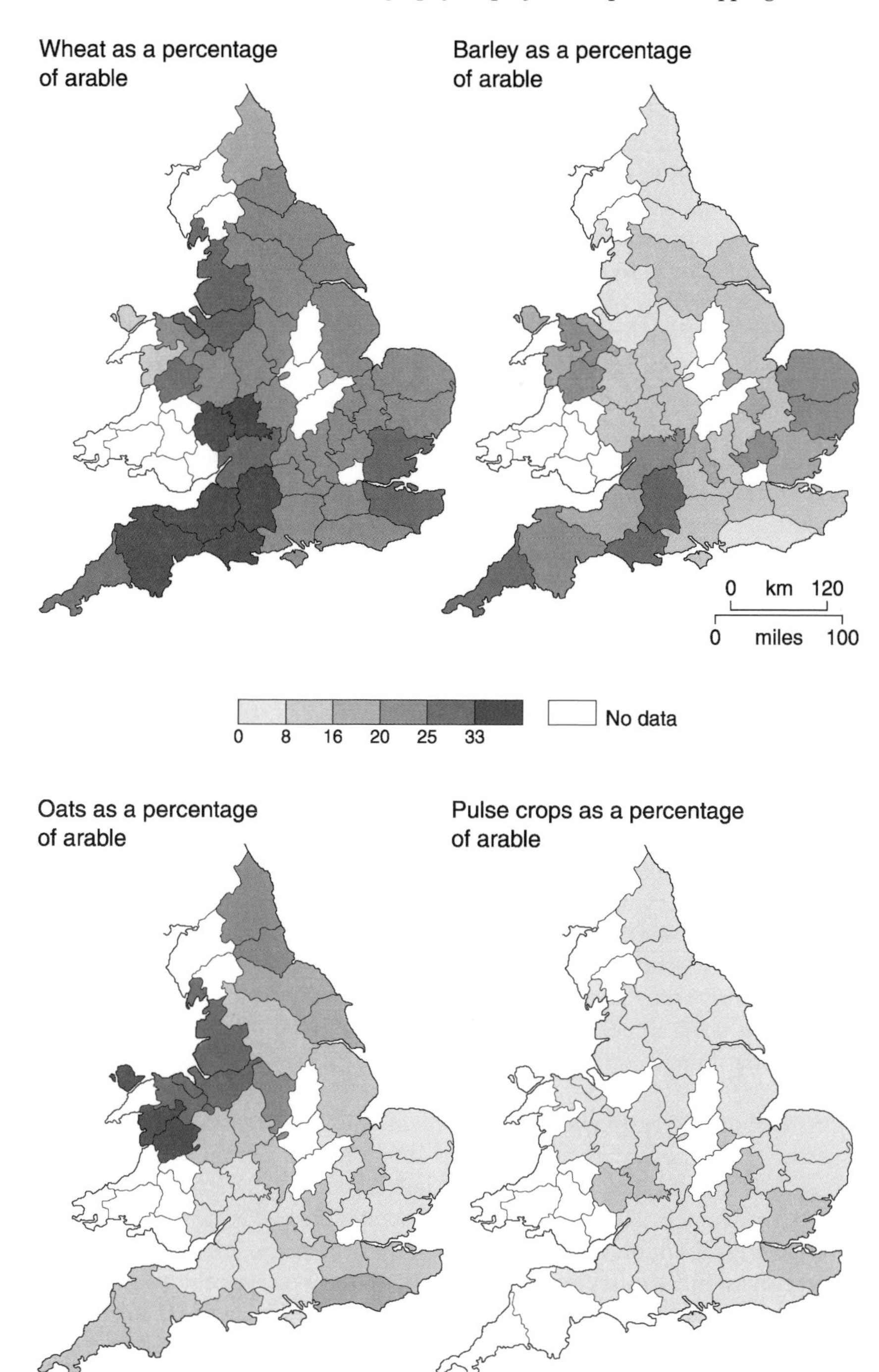

Fig. 11.13 Agriculture in England and Wales c. 1840, after Kain (1986).

(respectively Neave and Ellis, 1996; Dymond and Martin, 1999; Leslie and Short, 1999; Bennett and Bennett, 2001). In these atlases, priority is given to the map rather than the text as the prime method of communicating history. Geographers were much involved both as editors and contributors to these works, as they were in the production of the only English regional historical atlas to date: the *Historical Atlas of South-West England* was designed to highlight by means of a variety of maps, images, and text how the spatial distribution of features plotted out on maps aids the understanding of complex time–space relationships (Kain and Ravenhill, 1999). The analysis of the distribution of archaeological find spots is presented in relation to a detailed analysis of the physical environment. The distribution map of working mines in the nineteenth century is a reminder of the danger of reading too much from maps without accompanying documentation: a reduction of mining sites on the map, compared with earlier times, belies a dramatic increase in output, brought about through company amalgamations.

Maps as epistemology: new ways of reading maps

Faced with, as David Livingstone (1992, 333) puts it, 'statistics that don't bleed', and accepting that values as well as facts contribute to the behavioural environment, in the 1960s and 1970s some British geographers started to explore the way the human mind processes its relationship with the outside world (e.g. Wreford Watson, 1967; see also Kirk, 1952). The resultant mental map is the individual's perception, or belief, of the way that world is at a particular moment (Spencer and Weetman, 1981). The map may be (and typically is) factually incorrect but, as the British geographer Peter Gould (1966) pointed out in a seminal lecture given at the University of Michigan, the spatial image can be sufficiently powerful for economic and other administrative decisions to be taken according to it. Gould also drew attention to how little was known about the impact of these spatial images on people's minds (but see Lowenthal 1961; Wright, 1966).

In Britain, geographers also turned in the 1970s to the geographical aspects of psychology and cognitive development, eager to espouse a humanistic reaction against the mechanistic models of the quantitative revolution. In the cascade of examples that followed over the next two decades consideration of the implications of the term 'mental maps' and the cartographical aspects of their presentation was largely swept aside. Geographers were offered not only Gould's isopleth maps of residential desirability and other group preferences but also an outpouring of freehand sketch maps, many of which were applied to advertising, were used as cartoons, or celebrated the achievements of children (Pocock, 1976, 1981;

Board, 1979; Murray and Spencer, 1979). Many of the pictorial examples presented under the heading of cognitive mapping were to greater or lesser degree comical, such as Saul Steinberg's much-exploited New Yorker's view of the world (*New Yorker*, 29 March 1976; see also Holmes, 1992). Outside academic geography, the Parish Map Project was launched from a discreetly political platform to stimulate people towards the idea of creating their own parish map. A parish map, it was explained, 'need not be a map in the conventional sense but can be in any medium . . . [and] show the features of the parish that local people really value . . . to provide a focus for the community and encourage a sense of pride in, and responsibility for, their particular territory' (Greeves, 1987: inside cover) (Fig. 11.14). For published examples of such maps, see Mayfield and Clifford (*c.*1996); for a critique of the project, see Crouch and Matless (1996).

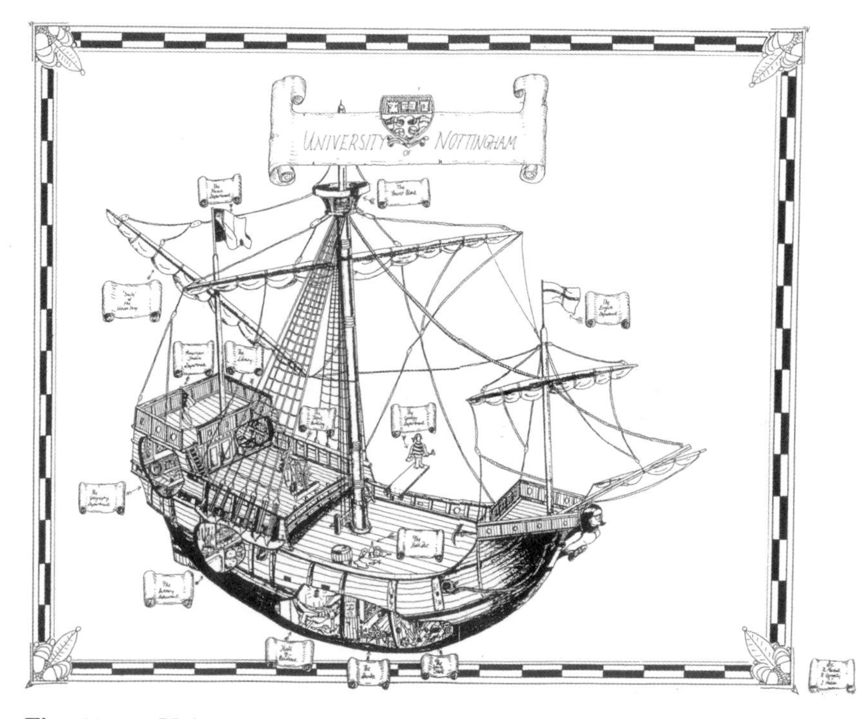

Fig. 11.14 University of Nottingham students' allegorical 'Parish Map', 1989. The Department of Geography, with its map, chart, atlases, and globe, is located on the bridge, next to the Library. Departments of English, French and American Studies are indicated by national flags. The Department of Geology, about to be closed, stands on the gang-plank. Departments in the Tower Block occupy the crow's nest. Halls of residence and the Buttery are in the hold. At the helm is the University administration.
Map compiled by R. Marshall, R. Geroghty, and J. Hallim.

If anything was said about the principles of cognitive mapping it was to suggest that these were identical to those of any other form of map-making, involving the purpose, perspective, and scale of the proposed map and the signs to be used on it (Downs and Stea, 1977, 64). In general, more attention was paid to the psychological and physiological aspects implied by the word 'mental' in the term 'mental map', and to the various methods used to extract, summarise, and represent spatial information held in the mind, than to the resultant artefact, the map itself (Downs, 1981; Graham, 1982). Whether the image is a map in the understood sense of the word—one that 'can be apprehended by researchers'—remained a problem (Johnston, 1997, 157), reflecting in some measure the behavioural geographer's unfamiliarity with the complex nature of maps. The debate rumbled on and in November 1999 half an issue of *The Professional Geographer* was devoted to mental maps and cognitive mapping. The *reductio ad absurdum* was reached when all maps were seen as mental maps, to the extent that the word 'map' is omitted from a geographical glossary containing more than 250 entries, to be replaced with a sole entry on mental (or cognitive) maps (Holt-Jensen, 1999). Behaviourism had reached the intellectual impasse foreseen by geographers such as Nigel Thrift (cited in Graham, 1982, 260).

Inasmuch as all maps are human constructs, behaviour as well as epistemology is a relevant factor in map-making as well as in the interpretation of maps. The debate amongst geographers, and between geographers and cartographers, over the subjectivity of maps reached new heights in the 1980s. Casting about for ways of dealing rationally with the interplay between individual and society, some human geographers found the place-based ideas of social theorists particularly attractive, especially those of Anthony Giddens and Michel Foucault (Giddens, 1979, 1981, 1984; Foucault, 1980; see also Livingstone, 1992, 340–1). Giddens called his ideas about that interplay 'structuration' and Foucault called them 'discourses', but both terms referred to the way a society's structures and institutions engage with individuals in specific locales or places and constrain (or manipulate, according to Foucault) their behaviour. Applied to maps by J. B. Harley, notions of 'strategies of power', 'control of territories', and 'knowledge/power' served to expose once and for all, as he saw it, the myth of cartographical 'objectivity' and the idea that the map is a 'perfect copy' of the landscape, the result of some impersonal process of selection and compilation, and mathematically accurate (Harley, 1988a, 1988b, 1989, 1997). Harley had already provided a methodological blueprint for the use of early maps as evidence in historical studies (Harley, 1969). Throughout the 1980s and up to his early death in 1991, he drew widely—some would say sometimes indiscriminately—on an eclectic range of authorities for his theoretical arsenal: literary philosophers and critics such as Michel Foucault, Roland Barthes, Jacques Derrida, Richard Rorty;

art historians such as Erwin Panofsky and Michael Baxandall; historians of science such as Joseph Rouse; sociologists such as Anthony Giddens; literary critics such as Edward Said; and economic historians such as Donald McCloskey. Harley's outstanding achievement was not only to pose questions about the nature of maps, early and modern, but in so doing also to close the gap between the history of cartography on the one hand, and history proper and the humanities and the social sciences on the other hand (Blakemore and Harley, 1980; Harley, 1983, 1991, 1997).

Harley also helped create what is proving to be probably one of the largest international collaborative projects in the humanities and social sciences conceived by geographers. In August 1981 he and David Woodward formally launched the 'History of Cartography Project', directed to the publication by the University of Chicago Press of a six-volume *History of Cartography* (Harley and Woodward, 1987, 1992, 1994; Woodward and Lewis, 1998; see Woodward, 2001a, for the story of the origins). The leitmotiv of this remarkable interdisciplinary venture is provided by the founding editors' rejection of the standard definition of a map (a scaled graphic representation, based on a projection, astronomically determined coordinates and mathematical measurement, which shows all of or part of the surface of the earth). Taking a more historical and a more liberal view, Harley and Woodward (1987, xvi) suggested that: 'Maps are graphic representations that facilitate a spatial understanding of things, concepts, conditions, processes, or events in the human world'. For some people, the *History*'s definition is too broad to have meaning, even when re-articulated for *The Dictionary of Human Geography* 'as codified images of all kinds by which humans formally articulate, represent, or construct their spatial knowledge of the world' (Woodward, 2000, 64). Others ignore the literature on the subject and continue with the old restrictive definition (e.g. Blaut, 1991, 64; Monmonier, 2000, 471). The overwhelming response in the wider arena of the humanities, however, has been to welcome the invitation to explore the historicity of all forms of spatial representation (see the reviews of the various volumes of the *History* listed in Woodward, 2001a, 25–6, n. 2). The result has been to expand the canon to include maps constructed differently from the products of modern (post-Enlightenment) geodetic cartography, such as maps drawn according to the principles of positional analysis (topology), maps depicting western society's own changing cosmological ideas, and maps relating to non-western religious beliefs. At the same time, modern cartography cannot but be better served by a holistic view of maps and mapping.

The new viewpoint has brought other advantages. There is now a conscious attempt to counter traditional Euro-centricity. Familiar aspects of maps are revisited and re-evaluated. The invitation to explore widely can bring, it has to be said, its own dangers, notably those of ahistoricism. The imposition of political correctness, for example, sometimes overtakes historical sensibility, with the result that over-simplistic expositions are arrived at. The line is a fine one between seeking to understand the 'silences and secrecy' of maps (Harley, 1988b) and the branding of all maps as purveyors of 'lies' (Monmonier, 1991; see also Crampton, 1994a). Even the punning title of a recent map exhibition, 'The Lie of the Land' (British Library, August 2001–April 2002), has more than a whiff of manipulation about it, however excusable that may have been in the interests of crowd-pulling. Away from the public eye, an important legacy of Harley's and the *History*'s stimulus to new ways of seeing maps will be the corpus of theoretical writing which, following on from Blakemore and Harley's (1980) 'review and perspective' of cartographical concepts, has been slowly building up. Critical discussion amongst British geographers of the assumptions underpinning the study of maps and map history have come from the pens of John Andrews (1994, 2001), Matthew Edney (1996), Catherine Delano-Smith (1996), Jeremy Crampton (1994b), and David Woodward (2001b). Empirical studies reflecting the new vision of maps as social documents, and not just mathematical compilations, include John Andrews on Ireland (1985, 1997), Delano-Smith's work on exegetical maps (1990; see also Delano-Smith and Ingram, 1991), Sarah Bendall's and David Fletcher's books on estate maps (1992 and 1995, respectively), David Woodward's Panizzi Lectures on Italian maps and prints (1996), Malcolm Lewis's investigations into the history of native North American and map-making (1998a), Catherine Delano-Smith and Roger Kain's overview of the history of English maps (1999), and John Rennie Short's account of the mapping of North America since colonial times (2001). A burgeoning enthusiasm for maps and the history of maps amongst the producers of television and radio programmes led to a veritable rash of broadcasts on maps and map history over the last decade. These included, most notably, the six-part BBC television series 'Tales from the Map Room' (1993), and a four-part radio programme 'The Medieval Ball' (2000). Contributors to the book *Tales from the Map Room*, which was co-edited by Christopher Board (Barber and Board, 1993), included 12 British geographers out of the total of 32 scholars and librarians.

Revisionism is discouragingly slow to filter through into the main geographical arena. Even the heavily updated fourth edition of the *Dictionary of Human Geography* (Johnston et al., 2000) gives an uncoordinated, and in places baffling, picture of the ideas supposedly influencing cartography and the study of maps by human geographers and others. The entry on cartography, oddly described by a cartographer as 'the study of maps and their use', says nothing about the crucial difference between past and present *methods* of map-making which, we would suggest, lies at the heart of the distinction between cartography *sensu stricto* and the making of maps. A 'cartogram' (the *only* entry for maps in Brian Goodall's *Penguin Dictionary of Human Geography*, 1987) is defined as 'a highly tailored map projection', without any explanation that similarly highly stylised maps (diagrammatic maps) are found throughout the past and in all map-making cultures as the most commonplace expression of spatial relationships. 'Mental map' is the term used in the *Dictionary of Human Geography* to refer 'to the psychological representation of space as shown by simple paper and pencil tests'. In the curiously entitled entry 'map image and map', the map is dismissed as 'a graphical representation of all or a portion of the earth or similarly vast environment' (Monmonier, 2000, 471) and the discussion focuses on map projections and on techniques of digital imaging. The entry on the history of cartography, in contrast, explains the subject as 'document[ing] and explain[ing] the motives for making maps, the agents and processes used to make them, their functions, and their role in forming society's views of space and place'.

Maps and metaphors rediscovered

A few British geographers, picking up the various batons dropped by Harley in his epistemological charge through the 1980s, engaged with the trope of power. Matthew Edney (1997), for instance, writing from North America, showed how, in imperial Britain, modern survey techniques were employed not only to produce the maps needed to administer British territories in India but also to legitimate its colonialist activities as triumphs of liberal, rational science. Maps in the service of propaganda have been reviewed (Atkinson, 1995, 2000). Other geographers in the 1980s were following a humanistic line and since then have been concerned with the human organisation of space as influenced by local or imposed systems of shared beliefs and values. Prominent in the cultural geography of the final decades of the twentieth century were notions of 'representation' and, to use John Berger's (1972) felicitous turn of phrase, new 'ways of seeing'. In 1988, *Word and Image*, the self-consciously interdisciplinary British journal of 'verbal/visual enquiry', devoted an entire issue of eight articles to 'maps and mapping'.

Taking their cue from the broad field of cultural studies, geographers explored new perspectives in the landscape itself and in landscapes represented in art or in cartography. To the geographer's traditional empirical concern with the economic history of landscape, articulated outstandingly by H. C. Darby in 1956, have been added studies of landscape in its social and cultural contexts. To the armoury of ways of reading maps has been added that of 'envisioning' maps. Where Harley explicitly re-read the map image in the light of iconology (Harley, 1983, 1985) or text (Harley, 1987–8, 1989, especially n. 41, and 1990a), Denis Cosgrove has been analysing a series of landscapes as embodiments of contemporary culture and cultural aspirations. Starting with the 'humanist vision' embedded in a relatively local landscape, that of the Venetian *Terrafirma* in the fifteenth century (Cosgrove, 1984), Cosgrove (1992) took a wider perspective on Italy, before embracing the globe as a whole, envisioned from antiquity to the present day (Cosgrove, 2001). Stephen Daniels has also led geographers to see the symbolism in landscape and representations of landscape (Cosgrove and Daniels, 1998; Daniels, 1999), although he specifically included maps only in the company of an art historian (Alfrey and Daniels, 1990).

There were others seeking to reach the 'message' in maps. David Harvey (1989, 239) considered the changing ways (as he saw it) that space and time—'never neutral [practices of power] in social affairs'—have been expressed in maps over time. Some cultural geographers engaged with the interpretation of landscape representation have omitted the cartographical format altogether. Douglas Pocock (1981), with 13 other British geographers, felt able to reflect on depictions of landscape and place in prose and poetry without so much as a nod in the direction of literary maps, as did Daniels and Rycroft (1993). In one of David Matless's explorations of the creation of the 'popular' view of landscape in the period between the two world wars, the word 'map' seems to be studiously avoided (e.g. Matless, 1992, but see pp. 471-2), although in others he does address, in particular, the role of the Ordnance Survey's Popular Editions (Matless, 1995, 1997).

For all its dynamism, the cultural turn in geography that characterised the penultimate decade of the twentieth century has enabled only a limited reclamation of geography's lost territory of the *study* of, as opposed to the *use* of, maps. The high ground of what a hundred years ago was the subject's heartland is, as we write, held not by geographers but by almost everybody else. In Britain, Jeremy Black (1997a) has claimed for history the first survey in English of how people in Europe and America depicted the space in which they lived, from Abraham Ortelius's *Theatrum Orbis Terrarum* (1570) to the latest digitised atlases, and he has written the first modern book-length study of maps and politics (Black, 1997b).

Less interest has so far been shown by British art historians in the way maps describe landscape apart from Nicholas Alfrey (Alfrey and Daniels, 1990) and Thomas Frangenberg (1994; see also Bann, 1988, 1994). None of the other main recent contributors to the fruitful marriage of art history with map history is British (e.g. Schulz, 1978, 1987; Alpers, 1983, 1987; Fiorani, 1995; Armstrong, 1996; Bourne, 1999; Kline, 2001). Irit Rogoff, an American currently head of the art history department at Goldsmith's College, London, takes a look at geography in general, as well as 'mapping' in particular, to see whether 'contemporary art can rewrite geography's relations with place and identity', geography having been found wanting 'as a source of authority in the modern world' (Rogoff, 2000, i). In contrast, British scholars from the general field of literature are contributing significantly to cultural and historical geography and to our understanding of maps. John Gillies's (1994) critique of Abraham Ortelius's atlas and other icons and emblems of late-sixteenth-century cartography is essential reading for map historians. In Lisa Jardine's reinterpretation of the European Renaissance and its materialist culture, maps are seen as key commodities (Jardine, 1996; Jardine and Brotton, 2000), as they are in Jerry Brotton's (1997) exploration of the ways maps and globes were used to mediate commercial and diplomatic disputes between the maritime empires of East and West in early modern times. Maps as creators of landscape rather than as mirrors is the theme of another book from an English literature department (King, 1997). In 1997, Andrew Gordon and Bernard Klein organised a conference at Queen Mary and Westfield College, University of London, on 'Paper Landscapes: Maps, Texts and the Construction of Space, 1500–1700'; of the 12 papers eventually collectively published, not one is by a geographer (Gordon and Klein, 2001). Of another 12 essays edited by Cosgrove in *Mappings* (1999b), only one author (besides the editor) is a geographer (Matless, 1999).

Looking further afield at recent English-language publications dealing with maps in the context of cultural geography (and leaving aside the ongoing *History of Cartography*), we again find many of the most substantive or challenging contributions coming not from geographers but from historians and students of literature. In America, Richard Helgerson (1986, 1992, 2001) led the literary field into cartographical pastures. More recently, Susan Schulten (2001) has shown how ideas and images there shaped popular understandings of world geography from the 1880s to 1950s. From Canada comes Lesley Cormack's (1997) charting of geography and maps at the English universities from 1580 to 1620, and from Germany comes Bernard Klein's (2001) account of the mental and material mapping of early modern England and its impact on Renaissance geographical thought and ideology of conquest. The gap in British writing on literary maps is being plugged by contributions from Italy (Moretti, 1998) and Austria (Reitinger, 1999).

And so we could continue; but the point has been made. Despite 'the startling explosion of academic, artistic and cultural interest in "cartography" as an object of critical attention' (Cosgrove, 1999a, 3), it has to be acknowledged that the impetus for English-language writing on maps had passed out of geography well before the end of the twentieth century. We also see, though, that the idea of the map is very much alive and well, both within and beyond the subject, albeit in sometimes unexpected ways. Search the catalogue of any of the British deposit libraries for book titles that include, for example, the word 'mapping': whereas all those listed for 1980 (11 in total) were concerned with actual mapping in one context or another (archaeology, agriculture, economic development, geology, and so on), the proportion had dropped to a quarter by 1990 (another 11 books) and has remained low, despite a doubling and more of the number of books with that word in their title. Even geographers have espoused the metaphor rather than the map. *Maps of Meaning*, a book aiming to combine 'some of the most important ideas from cultural studies with some recent developments in human geography', omits all reference to actual maps (Jackson, 1989). The map of part of Manhattan shown on the book's cover has as much, or as little, to do with the text inside as have the reproductions of Joseph Wright's *The Orrey* and the globe from William Hogarth's painting *Captain Coram* on the dust-jacket of Roy Porter's book on the Enlightenment, in which there is no mention of maps either (Porter, 2000).

An epitaph is premature

How should academic geographers regard the maps that were once their privileged territory? How far should they be concerned that the pages of their journals are all but mapless? A survey of the last ten years of *Transactions of the Institute of British Geographers* reveals that 3 per cent of all papers (six in total) are about maps, in 28 per cent maps are used as analytical or communication tools, and 69 per cent include no maps at all, other than location maps. What is the problem? As we hope to have shown, geographers still use maps. They are exploiters of the information carried in maps. Carto-literacy has spread with education for all to the opportunity for map-creation by all who have access to a desktop computer. Geographers are also producers of maps in the context of GIS, even if they are reluctant to admit to the common conceptual ground between the 'conventional' sheet map and their electronically created virtual maps. Geographers are no longer *studying* maps, however.

The study of maps and their history—by which we are referring to the study of their meaning rather than only their form, and their social rather than their mathematical construction—has never been exclusively geographical territory. But British geographers used to participate to a degree that they do not today. Since 1971 *Imago Mundi*, the flagship journal for the international history of cartography (and the only English-language scholarly journal dealing exclusively with non-current maps), has been edited in London by a British geographer (Eila M. J. Campbell to 1993, Catherine Delano-Smith since then). But whereas in the ten annual volumes published between 1964 and 1973 over a quarter (26 per cent) of the articles came from British geographers, the proportion for the decade 1992 to 2001 had dropped to 8 per cent.[13] The lack of interest from

[13] The figures for the last decade during which 104 articles were published are 9 British geographers out of 21 geographers in total.

geographers in the history of maps appears to be worldwide, to judge from the numbers who have applied for a J. B. Harley Fellowship in the History of Cartography since the Trust was established in his memory in 1992 (2 per cent of the applicants).[14]

Yet neither *Imago Mundi* nor, at a more modest level, the Harley Fellowships suffer from a shortage of students of map history. New books in English on almost any aspect of maps and their history arrive from scholarly presses in a steady stream. The fact that, to our knowledge, two major journals of literary review—the *New Yorker* and the *London Review of Books*—gave their lead pages to full-length reviews of Harley's *New Nature of Maps* (2001) and the 'new geography' (Daston, 2001; Lemann, 2001) signals the place of maps in the wider intellectual scene. Both reviewers equate geography with maps. Nicholas Lemann (2001, 131) remarked in the *New Yorker* that 'one of the main subspecialities of critical geography is critical cartography, the study of maps and mapmaking' (Plate 4). Lorraine Daston (2001, 3) in the *London Review of Books* explored the 'new geography and its claim to reveal that maps are texts to be read'. Harley was a historical geographer who had studied maps all his professional life. In the late 1980s and 1990s he was supported by a highly respected but small band of cultural geographers newly armed with theories from outside geography. In the first years of the twenty-first century, the upsurge of reactive humanism is showing no signs of flagging.

Our overview of maps in twentieth-century British geography suggests that maps as subjects for serious investigation have somewhere along the line been deemed irrelevant and unwelcome by all but a handful of human geographers. It is in fact not difficult to pin down the point of change. Writing in 1993 on the challenge facing geography in a changing world, Ronald Abler, one of the most influential of American quantitative geographers, confessed: 'My generation (I began graduate work in 1963) sold its intellectual birthright for a mass of theoretical and methodological pottage' (Abler, 1993). His diagnosis was that geography had distanced itself from area studies and regional geography to its detriment. Writing in the same book, Peter Taylor (1993) focused on what he sees as geography's 'full circle'—passing from a concern with world exploration in the era of imperial geography to the current preoccupation with the world's problems in an era of global geography. Maps, we have seen in this essay, were crucial adjuncts to geographical thinking at the turn into the twentieth century and central for the next five decades in regional geography. Paradoxically—or perhaps not so paradoxically?—maps now seem not to be part of the global geographer's professional equipment (whatever their usage elsewhere: Plate 5).

[14] It is true that many British geographers are barred from applying, as those resident within commuting distance of London are automatically disqualified.

We may no longer agree with Carl Sauer (1956, 289) when he says of maps, 'show me a geographer who does not need them constantly and want them about him, and I shall have my doubts as to whether he has made the right choice of life'; nor perhaps even with David Stoddart (1986, 56), who extends this sentiment to 'show me a geographer who cannot make a map, and I shall doubt his credentials'. We also doubt when we are confronted by a 'geography' that considers neither space, nor place, nor distributions. For such a 'geography', maps are wholly irrelevant.

In our view the clue to the present relationship between geography and maps is the word *use*. It is difficult to conceive of anyone interested in the study of spatial distributions not *using* a map at some point, but maps have become more than ever only one of a number of quantitative and qualitative tools used for geographical description and analysis. The modern geographer uses maps, but not exclusively maps; the modern geographer is not the exclusive user of maps. But at the end of the twentieth century most geographers had vacated the territory of the *study* of maps and their making, which they had so prominently occupied throughout its first 50 years.

Acknowledgements

We are particularly grateful to Jeremy Black, Christopher Board, Daniel Dorling, Andrew Gilg, Mike Heffernan, Janet Hooke, Ron Johnston, Paul Laxton, Bruce Webb, Michael Williams, and David Woodward for commenting on an early draft of this essay. For help on specific points we are pleased to thank Peter Chasseaud, Francis Herbert, Nick Millea, and Bruce Proudfoot. The line drawings are the work of Helen Jones and Andrew Teed prepared the photographic illustrations.

References

Abercrombie, P. (1938) Geography, the basis of planning. *Geography*, 23, 1–8.

Abler, R. F. (1993) Desiderata for geography: an institutional view. In R. J. Johnston, ed., *The Challenge for Geography. A Changing World: A Changing Discipline*. Oxford: Blackwell, 213–38.

Akerman, J. R. (1995) From books with maps to books as maps: the editor in the creation of the atlas idea. In J. Winnearls, ed., *Editing Early and Historical Atlases*. Toronto and London: University of Toronto Press, 3–48.

Alfrey, N. and Daniels, S., eds (1990) *Mapping the Landscape: Essays on Art and Cartography*. Nottingham: University Art Gallery and Castle Museum.

Alpers, S. (1983) *The Art of Describing: Dutch Art in the Seventeenth Century*. Chicago, IL and London: University of Chicago Press.

Alpers, S. (1987) The mapping impulse in Dutch art. In David Woodward, ed., *Art and Cartography: Six Historical Essays*. Chicago, IL and London: University of Chicago Press, 51–96.

Andrews, J. H. (1985) *Plantation Acres: An Historical Study of the Irish Land Surveyor and his Maps*. Belfast: Irish Historical Foundation.

Andrews, J. H. (1994) *Meaning and Power in the Map Philosophy of J. B. Harley*. Dublin: Department of Geography, Trinity College Dublin, Trinity Papers in Geography No. 6.

Andrews, J. H. (1997) *Shapes of Ireland: Maps and their Makers 1564–1839*. Dublin: Geography Publications.

Andrews, J. H. (2001) Introduction: meaning, knowledge, and power in the map philosophy of J. B. Harley. In J. B. Harley, ed. P. Laxton, *The New Nature of Maps: Essays in the History of Cartography*. Baltimore, MD and London: Johns Hopkins Press, 1–32.

Armstrong, L. (1996) Benedetto Bordon, *miniator*, and cartography in early sixteenth-century Venice. *Imago Mundi: The International Journal for the History of Cartography*, 48, 65–92.

Atkinson, D. (1995) Geopolitics, cartography and geographical knowledge: envisioning Africa from Fascist Italy. In M. Bell, R. Butlin, and M. Heffernan, eds, *Geography and Imperialism, 1820–1940*. Manchester: Manchester University Press, 265–97.

Atkinson, D. (2000) Geopolitical imaginations in modern Italy. In K. Dodds and D. Atkinson, eds, *Geopolitical Traditions: A Century of Geopolitical Thought*. London: Routledge, 93–117.

Baker, A. H. R. and Butlin, R. A., eds (1973) *Studies of Field Systems in the British Isles*. Cambridge: Cambridge University Press.

Baker, J. N. L. (1939) *Atlas of the War*. London. Oxford University Press.

Baker, J. N. L. (1959) A. G. Ogilvie and his place in British geography. In R. Miller and J. Wreford Watson, eds, *Geographical Essays in Memory of Alan G. Ogilvie*. London: Thomas Nelson, 1–6.

Balchin, W. G. V. (1976) Graphicacy. *The American Cartographer*, 3, 33–8.

Balchin, W. G. V. (1987) United Kingdom geographers in the Second World War. *Geographical Journal*, 153, 159–80.

Bales, K. (1991) Charles Booth's survey of life and labour of the people in London 1889–1903. In M. Bulmer, K. Bales, and K. Kish Sklar, eds, *The Social Survey in Historical Perspective*. Cambridge: Cambridge University Press, 66–110.

Bann, S. (1988) The truth in mapping. *Word and Image*, 4, 498–509.

Bann, S. (1994) The map as index of the real: Land Art and the authentication of travel. *Imago Mundi: The International Journal for the History of Cartography*, 46, 9–18.

Barber, P. and Board, C., eds (1993) *Tales from the Map Room: Fact and Fiction about Maps and their Makers*. London: BBC Books.

Barry, R. G. and Chorley, R. J. (1968) *Atmosphere, Weather and Climate*. London and New York: Methuen.

Bartholomew, J. G. (1914) *An Atlas of Economic Geography*. London: Oxford University Press.

Bartholomew, J. G., ed. (1922) *The Times Survey Atlas of the World*. London: The Times.

Batsford, H. (1940) *How to See the Country*. London: B. T. Batsford.

Bell, M., Butlin, R., and Heffernan, M. eds (1995) Introduction: geography and imperialism, 1820–1940. In *Geography and Imperialism 1820–1940*. Manchester: Manchester University Press, 1–11.

Bendall, A. S. (1992) *Maps, Land and Society: A History with a Cartobibliography of Cambridgeshire Estate Maps c.1600–1836.* Cambridge: Cambridge University Press.

Bennett, S. and Bennett, N., eds (2001) *An Historical Atlas of Lincolnshire.* Chichester: Phillimore (first published by University of Hull, 1993).

Berger, J. (1972) *Ways of Seeing.* Harmondsworth: Penguin.

Black, J. (1997a) *Maps and History: Constructing Images of the Past.* New Haven, CT and London: Yale University Press.

Black, J. (1997b) *Maps and Politics.* London: Reaktion Books.

Blakemore, M. J. and Harley, J. B. (1980) Concepts in the history of cartography: a review and perspective. *Cartographica*, 17/4, Monograph 26.

Blaut, J. (1991) Natural mapping. *Transactions, Institute of British Geographers*, NS16, 55–74.

Board, C. (1967) Maps as models. In R. J. Chorley and P. Haggett, eds, *Models in Geography.* London: Methuen, 671–725.

Board, C. (1968) Land use surveys: principles and practice. In Institute of British Geographers, *Land Use and Resources: A Memorial to Dudley Stamp.* London: Institute of British Geographers, Special Publication No. 1, 29–41.

Board, C. (1979) Maps in the mind's eye: maps on paper and maps in the mind. *Progress in Human Geography*, 3, 434–41.

Board, C. (1984) Higher order map-using tasks: geographical lessons in danger of being forgotten. In C. Board, ed., *New Insights in Cartographic Communication.* Toronto: Cartographica Monographs, 27, 85–97.

Board, C. (1995) The secret map of the County of London: 1926, and its sequels. *London Topographical Record*, 27, 257–80.

Board, C. (1999) Cartographic activities in the United Kingdom, 1995–1999. National report to the International Cartographic Association's 11th General Assembly, Ottawa, August 1999. *Cartographic Journal*, 36, 71–91.

Booth, C. (1902–3) *Life and Labour of the People in London*, 2nd edn. London: Macmillan and Co. (17 vols).

Bourne, M. (1999) Francesco II Gonzaga and maps as palace decoration in Renaissance Mantua. *Imago Mundi: The International Journal for the History of Cartography*, 51, 51–82.

Brace, C. (2001) Publishing and publishers: towards an historical geography of countryside writing, c.1930–1950. *Area*, 33, 287–96.

Brotton, J. (1997) *Trading Territories: Mapping the Early Modern World.* London: Reaktion Books.

Brown, E. H., ed. (1980) *Geography Yesterday and Tomorrow.* Oxford. Oxford University Press.

Browne, J. P. (1991) *Map Cover Art.* Southampton: Ordnance Survey.

Bryan, P. W. (1933) *Man's Adaptation to Nature: Studies in Cultural Geography.* London: University of London Press.

Bunge, B. (1962) *Theoretical Geography.* Lund: Gleerup.

Butlin, R. A. (1993) *Historical Geography Through the Gates of Space and Time.* London: Arnold.

Chasseaud, P. (1991) *Topography of Armageddon: A British Trench Map Atlas of the Western Front 1914–1919.* Lewes: Mapbooks.

Chasseaud, P. (1999) *Artillery's Astrologers: A History of British Survey and Mapping on the Western Front 1914–1918.* Lewes: Mapbooks.

Chisholm, M. (1971) Discussion of E. C. Willatts, 'Planning and geography in the last three decades'. *Geographical Journal*, 137, 335–7.

Chorley, R. J. and Haggett, P. (1965) Trend surface mapping in geographical research. *Transactions, Institute of British Geographers*, 37, 47–67.

Clarke, J. I. (1959) Statistical map-reading. *Geography*, 44, 96–104.

Cliff, A. D. and Haggett, P. (1988) *Atlas of Disease Distributions: Analytic Approaches to Epidemiological Data*. Oxford: Blackwell.

Coates, B. E., ed. (1974) *Census Atlas of South Yorkshire*. Sheffield: Department of Geography, University of Sheffield.

Coates, B. E. and Rawstron, E. M. (1971) *Regional Variations in Britain: Studies in Economic and Social Geography*. London: Batsford.

Coleman, A. (1961) The Second Land Use Survey: progress and prospect. *Geographical Journal*, 127, 168–86.

Coleman, A. (1964) Some cartographic aspects of the Second Series land use maps. *Geographical Journal*, 130, 167–70.

Coleman, A. (1980) Land-use survey today and tomorrow. In E. H. Brown, ed., *Geography Yesterday and Tomorrow*. Oxford: Oxford University Press, 216–28.

Conrad, J. (1928) Geography and some explorers. In *Last Essays*. London and Toronto: J. M. Dent, 1–21 (first published 1926).

Conzen, M. R. G. (1960) *Alnwick, Northumberland: A Study in Town Plan Analysis*. London: Institute of British Geographers, Publication No. 27.

Cooke, R. U. and Doornkamp, J. C. (1974) *Geomorphology in Environmental Management: An Introduction*. London. Oxford University Press.

Coones, P. (1987) *Mackinder's 'Scope and Methods of Geography' after a Hundred Years*. Oxford: University of Oxford, School of Geography.

Coppock, J. T. (1964) *An Agricultural Atlas of England and Wales*. London: Faber. New edition, with computer-generated maps, published in 1976.

Cormack, L. B. (1997) *Charting an Empire: Geography at the English Universities, 1580–1620*. London and Chicago, IL: University of Chicago Press.

Cosgrove, D. (1984) *Social Formation and Symbolic Landscape*. London: Croom Helm.

Cosgrove, D. (1992) *Palladian Landscape: Geographical Change and its Cultural Representation in Sixteenth-Century Italy*. Leicester: Leicester University Press.

Cosgrove, D. (1999a) Introduction: mapping meaning. In D. Cosgrove, ed., *Mappings*. London: Reaktion Books, 1–23.

Cosgrove, D., ed. (1999b) *Mappings*. London: Reaktion Books.

Cosgrove, D. (2001) *Apollo's Eye: A Cartographic Genealogy of the Earth in the Western Imagination*. Baltimore, MD and London: Johns Hopkins Press.

Cosgrove, D. and Daniels, S. eds (1998) *The Iconography of Landscape: Essays on the Symbolic Representation, Design and Use of Past Environments*. Cambridge: Cambridge University Press

Crampton, J. W. (1994a) Cartography's defining moment: the Peters projection controversy 1974–1990. *Cartographica*, 31, 16–32.

Crampton, J. W. (1994b) *Harley's Critical Cartography: In Search of a Language of Rhetoric*. Portsmouth: University of Portsmouth, Department of Geography, Working Paper 26.

Crouch, D. and Matless, D. (1996) Refiguring geography: parish maps of Common Ground. *Transactions, Institute of British Geographers*, NS21, 236–55.

Daniels, S. J. (1999) *Humphry Repton: Landscape Gardening and the Geography of Georgian England*. New Haven, CT and London: Yale University Press.

Daniels, S. and Rycroft, S. (1993) Mapping the modern city: Alan Sillitoe's Nottingham novels. *Transactions, Institute of British Geographers*, NS18, 460–80.

Darby, H. C. (1952) *The Domesday Geography of Eastern England*. Cambridge: Cambridge University Press.

Darby, H. C. (1956) The clearing of the woodland in Europe. In W. L. Thomas Jr, ed., *Man's Role in the Changing the Face of the Earth*. Chicago, IL: University of Chicago Press, 183–216.

Darby, H. C. (1977) *Domesday England*. Cambridge: Cambridge University Press.

Darby, H. C. and Campbell, E. M. J., eds (1962) *South-East England*. Cambridge: Cambridge University Press.

Darby, H. C. and Fullard, H. (1970) *The New Cambridge Modern History Atlas*. Cambridge: Cambridge University Press.

Darby, H. C. and Maxwell, I. C., eds (1962) *Northern England*. Cambridge: Cambridge University Press.

Darby, H. C. and Terrett, I. B., eds (1954) *Midland England*. Cambridge: Cambridge University Press.

Darby, H. C. and Versey, G. R. (1975) *Domesday Gazetteer*. Cambridge: Cambridge University Press.

Darby, H. C. and Welldon-Finn, R., eds (1967) *South-West England*. Cambridge: Cambridge University Press.

Daston, L. (2001) Language of power. *London Review of Books*, 1 November, 3, 6.

Debenham, F. (1936) *Map Making*. London and Glasgow: Blackie.

Debenham, F. (1950) *The Use of Geography*. London: English Universities Press.

de Boer, G. (1964) Spurn Head: its history and evolution. *Transactions, Institute of British Geographers*, 34, 71–89.

Delano-Smith, C. (1986) Changing environment and Roman landscape: the *Ager Lunensis*. In C. Delano-Smith, D. Gadd, N. Mills, and B. Ward-Perkins, Luni and the *Ager Lunensis*. The rise and fall of a Roman town and its territory. *Papers of the British School at Rome*, 54, 81–146 (at 123–41).

Delano-Smith, C. (1987) Cartography in the prehistoric period in the Old World: Europe, the Middle East, and North Africa. In J. B. Harley and D. Woodward, eds, *The History of Cartography, Vol. 1. Cartography in Prehistoric, Ancient, and Medieval Europe and the Mediterranean*. Chicago, IL: University of Chicago Press, 54–101.

Delano-Smith, C. (1990) Maps as art *and* science: maps in sixteenth-century Bibles. *Imago Mundi: The International Journal for the History of Cartography*, 42, 65–83.

Delano-Smith, C. (1996) Why theory in the history of cartography? *Imago Mundi: The International Journal for the History of Cartography*, 48, 98–203.

Delano-Smith, C. and Ingram, E. M. (1991) *Maps in Bibles 1500–1600: An Illustrated Catalogue*. Geneva: Droz.

Delano-Smith, C. and Kain, R. J. P. (1999) *English Maps: A History*. London and Toronto: The British Library and University of Toronto Press.

Doornkamp, J. C. and Gregory, K. J., eds (1980) *Atlas of Drought in Britain 1975–76*. London: Institute of British Geographers.

Dorling, D. (1995) *A New Social Atlas of Britain*. Chichester: Wiley.

Dorling, D. and Fairbairn, D. (1997) *Mapping: Ways of Representing the World.* London: Prentice Hall.

Downs, R. M. (1981) Maps and metaphors. *The Professional Geographer*, 33, 287–93.

Downs, R. M. and Stea, D. (1977) *Maps in Minds: Reflections on Cognitive Mapping*. New York: Harper & Row.

Dymond, D. and Martin, E. (1999) *An Historical Atlas of Suffolk*, revised and enlarged edition. Ipswich: Suffolk County Council and Suffolk Institute of Archaeology and History (first published 1988).

Edney, M. H. (1996) Theory and the history of cartography. *Imago Mundi: The International Journal for the History of Cartography*, 48, 185–90.

Edney, M. H. (1997) *Mapping an Empire: The Geographical Construction of British India, 1765–1843.* Chicago, IL and London: University of Chicago Press.

Fagg, C. C. and Hutchings, G. E. (1930) *An Introduction to Regional Surveying.* Cambridge: Cambridge University Press.

Fenneman, N. M. (1916) Physiographic divisions of the United States. *Annals of the Association of American Geographers* 6, 19–98. The map is Plate 1: Preliminary map of the physiographic division of the United States.

Fielding, A. J. (1993) *The Population of England and Wales in 1991: A Census Atlas.* Sheffield: Geographical Association.

Fiorani, F. (1995) The multimedia format of Renaissance maps. *Bulletin of the Society for Renaissance Studies*, 12, 7–12.

Fletcher, D. H. (1995) *The Emergence of Estate Maps: Christ Church Oxford Estate Management.* Oxford, Oxford University Press.

Forrest, R. and Gordon, D. (1993) *People and Place: A 1991 Census Atlas of England.* Bristol: School for Advanced Studies, University of Bristol.

Foucault, M. (1989) *Power/Knowledge: Selected Interviews and Other Writings, 1972–1977.* Brighton: Harvester Press.

Frangenberg, T. (1994) Chorographies of Florence: the use of city views and city plans in the sixteenth century. *Imago Mundi: The International Journal for the History of Cartography*, 46, 41–64.

Freeman, M. (1987) *Atlas of Nazi Germany*. London and Sydney: Croom Helm.

Fullard, H. and Darby, H. C., eds (1978) *The Library Atlas*, 13th edn. London: George Philip.

Gardiner, V. (1974) *Drainage Basin Morphometry*. London: British Geomorphological Research Group, Technical Paper No. 14.

Giddens, A. (1979) *Central Problems in Social Theory: Action, Structure, and Contradiction in Social Analysis*. Basingstoke: Macmillan.

Giddens, A. (1981) *The Contemporary Critique of Historical Materialism: Power, Property and the State*. Basingstoke: Macmillan.

Giddens, A. (1984) *The Constitution of Society: Outline of the Theory of Structuration.* Cambridge: Polity Press.

Gillies, J. (1994) *Shakespeare and the Geography of Difference.* Cambridge: Cambridge University Press.

Goodall, B. (1987) *Penguin Dictionary of Human Geography*. Harmondsworth: Penguin.

Gordon, A. and Klein, B. (2001) *Literature, Mapping and the Politics of Space in Early Modern Britain.* Cambridge: Cambridge University Press.

Gould, P. R. (1966) *On Mental Maps*. University of Michigan, Michigan Inter-University Community of Mathematical Geographers, Paper 9.
Gould, P. and White, R. (1974) *Mental Maps*. Boston, MA: Allen and Unwin.
Graham, E. (1982) Maps, metaphors and muddles. *The Professional Geographer*, 34, 251–9.
Green, D. R. (1995) Cartographic activities in the United Kingdom, 1991–1995. *Cartographic Journal*, 32, 50–73.
Greeves, T. (1987) *Parish Maps*. London: Common Ground.
Gregory, I. N. and Southall, H. R. (1998) Putting the past in its place: the Great Britain historical GIS. In S. Carver, ed., *Innovations in GIS 5*. London: Taylor and Francis, 210–21.
Gregory, I. N., Dorling, D., and Southall, H. (2001) A century of inequality in England and Wales using standardized geographical units. *Area*, 33, 297–311.
Gregory, K. J. (1985) *The Nature of Physical Geography*. London: Edward Arnold.
Gregory, K. J. and Walling, D. E. (1973) *Drainage Basin Form and Process: A Geomorphological Approach*. London: Edward Arnold.
Haggett, P. (1990) *The Geographer's Art*. Oxford: Blackwell.
Harley, J. B. (1969) The evaluation of early maps: towards a methodology. *Imago Mundi: The International Journal for the History of Cartography*, 22, 62–74.
Harley, J. B. (1983) Meaning and ambiguity in Tudor cartography. In S. Tyacke, ed., *English Map-Making, 1500–1650: Historical Essays*. London: The British Library, 29–38.
Harley, J. B. (1985) The iconology of early maps. In C. C. Marzoli, ed., *Imago et Mensura Mundi: Atti del IX Congresso Internazionale di Storia della Cartografia*. Rome: Istituto della Enciclopedia Italiana, Vol. 1, 29–38.
Harley, J. B. (1987–8) L'Histoire de la cartographie comme discours. In C. Jacob and H. Théry, eds, Dossier: la cartographie et ses méthodes, *Préfaces. Les idées et les sciences dans les bibliographies de la France*, 5, 66–114 (at 70–5).
Harley, J. B. (1988a) Maps, knowledge, and power. In D. Cosgrove and S. Daniels, eds, *The Iconography of Landscape: Essays on the Symbolic Representation, Design and Use of Past Environments*, Cambridge: Cambridge University Press. Reprinted in J. B. Harley (2001) *The New Nature of Maps*. Baltimore, MD: Johns Hopkins University Press.
Harley, J. B. (1988b) Silences and secrecy: the hidden agenda of cartography in early modern Europe. *Imago Mundi: The International Journal for the History of Cartography* 40, 57–76. Reprinted in J. B. Harley (2001) *The New Nature of Maps*. Baltimore, MD: Johns Hopkins University Press.
Harley, J. B. (1989) Deconstructing the Map. *Cartographica*, 26, 1–20. Reprinted in J. Agnew, D. N. Livingstone, and A. Rogers, eds (1996) *Human Geography: An Essential Anthology*. Oxford: Blackwell, 422–43. Also reprinted in J. B. Harley (2001) *The New Nature of Maps*. Baltimore, MD: Johns Hopkins University Press.
Harley, J. B. (1990a) Texts and contexts in the interpretation of early maps. In D. Buisseret, ed., *From Sea Charts to Satellite Images: Interpreting North American History through Maps*. Chicago, IL and London: University of Chicago Press, 3–15.
Harley, J. B. (1990b) Cartography, ethics and social theory. *Cartographica*, 27, 1–23.

Harley, J. B. (1991) Can there be a cartographic ethics? *Cartographic Perspectives*, 10, 9–16.

Harley, J. B. (1997) Power and legitimation in the English geographical atlases of the eighteenth century. In J. A. Wolter and R. E. Grim, eds, *Images of the World: The Atlas Through History*. New York: McGraw-Hill for the Library of Congress, 161–204. Reprinted in J. B. Harley (2001) *The New Nature of Maps*. Baltimore, MD: Johns Hopkins University Press.

Harley, J. B., ed. P. Laxton (2001) *The New Nature of Maps: Essays in the History of Cartography*. Baltimore, MD and London: Johns Hopkins University Press.

Harley J. B. and Woodward, D. (1987) Preface. The map and the development of the history of cartography. Concluding remarks. In J. B. Harley and D. Woodward, eds, *The History of Cartography, Vol. 1. Cartography in Prehistoric, Ancient, and Medieval Europe and the Mediterranean*. Chicago, IL: University of Chicago Press, xv–xxi, 1–42, and 502–9.

Harley, J. B. and Woodward, D. (1992) *The History of Cartography, Vol. 2, Book 1. Cartography in the Traditional Islamic and South Asian Societies*. Chicago, IL: University of Chicago Press.

Harley, J. B. and Woodward, D. (1994) *The History of Cartography, Vol. 2, Book 2. Cartography in the Traditional East and Southeast Asian Societies*. Chicago, IL: University of Chicago Press.

Hartshorne, R. (1939) *The Nature of Geography: A Critical Survey of Current Thought in the Light of the Past*. Lancaster, PA: Association of American Geographers.

Harvey, D. (1969) *Explanation in Geography*. London: Edward Arnold.

Harvey, D. (1989) *The Condition of Postmodernity: An Enquiry into the Origins of Cultural Change*. Oxford: Blackwell.

Heffernan, M. (1996) Geography, cartography and military intelligence: the Royal Geographical Society and the First World War. *Transactions, Institute of British Geographers*, NS21, 504–33.

Heffernan, M. (2002) The politics of the map in the early twentieth century. *Cartography and Geographic Information Science*, 29, 207–26. Special issue edited by M. Monmonier and D. Woodward, *Exploratory Essays: History of Cartography in the Twentieth Century*.

Helgerson, R. (1986) The land speaks: cartography, chorography, and subversion in Renaissance England. *Representations*, 16, 50–85.

Helgerson, R. (1992) *Forms of Nationhood: The Elizabethan Writing of England*. Chicago, IL and London: University of Chicago Press.

Helgerson, R. (2001) The folly of maps and modernity. In Andrew Gordon and Bernard Klein, eds, *Literature, Mapping and the Politics of Space in Early Modern Britain*. Cambridge: Cambridge University Press, 241–62.

Henderson, H. C. K. (1936) Our changing agriculture: the distribution of arable land in the Adur basin, Sussex, from 1780 to 1931. *Journal of the Ministry of Agriculture*, 43, 625–33.

Hodson, Y. (1999) *Popular Maps: The Ordnance Survey Popular Edition One-Inch Map of England and Wales 1919–1926*. London: Charles Close Society.

Holdich, T. H. (1902) Some geographical problems. *Geographical Journal*, 20, 411–27.

Holmes, N. (1992) *Pictorial Maps*. London: Herbert Press.

Holt-Jensen, A. (1999) *Geography, History and Concepts: A Student's Guide*, 3rd edn. London: Sage.

Hooke, J. M. (1977) An analysis of changes in river channel patterns. Unpublished PhD thesis, University of Exeter.

Hooke, J. M. (1995) Processes of channel planform change on meandering channels in the UK. In A. Gurnell and G. E. Petts, eds, *Changing River Channels*. Chichester: Wiley, 87–115.

Hooke, J. M. and Kain R. J. P. (1982) *Historical Change in the Physical Environment*. London: Butterworth.

Howe, G. M. (1970) *The National Atlas of Disease Mortality in the United Kingdom*, revised and enlarged edn. London: Thomas Nelson.

Hunt, A. J., ed. (1968) *Population Maps of the British Isles*. London: Institute of British Geographers, Special Publication No. 43.

Hyde, R. (1975) *Printed Maps of Victorian London: 1851–1900*. Folkestone: Dawson.

Jackson, P. (1989) *Maps of Meaning: An Introduction to Cultural Geography*. London: Unwin Hyman.

Jacob, C. (1992) *L'empire des cartes: Approche théorique de la cartographie à travers l'histoire*. Paris: Albin Michel. English translation in preparation as *The Sovereign Map: Theoretical Approaches in Cartography through History*. Chicago, IL and London: University of Chicago Press.

Jacob, C. (1996) Toward a cultural history of cartography. *Imago Mundi: The International Journal for the History of Cartography*, 48, 191–7.

Jardine, L. (1996) *Worldly Goods: A New History of the Renaissance*. London: Macmillan.

Jardine, L. and Brotton, J. (2000) *Global Interests: Renaissance Art between East and West*. London: Reaktion Books.

Johnston, R. J., ed. (1993) *The Challenge for Geography. A Changing World: A Changing Discipline*. Oxford: Blackwell.

Johnston, R. J. (1997) *Geography and Geographers: Anglo-American Human Geography Since 1945*, 5th edn. London: Arnold (1st edn 1979).

Johnston, R. J., Gregory, D., Pratt, G., and Watts, M., eds (2000) *Dictionary of Human Geography*, 4th edn. Oxford: Blackwell.

Kain, R. J. P. (1986) *An Atlas and Index of the Tithe Files of Mid-Nineteenth-Century England and Wales*. Cambridge: Cambridge University Press.

Kain, R. J. P. and Baigent, E. (1992) *The Cadastral Map in the Service of the State: a History of Property Mapping*. Chicago, IL: University of Chicago Press.

Kain, R. J. P. and Prince, H. C. (1985) *The Tithe Surveys of England and Wales*. Cambridge: Cambridge University Press.

Kain, R. J. P. and Prince, H. C. (2000) *Tithe Surveys for Historians*. Chichester: Phillimore.

Kain, R. J. P. and Ravenhill, W. eds (1999) *Historical Atlas of South-West England*. Exeter: University of Exeter Press.

Kearns, G. (1985) Sir Halford John Mackinder, 1861–1947. *Geographers Biobibliographical Studies*, 9, 71–86.

King, G. (1997) *Mapping Reality: An Exploration of Cultural Geography*. Basingstoke: Macmillan.

Kirk, W. (1952) Historical geography and the concept of the behavioural environment. In G. Kuriyan, ed., *The Indian Geographical Society Silver Jubilee*

Souvenir and N. Subrahmanyam Memorial Volume. Madras: Indian Geographical Society, 152–60.

Klein, B. (2001) *Maps and the Writing of Space in Early Modern England and Ireland*. Basingstoke: Palgrave.

Kline, N. R. (2001) *Maps of Medieval Thought: The Hereford Paradigm*. Woodbridge: The Boydell Press.

Knox, P. (1975) *Social Well-Being: A Spatial Perspective*. London. Oxford University Press.

Learmonth, A. T. A. (1972) Atlases in medical geography 1950–1970: a review. In N. D. McGlashan, ed., *Medical Geography: Techniques and Field Studies*. London: Methuen, 133–52.

Lemann, N. (2001) Atlas shrugs: the new geography argues that maps have shaped the world. *The New Yorker*, 9 April, 131–4.

Leslie, K. and Short, B. eds (1999) *An Historical Atlas of Sussex*. Chichester: Phillimore.

Lewis, G. M. (1968) Levels of living in the north-eastern United States *c.* 1960. *Transactions, Institute of British Geographers*, 45, 11–37.

Lewis, G. M., ed. (1998a) *Cartographic Encounters: Perspectives on Native North American Mapmaking and Map Use*. Chicago, IL and London: University of Chicago Press.

Lewis, G. M. (1998b) Maps, mapmaking and map use by native North Americans. In D. Woodward and G. M. Lewis, eds, *The History of Cartography, Vol. 2, Book 3. Cartography in the Traditional African, American, Arctic, Australian, and Pacific Societies.* Chicago, IL and London: University of Chicago Press, 51–182.

Linton, D. L. (1951) The delimitation of morphological regions. In L. D. Stamp and S. W. Wooldridge, eds, *London Essays in Geography: Rodwell Jones Memorial Volume*. London: Longmans, Green, 199–217.

Livingstone, D. (1992) *The Geographical Tradition: Episodes in the History of a Contested Enterprise*. Oxford: Blackwell.

Lowenthal, D. (1961) Geography, experience, and imagination: towards a geographical epistemology. *Annals of the Association of American Geographers*, 51, 241–60.

Lowenthal, D. and Bowden, M. J., eds (1976) *Geography of the Mind: Essays in Historical Geosophy in Honor of John Kirkland Wright.* New York. Oxford University Press.

Lyde, L. W. (1915) Types of political frontiers in Europe. *Geographical Journal*, 45, 126–45.

Mackinder, H. J. (1887) On the scope and methods of geography. *Proceedings of the Royal Geographical Society*, 9, 141–60, and discussion 160–74.

Mackinder, H. J. (1904) The geographical pivot of history. *Geographical Journal*, 23, 421–37.

Mackinder, H. J. (1919) *Democratic Ideals and Reality*. London: Holt.

Maitland, F. W. (1897) *Domesday Book and Beyond*. Cambridge: Cambridge University Press.

Martin, R. (2000) Editorial: in memory of maps. *Transactions, Institute of British Geographers*, NS25, 3–5.

Matless, D. (1992) Regional surveys and local knowledges: the geographical imagination in Britain, 1918–39. *Transactions, Institute of British Geographers*, NS17, 464–80.

Matless, D. (1995) 'The art of right living': landscape and citizenship, 1918–39. In S. Pile and N. Thrift, eds, *Mapping the Subject: Geographies of Cultural Transformation*. London: Routledge, 93–122.

Matless, D. (1997) Moral geographies of English landscape. *Journal of Historical Geography*, 22, 141–55.

Matless, D. (1999) The uses of cartographic literacy: mapping, survey and citizenship in twentieth-century Britain. In D. Cosgrove, ed. *Mappings*. London: Reaktion Books, 193–212.

Mayfield, B. and Clifford, S. (1996) *Parish Maps: Common Ground*. London: Common Ground (unbound folio).

Mill, H. R. (1930) *The Record of the Royal Geographical Society 1830–1930*. London: Royal Geographical Society.

Miller, A. A. (1951) Three new climatic maps. *Transactions, Institute of British Geographers*, 17, 13–20.

Millington, A. (1999) Editorial: cartography, geography and academia. *Geographical Journal*, 165, 253–4.

Mitchell, R., Dorling, D., and Shaw, M. (2000) *Inequalities in Life and Death: What if Britain Were More Equal?* Cambridge: Polity Press.

Ministère de l'Education Nationale (1971) *Atlas du Luxembourg*. Luxembourg: Ministère de l'Education Nationale.

Monkhouse, F. J. and Wilkinson, H. R. (1952) *Maps and Diagrams: Their Compilation and Construction*. London: Methuen (2nd edn 1964).

Monmonier, M. (1991) *How to Lie with Maps*. Chicago, IL and London: University of Chicago Press.

Monmonier, M. (2000) Map image and map. In R. J. Johnston, D. Gregory, G. Pratt, and M. Watts, eds, *Dictionary of Human Geography*, 4th edn. Oxford: Blackwell, 471–4.

Montague, C. E. (1924) *The Right Place: A Book of Pleasures*. London: Chatto and Windus.

Moretti, F. (1998) *Atlas of the European Novel, 1800–1900*. London and New York: Verso. First published 1997 as *Atlante del romanzo europeo 1800–1900*. Turin: Guilio Einaudi.

Muerhcke, P. (1981) Maps in geography. In L. Guelke, ed., *Maps in Modern Geography: Geographical Perspectives on the New Cartography*. Toronto, Cartographica Monograph, 27, 1–41.

Muir, R. (1911) *New School Atlas of Modern History*. London: G. Philip and Sons.

Murray, D. and Spencer, C. (1979) Individual differences in the drawing of cognitive maps: the effects of mental imagery and basic graphic ability. *Transactions, Institute of British Geographers*, NS4, 385–91.

Myres, J. L. (1933) The development of geography. *Geography*, 28, 69–75.

Neave, S. and Ellis, S., eds (1996) *An Historical Atlas of East Yorkshire*. Hull: University of Hull Press.

Nicholson, T. R. (1983) *Wheels on the Road: Maps of Britain for the Cyclist and Motorist 1870–1940*. Norwich: GeoBooks.

Nicolson, H. (1943) *Peacemaking 1919*, rev. edn. London: Methuen.

Ogilvie, A. G. (1922) *Some Aspects of Boundary Settlement at the Peace Conference*. London: Society for the Promotion of Christian Knowledge, Helps for Students of History Pamphlet No. 49.

Ogilvie, A. G. (1950) Isaiah Bowman: an appreciation. *Geographical Journal*, 115, 226–30.
Ordnance Survey (1982) *Ordnance Survey Atlas of Great Britain*. London: Ordnance Survey and Country Life Books.
Ortelius, A. (1570) *Theatrum Orbis Terrarum*.
Ortelius, A. (1579) *Parergon*.
Oxford University Press (1954) *Oxford Economic Atlas of the World*. London. Oxford University Press.
Oxford University Press (1960) *Oxford Regional Economic Atlas: The Middle East and North Africa*. London. Oxford University Press.
Pacione, M. (1995) The geography of rural deprivation in Scotland. *Transactions, Institute of British Geographers*, NS20, 173–92.
Palsky, G. (2002) Emmanuel de Martonne and the ethnographical cartography of Central Europe (1917–1920). *Imago Mundi: The International Journal for the History of Cartography*, 54, 111–19.
Pearson, A. W. (2002) Allied military model making during World War II. *Cartography and Geographic Information Science*, 29, 227–41. Special issue edited by M. Monmonier and D. Woodward, *Exploratory Essays: History of Cartography in the Twentieth Century*.
Petrie, G. (1977) Orthophotomaps. *Transactions, Institute of British Geographers*, NS2, 49–70.
Pocock, D. C. D. (1976) Some characteristics of mental maps: an empirical study. *Transactions, Institute of British Geographers*, NS1, 493–512.
Pocock, D. C. D., ed. (1981) *Humanistic Geography and Literature: Essays on the Experience of Place*. London: Croom Helm.
Poole, R. L. (1896–1902) *Historical Atlas of Modern Europe*. Oxford: Clarendon Press.
Porter, R. (2000) *Enlightenment: Britain and the Creation of the Modern World*. London: Allen Lane and Penguin Press.
Reeder, D. A. (1984) Introduction to *Charles Booth's Descriptive Map of London Poverty, 1889*. London: London Topographical Society, Publication No. 130.
Reitinger, F. (1999) Mapping relationships: allegory, gender and the cartographical image in eighteenth-century France and England. *Imago Mundi: The International Journal for the History of Cartography*, 51, 106–30.
Rhind, D. W. and Adams, T. A. (1980) Recent developments in survey and mapping. In E. H. Brown, ed., *Geography Yesterday and Tomorrow*. Oxford: Oxford University Press, 181–200.
Roberts, B. K. (1977) *Rural Settlement in Britain*. Folkestone: Dawson.
Robinson, A. H., Morrison, J. L. and Muehrcke, P. C. (1977) Cartography 1950–2000. *Transactions, Institute of British Geographers*, NS2, 3–18.
Rogoff, I. (2000) *Terra Infirma: Geography's Visual Culture*. London and New York: Routledge.
Rowley, G. (1984) *British Fire Insurance Plans*. Hatfield: Charles E. Goad.
Rowley, G. (1985) British fire insurance plans: the Goad productions, c.1885–c.1970. *Archives*, 17, 67–78.
Rycroft, S. and Cosgrove, D. (1995) Mapping the modern nation: Dudley Stamp and the Land Utilisation Survey. *History Workshop Journal*, 40, 91–105.
Sauer, C. O. (1941) Foreword to historical geography. *Annals of the Association of American Geographers*, 31, 1–24.

Sauer, C. O. (1956) The education of a geographer. *Annals of the Association of American Geographers*, 46, 287–99.

Scargill, D. I. (1976) The RGS and the foundations of geography at Oxford. *Geographical Journal*, 142, 438–61.

Schulten, S. (2001) *The Geographical Imagination in America, 1880–1950.* Chicago, IL and London: University of Chicago Press.

Schultz, J. (1978) Jacopo de' Barbari's view of Venice: map-making, city views and moralized geography before the year 1500. *Art Bulletin*, 60, 425–75.

Schultz, J. (1987) Maps as metaphors: mural map cycles in the Italian Renaissance. In David Woodward, ed., *Art and Cartography: Six Historical Essays.* Chicago, IL and London: University of Chicago Press, 97–122.

Short, B. (1997) *Land and Society in Edwardian Britain.* Cambridge: Cambridge University Press.

Short, J. R. (2001) *Representing the Republic: Mapping the United States, 1600–1900.* Chicago, IL and London: Chicago University Press.

Smith, D. M. (1977) *Human Geography: A Welfare Approach.* London: Edward Arnold.

Snow, J. (1855) *On the Mode of Communication of Cholera.* London (enlarged edn).

Spencer, C. and Weetman, M. (1981) The microgenesis of cognitive maps: a longitudinal study of new residents of an urban area. *Transactions, Institute of British Geographers*, NS6, 375–84.

Stamp, L. D. (1928) Regional survey exhibition. *Geography*, 14, 346–8.

Stamp, L. D. (1931) The Land Utilization Survey of Britain. *Geographical Journal*, 78, 40–7 and discussion 48–53.

Stamp, L. D. (1960) *Applied Geography.* Harmondsworth: Penguin.

Stamp, L. D. (1962) *The Land of Britain: Its Use and Misuse.* London: Longmans Green.

Steel, R. W. (1987) *British Geography 1918–1945.* Cambridge: Cambridge University Press, 1987.

Steers, J. A. (1926) Orford Ness: a study in coastal physiography. *Proceedings of the Geologists Association*, 37, 206–13.

Stoddart, D. R. (1986) *On Geography and its History.* Oxford: Blackwell.

Stoddart, D. R. (1992) Geography and war. The 'New Geography' and the 'New Army' in England, 1899–1914. *Political Geography*, 11, 87–99.

Stone, J. C. (1995) The cartography of colonialism and decolonisation: the case of Swaziland. In M. Bell, R. Butlin, and M. Heffernan, eds, *Geography and Imperialism 1820–1940.* Manchester: Manchester University Press, 298–324.

Taylor, E. G. R. (1921) *A Sketch-Map Geography: A Text Book of World Geography and Regional Geography for the Middle and Upper School.* London: Methuen (1950 onwards with E. M. J. Campbell).

Taylor, E. G. R. (1940) A national atlas of Britain. *Nature*, 145, 487–8.

Taylor, P. J. (1993) Full circle, or new meaning for the global?' in R. J. Johnston, ed., *The Challenge for Geography. A Changing World: A Changing Discipline.* Oxford: Blackwell, 181–97.

Waters, R. S. (1958) Morphological mapping. *Geography*, 43, 10–17.

Watson, J. W. (1967) *Mental Images and Geographical Reality in the Settlement of North America.* Nottingham: University of Nottingham, Cust Foundation Lecture.

Whittow, J. B. (1984) *The Penguin Dictionary of Physical Geography*. Harmondsworth: Penguin.

Willatts, E. C. (1933) Changes in land utilization in the south-west of the London Basin, 1840–1932. *Geographical Journal*, 82, 515–28.

Willatts, E. C. (1937) *Middlesex and the London Region*. London: Land Utilisation Survey of Great Britain, Report, Part 79.

Willatts, E. C. (1971) Planning and geography in the last three decades. *Geographical Journal*, 137, 311–30 and discussion 330–8.

Willatts, E. C. and Newson, M. G. C. (1953) The geographical pattern of population changes in England and Wales, 1921–1951. *Geographical Journal*, 119, 431–54.

Woods, R. and Shelton, N. (1997) *An Atlas of Victorian Mortality*. Liverpool: Liverpool University Press.

Woodward, D. (2000) Cartography. In R. J. Johnston, D. Gregory, G. Pratt, and M. Watts, eds, *Dictionary of Human Geography*, 4th edn. Oxford: Blackwell, 64–8.

Woodward, D. (2001a) Origin and history of *The History of Cartography*. In D. Woodward, C. Delano-Smith, and C. Yee, *Platejaments i objectuis d'una història universal de la cartografia. Approaches and Challenges in a Worldwide History of Cartography*. Barcelona: Institut Cartogràfic de Catalunya, Cicle de conferències sobra Història de la Cartogafia, 11è curs, 21–24 febrer 2000.

Woodward, D. (2001b) 'Theory' and *The History of Cartography*. In D. Woodward, C. Delano-Smith, and C. Yee, *Platejaments i objectuis d'una història universal de la cartografia. Approaches and Challenges in a Worldwide History of Cartography*, Barcelona: Institut Cartogràfic de Catalunya, Cicle de conferències sobra Història de la Cartogafia, 11è curs, 21–24 febrer 2000, 31–48.

Woodward, D. and Lewis, G. M., eds (1998) *The History of Cartography, Vol. 2, Book 3. Cartography in the Traditional African, American, Arctic, Australia, and Pacific Societies*. Chicago, IL and London: University of Chicago Press.

Wooldridge, S. W. and East, W. G. (1951) *The Spirit and Purpose of Geography*. London: Hutchinson (revised edn 1958).

Wooldridge, S. W. and Linton, D. L. (1939) *Structure, Surface and Drainage in South-East England*. London: Institute of British Geographers, Special Publication. Reprinted (1955) London: George Philip.

Wright, J. K. (1942) Map makers are human. Comments on the subjective in maps. *Geographical Review*, 32, 527–44. Reprinted in J. K Wright (1966) *Human Nature in Geography: Fourteen Papers 1925–1965*. Cambridge, MA: Harvard University Press, 33–52.

Wright, J. K. (1966) Notes on early American geopiety. In J. K Wright, *Human Nature in Geography: Fourteen Papers 1925–1965*. Cambridge: Harvard University Press, 250–85.

Young, A. E. (1920) *Some Investigations into the Theory of Projections*. London: Royal Geographical Society, Technical Series No. 1.

12

The geographical underpinning of society and its radical transition*

DAVID RHIND

Introduction

Maps and mapping are the manifestation of geography for the great bulk of the population. These play a key role in society and underpin many functions of the state. As Ratia (1999) said:

> When the European Commission invited representatives from the ministries in charge of mapping in member countries to a meeting in Luxembourg, at least the following ministries were represented: Ministry of Environment, Ministry of Agriculture and Forestry, Ministry of Housing and Physical Planning, Ministry of Finance, Ministry of the Interior, Ministry of Defence, Ministry of Justice. This shows how mapping and geographic information issues cover all the sectors of administration and it is in many cases a matter of taste which is the most natural ministry for these issues.

The situation is particularly marked in the UK, both in war and peace, where the Ordnance Survey (OS, Britain's national mapping agency) has been central to national mapping for over 200 years. Some 342 million maps were produced and/or printed for military use in the Second World War (Owen and Pilbeam, 1992) by the Ordnance Survey or its private contractors. Representatives of 168 major organisations in Britain in 1995 (Ordnance Survey, 1996) attested to peacetime dependency on detailed mapping. Many of these organisations had national responsibilities—for instance, the entire utilities industry and some 48 central government departments and agencies contributed to the study. On average, each organisation was able to identify about 15 applications of the mapping, usually spread across many different departments or units within the organisation as a whole. In addition, the annual sales of OS smaller-scale paper mapping (two million plus a year) for such purposes as leisure,

* This chapter is dedicated to the memory of Professor Terry Coppock FBA, pioneer in the computer-based analysis of geographic information.

motoring, and strategic planning make OS mapping central to the national 'perceived geography'. Finally, so far as the Ordnance Survey is concerned, the OXERA (1999) study showed that in 1996 between £79 and 136 billion of gross value added in the British economy was dependent to some extent on OS products and services. In addition to all this are the contributions of commercial and other government-based mapping (e.g. from the British Geological Survey). It is therefore no exaggeration to say that mapping underpins many of the activities of society, especially in the UK.

That said, professional geographers and the lay population have somewhat different views of maps—as Tobler (2000) has pointed out. He quotes Smith as summarising a common lay view of maps:

> Blair loved maps. He loved latitude, longitude and altitude . . . He loved the sense that with a sextant and a decent watch he could shoot the sun and determine his position anywhere on earth, and with a protractor and paper chart his position so that another man using his map could trace his steps to the exact same place . . . He loved topography, the twists and folds of the earth, the shelves that became mountains, the mountains that were islands. He loved the inconsistency of the planet . . . A map was, admittedly, no more than a moment in flux, but as a visualisation of time it was a work of art . . . People could no more resist maps of where they lived than they could portraits of themselves. (Smith, 1996, 39–40)

In contrast, according to Tobler, professional geographers use maps more as analytical tools, to help them understand and theorise about the earth and the phenomena distributed thereon, or to change and modify it. For Tobler (2000, 3) 'map-making consists of a selection and condensation from the immensities of reality to a depiction that presents aspects deemed important . . . cogent generalisation is a large part of the art of developing theoretical constructs or models from reality'. A good example of this is Dorling (1995).

In reality, these are two polar views of a spectrum since maps and mapping encompass a vast range of viewpoints and interests. The collation, visualisation, and analysis of geographical information through maps are intimately intertwined. Since British geography and British geographers have been involved in all aspects of mapping and its successor (see below), this chapter covers both academic and non-academic aspects of the subject area.

The transition from maps to geographic information

Maps have long been held to have at least four characteristics: as mechanisms for display of information[1] that varies across space, as information

[1] 'Information' is used in this chapter as described in Longley et al. (2001), as a class in a somewhat fuzzy hierarchy of descriptors ranging from data through information to evidence, knowledge and (perhaps) wisdom.

storage tools, as linkage mechanisms between different sources of information, and as objects of art. Roger Kain and Catherine Delano-Smith (Chapter 11, above) have elaborated many of the historical values of maps as display mechanisms and as information sources. One prime example of the latter is formed by Swedish cadastral maps, which give a detailed portrayal of the face of Sweden and land ownership therein from the sixteenth century onwards (Baigent, 1990).

This chapter is posited on three tenets. The first is that maps are increasingly only globally significant as information displays or as works of art; their information storage role and that of linkage (e.g. between descriptions of a particular soil set out in a monograph and supposedly homogeneous 'outcrops' of this soil on the surface of the earth) are simply historical artefacts (Rhind, 1994) that are being rendered redundant by use of digital storage. Thus the 230,000 'current-state' large-scale[2] OS maps of Britain were converted into digital form by 1995 and all 500,000 detailed historical OS maps of Britain dating from the 1860s were similarly encoded by 1998. All of these are now available on CD-ROMs or over computer networks and can be accessed by standard software on a personal computer. Similar situations are becoming the norm in many other countries. Major IT players like Microsoft are marketing mapping packages and Geographical Information Systems (GIS) used routinely by millions of customers: 'Do-It-Yourself cartography' is now commonplace, with literally millions of maps being made daily across the world.

The second tenet is even more fundamental: the new technology has permitted a separation of the geographic information base from the tools that enable its use (including cartography and individual GIS). It thus facilitates many different uses of the same information, whether for car guidance, missile tracking, re-districting of electoral boundaries, definition of statistical or policy-related zones (e.g. Housing Action Areas), location of the nearest cash machine, or calculations of cut-and-fill for earthworks (Raper, 2000). This is particularly significant because extracting information from maps for analysis has always been time-consuming. GIS enables a host of analyses, including the overlay of multiple distributions to detect spatial covariance, the seeking of clusters or other spatial regularity, distance-decay functions or 'optimal siting' (see Longley et al., 2001). This transition is of fundamental importance for lay users of maps and professional geographers alike: the encoded OS historical maps have been used both by academics for chronological analysis of settlements and by investment houses to ascertain whether the security on their loans is at risk because the land has previously been contaminated. Moreover, imagery

[2] Traditionally taken to be those originally published in paper form as 1:1,250, 1:2,500 or 1:10,000-scale maps. The use of map scale in a digital environment is of course a fraught and potentially misleading concept.

derived from satellites or aircraft has—for certain purposes—superseded conventional maps for some visualisation and certain analytical purposes; this is important for such information can now be collected largely automatically. The total volume of geographical information now available in digital form (especially that for 'state variables') and made available via these means is growing very rapidly (e.g. Curran et al., 1998). Though no quantification of it is known to the author, it seems likely that the volume of such geographical information has grown by an order of magnitude every few years in the last 15 years or so.[3]

The third tenet is that the era when technology drove the GIS and geographical information 'business' is over. Standard tools can cope with many routine tasks. The key issue is thus no longer one of inadequate technology seeking to simulate the artistic skills of highly trained artisans. Now the pressing policy and operational issues are those of finances to encode and maintain historical legacy information; international incompatibilities between information sets compiled by nation-states; use of common technical standards for inter-operability, plus information accuracy, reliability and 'believability'; access and charging policies; and information ownership (see below). Beyond this is the issue of the substantive contribution—for good or ill—that GIS brings to society, facilitating what are held by some to be undesirable trends (e.g. diminution of personal privacy) and fostering particular, positivist approaches to resolving issues over more inclusive ones.

For these reasons, this chapter concentrates on the transition from the cartographic world to the geographical information one; it also examines some of the societal consequences of the transition. This is a transition from a world where mapping was once carried out mainly by official governmental agencies and consumed largely by well-educated individuals in the pay of the state for professional purposes or for recreation—a world where analysis usually consisted of 'eye-balling' the map—to a world of many 'players' (including the commercial sector). In this new world, added value is created by linkage of geographical information and handling of the end result, including mapping, is now routinely (often unknowingly) carried out by the populace for innumerable individual purposes and GIS underpin visualisation and analysis of many kinds. The contemporary world is also one where the very concept of a simple 'map model' for the digital representation of the natural or human-created elements of the world has been found wanting: topological and thematic relationships between 'real-world entities', as well as purely geometric ones, need to be encoded if geographical information is to be useful for many different purposes. It is a world where, after a period of fierce academic debate

[3] It must be acknowledged that supply and demand are not perfectly matched, as Taylor's First Law of Geographical Information—that the need for such information tends to be greatest precisely where least is available (Taylor and Overton, 1991)—makes clear!

(largely in the United States), a complex and subtle form of geographical information science is evolving. Geography has thus in many senses been democratised and commercialised and its conceptualisation been re-thought in the course of this transition. Goodchild (2000) has termed this 'the digital transition'. The adjective has not been used here because the changes embrace far more than the (important) adoption of new technology (see below), as Pickles (2000) has argued.

This chapter highlights the contribution of British geographers to this radical transition—a situation unlike anything that had occurred in the previous 200 years. But, as indicated earlier, it would be extremely foolish to pretend that this revolution—for such it has been—was simply the work of professional or academic geographers, especially those in any one country. The transition has been enabled by technological change, itself fuelled by commercial and military factors, aided by key government policy actions. Academics *have* played some role, mainly in three ways: by periodic questioning of the contemporary wisdom and proposing alternative models; through 'salesmanship' of the concepts and potential of digitally stored geographical information to a wide audience (including policy-makers); and (probably most importantly) through the education of a generation of skilled individuals competent in understanding geographical information and handling it through such tools as GIS (see Coppock and Rhind, 1991; Longley et al., 1999, 2001).

GIS, geographical information and recorded geography as a sub-set of information technology

It has been argued that the advent of geographical information and GIS have simply been a reflection of the emerging computer technology and the declining cost of storage and transmission (see Plate 6). On this deterministic interpretation, GIS is nothing more than a small part of the information communication technology (ICT) industry. Such a causal explanation for what has become a £10+ billion turnover 'geographical industry' has some merits, notably in understanding the origins of the global homogeneity of concepts and tools used. But it is overly simplistic, for 'spatial is special' in certain respects, as Table 12.1 illustrates; handling geographical information is much less simple than, say, even the art form of creating annual accounts.

As indicated above, British geographers have contributed to the transition but often as parts of groups (e.g. with computer scientists) or in subtle and indirect ways. On occasions, significant contributions have also come from geographers operating in other disciplinary domains (e.g. Raper et al.'s 2002 information science perspective on the ontology, modelling, and system levels of geographical information). Many contributions have come from British individuals working outside academia as practitioners.

Table 12.1 The unusual characteristics of geographical information

- It is not like physical goods, for example, though it is typically expensive to collect and update, it is never used up and duplicating costs almost nothing (see Longley et al., 2001, section 16.4)
- Linking geographical data together has three potential benefits and at least one potential dis-benefit:
  - it permits infilling of missing data, thereby cutting the cost of data collection in some cases
  - it facilitates checking the quality of individual datasets through a check on their consistency
  - it provides added value, almost for nothing. The number of permutations rises very rapidly as the number of input information sets rises (see Longley et al., 2001, section 16.4.1). Hence many more applications can be tackled and new products and services created when datasets are linked together than when they are held separately.
  - it can lead to breaches of privacy and contravention of the UK Data Protection Act
- There are many different ways of representing the world (Longley et al., 2001, ch. 3) in geographical information form and how this is done influences what can be done and the results obtained
- Geographical information is typically sourced from many different origins and is fuzzy in that each and every element of it has associated imprecisions (see Longley et al., 2001, ch. 6); mixing data of different accuracies can lead to very significant problems
- Our techniques for describing 'data quality' are still primitive and our tools for analysing geographical information of different characteristics that has been combined are not good; it follows that deciding whether some information is 'fit for purpose' is often a matter of professional judgement—or irrelevant in the absence of any other (affordable) alternative.
- Some geographical information is in a form that makes it difficult to encode or combine with other, digitised geographical information (notably cultural properties and some qualitative historical information)
- Proof of ownership or provenance is sometimes difficult to demonstrate (e.g. when needed to demonstrate authenticity or fraud)
- The value of spatial data to a user often depends on the skills of the analyst
- The value of the data and analyses—and hence the price that can be charged—usually vary greatly between different applications
- The geographical information enterprise is more difficult for governments to regulate than, say, electricity, water, gas, or even telecommunications supply. The same geographical information may be manifested in many different forms by re-sampling or generalisation. There is no single constant unit of geographical information quantity (cf. water). Thus the capacity for gross profiteering or even fraud is greater than in some other industries

And much has been achieved by their counterparts elsewhere in the world. It follows from all this that the account of the transition is necessarily highly selective and involves citing the roles of only a few individuals—in the academic sector and without—from the many involved. It also needs to set the context in terms of national and other policies and military and commercial factors.

Non-trivial cartography

Finally, this focus on the transition from mapping to geographical information and to geographical information science does not imply that all map display—perhaps the one remaining important role for new cartography—is in any sense a trivial concern. Producing effective visualisation and communication through maps often requires much intelligence and skill (Hearnshaw and Unwin, 1994). Moreover, the results can often mislead or reinforce establishment views. Monmonier (1996) has shown convincingly how maps have long been used in both blatant and subtle propaganda. Harley (1989, 1990) demonstrated—with great force and clarity—that most official mapping has hitherto reflected the concepts of a 'ruling elite',[4] his work having been expanded by Wood (1993) and Pickles (2000, 9), who argued that 'maps do not simply represent territory, they also produce it'. This has been manifested in what is mapped, how it is categorised and depicted, and how the information is made available.[5] Limitations on space necessarily preclude further discussion of this aspect of maps and mapping, an important area for academic cartographers and geographers alike.

The geographical information context

Two aspects of the context are described here. The first is largely global (at least, it is increasingly if not universally so[6]) and is well known, whilst the second is very much specific to Britain.

[4] This point was reiterated by Simon Jenkins's article in *The Times* (3 August 2001, p. 16), celebrating the British Library's exhibition 'The lie of the land': 'Maps are indispensable. But each has its hidden agenda. It is selling a by-pass or excusing a bomber, allowing a tower block or defending a budget . . . the cartographer was trusted because he had a monopoly on truth.'

[5] The cost of detailed national re-survey following any change of elite view or user need ensures that such mapping often reflected the views of long-dead elites.

[6] 'If the world were reduced to a village of 1,000 people, there would be 584 Asians, 124 Africans, 136 from the Western Hemisphere (both North and South America), 95 Eastern/Western Europeans, and 55 Russians. 520 would be female. 650 would lack a telephone at home. 500 would never have used a telephone. 335 would be illiterate. 333 would lack access to safe, clean drinking water. 330 would be children. 70 would own automobiles. Ten would have a college degree. *Only one would own a computer.*' (Source: http://www.ntia.doc.gov/ntiahome/speeches/ntca120198.htm)

The technological revolution

The single most important technological factor has been the decrease in the cost of computing power, telecommunications, and data and information storage. Broadly speaking, Moore's Law describes this change: crudely expressed, it states that the cost of computing halves every 18 months. The practical manifestation of this is that in the 1960s and 1970s the largest mainframe computers were needed to carry out what are now regarded as trivial tasks for personal computers. The ready availability of microprocessors, allied to developments to meet military needs, has made a reality of Tobler's (1976) then-outrageous prediction of a wristwatch GPS and maps on hand-held displays. The free provision of signals from the US Global Positioning System (GPS), plus its Russian GLONASS equivalent, together with cheap hand-held receivers, have transformed the world of professional surveying, as well as both military and civilian leisure activities that require knowledge of position anywhere on the earth's surface. With suitable systems and processing, survey accuracies of a few centimetres or smaller may be obtained. As one consequence, the 6,000 OS trigonometric stations located on the peaks of British hills are now redundant, having been replaced by a series of active and passive GPS-based fixed points much nearer to centres of population (see Ordnance Survey, 2000). Satellites exploiting the panoply of digitally enabled technology are now able to image anywhere on the earth's surface. The 1-m resolution Ikonos imagery of Manhattan before and after 11 September 2001 was used worldwide. This, and images of Taliban soldiers on route marches in Afghanistan, were obtained by a commercial satellite orbiting at an altitude of 681 km and made available across the world wide web (see another example in Fig. 12.1) within a few hours of being collected. Behind all this, however, are many other important non-technological developments, such as the use of graph theory to develop topologically coherent maps by the US Bureau of Census.

The geographical-information-relevant policy context in the UK

Here we can distinguish three relevant areas: British government policy in regard to the role of public sector bodies in general and central government ones in particular; policy in regard to information accessibility, provision and pricing; and how both of these were mediated in terms of geographical information in Britain (for a review of international policies up to that point, see Masser, 1998; Rhind, 1999).

UK government policy on public sector bodies

The present and previous UK governments have regarded the improvement of public services as key policy objectives. From at least the late 1980s onwards, the Conservative government also saw the reduction of their costs

Fig. 12.1 An example of the impact of new technology on the collection and dissemination of geographical information. This shows an Ikonos satellite image of the Millennium Eye, Houses of Parliament, Whitehall, Waterloo Station and County Hall in central London, collected from an altitude of 681 km on 25 October 1999 at 1-m resolution and made available a few days later via the world wide web. In principle, such commercial imagery (based on technology originally created for the US military) can be produced for any area in the world.
Copyright Space Imaging Inc.

as being a parallel objective. The 1988 Ibbs report on the civil service argued that some 90 per cent of the work of central government was to provide services and that all customers—individuals, government bodies, businesses, and others—had a right to good service. Foster and Plowden (1996) urged government to focus even more on such customers. With the aim of reducing costs and improving quality, the 1979–97 party of government enforced the testing of services provided by the civil service against the best that the private sector could offer. Failure to succeed in such competitive 'market-testing' generally led to the service being 'contracted out' to the private sector. The policy applied across almost all of the state's remit. Without spelling out in detail what are the services that only the state can provide—unlike the US situation—the presumption has been that very few activities must *necessarily* or *solely* be carried out directly by the state. The views of the Labour government since 1997 in practice approximate to those of its predecessor, though a wider range of measures for defining efficiency and effectiveness are used and a major change in information policy has been introduced (see below).

To achieve better management of these services and greater value for taxpayers' money, some 75 per cent of all civil servants formerly working in government departments were transferred to about 140 Executive Agencies in the period 1989–95. Each agency has to create a corporate plan and meet annual targets set by ministers covering financial measures, operational efficiency, and quality of service. The targets and results are widely publicised and performance-related pay (and even jobs) depend upon their achievement. Recruitment of non-civil servants as chief executives occurred in nearly half of these agencies; open recruitment through advertising is the norm. To help it achieve its targets, each agency has somewhat greater freedoms than it had previously had in the public sector, though these do not generally include raising loan capital or competing aggressively with the private sector. Finally, each agency is reviewed every five years to see whether its function still needs to be met and, if so, whether it can best be met by the private or public sectors. Partnerships between the public and private sectors are universally held to be important, with contributions to meeting government policy and money inevitably being the prime factors in the determination of success of each partnership. It will be obvious that such operating circumstances have significant effects upon the actions of the executive agencies. In many respects their creation and operation has anticipated—and far extended—actions recommended in Osborne and Gaebler (1992).

Government policy on government information

Government is a large-scale acquirer of information to help frame policy and monitor its success. The great bulk of geographical information, in particular, in Britain is acquired by the state, through mechanisms like the population census and the work of government agencies like the Hydrographic Office, the Meteorological Office, the Office for National Statistics, and the Ordnance Survey. Openshaw and Goddard (1987) pointed out at an early stage the likelihood of such geographical information being 'commodified'.

It is largely true to say that there was no one information policy in Britain before 2000; rather there were many different policies, statutes, and administrative edicts and delegations that impacted on the availability of information, and geographical information in particular. All information collected by central government bodies was protected by Crown copyright and the Crown issued licenses where it deemed them appropriate for onward use of the material. Her Majesty's Stationery Office (HMSO) was the proxy owner of Crown copyright, though a number of organisations (like the Ordnance Survey) had delegated powers that effectively gave them freedom to practise locally set policies (such as 'the user pays') needed to meet the targets set by their ministers. To compound the situation, the Department of Trade and Industry issued guidance on how to provide information to the private sector and legal penalties potentially existed for any inconsistency of information supply policy (e.g. differential pricing for the same goods). The practical consequence of all this (and more) for any potential user were that he or she might well have to pay for (say) OS base map information but not for other information compiled by a different body using their copies of the OS map information. The separation of layers of information, which the digital environment permitted, was a crucial factor in enabling this policy inconsistency to operate.

From the mid-1990s onwards, the private sector argued forcefully that the government's aims to create a knowledge economy were being slowed enormously by the difficulty of unlocking the 'treasure troves' of government information. The need for individual licensing, a multiplicity of charging mechanisms, and highly variegated practices across government were all attacked. In addition, the European Commission formulated several directives which, when translated into British law, complicated the previous system. As a consequence, the Chancellor of the Exchequer carried out a review,[7] which concluded in December 2000 that it was highly desirable for the private sector to exploit government information. It was stipulated that no exclusive deals should be done with any one party, that in general the digital information should be made available at marginal cost (as in the US federal government) and that HMSO should become the government's information regulator. However, it was stipulated that the arrangements should be different for those agencies operating on a trading fund basis (i.e. those required to generate revenues from sale or licensing of information to meet their financial targets!). These bodies included the Hydrographic Office, HM Land Registry, the Meteorological Office, and the Ordnance Survey (but not the Office of National Statistics), and they were encouraged to employ differential pricing based on differentiated products for different markets (a practice the Ordnance Survey had employed for some years since completing its digital geography of the country—see below). In practice, then, it remains to be seen how much has actually changed, especially in regard to

[7] See http://www.hm-treasury.gov.uk/sr2000/associated/knowledge/index.html

geographical information. Further reviews of information pricing were promised in the report.

The evolution of government policy in relation to the Ordnance Survey

A summary of the policy context in earlier times is provided by Seymour (1980) and Owen and Pilbeam (1992). The remit under which the Ordnance Survey has operated since 1979 arose fundamentally from the report of the Ordnance Survey Review Committee (Department of the Environment, 1979),[8] which was set up by the Labour government in 1978 to examine and redefine the role of the Ordnance Survey after completion of the re-mapping of Britain begun in 1946. The report, produced at a time when digital processes in mapping were in their infancy, was a thorough and wide-ranging document, which re-affirmed the national need for country-wide paper map coverage at scales of 1:1,250 (in urban areas), 1:2,500 (in rural areas), and 1:10,000 (in mountain and moorland areas). It accepted the need for continuous revision of the maps rather than cyclic revision and urged the Ordnance Survey to proceed as rapidly as possible with its programme of converting its paper maps to computer form. Whilst generally supportive of the then management's plans, it was critical of an inward-looking culture and the lack of adequate investment in research and development. Most crucially, it argued that there were benefits in the Ordnance Survey obtaining as much revenue as it could from the sale of its maps: this would foster market focus and minimise the burden on the taxpayer. Since cost recovery in 1979 was about 40 per cent (see Fig. 12.2), however, the committee was clear that the Exchequer should continue to fund the difference between costs and revenue. The report's recommendations were eventually largely accepted by the Conservative government in 1984, the long delay in its acceptance reflecting the incoming government's distaste with the terms of reference and the largely public-sector composition of the committee. In practice, the recommendations had mostly been adopted by the Ordnance Survey before this formal acceptance.

The consequence of the funding mechanism came to pose a progressively greater problem for the Ordnance Survey over the next 18 years. Each year, ministers and Treasury officials looked for efficiency gains (i.e. cost-cutting) and improvements in revenues through improved marketing. The consequent effect of this was a progressive reduction in the vote of funds from Parliament: this dwindled from around £22 million around 1990 to £2 million eight years later. Whilst effective in minimising taxpayer funding, this has had several severe disadvantages. In the first instance, it perpetuated a misleading view that the Ordnance Survey was being subsidised for its ineptness. Even more seriously, it disguised the tension between a remit to map the whole country and yet find funds for

[8] Of which J. T. Coppock was a member.

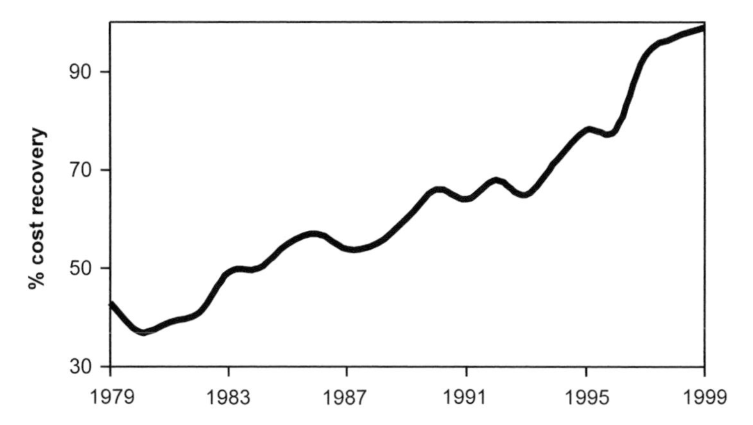

Fig. 12.2 The transition towards cost recovery by Britain's national mapping agency achieved by charging customers for geographical goods and services, creating new products and working with private-sector partners.
Source: Ordnance Survey Annual Reports.

that work from only the modest proportion of Britain that seems to be commercially viable (90 per cent of the population of Britain lives in about 10 per cent of the land area and many sales of mapping are population-related). This cross-subsidy is inequitable to customers and facilitates the entry of commercial competitors, who do not have to cover such a wide geographical overhead. Finally, the funding mechanism did not provide transparency and accountability: the taxpayer had no real idea what she or he was getting for the parliamentary funding provided.

Further influences on the Ordnance Survey remit and priorities came in the form of the report of the House of Lords' Select Committee on Science and Technology[9] in 1985 (Rhind, 1986), the report of the government's Committee of Inquiry into the Handling of Geographical Information (often called the Chorley Report after the committee's chairman[10]) in 1987 (Department of the Environment, 1987; Rhind and Mounsey, 1989; Heywood, 1997), and the report on charging principles for information by Coopers and Lybrand (1996). Amongst many other recommendation, the first two of these urged the Ordnance Survey to speed the rate of conversion of its 230,000 paper maps to computer form. The Chorley Report also urged the government and users to come together in creating a user forum for GIS and related matters (which became the thriving Association for Geographic Information) and urged universities to develop courses in GIS

[9] Chaired by Lord Shackleton, a former president of the Royal Geographical Society, and advised by the author.

[10] Lord Chorley, a former president of the Royal Geographical Society. The author and Professor J. Townshend, then of Reading University, were members of the Chorley Committee.

and research councils to fund research in this area. A joint funding programme by the Economic and Social Research Council (ESRC) and the Natural Environment Research Council (NERC) introduced substantial funding in GIS research at a critical time and, through the creation of Regional Research Laboratories, created something that has partially endured in self-sustaining form many years afterwards. After 1987, however, there was an almost total cessation of high-level policy debates about geographical information and GIS in Britain; this contrasts dramatically with the situation in the United States, where the National Research Council and congressionally affiliated bodies led major reviews (see, for instance, Mapping Science Committee, 1993, 1994, 1997; National Academy for Public Administration, 1998). What geographical information policy has emerged in Britain has come about through Ordnance-Survey-specific matters and general debates on charging for access to government information and the nature of government's role (see above).

So far as the Ordnance Survey itself is concerned, the manifestation of all the conflicting 'steers' on its activities is its 'license to operate' from government. This is set out in three succeeding framework documents (Ordnance Survey, 1990, 1995, 1999). Interpretation of these documents requires some skill (e.g. in identifying the significance of the low-key legitimation in the 1995 report for the Ordnance Survey to engage in all forms and dates of topographic mapping and GIS databases, rather than just contemporary mapping). Finally, a quinquennial review carried out in 2000–1 resulted in proposals to convert the Ordnance Survey from a government department and executive agency into a government-owned plc; this proposal was subsequently quashed by ministers (CMG, 2001).[11]

The reality of the transition in Britain

Policy issues and legacy aspects of the geographical framework, as well as the efforts of particular individuals, ensured that the early stages of the transition, particularly in Britain, had some distinctive features. In many ways what has happened differs significantly from the situation in the United States (Goodchild, 1988, 2000; Foresman, 1998).

Origins of the transition

The earliest known processing of geographical data by computer is probably the production of meteorological charts from about 1950 onwards. Simpson (1954) reported some early British work, using standard printing devices to construct embryonic raster maps. Within a few years, the first uses of cathode ray tubes for plotting maps were described in the literature (e.g. Sawyer

[11] See http://www.ordsvy.gov.uk/downloads/info_papers/QQR2001.pdf

1960). One surprising characteristic of all this is that geographers and cartographers appear not to have played any role in the earliest developments.

This situation changed very soon thereafter, especially initially in Sweden (Rhind, 1998). In Britain, however, the earliest machine-readable geographical database was probably the mechanical one produced by Perring and Walters (1958). They compiled information on the distribution by national grid square of 2,000 species throughout Britain, storing the results on 40-column punch cards. Initially, their plans were simply to use this for summarising and sorting the data but they quickly realised that the data could also be mapped using a mechanical tabulator. Another pioneer was Coppock, who first came to appreciate the need for electronic aids to analysis after attempting to analyse results in map form from his land use survey of the 1,600 square mile area of the Chiltern Hills in 1954. Subsequently he was involved in experiments with electronic area-measuring devices. At the end of the 1950s, however, he analysed about half a million records from the agricultural census using an early computer in the University of London. The programs summarised the data records and classified them, in readiness for mapping by hand. Though the potential value of computer mapping was clearly appreciated at the time, the limitations of machine performance and output devices rendered such automation impossible (Coppock, 1962). Coppock's work may be the earliest substantive 'GIS-based research'.

Elsewhere, a number of other important developments had occurred. By 1960, for instance, routine use was being made of computers in the Directorate of Overseas Surveys for projection change and other coordinate transformations, particularly for geodetic work in Africa. In addition, Bickmore had suffered so many problems in manually producing the mammoth *Atlas of Britain* that by 1960, at the latest, he had formulated plans for a sophisticated computer-based mapping system capable of matching the best quality that could be produced by hand. This prototype system was subsequently built as funds were obtained and was described by Bickmore and Boyle (1964). Amongst the advances achieved by this duo was the world's first free cursor digitising table.

By far the best-funded and most comprehensive work of the period was that by the Experimental Cartographic Unit (ECU), founded by Bickmore in 1967 and directed by him for about a decade. Rhind (1988) has described the work of the ECU in some detail. Its roots lay in work in the Clarendon Press of the University of Oxford from 1960 to 1965, though the Unit itself existed between 1967 and 1975, being based for part of this period in both Imperial College and the Royal College of Art as a deliberate attempt to bridge C. P. Snow's 'two cultures'. The ECU was funded by the NERC but its work spanned the full gamut of geographical activities. Thus, in 1969 for example, its staff created computer programs for converting digitised map coordinates to geographical ones, for projection change, for editing of cartographic features such as lines, for statistical summary of database characteristics, for the compression of data files, for generalisation of entities, for automated contouring, and for the production of three-dimensional anaglyph maps. They defined and published standards for the exchange of geographical data in a standard format. Simultaneously the ECU researchers were investigating digitising by automated line following and the accuracy of manual digitising of maps. They also created a 60,000-place gazetteer and produced various multicolour maps by computer. All this led in 1970 to the publication of a major work on the use of geographical data processing in planning (Experimental Cartographic Unit, 1971). The following year, ECU staff produced the first-ever multicolour map published as part of a standard map series—the geological map of Abingdon (Rhind, 1971)—and took the first tentative steps in planning global databases. At the same time, they discussed with commercial organisations the packaging and marketing of ECU's mapping software and planned an MSc course in the handling of spatial data.

By no means did the ECU represent the entirety of the British efforts in GIS-related fields. A branch of central government produced the LINMAP system, which provided statistical analysis and mapping capabilities within the Ministry of Local Housing and Government from about 1968 onwards (Gaits, 1969). But a particularly interesting and important development was launched by one individual, T. C. Waugh, who was mainly based in the University of Edinburgh. The author of a line-printer mapping system (CMS) whilst an undergraduate student around 1967, Waugh set out to create in the Geographic Information Mapping and Manipulation System (GIMMS) a portable, high-quality, vector mapping system with data manipulation and analysis capabilities (Waugh, 1980). The first work was carried out in 1969–70 whilst he was a graduate student at the Harvard Computer Graphics Laboratory and development continued for two decades afterwards. In many respects, the high point of GIMMS use came in the late 1970s to the early 1980s, when more than 100 sites worldwide ran it on mainframe and mini-computers. It was sold on a commercial basis from 1973 onwards (then at $250). Ultimately more than 300 sites in 23 countries ran GIMMS on a huge variety of computers, ranging from PCs and Macintoshs to a Cray supercomputer. Other than the much simpler, batch-operated SYMAP line-printer package emanating from Harvard University, GIMMS can be considered the first globally used GIS. It pioneered the use of topological data structures, user command languages for interactive operations, macro languages, and user control of high-quality graphics (including text placement and multiple fonts) in a widely used and wholly integrated system. In many respects, then, it is a prime antecedent of contemporary GIS and anticipated some key characteristics of the Harvard Odyssey system (Chrisman, 1998) by nearly five years and of the world's leading GIS (ARC/INFO and its descendants) by a decade. Other work by academics during this period included the first work on satellite remote sensing (see, for instance, Cooke and Harris, 1970).

In no sense does the preceding account exhaust the work going on in Britain at this period. Following a major project carried out with the ECU between 1969 and 1971—and sparked by constant public criticism by Bickmore—in 1973 the national mapping agency set up the world's first production line for converting its paper maps into computer form and the following year carried out widely publicised experiments into how to restructure simple vector map data into several alternative forms suited to different applications (Thompson, 1978). Until funding was withdrawn in 1974, a major central government initiative focused on the design and implementation of planning information systems to be used nationwide, fostered by the publication in 1972 of the report by local and central government on a General Information System for Planning (Department of the Environment, 1972). The use of population census data, both in academia and more generally, was transformed by the work of geographers, initially by a Durham group's work in mapping the data for 150,000 populated 1-km grid squares (Census Research Unit et al., 1980) and in a joint Edinburgh/Durham initiative creating the SASPAC software used universally to handle the 1981 census data. Baxter (1976) has a good claim—in competition with that by Nordbeck and Rystedt (1972)—to be the first real text on GIS. Members (many of them geographers) of BURISA—a British variant of the US Urban and Regional Information Systems Association—were catalysts in numerous attempts to build detailed descriptions of areas and their populations inside local government information systems in order to implement the 1960s and 1970s view of a planning-led society. Though successful and long-lived systems such as that for Merseyside resulted, most of this was thwarted by the poor technology of the time and by changing political ideologies. Yet between the late 1960s and mid-1970s it can fairly be said that the UK was at the forefront of the use of GIS, even if the systems were not thus described for many years thereafter.[12] Thereafter, Britain's relative domination of GIS in Europe has decreased (see Table 12.2). Even later, however, some world-leading work occurred, notably the creation of the first multimedia, hyper-linked database and display system—the Domesday system funded by the BBC, to which many British geographers contributed (Rhind, Armstrong

[12] The first use of the term 'GIS' in a European publication is impossible to determine with certainty: Unwin (1991) has suggested that it may be as the title of a small conference run by this author in 1976. It was certainly in use earlier in North America and is generally attributed to the English émigré Roger Tomlinson, who led the development of the Canada Geographic Information System from the mid-1960s (Tomlinson, 1998) and who subsequently obtained the first known PhD on GIS (from UCL).

Table 12.2 Geographical affiliations of first authors of papers given at the first three major, general-purpose GIS conferences held in Europe

Country	*Auto Carto London '86*	*EGIS '90 (Brussels)*	*EGIS/MARI '94 (Paris)*
France	6	6	45
Germany	3	8	11
Italy	1	6	23
Netherlands	3	44	26
UK	41	35	27
Rest of European Union	5	14	32
Rest of Greater Europe (20 countries)	6	13	48
Asia and Pacific	11	3	5
North America and Caribbean	37	8	7
Africa	1	0	10
Total papers	114	137	234

Note: The early dominance of UK authors is evident, as is the UK/US association in the strong presence of North Americans at the first conference.

and Openshaw, 1988). Globally significant contributions continue to be made in specific areas (such as the optimal location work of Wilson and Clarke and their colleagues at the University of Leeds, which built on two decades of modelling research: see for instance, Birkin et al., 1996). It is noteworthy that the great majority of those working in GIS in the formative period up to the early 1990s were geographers or cartographers.

A number of later British developments were also highly significant. The Geographical Algorithm Group's library in Britain, led from University College Geography Department and the Computer Science department in the University of East Anglia, was probably the first geographical software worldwide written to defined, professional standards of quality, consistency, and documentation (Geographical Algorithm Group, 1977). The first installation of a true commercial GIS in Europe was the US ARC/INFO v2 in Birkbeck College, University of London, which was funded by the ESRC as a national facility and launched in 1983–4. A decade later ARC/INFO became the *de facto* national standard in UK universities when its US creators ESRI won a national competition run by the Funding Councils' Joint Information System Committee.

One intermittent but passionately debated development of disciplinary significance centred on the nomenclature. In the 1980s there was a strong move by land surveyors and associated professions (e.g. Dale and McLaughlin, 1988) to make 'land information systems' the common parlance, with GIS relegated to cover small-scale mapping. This had limited success. Similarly, there has been from the mid-1990s onwards a geographically widespread adoption of the terms 'geomatics' or 'geomatics engineering' to replace land surveying and to sweep all aspects of the related subjects—including remote sensing, photogrammetry, GIS, and survey—under the same umbrella. Finally, surveyors in North America have been energetic in seeking to introduce professional certification in the field, though this has not taken root in the UK, where the much less formalised approach of the Association for Geographic Information has become accepted (Longley et al., 2001, 382–3).

Since the early days of the 1960s to 1980s, the range of GIS applications has widened and the number of systems has grown rapidly in Britain and the rest of Europe—as elsewhere in the world. Over 2000 universities are known to run GIS courses across the world and, even on a narrow definition of GIS, over a million systems are in current use, with the number growing at about 20 per cent per annum. This situation has been enabled by the technology but fostered by improvements in communications, both within and between countries and across and between disciplines. Several factors contributed to this, including the creation of a global GIS research journal: the *International Journal of GIS* grew from a meeting of British GIS workers but was consummated by marriage with colleagues in the United States and elsewhere; its founding editor was Terry Coppock. Alongside this academic development was the growth of a GIS book publishing industry—comprising such major players as Longman/GeoInformation International (whose GIS texts were subsequently absorbed by Wiley), Taylor and Francis, and Oxford University Press—and the popularity of trade magazines. Recent books by British authors (sometimes with others) include Atkinson and Tate (1999), Burrough and McDonnell (1998), Grimshaw (2000), Heywood, Cornelius and Carver (1998), Jones (1997), Longley et al. (1998, 1999, 2001), Martin (1996), Reeve and Petch (1999), and Worboys (1995). Also highly influential was the foundation of national GIS organisations, such as the Association for Geographic Information in the UK (most of whose senior officers have been geographers), national GIS research initiatives (funded by a joint research councils' initiative in the UK), pan-European initiatives such as the GISDATA project funded by the European Science Foundation, plus funding of multilateral GIS product development by the European Commission under its DRIVE, IMPACT and other initiatives.

One crucial factor fuelling all of the national and international developments was the growth of education in GIS. It is likely that the first GIS course in Europe comprised option classes in Edinburgh University in 1976 or some optional undergraduate courses and an MSc course in spatial data analysis run in the University of Durham in the same year. Certainly the Edinburgh MSc in GIS, run from 1984 to date, plus other undergraduate and postgraduate courses at City University and the universities of Leicester, London, and Kingston, allied to the British-led multinational

UNIGIS consortium, have had international as well as national ramifications. The particular nature of the education provided has mutated over time in line with market needs and concepts of the changing nature of GIS (see below); Forer and Unwin (1999) have reviewed the situation in the late 1990s and postulated newly emerging differentiated strands of GIS education. As already indicated above, Longley et al. (2001, 381–3) encapsulate this thinking and also summarise the developments in professional accreditation of GIS education.

All of the above is largely focused on the UK and, to some extent, on European developments. As indicated earlier, the UK played an influential role at an early stage in GIS, usually through the influence of a few individuals. For example, Stein Bie (a Norwegian) and Peter Burrough (a British national) gained their doctorates in the University of Oxford working in soil science under Dr Philip Beckett and all three came into contact with Bickmore (see above) and his activities. Both Bie and Burrough then went on to play central roles in the development of GIS in the Netherlands and far beyond. This historical account paints a very different picture to that commonly portrayed in US publications (see, for instance, Foresman 1998), which tend to imply that almost all significant developments in GIS and related fields originated in that continent. This is manifestly not true, even though the contributions of Ian McHarg, Roger Tomlinson, Jack Dangermond, the US Bureau of Census and its mathematicians like James Corbett and Marv White, plus many others, were hugely influential on a global basis. International interaction blossomed from the early 1970s—when less than a dozen UK academics were involved in such matters—through the 1980s and 1990s, when many research connections were made, up to the present day, when it is impossible to define a distinctive national approach to GIS. In part this evolution has come about through the movement of people: British émigrés like Michael Goodchild, Keith Clarke, John Pickles, and David Maguire have played a crucial role in US academic and related GIS developments. A distinctive British approach to geographic information is, however, still apparent and this arises largely from official policy and its influence on central and local government geographical information producers.

Ordnance Survey: a case study of the transition

In many respects Britain's national mapping agency forms a perfect case study to illustrate the admixture of technology effects, policy changes of various kinds, and the growth of commercialism in geographical information and mapping. Since it provides the 'topographic template' or geographical framework on which most other geographical information in Britain has long been collected, what has happened to the Ordnance Survey has ramifications for British geography far beyond topographic mapping.

The Ordnance Survey was essentially a factory producing paper and microfilm maps until the late 1980s (McMaster, 1991; Rhind, 1991). Its staff were largely artisans, recruited from school and retained until retirement: in 1979, only 23 of the 3,500 staff were graduates or had equivalent professional qualifications (HMSO, 1979). Planning was long term: the programme to re-map Britain as a consequence of under-investment in the 1920s and 1930s ran from 1946 until 1983. Within this long-term plan, annual planning was highly detailed and the methods by which work was carried out varied relatively slowly. Until 1974, the organisation's head was a major general in the British Army and military officers filled a number of key posts until 1984. Their task was to clone operations across the whole country and maintain the currency of 230,000 different maps at a wide variety of scales, producing these maps in paper form to assist the administration of the state. Smaller-scale maps were produced for walking and other purposes. Essentially, the Ordnance Survey now provides four services:

- national coverage of maps and data required for emergency purposes;
- mapping to meet the actions defined in various statutes and regulations passed by Parliament;
- products and services that can be sold to customers (including government departments) via commercial partners or directly; and
- expert geographical advice to government (the Director General is the official adviser on survey, mapping, and GIS to the whole of government).

The re-mapping of the whole of Britain was on a metric basis. These maps covered the entirety of the national territory at 1:10,000 scale. More importantly, some 70 per cent of Britain was also mapped in great detail at 1:1,250 and 1:2,500 scales. The national archive of mapping until the early 1970s consisted of lithographically printed copies of these maps—totalling some 40 million 'current maps' at any one time—and was held in one building at Southampton (see Plate 6), along with an incomplete collection of historical maps. In the 1970s the 'current maps' were stored on microfilm. From 1973 onwards, these 230,000 basic-scale[13] maps were converted into digital form, the pace accelerating dramatically in the early 1990s and the task being completed in 1995 (see Fig. 12.3). In the last few years of this process, over 90 per cent of the OS maps being digitised were encoded in the private sector under contract and to OS output specification. The way in which all this was done, however, was dictated by the technology available when the digitising began and by the legitimisation of

[13] The largest-scale or most detailed maps available in a particular area—either 1:1,250, 1:2,500 or 1:10,000-scale original mapping in Great Britain.

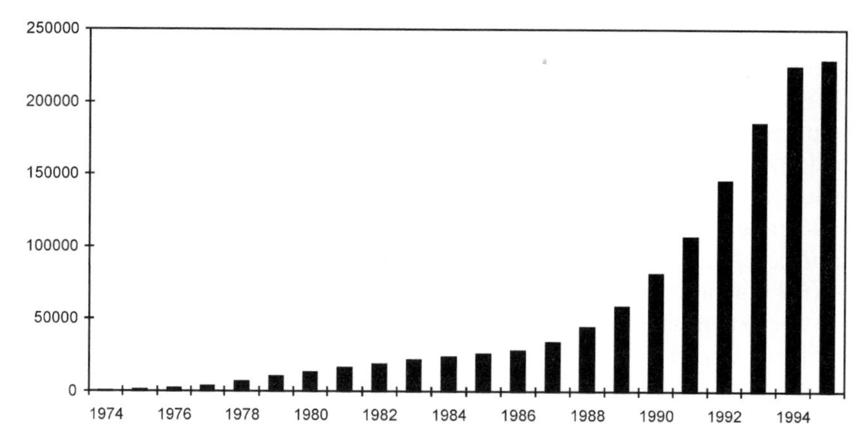

Fig. 12.3 Progress towards completion of digitising of the mapping of Britain: the number of basic-scale maps in computer form at the end of each financial year. *Source*: Rhind (1997).

this as an exercise to cut the cost of updating of the mapping. On this criterion, it was a significant failure for nearly 20 years, until national coverage was achieved and non-traditional mapping products were produced from 1995 onwards. Thereafter, the existence of the national topographic database enabled a rapid change in the Ordnance Survey's financial fortunes, as well as a major change in storage requirements (see Plate 6). Even more striking is the update regime: of the order of 1 per cent of the features on OS large-scale maps change annually and many of these are clustered in urban areas. Certain of these changes are of crucial importance to users such as the utility companies. In the 1960s, map updating was carried out on a cyclic timetable dictated by the Ordnance Survey: the average age of most detailed mapping was 20–30 years. By 1998, the update regime was such that major new features were incorporated in the database within six months of the change occurring and made available in new products within 24 hours later. One other consequence of technology changes was the reduction in staffing: in 1979 (at the time of the Ordnance Survey Review Committee: see p. 440), some 3,500 staff were employed; in 1992 this was 2,400 and by 1998 it had fallen to 1,880, including the arrival of 300 new staff with different skills and educational qualifications (see below).

The government policy of minimising costs to the taxpayer was, so far as the Ordnance Survey was concerned, achieved through generating revenues from the sale or leasing of information and the sale of services. This was facilitated through the creation of partnerships with a wide variety of private-sector bodies (e.g. to create and market a complete set of historical maps of Britain). Until 1992, the only such partnerships were in relation to a few paper atlases, but the numbers of partners and licensees rose rapidly thereafter and, in parallel, the Ordnance Survey's revenues from digitally held information rose to surpass that from traditional products. The need to protect investment and minimise arbitrage led the Ordnance Survey to take legal action against those whom it saw as infringing its copyright. The most notable case of this kind was that launched in 1995 against the Automobile Association: this involved substantial forensic cartography but was rewarded in 2001 by the payment of £20 million to the Ordnance Survey.

One aspect of the Ordnance Survey's progress in implementing this approach is illustrated in Table 12.3. The Ordnance Survey increased its revenue in 1997/8 to 150 per cent of its 1992/3 equivalent, despite falling prices for many products, and cut its costs in real terms over the previous period. Apart from paying staff salaries, much of the additional revenues was ploughed back into enhancement of services, creation of new products, and purchasing of goods and services from the private sector. In 1997/8, some 97 per cent of all costs were met from sales of goods and services to users; cost recovery levels achieved from sales of goods and services over a longer time period are shown in Fig. 12.2. All these figures are calculated on a commercial basis (i.e. they include interest on capital, accommodation charges, etc.). The Ordnance Survey skills mix is now very different to that of 20 years ago: the number of cartographers and surveyors has shrunk considerably but the number of staff with business and IT skills has multiplied.

The results were not, however, achieved through a reduction in product portfolio: indeed, they occurred whilst the Ordnance Survey was simultaneously providing a much greater range of geographical information products (including a national address database), completing digital coverage in the National Topographic Database, and enhancing the speed of delivery and quality of its goods and services. Central to this achievement was the separation of information and its representation in paper map form highlighted earlier. Following detailed scrutiny and user assessments, the Ordnance Survey was awarded a Charter Mark by the Prime Minister

Table 12.3 Key Ordnance Survey financial statistics over five financial years

	1992/3	*1997/8*
Total revenue (£m)	49.9	74.8
Total costs (£m)	74.8*	77.0
Cost recovery (%)	65.0	97.0

* Excludes exceptional item for voluntary early retirement programme.
Source: Ordnance Survey Annual Reports.

in 1993 in recognition of its 'excellence in the delivery of public services'; this was renewed in 1996 and in 2000.

The Ordnance Survey story in recent years has thus been one of some success. In a period of some turmoil, it has produced the first complete and detailed digital geographic framework in the world and its products have been adopted by every local government, all utility companies, most of central government, and many other organisations in Britain. The mapping is now much more up to date than ever before, thanks to imaginative use of the new technology, and many new products have been produced. It has pioneered public–private partnerships and established Service Level Agreements to provide bulk supply of information to particular groups tailored to their needs and purchasing power (the most difficult of which to negotiate was with academia, because of its diffuse power structure and the modest status of geographers therein across the country). As a result of Ordnance Survey efforts—aided by its consultative committees on which many British geographers sat—government ministers have recognised that not all activities can be run commercially and, through the National Interest in Mapping Service Agreement, agreed to pay for part of the cost of maintaining the framework in rural areas. Aside from that, the funding to do all this has largely been provided by customers, rather than parliamentary vote; as a consequence, the Ordnance Survey could plan its activities somewhat more freely than if it had been wholly dependent on the public purse at a time of substantial cuts.

But the policy drivers have had some downsides: Longley et al. (2001, 371–3) have set out these as well as the up-sides. They include the difficulty of pricing information in such a way that impecunious individual organisations like charities can exploit large-area 'map coverage'—a situation much simplified by the advent of the world wide web and high bandwidth, which enabled selling information by user-selected area as well as by theme. In addition, the structure and policies of British government have rendered very difficult the generation in the UK of a strong collaborative approach to the creation of a National Spatial Data Infrastructure such as that which (it is claimed) is now being forged in around 40 other countries (see Longley et al., 2001, 416). This was pioneered in the United States and fostered strongly by creation of an executive order by President Clinton and much continuing support by Vice-President Gore. In the UK, the only significant continuing manifestation of much work on the National Geospatial Data Framework is the metadata service (i.e. in effect a catalogue of government and other geographical information).[14]

[14] see http://www.agi.org.uk/public/gis-resources/index.htm

Reactions to technological imperialism

Thus far, much of the discussion has centred around either the impact of the new technologies or of policy frameworks set out by government and their influence on geographical information and GIS. But two other significant and rather different developments have occurred in recent years, both fired up by academic geographers in response to their perceptions of the ubiquity and shortcomings of existing technology. These are the growth of geocomputation and the antipathy to what is seen as the technological determinism and imperialism of standard GIS.

GIS has over the last 30 years evolved into a collection of tools that may be used in combination to assemble, check, analyse, and map much geographical information. Typically the analytical dimension has been relatively weak, though statistical analysis tools for spatial data are now widely available. The advent of geocomputation has, however, begun to address this long-standing weakness. It has been 'caricatured as uninformed pattern-seeking empiricism in the absence of clear theoretical guidance' (Longley et al., 1998, 8). Some important and highly innovative developments certainly made a virtue of avoiding preconceptions and theory by massive 'data dredging' (such as two decades of work by Openshaw and colleagues, perhaps best epitomised by Openshaw and Openshaw, 1997, and Openshaw, 1998). But others have argued that there is potentially much more to it than that, with the computer being an integral component of the modelling of complex systems and processes—rather than simply a solver of equations and a data handler—allied (potentially) to a sound theoretical base (Couclelis, 1998). Examples of early work in this putative discipline include that on remote (internet) browsing and interaction as avatars (Batty et al., 1998) and tracing and modelling the geography of disease transmission (Cliff and Haggett, 1998).

Perhaps rather more fundamental, however, is what has emerged from a decade-long period of fevered debate from the late 1980s. The earliest British manifestation of this was probably Taylor (1990), who saw GIS as the new imperialist geography, the emergence of which represented the positivists' revenge (to the demise of quantitative geography of the 1960s and 1970s) but which represented 'a return of the very worst kind of positivism, a most naïve empiricism' (pp. 211–12). This view was roundly rejected by Openshaw (1991, 1992), who saw it as based on reductionist devices and consisting solely of derogatory and confrontational language. Others saw the GIS developments as lacking in any intellectual challenge, simply a field for technicians. Behind all this collection of misunderstandings and posturings, however, was a more substantive issue. Pickles (1999) has summarised this concern clearly. From his summary, we may draw out at least three specific matters:

- concerns about the social engineering that GIS and geographical information make possible (e.g. the impact on personal privacy;[15] Curry 1999);
- the alleged lack of understanding of the societal context within which decision-making takes place and its replacement in a GIS world with a particular methodology and set of value systems. Johnston (1999) has also criticised GIS proponents for their lack of understanding of the multiple viewpoints in contemporary geography, such as the differing commitments of empiricist, hermeneutic or critical epistemologies; and
- the (apparently malevolent) role of the large institutional actors, notably the military and commercial enterprises, in steering the development of GIS.

Some of this seems untenable. For instance, it appears to assume a simplistic homogeneity of purpose and actions on the part of the leading actors. From personal experience within government departments and commercial enterprises, this is wholly unreasonable: what emerge as desired ends are often an almost chance product of internal or inter-organisational rivalry, partial information or lack of communications. It also seems posited on a view of GIS and geographical information that is a product of linear development, almost pre-ordained: on the contrary, the vast numbers of systems that have failed[16] testifies to the current situation being a matter of both Darwinian and serendipitous development. Moreover, not all key players take a utilitarian or short-term self-interested view: the most globally successful commercial GIS organisation (the Environmental Systems Research Institute) is one whose founder has always been a strong environmentalist, who has encouraged pluralist keynote views at his 11,000-attendee annual conference, and who has

[15] But much of this has little to do with geography as such if one accepts the *Economist* interpretation of 1 May 1999: 'Remember, they are always watching you. Use cash when you can. Do not give your phone number, social security number or address, unless you absolutely have to. Do not fill in questionnaires or respond to telemarketeers. Demand that credit and data marketing firms produce all information they have on you, correct errors and remove yourself from marketing lists. Check your medical records often. If you suspect a government agency has a file upon you, demand to see it. Block caller ID on your phone, and keep your number unlisted. Never use electronic toll-booths on roads. Never leave your mobile phone on—your movements can be traced. Do not use store credit or discount cards. If you must use the Internet, encrypt your email, reject all "cookies" and never give your real name when registering at web sites. Better still, use someone else's computer. At work, assume that calls, voice mail, email and computer use are all monitored.

. . . Anyone who took these precautions would merely be seeking a level of privacy available to all 20 years ago. . . . Yet . . . all these efforts to hold back the rising tide of electronic invasion will fail . . . Faced with the prospect of its [privacy] loss, many might prefer to eschew even the huge benefits that the new information economy promises. But they will not, in practice, be offered that choice.'

[16] Agreed that few of these have been written up, but see Openshaw et al. (1990).

supported education at all levels for 30 years. Even military developments have been made freely available to the civilian populace (notably the GPS signals). It is not, however, surprising that some academic critics have formulated the views they have. Understanding the dynamics of the evolution of GIS and geographical information and the rationale for many developments, and the transition generally, is almost impossible from any study of the literature because so little was written down at the time other than the success stories, usually composed long after the event and for purposes other than formal audit.

Yet, even if ill informed in part, such critiques and the subsequent debates have been almost entirely healthy ones in exposing apparently opposed views and leading to constructive dialogue. A particularly useful result was the creation of a new initiative (I 19) in the US National Center for Geographic Information and Analysis.[17] This has treated GIS as a set of social practices and institutions embedded in a particular discourse. Its aim was to create a theory of GIS and society in which 'description (of the development of particular logics, systems and uses of GIS), analysis (of the limits of access, range of diffusion, and effects of use) and critique (focused on the epistemological assumptions embedded in systems and use, conceptions of language in use, and logics and representations) are all present' (Pickles, 1999, 54; see also Schuurman, 2001).

Somewhat earlier, a number of individuals (e.g. Openshaw, 1991, 1992; Goodchild, 1992) began espousing the need for a mutation of GIS as Geographical Information Systems into Geographic Information Science. Initially this was conceptualised as something akin to a (relatively 'hard') science of space, but since the debate catalysed by Pickles's (1995) edited volume and the growing degree of constructive debate, this has widened considerably in scope (see for instance Goodchild, 2000; Pickles, 2000; Longley et al., 2001). Indeed the nature of this science is determinedly multidisciplinary: the first Geographical Information Science conference in 2000 deliberately sought (and obtained) those working in cognitive science, computer science, engineering, geography, information science, mathematics, philosophy, psychology, various social sciences, and statistics. Thus far, British geographers have played only a modest part in this expanded exploration of GIS and geographical information, but at least UK émigrés have been leaders in this development.

Conclusions

What is intended to be obvious from this inevitably superficial account is that British geography, in so far as it is manifested in maps, mapping,

[17] See http://www.geo.wvu.edu/i19/

geographical information systems, and geographic information, has evolved under many forces in the last 30 or so years. These forces include technological change, the evolution (including the separation and partial reharmonisation) of government policies, the commercialisation of many aspects of the field, globalisation effects of various kinds, and the role of individual personalities. This element of British geography has been claimed by technologists, accepted as part of the everyday world by much of the populace at large and by those in many other disciplines—to whom it often represents all they know of geography—and critiqued fiercely by some academic geographers and ignored by others. What seems clear at the time of writing, however, is that a GIS approach has permeated much of the workings of society—but also that the value systems and judgements that it embraces are being reviewed in a rather constructive way by geographers of many different persuasions. From this we are in the throes of building a new Geographic Information Science.

Acknowledgements

Some of the material in this chapter is drawn from previously published work by the author. Grateful thanks are due to Prentice-Hall and to Wiley and Sons for permission to reproduce it, albeit in updated form.

References

Atkinson, P. M. and Tate, N. J. (1999) *Advances in Remote Sensing and GIS Analysis*. Chichester: Wiley.

Baigent, E. (1990) Swedish cadastral mapping 1628–1700: a neglected legacy. *Geographical Journal*, 156(1), 62–9.

Batty, M., Dodge, M., Doyle, S., and Smith, M. (1998) Modelling virtual environments. In P. Longley et al., eds, *Geocomputation: A Primer*. Chichester: Wiley, 139–62.

Baxter, R. S. (1976) *Computer and Statistical Techniques for Planners*. Chichester: Wiley.

Bickmore, D. P. and Boyle, R. (1964) An automated system of cartography. In *Proceedings of the Technical Symposium of the International Cartographic Association*. Edinburgh.

Birkin, M., Clarke, G. P., Clarke, M., and Wilson, A. G. (1996) *Intelligent GIS: Location Decisions and Strategic Planning*. Cambridge: GeoInformation International.

Burrough, P. A. and McDonnell, R. A. (1998) *Principles of Geographical Information Systems*, 2nd edn. Oxford: Oxford University Press.

Census Research Unit, Office of Population Censuses and Surveys, and General Register Office (Scotland) (1980) *People in Britain*. London: HMSO.

Chrisman, N. (1998) Academic origins of GIS. In T. Foresman, ed., *The History of GIS*. Upper Saddle River, NJ: Prentice-Hall, 33–44.

Cliff, A. D. and Haggett, P. (1998) On complex geographical space: computing frameworks for spatial diffusion processes. In P. Longley et al., eds, *Geocomputation: A Primer*. Chichester: Wiley, 231–56.

CMG (2001) *Quinquennial Review of Ordnance Survey: Stage 1 Report*. Camberley: CMG Admiral

Cooke, R. U. and Harris, D. R. (1970) Remote sensing of the terrestrial environment: principles and progress. *Transactions, Institute of British Geographers*, 50, 1–27.

Coopers and Lybrand (1996) *Economic Aspects of the Collection, Dissemination and Integration of Government's Geospatial Information*. Southampton: Ordnance Survey.

Coppock, J. T. (1962) Electronic data processing in geographical research. *Professional Geographer*, 14, 1–4.

Coppock, J. T. and Rhind, D. W. (1991) The history of GIS. In D. J. Maguire, M. F. Goodchild, and D. W. Rhind, eds, *Geographical Information Systems*. Harlow: Longman, 21–43.

Couclelis, H. (1998) Geocomputation in context. In P. Longley et al., eds, *Geocomputation: A Primer*. Chichester: Wiley, 17–29.

Curran, P. J., Milton, E. J., Atkinson, P. M., and Foody, G. M. (1998) Remote sensing: from data to understanding. In P. Longley et al., eds, *Geocomputation: A Primer*. Chichester: Wiley, 33–59.

Curry, M. R. (1999) Rethinking privacy in a geocoded world. In P. A. Longley et al., eds, *Geographical Information Systems: Principles, Techniques, Management and Applications*, 2nd edn. New York: Wiley, 757–66.

Dale, P. F. and McLaughlin, J. D. (1988) *Land Information Management*. Oxford: Oxford University Press.

Department of the Environment (1972) *General Information System for Planning (GISP)*. London: Department of the Environment, Research Report 1.

Department of the Environment (1979) *Report of the Ordnance Survey Review Committee* (Chairman Sir David Serpell). London: HMSO.

Department of the Environment (1987) *The Report of the Government Committee of Enquiry on the Handling of Geographic Information*. London: Department of the Environment/HMSO.

Dorling, D. (1995) *A New Social Atlas of Britain*. Chichester: Wiley.

Experimental Cartography Unit (1971) *Automatic Cartography and Planning*. London: Architectural Press.

Forer, P. and Unwin, D. (1999) Enabling progress in GIS. and education. In P. A. Longley et al., eds, *Geographical Information Systems: Principles, Techniques, Management and Applications*, 2nd edn. New York: Wiley, 747–56.

Foresman, T. W., ed. (1998) *The History of GIS*. Upper Saddle River, NJ: Prentice-Hall.

Foster, C. D. and Plowden, F. J. (1996) *The State under Stress*. Buckingham: Open University Press.

Geographical Algorithms Group (1977) *Feasibility and Design Study for a Computer Algorithms Library*. London: Geographical Algorithms Group, University College London.

Gaits, G. (1969) Thematic mapping by computer. *Cartographic Journal*, 6, 50–8.

Goodchild, M. F. (1988) Stepping over the line: technological constraints and the new cartography. *American Cartographer*, 15, 311–22.

Goodchild, M. F. (1992) Geographical Information Science. *International Journal of Geographical Information Systems*, 6, 31–45

Goodchild, M. F. (2000) Cartographic futures on a digital earth. *Cartographic Perspectives*, 36, 3–11.

Grimshaw, D. J. (2000) *Bringing Geographical Information Systems into Business*, 2nd edn. Chichester: Wiley.

Harley, B. (1989) Deconstructing the map. *Cartographica*, 26(2), 1–20.

Harley, B. (1990) Cartography, ethics and social behavior. *Cartographica*, 27(2), 1–23

Hearnshaw, H. M. and Unwin, D. J., eds (1994) *Visualisation in Geographical Information Systems*. Chichester: Wiley.

Heywood, I., ed. (1997) *Beyond Chorley: Current Geographic Information Issues*. London: Association for Geographic Information.

Heywood, I., Cornelius, S., and Carver, S. (1998) *An Introduction to Geographical Information Systems*. Harlow: Longman.

HMSO (1979) *Report of the Ordnance Survey Review Committee*. London: HMSO.

Johnston, R. (1999) Geography and GIS. In P. A. Longley et al., eds, *Geographical Information Systems: Principles, Techniques, Management and Applications*, 2nd edn. New York: Wiley, 39–47.

Jones, C. (1997) *Geographical Information Systems and Computer Cartography*. London: Academic Press.

Longley, P., Brooks, S., McDonnell, R., and MacMillan, W., eds (1998) *Geocomputation: A Primer*. Chichester: Wiley.

Longley, P., Goodchild, M., Maguire, D., and Rhind, D., eds (1999) *Geographical Information Systems: Principles, Techniques, Applications and Management*, 2nd edn. Wiley: New York.

Longley, P., Goodchild, M., Maguire, D., and Rhind, D. (2001) *Geographic Information Systems and Science*. New York: Wiley.

Mapping Science Committee (1993) *Toward a Co-ordinated Spatial Data Infrastructure for the Nation*. Washington, DC: Mapping Science Committee, National Research Council, National Academy Press.

Mapping Science Committee (1994) *Promoting the National Spatial Data Infrastructure Through Partnerships*. Washington, DC: Mapping Science Committee, National Research Council, National Academy Press.

Mapping Science Committee (1997) *The Future of Spatial Data and Society*, Washington, DC: Mapping Science Committee, National Research Council, National Academy Press.

Martin, D. J. (1996) *Geographic Information Systems: Socio-Economic Applications*, 2nd edn. London: Routledge.

Masser, I. (1998) *Governments and Geographic Information*. London: Taylor and Francis.

McMaster, P. (1991) The Ordnance Survey: 200 years of mapping and on. *Journal of the Royal Society of Arts*, 139(5421), 581–93.

Monmonier, M. (1996) *How to Lie with Maps*, 2nd edn. Chicago, IL: University of Chicago Press.

National Academy for Public Administration (1998) *Geographic Information for the 21st Century: Building a Strategy for the Nation*. Washington, DC: NAPA.

Nordbeck, S. and Rystedt, B. (1972) *Computer Cartography*. Lund: Studentlitterateur.

Openshaw, S. (1991) A view on the GIS crisis in geography, or using GIS to put Humpty Dumpty back together again. *Environment and Planning A*, 23, 621–8

Openshaw S. (1992) Further thoughts on geography and GIS: a reply. *Environment and Planning A*, 24, 463–6.

Openshaw, S. (1998) Building automated Geographical Analysis and Explanation Machines. In P. Longley et al., eds, *Geocomputation: A Primer*. Chichester: Wiley, 95–116.

Openshaw, S. and Goddard, J. B. (1987) Some implications of the commodification of information and the emerging information economy for applied geographical analysis in the United Kingdom. *Environment and Planning A*, 19, 1423–39.

Openshaw, S. and Openshaw, C. (1997) *Artificial Intelligence in Geography*. Chichester: Wiley.

Openshaw, S., Cross, A., Charlton, M., Brunsden, C., and Lillie, J. (1990) Lessons learned from a post-mortem of a failed GIS. In *Proceedings of the Association for Geographic Information Conference*, Rickmansworth, Westrade Fairs, 2.3.1–2.3.5.

Ordnance Survey (1990) *Ordnance Survey Framework Document*. Southampton: Ordnance Survey.

Ordnance Survey (1995) *Ordnance Survey Framework Document*. Southampton: Ordnance Survey.

Ordnance Survey (1996) *Results of the Consultation Exercise on the 'National Interest in Mapping'*. Southampton: Ordnance Survey.

Ordnance Survey (1999) *Ordnance Survey Framework Document*. Southampton: Ordnance Survey

Ordnance Survey (2000) *Coordinate Positioning: Ordnance Survey Policy and Strategy*. Southampton: Ordnance Survey, Information Paper 1/2000.

Osborne, D. and Gaebler, T. (1992) *Re-inventing Government*. Reading, MA: Addison-Wesley.

Owen, T. and Pilbeam, E. (1992) *Ordnance Survey: Map-makers to Britain since 1791*. London: HMSO.

OXERA (1999) *The Economic Contribution of Ordnance Survey GB*. Oxford: Oxford Economic Research Associates (http://www/ordsvy/gov.uk/literatu/external.oxera99).

Perring, F. H. and Walters, S. M. (1958) Plant atlas of Great Britain. *Times Science Review*, 13.

Pickles, J. (1995) *Ground Truth: The Social Implications of GIS*. New York: Guilford Press.

Pickles, J. (1999) Arguments, debates and dialogues: the GIS–social theory debate and the concern for alternatives. In P. A. Longley et al., eds, *Geographical Information Systems: Principles, Techniques, Management and Applications*, 2nd edn. New York: Wiley, 49–60.

Pickles J. (2000) Cartography, digital transitions and questions of history. *Cartographic Perspectives*, 37, 4–18.

Raper, J. (2000) *Multidimensional Geographic Information Science*. London: Taylor and Francis.

Raper, J., Dykes, J., Wood, J., Mountain, D., Krause, K., and Rhind, D. (2002) A framework for evaluating geographical information. *Journal of Information Science*, 28(1), 51–62.

Ratia, J. (1999) Introduction by the chairman—Session 3: The policy framework. In *Proceedings of the Cambridge Conference of National Mapping Organisations*. Southampton: Ordnance Survey, Paper 3.0.

Reeve, D. E. and Petch, J. R. (1999) *GIS Organisations and People: A Socio-technic Approach*. London: Taylor and Francis.

Rhind, D. W. (1971) The production of a multi-colour geological map by automated means. *Nachrichten aus den Karten und Vermessungeswesen*, 52, 47–52.

Rhind, D. W. (1986) Remote sensing, digital mapping and Geographical Information Systems : the creation of government policy in the UK. *Environment and Planning C: Government and Policy*, 4, 91–102

Rhind, D. W. (1988) Personality as a factor in the development of a new discipline: the case of computer-assisted cartography. *American Cartographer*, 15(3), 277–89.

Rhind, D. W. (1991) The future role of Ordnance Survey. *Cartographic Journal*, 28(2), 188–99.

Rhind, D. W. (1994) Ordnance Survey and the history of the British landscape. In C. Board and P. Lawrence, eds, *Recording Our Changing Landscape*. London: British Academy and the Royal Society, 23–31.

Rhind, D. W. (1997) Facing the challenges: redesigning and rebuilding Ordnance Survey. In D. Rhind, *Framework for the World*, Cambridge: GeoInformation International (now handled by Wiley, Chichester and New York), 275–304.

Rhind, D. W. (1998) The incubation of GIS in Europe. In T. Foresman, ed., *The History of GIS*. Upper Saddle River, NJ: Prentice-Hall, 293–306.

Rhind, D. W. (1999) National and international geospatial data policies. In P. A. Longley et al., eds, *Geographical Information Systems: Principles, Techniques, Management and Applications*, 2nd edn. New York: Wiley, 767–87.

Rhind, D. W., Armstrong, P. A., and Openshaw, S. (1988) The Domesday machine : a nation-wide Geographical Information System. *Geographical Journal*, 154(1), 56–68.

Rhind, D. W. and Mounsey, H. M. (1989) The Chorley Committee and 'Handling Geographic Information'. *Environment and Planning A*, 21, 571–85.

Sawyer, J. S. (1960) Graphical output from computers and the production of numerically forecast or analysed synoptic charts. *Meteorological Magazine*, 89, 187–90.

Schuurman, N. (2001) Critical GIS: theorizing an emerging science. *Cartographica*, 53, 3–4.

Seymour, W. A. (1980) *A History of the Ordnance Survey*. Folkestone: Dawson.

Simpson, S. N. (1954) Least squares polynomial fitting to gravitation data and density plotting by digital computers. *Geophysics*, 19, 250–7.

Smith, M. C. (1996) *Rose*. New York: Ballentine Books.

Taylor, P. J. (1990) Editorial comment: GKS. *Political Geography Quarterly*, 9, 211–12.

Taylor, P. J. and Overton, M. (1991) Further thoughts on geography and GIS. *Environment and Planning A*, 23, 1087–90.

Thompson, C. N. (1978) Digital mapping in the Ordnance Survey 1968–78. Paper given to International Society of Photogrammetry Commission IV, Inter-Congress Symposium, Ottawa.

Tobler, W. (1976) Analytical cartography. *The American Cartographer*, 3, 21–31.

Tobler, W. (2000) The development of Analytical Cartography. *Cartography and Geographic Information Science*, 27(3), 189–94

Tomlinson, R. (1998) The Canada Geographic Information System. In T. Foresman, ed., *The History of GIS*. Upper Saddle River, NJ: Prentice-Hall, 21–32.

Unwin, D. (1991) The academic setting of GIS. In D. J. Maguire, M. F. Goodchild, and D. W. Rhind, eds, *Geographical Information Systems: Principles and Applications*. Harlow: Longman, 81–93.

Waugh, T. C. (1980) The development of the GIMMS computer mapping system. In D. R. F. Taylor, ed., *The Computer in Contemporary Cartography*. New York: Wiley, 219–34.

Wood, D. (1993) *The Power of Maps*. London: Routledge.

Worboys, M. F. (1995) *GIS: A Computing Perspective*. London: Taylor and Francis.

13

Geography applied

ROBERT J. BENNETT
ALAN G. WILSON

Introduction: charting the territory

Geography is by its nature *applied*. Geographers' systems of interest are of such direct concern to people, businesses, government, other organisations, and NGOs that any understanding achieved within the discipline is likely to be applicable. Since our understanding of geographical systems is encapsulated by the theory within the discipline, to survey applied geography we need to survey the development of geographical understanding, the theories that constitute geographical knowledge, how they have been applied, and developments in the broader realm of the relationships between scholarship and the users of knowledge.

The development of geography as an applied discipline spans many areas and interfaces with many other disciplines. Here we can only give an overview of the main trends and the most prominent focuses of research. Our subject matter is potentially vast. We restrict it in part by focusing on the contributions of British geographers, but draw in the lines of international influence where these are particularly relevant. We will primarily emphasise the applied developments in human geography, but it must be borne in mind that geography has developed using multidisciplinary perspectives from physical science as well as from the social sciences and humanities. The development of physical geography and earth sciences has been particularly influential on the development of applied geography at various stages, and we draw in discussion of these interrelationships where they are relevant.

We must also be explicit about what we treat as 'applied' geography. Much of the debate within geography itself about 'applications' of the discipline has tended to focus on contributions to public policy. We shall argue, however, that this debate has been to some extent falsely posed. The public policy of the state and its agencies is not the only focus of concern for applied geographical research. Also of crucial importance are the contributions of geography to understanding and decision-making in areas of the broader market dynamics of regions, location and business trading decisions, and spatial aspects of voluntary and other agencies. These aspects overlap with many other chapters in this volume. This means that if we focus on areas covered by our own expertise, we do so conscious that this presentation is complemented by other authors. A particular example of geographical application *par excellence* is the presentation of geographical data, particularly as maps, and, more recently, through geographical information systems. These topics are covered in the preceding two chapters and so here we focus on theory-based application, drawing out the general lines of applied work and the debates that have surrounded them. We first review the developments of the early part of the twentieth century and how the debate about 'applied' geography has developed. We then move on to examine the explosion in the volume and diversity of applied studies that occurred from the 1960s, concluding with an assessment of the challenges facing the discipline in the twenty-first century.

Early themes and disciplinary exegesis

An early broad conception of applied geography, termed 'applicable' geography, was propounded by Keltie (1908) as a discipline concerned with 'the intimate bearings of geographical conditions on collective humanity, on man in his striving after political, social and industrial development' (p. 4) and 'bearing on human interests, history, industry and commerce' (p. 3). This conception contained the possibility of broad contributions of an economic, social, and humanistic kind, as well as utilitarian public policy. A similarly broad call to arms derived from Mackinder (1919), who had a profound influence on much scholarship during the first half of the twentieth century (see Wise, 1973). In 1921, Stevens published a wide ranging book entitled *Applied Geography*, which had a very broad approach, including trade and commerce. However, neither Keltie nor Mackinder fully fleshed out their interpretations of applied geography to provide a blueprint for its development and Stevens's book appears to have been overlooked. The result has been that much of the breadth of these early conceptions became lost as individual scholars pursued applications under other labels.

There are many aspects of applied geography other than those elements that have been explicitly labelled as applied. For example, one of the most pervasive was Chisholm's (1889) *Commercial Geography*, which went through 20 editions up to 1980. Stevens's (1921) book also reflects this aspect. In contrast, the explicit label 'applied geography' became appended, chiefly from the 1930s, to a narrower range of geography mainly in the fields of regional planning, local land use planning, and a few other specific fields. As noted by Bennett (1981), the wider economic, social, and political elements of 'applied geography' in Keltie's conception of an applicable discipline were largely forgotten by those who debated applied geography. Even the founding editor of the journal *Applied Geography*, David Briggs (1981), argued that applied work had become dominated by a narrow conceptual view of land use planning, with little conceptual progress evident since the 1930s; he quotes in support of this view the contributions to the volumes by the Polish Academy of Sciences (1959), Beaver and Kosinski (1964), and the Institute of British Geographers (1968).

It would of course be inaccurate either to overemphasise the influence of polemical calls to arms by leading geographers, or to characterise the discipline as dominated by a single coherent structure. Early research on applied geography was, with a few exceptions, such as Stamp (1948, 1960, 1966), chiefly conducted on a limited scale, undertaken by a few individuals often in isolation from each other and from other parts of the discipline. In common with other subjects up to the 1920s or even 1950s, academic research geographers were small in number, they were frequently generalists, departments often consisted of a single lecturer or professor, team working was rare, and total staff numbers were meagre, with membership of the main academic society (the Institute of British Geographers) below 100 throughout the 1930s (see Wise, 1973; and Chapter 2, above). Indeed in the context of urban studies and planning, Hall (Chapter 16 in this volume) shows that, except for Geddes at the start of the twentieth century, the main research explicitly termed 'applied' began in the 1930s with the influence of Dudley Stamp's land use mapping. Indeed, in Briggs's (1981) review, Stamp is called the 'father' of 'applied geography'.

In our treatment here we first present these explicit contributions to 'applied geography', focusing on regional planning and policy, town and country planning, and other specific fields. More detailed reviews of this history are given elsewhere (e.g. House, 1973; Wise, 1973; Sheail, 1994). We then move on to discuss the debate about 'applied geography' that emerged in the 1970s, which leads into our review of more recent work in the rest of the chapter.

Regional planning and policy

Regional planning and policy, discussed at greater length by Ray Hudson (in Chapter 18, below), was born from the extremes of the 1920s depression. In the 1934 Special Areas Act, South Wales, Northeast England, West Cumberland, and Clydeside–Lanark were the chief focus of special measures providing government grants and aid. Although seen as temporary, eligibility for government aid was progressively extended to cover Dundee, Inverness, Northern Ireland, and Wigan and Northeast

Lancashire by 1945; and in the 'high period' of regional policy in the 1960s, more than half of the area of UK was eligible for special aid in two tiers of Development Areas and Special Development Areas (McCrone, 1969). The Barlow Committee (Royal Commission on the Distribution of the Industrial Population, 1940), which framed the post-war development of regional policy/planning, was influenced by, and was influential upon, the discipline's thinking at an institutional level. Eva Taylor (1938), both in her own contributions and as chair of a Royal Geographical Society (1938) committee that produced a response to consultations of the Barlow Committee, was a strong advocate and influence. Together with Stamp, she sought a more coherent approach by the discipline to opportunities for involvement with public policy (see review by Willatts, 1971, and Hebbert and Garside, 1989). However, the RGS's 1938 response failed to grasp the nettle, leaving open the answer to the difficult question of whether industrial development should follow old distribution patterns or adjust to the new.

Regional planning became embodied in national planning in the 1960s as part of the work of the National Economic Development Council (NEDC), the 1965 *National Plan* (Secretary of State for Economic Affairs, 1965), and the short-lived central government Department of Economic Affairs (DEA) in 1964–9. Within the DEA was a Regional Planning Group, which established Regional Economic Planning Boards (of civil servants) and Regional Economic Planning Councils (of local government and other institutions). The DEA, together with its separate focus for economic policy, was abolished in 1969, its functions returning to the Treasury. Only since 1997 has there been a re-emergence of debate about a separate focus for economic policy, as a result of Scottish and Welsh devolution and the establishment of Regional Development Agencies in England. However, the Regional Economic Planning Boards and Regional Economic Planning Councils did survive until abolished by Margaret Thatcher's Conservative government in 1979, and regional policy has continued in the form of Regional Selective Assistance up to the present day.

Many leading geographers were involved in development of this process, both in its advocacy and in its implementation: for example, Daysh (1949), Daysh and Symonds (1953), Smith (1949), Estall and Buchanan (1961), Wise (1963, 1965), and Caesar (1964) (see also reviews by Powell, 1970 and the economist Dennison, 1939). Later examples were Chisholm and Manners (1971), Keeble (1976), Moseley (1974), and contributors to *Regional Science and Urban Economics* (1976). Many geographers developed professional careers in the central ministries and the Regional Economic Planning Boards concerned with implementing regional planning. There is no full documentation of the geographer membership of Regional Economic Planning Councils, but at the least the following served at one time (see survey in House, 1973): Peter Haggett (South West),

David Keeble (East Anglia), Peter Hall and Gerald Manners (South East), Kenneth Edwards (East Midlands), John House and John Goddard (North). The work of geographers such as Peter Hall and Derek Diamond on the southeast and other early regional studies (Ministry of Housing and Local Government, 1963a, 1963b, 1964) was particularly influential (see Caesar, 1964; Powell, 1970; Willatts, 1971; House, 1973; Coppock, 1974; see also Hall, Chapter 16, below).

Land use planning

The early emphasis on land use studies in applied geography followed directly from the influence of Dudley Stamp's launch of the Land Utilisation Survey in 1930 (see Stamp, 1948), and this formed a large part of his own later text on *Applied Geography* (Stamp, 1960; previewed in Stamp, 1951). The opportunity for this to influence public policy was provided by the steady advance of planning legislation, begun in 1909, but developed in 1932 and especially in the 1947 Town and Country Planning Act. With the implementation by local authorities of the planning control regimes after 1947, a very large number of geographers were recruited into the planning profession from the 1950s through to the early 1970s.

Broader developments of agricultural and land use surveys were also made, examples including Coppock's (1964) *Agricultural Atlas* and the work of Best (1981). Although a second land utilisation survey instigated by Alice Coleman in the early 1960s was not completed, the Geographical Association managed a national coverage in 1996 through cooperation with schools (Walford, 1997). There was also significant use of Stamp's land use survey in the wartime 'plough-up' campaign and the work of the County War Agriculture Executive Committees (see e.g. Sheail, 1994), with Stamp acting as chief adviser on rural land use from 1942 to 1945 and as vice chair of the Scott Committee on Land Utilisation in Rural Areas throughout the war years (Stamp, 1943).

Sheail (1994) records the development of Stamp's influence on land use policy, and through the reports of the Scott and Barlow committees. A critical date was 1965, when the Ministry of Land and National Resources was set up, with a National Resources Advisory Committee, of which Stamp was chair. However, following the 1966 general election, the ministry was abolished and the advisory committee wound up. More long-lived was the Office of Science and Technology, established in 1965, together with the Natural Environment Research Council (NERC). It was apparently a 'grievous blow' to Stamp that the advisory committee was wound up and that he was not appointed to NERC (Sheail, 1994). But Kenneth Hare, Professor of Geography at King's College London from 1964 to 1968, was appointed (apparently chiefly as a meteorologist), and set up a Land Use Research Working Group with NERC members.

However, by 1967 NERC had developed in a different direction and had no specific land use focus, and in 1968 Hare moved to Canada. Stamp's call to see land use as the core of geography's public policy contribution, and a key way in which geographers could develop applied careers, then fell into the background; Stamp himself died in 1966. However, his influence perhaps found its strongest take-up in the planning system.

Willatts (1971) and Hebbert and Garside (1989) chronicle how the role of Dudley Stamp and Eva Taylor in regional policy also strongly influenced early planning surveys, such as those in Herefordshire in 1946 (Buchanan, 1947), Birmingham and the Black Country in 1948 (Buchanan and MacPherson, 1948), and the seven regional studies reported in Daysh (1949), of which the most widely cited is that by Daysh and Caesar (1949) for the northeast. Following the Barlow Report, which influenced regional planning, both Stamp and Taylor were members of the ministerial 'Reconstruction Group' that led to the development of the 1947 Act. There had also been precursors of these in local government cooperation studies that had developed sub-regional plans: for example, in the Doncaster area (in 1922) and South Wales (Ministry of Health, 1920; see also Alden and Morgan, 1974, 11–12). From the 1950s, therefore, geographers were at the centre of the development of a radical new policy agenda for planning and its implementation, a role that had earlier been accepted by Abercrombie (1938). As a result, many geographers followed careers in planning—administering the concept of balancing land uses, establishing green belts, and controlling the 'proper size' of settlements (see e.g. Smailes, 1941, 1944; Daysh and O'Dell, 1947; Dickinson, 1947; Best and Coppock, 1962). The considerable developments that have occurred through later work on the relationships between geography and planning are summarised in Diamond's (1995) presidential address to the 1994 conference of the Institute of British Geographers (IBG), and in Hall's chapter in this volume (Chapter 16).

There were also other convergences between regional policy/planning and town and country planning. Daysh (1949, xi) argued for regional planning of 'national needs', and Caesar (1964, 10) argued that because of the long life of the built environment, planning had to think of the future 'for 2064 or later'. Michael Wise, as chair of the influential Landscape Advisory Group, had a major influence on motorway development (Wise, 1963, 1965). Fawcett (1919) had developed the concept of economic regions, which was taken up by Dickinson (1960), Smailes (1944), and Carruthers (1957). Freeman (1958) sought to integrate regional concern with the sensitivity of physical landscape with concern for the human-made landscape, drawing on arguments from central place theory to order the scale of expansion allowed by planning in different cities and towns. Central place theory concepts also influenced the reform of local government, which became a major field for geographical input (e.g. Gilbert, 1939, 1948), both in terms of the 1960–70 reforms (Royal Commission on Local Government, 1969), discussed by Wise (1969) and House (1973), and through the work of Michael Chisholm as member of the Local Government Boundary Commission in the 1970s and 1980s and the Local Government Commission in the 1990s (Chisholm, 2000, 2001). This debate has continued. Later work on local government reform has suggested the need to balance the use of concepts from central place theory (to integrate government into functional regions) with the need to diffuse the tensions created by large units that undermine identity and sense of community. Bennett (1989, 1992), for example, argues for an approach to local government boundaries based on less static notions, which allow flexible aggregation using associations, bilateral agreements, and collectives of local government instead of requiring large-scale amalgamation. This alternative approach has been influential on reforms in Central Europe in the post-communist era since 1990, where the debates about the city region concept are still very much alive.

Other areas of applied geography

It would be unfair to characterise the early and mid-twentieth-century work of geography that was labelled as explicitly 'applied' as concerned only with regional and land use planning. A variety of other individual and group developments were made. Of particular note are the overlaps with physical geography through the works of Steers (1944, 1946) on coasts and wildlife conservation, begun in 1943–4, Balchin's (1958) water use survey, Wooldridge and Beaver's (1950) research on sand and gravel resources, which influenced the Waters Committee (1946–50) and which in turn was heavily used in the post-war rebuilding of British cities and the development of New Towns. The pervasive Dudley Stamp was also chair of the England Committee of the Nature Conservancy (see Stamp, 1969), whilst Wooldridge was one of the founders of the Council for the Promotion of Field Studies.

Broader developments of land use and resources planning were also reflected in Beaver's *Report on Derelict Land in the Black Country* (Ministry of Town and Country Planning, 1946), which was a model for replanning industrial areas, and in policy inputs to reclamation and re-afforestation of iron ore extraction areas in Northamptonshire (Beaver, 1944) and open-cast coal and iron areas (Evans, 1944). This was taken forward in subsequent research on international coal and oil trade by Manners (1964, 1981) and Odell (1963, 1970).

Much of this development was, therefore, also influenced by land use policy. In a similar vein, Stamp's influence as a member of the Royal Commission on Common Lands (1955–8) resulted in legislation to prepare common land registers, but second-round legislation has never followed this up (see Stamp, in Royal Commission on Common Lands, 1958, 185–267; Stamp and Hoskins, 1963). Michael Wise chaired the Departmental Committee of Inquiry into Statutory Small Holdings (1966, 1967), which produced recommendations, many of which were embodied in the 1970 Agriculture Act. House (1973) summarises the contributions of many other geographers to resource and planning boards in the early and middle years of the twentieth century.

The debate about 'applied' geography

In that field of geography explicitly labelled 'applied', the consequence of the emphasis on regional planning and town and country planning had a profound influence on subsequent debates. The sub-discipline of applied geography came to be seen as essentially analysis of, or contribution to, government policies and the development of the state. As stated by Caesar (1964, 1): 'Some measure of agreement on priorities and some coordination of (economic) activities is increasingly necessary—hence the need for planning in some form or another'. For Daysh (1949, ix), 'industry and planning go hand-in-hand'. Whilst other geographers developed many areas of real applications, the explicitly labelled applied sub-discipline tended to ignore wider areas of applications, such as economic geography, as well as Keltie's (1908) or Mackinder's (1919) early broad conceptions, at least at the polemical level.

The Second World War was a major influence on these changes. Wise describes the influence as one that shifted attention to 'working for the nation'; he does not view this as co-option by the state.[1] However, the appropriation of applied geography as the handmaiden of the state was accepted by others enthusiastically, and is perhaps clearest in John House's statements. In his inaugural lecture at Newcastle in 1965, he argued that: 'Applied Geography currently guides the geographer towards professional employment, on a world scale, in town, country and regional planning and in Government service in various ministries' (House, 1965, 14–15). He drew justification for this from the lineage of Daysh and Stamp, and from earlier theorising of classical geography. For example, he quoted Strabo, writing in the first century AD, that geography would 'help the Europeans to govern and administer the Roman Empire, and to find an appropriate policy' (p. 3, quoted from Hamilton and Falconer, 1892). He also quoted with approval the role of regional planning in Soviet countries, particularly the work of Gerasimov and colleagues (see summary in Gerasimov, 1962), and other state planning processes, including those of Napoleon.

House drew extensively on Daysh (1949, ix), who saw the 'distribution of industry problem', and town and country planning of land use, to be

[1] Personal communication April 2002.

key areas for geography, which needed greater 'inter-departmental coherency ... of administrative machinery [where] there can be agreement and synchronised action'. Daysh and O'Dell (1947) had indeed proposed an 'outline scheme' for regional planning surveys, whilst Daysh's (1949) collection of surveys was an attempt by a group of economic geographers, which included Caesar, not only to influence regional planning but to enmesh economic geography and planning within each other, in the same way that geography and town planning became enmeshed.

Subsequent developments of 'applied geography' as a term, or sub-discipline, tended to develop a similar agenda. Terry Coppock, for example, argued that public policy contribution was one of the chief ways in which geographers could increase their acceptability and influence as a discipline. He was critical of applied studies that were 'not designed to answer questions which are relevant to public policy', or research that focused on evaluating past policies, which 'generally failed to offer alternative strategies for the future' (Coppock, 1974, 4). He argued that filling this gap required identification of the areas to which geography can contribute, development of research projects that provided detailed assessment of these areas, opening of a dialogue with the relevant policy officers in government, and offering practical solutions to the problems identified. These were themes also echoed from international perspectives at the time of the plenary sessions at the 1974 IBG meeting by Hare (1974) and others. Hare, who had been influential as a founding member of NERC, argued that whilst there were 'policy directed minds at work [in geography] ... the group is a minority'; there was a need for more geographers to 'write, speak and lobby on behalf of formulated policy options' (Hare, 1974, 28). The subsequent development of this approach in Coppock and Sewell (1976, 257–9) suggests a process of problem identification, policy formulation, policy implementation, monitoring, and policy modification. They then relate policy processes to the levels of influence on policy and who should be influenced. Some of these ideas were subsequently developed in the control-systems approach to policy expounded in Bennett and Chorley (1978, chs 3 and 6).

This conception of 'applied' geography as public policy became broadly shared (see e.g. the summaries by House, 1973; Sant, 1982; Kenzer, 1989; and Hansom, 1992). Even in Johnston's (1981a) contribution to the journal *Applied Geography* it was possible to portray applied work as solely public-policy-focused: on impacts, policy variations, unintended consequences, prediction of future policy impact, and normative planning. The focus of applied geography on public policy was not universally shared, however. Even in the 1974 IBG meeting, where Coppock's (1974) presidential address was used as the touchstone of the plenary sessions, Peter Hall (1974) was advocating the need to bridge the gap between geography and political theory/political science. He recognised the danger of the specific bridging emerging from neo-Marxian concepts and called for a wider set of perspectives to develop an understanding of the spatial components within different groups of interests, and how alliances, coalitions, and conflicts could be analysed geographically and solutions resolved.

There was also no shortage of other contributions in the early 1970s that sought to shift the balance of study to a wider range of both public policy and other non-governmental fields of applications of geography: for example, Hägerstrand (1970), Thompson (1964), Massam (1974), Smith (1971, 1973), Chisholm (1971), Chisholm and Rodgers (1973), Berry (1972), and Chisholm and Manners (1971). However, none of these came close to satisfactorily defining what an applied geography should be, with the result that broad conceptions such as those of Keltie, Mackinder, or Stevens remained forgotten or undeveloped. Indeed, it was clear by the end of the 1960s that applied geographical work was in some form of crisis. The old style of much applied geography, focusing on description and classification with little theory, was ending; the fields of regional and land use planning were becoming isolated from the broader thrust of the discipline; and it was not clear where the cutting edge should be. The presidential address at the 1970 IBG meeting in Sussex was perhaps a turning point. Arthur Smailes' (1971) address, of a very traditional classificatory view of the urban hierarchy, was greeted by near silence when applause was called for. There were calls for presidential addresses to be no longer automatically published in the *Transactions*, and the IBG Council responded to members' criticisms by launching a wider range of plenary sessions at the annual conference, which were first implemented during Coppock's 1973–4 presidency.

However, these changes did not in themselves resolve the tensions for applied geography. The 1974 discussion of Terry Coppock's IBG presidential address led to a riposte from the emergent neo-Marxists. Framed within David Harvey's (1974) contribution to the same 1974 IBG conference, geography's perceived alliance with the corporate state was strongly challenged. Harvey argued that geography had to develop not out of the need for disciplinary imperialism or to ensure that its voice was heard in policy debates, but from 'social necessity'. Geographical 'research has become a commodity', and geography had 'been co-opted' by the state (Harvey, 1974, 21). The alternative was to develop geographers' *moral* obligation 'to subvert the ethos of the corporate state'. The task was to 'build a humanistic literature which collapses artificial . . . dualism between fact and nature, subject and object, man and nature, science and human interests' (p. 24). This debate did offer a 'wake-up call', but in practice it has had little practical impact on 'applied geography', although some have continued to argue that the key framework for applied geography is as a committed discipline (e.g. Pacione, 1999).

The 1974 IBG debate followed the 1971 Boston meeting of the Association of American Geographers which, for many, reflected a turning point in erecting a 'new' geography concerned with radicalism, relevance, or a 'revolution in social responsibility' (Smith, 1971, 153: see also Prince, 1971, Berry, 1972). Harvey and others were correct to observe, as we have seen, that applied geography could be perceived as largely co-opted by the state. He was also correct to observe that, in part at least, scientific geography, models, and theories were also part of this co-option (Harvey, 1974, 23). However, Harvey's call has not led to a specific focus for more recent work labelled 'applied geography'.

Analysing the specific critique of applied models and scientific geography as co-opted by the corporate state was particularly vulnerable, as argued by Bennett (1981, 387), because 'those elements of geography in Britain up to about 1974 which term themselves applied had neglected some of the most important aspects . . . [Instead, there was] an emphasis on centralized planning of the state at the aggregate level through regional policy, [and an] . . . overbearing concern with the physical structure and layout of settlements through land use planning'. Bennett called for Keltie's (1908) broad conception of applicable geography to be re-addressed as the answer to the criticisms of Harvey and others. He suggested that in Harvey's work itself, as well as in the work of D. M. Smith (1977, 1979), Johnston (1979, 1981b), and Bennett (1980), there was a natural set of focuses for a new 'applicable' geography, with the following attributes:

- construction of theories of how decisions should be made—normative study;
- a direct analysis of the workings of decision-making in terms of economic, social, and political impacts; and
- assessment of the direct and indirect impacts of decisions against criteria of economic efficiency, social equity, or political objectives, i.e. comparison of outcomes with goals.

Later (Bennett, 1989, 1997), this approach was enlarged to cover a broader conception of the need to study and improve the dynamics of markets and other fields, as well as the efficiency of government.

Indeed, the wider current interpretation of applied geography now appears to be well understood in practice, even if at the polemical level this is not yet fully enunciated. For example, analysis of the contributions to the journal *Applied Geography* shows a very broad range, although a shift occurs in the balance between physical and human geography from 41:59 in the years 1981–90, to 28:72 per cent in the years 1991–2001. Interestingly, very few *Applied Geography* papers have engaged with a particular legislative policy: only about 2 per cent between 1981 and 2001. Most papers instead focus on analysing policy implications, the approach criticised by Coppock (1974) and Hare (1974). Of course, any journal develops through editorial policy as well as its self-selection by authors as a target for their outputs. Many other geography journals also cover application areas, of most importance probably being *Regional Studies* and *Environment and Planning*, particularly Series C, *Government and Policy*. However, comparison with other geographical journals suggests that *Applied Geography* is fairly typical of the applied output of geography. The approaches of its founding editor, Briggs (1981), and second editor, Hansom (1992), were also explicit: that applied geography is problem oriented, focused on the society–environment interface, and on the methods to solve problems. However, despite criticism of the lack of theory (Hansom, 1992), few papers are theoretical.

The crisis of identity of the field labelled 'applied geography' has therefore been a continuing one. Looking back on the early developments, particularly Dudley Stamp's early work through the Scott and Barlow committees, Sheail (1994, 376) concludes that policy recommendations were often 'vague' and 'left a yawning gap between description as to what was wrong, and very general prescriptions as to what kind of body might be required to intervene'. The continuing gap between theory and geographical practice is reiterated by Kenzer (1992, 207), also analysing the content of the journal *Applied Geography*, who states that 'appallingly few authors writing in the journal have placed their work in the context of a larger body of applied studies'. Indeed Hansom (1992, 6), when taking over as editor, stated that applied geography was no more than 'simply geography applied' and is 'not theoretically or methodologically distinct'. If there is to be a modern concept of applied geography, therefore, it appears to lead to the integration of theory, studies of the dynamics of behaviour, and direct grounding in the needs of decision-makers (economic and social, as well as political agents), tracked against specific and explicit goals. We turn to examples of these developments in the next section.

Expansion and change from the 1960s

Systems, quantitative geography and the range of application

By the early 1960s, two major shifts away from static geographical concepts of description, land use survey mapping, and classification were evident: first, geography was seen as systematic—that is *functional-systems* focused, and second it was to an increasing extent using *quantitative* as well as other methods. Johnston (1986) saw this as 'the systematic fix' in which geographers focused on 'topical specialisms' rather than 'regions'. As well as accepting these shifts, Wise (1973) also argued that the post-1960s developments saw a move by geographers away from being generalists seeking to be comprehensive, towards becoming specialists and systematic. Each of these shifts has had significant implications for the volume and character of applied work: new techniques and focuses of interest have emerged, which themselves have been important influences on the subject matter studied and how it is structured.

One starting point for a review of developments, therefore, is go back to a broader view, in line with Keltie or Mackinder, to define those geographical systems that became the chief focus of most geographical applications. These 'topical specialisms' can be defined in a variety of ways and from different, overlapping, perspectives. Each possible perspective can be thought of as a *system of interest*, a system being defined as a set of interacting components and the associated infrastructure. The main range of examples to have occupied much of the post-1960 period have been regional systems such as cities, rural or urban regions, nations or groups of nations, such as the EU; urban systems—cities, disaggregated to show their structure, internal workings and external relationships; or functional systems, such as the economy, agriculture, resources, utilities, manufacturing, consumer services (retail, financial), public services (e.g. health, education), producer and business services, transport, social, housing, and the labour market. Or, as in Haggett (1965), there has been a more abstract focus on spatial structures involving point patterns (e.g. systems of cities), interactions (e.g. commuting flows), or networks (traffic, communications). This classification is certainly neither unique nor completely comprehensive, but it serves to illustrate the developing range of topics covered by 'applied' geography in the later part of the century.

It can be argued that geography has a similar relationship to such disciplines as town planning as physics has to engineering—supplying much of the underlying theory, though with the practice lying in a distinct profession (see Willatts, 1971 and Chapter 16 by Peter Hall in this volume; see also Hall, 1982, 1998). This kind of relationship has been typical: the geographer has often been one stage removed from practice, or has been a member of a multidisciplinary team. However, this is best illustrated by example and that is how we now proceed. We have seen that geographers' systems of interest may be defined by place, as with urban systems, or by function, as in retailing or health. However, in reviewing the new era, the capability to develop effective application depends on the underlying theory. A first step, therefore is to review the development of theory.

The development of geographical theory

Before the developments of the 1960s, as we have seen above and in Chapters 9 and 10, foundations were laid by those who worked on urban systems and regions, such as Stamp, Daysh, and Dickinson. And the functional-systematic theories were those of classical geography (interestingly, none of these were developed by British geographers): agriculture—von Thünen; industrial location—Weber; market areas and competing firms—Palander, Hoover, and Hotelling; central place theory—Christaller and Lösch; urban structure and development—Burgess, Hoyt, Harris, and Ullman; and interaction—the gravity model—Ravenstein and Reilly. The 1960s developments came from new directions, though by the end of the century a new synthesis was in sight, in part brought about by work in applied geography. This has meant that the applied focus on place intimately links the new methods and theory with 'regional' geography; and that classical theory can now be seen as being restated using the new ideas. We seek to articulate these ideas at the end of this section, following a review of the key developments.

In order to structure this account, we begin with a sketch of the methods and theories that underpin applicable geography and then show how these have been applied in a variety of contexts. The notions of 'methods' and 'theories' overlap, because new methods facilitate new—and more general and powerful—formulations of theories, and this will be reflected in the discussion that follows. The first stages of the quantitative revolution in geography (Burton, 1963) were characterised by the use of statistical methods. These included the use of standard statistical estimation and inferential statistics; factor analysis and principal components analysis were particularly developed for application to characterise regions and cities. These techniques were first introduced in sociology and factorial ecology and the first geographer to use them was David Herbert (1967, 1968; Williams and Herbert, 1962). However, Brian Berry, a US-based British geographer, became the leader in this area and his work influenced, and was taken up and extended by, notable scholars in the UK as well as the United States. Philip Rees (who was a PhD student of Berry's in Chicago), Nigel Spence, and Peter Taylor were examples (see Berry and Rees, 1969; Spence and Taylor, 1970; Rees, 1979). The parallel work in other fields such as sociology included authors like Duncan Timms (1971)—though Timms was trained as a geographer. These methods essentially involved taking a large number of variables associated with small areas within cities and regions and finding 'interesting' ways of characterising them. This underpinned—and still underpins—the major applied fields of geodemographics, of which more below.

A next step is to try to *model* the relationships between variables—seeking to predict dependent variables in terms of a set of independent variables using statistical methods such as regression analysis. There was a huge volume of work that connected geography to the most sophisticated of statistical techniques—and indeed opened up new areas of statistics through the challenge of spatial problems (Cliff and Ord, 1973, 1980; Bennett, 1979; Haining, 1990)—but also made explicit the task of articulating and developing theory by building *models* of relationships, and indeed of whole geographical systems of interacting variables. Here, the work of Haggett (1965) was of supreme importance—and see, for example, Haggett and Chorley (1969) and Haggett, Cliff, and Frey (1977).

Haggett (1965) articulated the concept of a model in geography: 'in model building we create an idealised representation of reality in order to demonstrate certain of its properties. Models are made necessary by the complexity of reality'. This was a critical development of the systems perspective: to build a model of a system in order to capture a manageable essence of a complex reality (Cliff et al., 1995). The model's hypotheses could be tested using statistical methods. The work of Gould was similarly important—as summarised, for example, in Gould (1985). Much development was influenced by thinking in physical geography, particularly through the work of Richard Chorley, in partnership with Peter Haggett. A further step in the development was to seek to build dynamic statistical geographical system models (Bennett and Chorley, 1978; Bennett, 1979), and this also embraced the methods of epidemiology and their application to geographical problems (see Chapter 15, below). In more recent times, the methods of neural nets—in effect using models of brain learning as a statistical learning technique—have been applied to the analysis of large geographical datasets (Openshaw, 1993; Reggiani et al., 1998). The work of Leung (1997) in Hong Kong has been particularly important here, and he and his colleagues (Leung, Gao, and Chen, 2001) have recently established the formal connection between neural network methods and spatial interaction models.

An important step in the evolution of quantitative geography was the development of *mathematical* models. The aim in this case was to represent *algebraically* the relationship between the variables of a theory relating to some system of interest and then, typically, to simulate the model on a computer. (This can be thought of as the geographical equivalent—say, for a city—of a flight simulator for training pilots.) Statistical techniques were used at a later stage to calibrate and validate the model. The impetus for the development of such models in geography largely came from outside the discipline in the first instance—with origins in transport planning in the United States from the mid 1950s—but were well incorporated in geography from the 1970s onwards (for an early review, see Carroll, 1955). They were the basis of much fruitful application. The initial American focus was on transport interaction and flows—because the aim was to model, and then to solve, the problem of traffic congestion. But it was almost immediately clear that transport flows depended on land use. At first, the land use was taken as independent within a transport model, but research programmes were soon launched to model the land use structure itself—indeed to seek to build a general urban model (Harris, 1962; Lowry, 1964, 1967). The aim was nothing less than to represent theories of urban functioning and development in a mathematical model—in turn represented as a computer model—and to use this model for effective city planning. This could, in principle, be used to address questions such as: what *forms* of a city (e.g. in terms of patterns of density) are consistent with the development of an efficient transport system? This is an interesting example of how research on a short-run applied problem led to theoretical developments that have had a profound effect on geographical theory and its further potential for application.

In the UK, these ideas were picked up by the Transport Research Group based in the Institute of Economics and Statistics in Oxford, led by Christopher Foster, one of the inventors of social cost–benefit analysis (see Foster and Beesley, 1963). The group, which included one of the present authors (AGW), was translated to the Ministry of Transport in London in 1966 on Foster's appointment as Director-General of Economic Planning. Modelling became a key element in its work through the Mathematical Advisory Unit, headed by Alan Wilson. In 1968, he became Assistant Director of the Centre for Environmental Studies (CES) and more opportunities were provided for the development of modelling, both in the CES itself and through CES funding of research in universities like Reading. There were recruits who went on to do distinguished work, illustrated by the following examples: Andrew Broadbent (1970, 1971, 1973), Martin Cordey-Hayes (1972, 1975), Shlomo Angel and Geoff Hyman (1976), and a researcher who then moved in a different direction, Doreen Massey—see Massey (1968) for what must now be an academic curiosity! This group was able to build on the earlier American experience and articulate mathematical and computer models for each of the key elements of urban and regional systems through a series of developments we describe below.

There were two main steps in the development of land-use transport models in the UK. The interaction element of the model, which calculated flows between locations, had long been based on a gravity model analogy: that any flow was proportional to the 'sizes' of the origin and destination areas (by analogy with Newtonian masses) and inversely proportional to a function of the distance between them. The first step in improving the situation was the recognition that the gravity analogy did not produce a satisfactory model. There was then a shift to a probability basis, initially using a statistical mechanics analogy, which gave rise to the notion of entropy-maximising models (Wilson, 1967, 1970). The entropy function provides a measure of the probability of a system state occurring; entropy-maximising is a means of finding the most probable state subject to constraints. The procedure was a statistical averaging one: by averaging over all the possible travel behaviours of a large population, stable and accurate predictions of flows could be made. In effect, given the spatial structure, it had been shown that the overwhelmingly most likely transport pattern could be determined by a model that contained only a small number of parameters. This created a new basis for building models that was theoretically well founded and fitted real patterns very well.

The second step was the recognition that spatial interaction models could also, in many cases, be made to function as *location* models. A family of models was constructed from which models could be selected for different applications (Wilson, 1971b). Retailing provided an archetypal example: flows from residences to retail centres could be summed to estimate total flows into each centre; and these totals were locational variables. (This idea can perhaps be attributed to the American town planner, Walter Hansen, 1959, and the American geographer David Huff, 1964, and was first effectively implemented by Lakshmanan and Hansen, 1965.) These ideas were used by a variety of authors to extend the original American models that had weaker theoretical underpinnings (Wilson, 1971a, 1971b; Batty, 1976). These models were given a variety of theoretical interpretations (as reviewed by Macgill and Wilson, 1979), of which the most important has been the development of economic behavioural interpretations.

However, there was one piece of the jigsaw missing: these models could not predict the urban structural variables, which had to be taken as independent. Again, take the retail model as an example to illustrate this issue. For a given pattern of residential location and densities, and for a given distribution of shopping centre sizes, it was possible to predict the person (or retail spending) flows into each centre (or even into each store within a centre). This supplied valuable commercial information. However, from a theoretical point of view, and for potential applications in city planning, it would be better to have a more comprehensive model that also predicted the evolution of residential patterns and the spatial structure of retail centres. These are the urban structural variables referred to above. There was a sense in which a classical theory like central place theory was tackling this problem, which the more recent models were failing to address. This situation began to be rectified in the 1970s when the mathematical methods of dynamic systems theory began to be available (Thom, 1975; Poston and Stewart, 1978). Harris and Wilson (1978) showed how to model retail structure (following some earlier, less satisfactory experiments: Poston and Wilson, 1977). This was done for retail structure, for example, by assuming a starting distribution about centre sizes and adding an hypothesis about change of size in each centre in relation to the profitability of the centre predicted by the flow model. In effect, this embedded the retail interaction model into a dynamic structural model.

These models have been tested in various ways (Wilson and Oulton, 1983; Clarke, Langley, and Caldwell, 1998). Although they have not yet reached a stage where they can contribute fully to applied work, they do offer valuable insights and represent important potential for future development and so are appropriately reported here: see Wilson (1981) for an early review and Wilson (2000) for a later one, which enables measures of progress to be assessed; an assessment of future prospects is given below (pp. 486–8).

Many of these models can be formulated as optimisation models (Wilson et al., 1981). In some cases, very simple assumptions were made about interaction (e.g. that service users went to the nearest facility) and this enabled the formulation of location problems, such as locating the optimum number of fire stations, as linear programming problems. These were the so-called location–allocation models (see Scott, 1971, for an early review). As we will see below, many of the applications in the public sector have been based on linear programming. Herbert and Stevens (1960) built a famous residential location model based on linear programming principles, in effect operationalising the economic model of Alonso (1960). The core entropy-maximising transport model—built on statistical averaging principles as described above—can be seen, as the name implies, as an optimising model—a non-linear programming problem. In this case, what is particularly interesting is that it can be linked, in a limit, to the transportation problem of linear programming (Evans, 1973; Senior and Wilson, 1974; Wilson and Senior, 1974). This is important theoretically, but also in terms of application—because this link gives a means of extending linear models in fruitful ways that can then be used in applications. We give specific examples below (pp. 483–4).

There were other functional geographical systems that demanded different modelling techniques—for example, the demographic and economic dimensions. The full set of methods for population modelling for small areas was set out by Rees and Wilson (1976) and interregional models were built on input–output principles, following the work of Leontief (1966) and Leontief and Strout (1963). The British economist-demographer, Richard Stone (1967, 1970), made important contributions in both fields.

The advent of increasingly powerful computers has had its own impact on available methods. Many models can now be represented in a micro-simulation mode. This method has a long history, being initiated in economics by Orcutt (1957; Orcutt et al., 1968), a tradition that has been continued by such authors as Hancock and Sutherland (1992). The basic idea can be described as follows. It has been seen that most urban models are based on interaction matrices. When the population is disaggregated, and there are many spatial zones, these matrices become very large and also sparse (i.e. with many empty cells). A more efficient way of storing the same information can be achieved by 'listing' in the computer a hypothetical population with the same properties as that represented (in effect probabilistically) in the matrices. These populations facilitate interpretation: they look and feel like real populations, but with a richer 'data base' than can be achieved, for example, from the census. This is because data from many sources can be combined in generating the populations. The method combines probabilities derived from mathematical models with computer-efficient data storage principles. The technique has been deployed in sociology by Gilbert (1995). It was first introduced into geography by Wilson and Pownall (1976) and has been developed by such authors as Clarke, Keys, and Williams (1981) and Duley and Rees (1991) (see Clarke, 1996, for a review).

In all these cases, it has been possible to build on the direct model outputs and to construct appropriate performance indicators. Williams (1977), for example (and see also Williams, Kim, and Martin, 1990), has laid the full and proper foundations for cost–benefit analysis—relating to measures such as consumers' surplus which are the basis of much transport planning. This field is reviewed and developed by Clarke and Wilson (1987a, 1987b). One example serves to illustrate the progress that has been made. Performance measures relating to a retail outlet will depend on the concept of a catchment population—sales per head of catchment population, for example. Traditionally, catchment populations have been measured by counting all who live within an arbitrary distance of the store. This leads to double counting and distorted answers. A precise measure can be obtained as a by-product of the models and more confidence can then be associated with the associated indicator. In this particular case, it became possible to use the indicator to identify underperforming outlets. This argument can also be turned round and indicators developed to represent effective delivery of services. This has been demonstrated in the context of food retailing by Clarke, Eyre, and Guy (2002), and on the basis of their analysis they have introduced the concept of *food deserts*.

There remains the task of combining all these elements into the general urban and regional model first heralded by the Lowry model and mentioned earlier. The Lowry model itself could be generalised (Wilson, 1971a) and this provided the framework for an ambitious programme of comprehensive model-building. The theory (and some of the practice) has been set out in a range of volumes (see Wilson, 1974; Wilson, Rees, and Leigh, 1977; Wilson and Bennett, 1985; Bertuglia et al., 1987; Bertuglia, Leonardi and Wilson, 1990; Bertuglia, Clarke, and Wilson, 1994). At the end of the next section, we briefly review the progress in building effective working models of this kind.

This brief account of the development of model-based theory which was then highly applicable should not be taken to disguise the fact that there was much controversy. This essentially stemmed from Harvey's (1973) book, *Social Justice and the City*, and a subsequent shift to an alternative 'radical' approach to geographical theory based on Marxist principles or, more broadly, on structuralist principles. There was a well-known explicit attack on the 'positivist' basis of modelling by Sayer (1976). At worst, these critiques labelled anything 'quantitative' as positivist and therefore suspect—conveniently ignoring the distinguished mathematical modelling work of, for example, Marxist economists. At best, they exposed weaknesses in the model-based theory's underpinnings. It has not yet proved possible to draw the two threads together, though this remains a legitimate ambition.

Examples of applicable geography

The development of applied modelling, sketched above, shows how a range of theory and methods interacted to influence the array of geographical problems addressed. But it also led to a wide range of other applications. It is not always easy to isolate the contribution of the geographer to these developments. Often, in urban and regional studies, the geographer, as we noted earlier, is part—but a crucial part—of an interdisciplinary team. The distinctive feature of the contribution, whether within the discipline or as part of a team, is the focus on spatial analysis, and, in particular, the addition of a fine-scale spatial dimension. It is not feasible in this review to consider the total range of all applications. We proceed by example, selecting from the list on p. 475 as follows: regional economies, population geography, urban 'description' (geodemographics and marketing), transport, retail, health and education, public services and performance, and integrated models for city and regional planning.

In regional economic analysis, a major tool, as we have seen, has been the input–output model. Economists usually work with a single region and the 'rest-of-the-world'. In providing the basis for regional and local economic planning, multiregional models are needed and, of course, this involves modelling economic flows across regional boundaries; this turns the economists' problem into one that is shared with geographers. The theoretical foundations were laid by Leontief and Strout (1963), as noted earlier, and developed by Wilson (1970, ch. 3). The field is now highly developed, mainly on an interdisciplinary basis within a regional science umbrella with geographers like Hewings (1971, 1977a, 1977b, for example), a British geographer working in Illinois, and Madden, Brown, and Batey in Liverpool (in the Department of Civic Design) prominent. (Two of the last three were originally geographers; see, for example, Batey and Madden, 1981a, 1981b.) Jin and Wilson (1993) tackled the problem of input–output modelling with very small zones. Further developments have been made of economic understanding of capital, industrial clustering, and agglomeration dynamics (Leyshon and Thrift, 1998; Bennett, Bratton, and Robson, 2000; Martin, 1999a, 1999b; Clark, 2000) and Martin's work has fed into government policy on regional cluster development (Department of Trade and Industry, 2001).

Population forecasting is an important component of many applied problems—for example, in estimating future demand for services or housing. Demographers, however, like economists, tend to work with large, even single, regions. Geographers have been prepared to work at very small spatial scales, such as postal districts, and also with the interrelations between spatial units at different scales (Openshaw, 1977). This is particularly important in the study of migration. A move that counts as migration at a fine scale may be intra-zonal at a coarser one. The works of Rees (1996), Stillwell and Congdon (1991), Alvanides et al. (1996), and Woods (1981) illustrate the possibility of applied population geography at fine spatial scales. The foundations were laid in a series of papers by Rees and Wilson, culminating in their 1976 book. Fine-scale population analysis also facilitates the study of communities that can only be identified at such scales—as in the studies of ethnic populations by Rees and Phillips (1996). These analyses have been valuable aids to local government for local forecasting for policy development and planning: the ability to forecast small area populations by type is critical for effective service provision.

The third set of examples are collected under the umbrella of 'urban description' and informed policy analysis. It has always been seen as an important geographical task to collect and map small-area data and there is a long tradition of informed geographical analysis (e.g. Gottmann, 1964; Carter, 1975). Indeed, it is important explicitly to decide which are the key features or entities that are the focus of complex systems of interest—what Chapman (1977) called 'entitation'. The presentation of such data is now facilitated by geographical information systems (see Chapter 12, above), enabling an almost infinite variety of maps to be generated (see Dorling, 1995). The next steps in the analysis involve deepening understanding of, and enhancing, the raw data through statistical analysis and modelling. Statistical methods have generated many applications. Factor analysis, for example, or its contemporary equivalents such as neural net analysis, underpins geodemographics, which are widely used in areas like marketing and targeted mailing through companies like Pinpoint (see Beaumont, 1991a, 1991b, 1991c), Experian (created and led by a geographer, Richard Webber) and CACI. (For reviews of the evolution of geodemographics, see Batey and Brown, 1995; Clarke, 1999.) Johnston and others (Johnston, Hay, and Taylor, 1982; Johnston et al., 1984) developed a programme for electoral redistricting which was used by the Boundary Commission to inform their work on European constituencies. The methods have been applied in the public sector in particular fields—particularly in health and in the characterisation of deprivation (see Brown, Hirschfield, and Batey, 1991). Epidemiological models (discussed in Chapter 15, below) have also offered fruitful applications. When data can be assembled in a comprehensive and systematic way, they can also form the basis for urban development policy work, exemplified by geographers such as Robson (1995) and Robson, Parkinson, and Bradford (1994), which has been very influential on urban neighbourhood regeneration policy since 1997 (see also Imrie and Thomas, 1999; and Chapter 16, below).

The widest range of application is probably through the deployment of interaction and location models and these can be illustrated through the transport, retail, and health examples. Many geographical systems are underpinned by interaction: residential and workplace location and the journey to work; retail turnover and the journey to shop; the use of schools and health facilities; the deployment of emergency services such as fire and ambulance; business services; and so on. It is often of direct interest simply to model the interaction matrices accurately. As we saw above (p. 478), they can then be totalled at each facility location to give an estimate of demand or usage; these totals are locational variables.

Geographical modellers have developed roles in transport planning, building on the earlier work of pioneers such as Hay (1973) and O'Sullivan (1978). The models have now been developed to a very high level of sophistication. Interaction models have been integrated with network assignment models, for example, within a mathematical programming framework (Boyce, 1984) and the model outputs have been fully integrated with the concepts of cost–benefit analysis—in principle facilitating applicable understanding. However, model application has rested largely with engineers; this is perhaps a missed opportunity for geographers. It would be possible to use models to ground transport planning—in the wider context and in city and regional planning—into something more realistic than has typically been the case (see Wilson, 1998, for a recent survey).

The second interaction-based example is in the analysis of 'retailing'. This should be interpreted very broadly, to include, for example, journey to shop, the purchase of motor vehicles through dealerships, or the use of banking and other businesses services. Early work building on classical geographical theory by authors such as Berry (1967) was later extended by authors like Wrigley and Lowe (1995, 2002) and Dawson and Broadbridge (1988). But in modelling terms, it has proved possible to estimate the flows of consumers to retailing facilities, or business clients to professional service suppliers, quite accurately—with errors of the order of 10 per cent or less. (For examples of different types of models, see Fotheringham, 1983; Breheny, 1988; Birkin et al., 1996; Bennett et al., 2000; Bennett and Smith, 2002.) The models estimate demand by zone by person type, and attractiveness of facilities by type and size, and then the matrices of interactions can be calculated with an appropriate interaction model as functions of average journey length. Relatively little consumer behavioural data is needed to calibrate such models (Batty, 1976; Birkin et al., 1996). However, substantial databases are needed to underpin them—to estimate demand and to characterise facilities. Both interaction and location (demand and facility attractivess) information has an obvious commercial value and has been exploited through companies like GMAP Ltd, which employs many geographers. A retailer, for example, can use the models in a '*What if?*' forecasting mode to plan an optimal distribution of outlets and such decision-support systems are now widely used.

Birkin et al. (1996) describe a wide range of applications. For example, GMAP has worked with two major car manufacturers, Toyota and Ford, to help plan optimum dealership networks for car sales. The interaction models are used to predict customer purchases of cars by model and manufacturer for a network of dealers—and so work for one company demands a model for the whole system—encompassing 8,000 dealers for the UK as a whole for example. The starting point is always to model the current system. It is then possible to explore changing configurations of dealerships and to work towards overall improvements. These systems have now been extensively used and tested (and continue to be used), as summarised in Longley et al. (2001) and Birkin, Clarke, and Clarke (2002b). The latter book in particular charts in detail the retail applications of geographical modelling. More detailed case studies can be found in Birkin, Boden, and Williams (2002a) on petrol forecourts, Birkin and Clarke (1998) on financial services, Birkin, Clarke, and Douglas (2002c) on spatial mergers, and Langston, Clarke, and Clarke (1997) on food retailing.

An interesting curiosity has been the application of the 'retail' model to the settlements of Ancient Greece, demonstrating their potential for productive application in archaeology. It was shown that by running location data through the model (and making some assumptions!) it was possible to reproduce the hierarchy of settlements indicated by archaeological digs (Rihll and Wilson, 1987a, 1987b, 1991).

The general analysis of public services and their finance has been given integrated treatment by Smith (1977, 1979), Bennett (1980, 1992, 1997), and Pinch (1985). There have been significant research monographs and papers on particular public service initiatives and programmes, some of which have been very influential in changing government policy, as for example on the spatial allocation of the rate support grant (Bennett, 1982), water privatisation (Rees, 1983), the role of Training and Enterprise Councils (Bennett, McCoshan, and Wilson, 1994), the 'New Deal' for unemployed people (Peck, 1999, 2001; Sunley, Martin, and Mativel, 2001), health care reforms (Mohan, 1998), education (e.g. Gordon, 1995), regional and local economic development (Bennett and Payne, 2000), and a range of local government boundary and finance issues (Chisholm, 1995, 2000, 2001).

We saw earlier (p. 481) that since the early days of modelling in the 1960s, there have been attempts to build integrated models—a general urban model. This enterprise has continued, though the applications have been mainly outside the UK—in the United States in the context of environmental pollution regulations, in South America and in Japan. Webster, Bly, and Paulley (1988) reviewed what has been achieved (and see also Paulley and Webster, 1991). Mackett (1980, 1990, 1993) and Echenique (a town planner; see Echenique et al., 1990) have been prominent British contributors in an international group that has included

Putman (1983, 1991) in the United States, Wegener (1986a, 1986b, 1994) in Germany, and Nakamura, Hayashi, and Miyamoto (1983) in Japan.

Prospects and Challenges

Geography has grown vastly in size, complexity, and breadth since Keltie's (1908) definition of applicable geography, Geddes' (1905) concept of civic life, culture, and city mapping, or Mackinder's (1919) 'Seven lamps of geography'. The suggestions of Keltie, Geddes, and Mackinder for a broadly based applied geography lay largely undeveloped for at least a generation. With the impact of 'national need' during the Second World War, and when the development of the state, especially the welfare state and regional planning, took off in the 1930s and 1940s, different ideas then came to the fore. Geographers became influential in many developments, but with the exception of the town planning profession, geography as a discipline was frequently marginal in its influence on theory. The theories on which policy were being built largely derived from elsewhere—particularly economics and sociology. Although Wise (1969) comments that there was close working with scholars from these other disciplines, Ackerman's statement (1963, 430) in his presidential address to the Association of American Geographers was equally applicable in Britain: geographers 'have not been on the forward salients of science, nor, until recently have we been associated closely with those that have'. Indeed, House (1973, 298) concludes that 'geography has rarely dominated in policy making'.

One effect of this evolution was the large scale co-option of much of the human geography discipline as a whole, and applied geography in particular, as an adjunct of state policy with scant theoretical strengths to defend its integrity of approach. Through the twentieth century, most polemical writers continued to interpret applied geography as chiefly serving public policy. We have shown how this follows from the way in which the leadership of Dudley Stamp and Eva Taylor was interpreted by others, and from the acquiescence or acceptance by many economic geographers (particularly House and Daysh) in the early post-war years, with its origins in the 1920s depression and the Second World War, and reiterated by Johnston (1981a, 1981b). Bennett (1989, 1997) has referred to this as the 'saturation' of the subject by static, idealist, and welfarist conceptions—a 'saturation' that also had a major influence on most areas of human geography, as well as on some of the developments and uses of modelling, quantitative analysis, and theory.

However, our discussion of developments in theory and methods since the 1960s has illustrated how in practice a very diverse range of concepts, subjects of study, theories, and methods have now come to characterise applied geography. Within these, public policy is but one of many research concerns. In this sense the diversity of the applied area of the discipline shows healthy pluralism and a grassroots response to a post-welfare agenda where co-option by the state has diminished. The scope of the discipline now to contribute across a wide range of fields is well established and significant. Hence, the rather different conceptions of Keltie, Geddes, and Mackinder may have been realised. Yet the clear enunciation of this emergent applied structure of the discipline is still largely lacking, and this chapter has demonstrated the dilemmas of trying to define or prescribe it too closely. In a real sense the definition of applied geography is tied up with the definition of the discipline as a whole: in Hansom's (1992, 6) words, applied geography is 'simply geography applied'. As a result, arguments about the distinctions of 'pure' and 'applied', so often discussed in the past (see e.g. House, 1965, 13–15; 1973), are no longer relevant.

In the space available here we focus on two main consequences. The first is the need to redress emerging imbalances. The burgeoning developments of the applied geography reviewed above are very largely the results of a cohort of quantitatively proficient geographers trained in the 1960s and early 1970s, as well as some distinguished later recruits. There is a very real problem of how far this cohort can and will be replaced. A recent survey of graduate studies by the British Academy (2001) identified within geography two endangered areas: geographical information systems and economic geography (meaning economics-proficient scholars within geography). This survey also identified a more general gap in the supply of postgraduates in quantitative social science as a whole, concerns that have also been noted by the ESRC. A challenge for the discipline is, therefore, how far it will replace its expertise in quantitatively and economically skilled scholars, and retain the strength of geography's scope to contribute at an applied level across a broad spectrum of fields. This challenge is not uncontested and it is, at the beginning of the twenty-first century, an open question whether the discipline will respond adequately with the training and appointment of those geographers with a wide range of applied skills and applicable knowledge. In this sense, then, the old debate about 'pure' and 'applied' has a new relevance and leaves the discipline with some awkward questions to address. For example, even in the 2001 Research Assessment Exercise (RAE) there was a relatively weak involvement of the user community (as a reference panel). There remains, therefore, a continuing gap between pure and applied work. Indeed this gap is embedded within UK institutions, as evident in the splits within government ministries such as that between the Department of Trade and Industry and the Office of Science and Technology.

A second area of concern is the extent to which the challenge of developing a fully integrated structure for the variety of models in applied geography will be possible—a 'final' theory (see Wilson, 1995) of geographical structure and its evolution. This would enable classical theory to be rewritten and would provide the basis for a new era in applied geography (Wilson, 2000). Model-based analysis could be used routinely to underpin policy development and planning in all the fields enumerated earlier (p. 475). As experience built up, this would be the basis of what is now commonly called evidence-based policy. Within geography, developments from classical models can be seen as components of a general model (see Birkin and Wilson, 1986a, 1986b, 1987, on industrial location and in relation to agriculture). The spatial interaction model used as a location paradigm fundamentally rewrites the classical notion of market areas and central place hierarchies—overtaking the authors mentioned earlier—Palander, Hoover, Hotelling, Christaller, Lösch, Burgess, Hoyt, Harris, and Ullman. We can also look forward to these models being applied to 'new' sectors such as telecommunications (Guldmann, 1999). Such developments can only be accomplished in a multidisciplinary framework. It would be important to embrace the ideas of economists, for example, who have been prepared to think outside the usual paradigms and to recognise that a key factor in urban evolution is the concept of positive returns to scale (Arthur, 1988, 1989; Romer, 1990)—an idea already implicit in models of urban retail evolution (Harris and Wilson, 1978). Herein lie many challenges for development in practice, and the development and growth of a community with the appropriate skills in quantitative geography.

Acknowledgements

We are most grateful for suggestions supplied by many colleagues, and for some specific comments on earlier drafts of this chapter. We particularly acknowledge the help of Michael Wise's critique, and the comments of Peter Hall, Derek Diamond, and Michael Chisholm.

References

Abercrombie, P. (1938) Geography, the basis of planning. *Geography*, 23(1), 1–8.

Ackerman, E. A. (1963) Where is the research frontier? *Annals of the Association of American Geographers*, 53, 429–40.

Alden, J. and Morgan, R. (1974) *Regional Planning: A Comprehensive View*. Leighton Buzzard: Leonard Hill.

Alonso, W. (1960) A theory of the urban land market. *Papers, Regional Science Association*, 6, 149–57.

Alvanides, S., Boyle, P., Duke-Williams, O., Openshaw, S., and Turton, I. (1996) Modelling migration in England and Wales at the ward level and the problem of estimating inter-ward distances. In *Proceedings, First International Conference on GeoComputation*, Leeds, 4–5.

Angel, S. and Hyman, G. M. (1976) *Urban Fields*. London: Pion.

Arthur, W. B. (1988) Urban systems and historical path dependence. In J. H. Ausubel and R. Herman, eds, *Cities and their Vital Systems: Infrastructure, Past, Present and Future*. Washington, DC: National Academy Press.

Arthur, W. B. (1989) Increasing returns and lock-ins by historical events. *The Economic Journal*, 99, 116.

Balchin, W. G. V. (1958) A water use survey. *Geographical Journal*, 124, 476–9.

Batey, P. W. J. and Brown, P. J. (1995) From human ecology to customer targeting: the evolution of geodemographics. In P. Longley and G. P. Clarke, eds, *GIS for Business and Service Planning*. Cambridge: Geoinformation, 77–103.

Batey, P. W. J. and Madden, M. (1981a) Demographic-economic forecasting within an activity-commodity framework: some theoretical considerations and empirical results. *Environment and Planning A*, 13, 1067–83.

Batey, P. W. J. and Madden, M. (1981b) An activity analysis approach to the integration of demographic-economic forecasts. In H. Voogdt, ed., *Strategic Planning in a Dynamic Society*. Delft: Delftsche Uitgevers Maatschappij, 143–53.

Batty, M. (1976) *Urban Modelling: Algorithms, Calibrations, Predictions*. Cambridge: Cambridge University Press.

Beaumont, J. R. (1991a) GIS and market analysis. In D. J. Maguire, M. F. Goodchild, and D. W. Rhind, eds, *Geographical Information Systems: Principles and Applications*. Harlow: Longman, 139–51.

Beaumont, J. R. (1991b) Managing information: getting to know your customers. *Mapping Awareness*, 5(1), 17-20.

Beaumont, J. R. (1991c) *An Introduction to Market Analysis*. Norwich: Geo-Abstracts, CATMOG 53.

Beaver, S. H. (1944) Minerals and planning. *Geographical Journal*, 104, 166–98.

Beaver, S. H. and Kosinski, L., eds (1964) Problems of applied geography II: Proceeding of the Anglo-Polish Seminar. *Geographica Polonica*, 3, 1–274.

Bennett, R. J. (1979) *Spatial Time Series*. London: Pion.

Bennett, R. J. (1980) *The Geography of Public Finance*. London: Methuen.

Bennett, R. J. (1981) Quantitative geography and public policy. In N. Wrigley and R. J. Bennett, eds, *Quantitative Geography*. London: Routledge and Kegan Paul, 387–96.

Bennett, R. J. (1982) *Central Grants to Local Governments: Political and Economic Impacts of the Rate Support Grant in England and Wales*. Cambridge: Cambridge University Press.

Bennett, R. J. (1989) Whither models and geography in a post-welfarist world? In W. D. Macmillan, ed., *Re-modelling Geography*. Oxford: Blackwell, 273–90.

Bennett, R. J., ed. (1992) *Local Government and Market Decentralization: Experiences of Industrial, Developing and Former Eastern Bloc Countries*. Tokyo: United Nations University Press.

Bennett, R. J. (1997) Administrative systems of economic spaces. *Regional Studies*, 31, 323–36.

Bennett, R. J. and Chorley, R. J. (1978) *Environmental Systems: Philosophy, Analysis and Control*. London: Methuen.

Bennett, R. J. and Payne, D. (2000) *Local and Regional Economic Development: Renegotiating Power under Labour*. Aldershot: Ashgate.

Bennett, R. J. and Smith, C. J. (2002) The influence of location and distance and the supply of business advice. *Environment and Planning A*, 34, 251–70.

Bennett, R. J., McCoshan, A., and Wicks, P. (1994) *Local Empowerment and Business Services: Britain's Experiment with TECs*. London: UCL Press.

Bennett, R. J., Bratton, W., and Robson, P. J. A. (2000). Business advice: the influence of distance. *Regional Studies*, 34, 813–28.

Berry, B. J. L. (1967) *Geography of Market Centers and Retail Distribution.* Englewood Cliffs, NJ: Prentice-Hall.

Berry, B. J. L. (1972) More on relevance and policy analysis. *Area*, 4, 77–80.

Berry, B. J. L. and Rees, P. H. (1969) The factorial ecology of Calcutta. *American Journal of Sociology*, 74, 445–91.

Bertuglia, C. S., Clarke, G. P., and Wilson, A. G., eds (1994) *Modelling the City: Performance, Policy and Planning.* London: Routledge.

Bertuglia, C. S., Leonardi, G., and Wilson, A. G. (1990) *Urban Dynamics: Designing an Integrated Model.* London: Routledge.

Bertuglia, C. S., Leonardi, G., Occelli, S., Rabino, G. A., Tadei, R., and Wilson, A. G., eds (1987) *Urban Systems: Contemporary Approaches to Modelling.* London: Croom Helm.

Best, R. (1981) *Land Use and Living Space.* London: Methuen.

Best, R. H. and Coppock, J. T. (1962) *The Changing Use of Land in Britain.* London: Faber and Faber.

Birkin, M. and Clarke, G. P. (1998) GIS, geodemographics and spatial modelling: an example within the UK financial services industry. *Journal of Housing Research*, 9(1), 87–112.

Birkin, M. and Wilson, A. G. (1986a) Industrial location models I: a review and an integrating framework. *Environment and Planning A*, 18, 175–205.

Birkin, M. and Wilson, A. G. (1986b) Industrial location models II: Weber, Palander, Hotelling and extensions in a new framework. *Environment and Planning A*, 18, 293–306.

Birkin, M. and Wilson, A. G. (1987) Dynamic models of agricultural location in a spatial interaction context. *Geographical Analysis*, 19(1), 31–56.

Birkin, M., Clarke, G. P., Clarke, M., and Wilson, A. G. (1996) *Intelligent GIS: Location Decisions and Strategic Planning.* Cambridge: Geoinformation International.

Birkin, M., Boden, P., and Williams, J. (2002a) Spatial decision support systems for petrol forecourts. In S. Geertman and J. Stillwell, eds, *Planning Support Systems in Practice.* Berlin: Springer.

Birkin, M., Clarke, G. P., and Clarke, M. (2002b) *Retail Geography and Intelligent Network Planning.* Chichester: Wiley.

Birkin, M., Clarke, G. P., and Douglas, L. (2002c) A model for optimising spatial mergers. *Progress in Planning*, 58, 229–318.

Boyce, D. E. (1984) Urban transportation network-equilibrium and design models: recent achievements and future prospects. *Environment and Planning A*, 16, 1445–74.

Breheny, M. J. (1988) Practical methods of retail location planning. In N. Wrigley, ed., *Store Choice, Store Location and Market Analysis.* London: Routledge.

Briggs, D. J. (1981) Editorial: the principles and practice of applied geography. *Applied Geography*, 1(1), 1–8.

British Academy (2001) *Review of Graduate Studies in the Humanities and Social Sciences* (Chair R. J. Bennett). London: British Academy.

Broadbent, T. A. (1970) Notes on the design of operational models. *Environment and Planning*, 2, 469–76.

Broadbent, T. A. (1971) A hierarchical interaction-allocation model for a two-level spatial system. *Regional Studies*, 5, 23–27.
Broadbent, T. A. (1973) Activity analysis of spatial allocation models. *Environment and Planning*, 5, 673–91.
Brown, P. J., Hirschfield, A., and Batey, P. (1991) Applications of geodemographic methods in the analysis of health conditions incidence data. *Papers in Regional Science*, 70, 329–44.
Buchanan, K. M. (1947) *Land Classification in the West Midlands*. London: Faber.
Buchanan, K. M. and MacPherson, A. W. (1948) *Conurbations: A Survey of Birmingham and the Black Country*. London: Architectural Press.
Burton, I. (1963) The quantitative revolution and theoretical geography. *Canadian Geographer*, 7, 151–62.
Caesar, A. A. L. (1964) Planning and the geography of Great Britain. *Advancement of Science*, 22, September, 1–11.
Carroll, J. D. (1955) Spatial interaction and the urban-metropolitan regional description. *Papers, Regional Science Association*, 1, 1–14.
Carruthers, W. I. (1957) A classification of service centres in England and Wales. *Geographical Journal*, 123, 371–85.
Carter, H. (1975) *The Study of Urban Geography*. London: Edward Arnold.
Chapman, G. T. (1977) *Human and Environmental Systems*. London: Academic Press.
Chisholm, G. G. (1889) *A Handbook of Commercial Geography*. London: Longman.
Chisholm, M. D. I. (1971) Geography and the question of relevance. *Area*, 3, 65–8.
Chisholm, M. D. I. (1995) Some lessons from the review of local government in England. *Regional Studies*, 29, 563–9.
Chisholm, M. D. I. (2000) *Structural Reorganisation of British Local Government: Rhetoric and Reality*. Manchester: Manchester University Press.
Chisholm, M. D. I. (2001) Economic doctrine, politics and local government finance. *Journal of Local Government Law*, 4, 18–23.
Chisholm, M. D. I. and Manners, G. (1971) *Spatial Policy Problems of the British Economy*. Cambridge: Cambridge University Press.
Chisholm, M. D. I. and Rodgers, B., eds (1973) *Studies in Human Geography*. London: Heinemann, for Social Science Research Council.
Clark, G. L. (2000) *Pension Fund Capitalism*. Oxford: Oxford University Press.
Clarke, G. P., ed. (1996) *Microsimulation for Urban and Regional Policy*. London: Pion.
Clarke, G. P. (1999) Geodemographics, marketing and retail location. In M. Pacione, ed., *Applied Geography*. London: Routledge, 577–92.
Clarke, G. P. and Wilson, A. G. (1987a) Performance indicators and model-based planning I: the indicator movement and the possibilities for urban planning. *Sistemi Urbani*, 9, 79–123.
Clarke, G. P. and Wilson, A. G. (1987b) Performance indicators and model-based planning II: model-based approaches. *Sistemi Urbani*, 9, 138–65.
Clarke, G. P., Langley, R., and Caldwell, W. (1998) Empirical applications of dynamic spatial interaction models. *Computer, Environmental and Spatial Systems*, 22, 157–84.
Clarke, G. P., Eyre, H., and Guy, C. (2002) Deriving indicators of access to food retail provision in British cities: studies of Cardiff, Leeds and Bradford. *Urban Studies*, 39, 2041–60.

Clarke, M., Keys, P., and Williams, H. C. W. L. (1981) Micro-analysis and simulation of socio-economic systems: progress and prospects. In N. Wrigley and R. J. Bennett, eds, *Quantitative Geography: A British View*. London: Routledge and Kegan Paul.

Cliff, A. D. and Ord, J. K. (1973) *Spatial Autocorrelation*. London: Pion.

Cliff, A. D. and Ord, J. K. (1980) *Spatial Processes: Models and Applications*. London: Pion.

Cliff, A. D., Gould, P. R., Hoare, A. G., and Thrift, N. J., eds (1995) *Diffusing Geography: Essays for Peter Haggett*. Oxford: Blackwell.

Coppock, J. T. (1964) *Agricultural Atlas of England and Wales*. London: Faber.

Coppock, J. T. (1974) Geography and public policy: challenges, opportunities and implications. *Transactions, Institute of British Geographers*, 63, 1–16.

Coppock, J. T. and Sewell, W. R. D., eds (1976), *Spatial Dimensions of Public Policy*. Oxford: Pergamon.

Cordey-Hayes, M. (1972) Dynamic framework for spatial models. *Socio-Economic Planning Sciences*, 6, 365–85.

Cordey-Hayes, M. (1975) Migration and the dynamics of multiregional population systems. *Environment and Planning A*, 7, 73–95.

Dawson, J. A. and Broadbridge, A. M. (1988) *Retailing in Scotland 2005*. Stirling: Institute for Retail Studies, University of Stirling.

Daysh, G. H. J., ed. (1949) *Studies in Regional Planning: Outline Surveys and Proposals for the Development of Certain Regions of England and Scotland*. London: George Philip.

Daysh, G. H. J. and Caesar, A. A. L. (1949) The North East of England. In G. H. J. Daysh, ed., *Studies in Regional Planning: Outline Surveys and Proposals for the Development of Certain Regions of England and Scotland*. London: George Philip, 77–108.

Daysh, G. H. J. and O'Dell, A. C. (1947) Geography and planning. *Geographical Journal*, 109, 1–3, 103–6.

Daysh, G. H. J. and Symonds, J. S. (1953) *West Durham: A Problem Area in North East England*. Oxford: Blackwell.

Department of Trade and Industry (2001) *Business Clusters in the UK*. London: DTI (2 vols).

Departmental Committee of Inquiry into Statutory Small Holdings (1966) *Statutory Smallholdings Provided by Local Authorities in England and Wales*. London: HMSO, first report of Departmental Committee of Inquiry into Statutory Smallholdings, Cmnd 2936.

Departmental Committee of Inquiry into Statutory Small Holdings (1967) *Statutory Smallholdings Provided by the Minister of Agriculture, Fisheries and Food*. London: HMSO, final report of Departmental Committee of Inquiry into Statutory Smallholdings, Cmnd 3303.

Dennison, S. R. (1939) The *Location of Industry and the Depressed Areas*. London: Oxford University Press.

Diamond, D. R. (1995) Geography and planning in the information age. *Transactions, Institute of British Geographers*, NS20, 131–8.

Dickinson, R. E. (1947) *City Region and Regionalism: A Geographical Contribution to Human Ecology*. London: Kegan Paul.

Dickinson, R. E. (1960) *City and Region: A Geographical Interpretation*. London: Routledge (1st edn 1947).

Doncaster Council (1922) *The Doncaster Regional Planning Scheme*, report prepared for Doncaster Council by P. Abercrombie and T. H. Johnson. Liverpool: University of Liverpool Press.

Dorling, D. (1995) *A New Social Atlas of Britain*. Chichester: Wiley.

Duley, C. and Rees, P. H. (1991) Incorporating migration into simulation models. In J. C. H. Stillwell and P. Congdon, eds, *Modelling Internal Migration*. London: Belhaven Press.

Echenique, M., Flowerdew, A. D. J., Hunt, J. D., Mayo, T. R., Skidmore, I. J., and Simmonds, D. C. (1990) The MEPLAN models of Bilbao, Leeds and Dortmund. *Transport Reviews*, 10, 309–22.

Estall, R. C. and Buchanan, R. O. (1961) *Industrial Activity and Economic Geography*. London: Hutchinson.

Evans, S. P. (1973) A relationship between the gravity model for trip distribution and the transportation model of linear programming. *Transportation Research*, 7, 39–61.

Evans, W. D. (1944) The opencast mining of ironstone and coal. *Geographical Journal*, 104, 102–19.

Fawcett, C. B. (1919) *Provinces of England: A Study of Some Geographical Aspects of Devolution*. London: Williams and Norgate.

Foster, C. D. and Beesley, M. E. (1963) Estimating the social benefit of constructing an underground railway in London. *Journal of the Royal Statistical Society A*, 126, 46–92.

Fotheringham, A. S. (1983) A new set of spatial interaction models: the theory of competing destinations. *Environment and Planning* A, 15, 1121–32.

Freeman, T. W. (1958) *Geography and Planning*. London: Hutchinson (4th edn 1974).

Geddes, P. (1905) Civics: as applied sociology. *Sociological Papers*, 1, 101–44.

Gerasimov, I. P., ed. (1962) *Soviet Geography: Accomplishments and Tasks. A Symposium*. New York: American Geographical Association, Occasional Publication No. 1 (English edition edited by C. D. Harris).

Gilbert, E. W. (1939) Practical regionalism in England and Wales. *Geographical Journal*, 94, 29–44.

Gilbert, E. W. (1948) The boundaries of local government areas. *Geographical Journal*, 111, 172–206.

Gilbert, N. (1995) Using computer simulation to study social phenomena. In R. M. Lee, ed., *Information Technology for the Social Scientist*. London: UCL Press, 208–20.

Gordon, I. (1995) Family structure, educational attainment and the inner city. *Urban Studies*, 33, 407–23.

Gottmann, J. (1964) *Megalopolis: The Urbanized Northeast Seaboard of the United States*. Cambridge, MA: MIT Press.

Gould, P. (1985) *The Geographer at Work*. London: Routledge and Kegan Paul.

Guldmann, J-M. (1999) Competing destinations and intervening opportunities interaction models of inter-city telecommunications flows. *Papers, Regional Science Association*, 78, 179–84.

Hägerstrand, T. (1970) What about people in regional science? *Papers, Regional Science Association*, 24, 7–21.

Haggett, P. (1965) *Locational Analysis in Human Geography*. London: Edward Arnold.

Haggett, P. and Chorley, R. J. (1969) *Network Analysis in Geography*. London: Edward Arnold.

Haggett, P., Cliff, A. D., and Frey, A. (1977) *Locational Analysis in Human Geography*, 2nd edn. London: Edward Arnold.

Haining, R. P. (1990) *Spatial Data Analysis in the Social and Environmental Sciences*. Cambridge: Cambridge University Press.

Hall, P. (1974) The new political geography. *Transactions, Institute of British Geographers* 63, 48–52.

Hall, P. (1982) *Urban and Regional Planning*. London: Allen and Unwin.

Hall, P. (1998) *Cities in Civilisation*. London: Weidenfeld and Nicholson.

Hamilton, H. C. and Falconer, W. (1892) *The Geography of Strabo*. London: George Bell.

Hancock, R. and Sutherland, H., eds (1992) *Microsimulation Models for Public Policy Analysis: New Frontiers*. London: London School of Economics.

Hansen, W. G. (1959) How accessibility shapes land use. *Journal of the American Institute of Planners*, 25, 73–6.

Hansom, J. D. (1992) Editorial. *Applied Geography*, 12, 5–6.

Hare, F. K. (1974) Geography and public policy: a Canadian view. *Transactions, Institute of British Geographers*, 63, 25–8.

Harris, B. (1962) *Linear Programming and the Projection of Land Uses*. Philadelphia, PA: Penn-Jersey Transportation Study, Paper 20.

Harris, B. and Wilson, A. G. (1978) Equilibrium values and dynamics of attractiveness terms in production-constrained spatial-interaction models. *Environment and Planning* A, 10, 371–88.

Harvey, D. W. (1973) *Social Justice and the City*. London: Edward Arnold.

Harvey, D. W. (1974) What kind of geography for what kind of public policy? *Transactions, Institute of British Geographers*, 63, 18–24.

Hay, A. M. (1973) *Transport and the Space Economy*. Basingstoke: Macmillan.

Hebbert, M. and Garside, P., eds (1989) *British Regionalism, 1900–2000*. London: Mansell.

Herbert, D. J. and Stevens, B. H. (1960) A model for the distribution of residential activity in an urban area. *Journal of Regional Science*, 2, 21–36.

Herbert, D. T. (1967) Social area analysis: a British study. *Urban Studies*, 4, 41–60.

Herbert, D. T. (1968) Principal components analysis and British studies of urban-social structure. *The Professional Geographer*, 20, 280–3.

Hewings, G. J. D. (1971) Regional input–output models in the UK: some problems and prospects for the use of non-survey techniques. *Regional Studies*, 5, 11–22.

Hewings, G. J. D. (1977a) *Regional Industrial Analysis and Development*. London: Methuen.

Hewings, G. J. D. (1977b) Evaluating the possibilities for exchanging regional input–output coefficients. *Environment and Planning A*, 9, 927–44.

House, J. W. (1965) *The Frontiers of Geography*. Newcastle: University of Newcastle upon Tyne (inaugural lecture, 8 March 1965).

House, J. W. (1973) Geographers, decision takers and policy makers. In M. D. I. Chisholm and B. Rodgers, eds, *Studies in Human Geography*. London: Heinemann, for Social Science Research Council, 272–305.

Huff, D. L. (1964) Defining and estimating a trading area. *Journal of Marketing*, 28, 34–8.

Imrie, R. and Thomas, H. (1999) *British Urban Policy*. London: Sage.

Institute of British Geographers (1968) *Land Use and Resources: Studies in Applied Geography*. London: Institute of British Geographers, Special Publication No. 1.

Jin, Y-X. and Wilson, A. G. (1993) Generation of integrated multispatial input–output models of cities. *Papers in Regional Science*, 72, 351–67.

Johnston, R. J. (1979) *Political, Electoral and Spatial Systems*. Oxford: Oxford University Press.

Johnston, R. J. (1981a) Applied geography, quantitative analysis and ideology. *Applied Geography*, 1, 213–19.

Johnston, R. J. (1981b) *Geography and the State*. Basingstoke: Macmillan.

Johnston, R. J., Hay, A. M., and Taylor, P. J. (1982) Estimating the sources of spatial change in election results: a multiproportional matrix approach. *Environment and Planning A*, 14, 951–61.

Johnston, R. J., Openshaw, S., Rhind, D. W., and Rossiter, D. J. (1984) Spatial scientists and representational democracy: the role of information processing technology in the design of parliamentary and other constituencies. *Environment and Planning C: Government and Policy*, 2, 57–66.

Johnston, R. J. (1986) *On Human Geography*. Oxford: Blackwell.

Keeble, D. (1976) *Industrial Location and Planning in the United Kingdom*. London: Methuen.

Keltie, J. S. (1908) *Applied Geography: A Preliminary Sketch*, 2nd edn. London: G. Philip.

Kenzer, M. S., ed. (1989) *Applied Geography: Issues, Questions, and Concerns*. Dordrecht: Kluwer.

Kenzer, M. S. (1992) Applied and academic geography and the remainder of the twentieth century. *Applied Geography*, 12, 207–10.

Lakshmanan, T. R. and Hansen, W. G. (1965) A retail market potential model. *Journal of the American Institute of Planners*, 31, 134–43.

Langston, P., Clarke, G. P., and Clarke, D. B. (1997) Retail saturation, retail location and retail competition: an analysis of British food retailing. *Environment and Planning A*, 29, 77–104.

Leontief, W. (1967) *Input–Output Analysis*. Oxford: Oxford University Press.

Leontief, W. and Strout, A. (1963) Multi-regional input-output analysis. In T. Barna, ed., *Structural Interdependence and Economic Development*. Basingstoke: Macmillan, 119–50.

Leung, Y. (1997) *Intelligent Spatial Decision Support Systems*. Berlin: Springer-Verlag.

Leung, Y., Gao, X-B., and Chen, K-Z. (2001) A dual neural network for solving entropy-maximising models. Unpublished manuscript, Department of Geography and Resource Management, Chinese University of Hong Kong.

Leyshon, A. and Thrift, N. (1998) *Money/Space: Geographies of Monetary Transformation*. London: Routledge.

Longley, P. A., Goodchild, M. F., MacGuire, D. J., and Rhind, D. W., eds (2001) *Geographic Information Systems Science*. Chichester: Wiley.

Lowry, I. S. (1964) *A Model of Metropolis*. Santa Monica, CA: The Rand Corporation, RM-4035-RC.

Lowry, I. S. (1967) *Seven Models of Urban Development: A Structural Comparison*. Santa Monica, CA: The Rand Corporation.

Macgill, S. M. and Wilson, A. G. (1979) Equivalences and similarities between some alternative urban and regional models. *Sistemi Urbani*, 1, 9–40.

Mackett, R. L. (1980) The relationship between transport and the viability of central and inner urban areas. *Journal of Transport Economics and Policy*, 14, 267–94.

Mackett, R. L. (1990) Comparative analysis of modelling land use–transport interaction at the micro and macro levels. *Environment and Planning* A, 22, 459–75.

Mackett, R. L. (1993) Structure of linkages between transport and land use. *Transportation Research* B, 27, 189–206.

Mackinder, H. (1919) *Democratic Ideas and Reality*. London: Constable & Co.

Manners, G. (1964) *The Geography of Energy*. London: Hutchinson.

Manners, G. (1981) *Coal in Britain*. London: Allen and Unwin.

Martin, R. L. (1999a) The new 'geographical turn' in economics: some critical reflections. *Cambridge Journal of Economics*, 23, 63–91.

Martin, R. L., ed. (1999b) *Money and the Space Economy*. Chichester: Wiley.

Massam, B. (1974) Political geography and the provision of public services. *Progress in Geography*, 6, 179–210.

Massey, D. B. (1968) *Problems of Location: Linear Programming*. London: Centre for Environmental Studies, Working Paper 14.

McCrone, G. (1969) *Regional Policy in Britain*. London: Allen & Unwin.

Ministry of Health (1920) *Report of South Wales Regional Survey*. London: HMSO.

Ministry of Housing and Local Government (1963a) *Central Scotland: A Programme for Development and Growth*. London: HMSO, Cmnd 2188.

Ministry of Housing and Local Government (1963b) *North East: A Programme for Development and Growth*. London: HMSO, Cmnd 2206.

Ministry of Housing and Local Government (1964) *The South East Study 1961–81*. London: HMSO.

Ministry of Town and Country Planning (1946) *Report on Derelict Land in the Black County*. London: HMSO.

Mohan, J. (1998) Uneven development, territorial politics and the British health care reforms. *Political Studies*, 46, 309–27.

Moseley, M. J. (1974) *Growth Centres in Spatial Planning*. Oxford: Pergamon.

Nakamura, H., Hayashi, Y., and Miyamoto, K. (1983) Land use transportation analysis system for a metropolitan area. *Transportation Research Record*, 931, 12–19.

National Plan (1965) London: HMSO, Cmnd 2764.

Odell, P. (1963) *Economic Geography of Oil*. London: Bell.

Odell, P. (1970) *Oil and World Power*. Harmondsworth: Penguin.

Openshaw, S. (1977) Optimal zoning systems for spatial interaction models. *Environment and Planning A*, 9, 169–84.

Openshaw, S. (1993) Modelling spatial interaction using a neural net. In M. M. Fischer and P. Nijkamp, eds, *GIS, Spatial Modelling and Policy Evaluation*. Berlin: Springer, 147–66.

Orcutt, G. H. (1957) A new type of socio-economic system. *Review of Economic Statistics*, 39, 116–23.

Orcutt, G. H., Watts, H. W., and Edwards, J. B. (1968) Data aggregation and information loss. *American Economic Review*, 58, 773–97.

O'Sullivan, P. (1978) Regions of a transport network. *Annals of the Association of American Geographers*, 68, 196–204.

Pacione, M. (1999) Applied geography: in pursuit of useful knowledge. *Applied Geography*, 19, 1–12.

Paulley, N. J. and Webster, F. V. (1991) Overview of an international study to compare models and evaluate land use and transport policies. *Transport Reviews*, 11, 197–222.

Peck, J. (1999) New labourers? Making a New Deal for the workless class. *Environment and Planning C: Government and Policy*, 17, 345–72.

Peck, J. (2001) *Workfare States*. New York: Guilford Press.

Pinch, S. (1985) *Cities and Services: The Geography of Collective Consumption*. London: Routledge and Kegan Paul.

Polish Academy of Sciences (1959) *Problems of Applied Geography I: Proceedings of the Anglo-Polish Seminar*. Warsaw: Polish Academy of Sciences (Institute of Geography), Geographical Studies 25.

Poston, T. and Stewart, I. (1978) *Catastrophe Theory and its Applications*. London: Pitman.

Poston, T. and Wilson, A. G. (1977) Facility size versus distance travelled: urban services and the fold catastrophe. *Environment and Planning A*, 9, 681–6.

Powell, A. G. (1970) The geographer in regional planning. In R. H. Osborne, F. A. Barnes, and J. C. Doornkamp, eds, *Geographical Essays in Honour of K. C. Edwards*. Nottingham: Department of Geography, University of Nottingham.

Prince, H. C. (1971) Questions of social relevance. *Area*, 3, 150–3.

Putnam, D. F. (1951) Geography is a practical subject. In G. Taylor, ed., *Geography in the Twentieth Century*. London: Methuen, 395–417 (3rd edn 1957).

Putman, S. H. (1983) *Integrated Urban Models: Policy Analysis of Transportation and Land Use*. London: Pion.

Putman, S. H. (1991) *Integrated Urban Models 2. New Research and Applications of Optimization and Dynamics*. London: Pion.

Rees, J. A. (1983) *Water Privatisation*. London: Department of Geography, London School of Economics, Discussion Paper No. 1.

Rees, P. H. (1979) *Residential Patterns in American Cities*. Chicago, IL: Department of Geography, University of Chicago, RP-189.

Rees, P. H. (1996) Projecting the national and regional populations of the European Union using migration information. In P. H. Rees, J. C. H. Stillwell, A. Convey, and M. Hupiszewski, *Population Migration in the Soviet Union*. Chichester: Wiley.

Rees, P. H. and Phillips, D. (1996) Geographical spread: the national picture. In P. Ratcliffe, ed., *Ethnicity in the 1991 Census*. London: HMSO, 23–109.

Rees, P. H. and Wilson, A. G. (1976) *Spatial Population Analysis*. London: Edward Arnold.

Reggiani, A., Tromanelli, R., Tritapepe, T., and Nijkamp, P. (1998) Neural networks: an overview and applications in the space economy. In V. Himanen, P. Nijkamp, and A. Reggiani, eds, *Neural Networks in Transport Applications*. Aldershot: Ashgate, 21–53.

Regional Science and Urban Economics (1976) Special issue on Public Economics and Planning in Space. *Regional Science and Urban Economics*, 6.

Rihll, T. E. and Wilson, A. G. (1987a) Spatial interaction and structural models in historical analysis: some possibilities and an example. *Histoire et Mesure*, II-1, 5–32.

Rihll, T. E. and Wilson, A. G. (1987b) Model-based approaches to the analysis of regional settlement structures: the case of ancient Greece. In P. Denley and D. Hopkin, eds, *History and Computing*. Manchester: Manchester University Press, 10–20.

Rihll, T. E. and Wilson, A. G. (1991) Settlement structures in Ancient Greece: new approaches to the polis. In J. Rich and A. Wallace-Hadrill, eds, *City and Country in the Ancient World.* London: Croom Helm, 58–95.

Robson, B. T. (1995) *The Index of Local Conditions: A Matrix of Results*. London: HMSO.

Robson, B. T., Parkinson, M., and Bradford, M. (1994) *Assessing the Impact of Urban Policy*. London: HMSO.

Romer, P. M. (1990) Indigenous technological change. *Journal of Political Economy*, 98, Supplement S71–102.

Royal Commission on Common Lands (1958) *Report of the Royal Commission on Common Lands 1955–58*. London: HMSO.

Royal Commission on the Distribution of the Industrial Population (1940) *Report of the Royal Commission on the Distribution of the Industrial Population*. London: HMSO, Cmd 6153 (Barlow Report).

Royal Commission on Local Government in England (1969) *Report of the Royal Commission on Local Government in England 1966–69*. London: HMSO, Cmnd 4040.

Royal Geographical Society (1938) Memorandum on the geographical factors relevant to the location of industry. *Geographical Journal*, 92, 499–526.

Sant, M. (1982) *Applied Geography: Practice, Problems and Prospects*. Harlow: Longman.

Sayer, R. A. (1976) A critique of urban modelling: from regional science to urban and regional political economy. *Progress in Planning*, 6, 187–254.

Scott, A. J. (1971) *Combinatorial Programming, Spatial Analysis and Planning*. London: Methuen.

Secretary of State for Economic Affairs (1965) *National Plan*. London: HMSO.

Senior, M. L. and Wilson, A. G. (1974) Explorations and syntheses of linear programming and spatial interaction models of residential location. *Geographical Analysis*, 6, 209–38.

Sheail, J. (1994) Geography and land use research: a UK historical perspective. *Applied Geography*, 14, 372–85.

Smailes, A. E. (1941) The re-distribution of settlement: the development of existing small towns. *Journal of the Town Planning Institute*, 27(6), 195–200.

Smailes, A. E. (1946) The urban mesh of England and Wales. *Transactions, Institute of British Geographers*, 11, 87–107.

Smailes, A. E. (1971) Urban systems. *Transactions, Institute of British Geographers*, 53, 1–14.

Smith, D. M. (1971) Radical geography: the next revolution. *Area*, 3, 153–7.

Smith, D. M. (1973) Alternative 'relevant' professional roles. *Area*, 5, 1–4.

Smith, D. M. (1977) *Human Geography: A Welfare Approach*. London: Arnold.

Smith, D. M. (1979) *Where the Grass is Greener*. London: Croom Helm.

Smith, W. (1949) *An Economic Geography of Great Britain*. London: Methuen.
Spence, N. J. and Taylor, P. J. (1970) Quantitative methods in regional taxonomy. *Progress in Geography*, 2, 1–64.
Stamp, L. D. (1943) The Scott Report. *Geographical Journal*, 101, 16–30.
Stamp, L. D. (1948) *The Land of Britain: Its Use and Misuse*. London: Longman.
Stamp, L. D. (1951) *Applied Geography: London Essays in Geography. Rodwell Jones Memorial Volume*. London: London School of Economics (later developed in Penguin edition, 1960).
Stamp, L. D. (1960) *Applied Geography*. Harmondsworth: Penguin.
Stamp, L. D. (1966) Ten years on. *Transactions, Institute of British Geographers*, 40, 11–20.
Stamp, L. D. (1969) *Nature Conservation in Britain*. Harlow: Longman.
Stamp, L. D. and Hoskins, W. G. (1963) *The Common Lands of England and Wales*. London: Collins.
Steers, J. A. (1944) Coastal preservation and planning. *Geographical Journal*, 104, 1–2, 7–27.
Steers, J. A. (1946) *The Coastline of England and Wales*. Cambridge: Cambridge University Press.
Stevens, A. (1921) *Applied Geography*. Glasgow: Blackie.
Stillwell, J. C. H. and Congdon, P., eds (1991) *Modelling Internal Migration*. London: Belhaven Press.
Stone, R. (1967) *Mathematics in the Social Sciences*. London: Chapman and Hall.
Stone, R. (1970) *Mathematical Models of the Economy*. London: Chapman and Hall.
Sunley, P., Martin, R. L., and Mativel, C. (2001) Mapping the New Deal: the local disparities in the performance of welfare-to-work. *Transactions, Institute of British Geographers*, NS26, 484–512.
Taylor, E. G. R. (1938) Discussion on the geographical distribution of industry. *Geographical Journal*, 92, 22-39.
Thom, R. (1975) *Structural Stability and Morphogenesis*. Reading, MA: Addison-Wesley.
Thompson, J. H. (1964) What about a geography of poverty? *Economic Geography*, 40, 283.
Timms, D. W. G. (1971) *The Urban Mosaic*. Cambridge: Cambridge University Press.
Walford, R. (1997) *Land-use UK*. Sheffield: Geographical Association.
Webster, F. V., Bly, P. H., and Paulley, N. J., eds (1988) *Urban Land Use and Transport Interaction*. Aldershot: Gower.
Wegener, M. (1986a) Transport network equilibrium and regional deconcentration. *Environment and Planning A*, 18, 437–56.
Wegener, M. (1986b) Integrated forecasting models of urban and regional systems. In *London Papers in Regional Science*, 15, 9–24.
Wegener, M. (1994) Operational urban models: the state of the art. *Journal of the American Institute of Planners*, 60, 17–29.
Willatts, E. C. (1971) Planning and geography in the last three decades. *Geographical Journal*, 137, 311–38.
Williams, H. C. W. L. (1977) On the formation of travel demand models and economic evaluation measures of user benefit. *Environment and Planning* A, 9, 285–344.

Williams, H. C. W. L., Kim, K. S., and Martin, D. (1990) Location-spatial interaction models: 1, 2 and 3. *Environment and Planning* A, 22, 1079–89, 1155–68 and 1281–90.

Williams, W. M. and Herbert, D. T. (1962) The social geography of Newcastle-under-Lyme. *North Staffordshire Journal of Field Studies*, 2, 108–26.

Wilson, A. G. (1967) A statistical theory of spatial distribution models. *Transportation Research*, 1, 253–69.

Wilson, A. G. (1970) *Entropy in Urban and Regional Modelling*. London: Pion.

Wilson, A. G. (1971a) Generalising the Lowry model. *London Papers in Regional Science*, 2, 121–34.

Wilson, A. G. (1971b) A family of spatial interaction models and associated developments. *Environment and Planning*, 3, 1–32. Reprinted in A. G. Wilson (1972) *Papers in Urban and Regional Analysis*. London: Pion, 170–201.

Wilson, A. G. (1974) *Urban and Regional Models in Geography and Planning*. Chichester and New York: Wiley.

Wilson, A. G. (1981) *Catastrophe Theory and Bifurcation: Applications to Urban and Regional Systems*. London: Croom Helm and Berkeley, CA: University of California Press.

Wilson, A. G. (1995) Simplicity, complexity and generality: dreams of a final theory in locational analysis. In A. D. Cliff, P. R. Gould, A. G. Hoare, and N. J. Thrift, eds *Diffusing Geography: Essays for Peter Haggett*. Oxford: Blackwell, 342–52.

Wilson, A. G. (1998) Land use transport interaction models: past and future. *Journal of Transport Economics and Policy*, 32, 3–26.

Wilson, A. G. (2000) *Complex Spatial Systems: The Modelling Foundations of Urban and Regional Analysis*. Harlow: Prentice Hall.

Wilson, A. G. and Bennett, R. J. (1985) *Mathematical Methods in Human Geography and Planning*. Chichester and New York: Wiley.

Wilson, A. G. and Oulton, M. J. (1983) The corner shop to supermarket transition in retailing: the beginnings of empirical evidence. *Environment and Planning* A, 15, 265–74.

Wilson, A. G. and Pownall, C. M. (1976) A new representation of the urban system for modelling and for the study of micro-level interdependence. *Area*, 8, 256–64.

Wilson, A. G. and Senior, M. L. (1974) Some relationships between entropy maximizing models, mathematical programming models and their duals. *Journal of Regional Science*, 14, 207–15.

Wilson, A. G., Rees, P. H., and Leigh, C. M., eds (1977) *Models of Cities and Regions*. Chichester and New York: Wiley.

Wilson, A. G., Coelho, J. D., Macgill, S. M., and Williams, H. C. W. L. (1981) *Optimization in Locational and Transport Analysis*. Chichester and New York: Wiley.

Wise, M. J. (1963) The South Wales motorway. *Geographical Magazine*, December.

Wise, M. J. (1965) The impact of a channel tunnel on the planning of south eastern England. *Geographical Journal*, 131, 167–79.

Wise, M. J. (1969) The future of local government in England: The 'Redcliffe-Maud Report'. *Geographical Journal*, 135, 583–7.

Wise, M. J. (1973) Introduction. In M. Chisholm and B. Rodgers, eds, *Studies in Human Geography*. London: Heinemann, for Social Science Research Council.
Woods, R. I. (1981) Spatiotemporal models of ethnic segregation and their implications for housing policy. *Environment and Planning A*, 13, 1415–33.
Wooldridge, S. W. and Beaver, S. H. (1950) The working of sand and gravel in Britain: a problem in land use. *Geographical Journal*, 115, 42–58.
Wrigley, N. and Lowe, M., eds (1995) *Retailing, Capital and Consumption: Towards the New Retail Geography*. Harlow: Longman.
Wrigley, N. and Lowe, M. (2002) *Reading Retail: A Geographical Perspective on Retailing and Consumption Spaces*. London: Arnold.

VI. Geography moving forwards

14

Geographers and environmental change

JOHN B. THORNES*

Within geography, physical geography is concerned with the characteristics of the natural environment (as outlined by Gregory in Chapter 3, above), the atmosphere, the lithosphere, and the biosphere; how they influence human activities and how they are affected by them across the face of the globe. It comprises geomorphology, climatology, and biogeography, and proceeds by monitoring, modelling, and managing environmental change. On the whole, the marine domain is excluded, mainly because it has attracted relatively little research interest from physical geographers. A notable exception is the department at Hull University, which has, over the years, contributed importantly to our knowledge of coastal processes and shoreline evolution under the guidance of Hardisty (1990). The lack of research input is partly due to the large and special resources needed for research in this area and the prominent role of national institutions.

One of the major global changes of the relatively recent past has been the cooling of climates that led to the glaciation of the northern hemisphere. Geographical research at first concentrated on the direct impacts of glaciation on the geomorphology of Britain, such as the glacial erosion of northern Britain and its indirect impacts, especially the effects of changing sea levels. David Linton (1949, 1951), one of the founding fathers of British geomorphology, carried out remarkable detective work to untangle the glacial history of highland Britain, whilst later Clayton (1960) established the glacial sequence of lowland Britain, notably in East Anglia and the London basin. This work continues to the present time in the British Isles under the leadership of D. Q. Bowen (1991). Of course the indirect impacts of

* The author completed this contribution whilst a Hugh Kelly Research Fellow at Rhodes University, Grahamstown, South Africa. He wishes to acknowledge the support of the Geography Department during his tenure and Rhodes University for providing the opportunity.

glaciation are worldwide and in former British colonies there is a legacy of geomorphological studies that show how glaciation changed the climate, the vegetation, and the rates and location of soil erosion in East and West Africa, Malaysia, Southeast Asia, Australia, and North America. Later the study of glacial processes overtook the chronological studies in contemporary glaciated environments from the Himalayas to the Alps and from Norway to Antarctica (Sugden and John, 1982). Biogeography has seen an important resurgence of activity in recent years, led by research by Ian Simmons at Durham University (e.g. Simmons et al., 1988) and encouraged by the establishment of the *Journal of Biogeography*, edited by Stott at SOAS.

Physical geographers in the last 100 years have taken some comfort from the knowledge that their skills are applied in matters of public interest and importance. Now the pace of global environmental change is such that these skills will be essential in the next 100 years, in solving some of the great contemporary environmental problems. Within the context of the newly emerged paradigms outlined later in this chapter, these changes are considered to be the springboard for future physical geography.

Global warming

Global environmental change mainly refers to climate change resulting from human actions in changing the gaseous composition of the atmosphere. These are the so-called 'greenhouse gases' that constrain the loss of heat back into the outer atmosphere by reducing the reflected heat from the surface. This global warming, caused by carbon dioxide, methane, and other gases, changes not only the temperature but also the circulation pattern and through it the rainfall, soil moisture, and type and amount of plant cover over large areas of the globe. It is the actual change and likely impacts that are the cause of international concern and that have led to the International Convention on Climate Change (ICCC), whose aim is to avoid dangerous interference in the climate system. Signatories to the convention agree to reduce greenhouse gas emissions and to report regularly on strategies to mitigate and adapt to climate change.

Although there have been two decades of doubt about the claims of disastrous consequences, evidence is quickly firming up that temperatures are rising faster than can be explained (for example by solar activity); already the impacts are beginning to show through, as the following list (published by the South African Institute of Botany, 2001) indicates.

- Temperature reconstruction since AD 1000 indicates that the twentieth century was unusually warm and the 1990s were the hottest decade on record.
- Global sea levels rose 10–15 cm in the last century.

- Glaciers in the European Alps have lost half their volumes since the 1850s.
- The Arctic ice cap has thinned by 40 per cent since the 1950s.
- The ranges of 63 per cent of non-migratory European butterfly species have shifted northwards by 35–240 km since 1990.
- The concentration of carbon dioxide in the atmosphere has increased more than 30 per cent since the dawn of the industrial revolution and is now higher than it has been in the past 430,000 years.

The UK Meteorological Office Hadley Centre, one of the world's leading centres for the study of climate change, predicts that carbon dioxide concentrations will rise from 370 ppm in 2000 to 550 ppm in 2050. Land areas will warm up faster than oceans and polar latitudes faster than temperate latitudes worldwide.

British geographers have contributed to the debate about the causes and impacts of global warming. Hare (1966), formerly at King's College London, clarified the dynamics of atmospheric circulation, leading the way to a much wider appreciation of the effects of changes on the circulation of the atmosphere. His *Restless Atmosphere* shifted the emphasis from the descriptive statistical climatology of the climate regions of the earth to a dynamic one. Hare's research also established the idea of permanency of the subtropical high pressure over the Sahel region, thereby confronting the over-simplified view of the 'advancing Sahara' that had gained currency just prior to the Second World War. The human potential for changing local climate in urban areas was recognised by Chandler (1995) in his seminal study, *The Climate of London*, in which he demonstrated that the city is a heat island compared with the surrounding countryside and that collecting climatic data in the world's great cities could give an exaggerated impression of global climate change. This result is found to be true of other world cities whose architecture has encouraged heat accumulation.

Geographers have been more directly involved in studying the *impacts* of global warming than either its causes and mechanisms or the policy implications developing therefrom. However, the study of impacts is a rather a difficult task, not least because of the uncertainty of predicting temperature changes. In fact future rainfall proves even more difficult than temperature to estimate, despite the continuing improvement of Global Circulation Models (GCMs). Constructed mainly by physicists, these have used the governing equations of atmospheric flow and the expected effects of aerosols on raindrop-forming processes. They are validated by running the models for the past 30 or 40 years to observe their capacity to reproduce known climate data. If found satisfactory, they can then be subject to the expected trends for the next 60 years or so, to provide the expected changes in climate for the middle of the current century.

If these are accepted as meaningful scenarios of the climate's response to changing greenhouse gases, the follow-up question is: what impact will they have on the geography of the global environment? Climate affects, most of all, the vegetation cover and, through this, the runoff from hillslopes into rivers and lakes. It also affects farming, which depends on bioproductivity and, above all, the capacity of agriculture to meet the food demands of European populations. In southern Europe, notably in the Mediterranean, reductions in rainfall will also bring about changes in the vegetation cover, leading to lower cereal production and increased soil erosion. The EU MEDALUS project (MEditerranean Desertification And Land USe) has shown that in areas such as southeast Spain, which have more than six months a year without rain and less than 300 mm of rainfall, the bush and grass cover will be severely affected by global warming. Teams of geographers from Leeds University and King's College London have created digital models to predict how this erosion will affect farming sustainability through erosion, soil salinisation, and lower moisture availability for growing crops. At a time when sustainable agriculture is high on the international agenda, it is ironic that natural forces on a global scale are constraining the potential for achieving sustainable agriculture in the Mediterranean area.

Parry has published extensively (e.g. 1990) on the effects of global warming on European agriculture. Essentially agricultural production is controlled by the length of the growing season, usually identified by the number of days that the temperature is above the threshold for growth, given that sufficient moisture is available. These studies indicate that European farmers will be able to extend the limits of cultivation northwards and grow a wider variety of crops. Worldwide, it is the drylands that are most likely to suffer from global warming, and Europe is no exception.

Another approach to defining the effects of changing climate is to seek contemporary analogues (places that have the same climate today) or temporal analogues (places that had the same climate in the past). The former was developed to good effect by geomorphologists, who used the methodology to establish how cold it was in the areas just beyond the ice sheets that occupied Britain in the Quaternary era. By comparing the processes with those in periglacial areas around existing glaciers, they were able to account for the landforms in upland areas of the Pennines, Scotland, and the southwest peninsula (Dartmoor and Exmoor). In Devon, Waters (1964) demonstrated that the thick blankets of periglacial sludge had smoothed off the crags and contours, leaving the more rounded hills we see today. These he attributed to soil flowage caused mainly by deep freezing of the soil mantle when glaciers were as far south as York. Since Waters' work, similar deposits have been reported from the Rockies and the Southern Alps of New Zealand, the Drakensbergs of the Eastern Cape in South Africa, and the mountains of southernmost China. Even more surprising, geomorphologists have inverted the procedure to argue for past climates on the basis of the landforms they observed. This seems very plausible when glacial landforms are used to infer glacial climates, but rather more tentative when extensive erosion surfaces are used to infer deep tropical weathering climates of the type found in, say, Nigeria today.

Spatial analogues are also used to infer the impacts of climate change along climate gradients, by looking for locations along the climate gradients that have conditions similar to those expected under global warming. In the Levant of Spain, from Barcelona to Almeria, the rainfall changes from about 1,000 mm in the north to 200 mm in the south; thus Valencia's landscape might provide a suitable analogue for a globally warmed Catalonian landscape. Unfortunately, the '*ceteris paribus*' argument defies this approach because geographically other factors—soil, topography and land use practices—also determine the response to warming. Moreover, it is assumed that the stable relationships between climate and surface characteristics will remain more or less constant. Unfortunately, the currently prevailing paradigm (see below) indicates that this is extremely unlikely in systems so close to thresholds of change.

Intellectually much more rigorous and satisfying was the careful analysis of both historical documentary sources and sedimentary records by J. M. Grove (1988), which cumulatively demonstrated the existence of a 'Little Ice Age' in the sixteenth/seventeenth centuries. This is now widely recognised to have affected most of Western Europe and to have been associated with environmental changes that affected crop growth, human health, the runoff of rivers, and the vegetation cover in British uplands, and even the creation of mini-glaciers in the Scottish islands. Goudie (1993) has accumulated evidence that the 'normal' condition of the earth's surface for the latest period of geological history was an arid one; the hotter earth of contemporary global warming is thus a reversion to the mean status over a much longer period of time.

Losing the forest

Another major environmental 'scare' at the end of the last millennium was the outcry about the global disappearance of the forest. Concern was expressed because of the potential feedback from forest to climate and the loss of biodiversity that the loss of forest (especially tropical rainforest) represents. Forests represent huge accumulations of carbon over centuries of time. As they are fired or cut down, this carbon is returned to the atmosphere, just as burning coal in the Industrial Revolution increased the greenhouse gases. Removal of forest also enhances soil erosion, as the interception of intensive tropical rainfall by the forest canopy reduces the energy of the rain reaching the ground surface. As the chemical constituents found at depth in tropical soils are exposed to the atmosphere, so they turn into hard carapaces at the surface; these duricrusts increase surface runoff and create difficulties for cultivation. Hence the notion of a transformation of 'Green Hell to Red Desert' and the exaggerated idea that the whole of Amazonia would be turned into a 'giant car park' if current rates of deforestation continue.

The global doom-mongers have been encouraged by the great increase in the capacity to observe environmental change as a result of the development of satellite-borne remote sensing and the techniques for analysing the resulting images. Geographers have made major contributions to this methodology. In the early days, the specialised computing equipment necessary to undertake the analysis restricted the research to a few departments: Reading, Bristol, and Nottingham must take much of the credit for this work. Using this technology to monitor changes of cover and at global scale enabled the evolution of the deforestation of Amazonia to be quantified more or less as it took place. As a result, Brazil became the apparent global culprit in environmental change circles—so much so that, by 2001, it had developed its own satellite, which is geostationary over Brazil to monitor the change in forested area. Brazil has long been a leader in capturing the effects of global change: its RADAM project, which provided photographic coverage of most of Amazonia by the 1970s, led the way to the assessment of forest transformation by remote sensing.

In a quite different application of this global technology, Barrett and his team at Bristol (Barrett and Martin, 1981) developed a method for estimating rainfall over the vast interior drylands of the Sahara–Sahel belt of Africa, where the density of rain gauges is so sparse as to provide little information of use to such vast areas. It was demonstrated that certain types of clouds at known temperatures have a given (higher) probability of providing rain. The Bristol group showed that the type and temperatures of clouds could indeed be determined from remotely sensed imagery. A logical extension of this is the capacity to predict outbreaks of locust swarms. Another innovative application of remote sensing was developed in the very early days of the technology by Cole (1980) of Bedford College, London. She showed that distinctive plant associations characteristic of metalliferous rocks and their surface soils could be identified and mapped using remote sensing techniques. This led to the discovery of new sources of wealth in Australia. Today the methodology is used to estimate soil erosion and its impact on sediment supply to Lake Tanganyika by Drake and his research group at King's College London (Drake et al., 1998). The leading UK geography departments in this field turn out undergraduates well trained in remote sensing, just as the masters' programmes enable agronomists, planners, and foresters from all over the world to make better use of their natural national resources.

It would be easy to forget that the basic lessons about forest clearance were largely learned through archival research and on foot. Darby's meticulous and detailed studies of the Domesday Survey of 1086 indicated the extent to which the forests of England, especially eastern England (e.g. Darby, 1952), were already decimated by the end of the first millennium, even though deforestation is widely assumed to be a scourge of the twentieth century. As the vegetation was changed and land freshly cultivated, heavy storms eroded the soil, producing large amounts of sediment, which silted up rivers and lakes. Across Europe these impacts can be traced by an almost universally redeposited soil, the *auhlehm*, which has generally been attributed to changed techniques of cultivation and the growth in demand for cultivated cereals. We must therefore conclude with Williams (1996) that most of the global forests had already disappeared in Europe, America, and Africa before the inroads in South America had hardly started.

Difficult though it has proved to be, estimating the amount and rate of loss of the global forest is not enough. As with climate, it is important to understand the impacts. It is here that the real academic contribution of the Royal Geographical Society (RGS) must be acknowledged. As it shifted from the unfashionable 'expedition' with its colonial image to the more acceptable 'project', it continued to undertake well-founded and scientifically worthwhile field campaigns in areas where global change and forest loss were critical: it has confidently led and inspired the bandwagon of environmental concern in the Matto Grosso, in the Guinean Shield, in the bushlands of Africa and Australia, and in tropical parts of Southeast Asia. The model that has evolved is of the 'field university', where visiting scientists of international reputation can cooperate with local scientists in the field. Above all, it is probably in the area of tropical forest change processes that it has achieved the most thorough and precise studies in small areas. The detailed study of change in the Isla do Maraca of northern Brazil involved a multidisciplinary team of geographers, botanists, hydrologists, and climatologists. The work of geographers such as Furley, Eden, McGregor, Cole, and Ross has illuminated the nature and the dynamics of the forest/savannah boundaries in Roraima, Brazil (Furley, Proctor, and Ratter, 1992). Although London based, the RGS established a network of regional branches that bring the nice mix of science and exploration to a much wider and generally much younger fellowship. Each year, through its corporate sponsors, the Society is able to support many field projects across the world for geographical scientists, young and old.

It has often proved difficult to separate out the effects of changes of climate from changes in human activity. By using the composition of the sediments in lakes, Walling (1999) and Foster (2000) have been able to identify the main areas in river basins that the eroded sediments come from, and hence infer the environmental changes in the catchment. Recently, the Exeter and Coventry departments have improved the methodology that uses radioactive fallout from weapons testing to date recent phases of erosion in the UK and around the Mediterranean. Walling was able to trace the redistribution of fallout from the Chernobyl disaster in Wales and the Welsh borderlands and, together with He (Walling and He, 1999), has provided new insights into the contemporary evolution of floodplain systems, using caesium-137. This stream of work was generally initiated by the work of Oldfield at Liverpool and Battarbee at University College London. Oldfield extensively explored and developed the measurement of remnant magnetism in lake sediments and showed that it could be used for dating the sediments. The strength of the earth's magnetic field varies through time, leaving a pattern in the minerals deposited, with deposits of the same age showing the same pattern (Oldfield, 1997). Oldfield currently serves as an executive director of the IGBP PAGES programme.

Acid rain was one of the first environmental change crises to attract international attention and concern. Large quantities of industrial atmospheric pollution were said to be acidifying rainwater, which in turn got into lakes, streams, and channels through the hydrological cycle and, it was claimed, were killing trees and fishes. Battarbee (2000) and his geography group at University College London were able to trace the history of atmospheric acid rainfall by recognising the sensitivity of assemblages of freshwater diatoms to the acid level in lake waters. These could then be compared with industrial emissions and power station production, to establish the main cause-and-effect relationships in a complex web of factors. It is also the case that lichens that live on trees are extremely sensitive to atmospheric pollution and F. Rose (1976) innovatively mapped the occurrence of indicator species in southeast England, thereby deciphering the spatial extent of atmospheric pollution caused by the city. This methodology is now widely used throughout Europe to map the effects of urban pollution on the environment.

Desertification

The third major environmental change that has attracted international attention, the problem of desertification, is global in scope and will loom large in the future agenda. Variously defined, desertification is essentially land degradation in arid and semi-arid regions, where 'land' is taken to includes soil and vegetation. Much of the confusion surrounding the term has arisen from the French usage that implies desertion—a loss of population. Often the results are quite similar, the one leading to the other. Almost a century ago, colonists in West Africa claimed to have demonstrated that the Saharan southern edge in the Sahel was moving southwards. Detailed examination suggested that the claim appears to have been mis-founded, the concept of a southward-moving east–west desert front being a rather gross distortion of reality. Despite the International Convention on Combating Desertification (ICCD), signed and ratified by 170 states (in 1994) and promoted by the

United Nations, British geographers are divided. Agnew and Chapell (2000) have demonstrated that the perception that the Sahara is drying out is probably an artefact of the changing distribution and density of rain gauges across this enormous continental area. Thomas and Middleton (1994) have argued that desertification is a myth perpetrated by the United Nations to rally world interest and attention around a contemporary environmental problem. A major difficulty here is that soil erosion varies greatly in time and space according to physiography, vegetation, agricultural practices and, of course, climate change. As a result, national claims about extent are, at best, 'guestimates', based more on hopes of compensation than observed erosion rates. Until the methods of estimating erosion described in the previous section are better developed and applied, the uncertainty will continue. An excellent review of global desertification and methods of combating it was provided by Grainger (1990) from Leeds University.

The case presented by Thomas and Middleton is compelling, but so too is the reality of falling well levels, dying vegetation, and declining agricultural productivity, eventually leading to rural depopulation and the rural welfare and poverty problems that are the consequences of desertification around the Mediterranean. As indicated earlier, geographers from Leeds University and King's College London have studied the phenomenon intensively, in conjunction with Spanish, Portuguese, Italian, and Greek colleagues. The results indicate that the future is likely to bring a 15 per cent reduction in annual rainfall and up to 56 per cent decline in bush and grass vegetation cover, with consequent accelerated losses in soil in the southeast of Spain, in mainland Greece, and in the Aegean islands (Diamond and Woodward, 1998).

Water

The future challenge at global scale will be concerned with freshwater resources. In all temperate, sub-humid and semi-arid areas, the demand for fresh water is outstripping the supply: populations continue to expand, rates of usage change with improved standards of living, and water quality is degraded by concentrated and diffuse runoff. For centuries, soil, rocks, and agricultural areas have been polluted through careless waste management. As transport facilities accelerate, dangerous chemicals run off from major highways. As agriculture develops and crop yields increase through the application of fertilisers and pesticides, residues accumulate in ponds, lakes, and reservoirs. The management of waste has become a major issue in both the developed and the developing world. A few years ago the primary problem was nitrate contamination of drinking water; now it is soil contamination in landfill sites and areas of mining waste.

Rivers have been undergoing human-induced changes for centuries, from the construction of mill weirs to the abstraction of river water to maintain public water supplies. Since the inception of agriculture in Neolithic times, the major catchments and rivers of Britain—the Severn, Trent, and Thames—have been substantially altered through changes in land use, draining, dredging, and straightening. Moreover, as the global environment changes, rivers respond in terms of discharge, morphology, and sediment transport, and there are complex unpredictable reactions. Fortunately, studies by Gregory, Lewin, and Thornes (1987) of palaeohydrology have enabled us to disentangle to some extent the human and climatic effects. Arnell (1999) has recently looked towards the future climatic effects on European rivers, showing that, in some cases, existing uses will quite definitely be unsustainable 50 to 60 years from now. Accepting the logic of this, the European Union has issued a Water Framework Directive that requires member states to address the issues of water quality, flood control, and over-exploitation by creating catchment management plans over the next 15 years.

Cambridge geographers Chorley and Kirkby (e.g. Kirkby, 1969) shifted the emphasis of hydrology from a preoccupation with rivers and river flow to an acceptance that it is on the hillside where rain is transformed into streamflow. The proper understanding of the relationship between climate change and river flow has to be based on this transformation, as it is affected by the soil and vegetation of the hillslopes. This hydrological revolution has completely changed the understanding of the environmental impacts of climate changes on rivers, because the hydrological regime is better appreciated when soil type and thickness and vegetation cover are taken into account. The very shape of the land, with its convergences and divergences, has a significant control on how the hillslope hydrology works, as shown by Anderson and Burt (1978) in the Bristol School of Geography. In short, the pathways taken by water through drainage basins are almost as important as the type and amount of rainfall.

At the end of the day, the drying up of rivers and other water-supply systems will have important economic and ecological impacts. Hollis (1998) has shown the threat to the great wetlands of Tunisia (Lake Ichkeul) implicit in current reductions of Mediterranean rainfall. In central Spain, the Ojos de Guadiana, a large wetland in Castilla la Mancha, famous as a stopover for birds migrating between Europe and Africa, has already dried up so much that the peat left behind has self-ignited and combusted. Wetlands are protected by an international convention and Maltby (1997) from Exeter University has played a major role in developing the science and social science of wetland conservation.

The main emerging problem of environmental regulation is the geographical mosaic onto which the regulatory mechanisms are imposed. These are so diverse that directives designed for the whole of Europe are invariably inappropriate in some areas and local regulatory flexibility has to be incorporated into the planning mechanisms. British geographers have been active in pressing home this message. O'Riordan and Voisey (1998) have demonstrated the difficulties in nations as environmentally diverse as Greece and Germany in meeting the Agenda 21 demands for sustainable planning at the local level. The European Environmental Agency has been structured along traditional lines into atmosphere, biosphere, and lithosphere, thus dividing environmental problems into process domains. A more satisfactory approach would have been to divide the administration of environmental change and protection into geographical domains—boreal forest, old industrial areas, mountain lands, and Mediterranean environments—to achieve better appreciation of the nature of the required regulation.

Three paradigm shifts

In the face of the great problems posed by environmental change and by environmental modification, more than physical description and explanation of the phenomena are required. There have been three important paradigm shifts in the last 20 years that have greatly altered the approaches to the problems and will determine the way they play out over the coming 60 years (the time horizon of climate change predictions).

First, it is now recognised through the work of Wilson (1981), and American geomorphologist S. A. Schumm (1979), that catastrophe theory and bifurcation are essential to understanding the location of thresholds and the rates of change in rural as well as urban economies. Quite small changes in environmental parameters may lead to profound and sometimes irreversible consequences for human activities, and even result in the elimination of entire settlement patterns (Wagstaff, 1978). This is well illustrated by soil erosion control. It is now established that an important bifurcation occurs when the vegetation cover (as viewed from above) is at about 30 per cent. Thornes (2002) has shown that small changes in rainfall may be responsible for significant shifts in the boundary between woody and grass vegetation under the pressure of heavy grazing. Below this value, erosion increases very rapidly, reaching a maximum when the soil is completely bare. An increase of cover above 30 per cent has less effect on restraining soil erosion. It has been demonstrated further that this relationship depends on the intensity of the rain. As the average rainfall is more concentrated in time, the bifurcation point shifts to a higher vegetation cover. The implication is clear: where the cover is close to 30 per cent (mainly in low rainfall areas), small changes in the vegetation cover to below 30 per cent will have big erosional impacts. This is why the erosion–grazing issue is so complex in the mathematical sense. The interaction between grazing and erosion, which is widely cited as cause and effect in the destruction of semi-arid environments, has been the subject of a deep and comprehensive review by Warren (1995). Brunsden and Thornes (1979) expressed the implicit non-linearity of geomorphological systems in their paper on landscape sensitivity and change, indicating that 'sensitive landforms' (those near a form threshold) would be highly susceptible to change and therefore somewhat uncommon in the landscape.

This paradigm of non-linear change has very deep implications for management and for modelling the response of systems to change, as was identified by Bennett and Chorley in 1978. They demonstrated that it is not enough to construct extensive box-and-arrow diagrams that connect everything to everything else in the environment. If the system is close to a threshold of change, then a dynamic system modelling approach is required. Quite small adjustments can create negligible or substantial effects. We need to know if a little push or a big push is needed to change the systems of concern and for management purposes we need to develop indicators that are sensitive to these thresholds.

The second main shift has been the recognition, more by those working in Africa than in Europe, that 'technical fixes' of environmental change that have their roots in social causes are simply insufficient as methodologies for addressing these acute problems. Rather what is called for is deeper community involvement in developing sustainable solutions. The issue has become closely linked to the democratic empowerment of those whose environment is being managed. In particular, the international agencies at the Rio Conference in 1992 pressed hard for sustainable development with a high level of local involvement. The human geographers Redclift (1992) and O'Riordan (1989) have been major contributors to this debate at the international level, whereas physical geographers, lacking the required skills, have tended to stand on the sidelines, whilst recognising the importance of the approach.

The third major paradigm change has been the recognition that environmental problems are multidisciplinary rather than simply interdisciplinary in scope. Given the social dimensions they can hardly be 'owned' by physicists, biologists or atmospheric chemists. One outcome has been the proliferation of degrees in the subject, but these are quickly becoming under-subscribed as the complexity of the problems is recognised. There has also been a steady growth of centres of research into environmental problems—the Environmental Change Unit at Oxford, the Climatic Research Unit at East Anglia, CSERGE at University College London, the Centre for Dryland Research at Sheffield, and so on.

Concluding remarks

The emerging messages of this chapter should be clear. They are that:

1 natural and human causes of local, regional and global environmental change are extremely difficult to separate;

2 almost invariably the symptoms are non-linear in behaviour;
3 the problems are usually multi-faceted and multidisciplinary in scope and they cannot be within the exclusive scope of one discipline; and
4 above all, they require a thorough understanding of the human dimensions in their management. Warren (1995, 199) expressed the problem succinctly:

> Different cultures have diverse systems of environmental science and these permit significantly different responses to the same environment.

The time when we can sit in ivory towers contemplating the origin of environmental changes taking place upwards of a million years ago is past. The responsibility is to engage with the present and the future of the globally changing environment. At least we can take some comfort in the fact that a first degree in geography inculcates a strong awareness of the human dimensions of environmental problems. Ten years ago, Redclift highlighted the problem as follows:

> The tortuous road to greater responsibility for environmental change will not, ultimately, be built on the uncertain predictions of natural scientists. Rather it is likely to be built on the daily lives of human subjects and the recognition that these lives involve choices of global proportions. (Redclift, 1992, 32)

Whilst the geographical environmental change community have taken this crucial message on board, the evidence is that other environmental scientists are still struggling with the concept.

References

Agnew, C. T. and Chappell, A. (2000) Desiccation in the Sahel. In S. J. Maclaren and D. R. Kniveton, eds, *Linking Climate Change to Land Surface Change*. The Hague: Kluwer, 27–48.

Anderson, M. J. and Burt, T. (1978) The role of topography in controlling throughflow generation. *Earth Surface Processes*, 3, 331–44.

Arnell, N. W. (1999) The effect of climate change on hydrological regimes in Europe: a continental perspective. *Global Environmental Change*, 9, 5–23.

Barrett, E. C. and Martin, D. W. (1981) *The Use of Satellite Data in Rainfall Monitoring*. London: Academic Press.

Battarbee, R. W. (2000) Palaeolimnological change approaches to climate change, with special regard to the biological record. *Quaternary Science Review*, 19, 107–24.

Bennett, R. J. and Chorley, R. J. (1978) *Environmental Systems, Philosophy, Analysis and Control*. London: Methuen.

Bowen, D. Q. (1991) Time and space in glacial sediment systems of the British Isles. In J. Ehlers, P. Gibbard, and J. Rose, eds, *Deposits in Great Britain and Ireland*. Rotterdam: Balkema, 213–23.

Brunsden, D. and Thornes, J. B. (1979) Landscape sensitivity and change. *Transactions, Institute of British Geographers*, NS4(4), 463–84.

Chandler, T. J. (1995) *The Climate of London.* London: Hutchinson.

Clayton, K. M. (1960) The landforms of part of southern Essex. *Transactions, Institute of British Geographers*, 28, 55–74.

Cole, M. M. (1980) The geobotanical expression of ore bodies. *Transactions of the Institution of Mining and Metallurgy (Section B: Applied Earth Science)*, 89, B73–143.

Darby, H. C. (1952) *The Domesday Geography of Eastern England.* Cambridge: Cambridge University Press.

Diamond, S. and Woodward F. I. (1998) Vegetation modelling. In P. Mairota, J. B. Thornes, and N. Geeson, eds, *An Atlas of Mediterranean Environments in Europe.* Chichester: Wiley, 68–9.

Drake, N. A., Zhang, X., Berkhart, E., Bonifacie, R., Grimes, D., Wainwright, J. and Mulligan, M. (1998) Modelling soil erosion at global and regional scales using RS and GIS techniques. In P. Atkinson, ed., *Spatial Analysis for RS and GIS.* Chichester: Wiley.

Foster, I. D. L., ed. (2000) *Tracers in Geomorphology.* Chichester: Wiley.

Furley, P. A., Proctor, J., and Ratter, J. A., eds (1992) *Nature and Dynamics of Forest–Savanna Boundaries.* London: Chapman and Hall.

Goudie, A. (1993) *The Nature of the Environment*, 3rd edn. Oxford: Blackwell.

Grainger, A. (1990) *The Threatening Desert: Controlling Desertification.* London: Earthscan.

Gregory, K. J., Lewin, J., and Thornes, J. B., eds (1987) *Palaeohydrology in Practice.* Chichester: Wiley.

Grove, J. M. (1988) *The Little Ice Age.* London: Methuen.

Hardisty, J. (1990) *Beaches: Form and Process.* London: Unwin-Hyman.

Hare, F. K. (1966) *The Restless Atmosphere.* London: Hutchinson.

Kirkby, M. J. (1969) Infiltration, throughflow and overland flow. In R. J. Chorley, ed., *Water, Earth and Man.* London and New York: Methuen, 215–27.

Linton, D. L. (1949 and 1951) Some Scottish river captures re-examined. *Scottish Geographical Magazine*, 65, 123–31 and 67, 31–44.

Maltby, E. M. and Lucas, E. (1997) Wetland restoration. In D. V. Brune, M. D. Gwynne, and J. M. Pacyna, eds, *The Global Environmental Science, Technology and Management.* Weinheim: VCH, vol. 2, 946–54.

Oldfield, F. (1997) Forward to the past: an update on PAGES. *Global Change Newsletter*, 31, 1–3.

O'Riordan, T. (1989) The challenge for environmentalism. In R. Peet and N. Thrift, eds, *New Models in Geography.* Boston, MA: UnwinHyman, vol. 1, 77–102.

O'Riordan, T. and Voisey, H., eds (1998) *The Transition to Sustainability: The Politics of Agenda 21 in Europe.* London: Earthscan.

Parry, M. L. (date?) *Climate Change, Agriculture and Settlement.* Folkestone: Dawson.

Parry, M. L. (1990) *Climate change and World Agriculture.* London: Earthscan.

Redclift, M. (1992) Sustainable development and global environmental change: implications of a changing agenda. *Global Environmental Change*, 2(1), 32–42.

Rose, F. (1976) Lichenological indicators of age and environmental continuity in woodlands. In D. H. Brown, D. L. Hawksworth, and R. H. Bailey, eds, *Lichenology: Progress and Problems.* New York: Academic Press, 279–307.

Schumm, S. A. (1979) Thresholds in geomorphology. *Transactions, Institute of British Geographers*, NS4(4), 440–62.

Simmons, I. G. (1989) *Changing the Face of the Earth: Culture, Environment, History*. Oxford: Blackwell.

Simmons, I. G. and Innes, J. B. (1988) Late Quaternary vegetation history of the North Yorkshire Moors VIII. *Journal of Biogeography*, 15, 249–72.

Sugden, D. E. and John, B. S. (1982) *Glaciers and Landscape*. London: Edward Arnold.

Thomas, D. S. G. and Middleton, N. J. (1994) *Desertification: Exploding the Myth*. Chichester: Wiley.

Thornes, J. B. (2001) Land degradation in the Mediterranean. In I. Douglas, ed., *Encyclopaedia of Earth Sciences, Vol 3. Causes and Consequences for Global Change*. Chichester: Wiley, 417–24.

Thornes, J. B. (2002) Emerging mosaics. In N. A. Geeson, J. B. Brandt, and J. B. Thornes, eds, *Mediterranean Desertification: A Mosaic of Processes and Responses*. Chichester: Wiley, 419–29.

Wagstaff, J. M. (1978) A possible interpretation of settlement pattern evolution in terms of 'catastrophe theory'. *Transactions, Institute of British Geographers*, NS3, 165–78.

Walling, D. E. and He, Q. (1999) Changing rates of overbank sedimentation in the floodplains of British rivers during the past 100 years. In A. G. Brown and T. A. Quine, eds, *Fluvial Processes and Environmental Change*. Chichester: Wiley, 207–22.

Warren, A. (*1*995) Changing understanding of African pastoralism and environmental paradigms. *Transactions, Institute of British Geographers*, NS20, 193–203.

Waters, R. S. (1964) Involutions and ice wedges in Devon. *Nature*, 189, 389–90.

Williams, M. (1996) European expansion and land cover transformation. In L. Douglas, R. Huggett, and M. Robinson, eds, *Companion Encyclopaedia of Geography: The Environment and Humankind*. London and New York: Routledge, 182–205.

Williams, M. (1997) Ecology, imperialism and deforestation. In T. Griffiths and L. Robin, eds, *Ecology and Empire: Environmental History of Settler Societies*. Edinburgh: Keele University Press, 169–84.

Wilson, A. (1981) *Catastrophe Theory and Bifurcation*. London: Croom Helm.

15

The geography of disease distributions

ANDREW CLIFF
PETER HAGGETT

Over the course of the last century, the confused landscape that lies on the marchland of two very ancient subjects—geography and medicine—has been explored from several directions. Occasionally, scientists and practitioners from the hugely powerful medical state have travelled confidently into geographical terrain. Less often and less confidently, a scholar or two from the smaller neighbour has wandered into medical country. This chapter describes some of the terrain explored, the body of knowledge that has grown up around these contacts, and the extraordinary growth of research activity that has occurred in the last couple of decades.

The chapter is confined in three ways other than the obvious one of length. First, it is restricted in time to the twentieth century. Secondly, it is constrained geographically to 'British' research, an appellation that has proved increasingly hard to hold in a century when ideas swept around the world ignoring national boundaries, and when an increasing number of scholars spent many years of their careers in more than one country. Thirdly, in concentrating on the geography of disease distributions, we survey only some small part of the wider field of overlap between geography and medicine. We refer readers to Frank Barrett's scholarly *Disease and Geography* (2000) for a critical review of the origin and usage of such much-debated terms as 'medical geography', 'geographical medicine', and 'geomedizin'.

Like the Walls of Jericho, the walls of academe have recently been tumbling down and disciplinary limits became ever more permeable as the century progressed. So we try here to pick out research in the field based on its intrinsic importance, rather than its disciplinary label. Given our backgrounds, most of the work discussed is by geographers but we also include work by demographers, epidemiologists, physiologists, statisticians, and health-service professionals where they are relevant to the narrative.

This review of British research over the twentieth century at the interface between geography and medicine follows an historical format.

It is divided into five phases each of roughly twenty years' duration. 'The opening decades' begins with Clemow's classic work on the *Geography of Disease* (1903) and considers the foundations for epidemiological modelling. 'The inter-war years' considers Stocks' (1928) work on cancer mapping, acclimatisation research, and the further development of epidemiological models. 'The Second World War and its aftermath' begins with the international movement for disease atlases originating in Germany and the United States and considers its implications for British research. 'The emergence of medical geography' mainly traces the impact of three geographers (Howe, Learmonth, and Prothero) on forging a recognisable school of British medical geography in the 1960s and 1970s. Finally, 'Expansion and diversity in the closing decades' considers the exponential growth of research in the last two decades on five disparate themes. The chapter concludes by trying to place the British century within a wider historical and geographical framework.

Phase I: The opening decades

Here we consider the 20-year period ending just after the close of the First World War. We pick out two themes: Clemow's work on the geography of disease and the progress in modelling the spread of epidemic diseases.

The geography of disease

The nineteenth century showed a wide range of approaches to what was variously called medical geography or geomedicine in France and Germany (Barrett, 2000, 175–370). Of critical importance was the work of August Hirsch (1817–94). His two-volume *Handbuch der Historische-Geographische Pathologie*, published between 1859 and 1864, was a monumental attempt to describe the world distribution of disease, drawing on more than 10,000 sources worldwide. Hirsch had close links with England and dedicated his book to the London Epidemiological Society. Twenty years later, a much revised version was translated by Charles Creighton as *Handbook of Geographical and Historical Pathology* (Hirsch, 1883–6). This time the adjectives in the title were reversed (a reversal approved by Hirsch) and the contents had swollen to three volumes, covering acute infectious diseases (1883), chronic and constitutional diseases (1885), and diseases of organs (1886). It was in that translated form that Hirsch's work became so well known and admired in the English-speaking world.

The strong influence of Hirsch can be seen in the work of Davidson (1892), Haviland (1892), and in the first major textbook on disease geography in the new century, Frank Clemow's *Geography of Disease* (1903). A physician with experience in Russia and the Ottoman empire, Clemow set out to study the way in which diseases were spread over the world. Diseases are organised alphabetically within three sections: a major section on general medicine and surgical diseases, followed by two briefer sections on diseases of the skin, along with the animal parasites and diseases associated with them. For each disease he gives (a) a brief aetiology followed by (b) a historical note. There follows two fuller sections on (c) recent geographical distribution of the disease and (d) factors governing that geographical distribution. For the most part Clemow relies on a listing to indicate geographical areas; only nine of the diseases have accompanying maps (see Plate 7) and there is little cross-reference to them in the text. No clue is given as to why the nine diseases (beriberi, blackwater fever, cholera, influenza, leprosy, Mediterranean or Malta fever, plague, yellow fever, and yaws) are chosen for mapping.

Clemow's section on cancer is influenced by Boudin's earlier work, 'law of geographical coincidence and antagonisms between diseases' (Barrett, 2000, 395). Clemow argues that diseases such as cancer and malaria may be antagonistic diseases in that they do not thrive in the same locations. Despite its medical thrust, the book was published as part of geographical series for the Cambridge University Press. It was edited by F. H. H. Guillemard, who had studied medicine at Gonville and Caius College, Cambridge, and was also the first lecturer in geography in the university (Stoddart, 1975, 8–9).

Epidemiological models

The nineteenth century was a rich period for epidemiological advance in Britain. John Snow showed in 1855 that cholera was being spread by the contamination of water supplies. Later, in 1873, William Budd established a similar manner of spread for typhoid. Parallel to these detailed studies in London and Bristol were the broader studies of statistical returns by William Farr, who tried to uncover empirical laws underlying the waxing and waning of epidemic outbreaks. In his *Second Report of the Registrar General of England and Wales*, Farr (1840) effectively fitted a normal curve to smoothed quarterly data on deaths from smallpox. Although both observed and predicted curves showed a distinctive bell-shaped form, the agreement was not very good, but similar models were used by Evans (1875) in a study of the 1871–2 smallpox outbreak.

The start of the new century saw a renewed attack on the prediction problem. In a series of papers published between 1906 and 1918, Brownlee (e.g. 1914) fitted various Pearson curves to epidemic data on many diseases. The curves were mainly approximations and represented small computational improvements on the Farr methods. Distinct improvements waited on a very different approach coming from disease modelling.

The improved bacteriological understanding that followed Koch and Pasteur's insights allowed Hamer (1906) to propose that the course of an epidemic must depend *inter alia* on the number of susceptibles and the contact rate between infectives and susceptible individuals. Hamer used simple mathematical assumptions not only to predict the course of individual outbreaks but also to deduce the existence of periodic recurrences. Parallel to these deterministic models was the fundamental work of Ross (1910) on the probability of malaria transmission. Ross's methods of '*a priori* pathometry' allowed for the first time deductions to be drawn from his models which could subsequently be tested out on actual malaria outbreaks.

Phase II: The inter-war years

While the 20 years from 1920 through to 1940 were difficult ones for the international and national economy, they were also years of major progress in the health sciences. At the international level, the League of Nations established a specialist health section and further international standard classifications of disease were published. Within the UK we take three themes: cancer mapping, climatic determinism, and later epidemiological modelling.

Cancer mapping

Percy Stocks, a medical statistician at the General Register Office, was one of the first to try to overcome the limitations of the crude death rate in mapping disease. His map of regional distribution by counties of cancer prevalence in England and Wales from 1919 to 1923 used corrections for differences in age, sex, and the urban/rural balance. He concluded that the mortalities 'vary over such wide limits and the counties group themselves into such definite regions of high and low prevalence, that there can be no question that geographical influences are in some way concerned' (Stocks, 1928, 518). More sophisticated adjustments were included by Stocks and Karn (1931) for both cancers and for tuberculosis. By the late thirties, the reports of the British Empire Cancer Campaign routinely included maps prepared by Stocks showing the spatial distribution for England and Wales of cancer of various organs in different age groups for each sex.

Climatic determinism

Environmental determinism was a critical part of geographical thinking in the early decades of the century. The publications of the Yale geographer Ellsworth Huntingdon, notably his *Civilization and Climate* (1915), focused attention on the contested hypothesis that a particular type of climate is necessary for the development of high civilisations. In Britain, the question that attracted both medical and geographic attention was the possibility of long-term settlement by white settlers in the humid tropics. Andrew Balfour, director of the London School of Hygiene and Tropical Medicine, was partly persuaded by Huntingdon's views and conducted a review of the physiology, psychology, and pathology of the tropics' different climatic zones. He argued in *The Lancet* that 'the hot and humid tropics are not suited to white colonization and never will be with our present knowledge' (Balfour, 1923, 245).

Balfour's views attracted further research on the physiology of heat regulation and metabolism by the human body under tropical conditions, but these studies were mostly carried out by groups outside Britain. Eijkman (1924) investigated European and Malay acclimatisation in the Dutch East Indies while Trewartha (1926) refined the taxonomic definitions of tropical climates and considered their relative suitability for white settlement. In Australia, the geographer Griffith Taylor entered the argument by contesting the Australian government's over-optimistic calculations for the settlement potential of the desert areas of that island continent. David Livingstone (1987) has written a scholarly perspective on this period.

Developments in epidemiological modelling

Following on from the work of Hamer and Ross noted above, the next two decades were rich ones for British modelling of epidemic diseases. Further work on deterministic models associated with measles was carried out by Soper (1929) to form the basis of a new set of Hamer–Soper models. The basic relationships were written as difference equations, which set off a train of harmonic waves whose timing could be matched with recurrent outbreaks. One problem was that the model produced a damped train (with oscillations reducing in amplitude), whereas observed disease outbreaks showed no such tendency. Soper argued (wrongly as was later shown) that allowing for an incubation period would remove such damping. Work by McKendrick (1926) followed on a genuinely stochastic version of an epidemic model. Alternative probability approaches by Reed and Frost in the United States and Greenwood (1931, 1932) in England used discrete stages or generations based on the chain-binomial theorem.

The most outstanding result during this period was the establishment of a threshold theorem by Kermack and McKendrick (1927). This demonstrated that the introduction of infectious cases into a community of susceptibles would not give rise to an epidemic outbreak if the density of susceptibles was below a certain critical value. If, on the other hand, the critical density was exceeded, there would be an epidemic of sufficient magnitude to drive the susceptible density as far below the threshold as it was above at the start of the epidemic. Thresholds were shown to be disease specific, varying with the infectiousness of the disease in question.

Phase III: The Second World War and its aftermath

The period from 1940 to 1960 was dominated by the Second World War and its repercussions. It saw the establishment of the World Health Organization (WHO) in 1947, with its headquarters at Geneva and regional offices worldwide. After a period of intensive research during the war years, a start was made on the wholesale expansion and rebuilding of universities, with its critical implications for the volume of research and for the growth of both the geographical and medical sciences.

Three significant mid-century atlases

The middle decades of the century were marked by the publication of three pioneering atlases of disease. Although appearing outside the UK (two in Germany and one in the United States), they were to have a profound effect on British thinking and each must be considered briefly here.

Seuchen Atlas

This atlas of epidemic disease (Zeiss, 1942–5; Anderson, 1947) was conceived by the German army as an adjunct to war, enhancing its ability to mount military campaigns. The atlas was produced as separate sheets over the years 1942–5, its distribution restricted to military institutes and to those German university institutes involved in training medical students. The scope of the atlas was not global but confined largely to those areas where the Army High Command expected to be fighting. As a consequence, it was divided into eight sections: (1) general maps, (2) Near East, (3) Trans-Caspian region, (4) Eastern Europe, (5) Baltic region, (6) Central Europe, (7) Mediterranean region, and (8) North and West Africa. Each map showed the way in which epidemic diseases 'nest' in a region, with the goal of the cartographic methods being to 'record on the geomedical map those disease-promoting geofactors which are crucial in determining the distribution or restriction of the relevant disease' (Barrett, 2000, 514). By geofactors, the atlas included animal reservoirs, vegetation zones, soil formations, or climatic areas.

American Geographical Society Atlas of Diseases

In 1944, the President of the American Geographical Society (himself a neurosurgeon) proposed to its Council that it should begin work on a world atlas of disease. This began formally four years later with the appointment of Jacques M. May as director of the Society's programme in medical geography. May was a French surgeon with an extensive background in tropical medicine through field experience in Thailand, French Indochina, Equatorial Africa, and the Caribbean (Thouez, 1983).

From 1950 to 1955, the Society released each year as supplements to the *Geographical Review* the sheets of its atlas of diseases (May, 1950–4) as they were completed. The first two plates showed the distribution of poliomyelitis 1900–50 and the distribution of cholera 1816–1950. The third map focused on the distribution of malaria vectors and the fourth on helminthiasis. Subsequent plates covered dengue and yellow fever, plague, leprosy, food sources, deficiency diseases, rickettsial diseases (three plates), and spirochetal diseases (three plates). Characteristic of the AGS maps were their innovative Breisemeister equal-area projections for the world base, the nesting of maps of critical areas at other scales, their dense thicket of detail on maps of all scales, and the use of the reverse side of the map sheet to cite the detailed references and sources on which the maps had been based. Given the limited resources and manpower available to the Society, the maps were easily the most advanced world disease atlas yet attempted.

World Atlas of Epidemic Diseases

At the end of the Second World War, with allied military occupation, Heidelberg fell within the US zone. Ernst Rodenwaldt, Professor of Hygiene at Heidelberg University, managed to secure American funding to rework and extend the maps produced for the original *Seuchen Atlas*. The bilingual global volume that emerged was entitled *World Atlas of Epidemic Diseases* (*Welt Seuchen Atlas*) (Rodenwaldt, 1952–61). In all, 120 maps of disease, climate, and population were produced in a three-volume folio.

British responses

Not for the first time in the history of geography in Britain, the two ancient universities and the Royal Geographical Society (RGS) played the leading part in taking up the international initiatives. At Oxford, E. W. Gilbert (1958) pursued his historical research on early maps of death and disease in England. At Cambridge, J. A. Steers attended the Washington Conference of the International Geographical Union (IGU) in 1952 and was strongly influenced by the first report there from the IGU Medical Geography Commission. On his return, he invited A. Leslie Banks, Professor of Human Ecology within the Cambridge Medical School, to give a course of lectures on medical geography as part of the geographical tripos.[1] Banks both accepted the invitation and gave keynote lectures to the RGS (Banks, 1956, 1959) at which leading figures from both the geographical and medical establishments engaged in discussion.

As a result, the RGS was led to set up a medical geography committee under the chairmanship of Lord Nathan in 1959 with a medical atlas

[1] Banks's course was attended by PH when a Cambridge undergraduate in the early 1950s.

of Britain (see below) as its first major research project. A member of that committee, Sir Dudley Stamp, pressed the importance of the emerging field through his Heath Clarke Lectures and in his *Geography of Life and Death* (Stamp, 1964). It is worth noting that informal links between the RGS, the Medical Research Council (MRC), and service interests (e.g. the Royal Army Medical Corps) predate this in the area of exploration and expedition medicine.

Epidemiological modelling

To some extent, the previous trends noted in epidemic modelling were continued, with deterministic models being further extended by Kendall (1956), and chain binomial theory by Greenwood (1946) and Bailey (1957). Important empirical results for the stochastic models were obtained by Bartlett (1957) using measles records for British towns. He showed that, in small communities, a complete fade-out of infection may occur if fresh cases are not introduced, whereas in communities above a certain critical size, infection reaches a low level but builds up again from a fresh outbreak. One feature of such empirical work was the extension from the usual disease model (measles) to a wider range of diseases: Bailey (1975) gives examples of modelling applied to malaria, typhoid fever, tetanus, cholera, and schistosomiasis.

Phase IV: The emergence of medical geography

Although the links between geography and medicine go back, in various complex forms, over a long historical period, it is in the second half of the century that medical geography appears to emerge in Britain as a distinct and self-conscious field of scholarship within geography. The period from the mid-1950s through to the late-1970s is one in which three remarkable geographers left their mark on the subject, laying a foundation on which the rich research variety of the end-of-century period has been founded. Each of the three brought different geographical focuses to their work.

Melvyn Howe and the British focus

Of all the British geography departments, Aberystwyth under E. G. Bowen's direction had, by mid-century, built the highest reputation in human ecology. It was therefore not surprising that the RGS medical committee should turn to an Aberystwyth scholar, G. Melvyn Howe, to lead its first research project. The publication of the *National Atlas of Disease Mortality in the United Kingdom* (Howe, 1963) represented, for the UK, its response to the international disease atlases noted above. Using data for the five-year period 1954–8 for each of 320 administrative units in the UK, it plotted standardised mortality ratios for 13 major causes of death, from heart disease to lung cancer. Unlike some countries, local areas could be used in Britain since its death statistics are published by the usual residence of the deceased rather than the hospital in which death occurred (unless the deceased had been there for more than six months). Like Stocks, Howe adjusted crude mortality ratios to allow for local age, sex, and degree of urbanisation.

A second edition of the national atlas (Howe, 1970) used data from a later five-year period, 1959-63. It covered an enlarged range of diseases and used different cartographic techniques. A demographic (rather than an area) base map was used to avoid giving undue prominence to mortality ratios for extensive and lightly populated areas of the country. A system of squares for urban populations and diamonds for rural population (each proportional to the at risk population in each area) was used. A stylistic coastline was added to aid in reading the maps. Howe's atlases were followed by a series of national atlases outside Britain—from Canada and the United States, through the Netherlands and West Germany to Japan and China. Within Britain itself, more specialised atlases on cancer (e.g. Gardner et al., 1983) and other selected diseases (Gardner, Winter, and Barker, 1984) were shortly to follow.

Towards the end of the period, Howe wrote a series of more general volumes giving structure to his broader work in medical geography. *Man, Environment, and Disease in Britain* (Howe, 1972) provided an historical view of the country's changing medical geography. His *World Geography of Human Diseases* (Howe, 1977), with contributions from both medical and geographical experts, reviewed the major human diseases, while his *Environmental Medicine* (Howe and Loraine, 1973) was a general synthesis of the field. Of special value was his chapter on medical mapping in Clarke's *Modern Methods in the History of Medicine* (Howe, 1971).

Learmonth and Prothero on tropical diseases

Parallel in time to Howe's work in Britain, two of his colleagues developed important studies of disease geography but with a tropical focus.

Andrew Learmonth and a South Asian focus

An Edinburgh graduate, Andrew Learmonth was foundation professor of geography at both the Australian National University and the Open University in the UK (see the memoir by Blunden, 1983). But it was during the 1950s and 1960s that he spent extended periods in India and Pakistan working on the medical geography of South and East Asia. His regional survey of survival, mortality, and disease in India–Pakistan, 1921–40, formed a milestone in third world medical geography. His sensitive handling of demographic and epidemiological data reflected his deep experience of India as well as his links to Mahalanobis and the Indian Statistical Institute. He was later to draw the strands together in two more general works, which drew heavily on his Indian experience—*Patterns of Disease and Hunger* (1978) and *Geography of Health* (1980).

Learmonth pioneered the use of cartographic methods to handle data of highly variable accuracy. His maps of cholera in India based upon data for 1921–40 plotted both the mean and the standard deviation of mortality ratios for the period for each area. By ranking areas into three groups on each of the two criteria, Learmonth arrived at a 3 x 3 matrix with nine cells, ranging from districts with low means but high variability (upper left) to regions with high means but low variability (lower right; see Fig. 15.1). He also made use of isopleth maps to show contours of urban infant mortality in India in the 1950s (Learmonth, 1965).

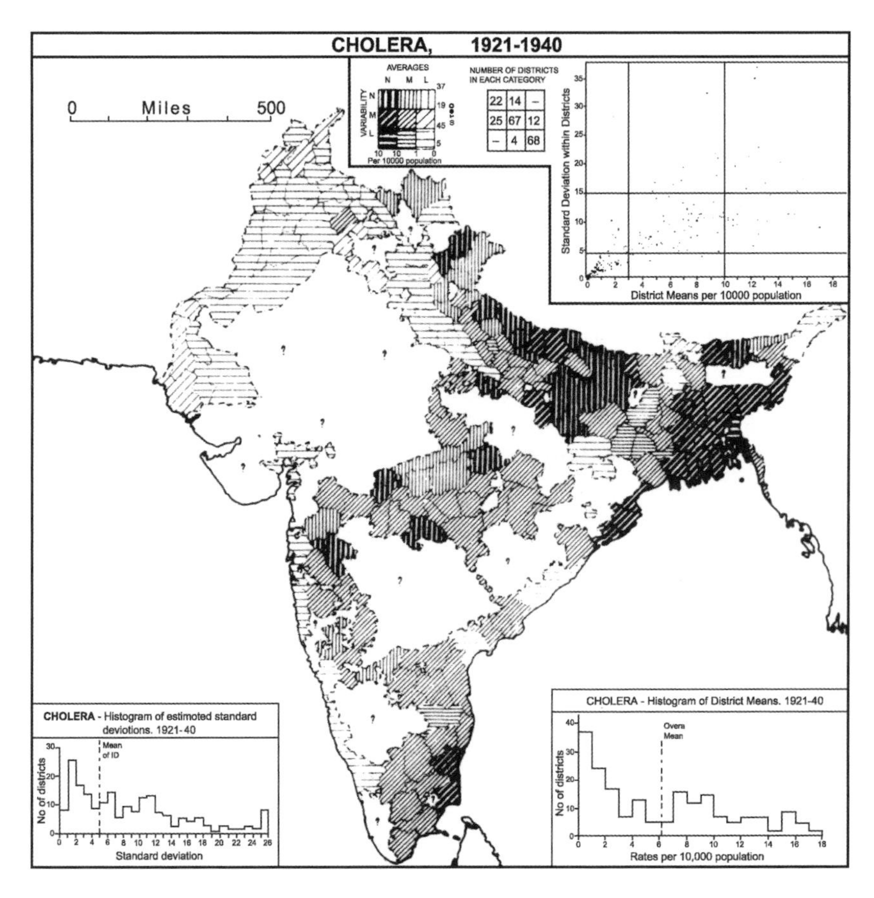

Fig. 15.1 Learmonth's map of cholera in India, 1921–1940.
Source: Reproduced in Pacione (1986, 45, fig. 2.3).

Mansell Prothero and an African focus

Like Learmonth, Mansell Prothero was an Edinburgh graduate and spent part of his academic career at Liverpool University, which had the benefit of both an outstanding geography department under the headship of a leading Africanist, Robert Steel, and the Liverpool School of Tropical Hygiene within the medical school. Prothero worked in Africa as the first WHO consultant geographer in 1960 and for further periods over the next two decades. Encouraged by the microbiologist Bruce-Chwatt (1980), he was particularly concerned with the role of rural population movements, which both contributed to malaria transmission and prejudiced programmes for its control and eradication (Prothero, 1961, 1977). Prothero's work was at several different spatial scales: tropical Africa, Morocco, Kano State (Nigeria), and the Horn of Africa. His *Migrants and Malaria* (1965) remains a classic account of the need to study the macro- and micro-geography of population movements if tropical diseases are to be understood and countered.

Other trends during the period

Most of the studies discussed so far emphasised the geographies of mortality and physical illness. However, geographies of mental illness have also received attention with the pioneering work of Giggs (1973) in urban contexts and Bain (1974) in rural contexts. Giggs used records of schizophrenia in Nottingham to determine the significance of spatial variations in its incidence within the city and their relationship to urban structure. Factor analysis was used to unravel the structure of the city and multiple regressions confirmed that there were statistically significant variations between (a) nine measures of schizophrenia and (b) 15 different kinds of socio-economic area. Giggs was also to use Nottingham as a test area for spatial variations in acute pancreatitis (Giggs, Bourke, and Ebdon, 1979).

A further extension during the period was from human to animal diseases. Gilg (1973) analysed the spatial spread of Newcastle disease (a virus disease affecting poultry) in Britain. Study of Gilg's maps suggests that Kendall Type I waves are characteristic of the central areas near the start of an outbreak but, as the epizootic spreads outwards, so the waveform changes towards Type II and eventually, at the outer edges of the outbreak, to Type III. Another geographical study of an animal disease was by Tinline (1970) who, in an analysis of the 1967–8 foot-and-mouth epizootic (a virus disease of cloven-hoofed animals) in central England, demonstrated both the role of lee-waves in accelerated spread of the disease and the relative efficiency of different types of ring-control mechanisms (see the discussion in Haggett, 2000, 117–19).

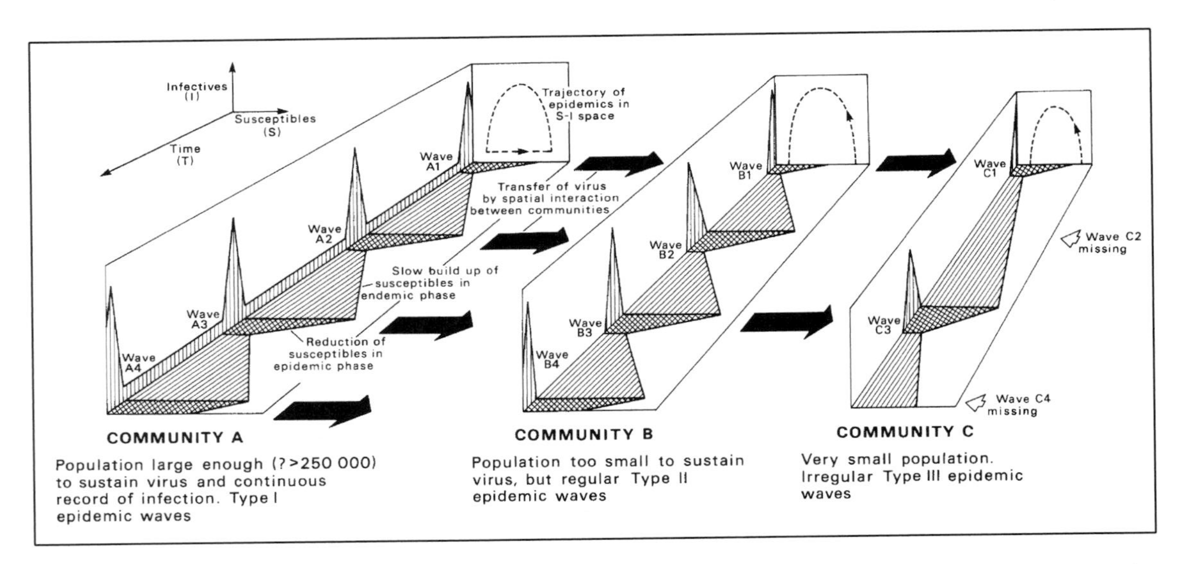

Fig. 15.2 Spatial modelling of disease. Cliff and Haggett's modification of the Bartlett model to show the relationship between epidemic cycles in communities of different sizes.
Source: Cliff and Haggett (1988, 246, fig. 6.5a).

Phase V: Expansion and diversity in the closing decades

By the start of the last quarter of the twentieth century, there was a sufficient body of British work on 'medical geography' to allow broad subject-wide surveys to be gathered together. Following McGlashan's pioneering 15-author volume on *Medical Geography: Techniques and Field Studies* (1972), a second volume of essays honouring Andrew Learmonth was published as *Geographical Aspects of Health* (McGlashan and Blunden, 1983). Two further reviews in the same genre were Pacione's *Medical Geography: Progress and Prospect* (1986) and Thomas's *Geomedical Systems* (1992). Surveys of this kind underlined how medical geography was now expanding rapidly and breaking out from a sole concern with diseases and their ecology to encompass the health services which delivered protection against disease. Valuable reviews of developments during this most recent period are provided in Mohan (2000a, 2000b). In this final section we illustrate the diversity of geographical work by selecting five areas for comment.

Spatial diffusion models

The ways in which the great plagues of the past and present have spread around the world continue to be a major focus of British research, with advances continuing to be made in the statistical modelling of diseases (Anderson and May, 1991). The significant contribution by British geographers has been the addition of spatial components to the aspatial models of disease incidence that were developed in biomathematics and for time-series forecasting of economic events. The models were systematically extended to cover multiple rather than single geographical areas, to handle the switching between epidemic and non-epidemic phases with recurrent epidemics, and to incorporate lead–lag relationships among geographical areas so that the spatial propagation of epidemics could be both modelled and forecast (Fig. 15.2).

Such models were tested across a spectrum of geographical environments from the sub-Arctic (Iceland: Cliff et al., 1981) to the sub-tropical (Fiji: Cliff and Haggett, 1985), at geographical scales from small island settings to continental areas (Cliff, Haggett, and Smallman-Raynor, 2000), for different diseases (e.g. influenza: Cliff, Haggett, and Ord, 1986; measles: Cliff, Haggett, and Smallman-Raynor, 1993; AIDS: Smallman-Raynor, Cliff, and Haggett, 1992; Thomas 1992), to examine past processes and to forecast future events for control purposes (Cliff et al., 1981).

Atlases and disease mapping

The trend towards national atlases of disease started by Howe continued apace into the final period, with the increasing power of computing now able to handle huge data arrays and GIS allowing sophisticated maps to be produced at will.

Howe himself led the way with a world atlas of human cancers: *Global Cancerology: A World Geography of Human Cancers* (1986). Cliff and Haggett produced an *Atlas of Disease Distributions* (1988) with a wide range of diseases mapped at different spatial scales. Its purpose was to provide worked examples of the widening range of analytical methods that could now be applied to epidemiological data. In contrast, their *Atlas of AIDS* (Smallman-Raynor et al., 1992) sought to catch the first decade of spread of a rapidly expanding world pandemic. A feature of the atlas is the heavy reliance on a unique database of attested movements of infectives and the use of a variety of spatial scales, from world maps down to spread within the individual districts of a city (e.g. San Francisco and New York City).

The process of providing a demographic rather than an areal base for disease mapping has been carried forward by Dorling (1995, 136–69) in the maps of health in his *New Social Atlas of Britain*. He uses maps of wards in Britain (there are over 10,000 of them) in which each ward is shown by a circle with area proportional to its population. The centre of each ward is first placed in its correct geographical location. A computer program is used to allow the circles to be expanded slowly and if they touch they push each other out of the way until their correct size is attained (Plate 8, facing p. 399). The shape of the map becomes distorted (with large cities occupying more of the map, empty rural areas less) but it is still possible to draw county and district boundaries, towns and villages upon it. Some maps (e.g. Permanently Sick, 1981) include Scotland, while mortality causes (e.g. Heart Disease, 1986–1989) relate to England and Wales only.

Testing for disease clusters

Studies of spatial clusters of disease have proceeded along two distinct tracks. The first is the use of ever-more sophisticated statistical modelling (Bailey and Gatrell, 1995). Growing out of nearest neighbour methods, this approach uses different mathematical functions to look for clusters in point distributions (e.g. individual cases of cancer around possible point causes) and to check for directional trends in the underlying causes of these patterns.

The second uses the power of computers to develop geographical information systems (GIS). Openshaw (1990; see also Openshaw, Charlton, and Craft, 1988) pioneered the use of a family of Geographical Analysis Machines (GAMs) to automate the analysis of spatial patterns of disease. He argued contentiously that in situations where we lack the prior knowledge and theories about the nature of the processes responsible for spatial clustering, sheer computer power might be used to explore the whole population at different scales of aggregation. In the simplest model

(GAM 1) he looked at the pattern of reported leukaemia cases across northern England falling within a circle of fixed radius and compared these with random expectations. If the count is significantly above the random expectation, the circle is drawn on the map (Fig. 15.3). The procedure is repeated for circles of the same radius until the entire study area has been tested and the test continues with larger and smaller radii. This 'trawling' method not only confirmed leukaemia clusters in the Sellafield area (previously known) but also a secondary cluster near Newcastle (not previously known). The GAM approach is one of a number of used to detect patterns in spatial distributions and it is critically reviewed in Bailey and Gatrell (1995).

Historical geography of disease

One major strand of historical research has been directed towards improving and standardising the sources from which epidemiological materials are drawn. Cliff and his co-workers made extensive use of the epidemiological data from Iceland; data that are outstanding both for their historical continuity, their spatial detail, and the level of contemporary epidemiological comment. Extensive epidemic data tables have been

Fig. 15.3 Openshaw's use of GAM in the search for significant cancer clusters. Maps of synthetic random data showing (a) apparent clusters and (b) the only significant cluster.
Source: Openshaw (1990, 69, fig. 1).

included in their monographs on Iceland (Cliff et al., 1981, 201–29). Islands in general have unusually rich source materials and an extensive review of such sources is given in the appendices to *Island Epidemics* (Cliff, Haggett, and Smallman-Raynor, 2000, 449–88). In another study, the unique weekly records for more than a hundred cities around the world from the US Public Health Service (the 'sanitary abstracts' that were later to become the *Morbidity and Mortality Weekly Reports*) were used to plot global trends in disease over a 25-year window from 1878 through to 1912. The outcome was published as *Deciphering Global Epidemics* (Cliff, Haggett, and Smallman-Raynor, 1998) and the underpinning data were also made available on disk to interested researchers.

Although the standard work on the history of medicine (Bynum and Porter, 1993) fails to include a chapter on historical geography, the decades were very productive in that field, with a second clearly identifiable strand of historical research focusing upon the demographic implications of disease. An outstanding contribution came from the Wellcome Unit in Oxford in Mary Dobson's (1997) *Contours of Death and Disease in Early Modern England*. A combination of demographic, epidemiological, and environmental history, it uses burial and baptism discrepancies in Sussex, Kent, and Essex to show dramatic geographical differences in health (Fig. 15.4). It was found to be largely a matter of contours: high ground had the lowest mortality, low ground the highest. Making use of both parish records and medical accounts, Dobson analyses the fevers that decimated the coastal marshes and confidently identifies malaria as the major culprit, with humid summers also accentuating the enteric fevers that increased infant deaths. While environmental factors were critical, Dobson shows how these were played out through social factors such as migration, wealth, class, and occupation. After 1750, marsh reclamation and drainage improved the situation, as did the widespread adoption of quinine against malaria in the following century.

Dobson's work reflects both her geographic origins but also the tradition of the French *Annales* school and the lead provided by the demographic researches of the Cambridge Population Group. The latter had been established by E. A. Wrigley in 1960 and developed to become the leading historical demographic research unit in the world. The two magisterial publications were the *Population History of England* (Wrigley and Schofield, 1981) and *English Population History from Family Reconstitution 1580–1837* (Wrigley et al., 1997). While the scope of its work falls largely outside this chapter, it is worth recalling that it was based upon a geographical sample of key parishes and that two of its directors (E. A. Wrigley and R. M. Smith) had begun academic life as historical geographers at Cambridge. The Cambridge work was paralleled by outstanding historical demographic work on the mortality of the Victorian period by Woods at Liverpool (Woods and Shelton, 1997; Woods, 2000).

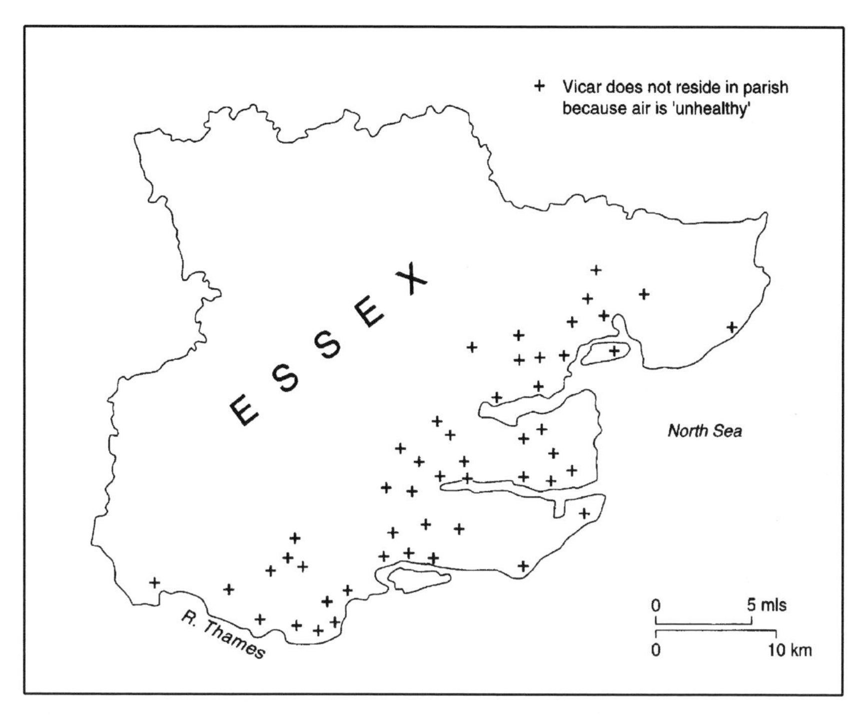

Fig. 15.4 Perception of malaria risk in eighteenth-century Essex based on vicars' perception of whether the air in the parish was regarded as 'unhealthy'.
Source: Dobson (1997, 296, fig. 6.2).

Health and health-care delivery systems: a social perspective

One of the strongest growths in research has been on the spatial structure of health-care delivery systems. These studies have ranged in scale from the micro-level mapping of attendance at general practitioner surgeries (Phillips, 1979) to global studies of health services using comparative international data (Phillips and Verhasselt, 1994). At the intermediate spatial level, Clarke and Wilson (1986) have used planning models in studying a health district in the north of England. They apply location-allocation models to (a) the spatial allocation of medical facilities, (b) the allocation of resources to hospital facilities, (c) the allocation of patients to hospitals, and (d) the allocation of patients to specialities and care types.

One of the trends in the late twentieth century that has attracted great interest and appears to be setting the pace in the new century is the combination of geographical and medical skills with behavioural and social perspectives. This is well illustrated by two influential texts that give shape and structure to this movement. Jones and Moon's (1987) *Health, Disease and Society* is significantly subtitled 'a critical medical geography'.

It illustrates the ideas, methods, and debates that inform the current approaches to medical geography, demonstrating the potential of a social and environmental approach to illness and health. With chapters on the social context of disease, concepts and issues in mental illness, inequalities in health care, planning heath services, and critical perspectives, it illustrates the progress made outside the epidemiological and cartographic constraints of the earlier decades.

A second influential volume was Curtis and Taket's (1996) *Health and Societies*. This traces on a worldwide basis the changing organisation of health-service delivery during a period of rapid change. They show how the health service professions are moving away from a disease-focused biomedical approach to one that emphasises health promotion and the prevention of illness. It covers equity issues and the social construction and contesting concepts of health, as well as more conventional health-planning topics. While both volumes retain strong geographical lines, with emphases on space, on scale, and on regional variation, they are also multidisciplinary in so far as they bring in anthropological, psychological, and political perspectives to show that the concept of health itself is changing and is a contested rather than a settled issue. With globalisation, and with spending on the health sector climbing steadily as a proportion on GDP, we can expect the contested nature of this approach to medical geography to rise higher up on the research agenda.

A consistent theme in such studies is that both the demand for and the supply of medical services have (a) distinctive spatial patterns and (b) improved efficiencies can come from meshing the two distributions together. Useful reviews of the very extensive research in this area are provided by Curtis and Taket (1996), Mohan (2000a), Phillips (1981), Whitelegg (1982) and Wilkinson (1996).

Conclusion

Looking back over our essay on the last century, we are aware of the partial nature of what we have been able to include in this brief survey. Much valuable work lies unmentioned but we are inevitably involved in what Mohan (1989) has termed 'competing diagnoses and prescriptions'. But all observers of research on disease distribution over the last century must be struck by the widening range and magnitude of the field. That the range of disciplines now involved (from statistics through epidemiology to operations research) has increased reflects the increasing maturity of those academic disciplines and the fact that the rise of computing and the social sciences is largely a late-century phenomenon. Given that disease mapping involves handling huge amounts of data, the improvement of computing power over the last quarter-century and of GIS over the last decade are critical.

Changes in the infrastructure of research have also proved important. The establishment of specialist medical geography groups (e.g. RGS group from 1959, and the IGU group from 1949) represents one element. The establishment of specialist journals has also played a part: *Social Science and Medicine* (from 1967) and *Health and Place* (from 1995). A third strand has been the establishment of specialists institutions (e.g. the Medical Research Council Environmental Epidemiology Unit at Southampton Hospital and the Institute of Health Studies at Lancaster University). The potential for further research in this area shows no sign of slackening. Indeed, the new knowledge arising from genetic studies (Cliff et al., 2000) and the impact of vaccination and climatic change upon disease distributions (Martens, 1998) open up wholly new possibilities for geomedical research. Meanwhile, the new studies summarised in Gatrell's comprehensive new *Geographies of Health* (2001) point the directions in which this field is now expanding. Certainly, our successors in the Academy at the end of the next century look like having a monumentally large field to try and summarise.

Acknowledgements

We are grateful for their recollections and comments to the three emeritus professors who played such a central role in establishing medical geography (Professors Melvyn Howe, Andrew Learmonth, and Mansell Prothero). We have also been helped by the work of a number of current researchers in the field, including Anthony Gatrell (Lancaster), Kelvyn Jones (Bristol), and Matthew Smallman-Raynor (Nottingham).

References

Anderson, G. W. (1947) A German atlas of epidemic diseases. *Geographical Review*, 37, 307–22.

Anderson, R. M. and May, R. M. (1991) *Infectious Diseases of Humans: Dynamics and Control*. Oxford: Oxford University Press.

Bailey, N. T. J. (1957) *The Mathematical Theory of Epidemics*. London: Griffin.

Bailey, N. T. J. (1975) *The Mathematical Theory of Infectious Diseases and its Applications*. London: Griffin.

Bailey, T. C. and Gatrell, A. C. (1995) *Interactive Spatial Data Analysis*. Harlow: Longman.

Bain, S. M. (1974) The geographical distribution of psychiatric disorders in the north-east region of Scotland. *Geographica Medica*, 2, 84–108.

Balfour, A. (1923) Problems of acclimatization. *Lancet*, 205, 84–5, 243–7.

Banks, A. L. (1956) Trends in the geographical pattern of disease. *Geographical Journal*, 122, 167–76.

Banks, A. L. (1959) The study of the geography of diseases. *Geographical Journal*, 125, 199–216.

Barrett, F. A. (2000) *Disease and Geography: The History of an Idea*. York: York University, Geographical Monographs No. 23.
Bartlett, M. S. (1957) Measles periodicity and community size. *Journal of the Royal Statistical Society, Series A*, 120, 48–70.
Blunden, J. R. (1983) Andrew Learmonth and the evolution of medical geography: a personal memoir of a career. In N. D. McGlashan and J. R. Blunden, eds, *Geographical Aspects of Health: Essays in Honour of Andrew Learmonth*. London: Academic Press, 15–32.
Brownlee, J. (1914) Periodicity in infectious disease. *Proceedings of the Royal Philosophical Society of Glasgow*, 45, 197–213.
Bruce-Chwatt, L. J. (1980) *Essential Malariology*. London: Heinemann.
Bynum, W. F. and Porter, R., eds (1993) *Companion Encyclopaedia of the History of Medicine*. London: Routledge (2 vols).
Clarke, M. and Wilson, A. G. (1986) Developing planning models for health care policy analysis. In M. Pacione, ed., *Medical Geography: Progress and Prospect*. London: Croom Helm, 248–83.
Clemow, F. G. (1903) *The Geography of Disease*. Cambridge: Cambridge University Press.
Cliff, A. D. and Haggett, P. (1985) *The Spread of Measles in Fiji and the Pacific*. Canberra: ANU Research School of Pacific Studies.
Cliff, A. D. and Haggett, P. (1988) *Atlas of Disease Distributions: Analytical Approaches to Epidemiological Data*. Oxford: Blackwell Reference.
Cliff, A. D., Haggett, P., Ord, J. K., and Versey, G. R. (1981) *Spatial Diffusion: An Historical Geography of Epidemics in an Island Community*. Cambridge: Cambridge University Press.
Cliff, A. D., Haggett, P., and Ord, J. K. (1986) *Spatial Aspects of Influenza Epidemics*. London: Pion.
Cliff, A. D., Haggett, P., and Smallman-Raynor, M. (1993) *Measles: An Historical Geography of a Major Human Viral Disease from Global Expansion to Local Retreat, 1840–1990*. Oxford: Blackwell Reference.
Cliff, A. D., Haggett, P., and Smallman-Raynor, M. (1998) *Deciphering Global Epidemics: Analytical Approaches to the Disease Records of World Cities, 1888–1912*. Cambridge: Cambridge University Press.
Cliff, A. D., Haggett, P., and Smallman-Raynor, M. (2000) *Island Epidemics*. Oxford: Oxford University Press.
Curtis, S. and Taket, A. (1996) *Health and Societies: Changing Perspectives*. London: Arnold.
Davidson, A. (1892) *Geographical Pathology: A Geographical Inquiry into the Geographical Distribution of Infectious and Climatic Diseases*. Edinburgh: Pentland.
Dobson, M. (1997) *Contours of Death and Disease in Early Modern England*. Cambridge: Cambridge University Press.
Dorling, D. (1995) *A New Social Atlas of Britain*. Chichester: Wiley.
Eijkman, C. (1924) Some questions concerning the influence of tropical climate on man. *Lancet*, 206, 887–93.
Evans, G. H. (1875) Some arithmetical considerations on the progress of epidemics. *Transactions of the Epidemiological Society of London*, 1875, 55–5.
Farr, W. (1840) Progress of epidemics. In *Second Report of the Registrar General of England and Wales*, 91–8.

Gardner, M. J., Winter, P. D., Taylor, C., and Acheson, E. D. (1983) *Atlas of Cancer Mortality in England and Wales, 1968–78*. Chichester: Wiley.

Gardner, M. J., Winter, P. D., and Barker, D. J. P. (1984) *Atlas of Mortality from Selective Diseases in England and Wales, 1968–78*. Chichester: Wiley.

Gatrell, A. C. (2001) *Geographies of Health*. Oxford: Blackwell.

Giggs, J. A. (1973) The distribution of schizophrenics in Nottingham. *Transactions, Institute of British Geographers*, 59, 55–76.

Giggs, J. A., Bourke, J. B., and Ebdon, D. S. (1979) Variations in the incidence and the spatial distribution of patients with acute pancreatitis in Nottingham, 1969–76. *Gut*, 20, 366.

Gilbert, E. W. (1958) Pioneer maps of health and disease in England. *Geographical Journal*, 124, 172–83.

Gilg, A. W. (1973) A study of agricultural disease diffusion: the case of the 1970–71 fowl pest epidemic. *Transactions, Institute of British Geographers*, 59, 77–97.

Greenwood, M. (1931) On the statistical measure of infectiousness. *Journal of Hygiene, Cambridge*, 31, 336–51.

Greenwood, M. (1932) *Epidemiology: Historical and Experimental*. Baltimore, MD: Johns Hopkins Press (The Herter Lectures for 1931).

Greenwood, M. (1946) The statistical study of infectious diseases. *Journal of the Royal Statistical Society*, Part II, 109, 85–103.

Haggett, P. (2000) *Geographical Structure of Epidemics*. Oxford: Oxford University Press.

Hamer, W. H. (1906) Epidemic disease in England. *Lancet*, 1, 733–9.

Haviland, A. (1892) *The Geographical Distribution of Disease in Great Britain*, 2nd edn. London: Sonenschein.

Hirsch, A. (1883–6) *Handbook of Geographical and Historical Pathology*. London: The New Sydenham Society (3 vols). Translation by Charles Creighton of *Handbuch der Geographische-Historische Pathologie*, 2nd German edn (1864).

Howe, G. M. (1963, 1970) *National Atlas of Disease Mortality in the United Kingdom*. London: Nelson, on behalf of the Royal Geographical Society.

Howe, G. M. (1971) The mapping of disease in history. In E. Clarke, ed., *Modern Methods: The History of Medicine*. London: Athlone Press, 335–57.

Howe, G. M. (1972) *Man, Environment, and Disease in Britain: A Medical Geography of Britain Through the Ages*. New York: Barnes & Noble. Also published by Penguin (1976) and in 1997 retitled as *People, Environment, Disease and Death: A Medical Geography of Britain Throughout the Ages*. Cardiff: University of Wales Press.

Howe, G. M., ed. (1977) *World Geography of Human Diseases*. London: Academic Press.

Howe, G. M., ed. (1986) *Global Cancerology: A World Geography of Human Cancers*. Edinburgh: Churchill Livingstone.

Howe, G. M. and Loraine, J. A. (1973) *Environmental Medicine*. London: Heinemann Medical.

Huntington, E. (1915) *Civilization and Climate*. New Haven, CT: Yale University Press.

Jones, K. and Moon, G. (1987) *Health, Disease, and Society: An Introduction to Medical Geography*. London: Routledge.

Kendall, D. G. (1956) Deterministic and stochastic epidemics in closed populations. *Proceedings of the Third Berkeley Symposium in Statistics and Probability*, 4, 149–65.

Kermack, W. O. and McKendrick, A. G. (1927) Contributions to the mathematical theory of epidemics. *Proceedings of the Royal Society Series A*, 115, 700–21.

Learmonth, A. T. A. (1965) *Health in the Indian Sub-continent, 1955–64: A Geographer's Review of Some Medical Literature.* Canberra: Department of Geography, ANU School of General Studies, Occasional Paper No. 2.

Learmonth, A. T. A. (1978) *Patterns of Disease and Hunger.* Newton Abbot: David and Charles.

Learmonth, A. T. A., ed. (1980) *The Geography of Health.* Oxford: Pergamon Press.

Livingstone, D. N. (1987) Human acclimatization: perspectives on a contested field of inquiry in science, medicine, and geography. *History of Science*, 25, 359–94.

McGlashan, N. D., ed. (1972) *Medical Geography: Techniques and Field Studies.* London: Methuen.

McGlashan, N. D. and Blunden, J. R., eds (1983) *Geographical Aspects of Health: Essays in Honour of Andrew Learmonth.* London: Academic Press.

McKendrick, A. G. (1926) Applications of mathematics to medical problems. *Proceedings of the Edinburgh Mathematical Society*, 14, 98–130.

Martens, P. (1998) *Health and Climate Change: Modelling the Impact of Global Warming and Ozone Depletion.* London: Earthscan.

May, J. M. (1950–4) Map of the world distribution of poliomyelitis [first of fifteen sheets of world diseases]. *Geographical Review*, 40, 646–8; 41, 272–3; 42, 98–101, 283–6; 43, 89–90, 408–10; 44, 133–6, 408–10, 583–4.

Mohan, J. F. (1989) Medical geography: competing diagnoses and prescriptions. *Antipode*, 20, 166–77.

Mohan, J. F. (2000a) Geography of health and health care. In R. J. Johnston, D. Gregory, G. Pratt, and M. Watts, eds, *Dictionary of Human Geography*, 4th edn. Oxford: Blackwell, 330–2.

Mohan, J. F. (2000b) Medical geography. In R. J. Johnston, D. Gregory, G. Pratt, and M. Watts, eds, *Dictionary of Human Geography*, 4th edn. Oxford: Blackwell, 494–6.

Openshaw, S. (1990) Automating the search for cancer clusters: a review of problems, progress and opportunities. In R. W. Thomas, ed., *Spatial Epidemiology*. London: Pion, 48–78.

Openshaw, S., Charlton, M., and Craft, A. W. (1988) Searching for leukaemia clusters using a geographical analysis machine. *Papers of the Regional Science Association*, 64, 95–106.

Pacione, M., ed. (1986) *Medical Geography: Progress and Prospect.* London: Croom Helm.

Phillips, D. and Verhasselt, Y., eds (1994) *Health and Development.* London: Routledge.

Phillips, D. R. (1979) Spatial variations in attendance at general practitioner services. *Social Science and Medicine*, 13D, 169–81.

Phillips, D. R. (1981) *Contemporary Issues in the Geography of Health Care.* Norwich: GeoBooks.

Prothero, R. M. (1961) Population movements and problems of malaria eradication in Africa. *Bulletin of the World Health Organization*, 24, 405–10.

Prothero, R. M. (1965) *Migrants and Malaria.* London: Longman.

Prothero, R. M. (1977) Disease and mobility: a neglected factor in epidemiology. *International Journal of Epidemiology*, 6, 259–67.

Rodenwaldt, E., ed. (1952-61) *World Atlas of Epidemic Disease: Welt Seuchen Atlas*. Hamburg: Flak (3 vols).

Ross, R. (1910) *The Prevention of Malaria*. London: John Murray.

Smallman-Raynor, M., Cliff, A. D., and Haggett, P. (1992) *International Atlas of AIDS*. Oxford: Blackwell Reference.

Soper, H. E. (1929) Interpretation of periodicity in disease-prevalence. *Journal of the Royal Statistical Society*, 92, 34–73.

Stamp, L. D. (1964) *The Geography of Life and Death*. London: Collins.

Stocks, P. (1928) On the evidence for a regional distribution of cancer prevalence in England and Wales. In *Report of the International Conference on Cancer, London, July 1928*. London: British Empire Cancer Campaign.

Stocks, P. and Karn, M. H. (1931) The distribution of cancer and tuberculosis in England and Wales. *Annals of Eugenics*, 4, 341–61.

Stoddart, D. R. (1975) The RGS and the foundations of geography at Cambridge. *Geographical Journal*, 141, 1–24.

Thomas, R. W., ed. (1990) *Spatial Epidemiology*. London: Pion.

Thomas, R. W. (1992) *Geomedical Systems: Intervention and Control*. London: Routledge.

Thouez, J. -P. (1983) Jacques M. May, 1896–1975. *Geographers: Biobibliographical Studies*, 7, 85–8.

Tinline, R. R. (1970) Lee wave hypothesis for the initial pattern of spread during the 1967–68 foot and mouth epizootic. *Nature*, 227, 860–2.

Trewartha, G. T. (1926) Recent thought on the problem of white acclimatization in the wet tropics. *Geographical Review*, 16, 467–78.

Whitelegg, J. (1982) *Inequalities in Health Care: Problems of Access and Provision*. Retford: Straw Branes.

Wilkinson, R. (1996) *Unhealthy Societies: The Afflictions of Inequality*. London: Routledge.

Woods, R. (2000) *The Demography of Victorian England and Wales*. Cambridge: Cambridge University Press.

Woods, R. and Shelton, N. (1997) *An Atlas of Victorian Mortality*. Liverpool: Liverpool University Press.

Wrigley, E. A. and Schofield, R. S. (1981) *The Population History of England, 1541–1981*. Cambridge: Cambridge University Press.

Wrigley, E. A., Davies, R. S., Oeppen, J. E., and Schofield, R. S. (1997) *English Population History from Family Reconstitution, 1580–1837*. Cambridge: Cambridge University Press.

Zeiss, H., ed. (1942–45) *Seuchen Atlas. Hrgs. im Aufrag des Chefs des Wehrmachtsanitätswesens*. Gotha: Perthes.

16

Geographers and the urban century

PETER HALL

British geographers only began to make a serious contribution to urban debates in the 1920s and 1930s. Their contributions fed actively into policy-making during and immediately after the Second World War, when geographers began to be recruited in substantial numbers into the new planning machinery at both central and local government levels, following the passage of the 1947 Town and Country Planning Act. But there was one notable if somewhat eccentric exception, who must form a preamble to the main story.

The precursor: Geddes

That exception made his mark at the very start of the twentieth century. He was Patrick Geddes, who was trained as a botanist but soon had 'begun on the idiosyncratic path which was to take him out of the mainstream of academic life, and eventually from the natural to the social sciences' (Meller, 1990, 19). He became heavily influenced by both the founding fathers of French geography, Elisé Reclus (1830–1905) and Paul Vidal de la Blache (1845–1918), and the French school of sociology founded by Frédéric Le Play (1806–82), whose new academic disciplines acquired respectability in France some years before they did in Britain or the United States (Weaver, 1984, 42, 47–8; Andrews 1986, 179).

Geddes had been introduced to France through his mentor T. H. Huxley, who arranged for him to study at the marine station at Roscoff. He soon mastered the language; 'the central and vital tradition of Scottish culture', he later argued, 'has always been wedded with that of France' (Defries, 1927, 251). From here, he visited the 1878 Paris Exhibition, where he encountered the ideas of Le Play—engineer, social scientist, and counsellor of Napoleon III—who had played a key role in the great 1867 Paris Exhibition, organising it around work and social life, and was repeating the themes in this Republican sequel. Very influential at that time, they emphasised the family as the basic social unit, in the context of its total environment, encapsulated in the phrase *Lieu, Travail, Famille* (place, work, folk) (Meller, 1990, 35).

Not long after, Geddes was introduced to the writings of the anarchist geographer Elisé Reclus, exiled from France for his association with the Communards in 1872; in the 1890s Reclus appeared at his Edinburgh summer school, and Geddes based his celebrated 'Valley Section' on his ideas (Meller, 1990, 31–2, 35, 40). Both Le Play and Reclus fed his idea of the natural region, which significantly he preferred to study in its purest form, far from the shadow of the metropolis:

> For our more general and comparative survey, then, simpler beginnings are preferable . . . the clear outlook, the more panoramic view of a definite geographic region such, for instance, as lies beneath us upon a mountain holiday . . . Such a river system is, as one geographer has pointed out, the essential unit for the student of cities and civilisations. Hence this simple geographical method must here be pled [sic] for as fundamental to any really orderly and comparative treatment of our subject. (Geddes, 1905, 105)

The 'Valley Section, as we commonly call it, makes vivid to us the range of climate, with its corresponding vegetation and animal life . . . the essential sectional outline of a geographer's "region," ready to be studied' (Geddes, 1925, 289). Geddes' greatest single contribution to planning, both in Britain and in the United States, was his stress on the need for basic geographical survey before plan-making. He owed a great deal here to the 'regional monographs' of Vidal de la Blache and his school; his famous Outlook Tower in the Old Town of Edinburgh was designed as a local survey centre for the study of Place–Work–Folk (Mairet, 1957, 216; Weaver, 1984, 47). Natural regions based on river basins, were, he argued, 'the soundest of introductions to the study of cities . . . it is useful for the student constantly to recover the elemental and naturalist-like point of view even in the greatest cities' (Geddes, 1905, 106).

The first link between geography and planning was thus based on a very traditional kind of geography, but it had a political point that was later to be seized upon by Lewis Mumford and his friends in the Regional Planning Association of America: 'the archaic quality of the regional survey, the emphasis on traditional occupations and on historic links, was no mere quirk: like Geddes' attempts to recapture past civic life through masques and pageants, it was a quite conscious celebration of what, for him, was the highest achievement of European culture': the old place-based, pre-industrial, peasant and handicraft culture (Hall, 2002, 149). He owed something here to the writings of another anarchist-in-exile, Peter Kropotkin, whom he met more than once, and who celebrated the life of the free city-states of the late middle ages (Kropotkin, 1920, 18–19). Here, like Ruskin and Morris before him, Raymond Unwin and Barry Parker in his own generation, and Lewis Mumford, Henry Wright, and Benton MacKaye after him, Geddes was contraposing this traditional culture to the life of the metropolis, though he never vilified the latter as many of these other writers did. Indeed, his other great contribution to geography-in-planning was his recognition—in his book *Cities in Evolution* (Geddes, 1915)—of the conurbation as an incipient urban form. Here too, it seems, he drew on Kropotkin, specifically his insight that new technologies—electric power and the motor car—would free humankind from the need to live in congested industrial cities.

> Some name, then, for these city-regions, these town aggregates, is wanted. Constellations we cannot call them; conglomerations is, alas! nearer the mark, at present, but it may sound unappreciative; what of 'Conurbations'? (Geddes, 1915, 34)

As examples, he identified Clyde–Forth, Tyne–Wear–Tees, 'Lancaston', the West Riding and 'South Riding', 'Midlandton', 'Waleston', and Greater London, as well as a score of foreign examples (Geddes, 1915, 41, 47, 48–9). He argued, in a crucial passage that hugely influenced Mumford, that the problem in planning them was to allow them to escape the constraints placed on them by the old 'palaeotechnic' era that was passing: henceforth, 'towns must now cease to spread like expanding ink-stains and grease-spots', but must grow organically, 'with green leaves set in alternation with its golden rays' (Geddes, 1915, 97); the country must be brought into the city. Here was a powerful argument for restraining the growth of cities and for preserving green wedges and green corridors within them, which enormously influenced British planning thought in subsequent decades.

Provinces, conurbations, and the coffin: a 1930s debate

The conurbations were defined more rigorously by an academic geographer, C. B. Fawcett, in 1932. Before that, in 1919, Fawcett had made one of the earliest and most robust attempts to devise a system of natural regions for England. His map is particularly interesting because it bears some resemblance to the Civil Defence regions of 1938, which in turn became the basis for the Standard Regions in the post-war period. In 1932 he defined the seven conurbations more rigorously, adding 30 minor conurbations; his article made it clear that 5 of the 7, and 15 out of the 30, were clustered in an axial belt stretching from Lancashire and Yorkshire, via the Midlands and London, to the south coast (Fawcett, 1932, 101). Subsequently he was more specific:

> There is an area of marked concentration of population in the form of a zone extending diagonally across England from south-east to north-west . . .

> In the last inter-Censal period (1921–31), three-fourths of the increase of population in Great Britain was in this zone . . . it had in 1921, 56 per cent of the total population, a proportion which rose to 58 per cent in 1931, when its population numbered 25,805,915. (Fawcett, 1934, 10)

This was the origin of the celebrated 'coffin', outlined by E. G. R. Taylor in her evidence on behalf of the Royal Geographical Society to the Barlow Commission in 1938 (Taylor et al., 1938, 23), but subsequently criticised by two other geographers, J. N. L. Baker and E. W. Gilbert, on two grounds: first, it was intersected by a zone of lower urban density from the Severn to the Wash; and secondly, it divided itself into a southern zone of high growth and a northern zone of much lower growth (Baker and Gilbert, 1944). In fact, the Royal Commission's report of 1940 (Royal Commission on the Distribution of Industrial Population, 1940) had already stressed this latter point as the relevant one (Hall et al., 1973, 89).

In this period, although there were many joint exercises in regional planning conducted by regional advisory committees, they seemed to have owed little to geographers. The major influence was George Pepler, principal planner to government from 1919 to 1946, as well as secretary and then president of the Town Planning Institute, who commissioned many of them; a few professional planners who had arrived via architecture or other routes, notably Thomas Adams and Patrick Abercrombie, conducted a disproportionate number of them (Cherry, 1981, 136; Wannop, 1995, 4–5).

Dudley Stamp and the Land Utilisation Survey

The outstanding geographical contribution to planning in this period came from quite another direction: in 1930, Lionel Dudley Stamp, who was appointed to the Sir Ernest Cassel Readership in Economic Geography at the London School of Economics in 1926—succeeding Sir Halford Mackinder, who had occupied the position continuously since his period as the School's director in 1904–8—single-handedly launched the Land Utilisation Survey of Great Britain. As Stamp's own account makes clear, it was an audacious step: he proposed to survey every piece of land in Britain with the support of a £500 grant from the LSE's own Rockefeller Research Fund, at a time when a set of the basic six-inch maps cost three times that amount. Though the work was carried out very largely on a voluntary basis by schoolchildren and their teachers up and down the country, and the professional work at the centre was supplemented over the next few years by further grants in cash and in kind, the Survey was in almost permanent financial difficulty; in 1936 the director of the LSE, Sir William Beveridge, announced its winding-up, and Stamp thence supported it personally through the sales of his profitable textbooks

(Stamp, 1962, 4–5, 12). In the event, of course, the Survey proved to be of inestimable use in the wartime food-production drive and at last received government support. But even before that, in 1938 a follow-up survey of agricultural land quality had been of value to the Barlow Commission. This led to Stamp's appointment in 1941–2 as vice-chairman of the Scott Committee on Land Utilisation in Rural Areas, whose report proved very influential in shaping post-war national land use policy (Stamp, 1962, 16).

The Land Utilisation Survey was thus without doubt the most considerable geographical contribution to planning in the key formative years of the Barlow Commission and the follow-up wartime reports, which provided the framework both for Abercrombie's London Plans and for the landmark 1947 Town and Country Planning Act. But it had an indirect effect too. E. C. (Christie) Willatts, a geographer, was appointed as organising secretary in autumn 1931 and took over the directorship of the Survey during Stamp's one-month absence in the United States in 1933–4. He became in effect controller on a day-to-day basis, until he was released for government service, first on a part-time and then on a full-time basis, in 1941–2, as maps officer to the new Research Section of the Ministry of Works, only to find his section incorporated into the new Ministry of Town and Country Planning as the Research Division in the spring of 1943.

The Ministry of Town and Country Planning Research Division

It was a key link, because the new division would play a key role in providing information for planning at the national level. It had the good fortune to be staffed by geographers of a very high calibre. As well as Willatts, it recruited J. R. ('Jimmy') James, Geoffrey Powell, Ian Carruthers, and Stanley Vince. All these had a completely academic cast of mind, albeit tuned to the policy needs of the day.

James's career was fairly typical. Born and educated in County Durham, he graduated in 1936 from the Joint School of Geography of King's College and the LSE, where he was taught by luminaries like Stamp, S. W. Wooldridge, and Stanley Beaver. He taught at Raynes Park County School in southwest London, where he cannot have failed to notice the new semi-detached housing as it spread along the Southern Electric lines. In 1941 he joined the Naval Intelligence Division where he wrote and edited three volumes of the celebrated Geographical Handbooks series on Greece (see Chapter 7, above). He returned to his native northeast as Research Officer in the Newcastle office of the Ministry of Town and Country Planning, in 1946, returning to London again in a more senior position in 1949;

thence he rose rapidly, to Deputy Chief Planner in 1958 and to Chief Planner in 1961. His obituary in the *Geographical Journal* noted that

> For a geographer to achieve this status in the face of competition from the professional town planners was a notable event, for hitherto the numerous geographers in the planning Ministry had been regarded largely as research workers, report writers and general providers of information, rather than high-level administrators. (Anon, 1981, 139)

That comment, which shows signs of inside knowledge, merely hints at considerable tensions within the ministry. Significantly, the Schuster report of 1950 had recommended that the entire basis of planning education be changed, away from the design basis that had characterised the first 30 years of the Town Planning Institute and towards a stronger basis in the social sciences (Committee on the Qualifications of Planners, 1950). Soon after the report was accepted, increasing numbers of geography graduates began to emerge with planning diplomas. But inside the ministry there was a tension between the older planners, with their design backgrounds, and the younger planning researchers. James and his colleagues received strong backing from Evelyn Sharp, the dynamic and unconventional permanent secretary, and from J. D. Jones, her deputy secretary, who later succeeded her as head of the Department of the Environment. One outcome was that the researcher-planners were given permission to cast off the cloak of anonymity with signed articles in academic journals. James himself delivered a paper, 'Land planning in an expanding economy' (1958), which earned him a silver medal from the Royal Society of Arts; Powell (1960) published a paper on 'The recent growth of Greater London', which excited considerable attention because it showed that the London region was growing despite official policies; and Carruthers (1957, 1967) published major research papers on the shopping hierarchy nationally and in Greater London. Another, less visible, sign was that as early as 1951 James, Jones, and others set up a private dining club, the Land Use Society, in order to meet and discuss planning issues with key academic and business figures on a confidential basis under Chatham House rules; highly influential in its early years, for obvious reasons its history has never been recorded in print.

The government, and Jones in particular, played two other crucially important roles. First, in 1966 the Ministry of Transport set up an Economic Planning Directorate, recruiting the economist Christopher Foster as director; Alan Wilson, a young mathematician who had worked with Foster in Oxford, was brought in to create a Mathematical Advisory Unit, there borrowing and further developing the new tradition of land use transportation modelling that had been developed in the United States. Secondly, as a result of a conference held in 1965 in Churchill College, Cambridge, attended both by the Minister of Housing and Local

Government (Richard Crossman) and the Permanent Secretary (Evelyn Sharp), in 1967 the government created an independent think-tank for urban research, the Centre for Environmental Studies. Headed first by Henry Chilver from Cranfield College of Technology, it recruited Alan Wilson as deputy and a young geography graduate then working for an advertising agency, Doreen Massey, as a member of the research team. It rapidly developed a formidable reputation, headed from 1969 by David Donnison and then by Christopher Foster. Long before that, in 1970, Wilson had moved to Leeds to take up a chair in geography: the youngest ever professor of geography in British history, and the first to have moved directly from mathematics into human geography. It was of course a sign of the times, because by then the quantitative revolution was in full swing. Later on still, Wilson achieved another rare feat in moving from his chair to take up the vice-chancellorship at Leeds, soon afterwards being elected as a human geographer to the British Academy. Geography was thus becoming steadily more fluid in its outlines, overlapping strongly with planning, which was similarly influenced by the new science of urban and regional modelling through the seminal works of Brian McLoughlin (1969) and George Chadwick (1971), and which generated new numerate graduates. One of the first was Michael Batty, who joined the Urban Systems Research Unit at Reading University's Department of Geography soon after graduating from Manchester University's Department of Planning.

Meanwhile, a steady stream of geography graduates began to gain postgraduate planning qualifications. An analysis of the biographies of 174 senior local authority planners, published in *Who's Who in Local Government* (Pead, 1996), shows that 153 gave relevant details of their education; of these, 47 (31 per cent) had first qualifications in planning, closely followed by 44 (29 per cent) with first degrees in geography, with 62 (40 per cent) listing a great variety of other first degrees or diplomas.

Academic geographers and planning

The direct impact of academic geography upon planning was less, but there were two conspicuous exceptions. At King's College of the University of Durham, G. H. J. Daysh took an early interest in regional development, which he parlayed into policy during the wartime years at the regional office of the Ministry of Town and Country Planning, returning to academia to publish some of the ideas he had developed there (Daysh et al., 1949). Alfred Augustus Levi Caesar, a young Cambridge geographer, worked under him in the wartime years, returning to Cambridge to teach generations of geographers a style of economic geography that was both academically rigorous and of practical application; among his students were several who became keen practitioners of applied geography, including Michael Chisholm, Peter Hall, Gerald Manners, Ron Martin,

Raymond Pahl, and Brian Robson, as well as a future chief planning inspector who would then retire and convert successfully to academic life, Stephen Crow. When a group of them produced a *Festschrift* for Caesar, they avoided calling it that in order not to embarrass him; the fact was buried deep in the preface (Chisholm and Manners, 1971). This was not a group in any formal sense, despite the huge influence that Caesar exercised on both the content and the style of their academic work. They specialised in diverse topics: Chisholm on rural development, Hall on urban growth and containment, Manners on the geography of energy, Martin on deindustrialisation, Pahl on social linkages and shifting urban–rural boundaries, and Robson on the decline of inner city economies. Though they were all economic geographers in whole or in major part, without exception their work also had important policy implications for planning in its widest sense.

In Birmingham a young geographer just returned from distinguished war service, Michael Wise, made a major contribution to *Conurbation*, a major planning survey of the West Midlands, which was published by the West Midland Group on Post-war Reconstruction and Planning (1948) as input to the regional plan then under preparation by Abercrombie and Jackson. Moving to the LSE, Wise was to exert a similar decisive influence on generations of students as did Caesar in Cambridge, and in very much the same field of applied economic geography.

The two streams came together briefly in the mid-1960s, when Hall joined Wise at the LSE to help set up (together with the economist Alan Day and the political scientist Peter Self) a new MSc in Regional and Urban Planning Studies, one of the first of the so-called ‘new planning courses’ that embodied the new analytical approaches coming from the quantitative revolution in geography and the development of urban and regional economics. This, and the similar course that Hall launched after taking the chair at Reading in 1968, was paralleled by the rapid appearance of new journals that embodied work in the newly emergent academic areas: *Urban Studies* (1964, edited by Barry Cullingworth and Derek Diamond), *Regional Studies* (1967, edited by Peter Hall), and *Environment and Planning* (1969, edited by Alan Wilson and Doreen Massey). The first and third of these were edited from the University of Glasgow and the Centre for Environmental Studies, respectively; the second was the house journal of the Regional Studies Association, founded in 1965 by a diverse group representing academic disciplines and professional interests; significantly, it did not affiliate with the US-based Regional Science Association because it took a deliberate decision to develop more specifically policy-relevant work.

Another important contribution came from geographers who studied urban fields and central place systems. This took place independently of—and to some extent in ignorance of—the pioneer work by Christaller in Germany, published there in 1933 but not translated until 1966.

One British geographer, Robert Dickinson, brought this work to the attention of colleagues in the 1950s, but it evoked little interest in the English-speaking world until a British émigré geographer working in the United States, Brian J. L. Berry, published his path-breaking work at the end of the decade (Berry and Garrison, 1958; Christaller, 1966; Johnston, 2001). Smailes's pioneer study of the urban hierarchy, published in 1944 and based on 1938 data, inspired a geographer-planner, Robin Smith, to extend and update his work, first for England and then for Wales and Scotland (Smith, 1968, 1970, 1978). F. H. W. Green's pioneering studies of bus services contributed powerfully to the emergent understanding of central place systems (Green, 1950). Simultaneously, David Thorpe published a major study of the retail hierarchy, work that was supplemented a little later by Russell Schiller (Thorpe, 1968; Schiller and Jarrett, 1985; Schiller and Reynolds, 1992). Later still, Clifford Guy (1996, 1997, 1998) published a number of studies that plotted the impacts of new forms of retailing, such as regional shopping centres and retail parks.

Academic geography came to the fore again in planning in the mid-1960s, significantly at a point when questions of regional strategic planning were being debated. Powell's 1960 paper, showing that despite controls the London region had added one million people in the 1950s, was the trigger. It led directly to his appointment as director of the South East Study from 1961 to 1981, an internal study (Ministry of Housing and Local Government, 1964) that contained proposals for major urban developments in south Hampshire, north Buckinghamshire, Peterborough, and east Kent. So controversial were these proposals, published as they were just before the 1964 general election, that the incoming Labour government immediately launched a review; but so firmly grounded were the proposals that the review led to no major change. Meanwhile, while the study was in preparation, Peter Hall had published his own proposals in his book *London 2000* (Hall, 1963), proposing some 30 new towns in the Outer Metropolitan Area around London. When the incoming government set up Economic Planning Councils for each region in 1965–6, geographers were appointed to several of them (then or subsequently). On the South East Economic Planning Council, Hall played a central role in drafting the *Strategy for the South East* (South East Economic Planning Council, 1967), based on a comparative international study of metropolitan planning strategies that he made in association with his book *The World Cities* (Hall, 1966, 1967). When the Strategy evoked controversy with the planning authorities in the region—partly, echoing earlier arguments within the ministry, based on the argument that this was not produced by professional planners—the government decided to set up a team, with both ministry and local government members and outside advisers, to produce a strategic plan. Reporting both to the Council and to the Standing Conference of Planning Authorities, the team was headed by James's successor Wilfred Burns, but much of the day-to-day direction was again in the hands of Geoffrey Powell; backed by an impressive volume of research, it appeared in 1970 (South East Joint Planning Team, 1970).

Spinning off these regional studies, in the southeast and elsewhere, the second half of the 1960s saw a series of feasibility studies and master plans for new towns, most of which were implemented. Geographers played a role here too. Notable was the publication *A New Town in Mid Wales* (Welsh Office, 1966), where the team was headed by Hall and also included Alan Wilson and a 21-year-old assistant, John Goddard, later director of the Centre for Urban and Regional Development Studies (CURDS) and professor of geography at Newcastle upon Tyne. The plan achieved another distinction: it was never properly implemented.

There was one major failure on the part of the geographer-planners in the 1960s: the attempt to reorganise local government in England on the basis of functional city regions. Originating in the group within the Ministry of Housing and Local Government, the idea was powerfully boosted by a symposium edited by the *Guardian's* planning correspondent (Senior, 1966), and by the adoption of a two-tier system of planning (structure plans and local plans), proposed by a Planning Advisory Group in 1965 and implemented in 1968. The group's proposals received strong support from Richard Crossman, as Minister for Housing and Local Government and thus responsible for planning. The Redcliffe-Maud Commission, duly appointed with Senior as key member, commissioned both in-house research from a team led by Stanley Vince as well as research from the LSE and other academic institutions. The LSE research on the structure of commuter and shopping flows in southeast England, which was later extended to the entire country, demonstrated that no system properly exhausted the territory in a system of mutually exclusive regions (Royal Commission on Local Government, 1968). Considerations like these caused the Commission to split, with Senior proposing a different and considerably more complex system than the majority. In the event, the report was overtaken by politics: at the 1970 election the Conservatives were returned, more than inclined to heed a noisy campaign by the rural localities to avoid what they saw as an urban take-over. Their piecemeal reform, preserving most of the historic counties, compromised the operation of the planning system for a generation or more. There was an ironic postscript: when reorganisation again appeared on the agenda in the late 1980s, Michael Chisholm, a distinguished geographer, was appointed to the commission. His work went unsupported by any serious research; afterwards, he wrote a paper berating the politicised nature of the whole exercise (Chisholm, 1997).

One final tradition of applied academic research was virtually the monopoly of a small department: Robin Best, a geographer, worked with the economist Gerald Wibberley at Wye College on problems of measuring and valuing the use of land. Their work first came to the fore in the 1950s in major contributions about the loss of agricultural land to proposed new towns, where they were able to show that the value of the lost agricultural land was infinitesimal in comparison with the costs of building higher and more densely in the cities. Later, Best extended his work to a comprehensive and exhaustive study of national land use and land use change in the UK; summarised in an overview volume, it showed that the extent of urban growth was far less than was popularly believed (Best, 1981). Unhappily, like so much excellent work of its kind, its message has been largely forgotten as shriller voices dominate the debate.

The containment of urban England

At the end of the 1960s, the Leverhulme Trust financed Political and Economic Planning (PEP) to carry out a major study of the 1947 planning system and its impacts. Published in 1973, the co-authored work concluded that planning had produced three outcomes, only one of which had been intended, and two of which were unfortunate and even perverse (Hall et al., 1973). The first was *containment*: planning had worked to physically constrain the suburban sprawl that had been so characteristic of the 1930s. The second was *suburbanisation*: suburban growth had not been stopped, but had been diverted into more distant locations, leading to more long-distance commuting than would otherwise have been the case. And the third was *rise in property values*: by restricting the supply of housing in areas of high demand, planning had made unsubsidised housing less affordable than it was in an unconstrained situation. This last conclusion was supported by a number of other empirical studies during the 1980s, concluding with a government-commissioned study, which confirmed that it was indeed the case, but that in order to reduce prices substantially it would be necessary to relax planning policies so much that it would effectively represent a return to the pre-1947 position (Department of the Environment, 1992).

Geography and planning in the 1970s: the great divorce

After 1970, with a profound shift in *Zeitgeist* following the *événements* of 1968 in Paris and elsewhere, geography, like other social sciences, took on a profoundly critical hue. UK geographers like Doreen Massey and David Harvey—whose 1973 book *Social Justice and the City* was hugely influential—were paralleled by their colleagues abroad, notably the English émigré Allen Scott, then teaching in Toronto, and by sociologists like Manuel Castells (Harvey, 1973; Castells, 1977; Dear and Scott, 1981; Massey, 1984). They were strongly represented at the Centre for

Environmental Studies (CES); ironically, this led to its immediate closure as soon as Margaret Thatcher assumed power in 1979, thus appearing to confirm the dire analysis of its practitioners. This tradition produced a great deal of very high-quality theoretical work as well as good empirical analyses: its problem was that it necessarily could not be greatly interested in offering practical advice to improve or ameliorate a system it saw as approaching the last stages of collapse.

That belief in the system's weaknesses received some support from events, as serious structural transformation occurred in the UK and other advanced economies at that time. From about 1970, urban economies began to suffer from major deindustrialisation, coupled with the loss of other basic industries like ports and warehousing; this deepened during the early years of the Thatcher government. This trend was first noticed by a few acute urban observers in the early 1970s (Donnison and Eversley, 1973), but in 1977 the publication of official consultants' reports on three inner cities (Liverpool, Birmingham, and Lambeth), in which geographer-planners played a role, confirmed the scale of the problem (Department of the Environment, 1977). It brought about a shift in policies and resources away from urban decentralisation and into inner-city regeneration, which the Thatcher government continued, albeit with different policy instruments. From the early 1980s, there was a flood of academic work analysing this problem, much of it from geographers (Hall, 1981; Boddy, Lovering, and Bassett, 1986; Lever and Moore, 1986; Martin and Rowthorn, 1986). The problem was that, like much work of the time, it was at one remove from the development of actual policy.

A sole exception was Peter Hall's somewhat whimsical 1977 proposal—in a speculative paper to the Royal Town Planning Institute's annual conference—for new-style plan-free zones in the most seriously impacted inner-city areas (Hall, 1977). Deliberately designed as a kite-flying exercise, it was ironically taken up by a leading Conservative politician, Geoffrey Howe, and became part of official policy after 1979, albeit in much-transmuted form.

The 1980s and 1990s: economic geography, technology, and clusters

One promising line of work did, however, begin to develop during the 1980s: anticipating work at the macro or national level by economists (Porter, 1990; Krugman, 1995), it was increasingly accepted that the clue to successful economic adaptation lay in endogenous growth through new and innovative firms and industries. Work emerged within the new neo-Schumpeterian paradigm in Reading (on London's Western Crescent), in Cambridge (on the so-called Cambridge phenomenon), and in Newcastle, where particularly important research was published on the role of telecommunications technology in the city (Oakey, Thwaites, and Nash, 1980; Oakey, 1981; Hall, 1985; Gillespie and Goddard, 1986; Hall et al., 1987; Gillespie and Hepworth, 1988; Keeble, 1988, 1989, 1990, 1997, 1999). A systematic worldwide study of so-called 'technopoles' (Castells and Hall, 1994) further concluded that it was apparently possible to generate such phenomena—as in the well-known case of the Cambridge phenomenon studied by Keeble—but that ill-judged state intervention could fail to generate the necessary synergies between research and development, and between the public and private sectors.

Generally, the direct and immediate effect of all this research on planning was limited, since it suggested that new industrial complexes—or 'clusters' as they came to be called in the late 1990s, following economist Paul Krugman's belated discovery of economic geography—developed slowly and cumulatively over long periods, sometimes as a result of quite fortuitous events. And this was given additional force by Peter Hall's (1998) work comparing cultural creativity and technical innovation, which concluded that both phenomena occurred over fairly short periods of history in a few selected places—the first in already-established centres of power and wealth, which could afford surplus resources to spend on the arts, the second in emergent places with an egalitarian social structure and a respect for technical education.

Geography and planning at the start of a new century

At the start of the new century, Brian Robson was warning that new economic divisions were emerging—between London and the northern cities, and between northern city centres and their surrounding inner-city rings. Narrowly focused regeneration programmes would fail, he argued, unless they addressed the wider problems of the regional economy. Writing in *The Guardian* in October 2002, just before the government's Urban Summit, he posed a stark distinction that was emerging in many British cities:

> . . . the growth of the doughnut city—jam in the middle; unleavened dough in the ring beyond. There is a deepening gulf between the city centres and the collar of decayed areas that surround them. Half a mile from the new penthouses, restaurants, offices and multiscreen cinemas, and you are in the land of the forgotten.
>
> The colonisation of city centres by young professionals living in new and refurbished apartments may have done wonders for the vitality of the cores of places like Manchester, Leeds and Birmingham. But a stone's throw away is the sad mix of decaying 19th-century terraces and post-war council housing

> which, in northern cities, has seen ever-increasing problems of collapsing housing markets. Residents are not only poor; many are trapped by negative equity.
>
> The environment of such areas is a mix of boarded-up houses, syringes in the alleyways, litter and dirt, abandoned shops, unruly packs of children 'bunking off' school. A new instability has been added through unregistered private landlords profiting from asylum seekers and refugees. (Robson, 2002)

There was hope, he suggested

> . . . that the combination of mainstreaming and developing cross-departmental programmes will help by improving services, and connecting training and job placements to physical improvements. Yet these are often the very areas that have been the target of the whole sequence of regeneration programmes—from the housing improvement areas of the 1960s and 70s, through Action for Cities in the 1980s and 90s, to other programmes of recent years. The only hope for sustainable regeneration must now rest with the focus on people as much as on property. (Robson, 2002)

This conclusion, ironically, could have equally applied a hundred years earlier. Then, too, as Charles Booth's maps of London so tellingly showed, there were stark contrasts within our cities. For decades after the Second World War, policy-makers fondly believed that they were being eliminated. Now it is evident that the claim was false and that there is a new urban agenda for geographers and all who labour with them in the study of the contemporary city.

References

Andrews, H. F. (1986) The early life of Paul Vidal de la Blache and the makings of modern geography. *Transactions, Institute of British Geographers*, NS11, 174–82.

Anon (1981) John Richings James, CB, OBE. *Geographical Journal*, 147, 139.

Baker, J. N. L. and Gilbert, E. W. (1944) The doctrine of an axial belt of industry in England. *Geographical Journal*, 103, 51.

Berry, B. J. L. and Garrison, W. L. (1958) The functional bases of the central-place hierarchy. *Economic Geography*, 34, 145–54. Reprinted in H. M. Mayer and C. F. Kohn, eds (1959) *Readings in Urban Geography*. Chicago, IL: University of Chicago Press, 218–27.

Best, R. H. (1981) *Land Use and Living Space*. London: Methuen.

Boddy, M., Lovering, J., and Bassett, K. (1986) *Sunbelt City? A Study of Economic Change in Britain's M4 Corridor*. Oxford: Oxford University Press.

Carruthers, W. I. (1957) A classification of service centres in England and Wales. *Geographical Journal*, 123, 371–85.

Carruthers, W. I. (1967) Major shopping centres in England and Wales, 1961. *Regional Studies*, 1, 65–81.

Castells, M. (1977) *The Urban Question: A Marxist Approach*. London: Edward Arnold.

Castells, M. and Hall, P. (1994) *Technopoles of the World: The Making of 21st Century Industrial Complexes*. London: Routledge.
Chadwick, G. (1971) *A Systems View of Planning: Towards a Theory of the Urban and Regional Planning Process*. Oxford: Pergamon.
Cherry, G. E. (1981) *Pioneers in British Planning*. London: Architectural Press.
Chisholm, M. (1997) Public management: independence under stress. *Public Administration*, 75, 97–107.
Chisholm, M. and Manners, G., eds (1971) *Spatial Policy Problems of the British Economy*. Cambridge: Cambridge University Press.
Christaller, W. (1966) *Central Places in Southern Germany*, translated by C. W. Baskin. Englewood Cliffs, NJ: Prentice-Hall (first published 1933).
Committee on the Qualifications of Planners (1950) *Report*. London: HMSO, Cmd 8059 (*BPP*, 1950, 14).
Daysh, G. H. J. et al. (1949) *Studies in Regional Planning: Outline Surveys and Proposals for the Development of Certain Regions of England and Scotland*. London: Philip.
Dear, M. and Scott, A. J. (1981) *Urbanization and Urban Planning in Capitalist Society*. London and New York: Methuen.
Defries, A. (1927) *The Interpreter Geddes: The Man and his Gospel*. London: George Routledge and Sons.
Department of the Environment (1977) *Inner Area Studies: Liverpool, Birmingham, and Lambeth. Summary of Consultants' Final Reports*. London: HMSO.
Department of the Environment (1992) *The Relationship between House Prices and Land Supply*. London: HMSO. Report prepared by Gerald Eve, Chartered Surveyors with the Department of Land Economy, University of Cambridge.
Donnison, D. and Eversley, D. E. C. (1973) *London: Urban Patterns, Problems and Policies*. London and Beverly Hills, CA: Sage.
Fawcett, C. B. (1919) *Provinces of England: A Study of some Geographical Aspects of Devolution*. London: Williams and Norgate.
Fawcett, C. B. (1932) Distribution of the urban population in Great Britain, 1931. *Geographical Journal*, 79, 100–16.
Fawcett, C. B. (1934) Areas of concentration of population in the English-speaking countries. *Population*, 1(3), 4–13.
Geddes, P. (1905) Civics: as applied sociology. *Sociological Papers*, 1, 101–44.
Geddes, P. (1915) *Cities in Evolution*. London: Williams and Norgate.
Geddes, P. (1925) The valley plan of civilization. *The Survey*, 54, 288–90, 322–5.
Gillespie, A. E. and Goddard, J. B. (1986) Advanced telecommunications and regional development. *Geographical Journal*, 132, 383–97.
Gillespie, A. E. and Hepworth, M. (1988) Telecommunications and the reconstruction of regional comparative advantage. *Environment and Planning A*, 20, 1311–21.
Green, F. H. W. (1950) Urban hinterlands in England and Wales: an analysis of bus services. *Geographical Journal*, 116, 64–88.
Guy, C. M. (1996) Corporate strategies in food retailing and their local impacts: a case study of Cardiff. *Environment and Planning A*, 28, 1575–1602.
Guy, C. M. (1997) Fixed assets or sunk costs? An examination of retailers' land and property investment in the United Kingdom. *Environment and Planning A*, 29, 1449–64.

Guy, C. M. (1998) Alternative-use valuation, open A1 planning consent, and the development of retail parks. *Environment and Planning A*, 30, 37–47.
Hall, P. (1963) *London 2000.* London: Faber.
Hall, P. (1966) *The World Cities.* London: Weidenfeld and Nicolson.
Hall, P. (1967) Planning for urban growth: metropolitan area plans and their implications for South-East England. *Regional Studies*, 1, 101–34.
Hall, P. (1977) Green fields and grey areas. In *Papers of the RTPI Annual Conference, Chester*. London: Royal Town Planning Institute.
Hall, P., ed. (1981) *The Inner City in Context: The Final Report of the Social Science Research Council Inner Cities Working Party*. London: Heinemann.
Hall, P. (1985) The geography of the fifth Kondratieff. In P. Hall and A. Markusen, eds, *Silicon Landscapes*. London: Allen & Unwin, 1–19.
Hall, P. (1998) *Cities in Civilization*. London: Weidenfeld and Nicolson.
Hall, P. (2002) *Cities of Tomorrow: An Intellectual History of Urban Planning and Design in the Twentieth Century*, 4th edn. Oxford: Blackwell.
Hall, P., Thomas, R., Gracey, H., and Drewett, R. (1973) *The Containment of Urban England*. London: Allen & Unwin (2 vols).
Hall, P., Breheny, M., McQuaid, R., and Hart, D. (1987) *Western Sunrise: The Genesis and Growth of Britain's Major High-Tech Corridor*. London: Allen & Unwin.
Harvey, D. (1973) *Social Justice and the City*. London: Edward Arnold.
James, J. R. (1958) Land planning in an expanding economy. *Journal of the Royal Society of Arts, Manufactures and Commerce*, 106, 589–604.
Johnston, R. (2001) Robert E. Dickinson and the growth of urban geography: an evaluation. *Urban Geography*, 22, 702–36.
Keeble, D. (1988) High-technology industry and local environments in the United Kingdom. In P. Aydalot and D. Keeble, eds, *High-Technology Industry and Innovative Environments: The European Experience*. London: Routledge and Kegan Paul, 65–98.
Keeble, D. (1989) High-technology industry and regional development in Britain: the case of the Cambridge phenomenon. *Environment and Planning C: Government and Policy*, 7, 153–72.
Keeble D. (1990) Small firms, new firms and uneven regional development in the United Kingdom. *Area*, 22, 234–45.
Keeble, D. (1997) Small firms, innovation and regional development in Britain in the 1990s. *Regional Studies, 31*, 281–93.
Keeble, D. (1999) Urban economic regeneration, SMEs and the urban–rural shift in the United Kingdom. In E. Weaver, ed., *Cities in Perspective, I: Economy, Planning and the Environment*. Assen: Van Gorcum, 29–47.
Kropotkin, P. (1920) *The State: Its Historic Role*, 5th edn. London: Freedom Press.
Krugman, P. (1995) *Development, Geography, and Economic Theory*. Cambridge, MA: MIT Press.
Lever, W. and Moore, C., eds (1986) *The City in Transition: Policies and Agencies for the Economic Regeneration of Clydeside*. Oxford: Oxford University Press
Mairet, P. (1957) *Pioneer of Sociology: The Life and Letters of Patrick Geddes*. London: Lund Humphries.
Martin, R. L. and Rowthorn, R., eds (1986) *The Geography of De-Industrialisation*. Basingstoke: Macmillan.

Massey, D. (1984) *Spatial Divisions of Labour: Social Structures and the Geography of Production*. Basingstoke: Macmillan.
McLoughlin, J. B. (1969) *Urban and Regional Planning: A Systems Approach*. London: Faber and Faber.
Meller, H. (1990) *Patrick Geddes: Social Evolutionist and City Planner*. London and New York: Routledge.
Ministry of Housing and Local Government (1964) *The South East Study 1961–1981*. London: HMSO.
Oakey, R. P. (1981) *High-Technology Industrial Location: The Instruments Example*. Aldershot: Gower.
Oakey, R. P., Thwaites, A. T., and Nash, P. A. (1980) The regional distribution of innovative manufacturing establishments in Britain. *Regional Studies*, 14, 235–53.
Pead, D., ed. (1996) *Who's Who in Local Government*. London: LGC Communications.
Porter, M. E. (1990) *The Competitive Advantage of Nations*. London: Collier Macmillan.
Powell, A. G. (1960) The recent development of Greater London. *The Advancement of Science*, 17, 76–86.
Robson, B. (2002) Filling the doughnut. *The Guardian (Society)*, 30 October.
Royal Commission on the Distribution of Industrial Population (1940) *Report of the Royal Commission on the Distribution of Industrial Population*. London: HMSO, Cmd 6153 (Barlow Report).
Royal Commission on Local Government in England (1968) *Research Studies 1: Local Government in South East England*, by the Greater London Group, the London School of Economics and Political Science. London: HMSO.
Schiller, R. K. and Jarrett, A. (1985) A ranking of shopping centres using multiple branch numbers. *Land Development Studies*, 2, 53–100.
Schiller, R. K. and Reynolds, J. (1992) A new classification of shopping centres in Great Britain using multiple branch numbers. *Journal of Property Research*, 9, 122–60.
Senior, D., ed. (1966) *The Regional City: An Anglo-American Discussion of Metropolitan Planning*. Harlow: Longman.
Smailes, A. E. (1944) The urban hierarchy in England and Wales. *Geography*, 29, 41–51.
Smith, R. D. P. (1968) The changing urban hierarchy. *Regional Studies*, 2, 1–19.
Smith, R. D. P. (1970) The changing urban hierarchy in Wales. *Regional Studies*, 4, 85–96.
Smith, R. D. P. (1978) The changing urban hierarchy in Scotland. *Regional Studies*, 12, 331–51.
South East Economic Planning Council (1967) *A Strategy for the South East: A First Report by the South East Planning Council*. London: HMSO.
South East Joint Planning Team (1970) *Strategic Plan for the South East: A Framework. Report by the South East Joint Planning Team*. London: HMSO.
Stamp, L. D. (1962) *The Land of Britain: Its Use and Misuse*. London: Longman.
Taylor, E. G. R. et al. (1938) Discussion on the geographical distribution of industry. *Geographical Journal*, 92, 22–39.
Thorpe, D. (1968) The main shopping centres of Great Britain in 1961: their location and structural characteristics. *Urban Studies*, 5, 165–206.

Wannop, U. (1995) *The Regional Imperative: Regional Planning and Governance in Britain, Europe and the USA*. London: Jessica Kingsley.
Weaver, C. (1984) *Regional Development and the Local Community: Planning, Politics and Social Context*. Chichester: Wiley.
Welsh Office (1966) *A New Town in Mid-Wales: Consultants' Proposals*. London: HMSO.
West Midland Group on Post-War Reconstruction and Planning (1948) *Conurbation: A Planning Survey of Birmingham and the Black Country*. London: Architectural Press.

17

Geographers and the fragmented city

CERI PEACH

Background

Whereas at the beginning of the twentieth century Mackinder could magisterially survey the world, declare it a closed system, and divide it into just three regions, by the end of the century geography's focus had narrowed, its content had become less disciplined by spatial concerns, and its subject matter had become fragmented. Geographers were writing about small, personal subjects: about identity and positionality, about statues and monuments.

How and why did this come about ? I think it is because both the urban geography of Britain underwent massive change and the way in which British geographers thought about cities was revolutionised. In reviewing this change in urban geography in Britain during the twentieth century, there are four reasons that emerge. The first is that the geography of British towns has changed significantly over the period. During the nineteenth and early twentieth centuries towns had sucked populations in and their boundaries expanded. In mid-century their girths were corseted by green belts, but in the latter part of the century, the conurbations spun population out to the New Towns and then, more diffusely, to villages and market towns in the wave of counter-urbanisation. The old cities were un-made through decay or re-made through gentrification and redevelopment, and their populations were transformed. Urban change gave geographers a large number of hares to chase and these hares were moving in different directions. Fragmentation was inevitable.

The second reason is that the major paradigms within which urban geography was written changed dramatically over the course of the twentieth century, first from ideographic to nomothetic, then to radical and humanistic and the new cultural turn. I discuss these changes in more detail below, but at this point it is sufficient to say that urban geography in Britain in the first half of the twentieth century was largely a-theoretic and informed more by empirical investigations of historical developments on geographical sites (see Hugh Clout's account in Chapter 7, above). Thereafter it tended to become a-historic, a-geographic, and driven by theory (as described by Ron Johnston in Chapter 9, above).

The third reason is that geography, in the post-1950 period, became internationalised. Many British urban geographers went overseas or were writing on non-British topics (e.g. Johnston, 1984) while some of those writing on British urban topics were not British. British urban geography became part of an international (but largely Anglo-American) literature rather than a discrete entity. Whereas Herbertson, at the end of the nineteenth century, had taken his PhD in Freiburg and reading German or French was almost compulsory for geography undergraduates in British universities up to the end of the 1950s, thereafter statistics replaced foreign languages and the United States replaced Europe as the source of many of the inspirations. (French philosophy, one would have to admit, has made a comeback since: see Bourdieu, 1977; Lefebvre, 1991.) Urban geography in this sense was not so much fragmented as fused into a much wider tradition from which a separate British identity became harder to discern.

Fourthly, in the first part of the twentieth century there were relatively few urban geographers, but they were writing about large topics; in the latter half, there were a larger number, but writing about smaller topics. As outlined by Ron Johnston in Chapter 2, geography departments expanded dramatically in Britain over the course of the twentieth century and the number of PhDs expanded likewise. In Herbertson's day, at the beginning of the twentieth century, a PhD had seemed a rather un-British qualification and he had taken his in Germany. After the 1960s, a PhD was a *sine qua non* for entering the academic geographical establishment. As the number of PhDs expanded, fields grew crowded and the topics tended to become smaller, more specialised, and fragmented.

Thus, to summarise, one could say that if we are talking of geographers and the fragmented city, part of that fragmentation is caused by the change in the nature of the cities themselves and part by the changes in the way in which geographers have come to view the city. The detailed history of regional change in Britain is well covered by Ray Hudson in Chapter 18, but a synoptic summary is given here to contextualise the story of geographers and the fragmented city. I start then with a brief account of the urban change and then move on to discuss the shifts within the philosophy of geography.

A brief account of the urban change

In the first two decades of the twentieth century, Britain was one of the greatest maritime imperial powers on the globe, with one of the largest merchant fleets and the greatest shipbuilding industry. London was the largest city in the world, but it was not a primate city within its own country as, say, Paris was within France. The Atlantic-facing ports of Liverpool, Glasgow, and Bristol, the textile cities of Lancashire and Yorkshire, the manufacturing cities of the Midlands, the steel-making cities of Scotland, Wales, and Yorkshire, the widespread and numerous coalfields (not least those of Durham and Northumberland, and the South Wales valleys) and coal-exporting ports—all gave a regional counterbalance to the metropolitan importance of the capital.

The great depression in the two decades following the First World War was regionally selective of the north and west and continued to the end of the century to leave these areas as the victims of the slash-and-burn industrialisation of the first two Kondratieff waves (Hall and Preston, 1988). The two decades following the Second World War saw a continuation of this decline, accentuated in the 1970s by the oil crisis. In the 1980s, Thatcherism not only worsened the problem but devastated the manufacturing base of the Midlands (the slash-and-burn industrial landscape of the third Kondratieff), leaving the south and east and a few pockets of the fourth Kondratieff prosperity elsewhere (Silicon Glen, for example) to carry the industrial torch for burnt-out urban landscapes.

The period from the late 1930s to 1980 can be regarded as the period of socialist planning. The 1930s saw national despair with the free market economy and between 1939 and 1945 there was coalition politics but rigid regulation. Following the war, the Labour governments of 1945–51 produced the most radical period of socialist planning experienced in Britain. Despite subsequent changes of government to the Conservatives, the post-war period into the 1960s was the high tide of state intervention and Butskellite consensual socialism. However, the 1970s saw a sharp break as the worldwide economic collapse of western industrial economies followed the 1973 and subsequent oil crises. The consequence in Britain was a disillusion with state intervention and the election in 1979 of the Thatcher government, dedicated to non-consensual politics, denationalisation, deregulation, and the elimination of social housing.

The first tentative steps in economic planning during the 1930s sought to sustain and create employment in the areas most devastated by the depression, with a policy of pinpointing Special Development Areas, which were extended in the next phase. By the 1960s, the government recognised the shadow effect that Development Areas were having on neighbouring areas: although nearly as badly affected by unemployment as the Development Areas, they could not attract industry because of the greater attraction of the Development Areas. Thus Intermediate Areas came into existence, and by 1979 much of the country was covered by special areas of one kind or another. In the 1960s, Labour governments had also recognised that the engine of economic growth was shifting from manufacturing to the service industries. The Location of Offices Bureau was set up to channel office development to the Development Areas and restrictions were placed on office expansion in London in particular.

The Thatcher government that came to power in 1979 scrapped the Development Areas and office construction controls, creating Enterprise Zones where planning restrictions were abolished. The Dockland Development Agency in London wrested the planning control of the dying docklands from its locally elected leaders and let the free market rip. It was an attempt to bring Houston (famous for having no zoning restrictions) to London or Hong Kong to Swansea. As the state was 'rolled back', subsidies to industries and services experiencing difficulties were abolished, the nationalised industries of coal, steel, electricity, railways and telephones were privatised, taxes on the rich were reduced, and polarisation of the population grew.

The new immigration

Another feature of the early post-war period was a dramatic shortage of labour in Britain, caused in part by the upward mobility of the workforce and the abandonment of poorly paid jobs with unsocial hours in the newly nationalised hospitals and transport services. In the metal-bashing industries of the Midlands and in the textile mills of west Yorkshire and in Lancashire, there was also a shortage of labour. The desperate attempts of the northern textile towns to compete with Third World industries produced a need for cheap labour. Population from the Caribbean and from South Asia moved in to fill the gaps in the labour market there and in the large conurbations. This movement of Commonwealth immigrants was closely related to British employment cycles.

Although Britain had a long history of non-European ethnic minority populations (Little, 1948; Richmond, 1954 ; Banton, 1955; Collins, 1957), the post-war scale of immigration was different in kind from the previous experience. The visible minority population of Britain grew from about 50,000 in 1951 (Peach, 1996a) to 4.6 million in 2001. From the 1950s to the 1970s, the movement of the Caribbean and to a lesser extent the South Asian population was largely controlled by the demand for labour (Robinson, 1986; Peach, 1965, 1997). Immigration controls instituted in the 1960s had the paradoxical effect of stimulating both the immigration and the settlement of the minority population, since they made it difficult for migrants returning home to re-enter the country. Choking off labour immigration through vouchers but leaving the family reunification route open meant that there were incentives to bring families in rather than for workers to return to their families.

The Caribbean and South Asian settlement pattern in Britain appeared like a barium meal in a body scan. Avoiding the dead areas of unemployment, they were impeded from settling in the areas of healthy circulation and concentrated in the areas unattractive to the white population. Minority settlement took place in the large conurbations in which there was demand for labour, but which the white population was leaving (Peach, 1966). They avoided altogether the areas of high unemployment (Northern Ireland, Scotland, Wales, and the north) and were blocked from substantial settlement in areas that were attractive to the white population (those places favoured by the tide of counter-urbanisation). The New Towns and the rural areas remained lily white, while immigrant settlement was greatest in the inner areas of the large cities. Visible minorities were not the only ones to arrive; as globalisation intensified, the movement of international executives such as those from Japan and the EU (White, 1998), young adventurers from the white Commonwealth, and refugees from Eastern Europe, Africa, the Middle East, and the Far East also increased, with a rather higher concentration on London and the west and east Midlands than had marked the early post-war flows.

Shifts within the philosophy of geography

The role of geographers in urban planning has been reported in Peter Hall's contribution to this volume (Chapter 16) and his *Cities of Tomorrow* (Hall, 1988) is a wonderful synopsis of urban movements in the world in general. The role of geographers in the new social geography of British cities is covered here. But whereas Hall records the active participation of geographers in the planning process, the role of urban and social geographers has been more one of commenting from the sidelines.

During the course of the twentieth century, Britain had changed from a raw-material, heavy-industry economy to a light-industrial manufacturing economy and then to a predominantly service-transactional economy. It had changed from a regionally diversified and developed economy to an economy of widespread dereliction and metropolitan dependence. Many of its cities had changed from being centres of production to sinks of welfare consumption. Thus while the urban geography of the 1950s was a confident geography of conurbations as a whole (Freeman, 1959), or nostalgic urban geographies such as E. W. Gilbert on Brighton (Gilbert, 1954) or on England and Germany's old university towns (Gilbert, 1961), or broad-ranging urban geographies such as R. E. Dickinson's *The West European City* (1951), the later geographies were more fragmented. The leapfrogging of the green belt had produced a new urban class in the countryside, charted by Ray Pahl in *Urbs in Rure* (1965). New Towns required new urban geographies and new urban geographers, although the foundations had been laid by Ebenezer Howard (1902).

The quantitative revolution

The quantitative revolution led to a radical change in British urban geography. Its most immediate impact came through Chorley and Haggett's *Frontiers in Geographical Teaching* (1965) and *Models in Geography* (1967), plus Haggett's *Locational Analysis in Human Geography* (1965). Barry Garner's chapter on 'Models of urban geography and settlement location' in *Models in Geography* (Garner, 1967) was one of the most influential pieces of writing in redirecting the literature of British urban geography. Christaller, Lösch, and Zipf were introduced to a stunned audience. A new generation discovered central place theory, rank-size rules, gravity models, internal models of city structure, and generalised land values surfaces. Brian Robson (1973) produced a new approach to British urban geography, treating cities not as individuals but as parts of a system. The fragmentation of urban geography is perhaps as apparent as anywhere in the study of retail structure. Studies of the CBD and retail hierarchies briefly flourished and then perished; they are no more.

To understand the extent of the quantitative revolution, one has only to compare the work on urban hinterlands, developed by A. E. Smailes (1947) or F. H. W. Green (1950), with the later work of retail geographers. Green's work was set in an historical but rather a-theoretic framework. Although it was relevant to the ideas of range, threshold, and hierarchy, it was only when the English-speaking world discovered the writings of Christaller and Lösch during the quantitative revolution that the theoretical relevance became apparent. Christaller's central place theory not only gave Green's work a theoretical framework, but immediately eclipsed it. Christaller's notions of range (how far people would travel to buy a certain good), threshold (how large a market was required to sustain the provision of that good), and hierarchy (the idea that there was a large number of places serving local low-value functions but a small number of places with high-order functions serving wide areas and large populations) were embraced by British urban geographies with enthusiasm.

Christaller's theory has central places developed on an isotropic surface. At each level of the hierarchy, places are uniformly distributed with tesselated hexagonal service areas around them. Geography was to be like a honeycomb. Range, threshold, and hierarchy were easily demonstrated, but isotropic surfaces proved elusive. The British cast envious eyes at the United States, where the mid-west looked monotonously flat and therefore exciting for the theorists. They also looked to the Netherlands, where some of the polders had actually been developed by applying Christaller's ideas. East Anglia and the Somerset Levels became places of interest in which to test the theories. Models were devised to measure the degree of evenness, clustering or randomness that settlements displayed. Geography almost succumbed to the danger of applying theories that mattered to regions that didn't—and forgetting the regions that did.

Having drawn a blank in the search for regular hexagonal spacing of settlements, geographers began to think that the failure lay not in the irregularity of geography but in their own perception. It was not that hexagons were not there but that we were looking for them in the wrong way. If one changed from a linear to a time metric, shapes changed (see also Haggett, 1965, 39). For example, if one could travel along good radial roads at 60 mph from a centre but at only 30 mph along the minor radial roads, then 10 minutes' driving along the good roads could carry you 6 miles, but only 3 miles along the minor, interstitial roads. Thus, a line representing 10 minutes' travelling time from that centre could appear petal-shaped on an Ordnance Survey map. However, in reality that petal shape was a circle if one's metric were time rather then linear distance.

From this kind of view developed the idea of mental maps and the way in which space is perceived. It soon became apparent that space was invested with meaning; distances to desirable places were seen as relatively shorter than distances to less desirable places. There was also a kind of logarithmic collapse of the perception of distance. Near and familiar places were perceived in detail, but with increasing distance, places collapsed into points. The Steinberg cover of the *New Yorker* in which Manhattan was seen in detail, with the Rockies and west coast in the distance and Asia as a dot, illustrates this view. Places also became invested with other kinds of meaning, such as danger (Valentine, 1989), as feminist authors also became interested in the perception of place (Valentine, 1993), complaining that urban planning was male dominated. Male ideas of landscape produced hazards for women, of which men were unaware. Bushes, doorways, and unlit spaces were the consequences of male insensitivity to women's needs.

Studies of crime and illness

Studies of crime and illness form another fragment within urban geography that received renewed and particular attention in the 1970s and 1980s. These studies tended to show that inner city areas were marked by social dystopia in the form of concentrated mental illness and criminal activity. The problem was to distinguish whether such concentrations were the cause or effect of inner city living. Early geographic investigation of mental illness was undertaken by Giggs (1973, 1975) and Gudgin (1975). In later work, Giggs and his associates (Giggs, Ebdon, and Bourke, 1980) were able to suggest an ecological basis for primary acute pancreatitis in the water supply in Nottingham. In crime, however, work by David Herbert

(1976) demonstrated, from Cardiff data, that there was no direct correlation between the ecology of crime and the ecology of social class or disadvantage. Taking groups with very similar socio-economic profiles, he showed that they differed significantly in terms of the degree of crime recorded. Part of the difficulty for urban geographers working on issues of crime was to do with data. The reporting of crime by victims is very incomplete and systems of classification are not always helpful. An alternative to police records of crime was the British Crime Survey, which questions a random sample of the population on their experiences of crime and of reporting incidents. This produced very different patterns from those shown by the police records. Susan Smith's work on these data was ground-breaking in this area (Smith, 1986).

After the dispassionate exuberance of the quantitative revolution, urban geographers became more socially conscious and the urban focus shifted in the 1970s from the city to inner cities (Boddy, 1976; Short and Bassett, 1981). Part of this repositioning originated from the social conditions themselves. The 1930s had focused on regional unemployment as the crisis; the 1970s focused on inner city dereliction. Part of the concern originated in the Marxist critique of urban geography (Harvey, 1973); part of the reorientation came from the new humanistic movement (Ley and Samuels, 1978).

Marxist approaches

The Marxist approach exemplifies the internationalisation of British urban geography. Its initial impact came principally from two books: Castells' *The Urban Question* (1977) and Harvey's *Social Justice and the City* (1973). Castells' argument, paradoxically, was that there was no *urban* question: the 'urban' was simply a particular form of the generalised capitalist model. There was nothing special that arose from urban concentration. Focusing on the urban form, according to Castells, represented a kind of geographical fetish, obscuring the critical issue of capitalist organisation and exploitation.

Harvey's attack in *Social Justice and the City* came from a different direction. Having written the ultimate geographical text in the quantitative revolutionary paradigm, *Explanation in Geography* (1969), Harvey changed direction in chapter 4 of *Social Justice*, which was entitled 'Revolutionary and counter-revolutionary theory and the problem of ghetto formation'. Since writing *Explanation*, Harvey had moved to Johns Hopkins University in Baltimore and been exposed to the anti-Vietnam involvement of the US student movements. He had experienced the massively segregated nature of American cities and he had discovered Marx and Engels.

Chapter 4 of *Social Justice and the City* explained to geographers that they had been looking at the world in the wrong way. He took as his analogy the work of Priestley and Lavoisier, both of whom had discovered oxygen in the eighteenth century. Priestley, however, had failed to understand his discovery. He had conceptualised it in terms of the old phlogiston theory of air; oxygen to him was dephlogisticated air. Lavoisier, on the other hand, understood that oxygen was something in its own right. Just as Priestley had identified oxygen but not understood his discovery, so, according to Harvey, Park, Burgess, and the Chicago School had discovered the concentric zone structure of cities but not understood their discovery. Engels, according to Harvey, had got it right a hundred years earlier, in Manchester. Harvey pointed out that, in the same way, geographers were interpreting segregation in terms of 'race'. In reality, according to Harvey, race was like dephlogisticated air. Economic class was the real explanation of the ghetto and to understand the ghetto in terms of race rather than class was to misunderstand the nature of the phenomenon. In a tour de force and also, one must say, by a legerdemain, Harvey's chapter discussed the problem of ghetto formation almost without mentioning race. The effect was seductive and a whole new literature of radical reinterpretation of urban geography came into being. What had been seen as problems turned out to be solutions. The ghetto was a solution for capitalist society, not a problem. Crime was a solution to inequality; police were the cause of riots (Keith, 1989, 1993) and increasing police numbers increased the amount of crime (Smith, 1989a).

Qualitative and humanistic approaches

There was, however, a counter-current to the Marxist revolution, partly nurtured in Oxford, which had been largely bypassed by the quantitative revolution sweeping Cambridge, Bristol, and Leeds; the qualitative and empirical tradition remained strong there. It is true that Chauncy Harris (one of the Chicago School of Geography and half progenitor of the Harris and Ullman model of urban development) was an Oxford alumnus. So, however, was Yi-Fu Tuan, the quintessential qualitative geographer (but he was a geomorphologist in his Oxford days). A new school of qualitative urban geographers developed from Oxford in the 1960s and 1970s. Denis Cosgrove, David Ley, Peter Jackson, and Susan Smith are perhaps its key figures and were instrumental in developing the new humanistic geography and the new cultural geography. North America, it must be said, played a large part in the development of all four.

David Ley moved to the United States for his graduate training and has spent his teaching career at the University of British Columbia. Interestingly, his classic book, *The Black Inner City* (1974), based on his Penn State PhD, was published shortly after David Harvey's *Social Justice*. Ley's book also dealt with ghettoisation, but his approach was diametrically opposed to Harvey's. Using participant observation, a methodology with its origins in anthropology, Ley deconstructed the external white representation of the black ghetto as an organised insurrectionary camp, a sort of Vietcong stronghold threatening the American state, and showed it to be a place of individuated despair, where status and respect came only as a result to someone else's loss. While Harvey's account was deeply theorised, Ley's was deeply personal. Ley nevertheless used quantitative techniques in his book, but they were applied to unexpected topics: obscene graffiti, elliptical walking routes to avoid danger, contour maps of the fear of place.

Peter Jackson's early work also dealt with issues of segregation (Jackson and Smith, 1981). In his case, it was the segregation of Puerto Ricans in New York City (Jackson, 1981). The early work, however, was quantitative, using indices of segregation to demonstrate that Puerto Ricans, despite being poorer and less well educated than blacks and later arrivals in New York, were nevertheless more accepted and less segregated than blacks; race and skin colour were more important in this respect than the expected socio-economic variables. Indeed, Jackson showed that differences in skin colour within the Puerto Rican population were as important as differences in skin colour between Puerto Ricans and African-Americans. The darker Puerto Ricans were concentrated on the edge of black areas while the white Puerto Ricans were far more widely distributed. Jackson, however, also had training in social anthropology and his work thereafter moved in a more qualitative direction. Like Harvey, but for different reasons, he eschewed the category of 'race' (Jackson, 1987) and worked on the social construction of identity (Jackson, 1989). Harvey, it may be noted parenthetically, briefly flirted with the humanistic approach, deconstructing the meaning of the Sacré Coeur basilica in Montmartre (Harvey, 1979) as not so much a tourist attraction as an attack on the working class.

Susan Smith was another of the Oxford geographers who began with a highly quantified approach (Jackson and Smith, 1981) Her book on crime, *Crime, Space and Society* (Smith, 1986), notably on the fear of crime, was in many ways the most innovative. Like Jackson, in *The Politics of 'Race' and Housing* (Smith, 1989b), she rejected the reality of 'race' as an explanatory construct, focusing instead on how and why racial inequality is constituted through economic, political and social activity. Although she maintained a quantitative expertise, notably in analyses of housing and health, her work moved progressively into cultural themes of procession (Smith, 1993) and sound (Smith, 1994).

Urban social segregation: race, ethnicity, gender, and sexuality

The Oxford School produced a significant number of scholars working in the field of ethnic relations and particularly on issues of segregation. Much of this impetus can be tracked back to Paul Paget, one of the old-style tutors who published little but who had an immense influence on his pupils. Colin Clarke, who took up Paget's interest in the Caribbean, produced a monumental study of the urban geography of Kingston (Clarke, 1975). Ceri Peach developed Paget's interests in West Indians in terms of their settlement in Britain (Peach, 1968) and thence moved into issues of ethnic segregation and social interaction more generally (Peach, 1996b). Anthony Lemon carried the work on segregation to a study of one of its heartlands, South Africa, with many studies of *apartheid* (Lemon, 1991). John Western's denunciation of the ethnic cleansing of District Six in *Outcast Cape Town* (Western, 1981) and his later, optimistic and humanistic account of Barbadian settlement in London, *A Passage to England* (Western, 1992), are also directly traceable to Paget's influence. This work was taken up by other graduate geographers in Oxford, Richard Black and David Simon. Work by Paget's pupils and, in turn, their pupils too can be found in Clarke, Ley, and Peach's *Geography and Ethnic Pluralism* (1984). Other Oxford graduates and graduate students who contributed to this more quantitative approach to issues of race and ethnicity were Robert Woods (1979) and Vaughan Robinson. Robinson (1986) wrote one of the classic works in the dissimilarlist mode (that is, work employing the index of dissimilarity). Lacking census data on ethnicity, he organised a whole team of young Asian men to chart the ethno-religio-linguistic distribution of the whole Asian population of Blackburn. From this he was able to demonstrate clearly that segregation was not simply a phenomenon of Asian segregation from whites, but of intra-Asian segregation by religion, linguistic group, and region of origin.

Robinson's findings were important to the debate between the urban social geographers and sociologists, represented by John Rex. In 1967, Rex and Moore published *Race, Community and Conflict*, a study of Pakistanis in the Sparkbrook district of Birmingham, in which they developed a theory of housing classes. In this, there was a hierarchy of housing tenure running from its apex of owner-occupiers without mortgages, through owner-occupiers with mortgages to council house tenants, owners of houses in which rooms were let out, and down to the private rental occupiers of such housing. Pakistanis were concentrated in this lowest housing class. Rex and Moore identified the racism of white society as the causal mechanism for this concentration in the multi-occupied sector and were easily able to demonstrate both racism and discrimination. The legislation against racial discrimination was not yet on the statute books and overtly discriminatory advertisements for the letting of property were the rule rather than the exception.

However, a Pakistani social anthropologist, Badr Dahya (1974), argued that, racism and discrimination notwithstanding, the reason for the concentration of Pakistanis in multi-occupied housing was a matter of choice rather than constraint. Pakistani immigration at that time was dominated by young, single men who came with the object of earning money and sending remittances to their families in Pakistan. They needed to minimise their expenses. They migrated along family, kinship, and village chains of contact. They did not speak English; many were employed in gangs with their landlord/kin acting as intermediaries with employers; and they were Muslims sharing the same needs of worship and *halal* dietary restrictions. Living together in multi-occupied houses was a desirable system of self-support, not a product of discrimination. Discrimination existed in plenty, but it was positive rather than negative factors that gave rise to the particular concentrations observed by Rex and Moore. Evidence of intra-Asian segregation and the way different ethnic groups avoided each other added evidence to the Dahya thesis.

There was, of course, much contrary evidence of ways in which discrimination operated to segregate ethnic minorities in British cities. Robert Woods's (1979) work demonstrated in Birmingham how, when policies of urban renewal, which were largely policies of slum clearance (Housing Action Areas), began to reach areas of ethnic minority rather than white slums, there was a policy change to one of Housing Improvement Areas. These did not require clearance and the rehousing of the occupants in council housing, but simply the patching up of existing dwellings. Work by Trevor Lee (1977) on London suggested that socio-economic factors made a major contribution to Caribbean segregation, but Henderson and Karn (1987) demonstrated how discrimination in local authority housing departments concentrated Caribbean tenants in the least desirable section of council housing.

The most important attempt to reconcile these differences between the choice and constraint schools of analysis of ethnic and racial segregation came from Philip Sarre and colleagues at the Open University. Using the rather unexpected example of Bedford as their Chicago, they applied Anthony Giddens's structuration theory to reconcile the differences between the apparent choice of ethnic minorities to concentrate together in unpromising housing areas and the white discrimination that confined them to such areas. Sarre, Phillips, and Skellington (1989) argued that minorities had internalised white constraints on their actions: when minorities chose housing in inner cities, it was because they had already discounted their ability to access housing elsewhere.

There is perhaps one ironic point to add to Rex and Moore's housing class argument. When they were writing *Race Community and Conflict*, it was in the era of the single male migrant. Immigration legislation put a stop to labour migration but allowed family reunification. Instead of returning to Pakistan, young men brought their families to Britain to join them. In an era of family reunification, multi-occupation of houses was no longer desirable; Indian and Pakistani households turned to owner-occupation, often buying outright. But owner-occupation, you may recall, was at the top of the housing class hierarchy. With a new co-writer (Tomlinson), Rex re-cast the housing class hierarchy so that Indian and Pakistani owner-occupiers of an inner city property, who had scaled the apex of the Rex and Moore schemata, were returned to the bottom of the housing class hierarchy (Rex and Tomlinson, 1979).

One of the most dramatic pieces of work was carried out at the University of Hull by Philip Jones (1970), who produced a series of 'contour maps' of the growth of ethnic minorities in Birmingham's inner city. The maps showed Birmingham's ethnic population concentrated into what looks like a volcanic caldera—a central crater with a high rim falling sharply towards the outer suburbs.

There was a significant consequence of this widespread engagement of British urban geographers with issues of race and ethnicity. When, from 1991, the census began to include an ethnic question, the Office of Population Censuses and Surveys and its successor, the Office for National Statistics, decided to commission work on the ethnic population; the majority of the analysts commissioned to write these official commentaries on the data were geographers rather than sociologists (Coleman and Salt, 1996; Peach, 1996b; Ratcliffe, 1996; Karn, 1997).

One outstanding contribution to urban social geography was Emrys Jones's 1960 classic, *A Social Geography of Belfast*. It was in some ways part of the old, confident school of holistic geography. However, in its use of statistical methods it foreshadowed the quantitative revolution. In its demonstration of segregation as a barometer of Catholic/Protestant struggle it gave a premonition of the future troubles. Another major set of work emerged from the troubles in Northern Ireland. Just before they began in earnest, Frederick Boal (1969) had demonstrated the behavioural impact of the segregation of the Catholic and Protestant communities in inner Belfast. The sharpest line of separation between the two traditions was in west Belfast between working-class Nationalists and Unionists along the Shankill/Falls divide. Boal selected a small area on the interface, demonstrated the almost complete absence of members of the rival communities from each other's areas, and then showed how even practices such as shopping patterns and catching buses were determined by the ethnicity of the area of shops and bus stops. The interface acted on social action like a Himalayan watershed. With the passage of time and intensification of the struggle, more sinister cartographies emerged (Boal, 1995). Murray and Boal (1979) analysed the distribution of doorstep murders in Belfast—political atrocities carried out on individuals as outrages against their communities. They identified geographical patterns to these murders: some of the victims were individuals were living as isolates in the wrong faith area; others were living on the edges of their faith communities. The most vulnerable, however, were individuals living on the edge of their faith communities where these areas were small and fragmented rather than large and consolidated. The small pockets made it easier for the gunmen to get in and out quickly.

It is already apparent, then, that whereas the urban literature of the first half of the century often dealt with large issues and in comprehensive ways (Howard, 1902; Geddes, 1915), those of the second half were first infused with the enthusiasm of planning and then with the social conscience of postcolonial injustice. Out of these preoccupations emerged the struggle against regional inequality, inner city inequality, racial injustice, gender injustice, and homophobia. Again following a lead from North America (Castells and Murphy, 1982), analyses of gay areas of cities began. The discovery of gay villages for men led to a search for lesbian areas—a more elusive goal since gay women seemed to lack the financial resources of gay men (Bell, 1991; Rose, 1993). Interest in areas of ethnic minority concentration led to an interest in the reversal of power relations in such areas (Keith, 1993), whereas interest in ethnic identity moved from race and colour to religion, especially to an interest in Islam as a more potent identifier than simply national origin. As the minority ethnic population in Britain grew from less than 50,000 in 1951 to 4.6 million in 2001, and the Muslim population from a few thousand to over 1.5 million, their impact on the cultural landscape of British cities shifted from the soft features of exotic clothes and dark skin to the hard features of ethnic shops and the dramatic features of mosques, temples and gurdwaras (Gale, 1999).

It is apparent, then, that the scale of interest has shifted from cities to inner cities (Winchester and White, 1995), and from an undifferentiated population to populations sharply defined by class (Bridge, 1995), race, ethnicity, religion (Peach, 1990; Dwyer, 1993), gender (McDowell, 1983, 2000; Bondi and Peake, 1988), and sexual orientation (Pile, 1994). Geography has also narrowed from the region to the body (Rodaway, 1994; Davidson, 2000; Hubbard, 2000). The geography of gender, in particular, emerged as a major source of interest and research from the 1970s to the 1990s. Geography had always been fairly sexist, unthinkingly using 'man' to represent women: *Man's Role in Changing the Face of the Earth* and man–environment relations, for example. Even the Marxist writers seemed to regard gender, particularly women's gender, as false consciousness. Everything was subordinate to class analysis. However, women geographers struck back and produced one of the great blossomings of the geographical literature. They formed the Women and Geography Study Group of the Institute of British Geographers and their 1984 publication *Geography and Gender* explored, among other issues, how urban spatial form has differential implications for women concerning their access to public facilities, employment, transport, retailing, and so on. Not all of their concerns had an urban focus, however.

(Feminist geography is discussed in more detail in Linda McDowell's contribution to this volume: Chapter 18.)

Conclusion

To summarise—the literature of British urban geography in the twentieth century falls into two parts, divided by the 1939–45 war. The first is one of certainty, whether the certainty of imperial grandeur or the certainty of the coming socialist revolution. The second half-century combined the longest period of economic prosperity with gathering self-doubt. In the second half, the Empire struck back. In rapid succession, faith in ideographic empiricism was replaced by nomothetic determinism, a revolutionary Marxist critique (undermined by the collapse of the socialist bloc) and postmodern, humanistic shoulder-shrugging. The first half of the century contributed a small number of large key works to the urban literature, but the second half produced an explosion of smaller works. Self-doubt is more fertile than certainty. Urban geography has splintered into a set of kaleidoscopic fragments, dazzling but with no stable pattern. The envelope has been pushed; the late twentieth century was much concerned with writing about the unmentionable. Some parts of urban geography have been colonised by cultural studies. Class, race, gender, sexuality, identity, and positionality have taken over from distance decay, and statues (Johnson, 1994) from gravity models. Geography has become less about seeing than about seeing through. The urban is seen less as a phenomenon in its own right than as a particular example of the way in which capitalism articulates itself. The modernist tendency of the twentieth century, with its belief in progress, economic growth, and social evolution has stalled in the face of the return of values that we thought had disappeared. Religion has failed to wither away and the spirit of Islam, at least, seems set to haunt the future of urban geography.

Although the story that I have told in this chapter emphasises the fragmentation of the study of urban geography during the course of the twentieth century (this was after all what the editors had requested!) there is also the story of integration. Many texts were written on urban geography in which the diverse research materials were integrated into a coherent account (e.g. Johnston, 1971; Herbert, 1972; Carter, 1981; Bridge and Watson, 2000). Successive editions of Carter's book illustrate the way in which the content of urban geography has changed. However, in the preface to the third edition, Carter (1981) stated that his approach had become outmoded because it dealt with form rather than the radical analysis of capitalism. Perhaps this is the point at which to end. Carter, like Castells, had come to the conclusion that there was no urban question: it was simply the capitalist system operating in an urban setting. Urban geography had not so much fragmented as disappeared.

References

Banton, M. (1955) *The Coloured Quarter*. London: Cape.

Bell, D. (1991) Insignificant Others: lesbian and gay geographies. *Area*, 23, 323–9.

Boal, F. W. (1969) Territoriality on the Shankill–Falls divide, Belfast. *Irish Geography*, 6(1), 30–50.

Boal, F. W. (1995) *Shaping a City: Belfast in the Late Twentieth Century*. Belfast: Belfast Institute of Irish Studies, Queen's University of Belfast.

Boddy, M. (1976) The structure of mortgage finance: building societies and British social formation. *Transactions, Institute of British Geographers*, NS1, 58–71.

Bondi, L. and Peake, L. (1988) Gender and the city: urban politics revisited. In J. Little, L. Peake, and J. Richardson, eds, *Women in Cities: Gender in the Urban Environment*. London: Hutchinson, 21–40.

Bourdieu, Pierre (1977) *Outline of a Theory of Practice*. Cambridge: Cambridge University Press.

Bridge, Gary (1995) Gentrification, class and community. In Alisdair Rogers and Steven Vertovec, eds, *The Urban Context: Ethnicity, Social Networks and Situational Analysis*. Oxford: Berg, 259–86.

Bridge, Gary and Watson, Sophie, eds (2000) *A Companion to the City*. Oxford: Blackwell.

Carter, Harold (1981) *The Study of Urban Geography*, 3rd edn. London: Methuen.

Castells, Manuel (1977) *The Urban Question*. London: Edward Arnold.

Castells, Manuel and Murphy, Karen (1982) Cultural identity and urban structure. In Norman I. Fainstein and Susan Fainstein, eds, *Urban Policy Under Capitalism*. London: Sage.

Chorley, Richard and Haggett, Peter (1965) *Frontiers in Geographical Teaching*. London: Methuen

Chorley, Richard and Haggett, Peter (1967) *Models in Geography*. London: Arnold.

Clarke, Colin (1975) *Kingston, Jamaica: Urban Growth and Social Change 1692–1962*. Berkeley, CA: University of California Press

Coleman, David and Salt, John, eds (1996) *Ethnicity in the 1991 Census, Vol. 1. Demographic Characteristics of the Ethnic Minority Populations*. London: Office of Population Censuses and Surveys.

Collins, Sydney (1957) *Coloured Minorities in Britain*. London: Lutterworth Press

Dahya, Badr (1974) The nature of Pakistani ethnicity in industrial cities in Britain. In Abner Cohen, ed., *Urban Ethnicity*. London: Tavistock, 77–118.

Davidson, J. (2000) '. . . the world was getting smaller': women, agoraphobia and bodily boundaries. *Area* 32(1), 31–40

Deakin, N. and Ungerson, C. (1977) *Leaving London: Planned Mobility and the Inner City*. London: Heinemann.

Dickinson, Robert E. (1951) *The West European City: A Geographical Interpretation*. London: Routledge and Kegan Paul.

Dwyer, Claire (1993) Constructions of Muslim identity and the contesting of power: the struggle over Muslim schools in the UK. In Peter Jackson and Jan Penrose, eds, *Constructions of Race, Place and Nation*. London: UCL Press, 143–59.

Freeman, T. W. (1959) *The Conurbations of Great Britain*. Manchester: Manchester University Press.

Gale, Richard (1999) *Pride of Places: South Asian Religious Groups and the City Planning Authority in Leicester.* Cardiff: University of Wales, Department of Planning, Planning Working Paper 124.

Garner, B. (1967) Models of urban geography and settlement location. In Richard J. Chorley and Peter Haggett, eds, *Models in Geography*. London: Methuen, 303–60.

Geddes, Patrick (1915) *Cities in Evolution*. London: Williams and Norgate.

Giggs, J. A. (1973) The distribution of schizophrenics in Nottingham. *Transactions, Institute of British Geographers*, 59, 55–76.

Giggs, J. A. (1975) The distribution of schizophrenics in Nottingham: a reply. *Transactions, Institute of British Geographers*, 64, 150–6

Giggs, J. A., Ebdon, G. S., and Bourke, J. B. (1980) The epidemiology of primary acute pancreatitis in the Nottingham defined population area. *Transactions, Institute of British Geographers*, NS5, 229–42.

Gilbert, E. W. (1954) *Brighton: Old Ocean's Bauble*. London: Methuen.

Gilbert, E. W. (1961) *The University Town in England and Germany*. Chicago, IL: Department of Geography, University of Chicago, Research Paper 71.

Green, F. H. W. (1950) Urban hinterlands in England and Wales: an analysis of bus services. *Geographical Journal*, 116, 64–88.

Gudgin, G. (1975) The distribution of schizophrenics in Nottingham: a comment. *Transactions, Institute of British Geographers*, 64, 148–9.

Haggett, Peter (1965) *Locational Analysis in Human Geography*. London: Arnold.

Hall, Peter (1988) *Cities of Tomorrow: An Intellectual History of Urban Planning in the Twentieth Century*. Oxford: Blackwell.

Hall, Peter and Preston, Paschal (1988) *The Carrier Wave: New Information Technology and the Geography of Innovation*. London: Unwin Hyman

Harvey, David (1969) *Explanation in Geography*. London: Edward Arnold.

Harvey, David (1973) *Social Justice and the City*. London: Arnold.

Harvey, David (1979) Monument and myth: the building of the Basilica of the Sacred Heart. *Annals of the Association of American Geographers*, 69, 363–81

Henderson, J. and Karn, V. (1987) *Race, Class and State Housing: Inequality and the Allocation of Public Housing in Britain*. Aldershot: Gower.

Herbert, David (1972) *Urban Geography: A Social Perspective*. Newton Abbot: David and Charles.

Herbert, David (1976) The study of delinquency areas: a social geographical approach. *Transactions, Institute of British Geographers*, NS1, 472–92.

Howard, Ebenezer (1902) *Garden Cities of Tomorrow*, being the 2nd edition of *To-Morrow a Peaceful Path to Real Reform*. London: S. Sonnenschein and Co.

Hubbard, P. (2000) Desire/disgust: mapping the moral contours of heterosexuality. *Progress in Human Geography*, 24(2), 191–217.

Jackson, Peter (1981) Paradoxes of Puerto Rican settlement in New York. In Ceri Peach, Vaughan Robinson, and Susan Smith, eds, *Ethnic Segregation in Cities*. London: Croom Helm, 109–26.

Jackson, Peter, ed. (1987) *Race and Racism*. London: Allen & Unwin.

Jackson, Peter (1989) *Maps of Meaning: An Introduction to Cultural Geography*. London: Unwin Hyman.

Jackson, Peter and Smith, Susan J., eds (1981) *Social Interaction and Ethnic Segregation*. London: Academic Press, for the Institute of British Geographers, Special Publication No. 12.

Johnson, Nuala (1994) Sculpting heroic histories: celebrating the centenary of the 1798 Rebellion in Ireland. *Transactions, Institute of British Geographers*, NS9, 78–93.

Johnston, R. J. (1971) *Urban Residential Patterns*. London: Bell.

Johnston, R. J. (1984) *Residential Segregation, the State and Constitutional Conflict in American Urban Areas*. London: Academic Press for the Institute of British Geographers, Special Publication No. 17.

Jones, Emrys (1960) *A Social Geography of Belfast*. Oxford: Oxford University Press.

Jones, Philip N. (1970) Some aspects of the changing distribution of coloured immigrants in Birmingham, 1961–1966. *Transactions, Institute of British Geographers*, 50, 199–219

Karn, Valerie, ed. (1997) *Ethnicity in the 1991 Census, Vol. 4. Employment, Education and Housing among the Ethnic Minority Populations of Britain*. London: Office for National Statistics.

Keith, Michael (1989) Riots as a social problem in British cities. In David T. Herbert and David M. Smith, eds, *Social Problems and the City*. Oxford: Oxford University Press, 289–306.

Keith, Michael (1993) *Race Riots and Policing: Lore and Disorder in a Multi-Racist Society*. London: UCL Press.

Lee, Trevor R. (1977) *Race and Residence: the Concentration and Dispersal of Immigrants in London*. Oxford: Clarendon Press.

Lefebvre, Henri (1991) *The Production of Space*. Oxford: Blackwell.

Lemon, Anthony, ed. (1991) *Homes Apart: South Africa's Segregated Cities*. London: Paul Chapman.

Ley, David (1974) *The Black Inner City as Frontier Outpost: Images and Behavior of a Philadelphia Neighborhood*. Washington, DC: Association of American Geographers, Monograph Series No. 7.

Ley, David and Samuels, Marwyn S. (1978) *Humanistic Geography: Prospects and Problems*. London: Croom Helm.

Little, K. (1948) *Negroes in Britain*. London: Kegan Paul.

Mackinder, H. J. (1904) The geographical pivot of history. *Geographical Journal*, 23(4), 421–37.

McDowell, L. (1983) Towards an understanding of the gender division of urban space. *Environment and Planning D: Society and Space*, 1, 59–72.

McDowell, L. (2000) The trouble with men? Young people, gender transformations and the crisis of masculinity. *International Journal of Urban Regional Research*, 24 (1) 201–9.

Murray, R. and Boal, F. W. (1979) The social ecology of urban violence. In David T. Herbert and David M. Smith, eds, *Social Problems and the City*. Oxford: Oxford University Press, 139–57.

Pahl, R. E. (1965) *Urbs in Rure: The Metropolitan Fringe in Hertfordshire*. London: London School of Economics, Geographical Papers No. 2.

Peach, Ceri (1965) West Indian migration to Britain: the economic factors. *Race, Journal of the Institute of Race Relations*, 7(1), 31–46.

Peach, Ceri (1966) Factors affecting the distribution of West Indians in Great Britain. *Transactions, Institute of British Geographers*, 38, 151–63.

Peach, Ceri (1968) *West Indian Migration to Britain : A Social Geography*. London: Oxford University Press.

Peach, Ceri (1990) The Muslim population of Great Britain. *Ethnic and Racial Studies*, 13(3) 414–19.
Peach, Ceri (1996a) Does Britain have ghettos? *Transactions, Institute of British Geographers*, NS21, 216–35.
Peach, Ceri, ed. (1996b) *Ethnicity in the 1991 Census, Vol. 2. The Ethnic Minority Populations of Great Britain*. London: Office for National Statistics.
Peach, Ceri (1997) Post-war migration to Europe: reflux, influx, refuge. *Social Science Quarterly*, 78(2), 269–83.
Pile, S. (1994) Masculinism, the use of dualistic epistemologies and 3rd spaces. *Antipode*, 26(3), 255–77.
Ratcliffe, Peter, ed. (1996) *Ethnicity in the 1991 Census, Vol. 3. Social Geography and Ethnicity in Britain: Geographical Spread, Spatial Concentration and Internal Migration*. London: Office for National Statistics.
Rex, John and Moore, Robert (1967) *Race, Community and Conflict: A Study of Sparkbrook*. London: Institute of Race Relations and Oxford University Press.
Rex, John and Tomlinson, Sally (1979) *Colonial Immigrants in a British City: A Class Analysis*. London: Routledge and Kegan Paul.
Richmond, Anthony H. (1954) *Colour Prejudice in Britain: A Study of West Indian Workers in Liverpool, 1941–1951*. London: Routledge and Kegan Paul
Robinson, Vaughan (1986) *Transients, Settlers and Refugees: Asians in Britain*. Oxford: Clarendon Press.
Robson, Brian T. (1973) *Urban Growth: An Approach*. London: Methuen
Rodaway, P. (1994) *Sensuous Geographies, Body, Sense and Place*. London: Routledge
Rose, G. (1993) *Feminism and Geography: The Limits of Geographical Knowledge*. Cambridge: Polity Press.
Sarre, P., Phillips, D., and Skellington, R. (1989) *Ethnic Minority Housing: Explanations and Policies*. Aldershot: Avebury.
Short, J. and Bassett, K. A. (1981) Housing policy in the inner city in the 1970s. *Transactions, Institute of British Geographers*, NS6, 293–312.
Smailes, A. E. (1947) The analysis and delimitation of urban fields. *Geography*, 32, 151–61.
Smailes, A. E. (1966) *The Geography of Towns*. London: Hutchinson (first published 1953).
Smith, Susan J. (1986) *Crime, Space and Society*. Cambridge: Cambridge University Press.
Smith, Susan J. (1989a) The challenge of urban crime. In David T. Herbert and David M. Smith, eds, *Social Problems and the City*. Oxford: Oxford University Press, 271–88.
Smith, Susan J. (1989b) *The Politics of 'Race' and Housing*. Cambridge: Polity Press.
Smith, Susan J. (1993) Bounding the borders: claiming space and making place in rural Scotland. *Transactions, Institute of British Geographers*, NS18(3), 291–308.
Smith, Susan J. (1994) Soundscape. *Area*, 26(3), 232–40.
Valentine, Gill (1989) The geography of women's fear. *Area*, 21(4), 385–90.
Valentine, Gill (1993) Negotiating and managing multiple sexual identities: lesbian time–space strategies. *Transactions, Institute of British Geographers*, NS8, 237–48.
Western, John (1981) *Outcast Cape Town*. London: Allen & Unwin.

Western, John (1992) *A Passage to England: Barbadian Londoners Speak of Home*. London: UCL Press (republished 1996).
White, Paul (1998) The settlement pattern of developed world migrants in London. *Urban Studies*, 35(10), 1725–44.
Winchester, H. P. M. and White, P. (1988) The location of marginalised groups in the inner city. *Environment and Planning D: Society and Space*, 6(1), 37–54
Women and Geography Study Group (1984) *Geography and Gender: An Introduction to Feminist Geography*. London: Hutchinson.
Woods, R. I. (1979) Ethnic segregation in Birmingham in the 1960s and 1970s. *Ethnic and Racial Studies*, 2, 455–77.

18

Geographers and the regional problem

RAY HUDSON

There is a long history of geographers in the UK analysing and engaging with 'the regional problem'. Despite periodic attempts to deny its significance, the regional problem has proved remarkably persistent and periodically re-emerges on the political agenda.[1] Furthermore, some geographers have argued that the 'regional problem' is genetically encoded within the social relations of a capitalist economy and as such the issue is not whether it exists but rather the particular form that it takes in given circumstances. It is, however, important to stress two points from the outset. First, there are no 'essential' problem regions: 'problem regions' (indeed, any regions) are social and political constructions by both geographers and public policy-makers. Such entities are always constructed in particular and typically contested ways, for specific purposes, in and as particular time/spaces, and it is through these processes of social construction that a society assumes its regional anatomy. Secondly, to acknowledge the persistence of the regional problem is not to deny that spatial inequalities are also constructed as existing at other spatial scales and in other domains (for example, in terms of the contrast of urban versus rural) in the UK. Indeed, a distinguishing feature of the way in which geographers have analysed spatial inequality in the UK has been an increasingly sophisticated and nuanced recognition of the multi-scalar and complex character of the map of spatial inequality.[2] In the remainder of this chapter, I summarise

[1] For example, in November 2000 Stephen Byers, Secretary of State for Trade and Industry, argued in a series of speeches that a 'North–South Divide' and, more generally, the regional problem were still present in the UK. This followed denials in 1999 by Prime Minister Tony Blair that there was a 'North–South Divide'.

[2] It is worth noting that Prime Minister Blair's argument for denying the existence of a North–South Divide is that the map of inequality is more complex and multi-scalar. This is hardly news to geographers, however, and does not justify a conclusion that the North–South Divide no longer exists, or that the regional problem is simply a subject of historical geography.

some of the main strands in the evolving ways in which geographers have analysed 'the regional problem' and associated regional policies in the light of two sorts of changes: first, changes in the map of regional uneven development and in government policies; and secondly, in terms of changing conceptions of human geography, and changes in geographical thought and practice. As a prelude to this, however, I consider some broader issues raised in recent debates about conceptualising and theorising regions, as they provide a contemporary reference point against which to view these issues and the 'regional problem' and 'problem regions' as objects of public policy.

(Re)-conceptualising the region

I want to focus upon two issues here: the conceptualisation of regions within geography, in the context of broader debates about (re)conceptualising place and space in relational terms; and the circumstances in which the regional problem became a policy issue (Hudson, 2001, ch. 8). Allen, Cochrane, and Massey (1998, 143, emphasis added) argue that an adequate understanding of regions (indeed, *any* 'place') can '*only* come through a conception [of them] as open, discontinuous, relational and internally diverse'. Thus '*in principle* the conception [of them] as bounded and undisturbed is incorrect' (Massey, 1995a, 64, emphasis in original). There is certainly great advantage in conceptualising regions as open, discontinuous, and permeable, but to claim that this is the only way is to overstate the case. It is one possible approach but not the only one. In some circumstances, conceptualising regions as closed, continuous, and bounded may offer greater analytic and/or political purchase. People who live in regions that are open and permeable have repeatedly sought discursively to construct them as closed and bounded as a way of defining and protecting their interests on a regional basis. The region thus becomes deployed rhetorically—the shared place around which otherwise divergent social interests for a time coalesce. However, this implies that 'regions' are always constructions in time/space, at best a temporary (albeit at times long-lived) settlement of social relations into a regional form. Regions are thus defined as spatio-temporal rather than simply as spatial constructions, conceptualised as discrete time/space envelopes (Hudson, 1990).

Some years ago, Doreen Massey (1978) presciently emphasised that regions should emerge from rather than be presupposed by analysis. This is an important point. Seen in this light, the degree to which regions are regarded as closed, continuous, and bounded, or as open, discontinuous, and permeable is a matter to be resolved *ex post facto* and empirically rather than *a priori* and theoretically. This is not simply an analytic point but one with considerable implications for policy and political debates about the 'regional problem'. Not least this is because there may be circumstances in which the region as object of analysis needs to be taken 'as given', as the socially constructed product of the analyses of others, which in turn requires analysis and comprehension. This is especially so in the context of understanding the 'regional problem'. It focuses attention on understanding why 'problem regions' are defined as discrete and bounded spaces characterised by common shared problems that must be addressed via public policies, and on understanding the changing conception of appropriate policies to address 'the regional problem' and 'problem regions'.

This is significant because it raises questions about two key issues. First, it problematises the circumstances in which people and/or governments seek to construct regions as problematic, as closed and bounded objects of analysis, and raises questions about the criteria that they use in so doing. Secondly, it foregrounds the issue of the circumstances in which regions may seek to become regions 'for themselves' (Lipietz, 1993). More precisely, particular groups of people within them may seek discursively to construct a 'regional interest', typically one that prioritises particular sectional interests while representing these as the interest of the 'region as a whole'. This highlights the circumstances in which 'regions' may seek in some way to become 'active subjects' in negotiating what constitutes the 'regional problem' and in defining appropriate ways of dealing with it, rather than just 'passive objects', targets of top-down central government regional policies. It is, however, important to emphasise that it is social groups that discursively construct and deploy particular concepts of regions and a 'regional interest', and so avoid simplistic notions of 'regional fetishism' and the suggestion that regions per se have agency.

Locating 'problem regions' and regional policies

Somewhat unusually in the context of policy analysis, the point at which 'the regional problem' emerged onto the policy agenda of central government in the UK can be identified quite precisely as 1928. This marks the moment of transition from regional uneven development being regarded as a 'natural' and unavoidable, if undesirable, feature of capitalist economic development to it being seen as a specific political problem that required explicit attention by the national state (McCrone, 1969; Parsons, 1986). This transition was informed by a variety of motives that can be placed into one of two broad categories. First, there are those concerned with economic efficiency—ensuring that regional imbalance remained within tolerable limits, as unbalanced regional growth became seen as a barrier to non-inflationary national economic growth. On this reading, regional policies were seen as necessary (though not necessarily sufficient) to ensure continuing capital accumulation within the national territory. Secondly, regional policies were seen as necessary for reasons of equity—and as a means of avoiding or defusing potential crises of legitimation and challenges to the authority of the government and the established order. Thus recognition of the regional problem and of the need for specific policies to address it involved an (albeit implicit) normative conception of the regional map of the UK, one with a greater degree of regional balance and more even socio-economic development. It acknowledged that central government had a responsibility in bringing this about. This amounted to a major revision to then-dominant liberal conceptions of economic policy and the relationships between economy and state.

Given this, it was understandable that the initial development of regional policies in the 1930s was halting and hesitant. Nevertheless, it led to the definition of the Special Areas (North East England, West Cumberland, South Wales, and Clydeside–North Lanarkshire in Scotland) and tentative new measures to entice new private sector investment to areas devastated by high unemployment. Regional industrial crisis and high levels of structural unemployment were consequences of the impacts of international recession on the 'staple' industries of northern Britain and south Wales. These regions had previously been created as core locations in the Industrial Revolution as capitalist social relationships revolutionised the what, how and where of production within the UK. At the same time as these now-peripheralised regions were subject to severe deindustrialisation, mass unemployment, and poverty, new forms and sectors of manufacturing growth were emerging in southeast England, further intensifying regional economic imbalances. Sir Malcolm Stewart, in his third report as Commissioner for the Special Areas, argued that further development of congested areas, notably London, must be controlled if the problems of the Special Areas were to be solved. A royal commission was established under the chairmanship of Sir Montague Barlow to inquire into the causes of and solutions to unbalanced regional growth. It produced its *Report on the Distribution of the Industrial Population* in 1940, the Royal Geographical Society (RGS) having been invited to submit a memorandum on the geographical aspects of the location of industry (Chetwode, 1939; Royal Commission on the Distribution of the Industrial Population, 1940). Reflecting the growing acceptance of Keynesian economic policies, it suggested government intervention to stimulate fresh industrial growth in regions other than the Midlands and the southeast. Thus it seemed that the efficiency and equity arguments for regional policy coincided and were complementary.

Rearmament and the war economy temporarily abolished 'the regional problem', but it was recognised that structural problems remained. While there was an expectation that these would be softened in the post-war years via the nationalisation of some key industries, especially coal mining, there was also recognition that regional economic diversification and modernisation remained necessary. This led to growing attention to issues of regional policy and land use planning during the wartime years, with a series of meetings at the RGS devoted to these issues of 'planning for the post-war' (Clerk, 1944). This helped shape a series of influential reports, two of which were published in 1942. The Scott Committee (the vice chairman of which was that prominent geographer, Dr (later Professor Sir) Dudley Stamp: see Stamp, 1943) reported on land utilisation in rural areas while the Utthwatt Committee reported on land management and land use planning. Shortly afterwards, in 1944, the Abercrombie Plan was published. This was a plan for Greater London which, echoing the earlier views of Sir Malcolm Stewart, suggested a policy of decentralisation to newly built satellite towns, separated from London via a green belt. This added a further dimension to the evolving debate on the need for inter-regionally and intra-regionally balanced growth, and the most appropriate mix of policies to attain this goal. These various reports, alongside the 1944 white paper on full employment, helped set the scene for the post-war radically reformist legislation to tackle—*inter alia*—the regional problem.

Consequently, strengthened regional policies became a central component of the post-war settlement and the Keynesian welfare state until the mid-1970s.[3] During this period geographers were active in a variety of ways in studying particular 'problem regions' and the regional problem, and, to a degree, in devising and implementing policies to tackle it. Much of this geographical work was cast in an empirical descriptive mould and produced some detailed and insightful analyses of particular regions. For example, Daysh and Symonds (1951) produced a classic study of north-west Durham. House and other colleagues at Newcastle carried out a series of valuable studies on northeast England (e.g. see House, 1969), a concern that continued with the formation of the Centre for Urban and Regional Development Studies (CURDS) there in the 1970s. Johnson and Wise (1950) and Wise and Johnson (1950), for example, carried out valuable work on Birmingham and the West Midlands, as did Smith (1953) and Smith and Lawton (1953) on Merseyside and northwest England. Dicken and Lloyd and colleagues at Manchester (e.g. see Lloyd and Mason, 1978) then took analysis of regional development issues in the northwest further, extending their concerns to pay particular attention to the effects of the changing place of the UK in the international division of labour on regional development patterns within it (e.g. Dicken and Lloyd, 1976, 1980). More generally, geographers retained a concern to analyse the processes leading to regional economic differentiation, to evaluate the impacts of regional policy, and to explore different conceptions of regional policy appropriate to changing circumstances (e.g. Chisholm and Manners, 1973; House, 1973; Chisholm, 1975,1976).

[3] In the 1950s their implementation was muted as post-war reconstruction and continuing wars outside Europe ensured high levels of effective demand for the outputs of the 'old' industrial regions (Hudson, 1986a).

By the early 1970s, there were already growing concerns with the 'costs per job' of seeking to tackle regional concentrations of unemployment via inducements to private capital to locate there. The limits to such policies were increasingly evident (see House of Commons Expenditure Committee (Trade and Industry Sub-Committee), 1973). This set the context for severe reductions in the scope of formal regional policy from the mid-1970s but especially as part of the political economy of Thatcherism in the 1980s (see Lewis and Townsend, 1989; Hudson and Williams, 1995).

Geographers and 'problem regions': from spatial science to Marxian political economy

As well as changes in the scope, content, and conception of regional policies and 'the regional problem', there have also been changes in the ways in which geographers have studied and analysed these issues. The emphasis has shifted from mapping the boundaries of problem regions and of the areas eligible for central government (and subsequently, post-1975, EU) regional policy assistance to seeking more powerful explanations of the causes of 'regional problems' and examining regional policies in relation to theories of the capitalist state.

Initially, the emphasis was upon descriptive studies of the regional policy framework of 'carrots and sticks' (incentives to private capital to invest in some areas and disincentives to invest in others), accounts of the legislation underpinning it, and of patterns of regional change in the context of intra-national and international competition for investment and jobs. From the 1960s human geographers began to use the then-novel methods of spatial science to try to explain regional uneven development and its relationship to regional policy. For example, there were attempts to use gravity models to analyse freight flows and the spatial structure of the economy (Chisholm and O'Sullivan, 1973) and account for branch plant movement and changes in regional employment patterns (Townsend and Gault, 1972; Keeble, 1976). There were, however, severe explanatory (and so policy) limitations to such approaches. This led to attempts to conceptualise and understand 'the regional problem' in fresh ways.

A fundamental problem of spatial science approaches is that they assume a seriously under-socialised conception of the processes that generate spatial patterns (Hudson, 2001). Recognition of this led human geographers to explore alternative epistemological and theoretical positions in explaining regional uneven development and 'the regional problem'. From the late 1960s, geographers increasingly turned to Marxian political economy in their search for more powerful explanations of regional inequality. Marxian political economy encompasses powerful concepts of structure and the social structural relations of capitalist societies and it offered a powerful challenge to the 'spatial fetishism' (Carney et al., 1976) of locational analysis and spatial science. This led geographers to emphasise regional uneven development and the 'regional problem' as integral to processes of capital accumulation (Carney, Hudson, and Lewis, 1977). This marked the beginning at Durham of a concerted period analysing the specific problems of northeast England and the regional problem more generally (e.g. Hudson, 1989, 1998).

The conceptual and theoretical advances in understanding the regional problem that were made from the 1970s cannot be understood without taking account of the seminal contributions of David Harvey and Doreen Massey. Their work profoundly influenced the ways in which geographers (and others) thought about the regional problem and, more generally, of processes of uneven development within capitalism. Harvey's (1982) re-conceptualisation of historical-geographical materialism revolutionised thinking about spatially uneven development, rigorously demonstrating why it was an integral and necessary feature of capitalist economies. Because it was constructed at a high level of theoretical abstraction, it necessarily left unanswered the critical question of *which* regions became problematic, and which successful, within an overall mosaic of regional uneven development. Analysing how the specific socially produced features of regions intersected with more general trajectories of capital accumulation and particular corporate strategies for production became the focus of the 'spatial divisions of labour' approach, particularly associated with the work of Massey (1984, 1995b). This encompasses a more spatially nuanced approach, building upon the critical insights of Marxian approaches in ways that recognise the central formative role of spatial differences in the constitution of capitalist societies. In particular, it sought to explain why different parts of the overall production process are located in different regions in response to variations in regional labour market conditions (see also Carney, Hudson, and Lewis, 1980). It drew attention to the strategies of capital, labour, national states, and local communities in seeking to shape geographies of economies and the anatomy of the regional problem and stimulated more subtle and sophisticated empirical research into formative processes. The work of Morgan and Sayer (1988), deploying a critical realist approach (Sayer, 1984), provides an excellent example of such theoretically informed sophisticated empirical analysis. Which regions grew and prospered, which declined and became problematic in various ways, is clearly an issue of considerable intellectual interest. In addition, it is also a question of great practical significance to people living in these regions and to policy-makers charged with responsibility for managing 'the regional problem'.

The varying fates of region raise issues of for whom and in what sense 'the regional problem' *is* a problem and the reasons as to why this is so, and of the criteria used to define regions as 'problematic' and 'non-problematic' (Massey, 1979). These questions are critical in terms of understanding the rationale for government involvement with 'the regional problem'. There is a permanent tension between the corporate and territorial development logics. The former defines regions as locations in which, for a time, to make profits, the latter seeks to ensure that regions remain locations in which people can find paid work to enable them to continue to live and learn there. As such, there is an unavoidable tension between regions as places to which their inhabitants are attached in complex ways and as locations in which capital seeks to make profits. Regional policies are one mechanism through which governments seek, for a time, to keep these tensions within tolerable limits, and at the same time preserve their own legitimacy and head off challenges to their authority.

Geographers also came to emphasise how other government policies can have significant and unintended regional consequences. This raises important questions about the links between regional policies seeking to contain 'the regional problem' and other national government and (post-1973) EU policies that may have 'unintended' and 'undesirable' regional effects. Such policies may be sectoral (such as defence policy, nationalised industry policies or welfare-state and public-service-sector policies) or spatial (e.g. urban or rural policies). This highlights the need to focus on the regionally differentiating effects of *all* government policies and public expenditure patterns, and not just those officially categorised as 'regional policies'. For example, it became recognised that defence policies, defined and implemented in pursuit of the national interest, had strongly regionally differentiated and differentiating effects, concentrating their R&D activities within the southeast. This emphasised the need to take account of qualitative as well as quantitative differences between regions in the economic activities located there (e.g. Lovering, 1988; Smith, 1988). The significance of the spatial consequences of 'aspatial' policies became even more clearly apparent with the rationalisation and privatisation of previously nationalised coal, steel, shipbuilding, and automobile industries (Hudson, 1986a, 1986b, 1989). These had been nationalised between the 1940s and 1970s, with the rationale for each nationalisation varying with timing: coal had been seen as critical to post-war recovery in a single-fuel economy in the late 1940s; steel had been seen as central to the strategy of modernising manufacturing in the 1960s; and shipbuilding and parts of the automobile industry had been identified as major sources of employment and strategic industries that could not be allowed to close in the 1970s.

In all cases, however, nationalisation reflected a recognition that vital industries were no longer sufficiently attractive to private capital to ensure the maintenance of required levels of capacity, output, and employment. Furthermore, there was a distinct regional geography to these industries, which were heavily concentrated in the 'old' industrial regions of northern Britain, the

English Midlands, south Wales, and Northern Ireland. Consequently, nationalisation was seen as a way of preserving capacity and employment in these industries and regions, of averting the emergence of 'problem regions' or avoiding further deterioration in the economic well-being of already 'problematic' regions, and so of further exacerbating 'the regional problem'.

The election of Thatcherite governments from 1979, however, heralded a dramatic change in the pattern of relations between economy, society, and government that had held on a cross-party basis for the preceding three decades (Hudson and Williams, 1995). Redefining the boundaries between state and economy involved a sharp reduction in state involvement and, as part of this, plans to return the nationalised industries to the private sector. At the same time, the extent and scope of regional policy was cut back sharply. However, before embarking upon the process of privatisation, it was necessary to restructure these industries to make them potentially profitable and attractive to private capital. Thus there was a period of severe rationalisation, leading to major regionally concentrated reductions in employment. Such instances of direct state involvement in creating new or exacerbating existing 'regional problems' generated both specific forms of protest (such as the 1984–5 miners' strike: Beynon, 1985) and specific government policy responses targeted on these areas. The latter typically took the form of small-area policies, sometimes linked to specific new agencies, such as British Steel (Industries) and British Coal (Enterprises). Such agencies were charged with job creation and retraining in what had become areas of high structural and long-term unemployment, and by the end of the 1990s to a task-force specifically for the coal districts (Coalfields Task Force, 1998; Bennett, Beynon, and Hudson, 2000). This was to be the first of several such task-forces charged with dealing with the consequences of regionally concentrated economic decline associated with industries such as automobiles and clothing.

New theoretical perspectives on the state and 'the regional problem'

Recognition of the unintended regional consequences of sectoral and nationalised industry policies emphasised the tangled and contradictory pattern of state involvement as both proximate cause of and solution to the 'regional problem'. In response, geographers turned their attention to theories of the state, regulation, and governance in an attempt to comprehend why state activity took these conflicting forms and to understand better the relations between the intended and unintended effects of state policies in relation to 'the regional problem'.

In particular, crisis theories of the capitalist state (O'Connor, 1973; Offe, 1975; Habermas, 1976) proved to be particularly helpful in this regard. These theories took their point of departure in the recognition that there were unavoidable economic crisis tendencies structurally encoded within capitalist relations of production and manifest—*inter alia*—in profitability crises. State policy interventions seek to address these problems but cannot abolish economic crisis tendencies; instead they displace them into the spheres of the state and civil society, from which they emerge in due course in new forms. In particular, they emerge as rationality crises, defined by a chronic disjunction between the stated aims and intended outcomes of state policies and their actual effects, many of which are 'unintended' and unwanted. For example, rather than ameliorating 'the regional problem' as intended, state policies may actually intensify it (as with the case of nationalised industry policies in the 1980s). Such disjunctions may in turn trigger legitimation crises, as the authority of the state to act in particular ways is called into question precisely because state policies fail to meet their stated objectives.[4] At the same time, growing pressures on state finances and the threat of a fiscal crisis of the state prevents expansion of the boundaries of state activities, because of the threat of capital flight in response to rising corporate taxation and inflation.

Thus geographers began to interpret privatisation and the retreat of central government from regional policy as part of the general reduction of the scope of state involvement in the national economy and society. Intensifying processes of globalisation weakened the capacity of the national state to steer the national economy (which to some came to be seen as 'a Keynesian myth': Radice, 1984). The dangers of fiscal crisis, with its implications for large-scale capital disinvestment from the UK, took precedence over those of a possible legitimation crisis. There was growing emphasis on national economic performance and securing the place of the national economy in dominant circuits of global capital. In the case of the UK post-1975, but especially post-1979, this above all meant securing the place of the City of London as a 'global city', as national economic policy became a regional policy for parts of southeast England, while 'traditional' regional policy was sharply reduced in scope.

The new international division of labour, the 'regional problem', and policy responses

For several decades, inward investment was regarded as central to 'solving the regional problem'. Initially, this was primarily seen in intra-national

[4] While such legitimation crises around the regional problem did not erupt in the UK, they did so in France in relation to regionally concentrated cutbacks in the steel industry in the late 1970s and 1980s (Hudson and Sadler, 1983). Even so, some geographers seemed to suggest that such regionalist challenges could and should emerge in the UK almost as an automatic response to regional deindustrialisation (Carney, 1980).

terms, involving the movement of industry from the southeast and the West Midlands to peripheral regions. As the West Midlands increasingly became seen as part *of* the regional problem, and as the economy became increasingly internationalised, the emphasis increasingly switched to inward investment by foreign multinationals (see Hudson, 1995b). However, such branch plant investment was seen as potentially problematic, creating new forms of externally dependent regional economies (Firn, 1975) composed of 'global outposts' (Hudson, 1995b) with little connection to the rest of the regional economy. Such factories were seen as vulnerable to closure due to corporate decisions made in their head offices in other countries.

By the mid-1970s branch plant disinvestment and 'capital flight' were increasingly regarded as a proximate cause of 'regional problems' rather than as necessarily part of their solution, especially as central government loosened controls on capital export from 1979. This provoked a sudden acceleration in regionally concentrated deindustrialisation, with the largest 40 manufacturing companies in the UK cutting manufacturing employment there by 415,000 jobs between 1980 and 1985 (Hudson and Williams, 1995). This was disproportionately concentrated in peripheral regions as companies switched the location of routine production activities in search of cheaper production locations and/or to establish capacity within foreign markets as part of emergent new international divisions of labour.

However, the changing configuration of the international economy and the emergence of new forms of corporate organisation also offered opportunities to rethink the character of 'the regional problem' and devise new ways of tackling it via inward investment. This led to renewed emphasis on inward investment, especially from Japan, as a way of reviving regional economies and of modernising managerial practices and working methods in large swathes of manufacturing via the demonstration effect and example of more-efficient 'Japanese' managerial practices (Hudson, 1995a). This repositioned many UK 'problem regions' in new ways within an evolving international division of labour while simultaneously seeking to reposition the UK economy within that division of labour, again raising questions as to how 'development' was to be understood.

This reassessment of 'development' was partly linked to analyses of new forms of inward investment and claims about its beneficial impacts upon 'problem regions'. In particular, it was claimed that there was now an alternative to the old 'low road' to regional development via branch plants offering only unskilled assembly work and with little connection to the surrounding regional economy. This alternative new 'high road' centred on 'embedded' performance plants, offering more skilled work in factories closely tied into the regional economy via their supply chains. However, while this became increasingly seen as central to a new regional development orthodoxy, encompassing ideas of clusters, there were evident dangers of recreating regional economies that would be very vulnerable to changes in demand for a narrow range of products. Furthermore, while many branch plant investments could not be simply classified as 'global outposts', neither did they correspond to the specification of performance plants. Instead, they constituted formative elements in a new form of enclave economy, linked in complex ways into both regional and global economies and wider corporate geographies that challenged the 'new orthodoxy' of cluster-based development (Hudson, 2003).

The reassessment of the nature of 'development' was also linked to an emphasis upon small and medium-sized enterprises (SMEs) as the new basis of regional policy and of regional economic well-being. SMEs were seen to offer a way of diversifying regional economies around a wider range of firms and products, spreading the risks of economic change and turbulence more widely within a region. This was not without its inconsistencies, however, especially as the interest in SMEs led to a fascination with industrial districts. These were understood as successful regions of SME growth, in which the traditional problems of small firm growth—marginalised companies with a very precarious existence—were largely overcome as cooperating networks of small firms flourished in particular niche markets, based on quality and flexibility of products. Furthermore, in some cases such districts transformed themselves into dynamic 'learning regions' (Morgan, 1995), or 'technology districts' (Storper, 1993), with competitive success based upon product innovation and strong Schumpeterian competitive strategies of market disturbance (Hudson, 2001).[5] However, it also became evident that many formerly successful industrial districts were 'hollowing out' in response to both the opportunities offered by and pressures of Europeanisation and globalisation, calling into question the validity of this particular developmental model (Hudson, 2003).

This focus on the rediscovery and reconstruction of regional economies was also linked with growing attention to non-economic relationships that were seen to underpin regional economic success. As well as the traded interdependencies of the supply chain, geographers also began to emphasise the significance of the untraded interdependencies of non-economic social and cultural ties within an associational economy (Storper, 1995, 1997). This was one element in a broader 'institutional turn' in analyses of regional development (Amin, 1998). Amin and Thrift (1994) had earlier introduced the important concept of 'institutional thickness' to explain variations in regional economic performance on the basis of the internal characteristics of regions. This denoted not simply the number and density of institutions within a region but the intensity and quality of the interactions between them. Many of the economically successful regions of

[5] Note, however that a concern with regional innovation and the creation of science parks can be traced to the political economy of Thatcherism in the 1980s.

Europe were seen to possess an appropriate institutional thickness, one that underpinned and supported economic activities located within the region, helping to reproduce regionally specific tacit knowledge and trust, seen as a critical determinants of continuing economic well-being. Conversely, however, an inappropriate institutional thickness, often a relict form from an earlier era when it *was* supportive of regional economic success, can act as a barrier to moving a regional economy onto a new and more promising developmental trajectory (Hudson, 1994), revealing 'the weakness of strong ties' (Grabher, 1993).

The recognition that some regions prospered whilst others declined, and that some formerly successful developmental models were becoming problematic (see above), led to intriguing intellectual questions with manifold practical implications as to the circumstances in which such developmental models were possible and successful. It raised questions about the possibility of transferring appropriate institutional arrangements and forms of 'institutional thickness' and 'best practice', enabling economically declining problem regions in the UK to learn from the success of more successful regions elsewhere (e.g. Dunford and Hudson, 1996). Such hopes proved false, however; they were based on a fundamental misunderstanding of the possibilities of transferring policies and 'best practice' from one regional context to another and of the 'limits to learning' as the basis for corporate and regional success (Hudson, 1999). As a result, SME policies generally reproduced precisely the sort of vulnerable small firm economy that industrial districts were seen to avoid. Equally, while giving rise to the occasional performance plant and a rather larger number of new forms of enclave development, as often as not inward investment policies continued to reproduce vulnerable economies based around 'global outposts', subject to corporate (dis)investment decisions taken in other continents.[6]

A corollary of the retreat of central government in the UK from direct engagement with the regional problem was the decentralisation of responsibility for regional economic well-being to the regional level. In comparison with much of Western Europe, this process of devolution was slow, despite the formation of the Scottish and Welsh Development Agencies in 1975 as part of central government's attempt to defuse nationalist sentiment in the Celtic fringes. The process of devolving responsibility for economic development was taken further in these areas in the late 1990s with the creation of the Welsh Assembly and the Scottish Parliament and

[6] This became very clear in the latter years of the 1990s. For example, both Fujitsu and Siemens closed very new 'state-of-the-art' factories making semi-conductors in northeast England (in Newton Aycliffe and North Tyneside, respectively) soon after they opened. In South Wales LG dramatically scaled back planned investment in a major production facility at Newport.

the re-instating of the Northern Ireland Assembly in 1999. It was not until the end of the 1990s, however, that Regional Development Agencies were established for the English regions, albeit with much more restricted domains of competencies, powers, and resources than was the case in Scotland and Wales.

Paralleling this hesitant and uneven process of devolution and the creation of new formal regional institutions responsible for regional economic development, the growing emphasis on regional institutions in explaining regional economic success and failure became closely linked to debates about new systems of multi-level governance and the 'hollowing out' and 're-structuring' of national states (MacLeod, 1998). These debates originated in claims that the national state was being emasculated by processes of globalisation and counter-claims that disputed that this heralded 'the death of the national state', even though its role was (again) changing (Anderson, 1995). Indeed, in the UK there was evidence of further centralisation of state power in central government ministries in the 1980s and 1990s (Hudson and Williams, 1995). Rather than the national state ceasing to matter in economic policy formation and in tackling 'the regional problem', the ways in which it did so were changing (as it took on an 'enabling' role alongside its existing market-facilitating and interventionist roles: Hudson, 2001). In this way the national state sought to keep 'the regional problem' within acceptable limits as part of a new architecture of governance and regulation. This encompassed the EU and sub-national institutions as well as national government and institutions in civil society.

In summary, there is no doubt that a radical conceptual break occurred in thinking about the regional problem and the reasons for regional uneven development in the 1990s. Previously, there had been considerable emphasis on external processes as both the causes of (for example, via transnational disinvestment) and solution to (notably by central government regional policies) regional problems. Comparatively little attention was paid to conditions within regions and to endogenous processes as explanatory factors.[7] The 'new' economic geography of the 1990s, in contrast, highlighted endogenous processes and conditions within regions. Reflecting broader cultural and institutional 'turns' in human geography, it also placed considerable emphasis upon non-economic processes and institutions that were claimed to underpin regional economic success and lie at the heart of regional economic failure and decline. While a useful corrective to the emphases of earlier approaches, it also proved problematic. It tended to focus attention on endogenous processes to the exclusion of

[7] This is understandable given the context of Anglo-American human geography more generally seeking to escape from the exceptionalist cul-de-sac of regional description from the 1950s onwards.

considerations of broader political-economic processes and the constraints that these impose upon the scope for regional autonomous action as the route to solving 'the regional problem'. One consequence of this was to encourage interregional competition as the solution to problems of regional uneven development, on the premise that all regions could pull themselves up by their bootstraps and engage in 'win–win' scenarios, if only they put in place appropriate intra-regional institutions. However, this effectively denied the character of capitalist development as unavoidably uneven and that there were necessarily 'losers' as well as 'winners' as a result of such competitive processes (and in all probability many more of the former than the latter). This is a point of immense significance in the context of future regional policies.

The regional problem and regional policies in the twenty-first century: what sort of future?

One strong conclusion to emerge from geographical analyses of 'the regional problem' is that regional uneven development is an unavoidable and necessary feature of the capitalist space economy, although its particular manifestations are contingent. It cannot be abolished by interregional competition, linked to endogenous development strategies that generate 'win–win' scenarios. Nor can it be abolished by state involvement in the economy and society. Indeed, the economic crisis tendencies inherent to capitalist production, and which lie at the heart of 'the regional problem', cannot be so abolished but instead are internalised within the operations of the state. In due course, they mature into crises of the state itself—that is, as fiscal, legitimation, or rationality crises. Consequently, questions about the future of 'the regional problem' and regional policies focus on two sets of issues: first, on the ways in which 'the problem' can be contained for particular periods of time, to produce a temporary spatialisation of the social relations of the economy within 'acceptable limits'; and secondly, on the relationships between this containment strategy and more general contradictions in the relations between economy, society, and state.

Within these parameters, some important questions remain unanswered. Will the future essentially reproduce the core–periphery, 'North–South' anatomy of 'the regional problem' that was dominant in the UK for most of the twentieth century, following the initial political recognition of the problem itself in the late 1920s? Or will new patterns of regional combined and uneven development emerge as an integral feature of a new 'post-industrial, knowledge economy'? In either case, there is no doubt that the map of spatial inequality will be complex and multi-scalar, with intra-regional as well as interregional contrasts—unavoidably so as the boundaries of regions are, and can only ever be, drawn according to specific criteria for particular purposes.

Finally, what are the implications of new relational concepts of regions—as porous, heterogeneous, and discontinuous—for defining 'problem regions' and policies to address their problems? The analytic advantage of relational thinking about regions is that it focuses attention precisely upon such issues and upon the social forces that seek to construct specific regions, the more general 'regional problem', and regional policies in particular ways. Relational thinking provides an analytic basis both for comprehending the material and discursive construction of regions and for a more realistic assessment of what sort of regional policies are possible and what sort of relationships they might have to 'the regional problem'. If regions are seen as sites through which global flows occur, and in which some aspects of corporate activity can be captured and tied down for a while, what sort of policies are required, and at what spatial scales (supra-national, national, sub-national)? How much can be done at regional level—what are the limits to the regional in this regard and what is the optimal configuration of regional, national, and EU policies, both sectoral and spatial? What is—and what should be—the relationship between formal mainstream regional economic regeneration policies and policies to encourage and foster the emergence of alternative regionally based community and social economies (Amin, Cameron, and Hudson, 2002)? In an era in which responsibilities for regional economic development are being further devolved to new regional institutions, these are important intellectual questions in terms of how regions are theorised. Perhaps even more important, however, they are critical practical questions of policy with profound implications for the living conditions and life chances of the majority of the population resident in the UK.

Acknowledgements

Ash Amin, Ron Johnston, Gordon MacLeod, and Joe Painter kindly commented on an earlier draft and made some valuable comments, drawing a number of points to my attention and to which I have tried to respond. The responsibility for the end product remains mine, however.

References

Allen, J., Cochrane, A., and Massey, D. (1998) *Re-thinking the Region.* London: Routledge.

Amin, A. (1998) An institutionalist perspective on regional economic development. Paper presented to the RGS Economic Geography Research Group Seminar on Institutions and Governance, 3 July, University College London.

Amin, A. and Thrift, N., eds (1994) *Globalization, Institutions and Regional Development in Europe.* Oxford: Oxford University Press.

Amin, A., Cameron, A., and Hudson, R. (2002) *Placing the Social Economy.* London: Routledge.

Anderson, J. (1995) The exaggerated death of the nation state. In J. Anderson, C. Brook, and A. Cochrane, eds, *A Global World.* Oxford: Oxford University Press, 65–112.

Bennett, K., Beynon, H., and Hudson, R. (2000) *Coalfields Regeneration: Dealing with the Consequences of Coalfields Decline.* Bristol: Policy Press.

Beynon, H., ed. (1985) *Digging Deeper: Issues in the Miners' Strike.* London: Verso.

Carney, J. (1980) Regions in crisis: accumulation, the regional problem and crisis. In J. Carney, R. Hudson, and J. Lewis, eds, *Regions in Crisis.* Beckenham: Croom Helm.

Carney, J., Hudson, R., Ive, G., and Lewis, J. (1976) Regional underdevelopment in the late capitalism. In I. Masser, ed., *London Papers in Regional Science 6.* London: Pion, 11–29.

Carney, J., Hudson, R., and Lewis, J. (1977) Coal combines and interregional uneven development in the UK. In P. Batey and D. Massey, eds, *London Papers in Regional Science 7.* London: Pion, 52–67.

Carney, J., Hudson, R., and Lewis, J., eds (1980) *Regions in Crisis: New Directions in European Regional Theory.* Beckenham: Croom Helm.

Chetwode, P. (1939) President's address to the Annual General Meeting of the Royal Geographical Society, 26 June 1939. *Geographical Journal*, 94(3), 189.

Chisholm, M. (1975) Regional policies for the 1970s. *Geographical Journal*, 140(2), 215–44.

Chisholm, M. (1976) Regional polices in an era of slow population growth and higher unemployment. *Regional Studies*, 10, 201–13.

Chisholm, M. and Manners, G., eds (1973) *Spatial Policy Problems of the British Economy.* Cambridge: Cambridge University Press.

Chisholm, M. and O'Sullivan, P. (1973) *Freight Flows and Spatial Aspects of the British Economy.* Cambridge: Cambridge University Press.

Clerk, G. (1944) President's address to the Annual General Meeting of the Royal Geographical Society, 19 June 1944. *Geographical Journal*, 104(1), 6–7.

Coalfields Task Force (1998) *Making the Difference: A New Start for England's Coalfield Communities.* London: Department for the Environment, Transport and Regions.

Daysh, G. H. J. and Symonds, J. S. (1951) *West Durham: A Problem Area in North-Eastern England.* Oxford: Blackwell.

Dicken, P. and Lloyd, P. E. (1976) Geographical perspectives on United States investment in the United Kingdom. *Environment and Planning A*, 8, 685–705.

Dicken, P. and Lloyd, P. E. (1980) Patterns and processes of change in the spatial distribution of foreign-controlled manufacturing plants in the United Kingdom, 1963 to 1975. *Environment and Planning A*, 12, 1405–26.

Dunford, M. and Hudson, R. (1996) *Successful European Regions: Northern Ireland Learning from Others.* Belfast: Northern Ireland Economic Council.

Firn, J. R. (1975) External control and regional development: the case of Scotland. *Environment and Planning A*, 7, 393–414.

Grabher, G. (1993) The weakness of strong ties: the lock-in of regional development in the Ruhr area. In G. Grabher, ed., *The Embedded Firm: On the Socio-Economics of Industrial Networks.* London: Routledge, 255–77.

Habermas, J. (1976) *Legitimation Crisis.* London: Heinemann.

Harvey, D. (1982) *The Limits to Capital.* London: Arnold.

House, J. (1969) *Industrial Britain: The North East.* Newton Abbott: David and Charles.

House, J. (1973) Geographers, decision-takers and policy makers. In M. Chisholm and B. Rodgers, eds, *Studies in Human Geography.* London: Heinemann, 272–305.

House of Commons Expenditure Committee (Trade and Industry Sub-Committee) (1973) *Regional Development Incentives: Report.* London: House of Commons Paper No. 85, Session 1973–4.

Hudson, R. (1986a) Producing an industrial wasteland: capital, labour and the state in north east England. In R. Martin and B. Rowthorne, eds, *The Geography of Deindustrialization.* Basingstoke: Macmillan, 169–213.

Hudson, R. (1986b) Nationalised industry policies and regional policies: the role of the state in the deindustrialisation and reindustrialization of regions. *Society and Space,* 4(1), 7–28.

Hudson, R. (1989) *Wrecking a Region: State Policies, Party Politics and Regional Change.* London: Pion.

Hudson, R. (1990) Re-thinking regions: some preliminary considerations on regions and social change. In R. J. Johnston, G. Hoekveld, and J. Hauer, eds, *Regional Geography: Current Developments and Future Prospects.* London: Routledge, 67–84.

Hudson, R. (1995a) The Japanese, the European market and the automobile industry in the United Kingdom. In R. Hudson and E. W. Schamp, eds, *New Production Concepts and Spatial Restructuring: Towards a New Map of Automobile Manufacturing in Europe?* Berlin: Springer, 63–92.

Hudson, R. (1995b) The role of foreign investment. In A. Darnell, L. Evans, P. Johnson, and B. Thomas, eds, *The Northern Region Economy: Progress and Prospects.* London: Mansell, 79–95.

Hudson, R. (1998) Restructuring region and state: the case of north east England. *Tijdschrift voor Economische and Sociale Geografie,* 89, 15–39.

Hudson, R. (1999) The learning economy, the learning firm and the learning region: a sympathetic critique of the limits to learning. *European Urban and Regional Studies,* 6(1), 69–72.

Hudson, R. (2001) *Producing Places.* New York: Guilford Press.

Hudson, R. (2003) Global production systems and European integration. In J. Peck and H. W.-C. Yeung, eds, *Global Connections.* London: Sage.

Hudson, R. and Sadler, D. (1983) Region, class and the politics of steel closures in the European Community. *Society and Space,* 1, 405–28.

Hudson, R. and Williams, A. (1995) *Divided Britain,* 2nd edn. Chichester: Wiley.

Johnson, B. L. C. and Wise, M. J. (1950) The Black Country 1800–1950. In R. H. Kinvig, J. G. Smith, and M. J. Wise, eds, *Birmingham and its Regional Setting: A Scientific Survey.* Birmingham: British Association for the Advancement of Science, 229–48.

Keeble, D. (1976) *Industrial Location and Planning in the United Kingdom.* London: Methuen.

Lewis, J. and Townsend, A., eds (1989) *The North–South Divide.* London: Paul Chapman.

Lipietz, A. (1993) The local and the global: regional individuality or interregionalism? *Transactions, Institute of British Geographers*, NS18(1), 6–18.

Lloyd, P. E. and Mason, C. M. (1978) Manufacturing industry in the inner city: a case study of Greater Manchester. *Transactions, Institute of British Geographers*, NS3, 66–90.

Lovering, J. (1988) Islands of prosperity: the spatial impact of high technology industry in Britain. In M. J. Breheny, ed., *Defence Expenditure and Regional Development.* London: Mansell, 29–48.

Massey, D. (1978) Regionalism: a review. *Capital and Class*, 6, 106–25.

Massey, D. (1979) In what sense a regional problem? *Regional Studies*, 13, 233–43.

Massey D. (1984) *Spatial Divisions of Labour.* Basingstoke: Macmillan.

Massey, D. (1995a) The conceptualization of place. In D. Massey and P. Jess, eds, *A Place in the World? Place, Culture and Globalization.* Oxford: Oxford University Press, 45–86.

Massey D. (1995b) *Spatial Divisions of Labour*, 2nd edn. Basingstoke: Macmillan.

MacLeod, G. (1998) Place, politics and 'scale dependence': exploring the structuration of Euro-regionalism. *European Urban and Regional Studies*, 6(3), 231–54.

McCrone G. (1969) *Regional Policy in Britain.* London: George Allen & Unwin.

Morgan, K. (1995) *The Learning Region: Institutions, Innovation and Regional Renewal.* Cardiff: Department of City and Regional Planning, Cardiff University, Papers in Planning Research No. 15.

Morgan, K. and Sayer, A. (1988) *Microcircuits of Capital: 'Sunrise' Industry and Uneven Development.* Cambridge: Polity Press.

O'Connor, J. (1973) *The Fiscal Crisis of the State.* New York: St Martins Press.

Offe C. (1975) The theory of the capitalist state and the problem of policy formation. In L. N. Lindberg, R. Alford, C. Crouch, and C. Offe, eds, *Stress and Contradiction in Modern Capitalism.* Lexington, MA: D. C. Heath, 125–44.

Parsons, D. W. (1986) *The Political Economy of Regional Policy.* Beckenham: Croom Helm.

Radice, H. (1984) The national economy: a Keynesian myth. *Capital and Class*, 22, 111–40.

Royal Commission on the Distribution of the Industrial Population (1940) *Report of the Royal Commission on the Distribution of the Industrial Population.* London: HMSO, Cmd 6153 (Barlow Report).

Sayer, A. (1984) *Method in Social Science: A Realist Approach.* London: Hutchinson.

Smith, R. P. (1988) The significance of defence expenditure in the US and UK national economies. In M. J. Breheny, ed., *Defence Expenditure and Regional Development.* London: Mansell, 29–48.

Smith, W. (1953) The location of industry. In W. Smith, ed., *Merseyside: A Scientific Survey.* Liverpool: University of Liverpool Press and British Association for the Advancement of Science, 170–80.

Smith, W. and Lawton, R. (1953) The West Lancashire coalfield. In W. Smith, ed., *Merseyside: A Scientific Survey*. Liverpool: University of Liverpool Press and British Association for the Advancement of Science, 268–77.

Stamp, L. D. (1943) The Scott Report. *Geographical Journal*, 101(1), 16–30.

Storper, M. (1993) Regional 'worlds' of production: learning and innovation in the technology districts of France, Italy and the USA. *Regional Studies*, 27(5) 433–56.

Storper, M. (1995) The resurgence of regional economies, ten years later: the region as a nexus of untraded interdependencies. *European Urban and Regional Studies*, 2(3), 191–222.

Storper, M. (1997) *The Regional World.* New York: Guilford Press.

Townsend, A. and Gault, F. (1972) A national model of factory movement and resultant employment. *Area*, 4, 92–8.

Wise, M. J. and Johnson, B. L. C. (1950) The changing regional pattern in the eighteenth century. In R. H. Kinvig, J. G. Smith, and M. J. Wise, eds, *Birmingham and its Regional Setting: A Scientific Survey.* Birmingham: British Association for the Advancement of Science, 161–86.

19

Geographers and sexual difference:

feminist contributions

LINDA McDOWELL

The aim of this chapter is to evaluate the challenges to geographical knowledge, and to the definition of knowledge more generally, that have arisen from critical debates about the meaning of difference and diversity in feminist scholarship. Although the history of feminist scholarship in the twentieth century is a comparatively short one, dating from some time in the late 1970s, it is also a complex one, and of huge significance for all areas of the discipline. Divisions based on the assumption that men and women are different from one another permeate all areas of social life as well as varying across space and between places. In the home and in the family, in the classroom or in the labour market, in politics, and in power relations, men and women are assumed to be different, to have distinct rights and obligations that affect their daily lives and their standard of living.

But 30 years ago, there were no courses about gender in British geography departments. Nor were there any published papers that explicitly addressed the centrality of gender divisions in shaping social life and in the differential participation by women and men in particular activities in different places, from the smallest to the largest spatial scale. Class and ethnicity were studied as central dimensions of socio-spatial differentiation but gender was ignored. Since then, however, three decades of pioneering work by feminist scholars, in geography as well as in other disciplines, have redressed this imbalance and gender is now widely recognised as a key axis of difference. As well as undertaking innovative research in virtually all the sub-disciplines that constitute human geography as a whole, and in a wide range of locations, feminist geographers have challenged the theoretical foundations of geographical knowledge. Together with other 'radical' and 'critical' scholars from the 1960s onwards (see Castree, 2000), whose academic work was inseparable from their political commitment to a range of radical social movements against inequality, feminists have argued that conventional notions of scientific knowledge as neutral or objective are, in fact, untenable. These knowledges instead represent the interests of the powerful. More recently, in the company of a diverse group of scholars, including postmodern and postcolonial theorists, these arguments have been expanded into a wide-ranging critique of the universal assumptions embodied in modernist social science and its view of a singular progressive political project. There are, however, also significant differences between the adherents of each of these different perspectives.

In this chapter, rather than presenting a chronological account of the exciting work carried out over the last three decades by feminist geographers and by other scholars interested in the connections and differences between places,[1] I have instead chosen to focus on a small number of significant conceptual ideas, previously often taken for granted by most geographers, where feminist arguments have challenged and disrupted conventional definitions, and so placed new issues and different ways of theorising and analysing spatial questions on the geographical agenda. The concepts that are discussed are: the public and the private; sex/gender/body; difference/identity/intersectionality; knowledge; and justice. It is here that some of the most innovative work has been undertaken.

The public and the private

It is now well established that the series of binary concepts that structured the development of knowledge in the social sciences and the humanities from the period of the western enlightenment until well into the twentieth century are a set of historically and temporally specific notions rather than universal social and political attributes (Pateman and Grosz, 1987). Even so, binaries or dichotomous divisions, and their mapping onto a gender division, which is the singular or categorical distinction between women and men, continue to exercise a dominant hold on the imagination and the social and political institutions of contemporary societies. The idea that women are one thing, men the opposite—women are emotional rather than rational, caring rather than instrumental, primarily mothers rather than workers—is deeply embedded in our sense of ourselves as individuals and in our daily interactions with others, as well as in the institutional structures of industrial societies and in western intellectual thought. Gillian Rose

[1] A number of useful surveys of feminist approaches and substantive contributions to geography were published in the 1990s, including Laurie et al. (1999), Jones, Nast, and Roberts (1997), McDowell (1992a, 1992b, 1999) and Women and Geography Study Group (1997). The Women and Geography Study Group of the Royal Geographical Society (with the Institute of British Geographers) began as a working party in 1979 and became a full study group in 1981. A further mark of the maturity of feminist geography was the establishment of a specialist journal, *Gender, Place and Culture*, in 1994.

(1993) has, for example, documented the ways in which geography from its origins as a university discipline was constituted as a masculinist discourse and series of practices, as well as being taught predominantly by men and, until recent decades, to men. Geography was, according to Denis Cosgrove (1993, 516), in a comment on the fieldwork tradition in geography, dominated by 'hairy-chested feats of scholarly endurance'.

In all cases, in the comparisons between women and men, the attributes associated with women and with femininity are constructed and perceived as inferior to those associated with men and with masculinity. Men, then, are seen as strong, as protectors of weaker women, as defenders of the feminine virtue of their wives, sisters and daughters, and as participators in the public sphere of work and politics rather than the private arena of the home. The home is defined as a feminine sphere of influence and as a haven or leisure space for men, as a respite from the more brutal world of the market and the state, where men labour to provide for their womenfolk (Davidoff and Hall, 1987). The fundamental institutions of modern societies—the family, the education system, the state, and the market—express and normalise these gender differences and in so doing produce and maintain social relations of inequality between men and women.

For human geographers, the spatial distinction between public and private spheres, and their association with gendered attributes and occupants, has had a profound influence on the structure and content of geographical research. Almost without exception, until the arguments of feminist scholarship began to influence the definition of appropriate matters for geographical investigation, it was the male-dominated public world of work and politics that was the focus of attention. Thus, in analyses of the city, for example, it was the public aspects of residential differentiation that captured attention, the methods of allocation and distribution that produced classed and raced patterns of difference in the city; the public world of urban politics, of boosterism or corruption; the use and control of public spaces, of urban leisure and sports. The internal world of the family, relations of power between men and women, adults and children, and the ways in which marriage was a key means of access to housing for women were all ignored. In economic analyses the definition of work included only public aspects of waged labour, and even here the sexual division of labour in workplaces was taken for granted, if commented on at all; the division of labour within the home, women's unpaid labour of biological and social reproduction, was ignored even though it was the basis for men's continued dominance in the public world of waged employment. And so Locke's separation of the public and private spheres and the association of affective social relations of caring among kin with the latter (compared to the instrumental and competitive relations among individuals in the state and the market) dominated the social sciences for centuries. Habermas (1989), for example, distinguished two separate institutional spheres, although in his work the family and the market were bracketed together in the private sphere and the state and the arena of public political debate between citizens constituted the public arena (Bowlby, 1999, 223). Castells (1977), in his influential work on the significance of state provision in cities to facilitate social reproduction—captured in the term 'collective consumption'—limited his definition of reproduction to the reproduction of capitalist social relations through public action, and so neglected the work undertaken by women within the family (Gamarnikow, 1978).

One of the most significant parts of the developing feminist critique of geographical work and that in the other social sciences was its challenge to this binary distinction between the public and the private. The absence of women in the urban landscape was pointed out, the ways in which idealised views of home and community reinforced women's subordination (Davidoff, L'Esperance, and Newby, 1976), as well as the ways in which the distribution and separation of different urban land uses divided, segregated, and confined women were investigated (McDowell, 1983). Feminist scholars also challenged the subject matter of economic geography, arguing that patterns of participation and the spatial variations in forms of work other than waged labour should be part of the discipline. Thus questions about domestic divisions of labour were raised, and the contribution of the domestic work and child care to total well-being was analysed and calculated (Tivers, 1985). Associations between women's domestic labour and their position in the labour market in different regions of the country were also explored in a major effort to make women's lives visible and to map and explain the ways in which hierarchical gender relations are both affected by and reflected in the spatial structure of society (Massey, 1979; Mackenzie and Rose, 1983; Foord and Lewis, 1984).

In political geography, the power relations between men and women and struggles to resist male domination were added to the agenda as a consequence of rethinking the meaning of so-called private issues. Thus questions about women's fear, in the home and in the public arena (Pain, 1991, 1997), about rape, about male violence, and about the normalisation of heterosexuality (Ettore, 1978; Valentine, 1989, 1992), as well as political movements and campaigns to improve the material conditions of women's lives, from local action to increase nursery provision (Mackenzie, 1989) to the establishment of women's committees by local authorities (Halford, 1988), became focuses for study under the redefinition of the political to include the 'personal'.

Building on this early work documenting the material inequalities between women and men in contemporary Britain, there has been great increase in the publication of feminist analyses of gender differences in all areas of life and in a great variety of circumstances, continuing to challenge the conceptual and empirical separation of the public and private arenas but also documenting in growing detail the significance of differences among women as well as between women and men. I shall turn to this issue about the recognition of difference and diversity in the third section of this chapter. First, however, I want to document the changing ways in which feminist scholars have distinguished and defined sex and gender.

Sex/gender/body

The second group of ideas that I want to discuss are those of sex, gender, and the body, as here too feminist redefinition of taken-for-granted terms has had a significant impact on the development of feminist scholarship. For much of the twentieth century the term 'sex' was used to describe sexual differences in a way that took for granted that social and psychological characteristics were based on the 'facts' of biology. Here, too, a simple binary categorisation was accepted. The sexes were divided into two—woman and man—each distinguishable by biological characteristics. A challenge faced feminist scholars: to disrupt what seemed like immutable sexual differences based in biology and so to undermine claims of absolute sexual difference between women and men, and, more significantly, women's 'natural' inferiority in physical and intellectual attributes. To do this the term 'gender' was adopted, theorised as the cultural or social elaboration of differences between women and men, and so mutable, variable, and open to change. This argument was prefigured in the classic claim by the great feminist existentialist Simone de Beauvoir in *The Second Sex* (first published in 1949) that: 'One is not born but rather becomes a woman. No biological, physiological or economic fate determines the figure that the human being presents in society: it is civilisation as a whole' (de Beauvoir, 1972, 525).

Thus, through institutional regulation, cultural practices, and everyday interactions 'sex' is transformed into 'gender'. For feminist geographers and anthropologists this conceptual separation was crucial as it opened the way for analyses of the multiple ways of becoming a woman. As the anthropologist Henrietta Moore (1988, 12) has argued, the aim of feminist scholarship is to explore 'what it is to be a woman, how cultural understandings of the category "woman" vary through space and time, and how those understandings relate to the position of women in different societies'. As she noted, the concepts of gender and gender relations were an essential aspect of this work, which documented the ways in which gender differences were constituted 'either as symbolic construction or as a social relationship' (p. 12), or as both. For feminist geographers, the recognition that gender is socially constructed and relational, created and maintained through that range of social practices and symbolic representations that attribute particular characteristics to women and different (usually superior) ones to men, was an impetus to research at all spatial scales, documenting the variety in the meaning of femininity, as well as the commonalities and differences in women's circumstances in particular places at particular times.

More recently, the stability of the construct gender has been deconstructed, followed by a more radical re-evaluation of the argument that sex and gender are separable, especially in new theorising about the body. Perhaps the key theorist here has been Judith Butler (1990, 1993), who has built on work by, among others, Michel Foucault, showing how the body is produced through discursive regulation in everyday interactions. Gender, rather than being a fixed binary category, is always provisional and fluid, in the process of becoming rather than fixed or stable. However, as Butler argues, the assumption of a binary division, the taken-for-granted view that there are distinct sexed bodies that are male or female, influences the ways in which gender is performed in bodily acts. Further, what Butler terms 'the regulatory fiction of heterosexuality' constrains gender performances to specific hegemonic versions of masculinity and femininity appropriate in particular contexts and circumstances. This binary division, she suggests, although dominant in contemporary societies, is open to challenge through transgressive and subversive acts such as drag. For geographers, Butler's arguments have been a great stimulus to investigations of how gender and place are co-constituted, how specific embodied performances both confirm and challenge conventional gender codings of different spaces and places (Bell et al., 1994; McDowell and Court, 1994; Walker, 1995; Nelson, 1999).

The work of theorists such as Butler (1990, 1993), Foucault (1978), and Grosz (1994, 1995) has also stimulated a provocative debate about the materiality of the body, which has been given a further impetus through the impact of new technologies and legal/social conventions that are not only challenging kinship relations, through, for example, *in vitro* fertilisation and surrogacy (Strathern, 1992), but also seem to have led to the decreasing relevance of bodily constraints. Viagra may be changing the associations between age and sexuality, post-menopausal births altering the links between age and fertility and exercise, but cosmetic surgery, in particular, means that bodies seem to be increasingly fluid and malleable, subjects of redefinition and reworking in ways that were previously not possible. In all cases, however, the potential limits of embodiment depend on material inequalities, as income levels largely determine access to the range of new services and provisions and significant class and spatial variations exist in the effects of the new technologies. There is a small but growing body of geographical work that is beginning to investigate the new questions about embodiment (Johnston, 1996; Nast and Pile, 1998; McCormack, 1999).

Difference/identity/intersectionality

If early work by feminist scholars in the 1970s and into the 1980s focused on the description and explanation of differences between women and men, as well as assessing policies to overturn the subordination of women, more recent work has turned much more explicitly to questions of difference. A significant stimulus to this was a powerful critique from women of colour in Britain, and especially in the United States, of the developing theory and politics of white feminism. Black women pointed out the multiple ways in which their lives differed from those of white women because of the intersection of their gender and their 'race', at the same time as geographers, anthropologists, and historians had been uncovering the dimensions of difference in women's lives across space and over time. As Kimmel (2000, 2) has noted in an introductory survey of gender differences 'what it means to be a man or a woman [varies] in four significant ways'.

First, the definitions and associated attributes of femininity and masculinity are spatially and culturally variable. As geographical research has demonstrated, what it means to be a woman in, say, the Australian outback (Jacobs, 1994) is different from what it means in Ireland (Nash, 1993; Johnson, 1995; Walters, 2001) or in the Andean states of Latin America (Radcliffe and Westwood, 1994). While in most societies the social construction of femininity is based on attributes that are regarded as inferior to those associated with masculinity, in certain cultures women are encouraged to be decisive and competitive, whereas in others women's 'natural' passivity, helplessness, and dependence are more highly valued. Regional and local differences are also apparent in the meaning of gender, based in local customs and practices and in established differences in women's and men's lives (McDowell and Massey, 1984). The traditional 'stroppiness' of Lancashire women, for example, compared to women in other regions of England, lies in part in their industrial experiences in the nineteenth and early twentieth centuries (Glucksmann, 2000), and was reflected in the unique role played by working-class women in this region in the struggle for suffrage before and after the First World War (Liddington, 1984).

These geographical differences, as this example suggests, are related to differences in historical development. Consequently, and secondly, the social construction, attributes, and meanings of femininity and masculinity vary historically as well as geographically. The social meanings of idealised femininity and masculinity have changed profoundly over the centuries. In his own research, Kimmel (1996) has documented how the meanings of manhood altered between the founding of the United States in 1776 and the end of the twentieth century. Thirdly, individuals' own senses of themselves change as they grow and mature. The issues confronting young people as they come to terms with themselves as a woman or as a man in different social circumstances and locales are different from the meanings of femininity or masculinity as people age. Glenda Laws (1994, 1997), for example, before her untimely death, began to explore the geographic meanings of ageing for women in the United States. Finally, recent work has convincingly demonstrated not only the connections between these three previous dimensions of difference, but also the complex ways in which the meaning of gender varies among different groups of women and men in any particular culture at any particular time. Gendered identities are structured by the intersection of class, race, ethnicity, age, sexuality, and location.

Questions about the fourth definition of difference, or what might be termed intersectionality, are at the heart of contemporary feminist scholarship, in which the ways in which women are differentially placed in what Collins (1990, 225) designated a 'matrix of domination' are being explored. In this work, notions of difference as diversity, pluralism, and hybridity have replaced simpler additive models of oppression (Brah, 1996). Thus, for example, poor women of colour are not seen as subjects of a threefold oppression (of race, class and gender) but instead their identity and social location are theorised as a consequence of the co-constitution of all these dimensions of difference. The increasing scale of migration within and between nations in recent decades has lent even greater significance to these questions about difference and identity, about the definition of gender as an ever-changing, complex, fractured, and fluid assemblage of traits, attitudes, and behaviours. As I noted in the discussion above, gender is often theorised as an ongoing performance (Butler, 1990), as something that is done and redone (West and Zimmerman, 1987) in particular ways in particular circumstances. At the very local spatial scale, for example, geographers have begun to explore the ways in which particular versions of femininity and masculinity either confirm or transgress the gendered meanings of different spaces and places.

For a discipline based on the exploration of spatial difference, constructed through the intersection of a range of social processes operating at different scales (captured, albeit as a simplification, in the common reference to global–local connections/processes), this contemporary theorisation of gender as context dependent, fluid, and variable is both persuasive and appealing, resonating with definitions of the relational construction of place (see Chapter 8 in this volume). Explorations of the co-constitution of place and gender relations is perhaps one of the most interesting and innovative areas of current work by geographers. Despite this recognition of the intersectionality of identity construction and the extent of variations across space and through time, it is also important not to lose sight of the relative fixity in definitions of masculinity and femininity and the ways in which they are constituted through social relations of power and inequality. Dominant notions of masculinity, for example, continue to be based on their differentiation from an inferior 'Other'. For men, women are the classic Other. To be masculine is to be not feminine, not a woman. But idealised, highly valued versions of both femininity and masculinity are also constructed through comparison with a range of other 'Others'—ethnic or sexual minorities for example. As Connell (1987, 183) argued, the hegemonic definition of masculinity is 'constructed in relation to various subordinated masculinities as well as in relation to women'. In contemporary western societies, hegemonic versions of both masculinity and femininity that stress whiteness, heterosexuality, good complexion, weight, and height continue to define men and women with different attributes as less valued (Young, 1990) and so, despite the growing theoretical recognition of cultural difference, there remain significant inequalities in the social structures and social relations that constitute, maintain, and reinforce such differences.

Feminist epistemologies and situated knowledges

In this fourth section I want to turn to a series of arguments about the nature of knowledge, which in part have developed out of the debates about difference but which also preceded them. Here I want to note the parallels between feminist questioning of orthodoxies and recent critical work in postmodern and postcolonial theorising. As I noted in the first section, a challenge to the authority of western conceptual frameworks has been a key part of the development of feminist scholarship since the 1970s, challenging the claims of rationality and objectivity in conventional scientific knowledge. Instead feminist theorists of science have argued that the systematic knowledge possessed by a culture or a sub-culture is a product of the kinds of interaction its knowledge producers have with their social and natural environment. In other words, cultures, or communities of knowledge producers, tend to gather bodies of knowledge that respond to their distinctively perceived values, needs, and interests. Knowledge and power are internally linked (Harding, 1986, 1991, 2001).

While these arguments about the social construction of knowledge are now widely accepted by philosophers and historians of science (Foucault, 1980; Quine, 1981; Bhaskar, 1991; Latour, 1999), feminist claims in the 1970s and 1980s that there is a difference between the knowledge derived from women's experiences and those of men (and later, a difference between knowledges of women from different cultures) were originally regarded by many geographers as an unacceptable challenge to conventional beliefs in western science and in its emancipatory potential. Feminist scholars argued not only that the knowledge of an oppressed group is different from that of the oppressors but also that it is more 'truthful', as the oppressed, in this case women, can see more clearly than those who are blinded by their own interests and involvement in the dominant ideology (Hartsock, 1985; Haraway, 1988, 1989). In this way feminist theorising has parallels with Marxian arguments about the biases of bourgeois science. However, it differed from Marxism in its denial of a singular truth and in its adherence to a belief in the significance of the everyday experiences of women in different locations and positions. Building on early practices of consciousness-raising in the women's movement, in which personal experiences were compared and reinterpreted as political issues, feminist scholars developed the concept of a standpoint epistemology—that is, knowledge based on the point of view of women—which, it was suggested, is more objective than pure theorising, unlocated in experience. As Harding (1991, 269), one of the key theorists in the development of standpoint epistemologies, has argued: 'when people speak from the opposite sides of power relations, the perspective from the lives of the less powerful can provide a more objective view than the perspective from the lives of the more powerful'.

In recognition of the arguments about difference and diversity outlined in the previous section, it has now become common to argue that there are multiple standpoints or multiple knowledges situated in relation to specific groups or positions. Women cannot be or hold to a single standpoint as they are divided by their class position, 'race', ethnicity, sexual orientation, generation, and physical capacity, as well as by their location within different societies and cultures. In this recognition of diversity and multiple knowledges, standpoint epistemology has similarities to postmodern and postcolonial epistemologies (Nicholson, 1990). Feminist standpoint theorists, however, reject accusations of relativism, insisting that 'the criteria for preferring some knowledges over others are ethical and political rather than purely "epistemological"' (Hartsock, 1997, 372). Thus, as Sylvia Walby (2001, 488) notes in a critical evaluation of standpoint theory, 'it [standpoint theory] is not relativism, in that there are criteria to decide between rival knowledges, but these criteria are aesthetics and values, rather than truth claims'. The values that are drawn on in the evaluation of knowledge claims are, as Sandra Harding (2001) has recently insisted, those based on pro-democratic ethics and politics. I return to the topic of ethics and social justice in the following section.

Standpoint epistemologies, with their insistence on the importance of location in the social construction of knowledge and knowledge claims, have an intrinsic appeal to geographers. Indeed, as Smith and Katz (1993) pointed out, spatial metaphors became a central feature of critical social theorising more generally in the 1990s. In the expansion of writing about identity that is taking place in the boundaries of geography, literary, and cultural studies, spatial concepts are key. For example, the literary critic Caren Kaplan (1996, 144) has noted that 'topography and geography now intersect literary and cultural criticism in a growing interdisciplinary inquiry into emergent identity formations and social practices'. Many feminist theorists have developed spatial metaphors to delineate new forms of identity: 'geometries' (Haraway, 1991) and 'geographies' (Pollock, 1996) of identity and 'topographies' of feminist engagement (Katz, 2001), defining identity as an 'historically embedded site, a positionality, a location, a standpoint, a terrain, an intersection, a cross roads of multiply situated knowledge' (Friedman, 1998, 19).

A new politics of writing and struggle is emerging from these spatial reconfigurations. A locational approach to understanding gendered identities demands an analysis of the ways in which the intersections of social processes of stratification and distinction and movements for social justice at different times and in different places result in different and changing gender identities and relations. Such a feminism is informed by recent geographical knowledge. It requires, and builds on, analyses of the ways in which local and global changes are mutually constituted and is enabled by new possibilities of making connections between heterogeneous feminist movements and struggles in specific places at particular times using new information technologies. Consequently, Mohanty (1991; Mohanty and Alexander, 1997), among others, has argued that feminist politics is no longer necessarily based on a settled or territorially based identity but rather on the development of networks between an imagined community of women with interests in common.

Standpoint theorising also became linked to the development of a long debate about the preferred research methods in feminist social sciences, including geography, in which qualitative methods were suggested as being the most appropriate way of keeping close to women's lives (*Professional Geographer*, 1995). More recently, however, the utility of a greater range of methods has been acknowledged. It has also been argued that the common emphasis on women's everyday lives as the key focus for research was based on a misunderstanding, or misreading, of feminist arguments by some researchers (Harding, 2001). While feminist research, by definition, must start with women's lives and their own understanding of them, it should not be limited or restricted to women's everyday experiences. Instead,

> . . . standpoint epistemologies direct researchers to 'study up', to start out from the everyday lives of people in oppressed groups in order to identify sources of their oppression to be found in the conceptual practices of power that are embedded in institutional cultures and practices and the epistemic standards that structure and sustain them. (Harding, 2001, 517)

Thus while questions about women's oppression are raised through studying women's lives, the answers to these questions lie elsewhere. As Harding continued: 'Oppressed groups, like all cultures, do possess many kinds of distinctive knowledge. But standpoint epistemologies are interested in how such knowledges can be used to identify otherwise obscured features of dominant institutions, their cultures and their practices' (p. 517).

Readings of feminist standpoint epistemologies that reject them as relativist and 'unscientific' misunderstand the aims and political purpose of such knowledges. Their aim is not to reject 'science' in the sense of objectivity, rationality, and good method but instead to extend and strengthen these characteristics in order to reveal the connections between power and knowledge and the ways in which systematic areas of ignorance are constructed as part of epistemological efforts. In common with postcolonial and postmodern theorists, feminist scholars are concerned to reveal not only the extent and effects of androcentric assumption but also the consequences of, *inter alia*, Eurocentrism, racism, and class-based values: 'Feminism does not introduce ethical and political considerations into heretofore politically neutral evaluation criteria but, rather, tries to redirect science to more scientifically effective as well as more politically progressive consideration' (Harding, 2001, 522).

Justice

I argued in the previous section that feminist epistemologies depend on pro-democratic values and ethics to judge between different knowledge claims. This argument in turn necessitates a consideration of feminist theories of social justice. Whose claims to inequality demand attention? Which inequalities matter? How might greater equality be achieved? Although the insistence on pro-democratic values in feminist theories of situated knowledge is helpful, it is unable to provide complete answers to these sorts of questions. My discussion here is relatively brief as the next chapter—by David Smith—is explicitly concerned with justice and ethics.

An influential feminist contribution to challenging conventional notions of distributive justice is to be found in the work of the political theorist Iris Marion Young (1990), and her work has had an important impact within geography (Harvey, 1996). One of the most significant aspects of the growth of research about the social constitution and effects of diversity and difference has been an acceptance that the bases of injustice in society are wider than purely distributive issues (i.e. unequal access to scarce resources). As well as material inequalities, in the main constituted in the economic sphere and so related to class position, injustice is also based on dimensions of 'Otherness', on attributes of identity that constitute an individual or a group as 'different from', as 'othered' compared to accepted norms. Thus a range of cultural inequalities, which Young has argued are based in differences associated with ethnicity, gender, and embodiment, parallel economic injustices. While not denying the great significance of economic inequality, Young has argued that these other forms of injustice may not be amenable to change through traditional redistributive politics and policies. Thus she identifies what she terms 'five faces of oppression'. Two of the five—economic exploitation (exploitation within the labour market—and Young adds exploitation on the basis of 'race' and gender to exploitation through the extraction of surplus value from the proletariat) and economic marginalisation (exclusion from the labour market, whether on the basis of skill, age, or infirmity)—are clearly material inequalities, and the third—powerlessness—is employment-related (the lack of power and autonomy experienced by non-professionals in the labour market). The other two—systemic violence (fear of random attack, in the main by men against women) and cultural imperialism (how groups are stereotyped as 'other' and discriminated against as deviant or inferior)—are cultural injustices based on forms of discrimination which, although related to, may also be distinguished from economic inequalities.

To differentiate these two forms of inequality it has become common to refer to political struggles to address and remedy the first set as 'the politics of redistribution' and to effect change in the second set as 'the politics of recognition'. But this division between economic and cultural injustice is, in my view, relatively unhelpful, leading as it has to bad-tempered, albeit explicable, debates among feminist scholars about the relative significance of economic versus cultural difference, in which the significance of the economic versus the 'merely cultural' were asserted (Fraser, 1995, 1997a; Phillips, 1997, 1999; Young, 1997; Butler, 1998). An insistence on the cultural dimensions of injustice was undoubtedly a necessary response to the long dominance of structuralist approaches within the social sciences, geography included, which insisted on the primacy of economic inequalities. Here David Harvey's series of books documenting the dimensions of injustice in capitalist economies and the ways in which spatial divisions and uneven development are a necessary part of capitalism justly deserve praise. Yet Harvey was reluctant to recognise the significance of sexual difference and gender relations as a key aspect of inequality and reacted angrily to feminist critiques (e.g. Deutsche, 1991; Massey, 1991; Harvey, 1992) of his best-selling book, *The Condition of Postmodernity* (Harvey, 1989). In the contemporary debates about the significance of economic inequities and cultural differences, Young's work has tended to be identified with the cultural axis. Her multiple, flexible, context-dependent, group-based notion of justice clearly has its origins in that wide range of work about difference and diversity. However, she herself has always insisted on the interdependence of the different bases of oppression. Indeed, in a forceful comment on David Harvey's work on justice, she made her position absolutely clear:

> It is both theoretically and politically counterproductive, it seems to me, to construct an account of our recent history as a fall away from the correct class-based universalist politics to a fragmented and relativist politics of

> difference and to suggest, however qualifiedly, that progressives ought to abandon the latter and return to an earlier and more correct road. We should not interpret our current theoretical and political situation as a choice between universal and particular, class unity and the recognition of social difference, but rather as a challenge to move beyond these oppositions. (Young, 1998, 37)

Young's insistence echoes the arguments of Nancy Fraser (although the two theorists have not always agreed), who also insists that 'critical theorists must rebut the claim that we must make an either/or choice between the politics of redistribution and the politics of recognition. We should aim instead to identify the emancipatory dimensions of both problematics and to integrate them into a single, comprehensive framework' (Fraser, 1997b, 4). In her work, Fraser has attempted to identify the different dimensions of current social policies to reduce inequality and injustice and their relative emphases on redistribution and recognition, and to address ways in which their aims might become compatible. Both affirmation of difference and the transformation of the material bases of oppression are essential parts of any programme to radically transform the bases of inequality, be they on the grounds of gender, race, sexuality, or other attributes. In the workplace, for example, feminists demand both recognition of differences, such as the impact of pregnancy on working lives, as well as a transformation of the current gender division of labour, which produces women's concentration in a limited range of positions in the labour market and the continuing pay gap between male and female employees (which is still evident in universities as well as in other workplaces).

Despite the undeniable significance of difference, however, long-standing systemic inequalities in life-chances continue to differentiate members of socially defined categories of persons, and historical and comparative analyses have illustrated the strength and persistence of certain of these categories—gender among them. And as Charles Tilly (1999) has argued in his interesting book *Durable Inequalities*, many of these categories are based on 'distinctly bounded pairs such as male/female, aristocrat/plebeian, citizen/foreigner, and more complex classifications based on religious affiliation, ethnic origin, or race' (p. 6). These bounded categories, or what I termed binaries earlier in the chapter, according to Tilly, 'deserve special attention because they provide clear evidence for the operation of durable inequality, because their boundaries do crucial organisational work, and because categorical differences actually account for much of what ordinary observers take to be the results of variation in individual talent or effort' (p. 6). While Tilly's arguments are compatible with much of the work that I have outlined above, they also remind us not to forget the political significance of these distinctions in arguments about the significance of difference.

There is also an interesting debate among feminist philosophers about the notion of universality. The joint work of the feminist theorist Martha Nussbaum and development economist Amartya Sen (e.g. Nussbaum and Sen, 1993; Nussbaum and Glover, 1995) in their interesting development of the notion of essential human capabilities is important here, as is the work of the Kantian philosopher Onora O'Neill (1995, 2000), all of whom continue to insist on the political significance of claiming universal human rights. Nussbaum defines capabilities as the 'rights' or entitlements that must be met to ensure a fulfilling human life. Because of their generality (including, for example, being able to think, to have attachments to things and persons outside oneself) but also their recognition of location, in, for example, the capability of living an independent life in one's own surroundings and context (see Nussbaum, 2000, 70–96), it seems that the claims of universalism inherent in the notion do not preclude the recognition of difference and diversity. It is clear that large theoretical and practical challenges still face feminist and other critical social theorists in thinking through their/our transformative projects and the political demands on which they rest. And geographers have a key part to play in this important work.

Concluding remarks

In this chapter, I have outlined the multiple ways in which feminist scholarship has been a key part of a wider recognition of the significance of difference and diversity. This recognition has led to an exhilarating and extensive re-evaluation of the nature of knowledge; the deconstruction of conventional binaries; the redefinition of both individual and group identity as fragmented, unstable, and differentiated; a general acceptance in the social sciences of the significance of location, and arguments for new forms of transformative politics: in short, a thorough, if still incomplete, challenge to almost all the assumptions on which the development of geography as a modern discipline in the twentieth century was based. As Harding (2001, 532–4) has noted, feminist contributions in the hugely important debates about the current epistemological crisis in the West have been extremely significant. 'Is it', she asks, 'too grandiose an evaluation to say that this crisis, and the changing social formations producing it, may well have effects as extensive as those that followed the Copernican, Darwinian and Freudian revolutions, and that feminist epistemology is fully part of this movement? Perhaps not'. Her claims may be ambitious but they seem entirely warranted to me.

I want, finally, to end with a comment on the role of feminism in the academy as a set of political practices as well as epistemological claims. For one of the key aims of feminist movements is to challenge the everyday practices and institutional arrangements that reinforce inequalities between men and women in a range of locations, including the universities. As Gillian Rose (1993), and many others, have argued, the masculinism of knowledge is inseparable from masculinist assumptions, practices, and behaviours in the academy that construct women as inferiors. From the way in which eye contact is made in seminars, through teaching styles, to forms of participation and assessment, it seems that what we might term masculine ways of doing have tended to be more highly valued. Similarly, the demands of the profession—in its commitment to after-hours social interactions, to seminars in the early evenings, to the model of the scholar as untrammelled by dependants or even by 'his' own bodily needs—place many women and most parents at a disadvantage. The most blatant aspects of sexist behaviour are no longer acceptable in universities as wider societal recognition of the debilitating aspects of sexist practices has filtered into the academy, and as the women therein challenge sexist assumptions and behaviours. Women scholars are now more numerous and have recently begun to challenge their own poorer promotion prospects and lower pay through membership of professional associations, through campaigns, and through the courts. In geography, where formerly the number of women academics was tiny—perhaps a sole woman on the academic staff in geography departments in most universities in the 1950s and 1960s—this position began to change with the post-Robbins expansion and accelerated towards the end of the twentieth century. In many departments there is now at least one women professor, a number of whom are feminist scholars. There has also been a marked rise in the participation rates of young women in university education and women are now a majority among undergraduates studying geography.

Despite these gains for women in the disciplinary structures of power in university geography, there is, however, still no doubt that men remain both numerically dominant and in the key positions of power—as heads of department, chairs of assessment panels, editors, examiners, and advisers. In the recent climate of growing regulation, continuous audits, inadequate funding, and what Katharyne Mitchell (1999) termed the corporatisation of the university, a number of geographers (Castree, 2000; Mitchell, 2000) have expressed the more general anxieties that seem to be common among academics about the nature of academic work in this changed environment. The general speed-up of demands to produce more, teach more and administer more, facilitated by new technologies such as email, are squeezing the time available for friends and family. The growing reliance on competitive rankings in the allocation of funding is exacerbating rivalries within and between departments and leaving less time for the 'unproductive activities' of, for example, student mentoring. New salary differentials are opening up as a rank of 'superstars' are appointed by the most market-oriented institutions, further exacerbating competition and reducing the collegiality that perhaps was once partial compensation for relatively low salaries. While the changing nature of academic life affects all scholars, and perhaps especially younger colleagues on the ladder to tenure and/or a permanent position, it may be that women academics are again in a relatively worse position. Given the continuing inequities in the distribution of domestic labour and child care, as well as the poor provision of child-care facilities in the workplace, even though universities have made some progress in this area over the past 20 years, maintaining that mythical work/life balance is problematic for many women geographers. The epistemological challenges of feminism may have changed ways of thinking, but many of the practical demands of the women's movement have yet to be met.

References

Bell, D., Binnie, J., Cream, J., and Valentine, G. (1994) All hyped up and no place to go. *Gender, Place and Culture*, 1, 31–47.

Bhaskar, R. (1991) *Philosophy and the Idea of Freedom*. Oxford: Blackwell.

Bowlby, S. (1999) Public–private division. In L. McDowell and J. Sharp, eds, *A Feminist Glossary of Human Geography*. London: Arnold, 222–4.

Brah, A. (1996) *Cartographies of Diaspora*. London: Routledge.

Butler, J. (1990) *Gender Trouble*. London: Routledge.

Butler, J. (1993) *Bodies that Matter: On the Discursive Limits of 'Sex'*. London: Routledge.

Butler, J. (1998) Merely cultural. *New Left Review*, 227, 33–44.

Castells, M. (1977) *The Urban Question*. London: Edward Arnold.

Castree, N. (2000) Professionalisation, activism and the university: whither 'critical geography'? *Environment and Planning A*, 32, 955–70.

Collins, P. H. (1990) *Black Feminist Thought*. London: Unwin Hyman.

Connell, R. W. (1987) *Gender and Power: Society, the Person and Sexual Politics*. Cambridge: Polity Press.

Cosgrove, D. (1993) On 'The reinvention of cultural geography' by Price and Lewis: commentary. *Annals of the Association of American Geographers*, 83, 515–16.

Davidoff, L. and Hall, C. (1987) *Family Fortunes: Men and Women of the English Middle Class*. London: Hutchinson.

Davidoff, L., L'Esperance, J., and Newby, H. (1976) Landscape with figures: home and community in English society. In J. Mitchell and A. Oakley, eds, *The Rights and Wrongs of Women*. Harmondsworth: Penguin, 142–70.

De Beauvoir, S. (1972) *The Second Sex*. Harmondsworth: Penguin (originally published 1949).

Deutsche, R. (1991) Boys town. *Environment and Planning D: Society and Space*, 9, 5–30.

Ettore, E. M. (1978) Women, urban social movements and the lesbian ghetto. *International Journal of Urban and Regional Research*, 2, 499–519.

Foord, J. and Lewis, J. (1984) Economic and urban change. In Women and Geography Study Group, *Geography and Gender*. London: Hutchinson.

Foucault, M. (1978) *The History of Sexuality, Volume 1. An Introduction*. London: Allen Lane.
Foucault, M. (1980) *Power/Knowledge: Selected Interviews and Other Writings, 1972–1977*. New York: Pantheon Books.
Fraser, N. (1995) Recognition of redistribution? A critical reading of Iris Young's *Justice and the Politics of Difference*. *Journal of Political Philosophy*, 3, 166–80.
Fraser, N. (1997a) A rejoinder to Iris Young. *New Left Review*, 223, 126–9.
Fraser, N. (1997b) *Justice Interruptus: Critical Reflections on the Postsocialist Condition*. London: Routledge.
Friedman, S. (1998) *Mappings: Feminism and the Cultural Geographies of Encounter*. Princeton, NJ: Princeton University Press.
Gamarnikow, E. (1978) Introduction: women and the city. *International Journal of Urban and Regional Research*, 2, 390–403.
Glucksmann, M. (2000) *Cottons and Casuals: The Gendered Organisation of Labour in Time and Space*. Durham: sociologypress.
Grosz, E. (1994) *Volatile Bodies: Towards a Corporeal Feminism*. Bloomington, IN: Indiana University Press.
Grosz, E. (1995) *Space, Time and Perversion: Essays on the Politics of Bodies*. London, Routledge.
Habermas, J. (1989) *The Structural Transformation of the Public Sphere*. Cambridge, MA: MIT Press (originally published in German in 1962).
Halford, S. (1988) Women's initiatives in local government . . . where do they come from and where do they go? *Policy and Politics*, 16, 251–9.
Haraway, D. (1988) Situated knowledges: the science question in feminism and the privilege of partial perspective. *Feminist Studies*, 14, 579–99.
Haraway, D. (1989) *Primate Visions: Gender, Race and Nature in the World of Modern Science*. London: Routledge.
Haraway, D. (1991) *Simians, Cyborgs and Women: The Reinvention of Nature*. London: Free Association Books.
Harding, S. (1986) *The Science Question in Feminism*. Ithaca, NY: Cornell University Press.
Harding, S. (1991) *Whose Science? Whose Knowledge? Thinking Women's Lives*. Milton Keynes: Open University Press.
Harding, S. (2001) A response to Walby's 'Against epistemological chasms': a standard misreading. *Signs*, 26, 511–25.
Hartsock, N. (1985) *Money, Sex and Power: Toward a Feminist Historical Materialism*. Boston, MA: Northeastern University Press.
Hartsock, N. (1997) Comment on Hekman's 'Truth and method: feminist standpoint revisited': truth or justice. *Signs*, 22, 367–74.
Harvey, D. (1989) *The Condition of Postmodernity*. Oxford: Blackwell.
Harvey, D. (1992) Postmodern morality plays. *Antipode*, 24, 300–26.
Harvey, D. (1996) *Justice, Nature and the Geography of Difference*. Oxford: Blackwell.
Jacobs, J. (1994) Earth honoring: western desires and indigenous knowledges. In A. Blunt and G. Rose, eds, *Writing Women and Space: Colonial and Postcolonial Geographies*. New York: Guilford Press, 169–96.
Johnson, N. (1995) Cast in stone: monuments, geography and nationalism. *Environment and Planning D: Society and Space*, 13, 51–65.

Johnston, L. (1996) Flexing femininity: female body builders reconfiguring 'the body'. *Gender, Place and Culture*, 3, 327–40.

Jones III, J. P., Nast, H., and Roberts, S., eds (1997) *Thresholds in Feminist Geography*. New York and Oxford: Rowman and Littlefield.

Kaplan, C. (1996) *Questions of Travel: Postmodern Discourses of Displacement*. Durham, NC and London: Duke University Press.

Katz, C. (2001) On the grounds of globalisation: a topography for feminist political engagement. *Signs*, 26, 1213–34.

Kimmel, M. S. (1996) *Manhood in America: A Cultural History*. New York: The Free Press.

Kimmel, M. S., with Aronson, A., eds (2000) *The Gendered Society Reader*. Oxford: Oxford University Press.

Latour, B. (1999) *Pandora's Hope: Essays on the Making of Science Studies*. Cambridge, MA: Harvard University Press.

Laurie, N., Dwyer, C., Holloway, S., and Smith, F. (1999) *Geographies of New Femininities*. Harlow: Longman.

Laws, G. (1994) Ageing, contested meanings and the built environment. *Environment and Planning A*, 26, 1787–1802.

Laws, G. (1997) Women's life course, spatial mobility and state policies. In J. P. Jones III, H. Nast, and S. Roberts, eds, *Thresholds in Feminist Geography*. New York: Rowman and Littlefield, 47–64.

Liddington, J. (1984) *The Making of a Respectable Rebel: Selina Cooper*. London: Virago.

McCormack, D. (1999) Body shopping: reconfiguring geographies of fitness. *Gender, Place and Culture*, 6, 157–77.

McDowell, L. (1983) Towards an understanding of the gender division of urban space. *Environment and Planning D: Society and Space*, 1, 59–72.

McDowell, L. (1992a) Space, place and gender relations. Part 1: Feminist empiricism and the geography of social relations. *Progress in Human Geography*, 17, 157–79.

McDowell, L. (1992b) Identity, difference, feminist geometries and geographies. *Progress in Human Geography*, 17, 305–18.

McDowell, L. (1999) *Gender, Identity and Place: Understanding Feminist Geographies*. Cambridge: Polity Press.

McDowell, L. and Court, G. (1994) Performing work: bodily representations in merchant banks. *Environment and Planning D: Society and Space*, 12, 727–50.

McDowell, L. and Massey, D. (1984) A woman's place. In D. Massey and J. Allen, eds, *Geography Matters!* Cambridge: Cambridge University Press.

Mackenzie, S. (1989) *Visible Histories: Women and Environments in a Post-war British City*. London and Montreal: McGill-Queens University Press.

Mackenzie, S. and Rose, D. (1983) Industrial change, the domestic economy and home life. In J. Anderson, S. Duncan, and R. Hudson, eds, *Redundant Spaces and Industrial Decline in Cities and Regions*. London: Academic Press, 155–99.

Massey, D. (1979) In what sense a regional problem? *Regional Studies*, 13, 233–43.

Massey, D. (1991) Flexible sexism. *Environment and Planning D: Society and Space*, 9, 31–57.

Mitchell, K (1999) Scholarship means dollarship, or, money in the bank is the best tenure. *Environment and Planning A*, 31, 381–8.

Mitchell, K. (2000) The value of academic labor: what the market has wrought. *Environment and Planning A*, 10, 1713–18.
Mohanty, C. T. (1991) Cartographies of struggle. In C. T. Mohanty, A. Russo, and L. Torres, eds, *Third World Women and the Politics of Feminism*. Bloomington, IN: Indiana University Press, 1–47.
Mohanty, C. T. and Alexander, J. (1997) *Feminist Genealogies, Colonial Legacies, Democratic Futures*. London: Routledge.
Moore, H. (1988) *Feminism and Anthropology*. Cambridge: Polity Press.
Nash, C. (1993) Remapping and renaming: new cartographies of identity, gender and landscape in Ireland. *Feminist Review*, 44, 39–57.
Nast, H. and Pile, S., eds (1998) *Places through the Body*. London: Routledge.
Nelson, L. (1999) Bodies (and spaces) do matter: the limits of performativity. *Gender, Place and Culture*, 6, 331–53.
Nicholson, L. (1990) *Feminism/Postmodernism*. London: Routledge.
Nussbaum, M. (2000) *Women and Human Development: The Capabilities Approach*. Cambridge: Cambridge University Press.
Nussbaum, M. and Glover, J., eds (1995) *Women, Culture and Development: A Study of Human Capabilities*. Oxford: Clarendon Press.
Nussbaum, M. and Sen, A., eds (1993) *The Quality of Life*. Oxford: Clarendon Press.
O'Neill, O. (1995) Justice, capabilities and vulnerabilities. In M. Nussbaum and J. Glover, eds, *Women, Culture and Development: A Study of Human Capabilities*. Oxford: Clarendon Press, 140–52.
O'Neill, O. (2000) *Bounds of Justice*. Cambridge: Cambridge University Press.
Pain, R. (1991) Space, sexual violence and social control: integrating geographical and feminist analyses of women's fear of crime. *Progress in Human Geography*, 15, 415–31.
Pain, R. (1997) Social geographies of women's fear of crime. *Transactions, Institute of British Geographers*, NS22, 231–44.
Pateman, C. and Grosz, E., eds (1987) *Feminist Challenges*. Boston, MA: North Eastern University Press.
Phillips, A. (1997) From inequality to difference: a severe case of displacement. *New Left Review*, 224, 143–53.
Phillips, A. (1999) *Which Equalities Matter?* Cambridge: Polity Press.
Pollock, G. (1996) *Generations and Geographies in the Visual Arts: Feminist Readings*. London: Routledge.
Professional Geographer (1995) Special issue on feminist research methods. *The Professional Geographer*, 48.
Quine, W. (1981) *Theories and Things*. Cambridge, MA: Harvard University Press.
Radcliffe, S. and Westwood, S., eds (1994) *Viva: Women and Popular Protest in Latin America*. London: Routledge.
Rose, G. (1993) *Feminism and Geography: The Limits of Geographical Knowledge*. Cambridge: Polity Press.
Smith, N. and Katz, C. (1993) Grounding metaphor: towards a spatialized politics. In M. Keith and S. Pile, eds, *Place and the Politics of Identity*. London: Routledge, 67–83
Strathern, M. (1992) *After Nature: English Kinship in the Late Twentieth Century*. Cambridge: Cambridge University Press.

Tilly, C. (1999) *Durable Inequality*. Berkeley, CA: University of California Press.
Tivers, J. (1985) *Women Attached: The Daily Activity Patterns of Women with Young Children*. London: Croom Helm.
Valentine, G. (1989) The geography of women's fear. *Area*, 21, 385–90.
Valentine, G. (1992) Images of danger: women's sources of information about the spatial distribution of male violence. *Area*, 24, 22–9.
Walby, S. (2001) Against epistemological chasms: the science question in feminism revisited. *Signs*, 22, 485–509.
Walker, L. (1995) More than just skin-deep: fem(me)ininity and the subversion of identity. *Gender, Place and Culture*, 2, 71–6.
Walters, B. (2001) *Outsiders Inside: Whiteness, Place and Irish Women*. London: Routledge.
West, C. and Zimmerman, D. (1987) Doing gender. *Gender and Society*, 1, 125–51.
Women and Geography Study Group (1997) *Feminist Geographies: Explorations in Diversity and Difference*. Harlow: Longman.
Young, I. M. (1990) *Justice and the Politics of Difference*. Princeton, NJ: Princeton University Press.
Young, I. M. (1997) Unruly categories: a critique of Nancy Fraser's dual systems theory. *New Left Review*, 222, 147–60.
Young, I. M. (1998) Harvey's complaint with race and gender struggles: a critical response. *Antipode*, 30, 36–42.

20

Geographers, ethics and social concern

DAVID M. SMITH

Social concern, or relevance, was one of the main themes in human geography during the last three decades of the twentieth century. Preoccupation with the areal differentiation of life on earth, which had dominated the discipline until the 1960s, gave way to an emerging sense of responsibility for improving the human condition. The previous chapters of this part of the book, on geography moving forward, have all considered aspects of the search for human betterment. Indeed, the very idea of moving forward, or progress, has normative connotations, shifting the focus from how things actually are to the way they ought to be. Thus the topics of environmental change, disease, urbanisation, the fragmented city, the regional problem, and the discourse of difference all address matters of social concern, the understanding and possible solution of which is considered amenable to geographical analysis.

Of course, normative issues have always been implicated in the discipline of geography, even when some of its practitioners claimed otherwise. The attempt to construct a value-free approach, associated with the location analysis or spatial science schools of human geography, was in fact highly selective in its methods and choice of material. Preference for the supposed rigour of quantification, statistics, and mathematical model-building was based on a particular conception of the human subject, dedicated to maximising place utility, minimising coverage of distance, or some similarly laudable goal. Qualitative research methods, based on more subtle notions of human agency, tended to be dismissed. And the dominance of economic geography left such conditions as social deprivation, ill-health, crime, and environmental pollution largely neglected, unless their particular manifestations invited positivist styles of description and interpretation.

It was this apparent lack of social concern on the part of the new numerical human geography that helped to provoke the 'radical' reaction of the 1970s. Inequality and social justice became central issues, as the role of values in geography was explicitly recognised. This prompted some restructuring of the discipline, including a revitalisation of social and cultural geography during the 1980s, accompanied by the introduction of more qualitative approaches. The 1990s saw a broader 'moral turn' (Smith, 1997), involving explorations of the interface between geography and ethics. So a century in which geographers consorted with a number of other disciplines—from geology, through economics, to political science, sociology, and cultural studies—ended in a new liaison, with moral philosophy.

British geography, and geographers, played a prominent part in the discipline's orientation towards ethics and social concern. But herein lies a difficulty for this chapter. Professional engagement with social problems involves responding to the societal context (economic, political, and cultural) within which they arise and are subject to debate. Elements of this context may be stronger motivating forces in some times and places than others, with the material conditions of particular times and places prompting individual geographers to work on particular issues. Thus the circumstances of the 1970s encouraged activism on the part of geographers, while particular problems (e.g. regional inequality in Britain and inner-city deprivation in the United States) stimulated particular responses. The difficulty for this chapter, against the book's brief to focus on British geographers working in the UK, is that some major contributions to the development of social relevance in geography were made by geographers of British birth but working elsewhere, notably in the United States. This is not a trivial matter, for it points to the increasing internationalisation of the profession in recent decades, as well as to how 'progress' in geography actually comes about. To cite obvious examples, David Harvey's conversion from the liberalism of the 1960s to the radicalism of the 1970s was stimulated by moving from Britain to the United States, as was my own shift from location theory to a welfare approach, and there is no one more closely associated with radical geography than another academic from Britain working in America—Richard Peet.

Geographers of British origin based in America also contributed to the quantitative revolution and to the development of geography as spatial science. But radical geography, at least in its early years, was more obviously grounded in the societal context of particularly turbulent times, experienced so vividly in the United States. Other cases of geographers from Britain working on issues of social concern elsewhere include underdeveloped countries, especially in Africa and Latin America. It could be that in some of these contexts the (British) outsider could engage social problems with greater detachment than could insiders, whose critical inclinations might have been limited by national loyalty. But the role of critical outsider makes special demands, on both ethics and prudence, which generate their own constraints. In any event, where research is done is as clearly implicated in how we work as is where we are from.

This chapter is therefore flexible about the meaning of British geography and geographers, in an account written by someone who's own sense of self is not strongly British. However, it must be recognised that adopting even the most inclusive definition of 'British' leaves out some geographers who played major parts in the developments described, adding to the inevitable selectivity of the work cited here. The significance of national affiliation and identity is considered again at the close, in the context of professional ethics.

Radical geography

British geography entered the second half of the twentieth century greatly influenced by the French school of regional synthesis. If there was any normative content to be detected, it was implicit approval of ways of life developed through close local interrelationships between man (sic) and environment. Another prominent strand was the 'commercial' geography associated with inventories of world resources, reflecting the far from neutral role of the discipline during Britain's era as a colonial power. More explicit value positions were revealed in the links between geography and the new profession of town and country planning, with its emphasis on the rational use of land. As David Matless (1994, 127) has pointed out, land use was a normative term in Dudley Stamp's famous Land Utilisation Survey of Britain (Stamp, 1948), the subtitle of which referred to 'use and misuse'. Avoiding misuse of land involved social evaluation, in which the preservation of a landscape unsullied by 'urban sprawl' and offensive structures featured prominently. Thus from the earliest days of geographical involvement with social policy, as in normative thinking more generally, ethics and aesthetics have been closely intertwined.

Engagement with problems of regional development, urbanisation, and planning has been an ongoing feature of British geography ever since (see Chapters 16 and 18). In the mid-1960s, the traditional preoccupation with unemployment as an indicator of spatial disparities was augmented by research on patterns of personal incomes, conducted by Eric Rawstron with Bryan Coates (e.g. Rawstron and Coates, 1965). This collaboration extended into a broader study of regional variations in Britain (Coates and Rawstron, 1971), motivated by the wish to measure economic and social conditions as thoroughly as the country's climate. In addition to income and employment, this book included chapters on population born overseas, medical and dental health services, mortality, and education. Compared with the traditional content of regional geography, the inventory of relevant conditions had been significantly enlarged.

The emergence of what became known as radical geography marked a major shift in the discipline's orientation, bringing a wide range of social concerns to the fore. Radical geography first surfaced in the United States, with the 1969 annual meeting of the Association of American Geographers (AAG), held in Ann Arbor, Michigan, enlivened by discussions of ethical issues, invasions of sessions by black activists, the launch of the radical journal *Antipode*, and William Bunge's famous 'expedition' to the poor ghetto neighbourhood of Fitzgerald in Detroit.[1] Radical geography consolidated its impact at the Boston AAG meeting in 1971. But, even in these early years, this movement was by no means exclusively American. For example, geographers of British origin working in the United States contributed to the Boston conference session on poverty and social well-being: one coined the term radical geography in a commentary on the meeting (Smith, 1971), while another was responsible for publishing the session's papers (Peet, 1972) and went on to edit a book on radical geography (Peet, 1977).

The radical movement called for greater social relevance, in the teaching of geography as well as in research. Inequality and social injustice became central concerns, as such issues as under-development, economic exploitation, political domination, social deprivation, and racial discrimination were added to the geographical agenda. There was also growing perception of environmental problems, as pollution and resource depletion were being recognised well beyond academia as costs of economic growth.

Radical geography focused attention on spatial variations in such hitherto neglected conditions as poverty, health, hunger, and crime. For example, criminal justice was a subject of sustained attention by another transatlantic migrant—Keith Harries (1974; Harries and Brunn, 1978). Attempts to identify a multivariate geography of social well-being or the quality of life added a spatial dimension to the contemporary social indicators movement, which sought to augment measures of human progress confined to economic criteria (Smith, 1973; Knox, 1975). Concern with the geographical expression of inequality at different scales prompted further research (Coates, Johnston, and Knox, 1977; Smith, 1979). And there was an attempt to restructure the discipline around the theme of human welfare, deploying the analytical apparatus of welfare economics to flesh out the notion of a territorial dimension to social justice (Smith, 1977). The growing significance of issues of social concern and public policy in Britain was reflected in contributions to *Area*, and in presidential addresses to both the Institute of British Geographers (IBG) (Coppock, 1974) and the Geographical Association (Steel, 1974).

A particularly important influence on both sides of the Atlantic was the work of David Harvey. A paper originally read at the Boston AAG meeting, outlining conditions that would have to be fulfilled for territorial social justice, was expanded into one of the seminal texts of the radical era—*Social Justice and the City* (Harvey, 1973). Harvey's move

[1] Thanks to Ron Johnston for these recollections.

from the liberal egalitarianism of John Rawls to the historical materialism of Karl Marx, laid out in his book, revealed the contested character of the values underpinning alternative ways of organising the societal process of production, exchange, and distribution. He also contributed to the growing debate on the implications of geography's links with public policy (e.g. Harvey, 1974), pointing to the limitations of what was likely to be achieved by the bourgeois state.

As radical geography gathered strength through the 1970s, the influence of Marxism increased. However, a mechanical form of structural determination, in which cultural and social forms followed from the economic base of society, allowed little scope for local expressions of individual volition. And the further elaboration of territorial social justice that Harvey had envisaged seemed unnecessary to those of a revolutionary inclination, so obvious were the injustices of capitalism and so convincing the superiority of socialism. However, as the roles of human agency and contingency began to penetrate Marxian metatheory, other intellectual movements came into play, including geography's dormant humanism, general social theory, and cultural studies. Radical geography was to be replaced by a reinvigorated social and cultural geography as the critical cutting edge of the discipline.

The new social and cultural geography

By the beginning of the 1980s, questions of social relevance had become an established part of mainstream human geography, not merely the preoccupation of a radical and politicised fringe. This was reflected in some reconsideration of the discipline's nature, history, and ideological content, recognising the complex influences on its development and underlining its affiliation with critical social science (Gregory, 1978; Stoddart, 1981). At a more practical level, new perspectives on a wide range of issues were being debated. For example, one collection (almost exclusively by British authors), *A World in Crisis,* included discussion of the international economy, energy, hunger, natural resources, population growth, the destruction of regional cultures, individual freedom, national congruence, world-power competition and local conflicts, the world-systems perspective, and war (Johnston and Taylor, 1986).

British geographers made substantial contributions to understanding parts of the world with distinctive issues of social concern. One of these was South Africa under apartheid, exemplified by the work of Anthony Lemon (1987, 1991; see also Smith, 1982, 1992), at a time when academic links with this country became an ethical question (see *Area* 20, 1988). Another was Eastern Europe, contact with which was facilitated by a series of British–Polish and British–Soviet seminars; of special note was work on the socialist city, which challenged western models of urban spatial structure

(French and Hamilton, 1979; see also Smith, 1989). These cases, and other examples elsewhere, underline the problem raised in the introduction to this chapter, of confining the discussion to British geographers working in the UK, especially when overseas research involved long periods in the field and attachment to local universities.

As to the more parochial social concerns of geographers actually working in and on Britain, their coverage in the other chapters of this book leaves only summary comments required here. Race and racism (Jackson, 1987), along with feminist or gender perspectives (Women and Geography Study Group, 1984), became defining topics of the new social and cultural geography. Work under the rubric of health and health care was added to the established field of medical geography (see Chapter 15). Mention may also be made of geographical perspectives on crime (Herbert, 1982), ethnic segregation (Peach, Robinson, and Smith, 1981), and social welfare services (Curtis, 1989; Pinch, 1985). There was a strong emphasis on social problems of the city (Herbert and Smith, 1979), but rural concerns were also recognised (Cloke, 1987). The ongoing preoccupation of many geographers with urban and regional development problems, and with the planning response, goes almost without saying.

The proliferation of issues of social concern prompted a rethinking of social geography. The normative content of so much of the new subject matter loosened this branch of the discipline from its empirical tradition of pattern recognition and challenged established positivist styles of interpretation. As was recognised by Peter Jackson and Susan Smith (1984), differences between such perspectives as humanism and structuralism, with their particular conceptions of human being, carry important ethical implications; ultimately, choice of explanatory framework can depend on moral criteria. Closely related to the changes in social geography was the emergence of what became known as the 'new' cultural geography, challenging the heritage of the Berkeley School and recognising culture as a site of power struggle and political conflict (Jackson, 1989).

These developments involved both sides of the Atlantic, with prominent American participation as well as Britons influenced by sojourns in the United States. But they were given particular impetus by the Social and Cultural Geography Study Group of the IBG. The earlier preoccupation with race and gender were extended to concerns about the disadvantage of various 'others', subject to unfair treatment by virtue of society's failure adequately to acknowledge and respond to their differences. These groups included cultural and ethnic minorities, the disabled, gay people, and postcolonial subjects. The emphasis on difference (see Chapter 19) reflected one of the prime concerns of the contemporary intellectual changes associated with postmodernism, with its hostility to the essentialising (homogenising and universalising) tendencies inherited from the Enlightenment. As the 1980s came to a close, the theory and practice of human geography could never again be separated from their ethical context.

While the radical geography of the 1970s had developed against an emerging societal consensus of opposition to discrimination, inequality, and injustice, along with expectations of ameliorative action on the part of government, the 1980s saw a change in context. The resurgence of a New Right political philosophy associated with Thatcherism, and with the Reagan presidency in the United States, was marked by greater respect for market outcomes and by growing hostility to state intervention. This was reflected by Robert Bennett (1989), who criticised much recent human geography for its political bias (to the left), arguing for a new orientation more in tune with the 'culture of the times'. This would respond to a move away from collective state provision for people's needs grounded in notions of rights towards what he termed 'post-welfarism', characterised by individual responsibility in a market-oriented society. Like David Harvey earlier, he challenged the basis on which the social relevance of research is defined in terms of assisting government policy, but from a quite different political perspective.

Bennett's paper aroused considerable interest (e.g. Johnston, 1993, 154–7), but more for its novelty than its conviction. Very few geographers in Britain, or anywhere else, responded positively to his lead. This may reflect the fact that the very changes that he welcomed appeared to be exacerbating the plight of the poor and marginalised, who have traditionally been the focus of social concern. The moral sentiments of British geographers have continued to be predominately egalitarian, whether underpinned by liberalism, Marxism, or a politics of difference, with little interest shown in either the authoritarian or the libertarian versions of conservatism. If we are out of tune with the prevailing culture and political ideology of mean times, it could be that our abiding practice of field inquiry brings such encounters with the reality of hard-earned lives as to make free-market fundamentalism an unappealing creed. It is difficult to work on, and with, the destitute, exploited, and oppressed without some feelings of empathy, grounded in mutual understanding of the human experience of suffering. As is increasingly recognised in ethics, feeling as well as reason is involved in the development of morality and a sense of justice (Smith, 2000b).

Moral geographies

During the era of radical geography and the consolidation of social concern that followed, there was little explicit consideration of ethics. Few human geographers read much moral philosophy; and hardly any ethicists knew anything about geography, so disconnected were these two fields. However, there was a growing recognition of a lacuna, the partial filling of which was to be a major preoccupation of the 1990s and the beginning of the present century.

The application of a 'moral lens' to human geography was anticipated by the IBG Social and Cultural Geography Study Group, in an argument for (re)connecting their inquiries to moral philosophy (Philo, 1991). They were interested in the everyday moralities of particular peoples, as they vary from place to place. These variations were thought to overlap with differences of social class, ethnic status, religious belief, political affiliation, and so on, with moral assumptions implicated in the social construction of such groups—with who is included and excluded. Attention was also drawn to the geographical content of moralities, pointing out that moral assumptions and arguments often incorporate thinking about space, place, environment, and landscape.

Thus, the relationship between local moral beliefs and practices, society, and culture was problematised. It was also recognised that recent debates about the relative merits of Marxism, humanism, and postmodernism had radically transformed the intellectual context of deliberations over the place of morality in geography (Driver, 1991). Power, representation, and resistance became keywords in the new social and cultural geography, suspicious of any imposed codes of morality as possible instruments of domination. There had already been an impact on the long-established practice of landscape interpretation, with landscape now taken as a text to be read critically, much as a written document or painting might be (Cosgrove and Daniels, 1988). Landscapes came to be seen very much as reflections of economic, political, and social systems, with their prevailing ideologies or 'moral order' symbolised at the scale of settlement design as well as of individual structures. Landscapes could be viewed as collages of changing and competing discourses, each deploying aesthetics to convey something of their moral content and strength.

The term 'moral geography' was adopted to cover a variety of moral readings of geographical situations (Smith, 1998a, 14–18; 2000b, 45–53). Some writers preferred the terms 'moral landscape', 'moral location', 'moral terrain', or 'moral topography'. The earliest usage of 'moral' in this context appears to be by Peter Jackson and Susan Smith (Jackson, 1984; Jackson and Smith, 1984) in their examinations of the 'moral order' of the city as revealed by the Chicago School of urban sociology in the 1920s and 1930s. This notion was also applied to the Victorian city in Britain (Jackson, 1984; Driver, 1988), with space deeply implicated in class struggle and moral order underpinned by residential segregation (Jackson, 1989). There were links with the work of David Sibley (1995), who suggested that nineteenth-century schemes to reshape the city could be seen as a process of purification designed to exclude such 'polluting' groups as the poor, racial minorities, and prostitutes. Faith in the possibility of amelioration through environmental reform was reflected not only in building model housing, but also in provision of such facilities as reformatories, hospitals and public parks. Case studies included the location of facilities for servicemen in nineteenth-century Portsmouth well away from sources of vice and disease (Ogborn and Philo, 1994), and of reformatories in rural settings, which combined an environment considered favourable to rehabilitation with the virtue of hard labour (Ploszajska, 1994).

A variation on the link between environment and morality was provided by the notions of the 'moral discourse of climate' and 'climate's moral economy' in the work of David Livingstone (1991, 1992). He explained that discussions of climate by geographers throughout the nineteenth century and well into the twentieth century were frequently cast in language involving moralistic appraisals of both people and places. Thus, in the prevailing spirit of environmental determinism, native inhabitants of tropical lands could be described as in a low state of morality: sensual, unseemly, lazy, and so on. Such behaviour as excessive use of alcohol and sexual indulgence characterised the 'moral topography of the white tropical experience' (Livingstone, 1992, 236), with advice to the colonisers on how to survive in the tropics couched in terms of prudence, abstemiousness, circumspection, and hygienic discipline.

Gender roles and sex have provided fertile ground for investigations of contemporary geographical expressions of morality. For example, a study of local child-care cultures in Sheffield identified a 'moral geography of mothering', involving a localised discourse concerning what is right and wrong in raising children (Holloway, 1998). Similarly, suburban differences around moral codes of mothering and participation in paid work have been observed (Dowling, 1998). Spatial expressions of sexuality that have attracted moral readings include presence of prostitutes (Hubbard, 1999), the 'moral contours' of heterosexuality (Hubbard, 2000), and the spaces of gay men and lesbians (Valentine, 1993; Bell and Valentine, 1995; Binnie and Valentine, 1999). All this, as well as research on children (Valentine, 1996), raises the question of the 'right' place for certain kinds of persons and practices, further elaborated by Tim Cresswell (1996) using the notion of transgression.

Few writers on moral geography and similar concepts have given much direct attention to the meaning of 'moral'. An exception is David Matless (1994), who has suggested that moral geographies work around Michel Foucault's three senses of morality: as moral codes, as the exercise of behaviour in transgression or obedience of a code, and as the way in which individuals act in relation to a code. He has also underlined the role of environmentalism as moral improvement, showing how leisure pursuits in the English countryside have been presented in terms of the development of good citizenship (Matless, 1997).

A number of important themes may be found in studies of moral geographies and the like. The most obvious is the identification of spatial variations in moral values and practices, which challenges the notion of universal moral truth. Another is the role of environment in moulding morality, reflecting the discipline's traditional preoccupation with behavioural responses to both nature and built form. There is also a focus on power in the spatial practices of social control and exclusion. Such studies might be regarded as exercises in 'thick' descriptive ethics (Proctor, 1998), in which empirical observation fuses with contextual interpretation. They add substance to the abstractions of moral philosophy, helping to reveal different ways in which the right and the good is understood and acted upon, by real people in actual situations.

Geography and ethics

By the beginning of the 1990s the concern with ethical issues in geography was being augmented from a number of directions. This prompted sessions at conferences, and special issues of journals on social (in)justice (*Urban Geography*, 1994), environmental ethics (*Antipode*, 1996), and normative theory (*Society and Space*, 1997). New books on social justice appeared, with a review of alternative theoretical perspectives and case studies (Smith, 1994), and the reassertion of a materialist perspective exploring 'the just production of just geographical differences' (Harvey, 1996). Moral content also permeates the notions of 'geographies of resistance' (Pile and Keith, 1997) or 'dissident geographies' (Blunt and Wills, 2000), and the utopian vision of 'spaces of hope' imagined by David Harvey (2000). Other indications are found in the journal *Ethics, Place and Environment*, and in reviews of the growing literature in *Progress in Human Geography*.

Again, these developments have been very much a transatlantic initiative. Indeed, it would be remiss not to recognise the role of distinguished American geographers in promoting interest in ethics and morality, especially Robert Sack and Yi-Fu Tuan (see Smith, 2000b). It was an international collaborative venture: the Geography/Ethics Project convened at conferences in Britain and the United States, which generated the first edited collection of papers on a range of topics linking ethics with space, place, nature, and the construction of geographical knowledge (Proctor and Smith, 1999).

The first integrated text to explore the interface of geography with ethics shows how those familiar geographical concepts of landscape, location and place, proximity, distance and community, space and territory, justice, development, and nature can all be elucidated by moral readings (Smith, 2000b). Furthermore, the geographer can make a significant contribution to ethics, with its abiding predilection towards philosophical abstraction, by arguing and illustrating the importance of context to the way in which real people act, in actual geographical and historical circumstances, in pursuit of the right and the good as they understand morality. Particular interest has been shown in links with a feminist-inspired ethic of care, manifest in the question of the spatial scope of beneficence (Silk, 1998; Smith, 1998b), which links with crucial aspects of development studies (Corbridge, 1993) as well as with moral philosophy.

Geographers' concern with ethics in recent years may be related to the intellectual context of the times, including a broader normative turn in social theory (Sayer and Storper, 1997). Just as the era of radical geography reflected the emergence of a critical social science in response to particular problems, so the contemporary engagement with normative issues is associated with moral as well as practical concern over increasing economic inequality, social instability, political alienation, and environmental despoliation in the post-socialist world of globalising capital. And unease is by no means confined to the political left, for there are powerful arguments from the right pointing to the erosion of traditional values and sources of moral authority. Postmodern scepticism about the very idea of moral truth encourages relativism, at a time when such evident evils as starvation, genocide, and ethnic cleansing invite the assertion of some moral certainties. Hence the importance of working towards a universal theory of the right and good, however thin it might be, in a dialectical relationship with understanding actual human conduct in context.

The growing interest in the interface between geography and ethics has stimulated contributions to some interdisciplinary normative concerns. Example are development ethics, including Stuart Corbridge (1993, 1998) on responsibility to distant strangers, and related aspects of environmental ethics (Pepper, 1993; Peet and Watts, 1996). Research on the 'moral economic geographies' formed by participants in local exchange, employment, and trading systems (LETS) shows how new forms of economic and social relations mediated by 'virtual currencies' might challenge the customary preoccupations of economics (Lee, 1996). And so on. Thus, geographers are beginning to participate in a more general, multi-disciplinary search for ways of engaging with ethics, from which moral philosophers have as much to learn as we have.

The ethics of 'British Geography'

It remains to make brief reference to the ethics of professional geography. Vigorous debates on the meaning and practice of social relevance were initiated in the early days of radical geography, but it was some years before the arguments were brought together systematically by Bruce Mitchell and Dianne Draper (1982). They also provided a sustained examination of research ethics, including responsibility to and treatment of the human subjects of geographical inquiry. This greatly extended more familiar ethical conventions, such as not falsifying data or plagiarising the work of others.

Interest in professional ethics has grown considerably in recent years. A wide range of issues have been raised, including ethics and power in higher education (Kirby, 1991), multiple voices and pedagogic authority (McDowell, 1994), and the positionality or situatedness of the researcher (Jackson, 1993; Rose, 1997). The construction of geographical knowledge has become part of the discipline's problematic. Among the issues discussed by contributors to the collection edited by Proctor and Smith (1999) are the importance of a communicative ethics in participatory research, the legitimacy of persons writing about a group of which they are not members, the conduct of cross-cultural research involving encounters with alternative views of the world, and the relationship between research students and supervisors with different agendas. The widespread adoption of qualitative methods has generated a growing recognition of a relational ethics, emphasising reciprocity between researchers and researched. Reflections on sensitive aspects of professional ethics has been stimulated by continuing research on some groups vulnerable to their particular difference; for example, by Rob Kitchin (1999) and others on disability and by Paul Cloke and his collaborators (2000) on rural homelessness.

Some of the interest in professional ethics has been prompted by the changing context of academic activity, including increasing commodification of knowledge and the pressure of performance assessment. The importance of associating a personal or institutional name with research findings, which may generate commercial value as well as professional prestige, raises questions of intellectual property rights. Innovations in the collection and dissemination of information, such as remote sensing, geographical information systems and the internet, pose such issues as unequal access and intrusions on privacy.

Of particular importance in British universities has been the imposition of formal assessment of the quality of research and teaching. The Research Assessment Exercise (RAE) has generated an ongoing debate in *Area* (see also Smith, 2000a, 2001), questioning among other things the ethics of maintaining the dubious technical claims that quality of departmental research can be reliably placed on a numerical scale, and that resources should be allocated accordingly to achieve value for money. Like Teaching Quality Assessment (TQA), which appears to prioritise a particular pedagogic style, the RAE raises fundamental questions of academic freedom.

While much of the debate on professional ethics concerns the conduct of individual scholarship, there are some wider issues. One is that the increasingly competitive UK academic scene, which pits department against department in pursuit of limited financial resources and students, encourages parochialism in the form of an exaggerated ethic of care for one's own. This raises the question of relationships with professional colleagues elsewhere, who may depend in various ways on the possibility of collaboration with geographers from Britain whose professional standing may be better advanced by work at home.[2] Such issues have been debated within the broader context of responsibility to distant others (Smith, 1994a; see also Cline-Cole, 1999; Mohan, 1999), as a case of general ethical argument applied to a specific professional problem, or metaethics informing normative ethics.

Ethics are often sharpened by hard cases. One of the most difficult professional issues in recent years, at least for some British geographers, was a special case of responsibility to distant others that arose in the context of institutional change. The merger of the IBG with the Royal Geographical Society (RGS) became complicated by the fact that one of the corporate patrons of the RGS was the Shell oil company. In addition to questioning Shell's environmental record, it was claimed that the company was implicated in the fate of the Nigerian writer Ken Saro-Wiwa, who had criticised the pollution of land in the Ogoni region of the Niger delta and was executed in 1995 by the ruling military regime. This stimulated a debate on business sponsorship, academic independence, and critical engagement, at just the time when institutional as well as personal ethics were becoming an issue (see *Ethics, Place and Environment* 2, 1999). But this was not all. The designation of 'Fellow' for members of the RGS, and of the merged institution, was regarded with hostility in some circles sensitive to gender-specific labels. And the 'Royal' association raised difficulties for those whose political philosophy and sense of justice exclude recognition of a monarchy. The combination of business patronage, patriarchy, and inherited privilege invoked by the RGS was too much for some members of the IBG, who declined to become Fellows. Some feel more at home in the Critical Geography Forum, in which the internet connects over 600 geographers from various parts of the world, sharing mutual interests in the kind of issues raised by engagement with ethics and social concern. A case of moving forward by moving out(wards)?

This episode re-introduces the issue of national identity, raised in the introduction of this chapter. It helps to explain that affiliation with 'British Geography' is not only difficult to grasp in a practical sense, given the increasingly international scope of research collaboration and personal responsibility, but also raises ethical issues of association with the values of a national institution. If our professional practice is now open to rigorous ethical scrutiny, then nothing is beyond critique, not even the very idea of a 'British Geography'.

[2] For example, on the day I planned to finish this chapter I broke off to respond to a request from a Polish colleague to correct the English in a synopsis of her book, work which contributes to no 'performance indicators' but is surely the right thing to do.

Acknowledgements

Parts of this chapter draw on some of my other publications (see below). My research on moral issues in human geography has been supported by a Leverhulme Fellowship. Thanks to Ron Johnston for some helpful suggestions.

References

Antipode (1996) Special issue on environmental ethics. *Antipode*, 28(2).

Bell, D. and Valentine, G., eds (1995) *Mapping Desires: Geographies of Sexuality.* London: Routledge.

Bennett, R. J. (1989) Whither models and geography in a post-welfarist world? In B. Macmillan, ed., *Remodelling Geography.* Oxford: Blackwell, 273–70.

Binnie, J. and Valentine, G. (1999) Geographies of sexuality: a review of progress. *Progress in Human Geography*, 23, 175–87.

Blunt, A. and Wills, J. (2000) *Dissident Geographies: An Introduction to Radical Ideas and Practices.* Harlow: Prentice-Hall.

Cline-Cole, R. (1999) Contextualising professional interaction in Anglo-(American) African(ist) geographies. In D. Simon and A. Norman, eds, *Development as Theory and Practice: Current Perspectives on Development and Development Cooperation.* Harlow: Longman, 95–133.

Cloke, P., ed. (1987) *Rural Planning: Policy into Action?* London: Harper & Row.

Cloke, P., Cooke, P., Cursons, J., Milbourne, P., and Widdowfield, R. (2000) Ethics, reflexivity and research: encounters with homeless people. *Ethics, Place and Environment*, 3, 133–54.

Coates, B. E. and Rawstron, E. M. (1971) *Regional Variations in Britain: Studies in Economic and Social Geography.* London: Batsford.

Coates, B. E., Johnston, R. J., and Knox, P. L. (1977) *Geography and Inequality.* Oxford: Oxford University Press.

Coppock, J. T. (1974) Geography and public policy: challenges, opportunities and implications. *Transactions, Institute of British Geographers*, 63, 1–16.

Corbridge, S. (1993) Marxisms, modernities and moralities: development praxis and the claims of distant strangers. *Environment and Planning D: Society and Space*, 11, 449–72.

Corbridge, S. (1998) Development ethics: distance, difference, plausibility. *Ethics, Place and Environment*, 1, 35–53.

Cosgrove, D. and Daniels, S. (1988) *The Iconography of Landscape.* Cambridge: Cambridge University Press.

Cresswell, T. (1996) *In Place/Out of Place: Geography, Ideology and Transgression.* Minneapolis, MN: University of Minnesota Press.

Curtis, S. (1989) *The Geography of Public Welfare Provision.* London: Routledge.

Dowling, R. (1998) Suburban stories, gendered lives: thinking through difference. In R. Fincher and J. M. Jacobs, eds, *Cities of Difference.* London: Guilford Press, 69–88.

Driver, F. (1988) Moral geographies: social science and the urban environment in mid-nineteenth century England. *Transactions, Institute of British Geographers*, NS13, 275–87.

Driver, F. (1991) Morality, politics, geography: brave new words. In C. Philo, compiler, *New Words, New Worlds: Reconceptualising Social and Cultural Geography*. Lampeter: Department of Geography, St David's University College, 61–4.

French, R. A. and Hamilton, F. E. I., eds (1979) *The Socialist City: Spatial Structure and Urban Policy*. Chichester: Wiley.

Gregory, D. (1978) *Ideology, Science and Human Geography*. London: Hutchinson.

Harries, K. D. (1974) *The Geography of Crime and Justice*. New York: McGraw-Hill.

Harries, K. D. and Brunn, S. D. (1978) *The Geography of Laws and Justice: Spatial Perspectives on the Criminal Justice System*. New York: Praeger.

Harvey, D. W. (1973) *Social Justice and the City*. London: Edward Arnold.

Harvey, D. W. (1974) What kind of geography for what kind of public policy? *Transactions, Institute of British Geographers*, 63, 18–24.

Harvey, D. (1996) *Justice, Nature and the Geography of Difference*. Oxford: Blackwell.

Harvey, D. (2000) *Spaces of Hope*. Edinburgh: Edinburgh University Press.

Herbert, D. T. (1982) *The Geography of Urban Crime*. Harlow: Longman.

Herbert, D. T. and Smith, D. M., eds (1979) *Social Problems and the City*. Oxford: Oxford University Press.

Holloway, S. H. (1998) Local childcare cultures: moral geographies of mothering and the social organisation of pre-school education. *Gender, Place and Culture*, 5, 29–53.

Hubbard, P. (1999) *Sex and the City: Geographies of Prostitution in the Urban West*. Aldershot: Ashgate.

Hubbard, P. (2000) Desire/disgust: mapping the moral contours of heterosexuality. *Progress in Human Geography*, 24, 191–217.

Jackson, P. (1984) Social disorganization and moral order in the city. *Transactions, Institute of British Geographers*, NS9, 168–80.

Jackson, P., ed. (1987) *Race and Racism*. London: Allen & Unwin.

Jackson, P. (1989) *Maps of Meaning: An Introduction to Cultural Geography*. London: Unwin Hyman.

Jackson, P. (1993) Changing ourselves: a geography of position. In R. J. Johnston, ed., *The Challenge for Geography. A Changing World: A Changing Discipline*. Oxford: Blackwell, 198–214.

Jackson, P. and Smith, S. (1984) *Exploring Social Geography*. London: Allen & Unwin.

Johnston, R. J., ed. (1993) *The Challenge for Geography. A Changing World: A Changing Discipline*. Oxford: Blackwell.

Johnston, R. J. and Taylor, P. J. (1986) *A World in Crisis? Geographical Perspectives*. Oxford: Blackwell.

Kitchin, R. (1999) Morals and ethics in geographical studies of disability. In J. D. Proctor and D. M. Smith, eds, *Geography and Ethics: Journeys in a Moral Terrain*. London: Routledge, 223–36.

Kirby, A. (1991) On ethics and power in higher education. *Journal of Geography in Higher Education*, 15, 75–7.

Knox, P. L. (1975) *Social Well-being: A Spatial Perspective*. Oxford: Oxford University Press.

Lee, R. (1996) Moral money? LETS and the social construction of local economic geographies in southeast England. *Environment and Planning A*, 28, 1377–94.

Lemon, A.(1987) *Apartheid in Transition.* Aldershot: Gower.
Lemon, A., ed. (1991) *Homes Apart: South Africa's Segregated Cities.* London: Paul Chapman.
Livingstone, D. N. (1991) The moral discourse of climate: historical considerations on race, place and virtue. *Journal of Historical Geography*, 17, 413–34.
Livingstone, D. N. (1992) *The Geographical Tradition.* Oxford: Blackwell.
McDowell, L. (1994) Polyphony and pedagogic authority. *Area*, 26, 41–8.
Matless, D. (1994) Moral geographies in Broadland. *Ecumene*, 1(2), 127–56.
Matless, D. (1997) Moral geographies of English landscape. *Landscape Research*, 22, 141–55.
Mitchell, B. and Draper, D. (1982) *Relevance and Ethics in Geography.* Harlow: Longman.
Mohan, G. (1999) Not so distant, not so strange: the personal and the political in participatory research. *Ethics, Place and Environment*, 2, 41–54.
Ogborn, M. and Philo, C. (1994) Soldiers, sailors and moral locations in nineteenth-century Portsmouth. *Area*, 26, 221–31.
Peach, H, C., Robinson, V., and Smith, S., eds (1981) *Ethnic Segregation in Cities.* London: Croom Helm.
Peet, R., ed. (1972) *Geographical Perspectives on American Poverty.* Worcester, MA: Antipode.
Peet, R., ed. (1977) *Radical Geography: Alternative Viewpoints on Contemporary Social Issues.* Chicago, IL: Maaroufa Press.
Peet, R. and Watts, M., eds (1996) *Liberation Ecologies: Environment, Development, Social Movements.* London: Routledge.
Pepper, D. (1993) *Eco-socialism: From Deep Ecology to Social Justice.* London: Routledge.
Philo, C., compiler (1991) *New Words, New Worlds: Reconceptualising Social and Cultural Geography.* Lampeter: Department of Geography, St David's University College.
Pile, S. and Keith, M., eds (1997) *Geographies of Resistance.* London: Routledge.
Pinch, S. (1985) *Cities and Services.* London: Routledge.
Ploszajska, T. (1994) Moral landscapes and manipulated spaces: gender, class and space in Victorian reformatory schools. *Journal of Historical Geography*, 20, 413–29.
Proctor, J. D. (1998) Ethics in geography: giving moral form to the geographical imagination. *Area*, 30, 8–18.
Proctor, J. D. and Smith, D. M., eds (1999) *Geography and Ethics: Journeys in a Moral Terrain.* London: Routledge.
Rawstron, E. M. and Coates, B. E. (1965) Opportunity and affluence. *Geography*, 51, 1–15.
Rose, G. (1997) Situating knowledge, positionality, reflexivities and other tactics. *Progress in Human Geography*, 21, 305–20.
Sayer, A. and Storper, M. (1997) Ethics unbound: for a normative turn in social theory. *Environment and Planning D: Society and Space*, 15, 1–17.
Sibley, D. (1995) *Geographies of Exclusion: Society and Difference in the West.* London: Routledge.
Silk, J. (1998) Caring at a distance. *Ethics, Place and Environment*, 1, 165–82.
Smith, D. M. (1971) Radical geography: the next revolution? *Area* 3, 53–7.

Smith, D. M. (1973) *The Geography of Social Well-being in the United States: An Introduction to Territorial Social Indicators.* New York: McGraw-Hill.
Smith, D. M. (1977) *Human Geography: A Welfare Approach.* London: Edward Arnold.
Smith, D. M. (1979) *Where the Grass in Greener: Living in an Unequal World.* Harmondsworth: Penguin.
Smith, D. M., ed. (1982) *Living Under Apartheid: Aspects of Urbanization and Social Change in South Africa.* London: Allen & Unwin.
Smith, D. M. (1989) *Urban Inequality Under Socialism: Case Studies from Eastern Europe and the Soviet Union.* Cambridge: Cambridge University Press.
Smith, D. M., ed. (1992) *The Apartheid City and Beyond: Urbanization and Social Change in South Africa.* London: Routledge.
Smith, D. M. (1994a) *Geography and Social Justice.* Oxford: Blackwell.
Smith, D. M. (1994b) On professional responsibility to distant others. *Area*, 26, 359–67.
Smith, D. M. (1997) Geography and ethics: a moral turn? *Progress in Human Geography*, 21, 596–603.
Smith, D. M. (1998a) Geography and moral philosophy: some common ground. *Ethics, Place and Environment*, 1, 7–34.
Smith, D. M. (1998b) How far should we care? On the spatial scope of beneficence. *Progress in Human Geography*, 22, 15–38.
Smith, D. M. (2000a) Moral progress in human geography: transcending the place of good fortune. *Progress in Human Geography*, 24, 1–18.
Smith, D. M. (2000b) *Moral Geographies: Ethics in a World of Difference.* Edinburgh: Edinburgh University Press.
Smith, D. M. (2001) On performing geography. *Antipode*, 33, 142–6.
Society and Space (1997) Special issue on normative theory. *Society and Space*, 15(1).
Stamp, L. D. (1948) *The Land of Britain: Its Use and Misuse.* London: Longman.
Steel, R. W. (1974) The Third World: geography in practice. *Geography*, 59, 189–207.
Stoddart, D. R. (1981) *Geography, Ideology and Social Concern.* Oxford: Blackwell.
Urban Geography (1994) Special issue on social justice. *Urban Geography*, 15(7).
Valentine, G. (1993) (Hetero)sexing space: lesbian perceptions and experiences of everyday spaces. *Environment and Planning D: Society and Space*, 9, 395–413.
Valentine, G. (1996) Angels and devils: moral landscapes of childhood. *Environment and Planning D: Society and Space*, 14, 581–99.
Women and Geography Study Group (1984) *Geography and Gender.* London: Hutchinson.

Index

Note: Alphabetical arrangement is word-by-word. Maps and diagrams are denoted by page references in *italics*, and colour plates as *pl. 1*, etc. (preceding any page numbers). Material in footnotes is indicated by the abbreviation 'n', following the page number. Name entries are restricted to those individuals whose work is discussed or quoted in the text; passing references are excluded. For further details of authors cited in the text, the reference lists at the end of each chapter should be consulted.